Electrical Engineering
Concepts and Applications

AF889515

P.V. Prasad

Department of Electrical and Electronics Engineering
Chaitanya Bharathi Institute of Technology, Hyderabad

S. Sivanagaraju

Department of Electrical and Electronics Engineering
University College of Engineering
JNTU, Kakinada

Cengage

Australia • Brazil • Canada • India • Mexico • Singapore • United Kingdom • United States

Cengage

Electrical Engineering: Concepts and Applications

P.V. Prasad
S. Sivanagaraju

© 2012 Cengage Learning India Pvt. Ltd.

ALL RIGHTS RESERVED. No part of this work covered by the copyright herein may be reproduced or distributed in any form or by any means without the prior written permission of the copyright owner.

For product information and technology assistance, contact us at
asia.infoindia@cengage.com

For permission to use material from this text or product,
submit all requests online at
www.cengage.com/permissions

ISBN-13: 978-81-315-1787-1
ISBN-10: 81-315-1787-X

Cengage Learning India Pvt. Ltd.
T-303, 3rd Floor
DAV Complex, Plot 3
Mayur Vihar Phase 1
Delhi-110091

Cengage is a leading provider of customized learning solutions with employees residing in nearly 40 different countries and sales in more than 125 countries around the world. Find your local representative at **www.cengage.com**

Cengage is committed to creating inclusive and diverse products, including perspectives that represent the diversity of people across all backgrounds, identities, opinions, and experiences. If you have any feedback related to this, please contact us at **support.cengage.com**

To learn more about Cengage products, platforms and services, visit at **www.cengage.co.in**

Printed by:
IBH Print and Bind Private Limited
Plot No. 81, Ecotech 12, Greater Noida, Gautam Budh Nagar,
Uttar Pradesh 201308

Printed in India
Print Number: 05 Print Year: 2025

Dedicated to the foremost candles in existence
and our parents

Preface

The purpose of the *Electrical Engineering: Concepts and Applications* is to provide a comprehensive reference work covering the broad spectrum of electrical engineering. This book is written primarily as an introduction to electrical engineering. It is designed to cater to the needs of students in electrical and other engineering disciplines. The coverage of the book is designed to allow its use in a number of ways including service courses taught to non-electrical majors. The organization and details of the material in this book allow maximum flexibility for the instructor to select topics to include in courses within the curriculum.

The salient features of this book are a number of solved problems in each chapter for better learning of concepts. The presentation is easy to understand. "Chapter at a glance" and "short answer questions" are given at the end of each chapter. "Review questions" are given to help the students evaluate themselves. Since most of the competitive examinations nowadays use multiple choice questions, the importance for this segment is increasing; hence, multiple choice questions are given at the end of each chapter.

The information covers both circuits and electric machinery fundamentals and is organized into 11 chapters. The first six chapters discuss electrical circuits; the next four chapters explain the electrical machinery; and the last chapter discusses measuring instruments.

Chapter 1 deals with the basic definitions of various electrical quantities, types of sources, and Kirchhoff's laws. It also provides the techniques to simplify electrical networks.

Chapter 2 comprises dc circuits. This chapter explains comprehensively source transformation techniques and mesh and nodal analyses. It also explains the star–delta transformation techniques with many examples.

Chapter 3 presents magnetic circuits. This chapter explains the concepts of electromagnetism, basic definitions, magnetic effect of electromagnets, leakage flux and fringing, Faraday's laws of electromagnetic induction, and various laws of electromagnetism.

Chapter 4 discusses the ac circuit analysis. This chapter explains the generation of ac voltages and expression of emf, terms related to alternating quantities, rms, and average values. It also describes phasor analysis.

Chapter 5 describes power calculations in ac circuits, series and parallel circuit analyses, and series and parallel resonances. It also provides network theorems with a number of worked-out examples.

Chapter 6 discusses the comparison of single-phase and three-phase systems, three-phase voltage generation, connections of three-phase supply and loads, different configurations of star and delta combinations, and method to measure the three-phase power.

Chapter 7 presents transformers. In this chapter the importance of transformer, types of transformers, equivalent circuit of transformer, performance of transformer, and special transformers are presented in detail.

Chapter 8 explains dc machines. The basic principle of the operation of dc machines when working as generator and motor, types of generator, generator characteristics, motor types, motor characteristics, and performance of motor explained with numerical examples.

Chapter 9 focuses on three-phase induction motor. This chapter provides constructional details, principle of operation, slip–torque characteristics, and losses and efficiency calculations of induction motor. It also explains testing and starting of induction motors, speed control and applications of induction motors, and comparison between synchronous and induction motors.

Chapter 10 discusses synchronous machines. This chapter explains construction and principle of the operation of synchronous machine, types of windings, emf equation, etc. It also provides the behavior of synchronous machine both as generator and motor, regulation of an alternator and the hunting phenomenon.

Chapter 11 describes measuring instruments. This chapter also describes the essential features of the indicating instruments, the principle of operation of moving coil, moving iron-type instruments, extension of meter ranges, and the principle of operation of single-phase energy meter and megger.

We are thankful to the management of Cengage Learning for their excellent contributions toward production values of the book. Heartfelt appreciation is also due to the editorial and production staff of Cengage Learning.

P. Venkata Prasad is also grateful to the management and Principal of CBIT, Hyderabad, for their encouragement and support provided in this effort. He also places on record his indebtedness to his wife Usha and son Saket Reddy without whose support and understanding this effort would never been fruitful.

S. Sivanagaraju sincerely thanks the higher authorities at JNTU, Kakinada, for help given. He wishes to thank his wife Vijaya Lakshmi, son Teja, and daughter Sharmila for their continued support, understanding, and patience.

P. Venkata Prasad
S. Sivanagaraju

Contents

Chapter 1

Introduction to Networks

Chapter Objectives

- To explain basic definitions of electrical quantities
- To give various types of sources
- To discuss Kirchhoff's current law and Kirchhoff's voltage law
- To explain how to simplify the given networks

1.1 INTRODUCTION

Electricity is a naturally occurring energy that exists in nature. Humans have been conscious of this energy since ages. It is an expedient form of energy which has numerous applications. Today, it is difficult to imagine human civilization without electricity. It is considered the backbone of economic development of any country.

To understand the flow of electrical energy, it is necessary to have a basic knowledge of the various elements through which it flows. The path of energy flow is called *electrical circuit*, and the electrical elements we come across are voltage source, current source, resistors, inductors, and capacitors. Based on the elements in the circuit, an electrical circuit can transmit, convert, absorb, and store energy. An interconnection of electrical elements is called a *circuit*. The circuit elements are divided into two groups: passive elements and active elements. The passive elements are resistors, capacitors, and inductors. The active elements are sources (or driving forces). The sources or excitations are of two types: independent sources and controlled sources.

An electric network is defined as an interconnection of two or more circuit elements such as sources, resistors, inductors, and capacitors. If a network contains at least one closed path, it is called as an electric circuit. Analogous to an electric circuit is water pipe. The flow of water through the pipe is similar to the flow of charge through wires. If a pipe breaks, water cannot reach the destination point. Similarly, if a circuit opens, there will not be any flow of charge. Further, it is important that all connections are made with possible charge carriers.

1.2 STRUCTURE OF MATTER

The knowledge of the structure of matter is fundamental to understanding the basics of electricity. In order to understand electricity, we need to look closely at the tiny particle components that the matter is composed of. All matter, such as solids, liquids, and gases, is composed of atoms. Therefore, atoms are considered to be the basic building block of matter. However, atoms are almost always grouped together with other atoms to form a molecule. Only a few gases, such as helium, are composed of individual atoms as the structural unit. Any material that is composed of only one type of atom is called a basic element, or just an element. A compound consists of more than one type of atom. Thus, every component has a distinctive atomic structure. Every matter is composed of combinations of one or more of these components.

1.2.1 Structure of Atom

An atom is composed of three particles, namely protons, electrons, and neutrons. Protons have a positive electrical charge, electrons have a negative charge, and neutrons have no charge at all. Generally, atoms are neutrally charged because they have the same number of protons and electrons or the same number of positive and negative charges. Atoms become ions when they are charged. An atom is detained by forces of attraction between the electrons and the protons. The neutrons help to grip the protons together. Niels Bohr was a Danish physicist who introduced the model of an atom and also contributed in understanding the atomic structure. The atomic model consists of a central nucleus surrounded by minute particles called electrons. These electrons revolve around the nucleus of the atom much like the planets around the Sun in our solar system. But electrons do not really orbit in the same path. They actually vary their orbit with each revolution. Figure 1.1 shows the atomic structure and electronic path.

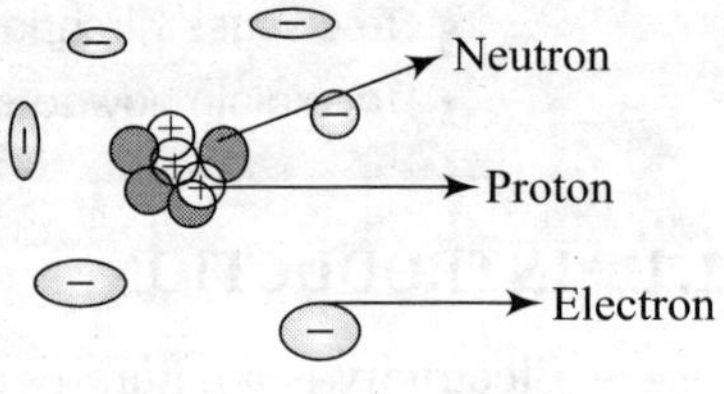

Figure 1.1 Atomic structure and electronic path

1.2.2 Concept of Charge

The nucleus of an atom consists of protons and neutrons. Protons have a positive charge and electrons have a negative charge but their magnitudes are equal. However, protons are much larger and heavier than electrons. Neutrons do not have any charge. Generally the number of protons is equal to the number of electrons for a given atom. Hence, an atom is electrically neutral. If an atom gains an electron it becomes a negative ion. If an atom loses an electron it becomes a positive ion. Electrons appear to be surrounded by a kind of invisible force field, called an electrostatic field. The positively charged electrostatic field in protons is exactly the same strength as the electrostatic field in an electron, but it is opposite in polarity.

The sign conventions of electron and proton are shown in Figure 1.2. The electron sign is marked with negative as shown in Figure 1.2(a) and the proton sign is marked with positive as shown in Figure 1.2(b).

Two electrons will repel each other because they both have a negative electrical charge. Similarly, two protons will also repel each other because they both have a positive charge. On the other hand, protons and electrons will be attracted to each other because of their different charges.

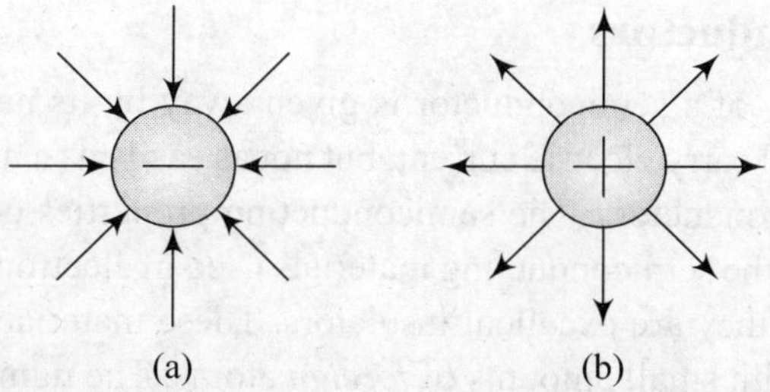

Figure 1.2 Sign conventions of (a) electron and (b) proton

Since electron mass is smaller than that of proton, when the forces of attraction are in play, it is the electrons that do most of the moving. Another important point about the electrical charges of protons and electrons is that if the distance between these is very high then they have less force on each other. Similarly, the closer they are to each other, the more force they will experience from each other due to an electric field.

Unit of charge

Electrical charge is measured in coulombs (C). One electron has a charge of -1.602×10^{-19} C and one proton has a charge of 1.602×10^{-19} C. One coulomb charge is defined as the charge owned by 6.24×10^{18} number of electrons.

1.3 CONDUCTORS, SEMICONDUCTORS, AND INSULATORS

The outermost shell of an atom is called the valence shell. When the valence electron in any atom gains sufficient energy from some outside force, it can break away from the parent atom and become what is called a *free electron*.

Atoms with few electrons in their valence shell tend to have more free electrons since these valence electrons are more loosely bound to the nucleus. In some materials, such as copper, the electrons are so loosely held by the atom and so close to the neighboring atoms that it is difficult to determine which electron belongs to which atom. Under these conditions, the valence or free electrons tend to drift randomly from one atom to its neighboring atoms. Under normal conditions the movement of the electrons is truly random, meaning they are moving in all directions by the same amount. However, if some outside force acts upon the material, this flow of electrons can be directed through materials and this flow is called *electrical current*. Based on how freely current can flow through them, materials are classified as conductors, semiconductors, and insulators.

1.3.1 Conductors

Materials that have free electrons allow electrical current to flow easily are called *conductors*. Most of the conductors used in electronics are metals such as copper, aluminum, and steel. Conductors are materials that obey Ohm's law and have very low resistance. They can therefore carry electric currents from place to place without dissipating a lot of power. As a result, metals are useful as connecting wires to carry electrical signals from place to place. They help ensure that most of the signal's power reaches its destination instead of warming up the wires in between.

1.3.2 Semiconductors

The basic property of a semiconductor is given away by its name—it "conducts a little bit." A semiconductor will carry electric current, but not as easily as a normal conductor. Some materials are *intrinsic* semiconductors. The semiconducting properties occur in these materials naturally. However, most of the semiconducting materials used in electronics are *extrinsic*. This means that left to themselves they are excellent insulators. These materials are turned into semiconductors by *doping* them with small amounts of *foreign* atoms. The number of doping atoms you need to add is very small. If you leave all the doping atoms inside the package of a transistor and remove the bulk of the material you will be left with a vacuum better than that exists between the planets of the solar system.

1.3.3 Insulators

Insulators are materials that have just the opposite effect on the flow of electrons. They do not let electrons flow very easily from one atom to another. Insulators are materials whose atoms have tightly bound electrons. These electrons are not free to roam around and be shared by neighboring atoms.

Some common insulator materials are glass, plastic, rubber, air, and wood. These are materials which will refuse to carry an electric current. They are useful for functions such as coating electric wires to prevent them from "shorting together" or giving you a shock. Silk and cotton are also good insulators (when they are dry) and some of the mains wiring in very old houses once used them. But by modern standards this was pretty dangerous since one could get a shock when wet or a spark would set them alight when dry!

Modern insulators such as PVC (poly vinyl chloride) are much better and safer. Insulators are also very useful to fill the "gap" in between the metal plates of a capacitor. Very old capacitors (and some modern radio ones) use air gaps since air is a good insulator.

1.4 ELECTRICAL QUANTITIES

The value of a parameter characterizes the relationship between two variables. In electrical terms, a parameter is usually regarded as a circuit characteristic which defines circuit's behavior.

Electric circuit is made up of electrical elements. These electrical elements are classified as follows.

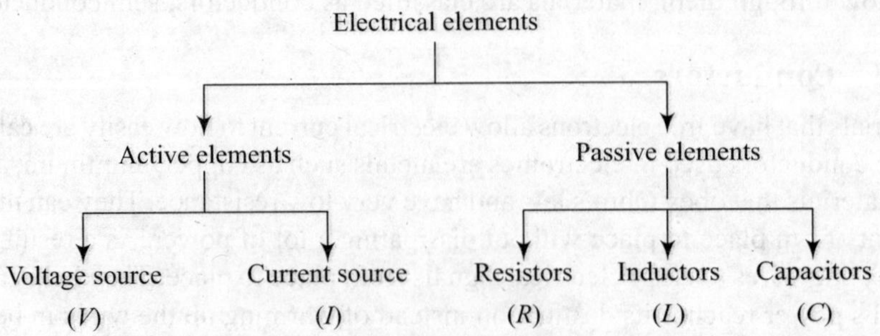

1.4.1 Electric Field

In physics, the space surrounding an electric charge, or in the presence of a time-varying magnetic field, has a property called an *electric field.* This electric field exerts a force on other electrically charged objects. The concept of an electric field was introduced by Michael Faraday. The strength of an electric field, E, is measured in volts per meter (V/m). It is a type of electromagnetic field. An electric field is formed between two metal plates that are parallel to each other and connected to a voltage supply. An electron beam moving through these plates will be deflected slightly towards the positive plate.

1.4.2 Electric Current

In all semiconductors and conductive materials, free electrons are present. These free electrons move at random in all directions within the structure in the absence of external pressure or voltage. If a certain amount of voltage is applied, then all the free electrons will move in one direction depending on the polarity of the applied voltage as shown in Figure 1.3.

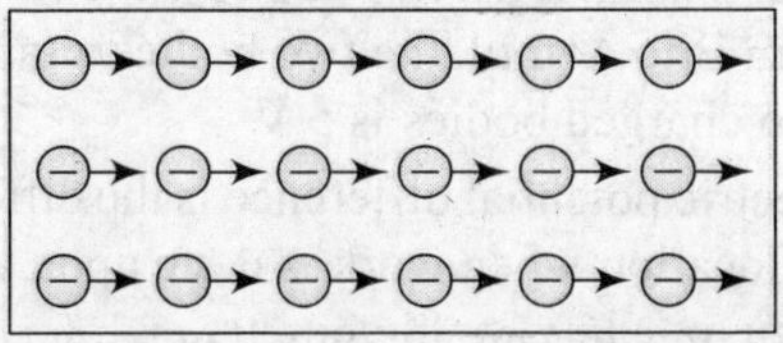

Figure 1.3 Movement of free electrons on application of voltage

This movement of electrons from one end of the material to the other end is known as *electric current,* which is denoted by I or i. It is the rate at which charge flows past a point on a circuit. The current in a circuit can be determined if the quantity of a charge Q passing through a cross section of a wire in a time t can be measured. Thus, current is simply the ratio of the quantity of charge and time.

When one coulomb of electric charge continuously passes a given point in a second, the electric current is said to be one ampere.

$$I = \frac{Q}{t} \tag{1.1}$$

where I is the current in amperes, Q is the charge in coulombs, and t is the time in seconds.

1.4.3 Electric Potential

At a point in space, the electric potential (also called the *electrostatic potential*) is the potential energy per unit of charge that is associated with a static (time-invariant) electric field. It is a scalar quantity, typically measured in volts.

1.4.4 Potential Difference

According to the structure of an atom, there are two types of charges—positive and negative. A force of attraction exists between these charges. A certain amount of energy is required to overcome that force and move the charge through a specific distance. All opposite charges possess a certain amount of potential energy because of the separation between them. Electric potential is the amount of electric potential energy per unit of charge that would be possessed by a charged object if placed within an electric field at a given location.

The difference in the potentials (i.e., power or energy) of two charged bodies is called *potential difference*.

Consider two charged bodies A and B having potentials of +5 V and +10 V, respectively, as shown in Figure 1.4.

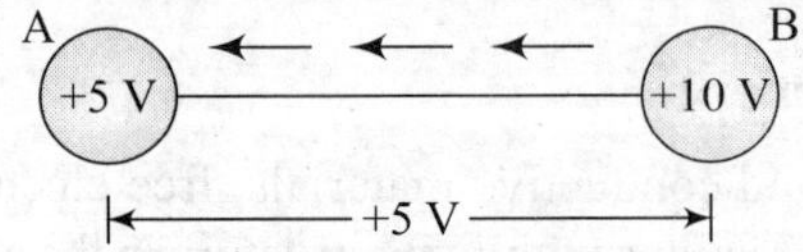

Figure 1.4 Two charged bodies

B is at higher potential than A. If the two bodies are joined through a conductor then the electrons would start flowing from B to A until the two bodies attain the same potential. The difference in the potentials of the two charged bodies is 5 V.

By definition, the electric potential difference is the difference in the electric potential between the final and the initial location when work is done upon a charge to change its potential energy.

Note: The current will flow in a circuit only if potential difference exists. If there is no potential difference then no current flows. It is similar to how water flows in a closed pipe driven by a mechanical pump. If there is no pressure difference between the two ends in the pipe, no water flows. Similarly, if there is no voltage difference no current flows.

1.4.5 Electromotive Force

Electromotive force (emf) represents the electric pressure or potential difference between two ends of a conductor that tends to create electron flow. Electromotive force can be developed in a battery or in a generator. The two interpretations of voltage can be given as follows:

(i) Electromotive force is a voltage associated with generated energy.

(ii) Potential difference is a voltage associated with dissipated energy.

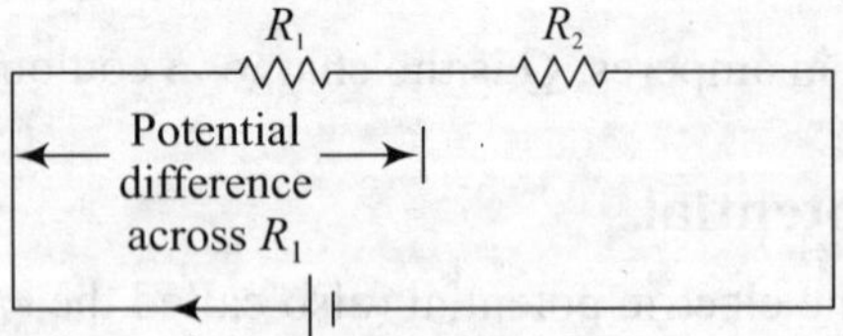

Figure 1.5 Interpretation of emf and potential difference

The general symbol for voltage is V. The symbol V may be used whenever it is unnecessary to distinguish between the voltages of generation and dissipation. The specific symbol for emf is E. The symbol E only refers to voltage of generation. The specific symbol for potential difference is V. This symbol is used to refer to the voltage of dissipation.

The unit for either type of voltage, emf or potential difference, is volt (V). The voltage is defined by the following statement. If between two points in a circuit 1 J of energy conversion is associated with the movement of 1 C of charge, a potential difference of 1 V exists between those two points. This definition gives rise to the following equation:

$$1\ \text{V} = 1\ \text{joule/1 coulomb}$$

$$V = \frac{dW}{dQ} \tag{1.2}$$

where V is the voltage (V), W the energy of work (J), and Q the charge (C).

1.5 WORK, POWER, AND ENERGY

1.5.1 Work

Electrical work refers to the work which is done on an electrically charged particle by an electric field. This form of energy is expressed on similar lines to the mechanical work that is done by a force. The basic equation for work done due to an electric field is

$$W = q\int E\,dr \tag{1.3}$$

where q is the charge of the particle, E the strength of the electric field, and dr the distance involved in the exertion of the electric field.

1.5.2 Power

Power is defined as the rate at which electrical energy is transferred by the circuit or as the rate of doing work. The symbol for power is P. The SI unit is watt. The unit of electric power is defined in terms of joule per second. One joule per second is the work done when one coulomb of electricity is moved through a potential difference of one volt in one second. As one coulomb per second is one ampere, one joule per second is one watt, and this is the product of E (volt) and I (current). The power in an electric circuit is

$$P = \frac{dW}{dt} \tag{1.4}$$

Equation (1.4) can be written as

$$\frac{dW}{dt} = \frac{dW}{dq} \times \frac{dq}{dt} \tag{1.5}$$

Therefore,

$$P = V \times I \tag{1.6}$$

Hence, power is the product of voltage and current.

1.5.3 Energy

Electrical energy is defined as the total amount of electrical work done in an electric circuit. The unit of energy is joules or watt-second.

Electrical energy, E = power (P) × time (t)

i.e.,

$$E = VIt \text{ joules or watt-second} \tag{1.7}$$

EXAMPLE 1.1 **For every 40 C of charge, energy available is 100 J. What is the voltage?**

Solution Voltage, $V = \dfrac{W}{Q} = \dfrac{100}{40} = 2.5\ \text{V}$

EXAMPLE 1.2 **What is the current flowing if 10 C of charge flows through a point in a conductor for 2 s?**

Solution Current, $I = \dfrac{Q}{t} = \dfrac{10}{2} = 5\ \text{A}$

EXAMPLE 1.3 **If energy consumed by load is 100 J in 10 s, what is the power in watts?**

Solution Power, $P = \dfrac{E}{t} = \dfrac{100}{10} = 10\ \text{A}$

EXAMPLE 1.4 **A heater element draws 20 A when connected to a 230-V line. How long does it take to consume 60 kJ (1 J = 1 W-s)?**

Solution Energy, $E = V \times I \times t$

Time taken to consume energy, $t = \dfrac{E}{VI} = \dfrac{60000}{230 \times 20} = 13.04\ \text{s}$

1.6 BASIC CIRCUIT ELEMENTS

The basic circuit elements are resistor, inductor, and capacitor. In this section, we will discuss the characteristics of these elements.

1.6.1 Resistor

As its name suggests, a resistor resists or opposes the flow of current through it. Resistance in a circuit is the parameter that dissipates energy. Resistance is the hindrance to the flow of charge. The flow of charge through wires is often compared to the flow of water through pipes. The resistance to the flow of charge in an electric circuit is analogous to the frictional affects between water and the pipe surfaces as well as the resistance offered by obstacles which are present in its path. It is this resistance which hinders the water flow and reduces both its flow rate and its *drift* speed. Like the resistance to water flow, the total amount of resistance to charge flow within a wire of an electric circuit is affected by some clearly identifiable variables.

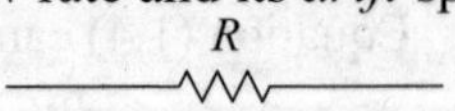

Figure 1.6 Symbol for resistance

The symbol for resistor is R as shown in Figure 1.6. The SI unit of resistance is ohm, symbol Ω (capital Greek letter omega). Typical examples of resistance are the heating elements in stoves and filament bulbs.

Factors affecting resistance

The resistance of a conductor depends upon

(i) the material of the conductor, i.e., resistivity (ρ) in ohm-m

(ii) the length (l) in meters

(iii) the cross-sectional area (a) in square meters
(iv) the temperature

The resistance of electrical materials is usually given per unit cross-sectional area and per unit length. This is called *specific resistance* or *resisitivity* of the material represented by symbol ρ.

$$R = \rho \frac{l}{a} \tag{1.8}$$

Temperature coefficient of resistance

Since the electrical resistance of a conductor such as a copper wire is dependent upon the collisional processes within the wire, the resistance could be expected to increase with temperature since there will be more collisions. The temperature dependence of conductors is linear to a great degree, but that of a semiconductor is exponential. An intuitive approach to temperature dependence leads one to expect a fractional change in resistance which is proportional to the temperature change,

$$\frac{\Delta R}{R_0} = \alpha\,(\Delta T)$$

or expressed in terms of the resistance at some standard temperature,

$$\frac{R - R_0}{R_0} = \alpha\,(T - T_0) \quad \text{or} \quad R = R_0\,[1 + \alpha\,(T - T_0)]$$

where R is the conductor resistance at temperature T, R_0 is the conductor resistance at reference temperature T_0 (either room temperature or 0°C), and α is the temperature coefficient of resistance having the dimension of an inverse temperature (1/K or K^{-1}).

Conductance and conductivity

The reciprocal of resistance is called conductance. It is represented by letter "G" and its unit is mho (℧) or siemens (S).

$$G = \frac{1}{R} \quad \text{or} \quad R = \frac{1}{G} \tag{1.9}$$

The reciprocal of resistivity is called conductivity and is represented by the symbol g.

$$g = \frac{1}{\rho} \tag{1.10}$$

1.6.2 Inductor

A wire of certain length, when twisted into a coil becomes a basic inductor. When the current is passed through an inductor an electromagnetic field is formed.

A change in the magnitude of the current changes the electromagnetic field, i.e., increase in current increases the field and decrease in current reduces the field. Therefore, a change in the current produces a change in the electromagnetic field, which induces a voltage across the coil according to Faraday's law of electromagnetic induction. The practical unit of inductance (L) is henry or H. Figure 1.7 shows the depiction of inductance in electrical diagrams.

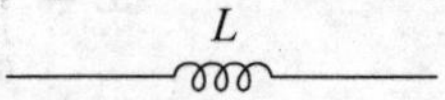

Figure 1.7 Symbol for inductance

The current–voltage relation in an inductor is given by the following equation:

$$V = L\frac{di}{dt} \tag{1.11}$$

where V is the voltage across inductor in volts, i is the current through inductor in amperes, and L is the inductance of the coil in henry.

Assuming that current i is flowing through the coil, the energy stored in the time interval is given by

$$E = \int_0^t vi\,dt \tag{1.12}$$

$$= \int_0^t \left(L\frac{di}{dt}\right) i\,dt$$

$$= \int_0^t L\,i\,dt$$

$$= L\int_0^t i\,dt = L\left.\frac{i^2}{2}\right|_0^t = \frac{1}{2}Li^2 \text{ (assuming zero initial current)}$$

$$\text{Energy stored in the inductor} = \frac{1}{2}Li^2 \text{ joules} \tag{1.13}$$

1.6.3 Capacitor

A *capacitor* is a device that is capable of storing charge. It consists of two conducting surfaces separated by an insulating material. The conducting surface is called plates of capacitor and the insulating material is called dielectric, for instance, air and mica. The symbol for capacitance is C (Figure 1.8) and the SI unit is farad or F.

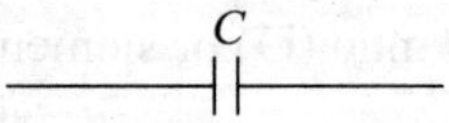

Figure 1.8 Symbol for capacitor

A capacitor is used to store electrical energy in dielectric.

$$C = \frac{Q}{V} \tag{1.14}$$

where C is the capacitance in farads, Q is the charge in coulombs, and V is the voltage in volts.

The capacitance opposes sudden changes in circuit voltages by the releasing part of the stored energy.

$$i = C\frac{dV}{dt} \tag{1.15}$$

The energy stored in a capacitor is given by

$$E = \int_0^t vi\,dt \tag{1.16}$$

$$= \int_0^t vC\frac{dV}{dt}\,dt$$

$$= \int_0^t vC\,dV$$

$$= C\int_0^t v\,dV = \frac{1}{2}CV^2 \text{ (assuming zero initial voltage across the capacitor)}$$

Therefore, the energy stored,

$$E = \frac{1}{2}CV^2 \text{ joules} \qquad (1.17)$$

EXAMPLE 1.5 **A 20-Ω resistor is connected to a 100-V voltage source. What is the power consumed by the resistor?**

Solution Power, $P = \dfrac{V^2}{R} = \dfrac{100^2}{20} = 500$ W

EXAMPLE 1.6 **How much energy is stored by a 100-mH inductance with a current of 1 A?**

Solution Energy stored in inductor, $E = \dfrac{1}{2}LI^2$ joules $= \dfrac{1}{2} \times 100 \times 10^{-3} \times 1 = 0.05$ J

EXAMPLE 1.7 **If a resistor has 6 V across it and 5 mA flowing through it, what is the power?**

Solution Power, $P = V \times I = 6 \times 5 \times 10^{-3} = 0.03$ W

EXAMPLE 1.8 **Calculate the energy stored by a 0.05-μF capacitor with a voltage of 200 V.**

Solution Energy stored in the capacitor $= \dfrac{1}{2}CV^2 = \dfrac{1}{2} \times 0.05 \times 10^{-6} \times (200)^2 = 1$ mJ

EXAMPLE 1.9 **The rate of change of current in a 4-H inductor is 2 A/s. Find the voltage across the inductor.**

Solution Voltage across the inductor, $V = L\dfrac{di}{dt} = 4 \times 2 = 8$ V

EXAMPLE 1.10 **What is the charge when the capacitance is 0.001 μF and the voltage is 2 kV?**

Solution Charge, $Q = CV = 0.001\ \mu\text{F} \times 2\ \text{kV} = 2\ \mu\text{F}$

EXAMPLE 1.11 **Calculate the voltage of a battery that uses 1000 J of energy to move 50 C of charge through a conductor.**

Solution Voltage, $V = \dfrac{E}{Q} = \dfrac{1000}{50} = 20$ V

EXAMPLE 1.12 **Find the power delivered to an element at $t = 4$ ms if the current entering its positive terminal is $i = 10 \sin 100\pi t$ and the voltage is $V = 5i$?**

Solution Power delivered, $P = VI = 5i \times i = 5i^2$

$$= 5(10 \sin 100\pi t)^2$$

At $t = 4$ ms,

$$P = 5\,[10 \sin (100\pi \times 4 \times 10^{-3})]^2 = 452.25 \text{ W}$$

1.7 NETWORK TERMINOLOGY

By using a network shown in Figure 1.9, the following terms are defined:

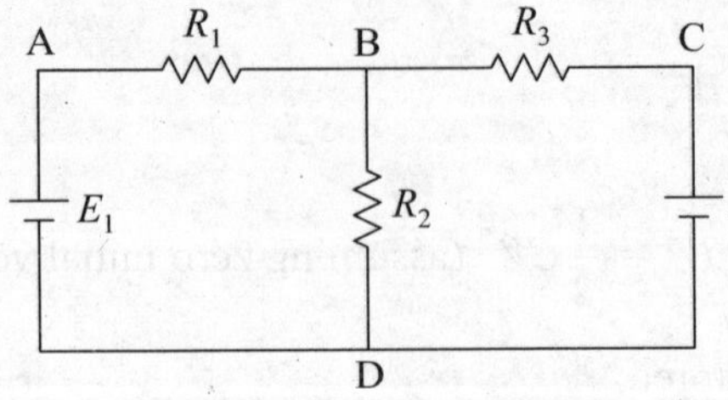

Figure 1.9 Electrical circuit

Node A node of a network is an equipotential surface at which two or more elements are joined. In Figure 1.9, the circuit elements R_1 and E_1 are joined at point A, hence A is the node. Similarly, points B, C, and D are nodes.

Junction A junction is that point in a network where three or more circuit elements are joined. In Figure 1.9, there are only two junction points B and D.

Branch A branch is that part of a network which lies between two junction points. In Figure 1.9, there are three branches BAD, BCD, and BD.

Loop A loop is any closed path of a network. In Figure 1.9, ABDA, BCDB, and ABCDA are loops.

Mesh A mesh is the most elementary form of a loop and cannot be further divided into other loops. Both loops ABDA and BCDB qualify as meshes because they cannot be further divided into other loops. However, loop ABCDA cannot be called as mesh, because it encloses two loops ABDA and BCDB.

1.8 CLASSIFICATION OF ELECTRICAL NETWORKS

Electrical networks are classified as follows:

(i) Linear and nonlinear
(ii) Unilateral and bilateral
(iii) Active and passive
(iv) Lumped and distributed

1.8.1 Linear and Nonlinear

A linear network is one whose values are constant, i.e., they do not change with voltage, time, or temperature. Ohm's law can be applied to linear networks.

A nonlinear network is one whose values are not constant, i.e., they change with voltage and current. Ohm's law is not applicable to nonlinear networks. One of the examples is the circuit which consists of a diode or transistor.

1.8.2 Unilateral and Bilateral

Unilateral elements offer a low value of impedance for the flow of current in one direction and offer a high value of impedance for the flow of current in another direction, i.e., the magnitude of current flowing through the element changes with the direction of current flow. For example, diodes and semiconductor devices.

A network which consists of unilateral elements is called a *unilateral network*. Bilateral elements offer the same value of impedance for the flow of current in either directions, i.e., the magnitude of current flowing through the element does not change with its direction. Such networks are called *bilateral networks*, for example, resistors, inductors, and capacitors.

1.8.3 Active and Passive

Voltage and current sources are active elements. They are capable of delivering energy to some external devices. A circuit which consists of at least one source of energy is called an *active network*.

Passive elements are capable only of receiving energy, for example, resistors, inductors, and capacitors. Inductors and capacitors are also capable of storing some finite amount of energy. A circuit which contains no energy source is called a *passive network*.

1.8.4 Lumped and Distributed

In a *lumped network*, all the network elements are physically separable. Most of the electric networks are lumped in nature as they comprise resistors, inductors, and capacitors.

A network in which the network elements cannot be separated is called a *distributed network*. The best example of such a network is a transmission line. In this, resistance, inductance, and capacitance are distributed all along its length. These cannot be separated anywhere in the circuit.

1.9 ENERGY SOURCES

The energy sources are divided into two categories: (i) independent sources and (ii) dependent sources.

1.9.1 Independent Sources

There are two types of independent sources. One is voltage source and the other is current source.

Independent voltage source

An independent voltage source is an energy source that pushes electrons with a constant force or pressure regardless of the load presented to it. In other words, a 1-V source will maintain 1 V between its terminals regardless of the load. The symbol for the voltage source and its *V–I* curve is shown in Figure 1.10. The voltage source symbol must be identified with its positive or negative terminal marked and a value or variable for its voltage.

The independent voltage source will supply whatever current is necessary to maintain a constant terminal voltage.

A circle is used to indicate the independent sources as shown in Figure 1.10(a). The strength of voltage or current is not changed by any variation in the connected network. The *V–I* characteristics are shown in Figure 1.10(b).

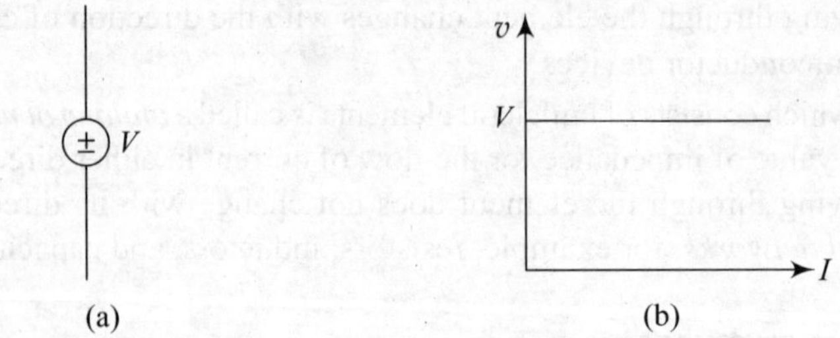

Figure 1.10 (a) Independent voltage source representation and (b) *V–I* characteristics

Independent current source

The source which gives constant current at its terminals irrespective of the voltage appearing across its terminals is called *ideal current source*. The symbol for ideal current source is shown in Figure 1.11(a) and its *V–I* characteristics are shown in Figure 1.11(b).

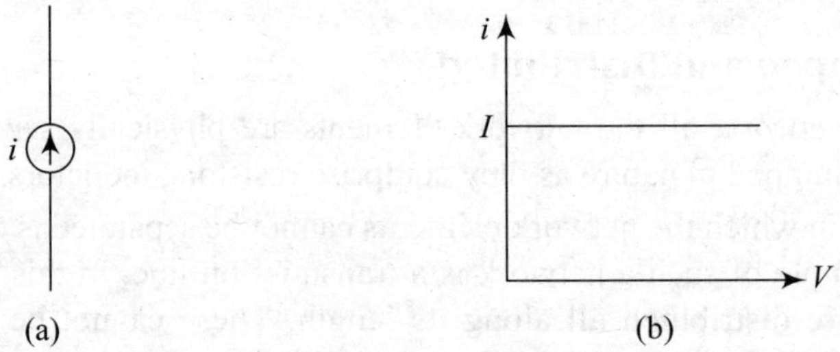

Figure 1.11 (a) Symbol for independent current source and (b) *V–I* characteristics

1.9.2 Dependent Sources

The source whose output voltage or current is dependent on the voltage or current in another part of a circuit is called a *dependent source*. It is classified as follows:

(i) Voltage-controlled voltage source
(ii) Voltage-controlled current source
(iii) Current-controlled voltage source
(iv) Current-controlled current source

A diamond shape is used to indicate the dependent sources as shown in Figure 1.12. The strength of voltage or current changes by any variation in the connected network.

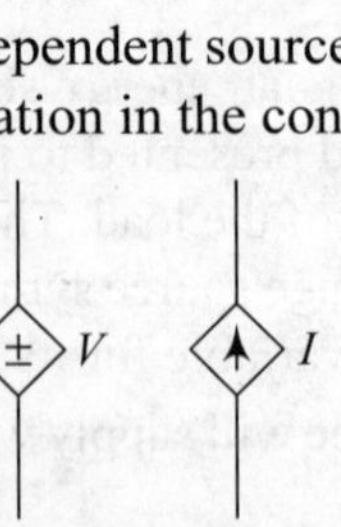

Figure 1.12 Representation of dependent source

Voltage-controlled voltage source

It is a voltage source dependent on voltage elsewhere in the circuit. Its diagram is shown in Figure 1.13.

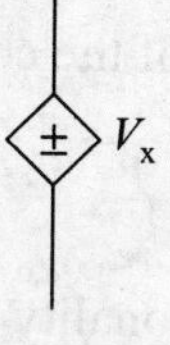

Figure 1.13 Voltage-controlled voltage source

Voltage-controlled current source

It is a current source dependent on voltage elsewhere in the circuit. Its diagram is shown in Figure 1.14.

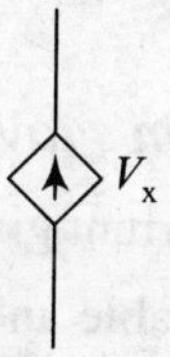

Figure 1.14 Voltage-controlled current source

Current-controlled voltage source

It is a voltage source dependent on the current elsewhere in the circuit. Its diagram is shown in Figure 1.15.

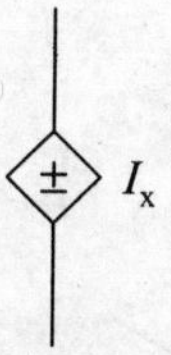

Figure 1.15 Current-controlled voltage source

Current-controlled current source

It is a current source dependent on the current elsewhere in the circuit. Its diagram is shown in Figure 1.16.

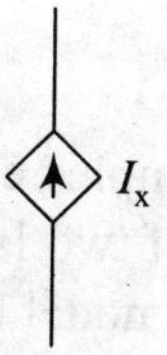

Figure 1.16 Current-controlled current source

1.10 BASIC CIRCUIT LAWS

1.10.1 Ohm's Law

Ohm's law states that the electric current in a conductor between two points is directly proportional to the voltage across the two points of the conductor. That means

$$i \propto V$$

$$i = GV \tag{1.18}$$

where G is the constant of proportionality and is called the conductance of the element. The conductance of an element suggests its ability to conduct electric current. The above equation can be written as

$$V = RI \tag{1.19}$$

where $R = 1/G$ is the resistance of the conductor. This law is subject to temperature remaining constant as R tends to vary with temperature.

Graphical representation of Ohm's law

Plotting values of voltage against current gives a graph of conductance and resistance.

If voltage is the independent variable and current is the dependent variable, the slope of the curve will represent the conductance.

From Figure 1.17(a), $I = VG$

$\therefore \quad I \propto G$

If current is the independent variable and voltage is the dependent variable, the slope of the curve will represent resistance.

From Figure 1.17(b), $V = IR$

$\therefore \quad V \propto R$

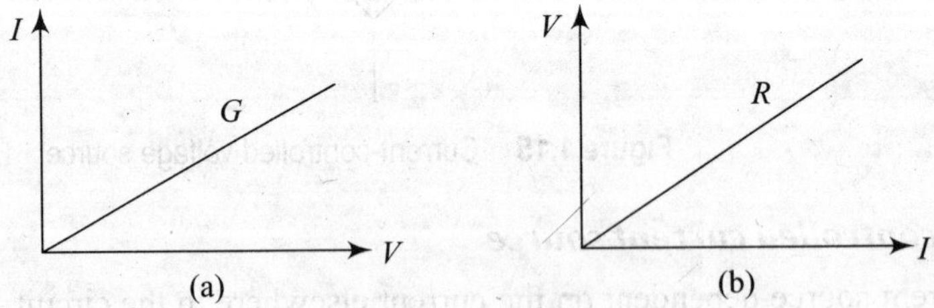

Figure 1.17 Ohm's law

1.10.2 Kirchhoff's Laws

In order to analyze comparatively complex electrical networks comprising more than one voltage source, it is necessary to make use of two laws known as Kirchhoff's laws. These two laws are (i) Kirchhoff's current law (KCL) or nodal law and (ii) Kirchhoff's voltage law (KVL) or mesh law.

KCL or nodal law

The nodal law states that in any electrical network, the algebraic sum of the currents meeting at a point or junction is zero.

It simply means that the total current leaving a junction is equal to the total current entering that junction.

Consider Figure 1.18. According to KCL

Incoming current = Outgoing current

$$i_1 + i_3 + i_4 = i_2 + i_5$$

$$i_1 + i_3 + i_4 - i_2 - i_5 = 0 \quad (1.20)$$

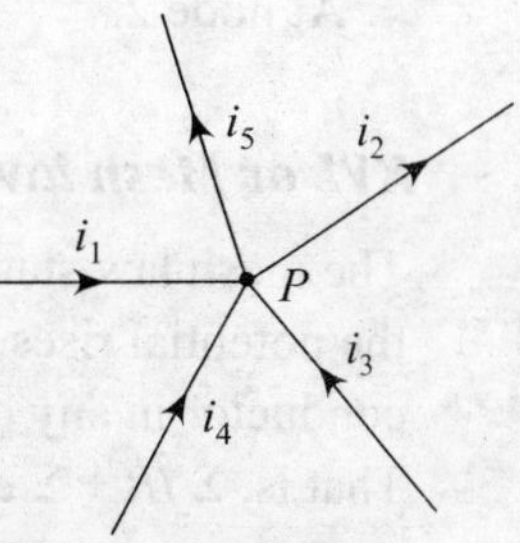

Figure 1.18 Illustration of KCL

Procedure to write node equation The following procedure is followed to write the equation for a node. The term "local node" is applied to the node for which the equation is being written. The term "remote node" implies a node connected to the local node by a resistance.

(i) Place element currents on the left side of the equation.

(ii) Divide the local node voltage (with respect to the reference) by the sum of the resistances connected to the node. Show the product as a positive quantity. Ignore elements in series with current sources, as they cannot affect circuit currents.

(iii) Divide each remote node voltage by the resistance of the element connecting it to the local node. Show this as a negative quantity.

(iv) Place source currents on the right side of the equation.

(v) Show currents entering the local node as negative quantities.

(vi) Show currents leaving the local node as positive quantities.

(vii) If there are no source currents at the local node, show as zero.

For example, consider the circuit shown in Figure 1.19. The node equations for the diagram are written as follows:

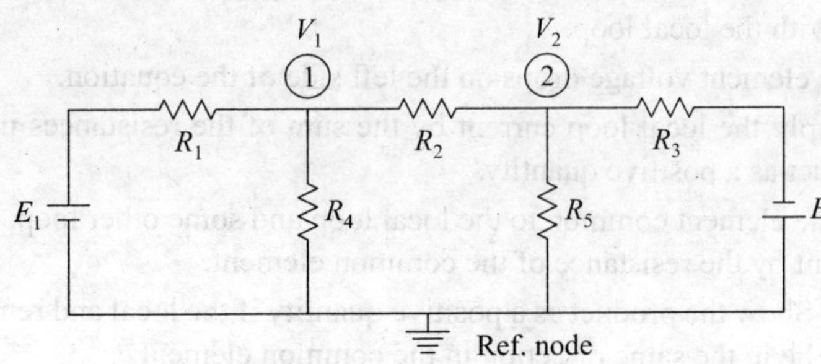

Figure 1.19 Electrical circuit

At node 1, $\frac{V_1 - E_1}{R_1} + \frac{V_1}{R_4} + \frac{V_1 - V_2}{R_2} = 0$

At node 2, $$\frac{V_2+E_2}{R_3}+\frac{V_2-V_1}{R_2}+\frac{V_2}{R_5}=0$$

KVL or Mesh law

The mesh law states that in a closed circuit the sum of the potential drops is equal to the sum of the potential rises, or the algebraic sum of the product of currents and resistances in each of the conductor in any closed path in a network plus the algebraic sum of the emf in that path is zero. That is, $\Sigma IR + \Sigma$ emf $= 0$ (in a closed path).

Consider Figure 1.20. If we start from a particular junction and round the mesh till we come back to the starting point, then we must be at the same potential with which we started.

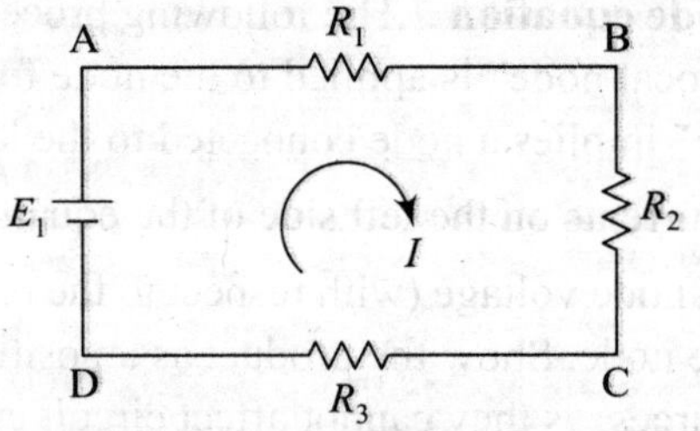

Figure 1.20 Illustration of KVL

Consider loop ABCDA, the loop equation is

$$IR_1 + IR_2 + IR_3 - E = 0$$

or

$$IR_1 + IR_2 + IR_3 = E$$

Potential drop = Potential rise

Procedure to write loop equation The following procedure is followed to write the equation for a loop. "Local" refers the loop being considered and "remote" refers to a loop which shares an element with the local loop.

(i) Place element voltage drops on the left side of the equation.

(ii) Multiply the local loop current by the sum of the resistances in that loop and show the product as a positive quantity.

(iii) For the element common to the local loop and some other loop, multiply the remote loop current by the resistance of the common element.

  (a) Show the product as a positive quantity if the local and remote loop currents tend to be in the same direction in the common element.

  (b) Show the product as a negative quantity if the current directions are opposite.

(iv) Place the source emf on the right side of the equation.

(v) Show emfs supporting the direction of the local loop current as positive quantities.

(vi) Show opposing emfs as negative quantities.

(vii) If there are no voltage sources in the loop, show as zero.

Consider the example shown in Figure 1.21.

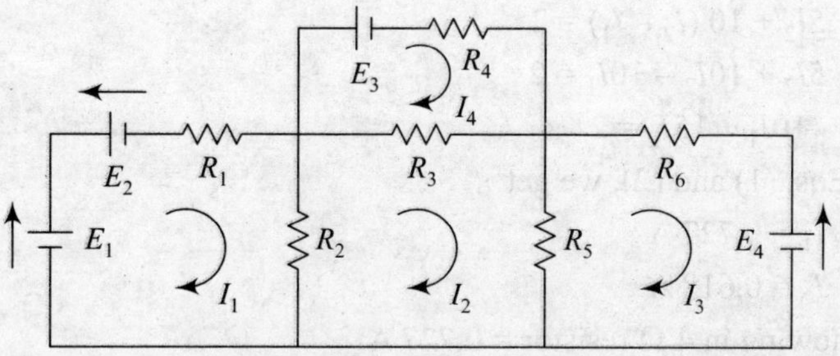

Figure 1.21 Electrical circuit

The loop equations are as follows:

Loop 1: $(R_1 + R_2) I_1 - R_2 I_2 = E_1 - E_2$

Loop 2: $(R_2 + R_3 + R_5) I_2 - R_2 I_1 - R_3 I_4 - R_5 I_3 = 0$

Loop 3: $(R_5 + R_6) I_3 - R_5 I_2 = - E_4$

Loop 4: $(R_3 + R_4) I_4 - R_3 I_2 = - E_3$

EXAMPLE 1.13 **In the circuit shown in Figure 1.22, find the magnitude and direction of the current in each resistor. Internal resistance of the battery is negligible.**

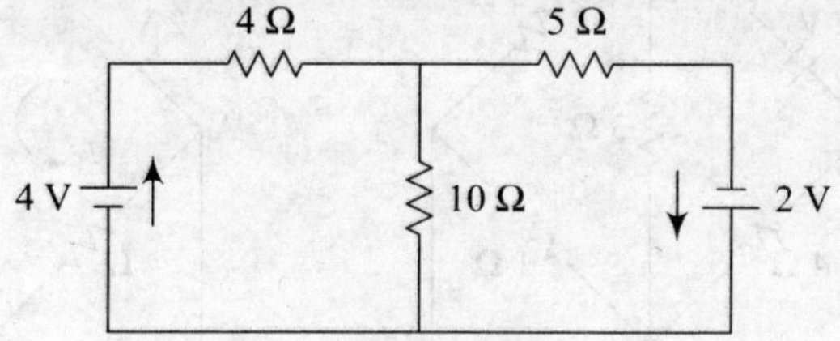

Figure 1.22

Solution Assuming current directions, the circuit is redrawn as shown in Figure 1.23.

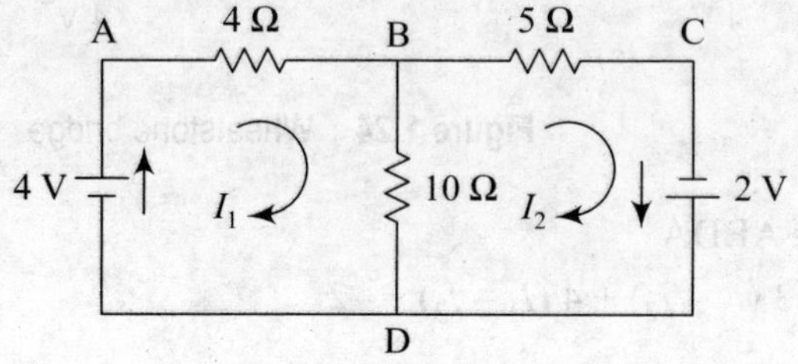

Figure 1.23

Consider loop ABDA,

$$4I_1 + 10 (I_1 - I_2) = 4$$

$$4I_1 + 10I_1 - 10I_2 = 4$$

$$14I_1 - 10I_2 = 4 \qquad (1)$$

For loop BCDB,

$$5I_2 + 10\,(I_2 - I_1) = 2$$
$$5I_2 + 10I_2 - 10I_1 = 2$$
$$-10I_1 + 15I_2 = 2 \qquad (2)$$

Solving Eqs. (1) and (2), we get

$$I_1 = 0.727 \text{ A}$$
$$I_2 = 0.618 \text{ A}$$

Current flowing in 4 Ω resistor = 0.727 A

Current flowing in 5 Ω resistor = 0.618 A

Current flowing in 10 Ω resistor = $I_1 - I_2$ (because $I_1 > I_2$)

= 0.727 – 0.618 = 0.109 A

The direction of current is B to D.

EXAMPLE 1.14 In the Wheatstone bridge network shown in Figure 1.24, find the current in each resistor.

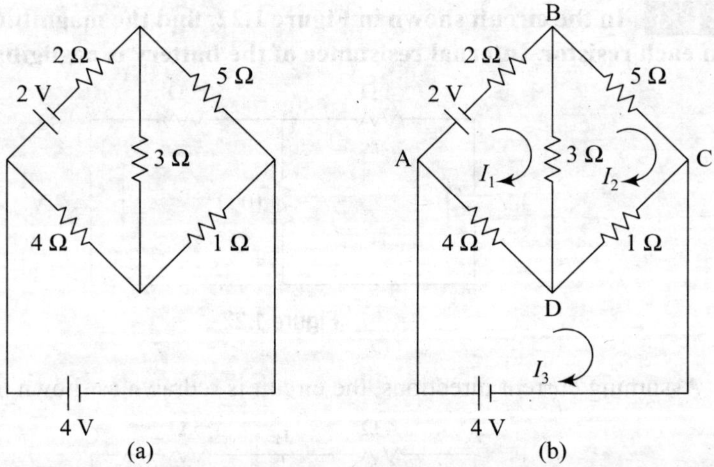

Figure 1.24 Wheatstone bridge

Solution Loop 1: ABDA

$$2I_1 + 3\,(I_1 - I_2) + 4\,(I_1 - I_3) = -2 \qquad (1)$$

Loop 2: BCDB

$$5I_2 + 1\,(I_2 - I_3) + 3\,(I_2 - I_1) = 0 \qquad (2)$$

Loop 3: ADCA

$$4\,(I_3 - I_1) + 1\,(I_3 - I_2) = 4 \qquad (3)$$

Solving Eqs. (1), (2), and (3), we get

$$I_1 = 0.3715 \text{ A};\ I_2 = 0.2514 \text{ A};\ I_3 = 1.147 \text{ A}$$

EXAMPLE 1.15 **Determine the current "x" in the 4-Ω resistance of the circuit shown in Figure 1.25.**

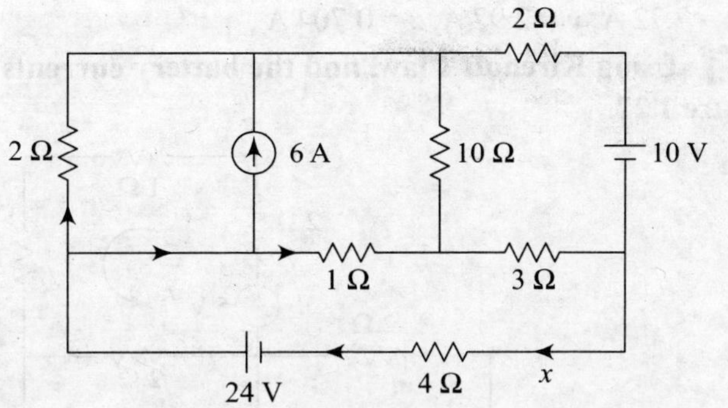

Figure 1.25

Solution We can use current division method to solve the given network. The same circuit can also be solved by using the loop analysis.

Indicate the current directions in the network taking care that the current division should start with some variable and should end with the same variable.

We started with "x" variable at 24 V source and we ended up with x variable at same 24 V source.

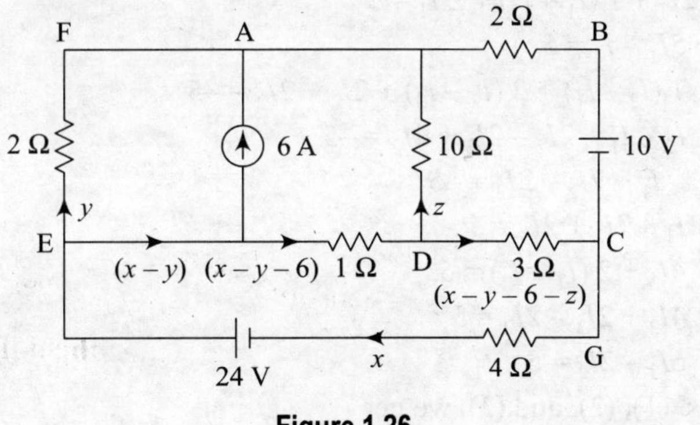

Figure 1.26

Loop 1 EFADE:

$$-2y + 10(z) + 1(x - y - 6) = 0$$

$$-2y + 10z + x - y - 6 = 0$$

$$x - 3y + 10z = 6 \quad (1)$$

Loop 2 ABCDA:

$$2(y + 6 + z) - 3(x - y - 6 - z) + 10z = -10$$

$$-3x + 5y + 15z = 40 \quad (2)$$

Loop 3 EDCGE:

$$1(x - y - 6) + 3(x - y - 6 - z) + 4x = 24$$

$$8x - 4y - 3z = 48 \qquad (3)$$

Solving Eqs. (1), (2), and (3), we get

$$x = 7.72 \text{ A}, y = 2.92 \text{ A}, z = 0.704 \text{ A}$$

EXAMPLE 1.16 Using Kirchoff's law, find the battery currents in all sources of the circuit shown in Figure 1.27.

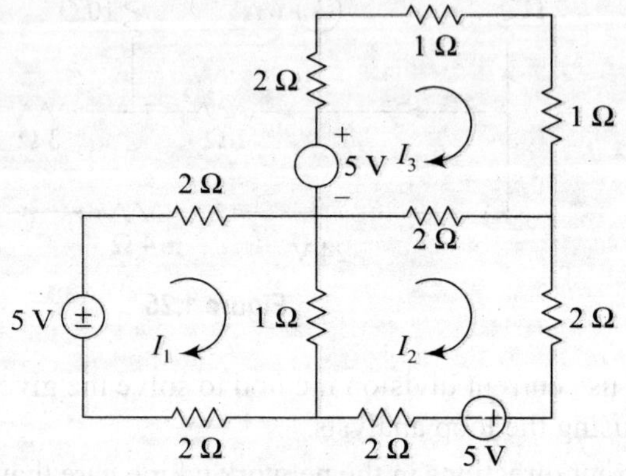

Figure 1.27

Solution

For loop 1: $2I_1 + 1\,(I_1 - I_2) + 2\,I_1 = 5$

$$5I_1 - I_2 = 5 \qquad (1)$$

For loop 2: $1\,(I_2 - I_1) + 2\,(I_2 - I_3) + 2I_2 + 2I_2 = -5$

$$I_2 - I_1 + 2I_2 - 2I_3 + 4I_2 = -5$$

$$-I_1 + 7I_2 - 2I_3 = -5$$

$$I_1 - 7I_2 + 2I_3 = 5 \qquad (2)$$

For loop 3: $4I_3 + 2\,(I_3 - I_2) = 5$

$$4I_3 + 2I_3 - 2I_2 = 5$$

$$6I_3 - 2I_2 = 5 \qquad (3)$$

Solving Eqs. (1), (2), and (3), we get

$$I_1 = 0.92\text{A}, I_2 = -0.38 \text{ A}, I_3 = 0.70 \text{ A}$$

EXAMPLE 1.17 Solve the network by using KCL and find the power supplied by the source for the circuit shown in Figure 1.28.

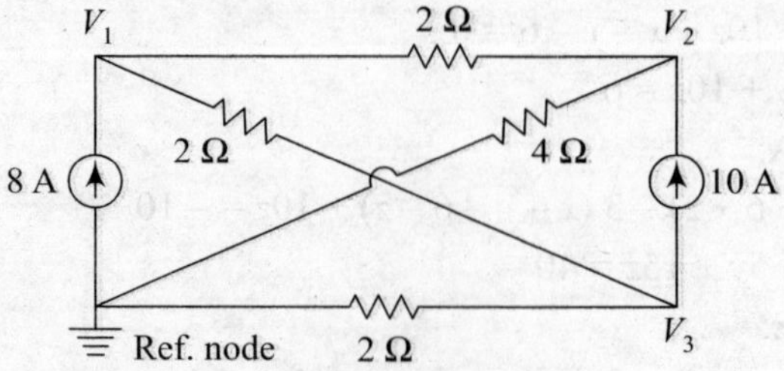

Figure 1.28

Solution At node 1,

$$\frac{V_1 - V_2}{2} + \frac{V_1 - V_3}{2} - 8 = 0$$

$$\frac{V_1}{2} - \frac{V_2}{2} + \frac{V_1}{2} - \frac{V_3}{2} = 8$$

$$V_1\left(\frac{1}{2} + \frac{1}{2}\right) - \frac{V_2}{2} - \frac{V_3}{2} = 8$$

$$V_1 - \frac{V_2}{2} - \frac{V_3}{2} = 8 \tag{1}$$

At node 2,

$$\frac{V_2 - V_1}{2} + \frac{V_2}{4} - 10 = 0$$

$$-\frac{V_1}{2} + V_2\left(\frac{1}{4} + \frac{1}{2}\right) = 10 \tag{2}$$

At node 3,

$$\frac{V_3 - V_1}{2} + \frac{V_3}{2} + 10 = 0$$

$$-\frac{V_1}{2} + V_3\left(\frac{1}{2} + \frac{1}{2}\right) = -10 \tag{3}$$

Solving Eqs. (1), (2), and (3), we get

$V_1 = 23.2$ V, $V_2 = 28.8$ V, $V_3 = 1.6$ V

Power supplied by 8-A source, connected in between node V_1 and the reference node,

$$P_8 = V_1 \times 8 = 23.2 \times 8 = 185.6 \text{ W}$$

Power supplied by 10-A source, connected in between nodes V_2 and V_3,

$$P_{10} = (V_2 - V_3)\,10 = (28.8 - 1.6) \times 10$$
$$= 272 \text{ W}$$

EXAMPLE 1.18 **Find the value of the dependent source for the network given in Figure 1.29.**

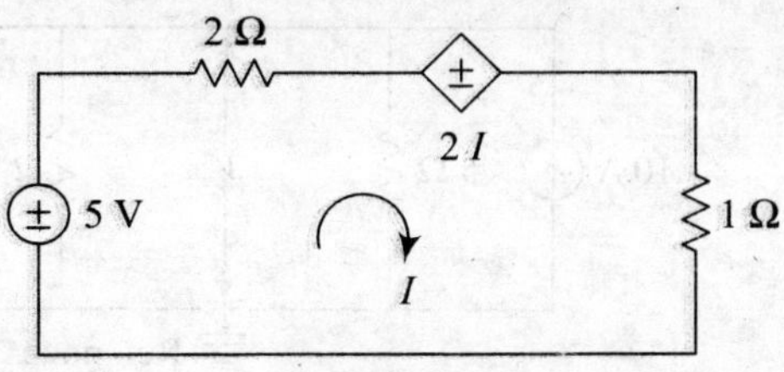

Figure 1.29

Solution In the given circuit, there is one independent 5-V source and one current-dependent voltage source. This dependent voltage source is dependent upon the circuit current I.

By writing the loop equation for the above network,

$$2I - 2I + I = 5$$

$$I = 5$$

The value of dependent source, i.e., $2I = 2 \times 5 = 10$ A.

EXAMPLE 1.19 **Determine the power absorbed by each element for the circuit shown in Figure 1.30.**

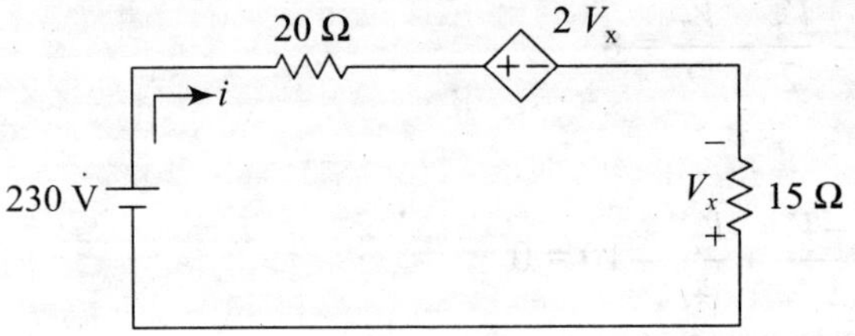

Figure 1.30

Solution Applying KVL,

$$-230 + 20i + 2V_x - V_x = 0$$

$$V_x + 20i = 230 \qquad (1)$$

Applying Ohm's law at the 15-Ω resistor, we get

$$V_x = -15i \qquad (2)$$

Substituting Eq. (2) into Eq. (1), we get

$$-15i + 20i = 230$$

$$i = 46 \text{ A}$$

Power supplied by 230 V battery, $P = 230 \times 46 = 10.58$ kW

Power supplied by dependent source, $P = 2 \times 690 \times 46 = 63.48$ kW

Power dissipated in the two resistors together, $P = 46^2(20 + 15) = 74.06$ kW

Hence, power supplied is equal to power consumed.

EXAMPLE 1.20 **Obtain the numerical value of the voltage-dependent current source given in the circuit shown in Figure 1.31.**

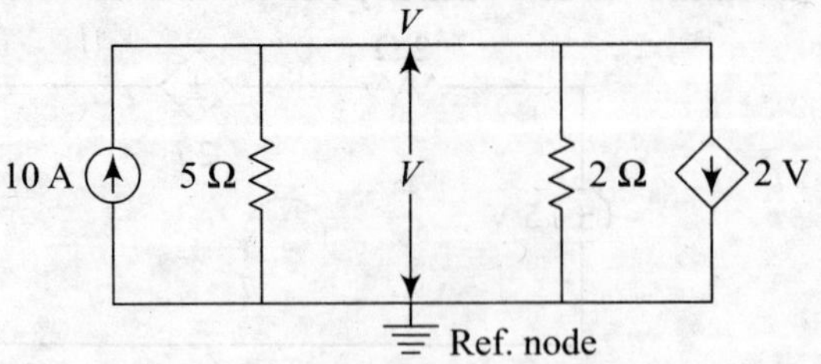

Figure 1.31

Solution For the given circuit, there are two nodes: V and the reference node.

While writing the nodal equation, consider the voltage-dependent current source as normal current source and apply normal procedure to write the equation.

Nodal equation is

$$\frac{V}{5}+\frac{V}{2}-10+2V=0$$

$$V\left(\frac{1}{5}+\frac{1}{2}+2\right)=10$$

$$V=\frac{10}{2.7}=3.7\text{ V}$$

The value of voltage-dependent current source is

$$2\text{ V}=2\times 3.7=7.4\text{ A}$$

EXAMPLE 1.21 **Calculate the value of resistance *R* for the following cases:**

(i) A voltmeter *V* of 2000-Ω resistance connected across *R* reads 200 V, while the total current supplied to *V* and *R* is 0.5 A.

(ii) A voltage of 10 V is applied to *R* in series with an ammeter A of 0.1 Ω resistance, while A reads 50 A.

Solution

(i) $I_1=\frac{200}{2000}=0.1\text{ A}$

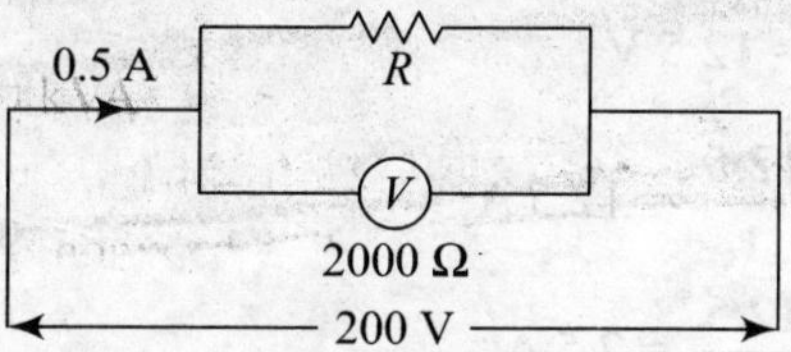

Figure 1.32

$$I_2=0.5-0.1=0.4\text{ A}$$

$$R=\frac{200}{0.4}=500\ \Omega$$

(ii) $R+0.1=\frac{10}{50}=0.2$

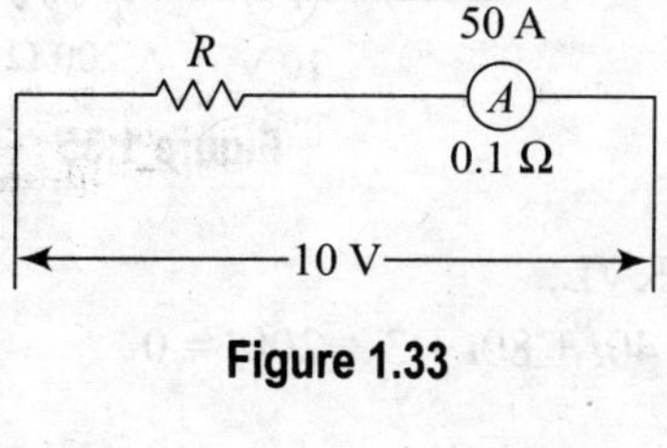

Figure 1.33

$$R=0.2-0.1=0.1\ \Omega$$

EXAMPLE 1.22 **Find i_1 and i_2 in the circuit shown in Figure 1.34.**

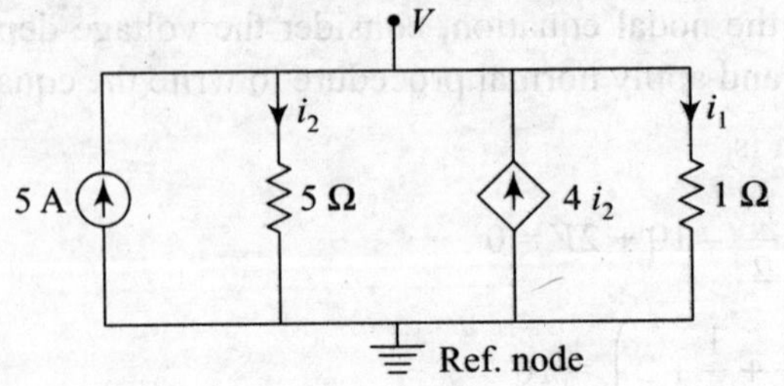

Figure 1.34

Solution Given that the i_1 and i_2 directions are flowing from node V to reference node.

By writing the nodal equation,

$$\frac{V}{5}+\frac{V}{1}-5-4i_2=0$$

where $i_2 = V/5$, i.e., i_2 is an outgoing source, so consider positive value substituting the i_2 value in the above equation.

$$V\left(\frac{1}{5}+1\right)-4\times\frac{V}{5}=5$$

$$V\left(1.2-\frac{4}{5}\right)=5$$

$$V=\frac{5}{0.4}=12.5\text{ V}$$

$$i_1=\frac{V}{1}=\frac{12.5}{1}=12.5\text{ A}$$

$$i_2=\frac{V}{5}=\frac{12.5}{5}=2.5\text{ A}$$

EXAMPLE 1.23 **For the circuit shown in Figure 1.35, determine the current *i*.**

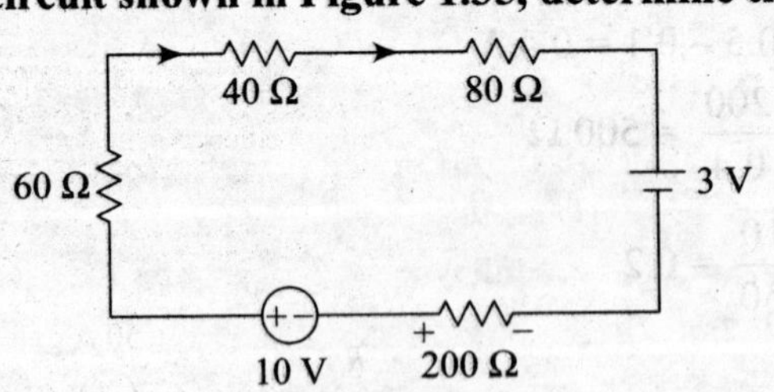

Figure 1.35

Solution Applying the KVL,

$$-10+60i+40i+80i+3+200i=0$$

$$380i=7$$

$$i=18.42\text{ mA}$$

EXAMPLE 1.24 **Determine the number of branches and nodes for the circuit given in Figure 1.36.**

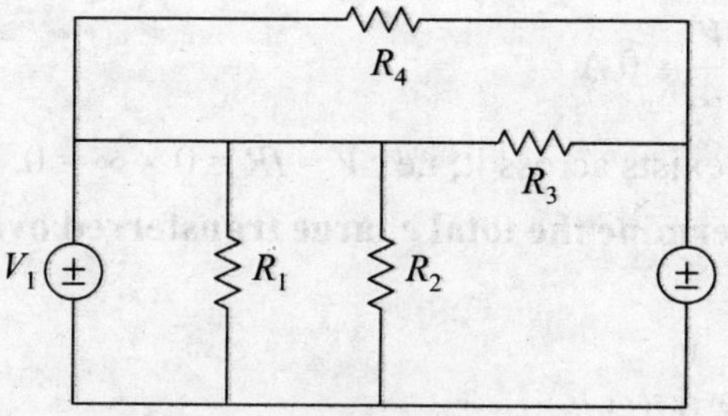

Figure 1.36

Solution The number of branches in the given circuit is 6 and the number of nodes is 3.

1.11 SHORT CIRCUIT AND OPEN CIRCUIT

1.11.1 Short Circuit

When two points are connected together by a metallic wire as shown in Figure 1.37, they are said to be short circuited. The characteristics of a short circuit are as follows:

(i) It has zero resistance.

(ii) No voltage exists across it, i.e., $V = IR = 1 \times 0 = 0$ V.

(iii) Current through it is very high (theoretically, it is infinite), i.e., $I = V/R = V/0 = \infty$.

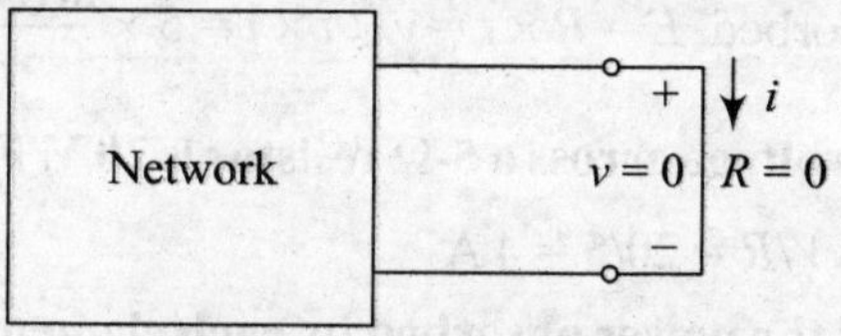

Figure 1.37 Short circuit

1.11.2 Open Circuit

Two points are said to be open circuited, when there is no continuity in the circuit or when there is a break in the circuit as shown in Figure 1.38. The characteristics of an open circuit are as follows:

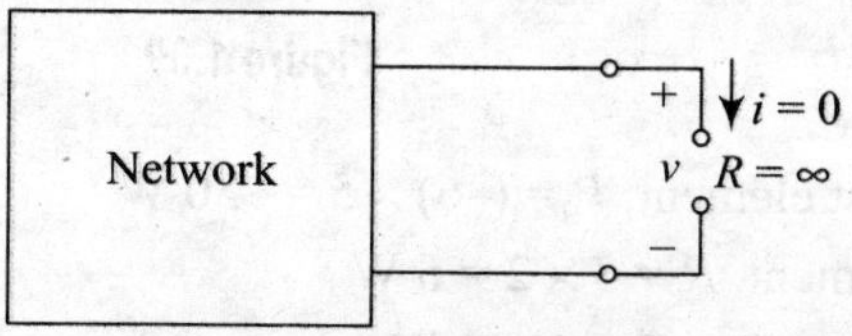

Figure 1.38 Open circuit

(i) It has infinite resistance.

(ii) No flow of current between the points, i.e.,

$$I = \frac{V}{R} = \frac{V}{\infty} = 0 \text{ A}$$

(iii) Finite voltage exists across it, i.e., $V = IR = 0 \times \infty = 0$.

EXAMPLE 1.25 **Determine the total charge transferred over the time interval of $0 \le t \le 10$ s when $i(t) = 0.5t$ A.**

Solution Charge $q = \int_0^{10} i(t)dt$

$$= \int_0^{10} 0.5\, t dt = 0.5 \left.\frac{t^2}{2}\right|_0^{10} = 25 \text{ C}$$

EXAMPLE 1.26 **How much energy does a 10-hp motor deliver in 30 min? Assume that 1 hp = 746 W.**

Solution $P = 10 \text{ hp} = 10 \times 746 = 7460$ W

$T = 30$ min

Energy $= P \times T = 7460 \times 30 \times 60 = 13.4$ MW-s

EXAMPLE 1.27 **Current entering the positive terminal of a device is $i(t) = 3e^{-2t}$ amperes and the voltage across the device is $v(t) = 5\ (di/dt)$ volts. Calculate the energy absorbed in 3 s.**

Solution Energy absorbed, $E = P \times t = v \times i \times t = 5 \times \frac{d(3e^{-2t})}{dt}\ 3e^{-2t} = -1.66$ mJ

EXAMPLE 1.28 **The voltage across a 5-Ω resistor is 20 V. Find the current through the resistor.**

Solution Current $i = V/R = 20/5 = 4$ A

EXAMPLE 1.29 **Find the power absorbed by each element in the circuit shown in Figure 1.39.**

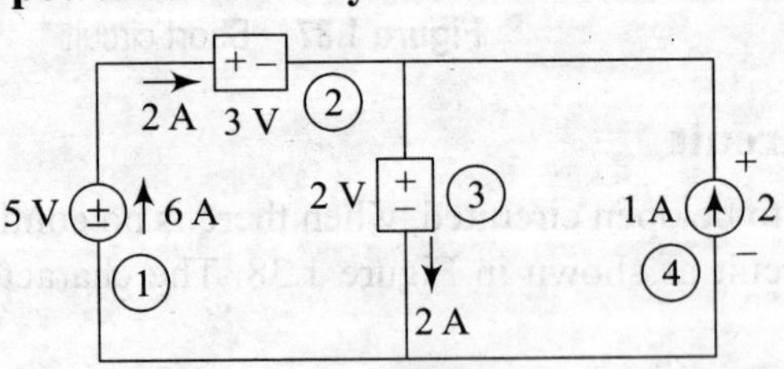

Figure 1.39

Solution For the first element, $P_1 = (-6) \times 5 = -30$ W

For the second element, $P_2 = 3 \times 2 = 6$ W

For the third element, $P_3 = 2 \times 2 = 4$ W

For the fourth element, $P_4 = 2 \times (-1) = -2$ W

EXAMPLE 1.30 **Determine the currents i_A, i_B, i_C, and i_D for the circuit shown in Figure 1.40.**

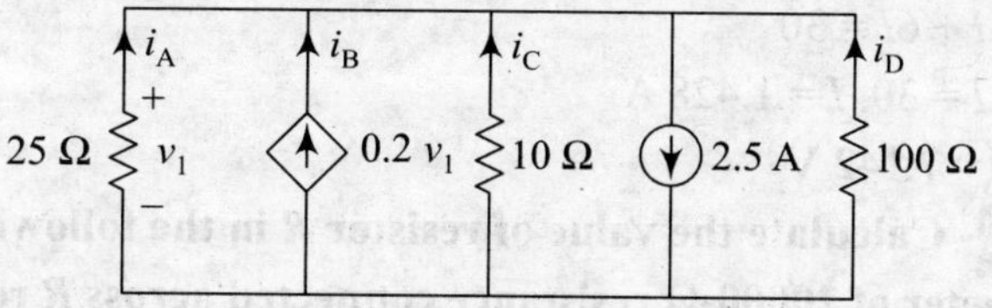

Figure 1.40

Solution Applying KCL for the given circuit,

$$i_A + i_B + i_C + i_D = 2.5 \quad (1)$$

$$i_B = 0.2\, v_1 = 0.2 \times (-25 i_A) = -5 i_A \quad (2)$$

Substituting Eq. (2) into Eq. (1), we get

$$i_A - 5i_A + i_C + i_D = 2.5$$

$$-4i_A + i_C + i_D = 2.5 \quad (3)$$

Assuming node voltage as v_1 and writing Eq. (3) into node voltage, we get

$$-4\frac{v_1}{25} + \frac{v_1}{10} + \frac{v_1}{100} = 2.5 \quad (4)$$

Solving Eq. (4), $v_1 = -50$ V

$$i_A = \frac{-50}{25} = -2 \text{ A}$$

$$i_B = 0.2 \times 50 = 10 \text{ A}$$

$$i_C = \frac{-50}{10} = -5 \text{ A}$$

$$i_D = \frac{-50}{100} = -0.5 \text{ A}$$

EXAMPLE 1.31 **Find V_A in the circuit shown in Figure 1.41.**

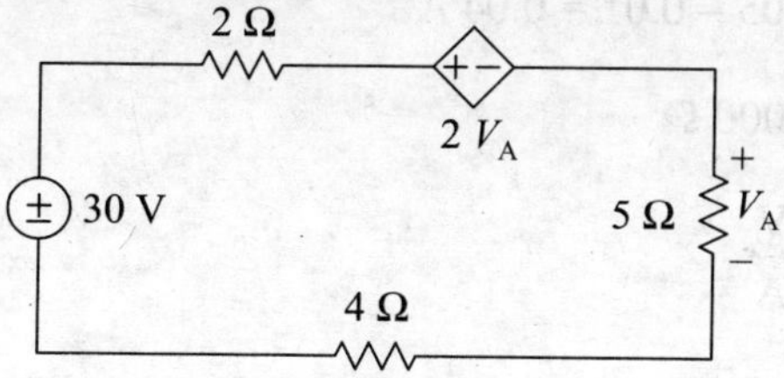

Figure 1.41

Solution Applying KVL,

$$-30 + 2I + 2V_A + V_A + 4I = 0 \quad (1)$$

By applying Ohm's law, $V_A = 5I$ (2)

Substituting Eq. (2) into Eq. (1), we get

$$15I + 6I = 30$$

$$21I = 30;\ I = 1.428\ \text{A}$$

Voltage, $V_A = 7.142$ V

EXAMPLE 1.32 **Calculate the value of resistor R in the following cases:**

(i) A voltmeter of 20000-Ω resistance connected across R reads 200 V, while the current supplied to V and R is 0.05 A.

(ii) A voltage of 10 V is applied to R in series with an ammeter of 0.1-Ω resistance, while A reads 50 A.

(iii) A voltmeter of 5000-Ω resistance in series with R across a 250-V supply reads 50 V.

Solution

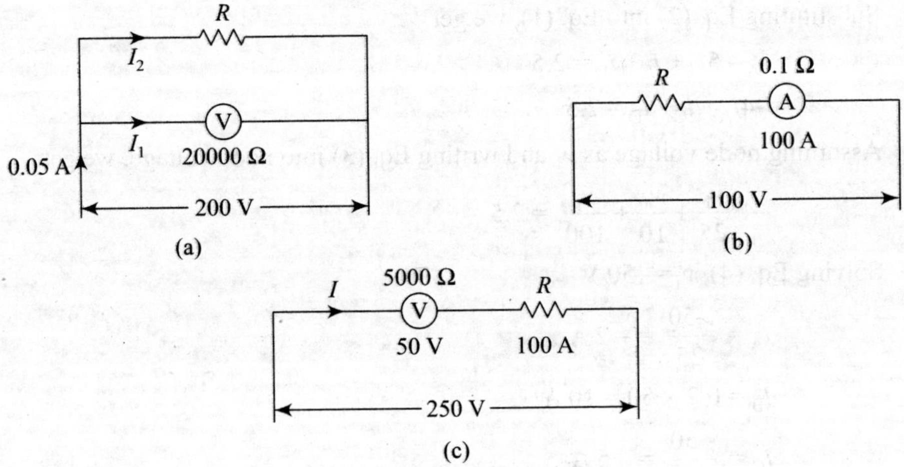

Figure 1.42

(i) $I_1 = \dfrac{200}{20000} = 0.01$ A

$I_2 = I - I_1 = 0.05 - 0.01 = 0.04$ A

$R = \dfrac{200}{0.04} = 5000\ \Omega$

(ii) $R + 0.1 = \dfrac{10\ \text{V}}{50\ \text{A}}$

$R = 0.1\ \Omega$

(iii) $I = \dfrac{50}{5000} = 0.01$ A

Voltage drop across $R = 250 - 50 = 200$ V

$R = \dfrac{200}{0.01} = 20000\ \Omega$

EXAMPLE 1.33 Find the current and power supplied by the battery to the circuit shown in Figure 1.43 (i) under normal conditions and (ii) when a short circuit occurs across terminals A and B.

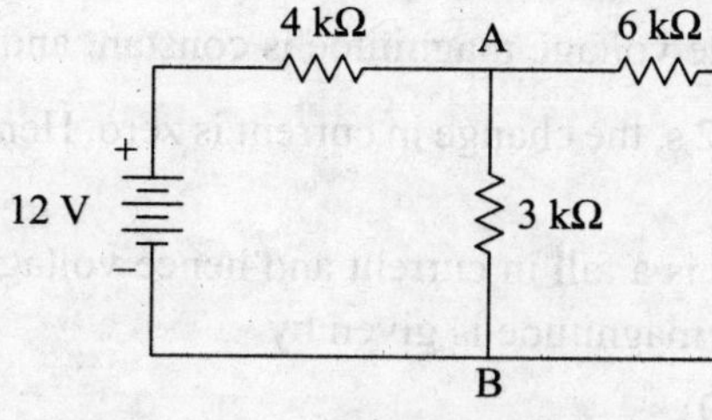

Figure 1.43

Solution

(i) Under normal conditions,

The equivalent resistance of the circuit, $R_{eq} = (3\ \text{k}\Omega \parallel 6\ \text{k}\Omega) + 4\ \text{k}\Omega$
$= 6\ \text{k}\Omega$

Current supplied by source, $I = \dfrac{12}{6\ \text{k}\Omega} = 2\ \text{mA}$

Power supplied, $P = VI = 12 \times 2\ \text{mA} = 24\ \text{mW}$

(ii) When AB is shorted,

The equivalent resistance $R_{eq} = (0 \parallel 6\ \text{k}\Omega) + 4\ \text{k}\Omega$
$= 4\ \text{k}\Omega$

Current supplied by source $I = \dfrac{12}{4\ \text{k}\Omega} = 3\ \text{mA}$

Power supplied $P = VI = 12 \times 3\ \text{mA} = 36\ \text{mW}$

EXAMPLE 1.34 A current waveform shown in Figure 1.44 is applied to an inductor of 4 H. Find the voltage waveform.

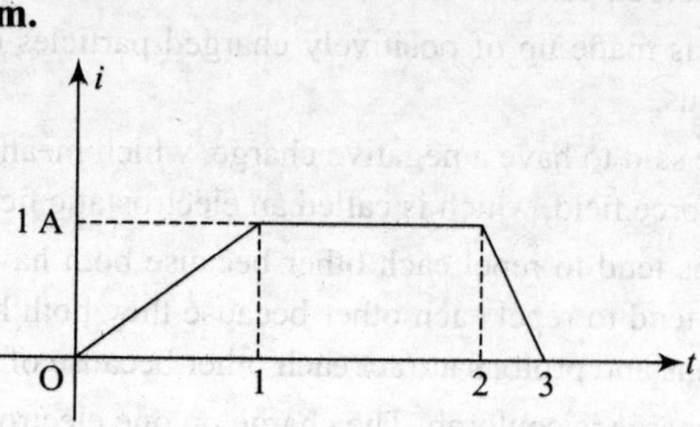

Figure 1.44 Current waveform

Solution The voltage across the inductor is given by $v = L\ (di/dt)$.
From Figure 1.44, it is observed that there is change in current from 0 A to 1 A in time 0 s to 1 s.

Hence, the voltage

$$v = 4 \times \frac{(1-0)}{(1-0)} = 4\text{ V}$$

That is, from 0 to 1 s the voltage magnitude is constant and is equal to 4 V.

Similarly, from $t = 1$ to 2 s, the change in current is zero. Hence, the voltage across the inductor is also equal to zero.

From $t = 2$ to 3 s, there is a fall in current and hence voltage will go into the negative side of the axis, and therefore, magnitude is given by

$$v = 4 \times \frac{(1-0)}{(2-3)} = -4\text{ V}$$

The diagram of the voltage waveform is shown in Figure 1.45.

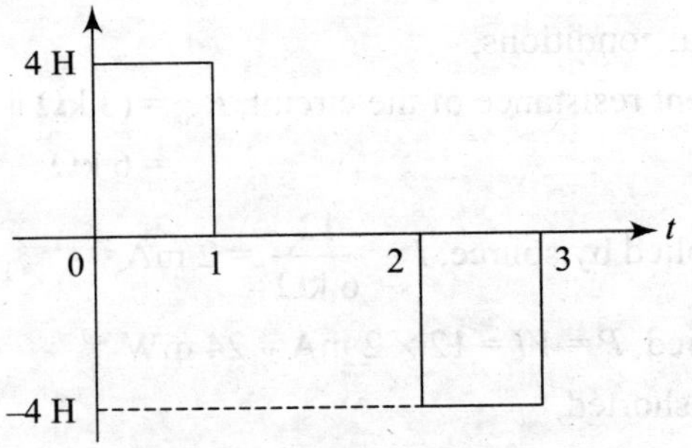

Figure 1.45 Voltage waveform

CHAPTER AT A GLANCE

- An atom is the smallest part of any element that retains the characteristics of that element.
- Atoms consist of a central, dense nucleus that is surrounded by one or more lightweight negatively charged particles called electrons.
- The nucleus is made up of positively charged particles called protons and neutral particles called neutrons.
- Electrons are said to have a negative charge, which means that they are surrounded by a kind of invisible force field, which is called an electrostatic field.
- Two electrons tend to repel each other because both have a negative electrical charge. Two protons also tend to repel each other because they both have a positive charge. On the other hand, electrons and protons attract each other because of their unlike charges.
- The unit of charge is coulomb. The charge on one electron is 1.602×10^{-19} C.
- Based on the values of the forbidden energy gap, the materials are classified as conductors, semiconductors, and insulators.
- Electromotive force is defined as the force required to drift the free electrons in a particular direction in a conductor.

- Current is defined as the rate of flow of charge in an electric circuit.
- Electric potential is defined as the ability of a charged particle to do work.
- The difference between the electric potentials at any two given points in a circuit is known as potential difference.
- Power is defined as the rate at which electrical energy is transferred by the circuit or the rate of doing work.
- Energy is defined as the total amount of electrical work done in an electric circuit. The unit of energy is joules or watt-sec.
- Resistance is the hindrance to the flow of charge.
- The SI unit of resistance is ohm, symbol Ω.
- The reciprocal of resistance is called conductance.
- A wire of a certain length, when twisted into a coil becomes a basic inductor.
- Energy stored in inductor, $\omega = (1/2)\ LI^2$ joules.
- Capacitor consists of two conducting surfaces separated by an insulating material.
- Energy stored in capacitor, $\omega = (1/2) = CV^2$ joules.
- A node of a network is an equipotential surface at which two or more elements are joined.
- A junction is that point in a network where three or more circuit elements are joined.
- A branch is that part of a network which lies between two junction points.
- A loop is any closed path of a network.
- Electrical networks are classified as (i) active and passive, (ii) unilateral and bilateral, (iii) linear and nonlinear, and (iv) lumped and distributed.
- The energy sources are divided into two categories: (i) independent sources and (ii) dependent sources.
- Ohm's law states that the electric current in a conductor between two points is directly proportional to the voltage across the two points and inversely proportional to the resistance between them.
- KCL states that in any electrical network, the algebraic sum of the currents meeting at a point or junction is zero.
- KVL states that the algebraic sum of the product of currents and resistances in each of the conductor in any closed path in a network plus the algebraic sum of the emf in that path is zero.
- When two points are connected together by a metallic wire, they are said to be short circuited. Two points are said to be open circuited, when there is no continuity in the circuit or when there is a break in the circuit.

MULTIPLE CHOICE QUESTIONS

1. A current is said to exist whenever

   (a) a wire is charged (b) a battery is present
   (c) electric charges are unbalanced (d) electric charges move in a loop

2. If an electric circuit could be compared to a water circuit at a water park, then the current would be analogous to the
   (a) water pressure
   (b) gallons of water flowing down slide per minute
   (c) water
   (d) water pump
3. Which of the following will cause the current through an electrical circuit to decrease?
   (a) Decrease the voltage (b) Decrease the resistance
   (c) Increase the voltage (d) Increase the resistance
4. A certain electrical circuit contains a battery with three cells, wires, and a light bulb. Which of the following would cause the bulb to shine less brightly?
   (a) Increase the voltage of the battery (add another cell)
   (b) Decrease the voltage of the battery (remove a cell)
   (c) Decrease the resistance of the circuit
   (d) Increase the resistance of the circuit
5. You have likely been warned to avoid contact with electrical appliances or even electrical outlets when your hands are wet. Such contact is more dangerous when your hands are wet (vs. dry) because wet hands cause
   (a) the voltage of the circuit to be higher (b) the voltage of the circuit to be lower
   (c) your resistance to be higher (d) your resistance to be lower
   (e) the current passing through you to be lower
6. If the resistance of a circuit is tripled, then the current through the circuit would be
   (a) one-third as much
   (b) three times as much
   (c) unchanged
   (d) ... nonsense! There would be no way to make such a prediction
7. If the voltage across a circuit is quadrupled, then the current through the circuit would be
   (a) one-fourth as much
   (b) four times as much
   (c) unchanged
   (d) ... nonsense! There would be no way to make such a prediction
8. Which of the following material is a semiconductor?
   (a) Germanium (b) Wood
   (c) Iron (d) Mica
9. What is the unit of resistance?
   (a) Coulomb (b) Ohm
   (c) Ampere (d) Henry
10. The unit of electrical energy is
   (a) newton meter (b) watt
   (c) joule (d) watt hour

11. Ohm's law is applicable to
(a) semiconductors (b) vacuum tubes
(c) electrolytes (d) resistors

12. A linear resistor is one which obeys
(a) Ampere's law (b) Lenz's law
(c) Ohm's law (d) Kirchhoff's law

13. A wire is stretched to double its length. The new resistance would be
(a) four-times the original resistance (b) double the original resistance
(c) half of the original resistance (d) one-fourth of the original resistance

14. For the same voltage, the ratio of resistances of 25 W and 100 W lamps will be
(a) 1:1 (b) 2:1
(c) 4:1 (d) 1:4

15. Constant voltage source is
(a) active and bilateral (b) passive and bilateral
(c) passive and unilateral (d) active and unilateral

16. A terminal where more than two branches meet is known as
(a) node (b) junction
(c) (a) and (b) (d) anode

17. For determining the polarity of the voltage drop across a resistor, it is necessary to know the
(a) value of resistor
(b) value of current
(c) value of emf in the circuit
(d) direction of current flowing through the resistor

18. Electric pressure is also called as
(a) resistance (b) power
(c) voltage (d) energy

19. Resistance of a human body is about
(a) 10 Ω (b) 1000 Ω
(c) 100 Ω (d) 10000 Ω

20. Ampere-second could be the unit of
(a) emf (b) charge
(c) power (d) energy

SHORT ANSWER QUESTIONS

1. Find the voltage drop impressed across an electrical device with a resistance of 3.0 Ω that will allow a current of 4.0 A to flow through it.
2. A voltage of 120 V is impressed across an electric heater. Determine the resistance when a current of 10.0 A will flow through the heater.

3. Find the current if a flashlight is powered by 3 V and uses a bulb with a resistance of 60 Ω.
4. Which would be thicker (wider)—the filament of a 60-W light bulb or the filament of a 120-W light bulb? Explain.
5. The sticker on a compact disc player says that it draws 288 mA of current when powered by a 9-V battery. What is the power (in W) of the CD player?
6. What is the reason for temperature rise when the current flows through a conductor?
7. What are the various types of dependent sources?
8. Define a network.
9. Define a node.
10. Define loop and mesh.
11. A voltage source of 20 V and 2 Ω is connected to a load resistance of 8 Ω. Find the load current.

REVIEW QUESTIONS

1. What is resistance and what are the factors affecting it?
2. State and explain Kirchoff's laws.
3. Explain Ohm's law.
4. What is the difference between emf and potential difference?
5. What are the characteristics of an open circuit?
6. What are the characteristics of a short circuit?
7. How is voltage divided in a series circuit?
8. How is current divided in a parallel circuit?
9. How do you classify the sources?
10. What is the difference between linear and nonlinear elements?
11. What is the difference between unilateral and bilateral elements?
12. Describe independent and dependent sources.

EXERCISE PROBLEMS

1. Two wires—A and B—with circular cross sections have identical lengths and are made of the same material. Yet, wire A has four times the resistance of wire B. How many times greater is the diameter of wire B than wire A?
2. Calculate the resistance and the current of a 7.5-W light bulb plugged into a household outlet (230 V).
3. A color TV has a current of 1.99 A when connected to a 230-V household circuit. What is the resistance (in ohms) of the TV set?
4. Find the currents i_1, i_2, and i_3 for the network shown in Figure 1.46.

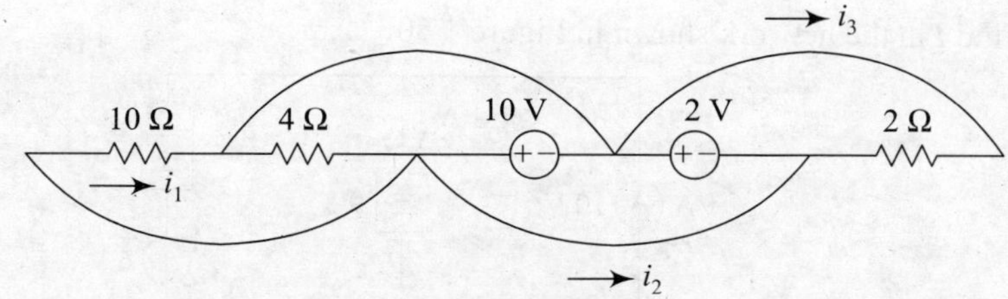

Figure 1.46

5. Determine R_L for the circuit shown in Figure 1.47.

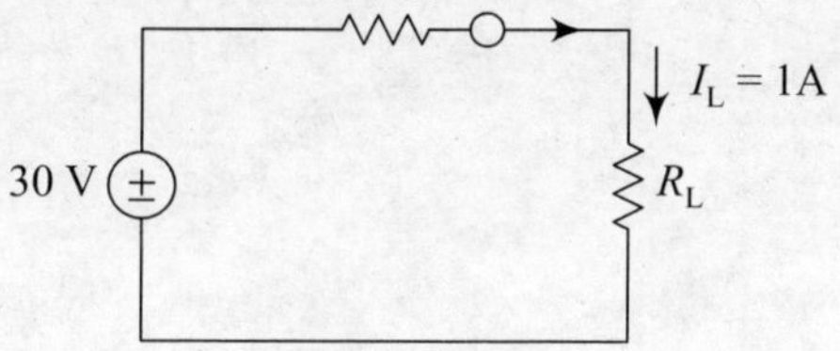

Figure 1.47

6. A 1.2-kW toaster takes roughly 5 min to heat four pieces of bread. Find the cost of operating the toaster once per day for 15 days. Assume the cost of energy ₹ 2/kWh.

7. The current flowing through an element is $i = 2$ A, $0 < t < 1$; $i = 2t^2$, $t > 1$. Calculate the charge entering the element from $t = 0$ to $t = 2$ s.

8. When a dc voltage is applied to a capacitor, the voltage across its terminals is found in accordance with $V_c = 100\,(1 - e^{-50t})$. After the time interval of 0.1 s, the current flow is equal to 1 A. Determine the value of capacitance and energy stored in the electric field at this time.

9. Determine the number of nodes and branches for the circuit shown in Figure 1.48.

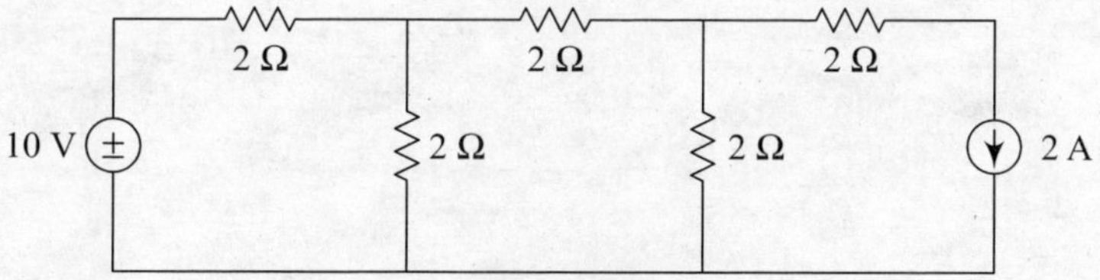

Figure 1.48

10. Find i_1, i_2, and i_3 for the circuit shown in Figure 1.49.

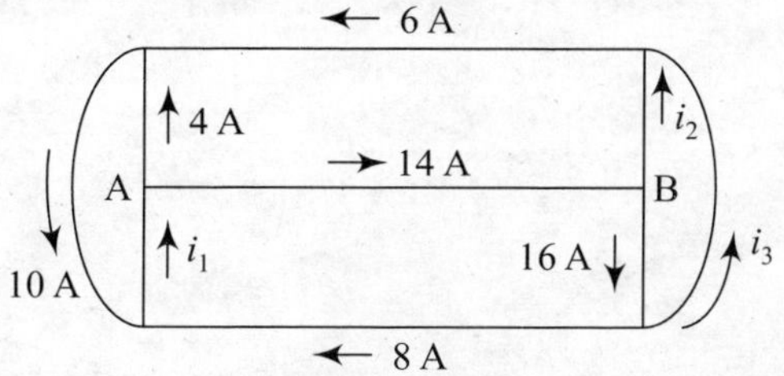

Figure 1.49

11. Find I in the network shown in Figure 1.50.

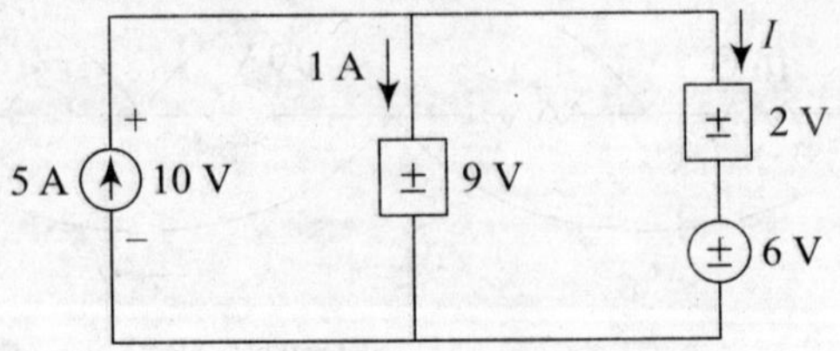

Figure 1.50

Chapter 2

DC Circuit Analysis

Chapter Objectives

- To explain voltage and current division rules
- To discuss source transformation techniques
- To describe mesh analysis and nodal analysis
- To explain star–delta transformation

2.1 INTRODUCTION

An electrical circuit consists of one or more sources of energy and one or more number of basic elements such as resistors, inductors, and capacitors. These elements and sources may be connected in a desired manner. The interconnection of such elements gives rise to complexity. In order to find the current flowing through various elements in the network, and to find the voltage across the elements the circuit must be solved. To solve such circuits, various techniques have been proposed in this chapter. Here, only dc circuit analysis has been considered.

2.2 SERIES AND PARALLEL NETWORKS

A network in which two or more elements (i.e., resistors or inductors or capacitors) are connected one after the other in series is called *series network*. A network in which each element is connected across power source is called *parallel network*. These are basics of all the network analysis and are discussed in the following sections.

2.2.1 Resistances in Series

The characteristics of a series circuit are as follows:

(i) There is only one current in all parts of the circuit.

(ii) Elements have individual voltage drops.

(iii) Voltage drops are additive.
(iv) Applied emf equals the sum of element voltage drops.
(v) Resistances are additive.
(vi) Powers and energies are additive.

Consider two resistors of ohmic values R_1 and R_2 connected in series as shown in Figure 2.1. The current flowing in these resistors is same, say I amperes. The voltages are

$$V_1 = IR_1 \tag{2.1}$$

$$V_2 = IR_2 \tag{2.2}$$

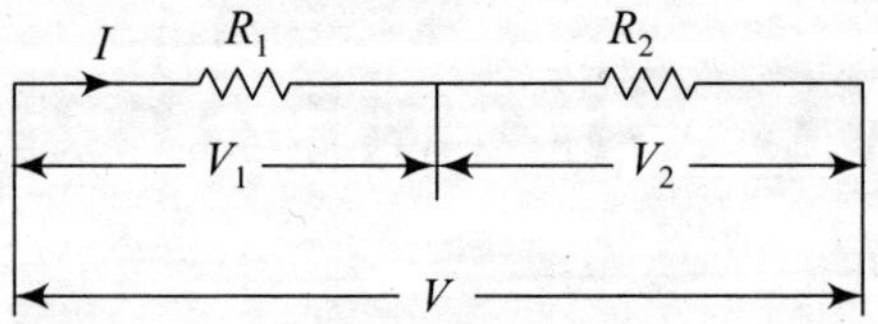

Figure 2.1 Two resistors connected in series

The applied emf equals the sum of element voltage drops

$$V = V_1 + V_2 \tag{2.3}$$

$$= IR_1 + IR_2$$

$$= I(R_1 + R_2)$$

Since $V = I\,R_{eq}$

$$\therefore \quad R_{eq} = R_1 + R_2 \tag{2.4}$$

where R_{eq} is the equivalent resistance of the series combination.

2.2.2 Resistances in Parallel

The characteristics of a parallel circuit are as follows:

(i) Same voltage acts across all parts of the circuit.
(ii) Elements have individual currents.
(iii) Branch currents are additive.
(iv) Conductances are additive.
(v) Powers and energies are additive.

Consider two resistors of ohmic values R_1 and R_2 connected in parallel as shown in Figure 2.2.

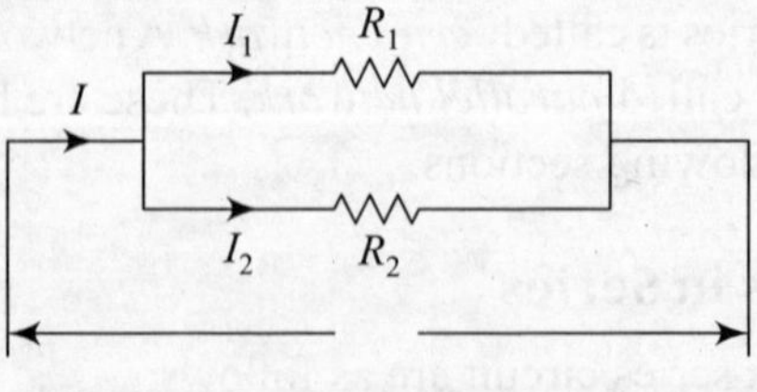

Figure 2.2 Two resistors connected in parallel

The voltage across each one of them is same, say V volts, hence the currents are

$$I_1 = \frac{V}{R_1} \quad \text{and} \quad I_2 = \frac{V}{R_2} \tag{2.5}$$

The total current of the parallel combination,

$$I = I_1 + I_2 \tag{2.6}$$

$$= \frac{V}{R_1} + \frac{V}{R_2} + V\left(\frac{1}{R_1} + \frac{1}{R_2}\right)$$

If R_{eq} is the equivalent resistance of the parallel combination, then

$$\text{I} = \frac{V}{R_{eq}}, \quad \frac{1}{R_{eq}} = \frac{1}{R_1} + \frac{1}{R_2} \quad \text{or} \quad R_{eq} = \frac{R_1 R_2}{R_1 + R_2} \tag{2.7}$$

$$G = G_1 + G_2$$

If three resistors are in parallel, then

$$\frac{1}{R_{eq}} = \frac{1}{R_1} + \frac{1}{R_2} + \frac{1}{R_3} \tag{2.8}$$

$$R_{eq} = \frac{R_1 R_2 R_3}{R_1 R_2 + R_2 R_3 + R_3 R_1}$$

If four resistors are in parallel, then

$$\frac{1}{R_{eq}} = \frac{1}{R_1} + \frac{1}{R_2} + \frac{1}{R_3} + \frac{1}{R_4} \tag{2.9}$$

$$R_{eq} = \frac{R_1 R_2 R_3 R_4}{R_1 R_2 R_3 + R_2 R_3 R_4 + R_3 R_4 R_1 + R_4 R_1 R_2}$$

2.2.3 Inductances in Series

By neglecting mutual inductance, we can consider the same formulas which will be implemented in series and parallel combinations of resistances. We can use the same formulas for inductance in series also. Figure 2.3 shows the inductances connected in series.

Equivalent inductance, $L_{eq} = L_1 + L_2$ (2.10)

2.2.4 Inductances in Parallel

When inductances are connected in parallel, as shown in Figure 2.4, their equivalent inductance is given by

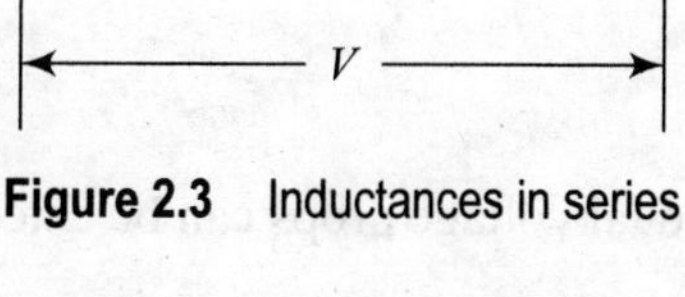

Figure 2.3 Inductances in series

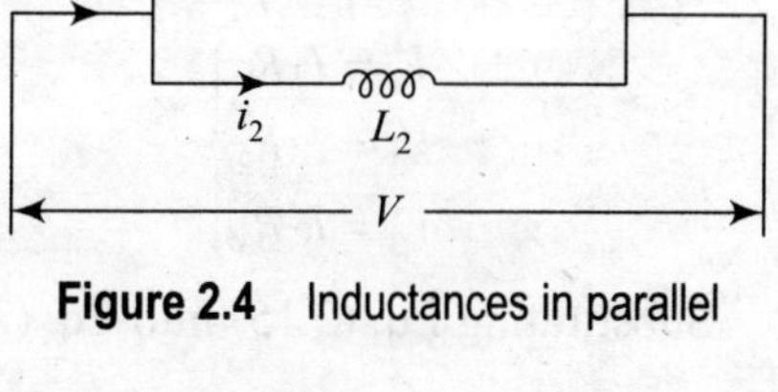

Figure 2.4 Inductances in parallel

$$\text{Equivalent inductance, } L_{eq} = \frac{L_1 L_2}{L_1 + L_2} \tag{2.11}$$

2.2.5 Capacitances in Series

The two series capacitances as shown in Figure 2.5 have a common current I. The charge on all capacitors is same, but potential difference across each is different. The equivalent capacitance C_{eq} is given by

$$\frac{1}{C_{eq}} = \frac{1}{C_1} + \frac{1}{C_2} \tag{2.12}$$

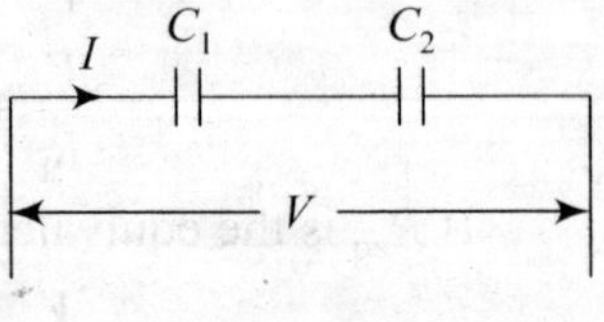

Figure 2.5 Capacitors connected in series

2.2.6 Capacitances in Parallel

Figure 2.6 shows the parallel connection of two capacitances. The charge on each capacitance is different, but the potential differences across each are the same. The equivalent capacitance C_{eq} is given by

$$C_{eq} = C_1 + C_2 \tag{2.13}$$

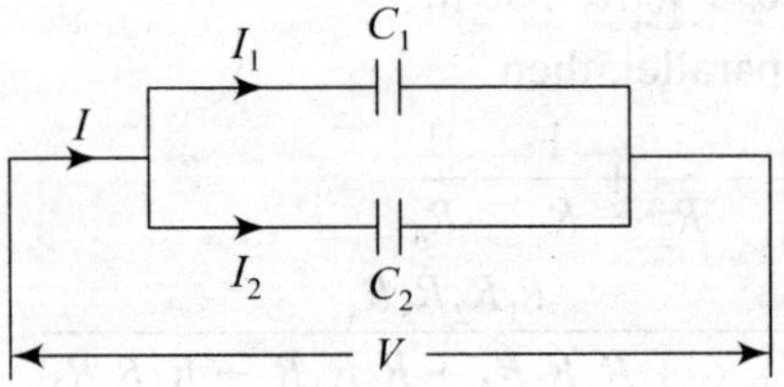

Figure 2.6 Capacitors connected in parallel

2.3 VOLTAGE DIVISION RULE

A practical application of the voltage division rule is when the resistors are connected in series across a voltage source the voltage drop across each resistor can be calculated easily.

Consider a series circuit as shown in Figure 2.7. The current I flowing through all the series-connected resistors is same. Applying KVL, we get

$$V = I(R_1 + R_2 + R_3) \tag{2.14}$$

$$\therefore \quad I = \frac{V}{R_1 + R_2 + R_3} \tag{2.15}$$

Total voltage drop applied must be equal to the sum of voltage drops across each resistor. Hence,

$$\left.\begin{aligned} V_1 &= I \cdot R_1 \\ V_2 &= I \cdot R_2 \\ V_3 &= I \cdot R_3 \end{aligned}\right\} \tag{2.16}$$

Substituting Eq. (2.15) into Eq. (2.16), the individual voltage drops can be calculated as

$$V_1 = V \cdot \frac{R_1}{R_1 + R_2 + R_3} \tag{2.17}$$

$$V_2 = V \cdot \frac{R_2}{R_1 + R_2 + R_3} \tag{2.18}$$

$$V_3 = V \cdot \frac{R_3}{R_1 + R_2 + R_3} \tag{2.19}$$

Note: Voltage across a resistor in a series circuit = $\dfrac{\text{Applied voltage} \times \text{Value of that resistor}}{\text{Sum of all series-connected resistors}}$

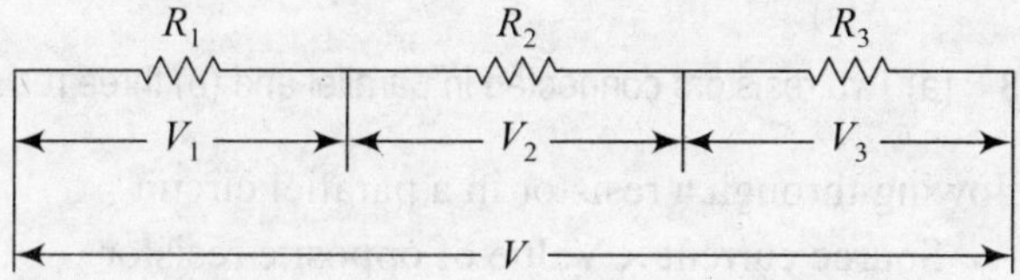

Figure 2.7 Series circuit to illustrate voltage division rule

2.4 CURRENT DIVISION RULE

The current division rule is useful in finding the currents when resistors are connected in parallel. With this rule, the reduction of network can be avoided.

Consider a parallel circuit as shown in Figures 2.8(a) and (b). The voltage applied across a parallel circuit is same. The current supplied by the source is I. The current flowing through parallel resistors is I_1 and I_2. Applying KCL, we get

$$I = I_1 + I_2 \tag{2.20}$$

But

$$I_1 = \frac{V}{R_1}; \quad I_2 = \frac{V}{R_2} \tag{2.21}$$

From Eq. (2.21), we get

$$V = I_1 R_1 = I_2 R_2 \tag{2.22}$$

From Eq. (2.22), we get

$$I_2 = I_1 \frac{R_1}{R_2} \tag{2.23}$$

Substituting Eq. (2.23) into Eq. (2.20), we get

$$I = I_1 + I_1 \frac{R_1}{R_2} = \left(1 + \frac{R_1}{R_2}\right) I_1 \tag{2.24}$$

From Eq. (2.24), we get

$$I_1 = \frac{R_2}{R_1 + R_2} I \tag{2.25}$$

Similarly,

$$I_2 = \frac{R_1}{R_1 + R_2} I \tag{2.26}$$

$$\Rightarrow \qquad I_1 = I \cdot \frac{R_2}{R_1 + R_2}\ ; \quad I_2 = I \cdot \frac{R_1}{R_1 + R_2}$$

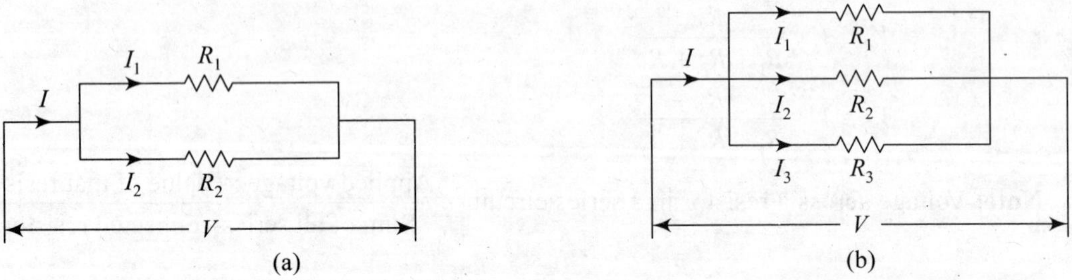

Figure 2.8 (a) Two resistors connected in parallel and (b) three resistors connected in parallel

Note: Current flowing through a resistor in a parallel circuit

$$= \frac{\text{Source current} \times \text{Value of opposite resistor}}{\text{Sum of all parallel-connected resistors}}$$

If three resistors are connected in parallel then the individual branch current can be calculated as

$$I_1 = I\left(\frac{R_2R_3}{R_1R_2 + R_2R_3 + R_3R_1}\right) \tag{2.27}$$

$$I_2 = I\left(\frac{R_1R_3}{R_1R_2 + R_2R_3 + R_3R_1}\right) \tag{2.28}$$

$$I_3 = I\left(\frac{R_1R_2}{R_1R_2 + R_2R_3 + R_3R_1}\right) \tag{2.29}$$

EXAMPLE 2.1 **Three resistors of 8.4 Ω, 6.8 Ω, and 4.8 Ω are connected in a series across 100-V supply. Find**

(i) Total equivalent resistance

(ii) Current

(iii) Voltage across each resistor

Solution Given data

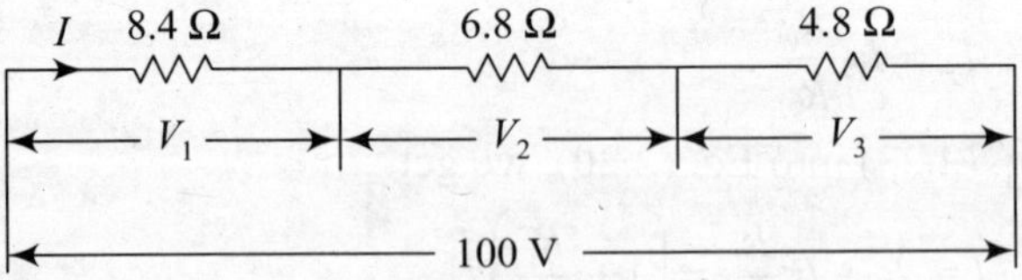

Figure 2.9

$$R_1 = 8.4\ \Omega, \quad R_2 = 6.8\ \Omega, \quad R_3 = 4.8\ \Omega, \quad V = 100\ \text{V}$$

(i) $R_{eq} = R_1 + R_2 + R_3 = 8.4 + 6.8 + 4.8 = 20\ \Omega$

(ii) Current $I = V/R_{eq} = 100\ \text{V}/\ 20\ \Omega = 5\ \text{A}$

(iii) Voltage across each resistor

$$V_1 = IR_1 = 5 \times 8.4 = 42 \text{ V}$$
$$V_2 = IR_2 = 5 \times 6.8 = 34 \text{ V}$$
$$V_3 = IR_3 = 5 \times 4.8 = 24 \text{ V}$$

Supply voltage,

$$V = V_1 + V_2 + V_3$$
$$= 42 + 34 + 24$$
$$= 100 \text{ V}$$

EXAMPLE 2.2 **Two resistors 4 Ω and 6 Ω are connected in parallel. If the current supplied by source is 30 A. Find the equivalent resistance and current through each branch.**

Solution Given data

$$R_1 = 4\ \Omega, \quad R_2 = 6\ \Omega, \quad I = 30 \text{ A}$$

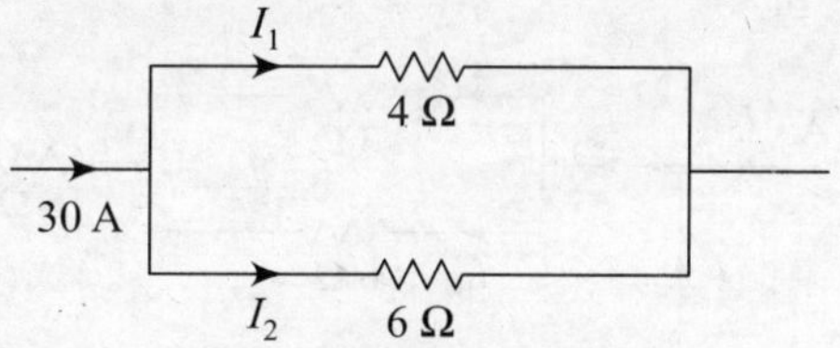

Figure 2.10

Equivalent resistance,

$$R_{eq} = \frac{R_1 R_2}{R_1 + R_2} = \frac{24}{10} = 2.4\ \Omega$$

By using the current division rule, the current flowing in each branch is given by

$$I_1 = I \cdot \frac{R_2}{R_1 + R_2} = 30 \times \frac{6}{4+6} = 18 \text{ A}$$

$$I_2 = I \cdot \frac{R_1}{R_1 + R_2} = 30 \times \frac{4}{4+6} = 12 \text{ A}$$

Total current,

$$I = 18 + 12 = 30 \text{ A}$$

EXAMPLE 2.3 **Calculate the effective resistance of the following combination of resistances and the voltage drops across each resistance, when a voltage of 60 V is applied between points A and B as shown in Figure 2.11.**

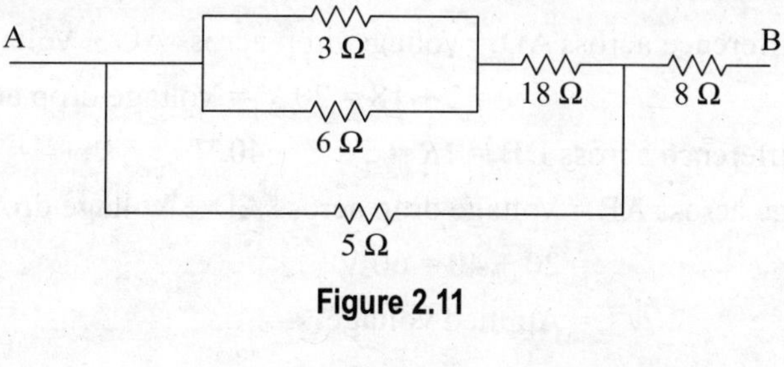

Figure 2.11

Solution Given data

$V = 60$ V

3 Ω and 6 Ω resistances are parallel $= \dfrac{6 \times 3}{6+3} = 2\ \Omega$

This equivalent resistance is in series with 18 Ω = 2 + 18 + 20 Ω

20 Ω is parallel to 5 Ω $= \dfrac{20 \times 5}{20+5} = 4\ \Omega$

This 4-Ω resistance is in series with 8 Ω = 4 + 8 = 12 Ω

Therefore, equivalent resistance of the given circuit = 12 Ω

Total current, $I = \dfrac{V}{R_{eq}} = \dfrac{60}{12} = = 5$ A

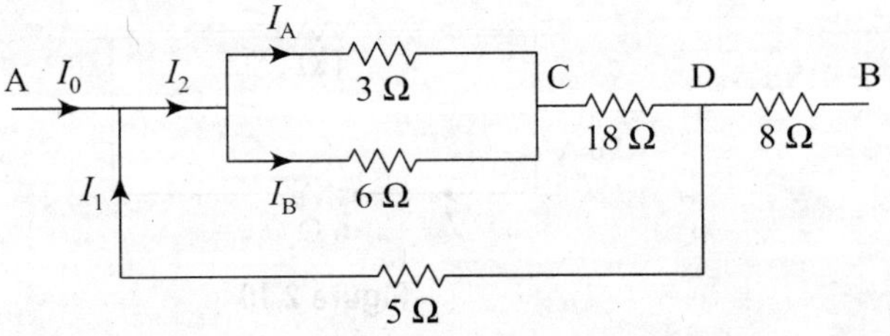

Figure 2.12

As shown in Figure 2.12, by assuming the current divisions, we get

$$I_1 = 5 \times \frac{20}{25} = 4 \text{ A}$$

$$I_2 = 5 \times \frac{5}{25} = 1 \text{ A}$$

$$I_A = 1 \times \frac{6}{9} = \frac{2}{3} \text{ A}$$

$$I_B = 1 \times \frac{3}{9} = \frac{1}{3} \text{ A}$$

Potential difference across AC = 1R = 1 × 2 = 2 V

Potential difference across CD = 1R = 1 × 18 = 18 V

Potential difference across AD = Voltage drop across AC + Voltage drop across CD

= 2 + 18 = 20 V = Voltage drop across 5 Ω

Potential difference across DB = 1R = 5 × 8 = 40 V

Total voltage across AB = Voltage drop across AD + Voltage drop across DB

= 20 + 40 = 60 V

= Applied voltage

EXAMPLE 2.4 **For the network shown in Figure 2.13, determine the currents in all the branches and also find the power loss in all the branches.**

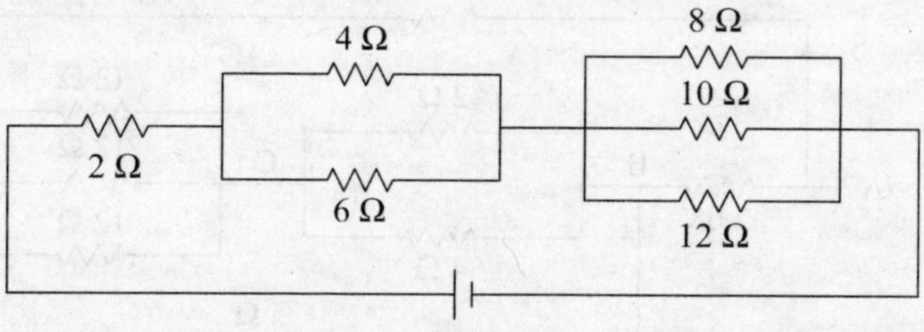

Figure 2.13

Solution Given data

$$4\ \Omega \text{ and } 6\ \Omega \text{ resistors are parallel} = \frac{6\times 4}{6+4} = 2.4\ \Omega$$

8 Ω, 10 Ω, and 12 Ω resistors are parallel = 3.24 Ω

Total equivalent resistance, $R_{eq} = 2 + 2.4 + 3.24 = 7.64\ \Omega$

The current drawn from the battery is

$$I = \frac{V}{R} = \frac{24}{7.64} = 3.14\text{ A}$$

Current in 2 Ω resistor, $I_2 = 3.14$ A

$$\text{Current in } 4\ \Omega \text{ resistor, } I_4 = 3.14 \times \frac{6}{4+6} = 1.884\text{ A}$$

$$\text{Current in } 6\ \Omega \text{ resistor, } I_6 = 3.14 \times \frac{4}{4+6} = 1.256\text{ A}$$

$$\text{Current in } 8\ \Omega \text{ resistor, } I_8 = 3.14 \times \frac{10\times 12}{8\times 10 + 10\times 12 + 12\times 8} = 1.274\text{ A}$$

$$\text{Current in } 10\ \Omega \text{ resistor, } I_{10} = 3.14 \times \frac{12\times 8}{8\times 10 + 10\times 12 + 12\times 8} = 1.018\text{ A}$$

The power drawn from the battery is $P = VI = 24 \times 3.14 = 75.40$ W

This power is equal to the sum of power losses in all the branches whose values can be computed by using the formula I^2R

$$P_2 = 3.14^2 \times 2 = 19.7\text{ W}$$
$$P_4 = 1.884^2 \times 4 = 14.19\text{ W}$$
$$P_6 = 1.256^2 \times 6 = 9.46\text{ W}$$
$$P_8 = 1.274^2 \times 8 = 12.98\text{ W}$$
$$P_{10} = 1.019^2 \times 10 = 10.38\text{ W}$$
$$P_{12} = 0.849^2 \times 12 = 8.649\text{ W}$$

$$\text{Total power loss} = P_2 + P_4 + P_6 + P_8 + P_{10} + P_{12}$$
$$= 75.40\text{ W}$$
$$= \text{Power drawn from the battery}$$

EXAMPLE 2.5 **Find the equivalent resistance of the circuit shown in Figure 2.14.**

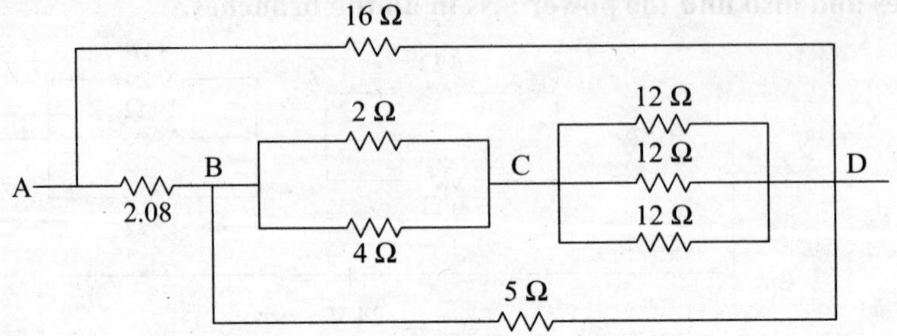

Figure 2.14

Solution Three 12-Ω resistors are in parallel, $R_{CD} = 4\ \Omega$

2 Ω and 4 Ω resistors are in parallel, $R_{BC} = \dfrac{4 \times 2}{4+2} = 1.33\ \Omega$

R_{BC} and R_{CD} are in series, $R_{BD} = 4 + 1.33 = 5.33\ \Omega$

5-Ω resistor is parallel to, $R_{BD} = \dfrac{5.33 \times 5}{5.33+5} = 2.579\ \Omega$

R_{CD} is parallel to the above combination

$$R_{AD} = \frac{16 \times 4.659}{16 + 4.659} = \frac{74.55}{20.65} = 3.62\ \Omega$$

Equivalent resistance of given circuit, $R_{AD} = 3.62\ \Omega$

EXAMPLE 2.6 **Three resistors A, B, and C are connected in parallel as shown in Figure 2.15 and take a total current of 7.9 A. $R_A = 48\ \Omega$ and takes 2.5 A, $I_B = 2I_C$. Calculate (i) I_B and I_C, (ii) the line voltage, and (iii) R_B and R_C.**

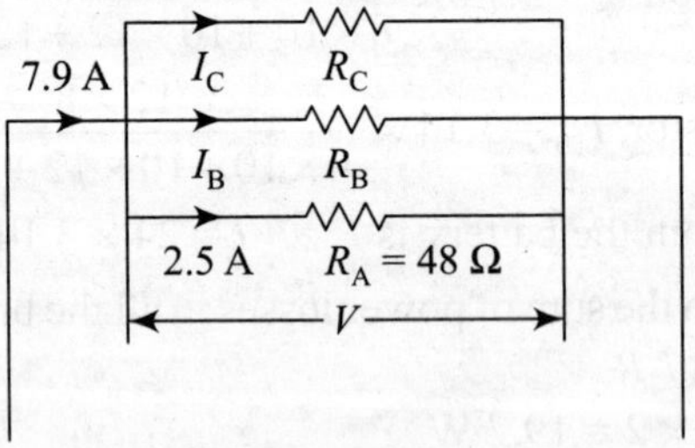

Figure 2.15

Solution $V = 1, \quad R = 2.5 \times 48 = 120$ V

$$I_B = \frac{120}{R_B}$$

$$I_C = \frac{120}{R_C}$$

$$I_B = 2I_C$$

$$\frac{120}{R_B} = 2 \cdot \frac{120}{R_C}$$

$$\therefore \quad R_B = \frac{R_C}{2}$$

$$I_A + I_B + I_C = 7.9 \text{ A}$$

$$I_B + I_C = 7.9 - 2.5$$

$$I_B + I_C = 5.4 \text{ A}$$

Given, $I_B = 2I_C$

$$\therefore \quad 3I_C = 5.4 \text{ A}$$

$$I_C = \frac{5.4}{3} = 1.8 \text{ A}$$

$$I_B = 3.6 \text{ A}$$

$$R_B = \frac{120}{3.6} = 33.33\ \Omega$$

$$R_C = \frac{120}{1.8} = 66.67\ \Omega$$

EXAMPLE 2.7 **What is the value of the unknown resistance R in the circuit shown in Figure 2.16, if the voltage across 500-Ω resistors is 2.5 V?**

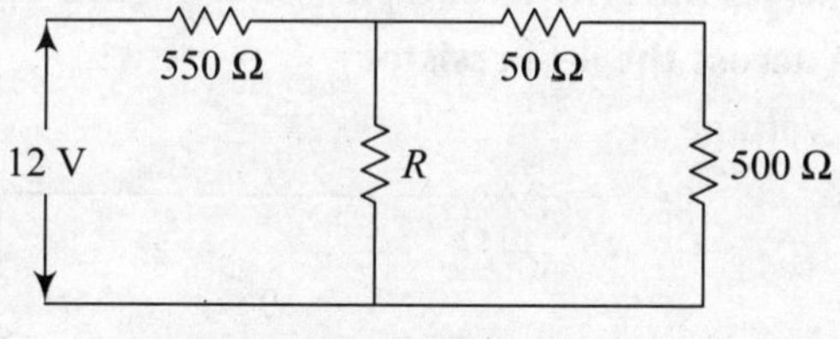

Figure 2.16 Electrical circuit

Solution Voltage drop across 500 Ω is 2.5 V.

$$\therefore \quad I_2 = \frac{2.5}{500} = 0.005 \text{ A}$$

Voltage drop across 50 Ω is $I_2 \times 50 = 0.005 \times 50 = 0.25$ V

Total voltage drop across 50- and 500-Ω resistors = 0.25 V + 2.5 V = 2.75 V. This voltage drop is applied across resistor R because both are connected parallel. Therefore, the voltage drop across resistor R is 2.75 V.

Voltage drop across the 550-Ω resistor = 12 – 2.75 = 9.25 V

$$I = \frac{9.25}{550} = 0.0168 \text{ A}$$

$$I_1 = I - I_2 = 0.0168 - 0.005 = 0.0118 \text{ A}$$

$$I_1 = \frac{V}{R}$$

$$0.0118 = \frac{2.75}{R}$$

$$R = 233\ \Omega$$

EXAMPLE 2.8 **In the network shown in Figure 2.17, find *R* if R_{eq} = 200 Ω.**

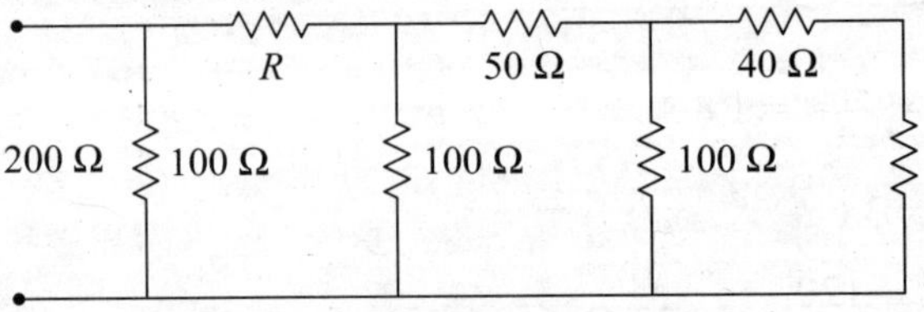

Figure 2.17

Solution $$\frac{(50 + R)100}{50 + R + 100} = 200$$

$$R = -250\ \Omega$$

EXAMPLE 2.9 **In series parallel circuit shown in Figure 2.18, find**

(i) The voltage across the 4-Ω resistor
(ii) The supply voltage

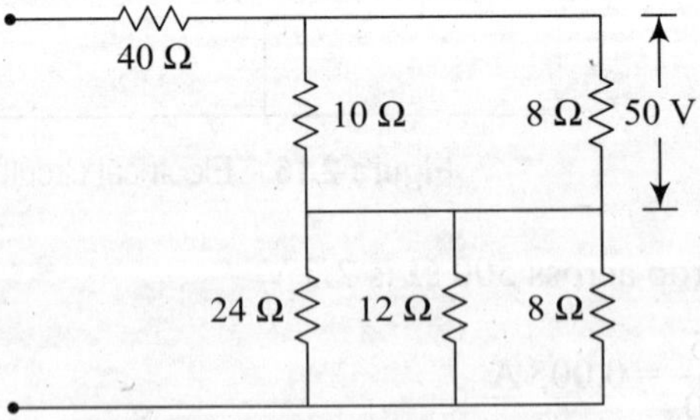

Figure 2.18 Series parallel circuit

Solution

(i) Voltage across 8 Ω and 10 Ω is given as 50 V.
Current through 8-Ω resistor, I_8 = 50/8 = 6.25 A

Current through 10-Ω resistor, $I_{10} = 50/10 = 5$ A

Current through 4 Ω is the sum of currents I_8 and I_{10}

$I_4 = 6.25 + 5 = 11.25$ A

So, voltage drop across 4 Ω is $11.25 \times 4 = 45$ V

(ii) For the sake of clarity, we can determine the above circuit as follows:

24 Ω, 12 Ω, and 8 Ω are parallel = 4 Ω

Voltage across this combination = $1R = 11.25 \times 4 = 45$ V

Therefore, the supply voltage = 45 + 50 + 45 = 140 V

2.5 SOURCE TRANSFORMATION TECHNIQUE

For the sake of numerical calculations, it is possible to replace a current source with an equivalent voltage source and vice versa without affecting the response in the rest of the circuit.

A given voltage source with a series resistance can be converted into an equivalent current source with a parallel resistance. Similarly, a current source with a parallel resistance can be converted into a voltage source with a series resistance.

It should be kept in mind that a voltage source series resistance combination is equivalent to a current source parallel resistance combination if and only if their (i) respective open-circuit voltages are equal and (ii) respective short-circuit currents are equal.

The two sources are said to be equivalent, if they supply equal load current to the load, with same load connected across its terminals. How the sources are transformed is explained below with the help of circuit diagram.

Consider a practical voltage source as shown in Figure 2.19 in series with internal resistance R_S and load resistance R_L. Similarly, also consider the current source as shown in Figure 2.20.

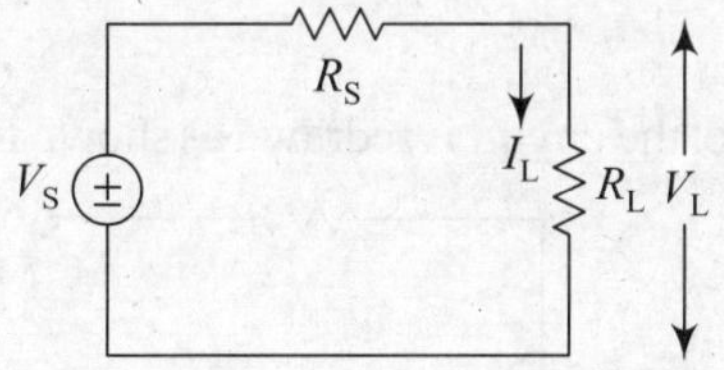

Figure 2.19 Practical voltage source

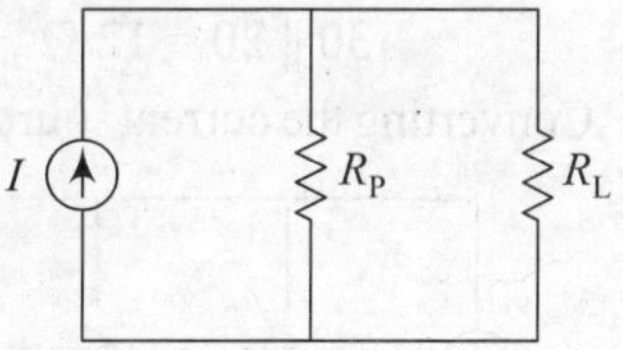

Figure 2.20 Practical current source

As shown in Figure 2.19, the load voltage can be calculated as

$$V_L = V_S - I_L R_S \tag{2.30}$$

The open-circuit voltage $V_{OC} = V_S$ and the short-circuit current $I_{SC} = V_S/R_S$. From Figure 2.20,

$$I_L = I_S - I = I_S - \frac{V_L}{R_P} \tag{2.31}$$

The open-circuit voltage $V_{OC} = I_S R_P$; the short-circuit current $I_{SC} = I_S$. The two sources are said to be equal, when they produce equal amount of current and voltage across the load resistance.

Therefore, by equating the open-circuit voltages and short-circuit currents of the above two sources, we obtain

$$\left.\begin{aligned} V_{\infty} &= I_S R_P = V_S \\ I_{SC} &= I_S = \frac{V_S}{R_S} \end{aligned}\right\} \tag{2.32}$$

From Eq. (2.32), $R_P = R_S = R_I$. Therefore, we can write $V_S = I_S = R_I$, where R_I is the internal resistance of the voltage or current source. Thus, any practical voltage source can be converted into current source and vice versa.

Note: A voltage source with voltage V_S and internal series resistance R_I can be replaced by a current source $I_S = V_S/R_I$ in parallel with internal resistance R_I. Similarly, a current source can be replaced with a voltage source $V_S = I_S R_I$ in series with an internal resistance R_I.

EXAMPLE 2.10 **Calculate V_X in the circuit shown in Figure 2.21 using the method of source transformation.**

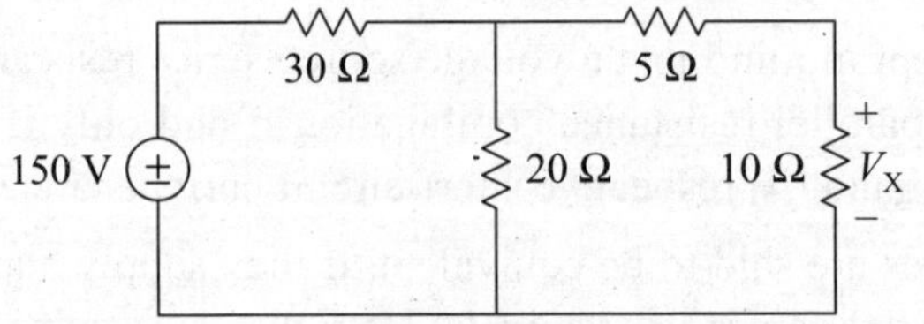

Figure 2.21

Solution Transforming the voltage source into current source, the circuit is redrawn as shown in Figure 2.22(a).

$$30 \parallel 20 = 12\ \Omega$$

Converting the current source into voltage source, the circuit is redrawn as shown in Figure 2.22(b).

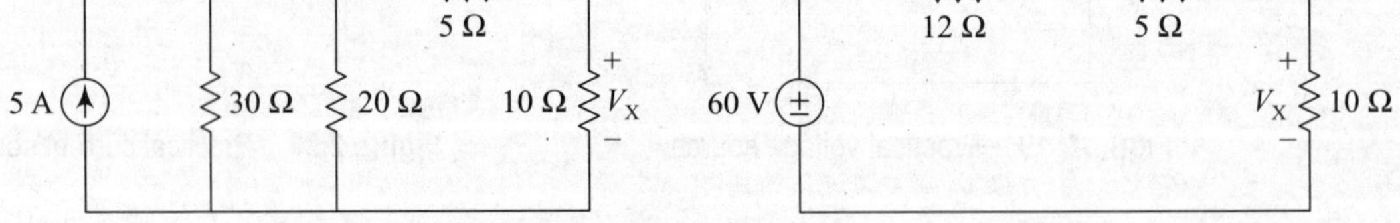

(a) (b)

Figure 2.22

Using the voltage division rule

$$V_X = V \times \frac{10}{10 + 12 + 5} = 60 \times \frac{10}{10 + 12 + 5} = 22.22\ \text{V}$$

EXAMPLE 2.11 For the circuit shown in Figure 2.23, determine the power dissipated in the 6-Ω resistor using source transformation.

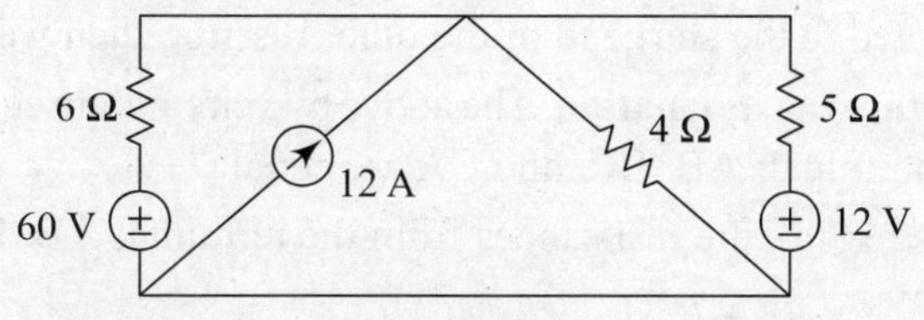

Figure 2.23

Solution Converting the given current source into voltage source and circuit is redrawn as shown in Figure 2.24.

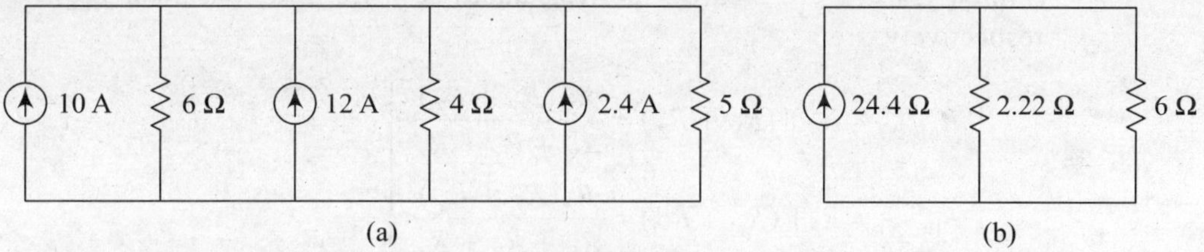

Figure 2.24

Current supplied by the voltage source, $I = \dfrac{50}{10.5} = 4.76$ A

Current through the 6-Ω resistor, $I_6 = 4.76 \times \dfrac{18}{24} = 3.57$ A

Power dissipated in the 6-Ω resistor, $P_6 = \; I_6^2 R = 3.57^2 \times 6 = 76.5$ W

2.6 STAR–DELTA TRANSFORMATIONS

Like series and parallel connections, resistors may be connected in "star" or Y and "delta" or Δ connection as shown in Figures 2.25(a) and (b). For simplification of networks, the networks may be converted from star to delta and vice versa. Three resistors R_A, R_B, and R_C can be connected in either star or delta.

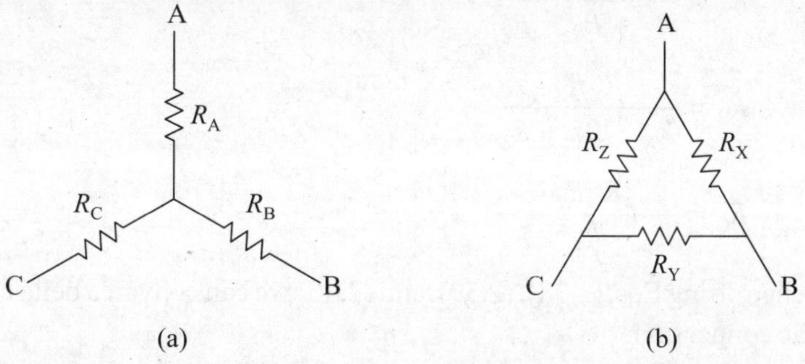

Figure 2.25 Star–Delta transformation: (a) star connection and (b) delta connection

Each resistor will have two ends, say the start end and the finish end. If all the finish ends are connected together to form a junction, then we call it as a star connection. If the finish end of one resistor is connected to the start end of the other resistor, then we call it as a delta connection.

(i) Delta to star transformation. The above two circuits are equal if their respective resistances from the terminals AB, BC, and CA are equal.

In Figure 2.25(a), the resistances from the terminals AB, BC, and CA, respectively, are

$$\left.\begin{aligned} R_{AB(Y)} &= R_A + R_B \\ R_{BC(Y)} &= R_B + R_C \\ R_{CA(Y)} &= R_C + R_A \end{aligned}\right\} \quad (2.33)$$

Similarly, in Figure 2.25(b), the resistances seen from the terminals AB, BC, and CA, respectively, are

$$\left.\begin{aligned} R_{AB(\Delta)} &= R_X \parallel (R_Y + R_Z) = \frac{R_X(R_Y + R_Z)}{R_X + R_Y + R_Z} \\ R_{BC(\Delta)} &= R_Z \parallel (R_X + R_Y) = \frac{R_Z(R_X + R_Y)}{R_X + R_Y + R_Z} \\ R_{CA(\Delta)} &= R_Y \parallel (R_X + R_Z) = \frac{R_Y(R_X + R_Z)}{R_X + R_Y + R_Z} \end{aligned}\right\} \quad (2.34)$$

Now, equate the resistances of star and delta circuits, i.e., Eqs. (2.33) and (2.34), we get

$$R_A + R_B = \frac{R_X(R_Y + R_Z)}{R_X + R_Y + R_Z} \quad (2.35)$$

$$R_B + R_C = \frac{R_Z(R_X + R_Y)}{R_X + R_Y + R_Z} \quad (2.36)$$

$$R_C + R_A = \frac{R_Y(R_X + R_Z)}{R_X + R_Y + R_Z} \quad (2.37)$$

Subtracting Eq. (2.36) from Eq. (2.35) and adding resultant to Eq. (2.37), we get

$$R_A = \frac{R_X R_Y}{R_X + R_Y + R_Z} \quad (2.38)$$

$$R_B = \frac{R_X R_Z}{R_X + R_Y + R_Z} \quad (2.39)$$

$$R_C = \frac{R_Y R_Z}{R_X + R_Y + R_Z} \quad (2.40)$$

Hence, using Eqs. (2.38), (2.39), and (2.40) we can convert a delta connection into equivalent star connection.

Note: The equivalent star resistance between any terminal and star point is equal to the product of the two resistances in delta, which are connected to the same terminal, divided by the sum of all three delta-connected resistances.

(ii) Star to delta transformation. Consider Eqs. (2.38), (2.39), and (2.40). Now multiplying Eqs. (2.38) and (2.39), (2.39) and (2.40), and (2.40) and (2.38), we get

$$R_A R_B = \frac{{R_X}^2 R_Y R_Z}{(R_X + R_Y + R_Z)^2} \tag{2.41}$$

$$R_B R_C = \frac{{R_Z}^2 R_X R_Y}{(R_X + R_Y + R_Z)^2} \tag{2.42}$$

$$R_C R_A = \frac{R_Y^2 R_X R_Y}{(R_X + R_Y + R_Z)^2} \tag{2.43}$$

Now, adding Eqs. (2.41), (2.42), and (2.43) and simplifying, we get

$$R_A R_B + R_B R_C + R_C R_A = \frac{R_X R_Y R_Z}{R_X + R_Y + R_Z} \tag{2.44}$$

Substituting Eq. (2.38) in the RHS of Eq. (2.44), we get

$$R_A R_B + R_B R_C + R_C R_A = R_A R_Z \tag{2.45}$$

Dividing Eq. (2.45) by R_A, we get

$$R_Z = R_B + R_C + \frac{R_B R_C}{R_A} \tag{2.46}$$

Similarly, substituting the remaining values, in the RHS of Eq. (2.44), we can write expressions for the remaining two resistances.

$$R_Y = R_A + R_B + \frac{R_A R_B}{R_C} \tag{2.47}$$

$$R_X = R_C + R_A + \frac{R_C R_A}{R_B} \tag{2.48}$$

Thus, using Eqs. (2.46), (2.47), and (2.48), we can convert the star into delta connection.

Note: The equivalent delta-connected resistance between any two terminals is the sum of the two resistances connected between the same two terminals, and star point, respectively, in star, plus the product of the same two star resistances divided by the third star resistance.

EXAMPLE 2.12 **By using star–delta transformation for Figure 2.26(a), find current *I* supplied by the battery.**

Solution Consider delta ABC. Convert above delta into star, by using delta/star transformation equations

$$R_A = \frac{50 \times 20}{50 + 30 + 20} = 10\ \Omega$$

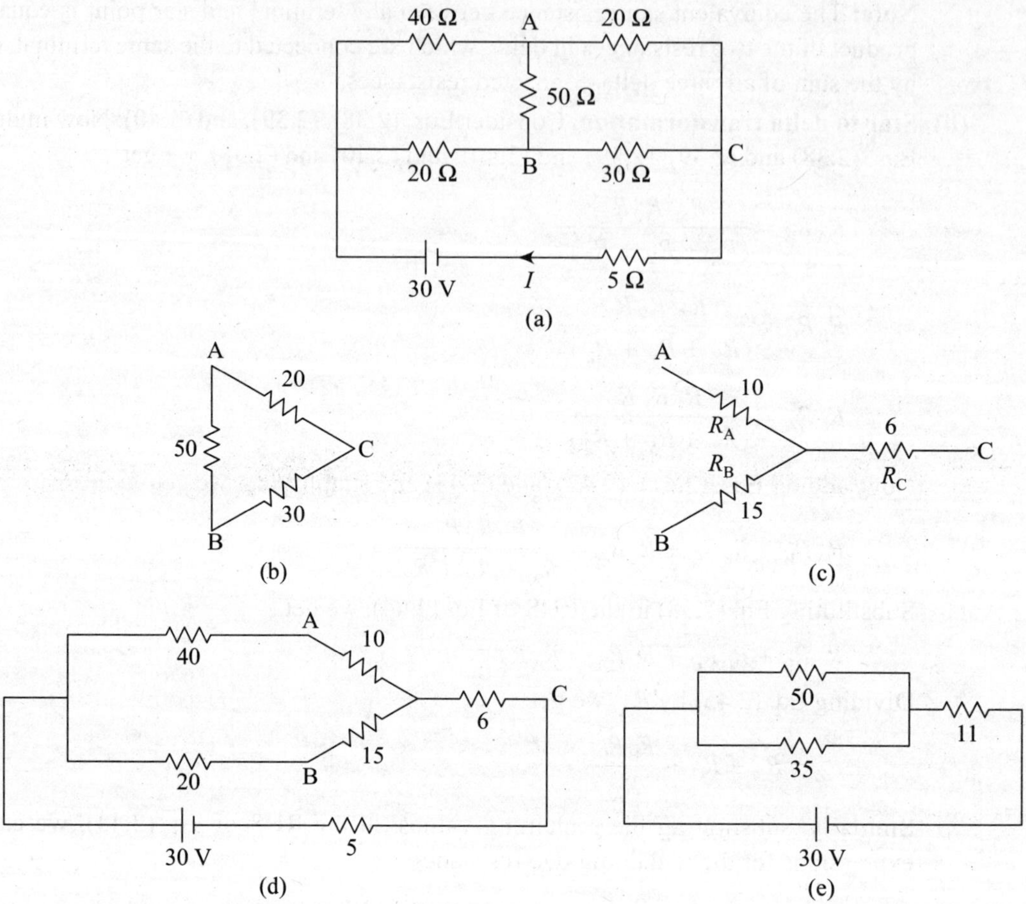

Figure 2.26

$$R_B = \frac{50 \times 30}{50 + 30 + 20} = 15\ \Omega$$

$$R_C = \frac{30 \times 20}{50 + 30 + 20} = 6\ \Omega$$

Replacing the above network into original network, we get

40 Ω and 10 Ω connected in series = 40 + 10 = 50 Ω

20 Ω and 15 Ω connected in series = 20 + 15 = 35 Ω

6 Ω and 5 Ω connected in series = 6 + 5 = 11 Ω

$$50\ \Omega \text{ and } 35\ \Omega \text{ connected in parallel} = \frac{35 \times 50}{35 + 50} = 20.58\ \Omega$$

11 Ω resistance in series with above combination = 11 + 20.58 = 31.58 Ω

$$I = \frac{V}{R} = \frac{30}{31.5} = 0.95 \text{ A}$$

EXAMPLE 2.13 **Find the power supplied by the battery in the network shown in Figure 2.27(a).**

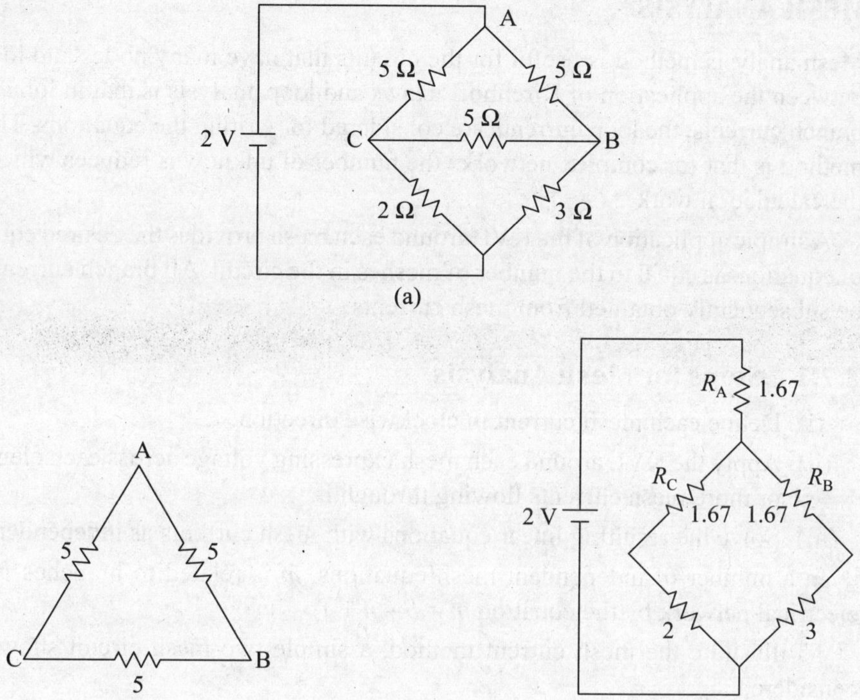

Figure 2.27

Solution Consider delta ABC

$$R_A = \frac{5 \times 5}{5 + 5 + 5} = 1.67\ \Omega$$

$$R_B = 1.67\ \Omega$$

$$R_C = 1.67\ \Omega$$

1.67 Ω and 3 Ω connected in series = 1.67 + 3 = 4.67 Ω

1.67 Ω and 2 Ω connected in series = 1.67 + 2 = 3.67 Ω

4.67 Ω and 3.67 Ω connected in parallel $= \frac{4.67 \times 3.67}{4.67 + 3.67} = \frac{17.138}{8.34} = 2.054\ \Omega$

2.054 Ω connected in series to 1.67 Ω = 2.054+ 1.67 = 3.72 Ω

$$I = \frac{2}{3.72} = 0.536 \text{ A}$$

$$P = VI = 2 \times 0.536 = 1.073 \text{ W}$$

Therefore, power supplied by the battery is given by 1.073 W.

2.7 MESH ANALYSIS

Mesh analysis method is useful for the circuits that have many nodes and loops. The difference between the application of Kirchhoff's laws and loop analysis is that in loop analysis, instead of branch currents, the loop currents are considered for writing the equations. The advantage of this method is that for complex networks the number of unknowns reduces which greatly simplifies the calculation work.

A simple application of the KVL around each mesh provides the desired equations. The number of equations is equal to the number of meshes in the circuit. All branch currents and voltages may be subsequently obtained from mesh currents.

2.7.1 Steps for Mesh Analysis

(i) Define each mesh current in clockwise direction.

(ii) Apply the KVL around each mesh expressing voltage across each element in terms of one or more mesh currents flowing through it.

(iii) Solve the resulting linear equations with mesh currents as independent variables.

In a number of independent mesh equations, m is related to branches b and nodes n in an electrical network by the equation $m = b - n + 1$.

To illustrate the mesh current method, a simple two-mesh circuit shown in Figure 2.28 is considered.

From Figure 2.28, it is observed that there are two meshes. Let us assume the mesh currents I_1 and I_2 with directions as shown in Figure 2.28.

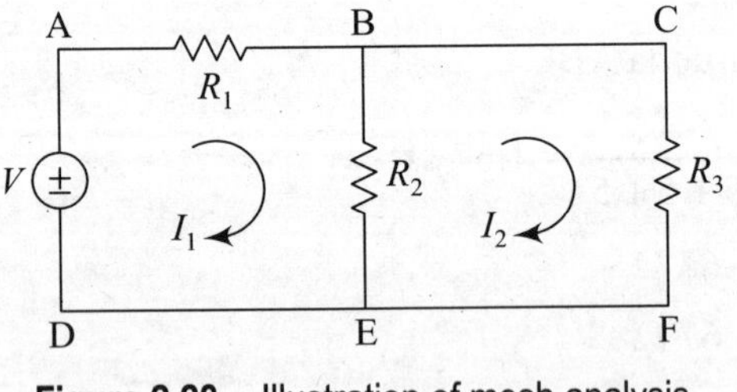

Figure 2.28 Illustration of mesh analysis

By applying KVL for mesh 1,

$$V = I_1R_1 + R_2(I_1 - I_2) \tag{2.49}$$

Similarly, if we consider mesh 2 and apply the KVL, we get

$$R_2(I_2 - I_1) + R_3I_2 = 0 \tag{2.50}$$

Rewriting Eqs. (2.49) and (2.50) and the corresponding mesh current equations are

$$I_1(R_1 + R_2) - I_2R_2 = V \tag{2.51}$$

$$-I_1R_2 + (R_2 + R_3)I_2 = 0 \tag{2.52}$$

By solving Eqs. (2.51) and (2.52), we can get the currents I_1 and I_2.

EXAMPLE 2.14 Determine the currents in various elements of the circuit shown in Figure 2.29 using mesh analysis.

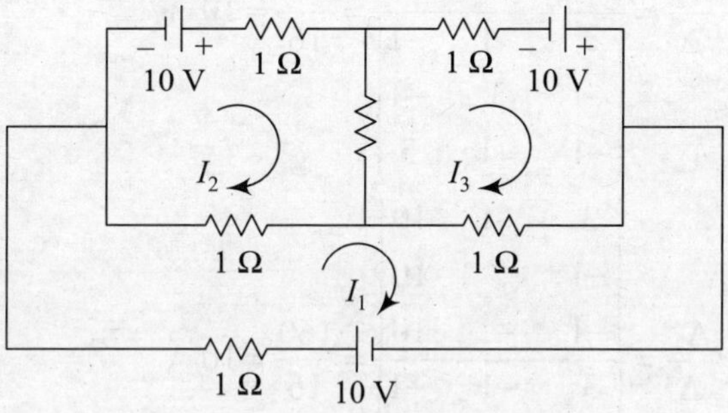

Figure 2.29 Electrical circuit

Solution By applying KVL for mesh 1,

$$I_1 + (I_1 - I_2) + (I_1 - I_2) = 10$$

Similarly, by applying KVL for mesh 2,

$$I_2 + (I_2 - I_3) + (I_2 - I_1) = 10$$

For mesh 3,

$$I_3 + (I_3 - I_1) + (I_3 - I_2) = 10$$

Rewriting all the mesh equations

$$3I_1 - I_2 - I_3 = 10$$

$$-I_1 + 3I_2 - I_3 = 10$$

$$-I_1 + 3I_3 - I_2 = 10$$

Solving the above equations, we can get I_1, I_2, and I_3.

These equations can be solved using Crammer's rule.

$$\begin{bmatrix} 3 & -1 & -1 \\ -1 & 3 & -1 \\ -1 & -1 & 3 \end{bmatrix} \begin{bmatrix} I_1 \\ I_2 \\ I_3 \end{bmatrix} = \begin{bmatrix} 10 \\ 10 \\ 10 \end{bmatrix}$$

$$I_1 = \frac{\Delta_1}{\Delta} = \frac{\begin{vmatrix} 10 & -1 & -1 \\ 10 & 3 & -1 \\ 10 & -1 & 3 \end{vmatrix}}{\begin{vmatrix} 3 & -1 & -1 \\ -1 & 3 & -1 \\ -1 & -1 & 3 \end{vmatrix}} = \frac{160}{16} = 10 \text{ A}$$

Similarly,

$$I_2 = \frac{\Delta_2}{\Delta} = \frac{\begin{vmatrix} 3 & 10 & -1 \\ -1 & 10 & -1 \\ -1 & 10 & 3 \end{vmatrix}}{\begin{vmatrix} 3 & -1 & -1 \\ -1 & 3 & -1 \\ -1 & -1 & 3 \end{vmatrix}} = \frac{160}{16} = 10\text{ A}$$

$$I_3 = \frac{\Delta_3}{\Delta} = \frac{\begin{vmatrix} 3 & -1 & 10 \\ -1 & 3 & 10 \\ -1 & -1 & 10 \end{vmatrix}}{\begin{vmatrix} 3 & -1 & -1 \\ -1 & 3 & -1 \\ -1 & -1 & 3 \end{vmatrix}} = \frac{160}{16} = 10\text{ A}$$

Therefore, the currents are $I_1 = I_2 = I_3 = 10$ A.

2.7.2 Supermesh

Suppose any of the branches in the network has a current source, then it is slightly difficult to apply mesh analysis straightforward because a loop cannot be defined through the current source as the voltage drop across the current source is unknown from the KVL point of view. One way to overcome this difficulty is by applying the supermesh technique. A supermesh is constituted by two adjacent loops that have a common current source. This is illustrated with the following example.

Consider the circuit shown in Figure 2.30.

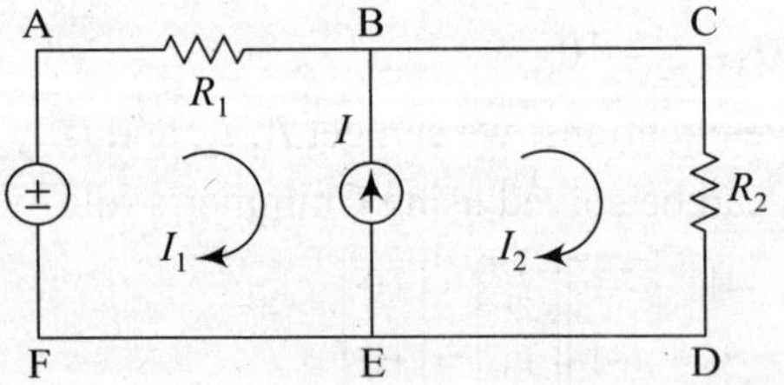

Figure 2.30 Illustration of supermesh analysis

In this circuit, a current source is present in branch BE, hence we cannot directly write the KVL equation. This forms a supermesh along the outer periphery of the circuit, "ABCDEF" which forms the supermesh. By applying the KVL for the supermesh, we can get one equation and by establishing the relationship between two mesh currents and the current source, we can get one more equation. By solving these two equations, the mesh currents I_1 and I_2 can be determined.

Writing the supermesh equation, i.e., applying the KVL for "ABCDEF," we get

$$-V_S + I_1R_1 + I_2R_2 = 0 \tag{2.53}$$

Relating the current source between the two mesh currents,

$$I = I_2 - I_1 \tag{2.54}$$

By solving the above equations, we can get I_1 and I_2.

EXAMPLE 2.15 **Evaluate the mesh currents using the mesh analysis for the circuit shown in Figure 2.31.**

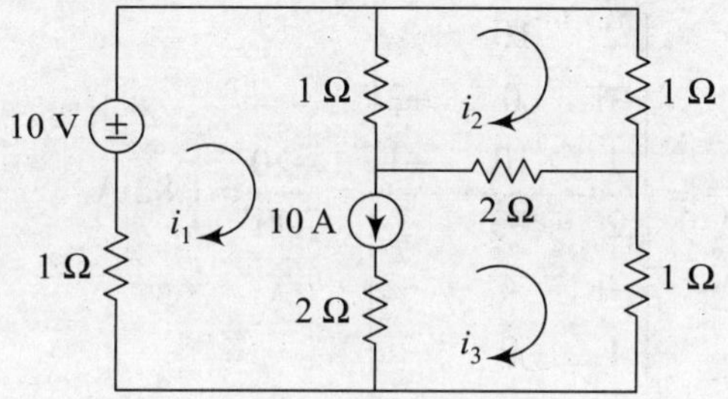

Figure 2.31

Solution Here, 10-A current source is common to two meshes. Hence, it creates a supermesh along the boundaries of mesh 1 and mesh 3.

Writing the KVL for supermesh,

$$-10 + I_1 + 1(I_1 - I_2) + 2(I_3 - I_2) + I_3 = 0$$

Writing the KVL for mesh 2,

$$1(I_2 - I_1) + 1(I_2) + 2(I_2 - I_3) = 0$$

Relating the source current with assumed mesh currents,

$$I_1 - I_3 = 10$$

Rewriting the above equations, we get

$$2I_1 - 3I_2 + 3I_3 = 10$$

$$-I_1 + 4I_2 - 2I_3 = 0$$

$$I_1 - I_3 = 10$$

Solving the above equations using Crammer's rule, we get

$$\begin{bmatrix} 2 & -3 & 3 \\ -1 & 4 & -2 \\ 1 & 0 & -1 \end{bmatrix}\begin{bmatrix} I_1 \\ I_2 \\ I_3 \end{bmatrix} = \begin{bmatrix} 10 \\ 0 \\ 10 \end{bmatrix}$$

$$I_1 = \frac{\Delta_1}{\Delta} = \frac{\begin{vmatrix} 10 & -3 & 3 \\ 0 & 4 & -2 \\ 10 & 0 & -1 \end{vmatrix}}{\begin{vmatrix} 2 & -3 & 3 \\ -1 & 4 & -2 \\ 1 & 0 & -1 \end{vmatrix}} = \frac{-100}{-11} = 9.09 \text{ A}$$

$$I_2 = \frac{\Delta_2}{\Delta} = \frac{\begin{vmatrix} 2 & 10 & 3 \\ -1 & 0 & -2 \\ 1 & 10 & -1 \end{vmatrix}}{\begin{vmatrix} 2 & -3 & 3 \\ -1 & 4 & -2 \\ 1 & 0 & -1 \end{vmatrix}} = \frac{-20}{-11} = 1.82 \text{ A}$$

$$I_3 = \frac{\Delta_3}{\Delta} = \frac{\begin{vmatrix} 2 & -3 & 10 \\ -1 & 4 & 0 \\ 1 & 0 & 10 \end{vmatrix}}{\begin{vmatrix} 2 & -3 & 3 \\ -1 & 4 & -2 \\ 1 & 0 & -1 \end{vmatrix}} = \frac{10}{-11} = -0.909 \text{ A}$$

Hence,

$$I_1 = 9.09 \text{ A}; \quad I_2 = 1.82 \text{ A}; \quad I_3 = -0.909 \text{ A}$$

EXAMPLE 2.16 **Determine the currents I_1, I_2, and I_3 for the circuit shown in Figure 2.32 by writing mesh equations.**

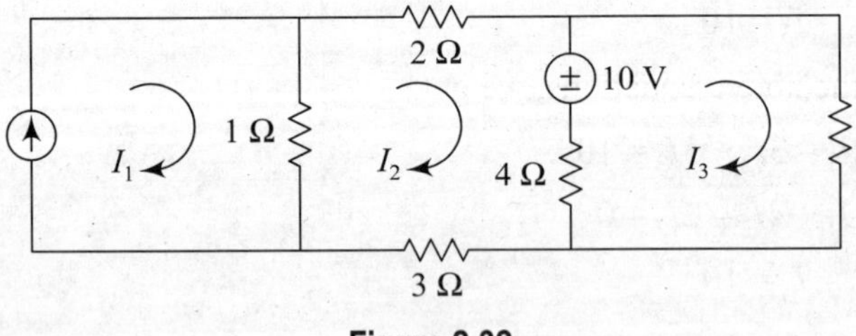

Figure 2.32

Solution In Figure 2.32, current source lies on the boundary of the circuit, hence we can ignore the first mesh. The KVL is applied for second and third meshes only.

From the first mesh, I_1= 5A

Writing the KVL for mesh 2,

$$1(I_2 - I_1) + 2(I_2) + 10 + 4(I_2 - I_3) + 3I_2 = 0$$

Writing the KVL for mesh 3,

$$2I_3 + 4(I_3 - I_2) = 10$$

Rewriting the above equations, we get

$$-I_1 + 10I_2 - 4I_3 = -10$$

Substituting $I_1 = 5\text{A}$ and simplifying, we get

$$-10I_2 + 4I_3 = 5$$

For mesh 3,

$$-4I_2 + 6I_3 = 10$$

Solving the above two equations, we get

$$I_2 = 0.227 \text{ A}; \quad I_3 = 1.818 \text{ A}$$

2.8 NODAL ANALYSIS

Nodal analysis method is mainly based on the KCL. Node voltage analysis is one of the general methods for the analysis of electrical networks. The node voltage method is based on defining the voltage at each node as an independent variable. One of the nodes is selected as reference node (usually ground) and each of the other nodes is referenced to this node. The KCL is applied at each of the nodes except the reference node. If a circuit consists of n nodes, there will be $n - 1$ equations to solve. However, one of the node voltages is the reference voltage and is therefore already known, since it is usually assumed to be zero. Thus, the $n - 1$ independent linear equations may be written in terms of the $n - 1$ independent node voltage variables. After solving these equations we can get the $n - 1$ node voltages. The branch currents and voltages may be determined from nodal voltages.

2.8.1 Steps for Nodal Analysis

(i) Select a reference node, i.e., usually ground.

(ii) Define the remaining $n - 1$ nodal voltages with respect to this reference node as independent variables.

(iii) Express each current in terms of the adjacent node voltages. Apply the KCL at each of the $n - 1$ nodes.

(iv) Solve the linear system of $n - 1$ equations in $n - 1$ unknown node voltages.

To illustrate the nodal analysis, a simple two-node circuit is considered which is shown in Figure 2.33(a).

Let the voltages at nodes 1 and 2 are V_1 and V_2. Mark various branch currents as shown in Figure 2.33(b). Ground node is taken as reference node which is at zero potential.

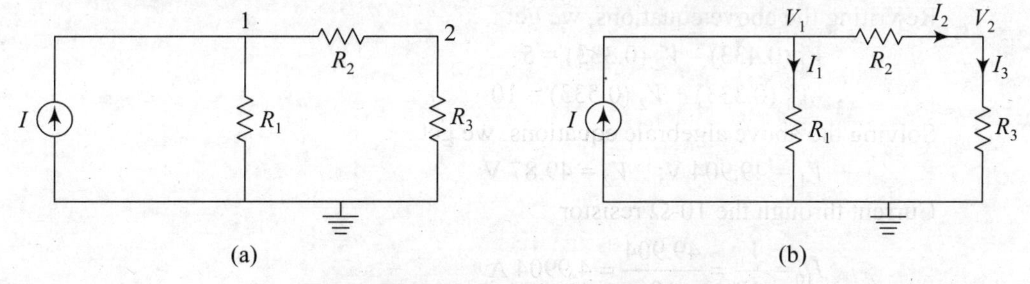

Figure 2.33 Illustration of nodal analysis

Writing a KCL equation for node 1: Current I enters node 1, and I_1 and I_2 leave node 1; hence, entering currents are assigned with positive direction and leaving currents are assigned with negative direction, accordingly

$$I - I_1 - I_2 = 0 \tag{2.55}$$

KCL at node 2,

$$I_2 - I_3 = 0 \tag{2.56}$$

Currents in the above equations can be expressed in terms of node voltages as

$$I - \frac{V_1}{R_1} - \frac{V_1 - V_2}{R_2} = 0 \tag{2.57}$$

$$\frac{V_1 - V_2}{R_2} = \frac{V_2}{R_3} \tag{2.58}$$

Solving the above equations, we can get the node voltages V_1 and V_2 and can calculate the branch currents.

EXAMPLE 2.17 Determine the currents in each branch using the nodal analysis for the circuit shown in Figure 2.34.

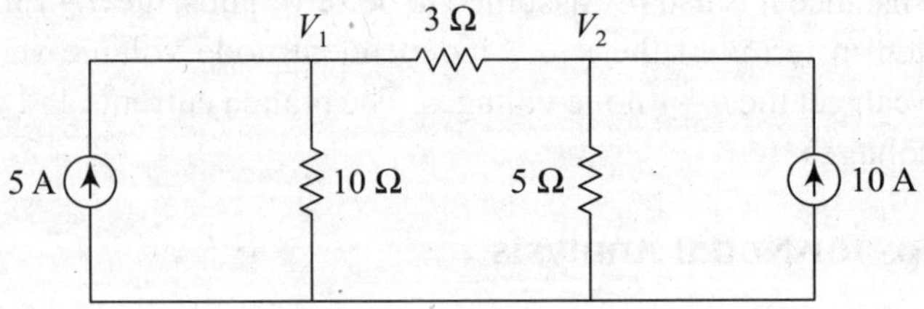

Figure 2.34 Electrical circuit

Solution There are two nodes in the given circuit. The third node is taken as reference node.

Writing the KCL equation for node 1,

$$5 = \frac{V_1}{10} + \frac{V_1 - V_2}{3}$$

Writing the KCL equation for node 2,

$$10 = \frac{V_2}{5} + \frac{V_2 - V_1}{3}$$

Rewriting the above equations, we get

$$V_1 (0.433) - V_2 (0.333) = 5$$

$$-V_1 (0.333) + V_2 (0.533) = 10$$

Solving the above algebraic equations, we get

$$V_1 = 49.904 \text{ V}; \quad V_2 = 49.87 \text{ V}$$

Current through the 10-Ω resistor

$$I_{10} = \frac{V_1}{10} = \frac{49.904}{10} = 4.9904 \text{ A}$$

Current through the 3-Ω resistor

$$I_3 = \frac{49.904 - 49.87}{3} = 0.011 \text{ A}$$

Current through the 5-Ω resistor

$$I_5 = \frac{V_2}{5} = \frac{49.87}{5} = 9.974 \text{ A}$$

2.8.2 Supernode

Suppose any one of the branches in the network has a voltage source alone, then it becomes difficult to apply the nodal analysis. One way to overcome this difficulty is the supernode technique. In this method, the two adjacent nodes that are connected by a voltage source are reduced to a single node and then equations are formed by applying the KCL as usual. This is explained with the following example as shown in Figure 2.35.

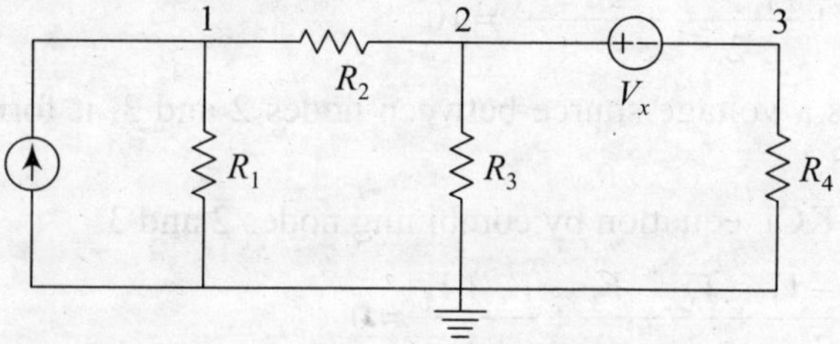

Figure 2.35 Illustration of supernodal analysis

In this circuit there are four nodes including the ground node. The three node voltages are represented by V_1, V_2, and V_3.

There is a voltage source alone between nodes 2 and 3. The area connected to the voltage source between the two nodes is called *supernode*. Writing all the node equations separately, relating the voltage source connected between two node voltages separately, and solving all the equations give node voltages.

Writing the KCL equation for node 1,

$$I = \frac{V_1}{R_1} + \frac{V_1 - V_2}{R_2} \tag{2.59}$$

Due to the presence of voltage source between nodes 2 and 3, it becomes difficult to apply KCL. According to the supernode technique, we can write a combined equation for nodes 2 and 3 as given below

$$\frac{V_2 - V_1}{R_2} + \frac{V_2}{R_3} + \frac{V_3}{R_4} = 0 \tag{2.60}$$

The voltage source equation is

$$V_2 - V_3 = V_x \tag{2.61}$$

Solving the above equations, we can get the node voltages V_1, V_2, and V_3.

EXAMPLE 2.18 **Determine the current in the 3-Ω resistor for the circuit shown in Figure 2.36 using the nodal analysis.**

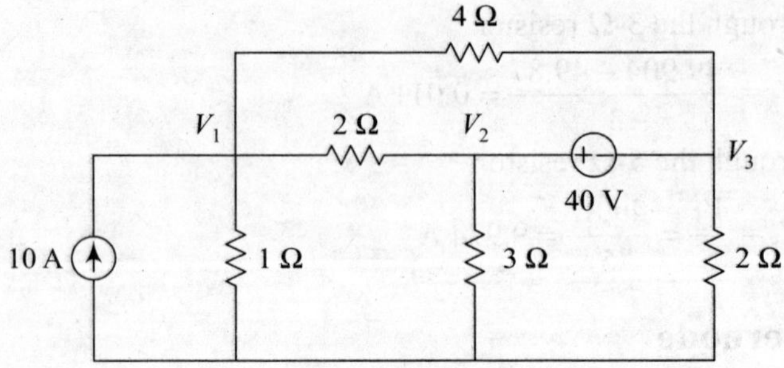

Figure 2.36

Solution Let the voltages be V_1, V_2, and V_3 for nodes 1, 2, and 3, respectively.

Writing the KCL equation for node 1,

$$\frac{V_1}{1}+\frac{V_1-V_2}{2}+\frac{V_1-V_3}{4}=10$$

Since there is a voltage source between nodes 2 and 3, it forms a supernode by combining nodes 2 and 3.

We write the KCL equation by combining nodes 2 and 3.

$$\frac{V_2-V_1}{2}+\frac{V_2}{3}+\frac{V_3}{2}+\frac{V_3-V_1}{4}=0$$

Relating the source voltage with V_2 and V_3,

$$V_2-V_3=40$$

Rearranging the above equations, we get

$$1.75V_1-0.5V_2-0.25V_3=10$$
$$-0.75V_1+0.833V_2+0.75V_3=0$$
$$V_2-V_3=40$$

Solving the above equations using Crammer's rule, we get

$$\begin{bmatrix} 1.75 & -0.5 & -0.25 \\ -0.75 & 0.833 & 0.75 \\ 0 & 1 & -1 \end{bmatrix}\begin{bmatrix} V_1 \\ V_2 \\ V_3 \end{bmatrix}=\begin{bmatrix} 10 \\ 0 \\ 40 \end{bmatrix}$$

$$V_1=\frac{\Delta_1}{\Delta}=\frac{\begin{vmatrix} 10 & -0.5 & -0.25 \\ 0 & 0.833 & 0.75 \\ 40 & 1 & -1 \end{vmatrix}}{\begin{vmatrix} 1.75 & -0.5 & -0.25 \\ -0.75 & 0.833 & 0.75 \\ 0 & 1 & -1 \end{vmatrix}}=\frac{-22.5}{-2.21}=10.18$$

$$V_2 = \frac{\Delta_2}{\Delta} = \frac{\begin{vmatrix} 1.75 & 10 & -0.25 \\ -0.75 & 0 & 0.75 \\ 0 & 40 & -1 \end{vmatrix}}{\begin{vmatrix} 1.75 & -0.5 & -0.25 \\ -0.75 & 0.833 & 0.75 \\ 0 & 1 & -1 \end{vmatrix}} = \frac{-52.5}{-2.21} = 23.75$$

$$V_3 = \frac{\Delta_3}{\Delta} = \frac{\begin{vmatrix} 1.75 & -0.5 & 10 \\ -0.75 & 0.833 & 0 \\ 0 & 1 & 40 \end{vmatrix}}{\begin{vmatrix} 1.75 & -0.5 & -0.25 \\ -0.75 & 0.833 & 0.75 \\ 0 & 1 & -1 \end{vmatrix}} = \frac{35.8}{-2.21} = -16.2 \text{ A}$$

Current through the 3-Ω resistor, $I_3 = \frac{V_2}{3} = \frac{23.75}{3} = 7.91\text{A}$

EXAMPLE 2.19 **Using the nodal analysis, determine the node voltages for the circuit shown in Figure 2.37.**

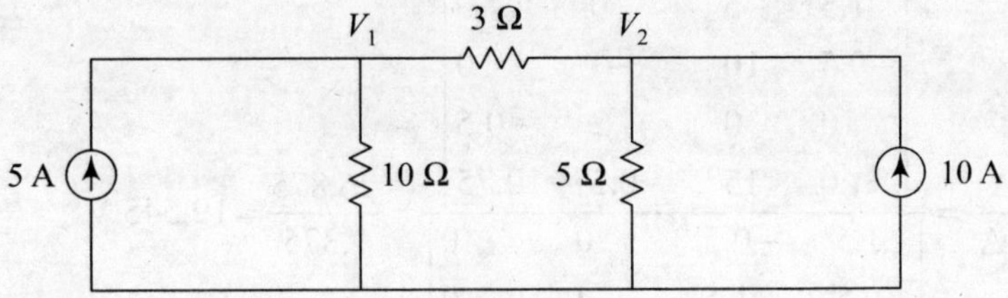

Figure 2.37

Solution Writing the KCL equation for node 1,

$$\frac{V_1}{1} + \frac{V_1 - V_2}{2} + 15 = 0$$

Writing the KCL equation for node 2,

$$\frac{V_2 - V_1}{2} + \frac{V_2 - V_3}{1} = 10$$

Writing the KCL equation for node 3,

$$\frac{V_4 - V_3}{2} + \frac{V_4}{4} = 15$$

Rewriting the above nodal equations, we get

$$1.5V_1 - 0.5V_2 = -5$$

$$-0.5V_1 + 1.5V_2 - V_3 = 10$$
$$-V_2 + 2V_3 - 0.5V_4 = 0$$
$$-0.5V_3 + 0.75V_4 = 15$$

Solving these equations using Crammer's rule, we get

$$\begin{bmatrix} 1.5 & -0.5 & 0 & 0 \\ -0.5 & 1.5 & -1 & 0 \\ 0 & -1 & 2 & -0.5 \\ 0 & 0 & -0.5 & 0.75 \end{bmatrix} \begin{bmatrix} V_1 \\ V_2 \\ V_3 \\ V_4 \end{bmatrix} = \begin{bmatrix} -5 \\ 10 \\ 0 \\ 15 \end{bmatrix}$$

$$V_1 = \frac{\Delta_1}{\Delta} = \frac{\begin{vmatrix} -5 & -0.5 & 0 & 0 \\ 10 & 1.5 & -1 & 0 \\ 0 & -1 & 2 & -0.5 \\ 15 & 0 & -0.5 & 0.75 \end{vmatrix}}{\begin{vmatrix} 1.5 & -0.5 & 0 & 0 \\ -0.5 & 1.5 & -1 & 0 \\ 0 & -1 & 2 & -0.5 \\ 0 & 0 & -0.5 & 0.75 \end{vmatrix}} = \frac{4.375}{1.375} = 3.182 \text{ V}$$

$$V_2 = \frac{\Delta_2}{\Delta} = \frac{\begin{vmatrix} 1.5 & -5 & 0 & 0 \\ -0.5 & 10 & -1 & 0 \\ 0 & 0 & 2 & -0.5 \\ 0 & 15 & -0.5 & 0.75 \end{vmatrix}}{\begin{vmatrix} 1.5 & -0.5 & 0 & 0 \\ -0.5 & 1.5 & -1 & 0 \\ 0 & -1 & 2 & -0.5 \\ 0 & 0 & -0.5 & 0.75 \end{vmatrix}} = \frac{26.875}{1.375} = 19.545 \text{ V}$$

$$V_3 = \frac{\Delta_3}{\Delta} = \frac{\begin{vmatrix} 1.5 & -0.5 & -5 & 0 \\ -0.5 & 1.5 & 10 & 0 \\ 0 & -1 & 0 & -0.5 \\ 0 & 0 & 15 & 0.75 \end{vmatrix}}{\begin{vmatrix} 1.5 & -0.5 & 0 & 0 \\ -0.5 & 1.5 & -1 & 0 \\ 0 & -1 & 2 & -0.5 \\ 0 & 0 & -0.5 & 0.75 \end{vmatrix}} = \frac{24.375}{1.375} = 17.72 \text{ V}$$

$$V_4 = \frac{\Delta_4}{\Delta} = \frac{\begin{vmatrix} 1.5 & -0.5 & 0 & -5 \\ -0.5 & 1.5 & -1 & 10 \\ 0 & -1 & 2 & 0 \\ 0 & 0 & -0.5 & 15 \end{vmatrix}}{\begin{vmatrix} 1.5 & -0.5 & 0 & 0 \\ -0.5 & 1.5 & -1 & 0 \\ 0 & -1 & 2 & -0.5 \\ 0 & 0 & -0.5 & 0.75 \end{vmatrix}} = \frac{43.75}{1.375} = 31.82\text{ V}$$

Therefore, the node voltages are $V_1 = 3.182$ V; $V_2 = 19.545$ V; $V_3 = 17.72$ V; and $V_4 = 31.82$ V.

2.9 NETWORK THEOREMS

Kirchhoff's laws can be used to solve all network problems. By using Kirchhoff's law it is possible to determine all the branch currents and voltages. Sometimes it is necessary to determine a particular branch current or voltage only (like load voltage drop or load current). In such cases, network theorems are useful.

2.9.1 Superposition Theorem

The superposition theorem may be stated as: "In any linear network containing more than one source (voltage or current), the total current in any branch is the algebraic sum of the individual currents produced by each source acting alone. Meanwhile all other sources are replaced by their respective internal resistances."

Explanation of superposition theorem

Many electrical circuits may contain more than one source of either voltage or current. In such a case, it becomes more convenient to solve the circuit for the desired response (voltage or current) produced by each source separately and then combine the results. This is done with the help of the superposition theorem. The superposition theorem is valid only for linear systems and is not applicable to power. This theorem is applicable to only voltages and currents. What follows is a list of steps to apply.

(i) While applying the superposition theorem only one source is kept in the circuit, making the other sources equal to zero.

(ii) Replace the voltage source by a short circuit by retaining their internal resistances.

(iii) Replace the current source by an open circuit by retaining their internal resistances.

(iv) In obtaining the algebraic sum, if the currents flow in the same direction they add to each other; however, if they are in opposite directions, they subtract from one another.

When dealing with the networks of more than one active source, it is possible to use a method that avoids solving a set of simultaneous linear equations. This is accomplished by obtaining the effects of each source separately.

Consider a circuit shown in Figure 2.38(a) as a combination of the voltage sources of 2 V and 4 V. The total voltage between terminals "AB" of circuit in Figure 2.38(a) is 6 V. Therefore, the total current flowing in circuit in Figure 2.38(a) is

$$I_T = \frac{2 + 4\ \text{V}}{1\ \text{k}\Omega} = 6\ \text{mA}$$

In the circuit in Figure 2.38(b), the 2-V source is acting alone, and the current I_1 is given by

$$I_1 = \frac{2\ \text{V}}{1\ \text{k}\Omega} = 2\ \text{mA}$$

In the circuit in Figure 2.38(c), the 4-V source is acting alone and the current I_2 is given by

$$I_2 = \frac{4\ \text{V}}{1\ \text{k}\Omega} = 4\ \text{mA}$$

The summation of individual currents I_1 and I_2 yields 6 mA, which is equal to the total current I_T.

This procedure can be applied to any passive network containing any number of voltage sources, current sources, or combination of both.

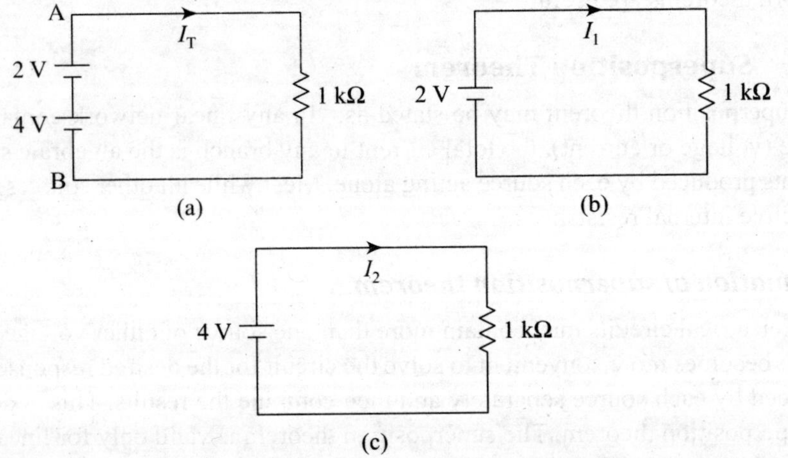

Figure 2.38 Circuit illustrating the method of superposition: (a) complete circuit; (b) considering a 2-V source only; and (c) considering a 4-V source only

EXAMPLE 2.20 **Using the method of superposition, determine the current through the 5-kΩ resistors for the circuit shown in Figure 2.39.**

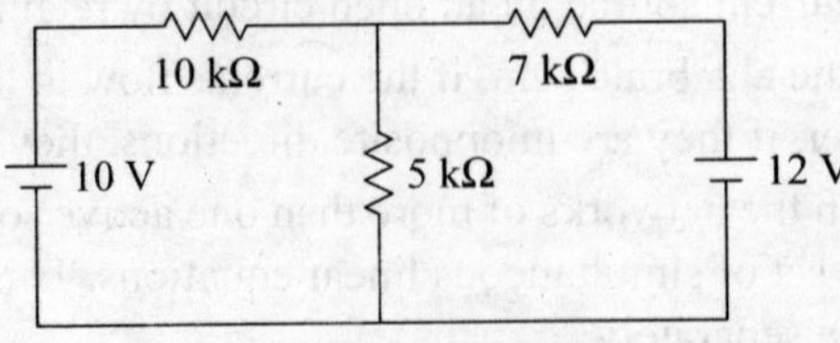

Figure 2.39

Solution **Case 1:** A 10-V source is acting alone by replacing voltage E_2 with a short circuit. In Figure 2.40, R_2 and R_3 are connected in parallel.

$$R_A = \frac{R_2 R_3}{R_2 + R_3} = \frac{5 \times 7}{5+7} = 2.92\ \text{k}\Omega$$

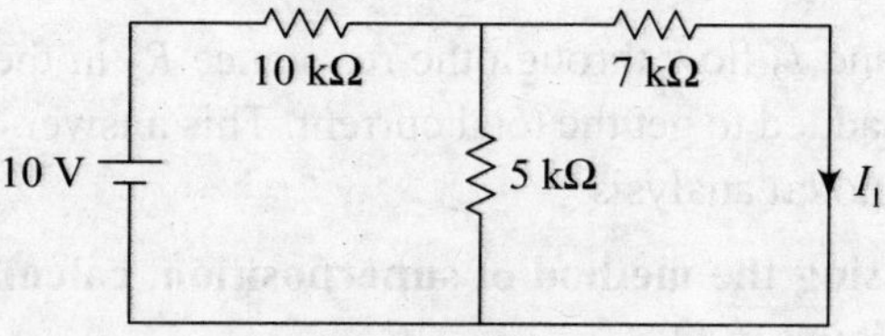

Figure 2.40

R_1 is in series with R_A and the total equivalent resistance is

$$R_T = R_1 + R_A = 10 + 2.92 = 12.92\ \text{k}\Omega$$

The current drawn from E_1 is

$$I_T = \frac{E_1}{R_T} = \frac{10}{12.92\ \text{k}\Omega} = 0.775\ \text{mA}$$

Using the current division rule, we get

$$I_1 = I_T \times \frac{R_3}{R_2 + R_3} = 0.775\ \text{wA} \times \frac{7}{5+7} = 0.45\ \text{mA}$$

$$I_1 = 0.45\ \text{mA}$$

Case 2: A 12-V source is acting alone by replacing voltage E_1 with a short circuit. In Figure 2.41, R_1 and R_2 are connected in parallel.

$$R_B = \frac{R_1 R_2}{R_1 + R_2} = \frac{10 \times 5}{10+5} = 3.33\ \text{k}\Omega$$

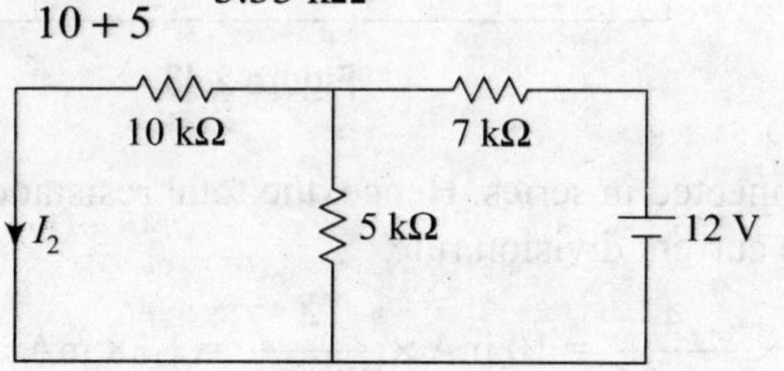

Figure 2.41

R_3 is in series with R_B and the total equivalent resistance is

$$R_T = R_3 + R_B = 7 + 3.33 = 10.33\ \text{k}\Omega$$

The current drawn from E_2 is

$$I_T = \frac{E_2}{R_T} = \frac{12}{10.33} = 1.16\ \text{mA}$$

Using the current division rule, we get

$$I_2 = I_T \times \frac{R_1}{R_1 + R_2} = 1.16 \times \frac{10}{10+5} = 0.775 \text{ mA}$$

The algebraic sum of the current which is flowing through the 5-kΩ resistance is

$$I_5 = I_1 + I_2 = 0.45 \text{ mA} + 0.775 \text{ mA} = 1.225 \text{ mA}$$

The currents I_1 and I_2 flow through the resistance R_2 in the same direction. Therefore, these two currents are added to get the total current. This answer can be checked by using either the loop analysis or nodal analysis.

EXAMPLE 2.21 Using the method of superposition, calculate the branch current I_2 in the circuit shown in Figure 2.42.

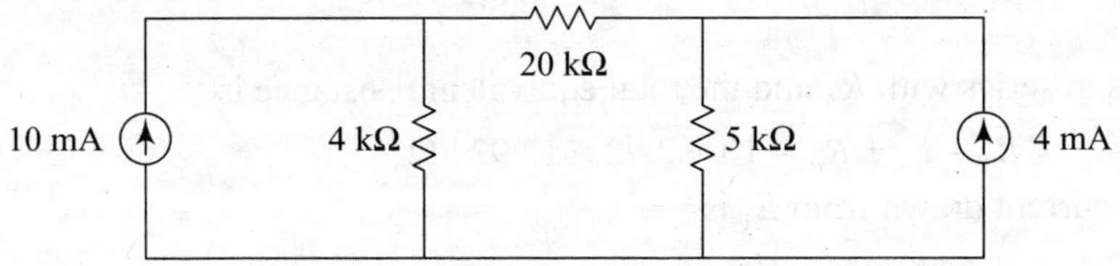

Figure 2.42

Solution **Case 1:** I_1, i.e., the 10-mA current source, is acting alone by replacing I_2 with an open circuit as shown in Figure 2.43.

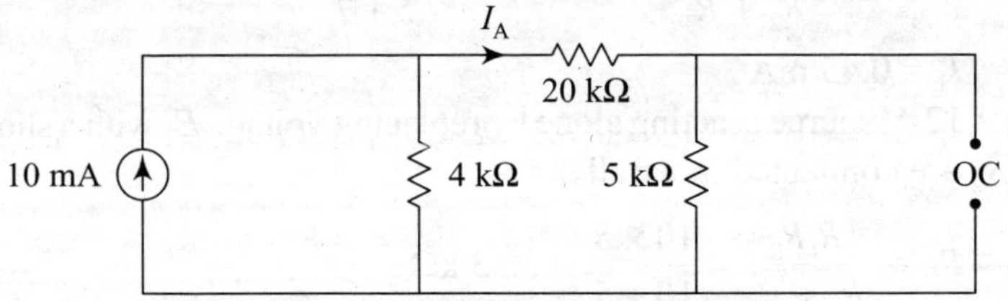

Figure 2.43

R_2 and R_3 are connected in series. Hence, the total resistance is $R_2 + R_3 = 20 + 5 = 25$ kΩ. To find I_A, apply the current division rule.

$$I_A = I_1 \times \frac{4}{4+25} = 10 \text{ mA} \times \frac{4}{4+25} = 1.38 \text{ mA}$$

Case 2: I_2, i.e., the 4-mA current source, is acting alone by replacing I_1 with an open circuit as shown in Figure 2.44.

R_1 and R_2 are connected in series. Hence, the total resistance is $R_1 + R_2 = 4 + 20 = 24$ kΩ. To find I_B, apply the current division rule.

$$I_B = I_2 \times \frac{5}{5+24} = 4 \times \frac{5}{5+24} = 0.69 \text{ mA}$$

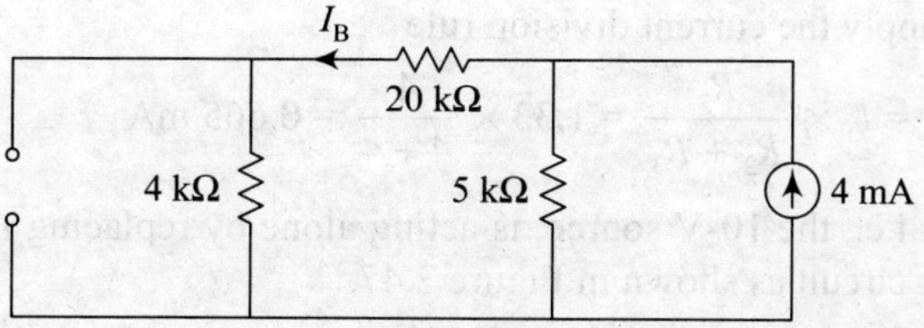

Figure 2.44

By considering cases 1 and 2, the currents I_A and I_B are opposite in direction. Hence, the algebraic sum of currents which is flowing through the 20-kΩ resistance is

$$I_2 = I_A - I_B = 1.38 \text{ mA} - 0.69 \text{ mA} = 0.69 \text{ mA}$$

EXAMPLE 2.22 **Determine the current through and the power dissipated by R_2 in the circuit shown in Figure 2.45.**

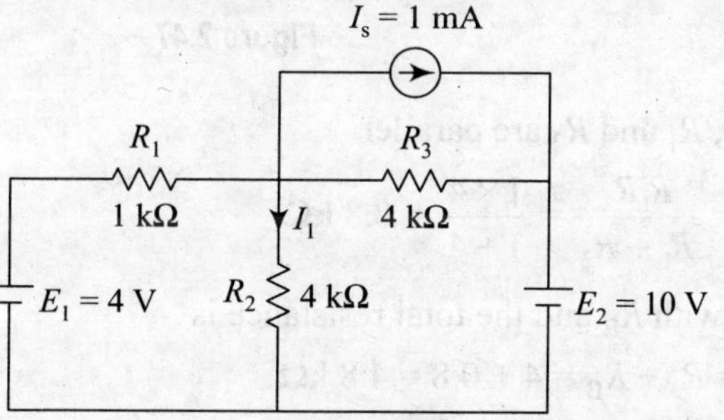

Figure 2.45

Solution **Case 1:** E_1, i.e., the 4-V source, is acting alone by replacing I_s with an open circuit and E_2 with a short circuit as shown in Figure 2.46.

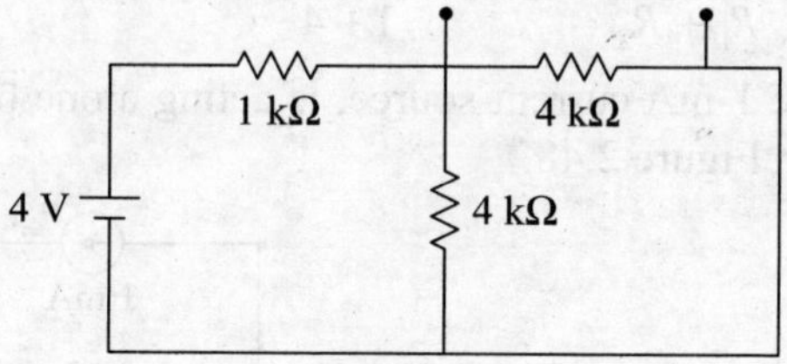

Figure 2.46

$$R_A = \frac{R_2 R_3}{R_2 + R_3} = \frac{4 \times 4}{4+4} = 2 \text{ k}\Omega$$

R_1 is in series with R_A and the total resistance is

$$R_T = R_1 + R_A = 1 + 2 = 3 \text{ k}\Omega$$

$$I_T = \frac{4}{R_T} = \frac{4}{3 \text{ k}\Omega} = 1.33 \text{ mA}$$

To find I_1, apply the current division rule

$$I_1 = I_T \times \frac{R_3}{R_2 + R_3} = 1.33 \times \frac{4}{4+4} = 0.665 \text{ mA}$$

Case 2: E_1, i.e., the 10-V source, is acting alone by replacing I_s with an open circuit and E_1 with a short circuit as shown in Figure 2.47.

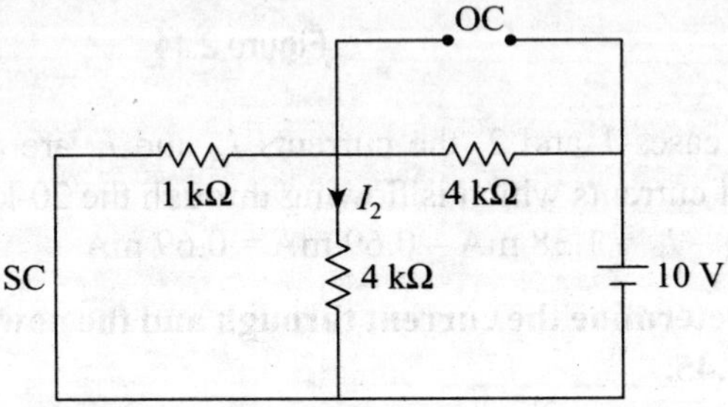

Figure 2.47

In Figure 2.47, R_1 and R_2 are parallel

$$R_B = \frac{R_1 R_2}{R_1 + R_2} = \frac{1 \times 4}{1+4} = 0.8 \text{ k}\Omega$$

R_3 is in series with R_B and the total resistance is

$$R_T = R_3 + R_B = 4 + 0.8 = 4.8 \text{ k}\Omega$$

$$\text{Total current, } I_T = \frac{10}{4.8 \text{ k}\Omega} = 2.08 \text{ mA}$$

To workout I_2, apply the current division rule

$$I_2 = I_T \times \frac{R_1}{R_1 + R_2} = 2.08 \times \frac{1}{1+4} = 0.416 \text{ mA}$$

Case 3: I_s, i.e., the 1-mA current source, is acting alone by replacing E_1 and E_2 with a short circuit as shown in Figure 2.48.

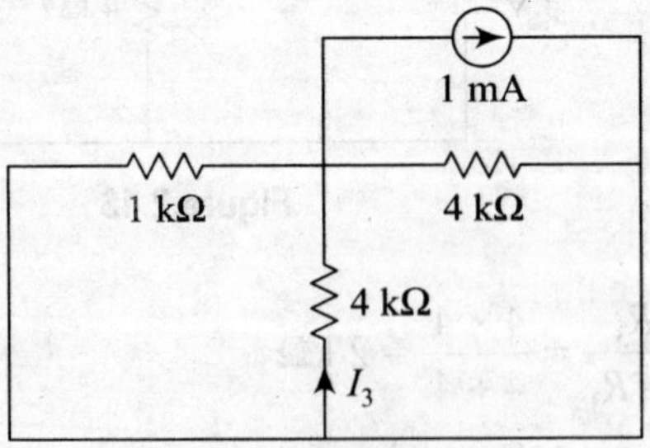

Figure 2.48

We can draw the above circuit as follows:

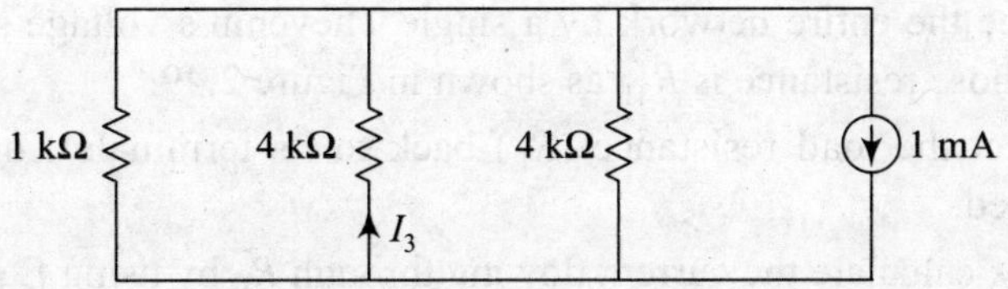

R_1 and R_3 are parallel

$$R_C = \frac{R_1 R_3}{R_1 + R_3} = \frac{1 \times 4}{1+4} = 0.8 \text{ k}\Omega$$

Using the current division rule,

$$I_3 = I_s \times \frac{0.8}{0.8+4} = 1 \text{ mA} \times \frac{0.8}{0.8+4} = 0.166 \text{ mA}$$

By considering cases 1, 2, and 3, the currents I_1 and I_2 are flowing in the same direction and I_3 is flowing in the opposite direction. Hence, the current through R_2 is

$$I_1 + I_2 - I_3 = 0.665 \text{ mA} + 0.416 \text{ mA} - 0.166 \text{ mA}$$

$$I_{R_2} = 0.915 \text{ mA}$$

The power dissipated by R_2 is

$$P_{R_2} = I_{R_2}^2 \; R_2$$

$$= 0.915^2 \text{ mA} \times 4 \text{ k}\Omega$$

$$= 3.35 \text{ mW}$$

2.9.2 Thevenin's Theorem

Thevenin's theorem may be stated as: "Any two terminals of an active network may be represented by a series combination of a voltage source and a resistor. The value of this equivalent source is the voltage appearing at the open-circuited terminals. The value of the equivalent resistance is equal to the resistance seen looking back into the network with all energy sources replaced by their respective internal resistances."

Using Thevenin's theorem we are able to mathematically analyze any network to obtain an equivalent circuit which contains a voltage source V_{Th} and in series with internal resistance R_{Th}.

Steps to obtain Thevenin's equivalent circuit

(i) Temporarily remove the load resistance whose current is required.

(ii) Find the open-circuit voltage V_{OC} which appears across the two terminals from where the load resistance has been removed. It is also called Thevenin's voltage, V_{Th}.

(iii) Calculate the resistance of the whole network as looked into from these two terminals, after all voltage sources are replaced by a short circuit and all current sources are replaced by an open circuit leaving the internal resistance (if any). It is also called Thevenin's resistance, R_{Th}.

(iv) Replace the entire network by a single Thevenin's voltage source whose voltage is V_{Th} and whose resistance is R_{Th} as shown in Figure 2.49.

(v) Connect the load resistance (R_L) back to its terminals from where it was previously removed.

(vi) Finally, calculate the current flowing through R_L by using Eq. (2.62).

$$I_L = \frac{V_{Th}}{R_{Th} + R_L} \tag{2.62}$$

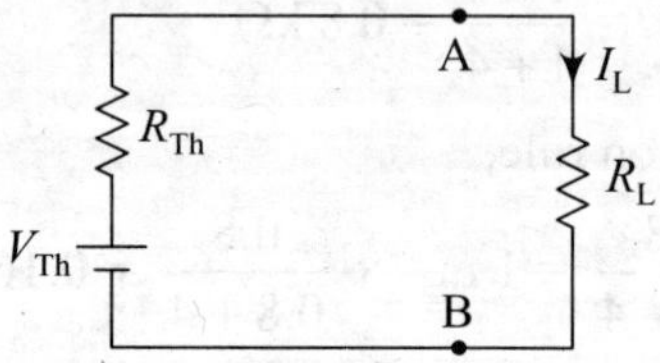

Figure 2.49 Thevenin's equivalent network

EXAMPLE 2.23 **Consider the circuit shown in Figure 2.50 and determine Thevenin's equivalent circuit for the load resistance R_L. Then determine I_L if R_L is 4, 8, and 10 kΩ.**

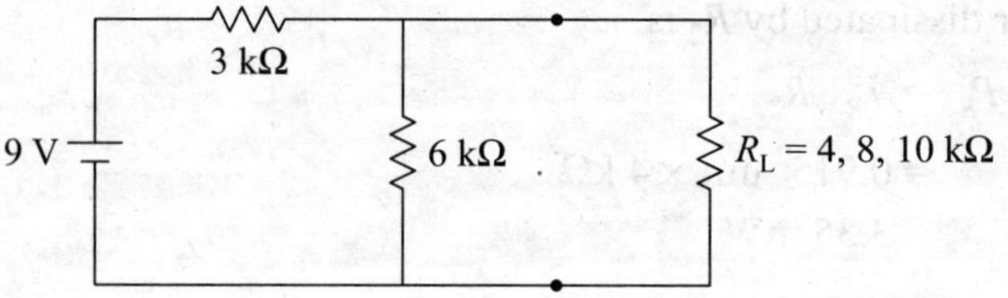

Figure 2.50

Solution First, remove the load resistance and draw the circuit diagram as shown in Figure 2.51(a) to find the open-circuit voltage, i.e., V_{Th}.

$$I = \frac{V}{R_1 + R_2} = \frac{9}{3+6} = 1 \text{ mA}$$

The current flowing through R_2 is 1 mA. Hence, the voltage drop across R_2 is $1_2 = 1 \text{ mA} \times 6 \text{ k}\Omega = 6$ V; $V_{Th} = 6$ V. Thevenin's resistance can be determined from Figure 2.51(b). Thevenin's resistance R_{Th} is the equivalent resistance looking from terminals A and B, i.e., where the load resistance has been removed. Then, R_1 is parallel with R_2.

$$R_{Th} = \frac{R_1 R_2}{R_1 + R_2} = \frac{3 \times 6}{3+6} = 2 \text{ k}\Omega$$

Replace the entire network by a single equivalent resistance, i.e., R_{Th}, and a single voltage source, i.e., V_{Th}, as shown in Figure 2.51(c).

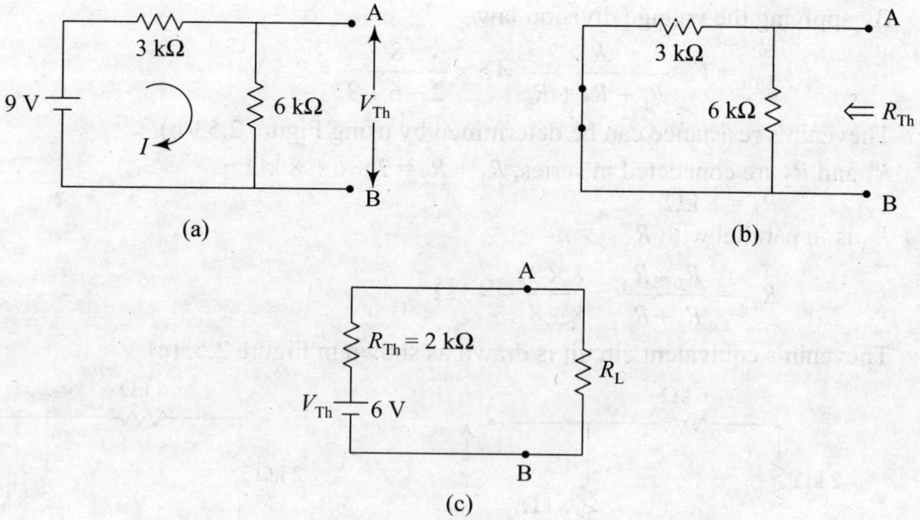

Figure 2.51

Connect R_L back and calculate the load current. In this case there are three load resistances. We may determine I_L for different values of R_L.

If $R_L = 4\ k\Omega$, then

$$I_L = \frac{V_{Th}}{R_{Th} + R_L} = \frac{6}{2+4} = 1\ mA$$

If $R_L = 8\ k\Omega$, then

$$I_L = \frac{V_{Th}}{R_{Th} + R_L} = \frac{6}{2+8} = 0.6\ mA$$

If $R_L = 10\ k\Omega$, then

$$I_L = \frac{V_{Th}}{R_{Th} + R_L} = \frac{6}{2+10} = 0.5\ mA$$

EXAMPLE 2.24 **Determine Thevenin's equivalent circuit between terminals A and B for the circuit shown in Figure 2.52.**

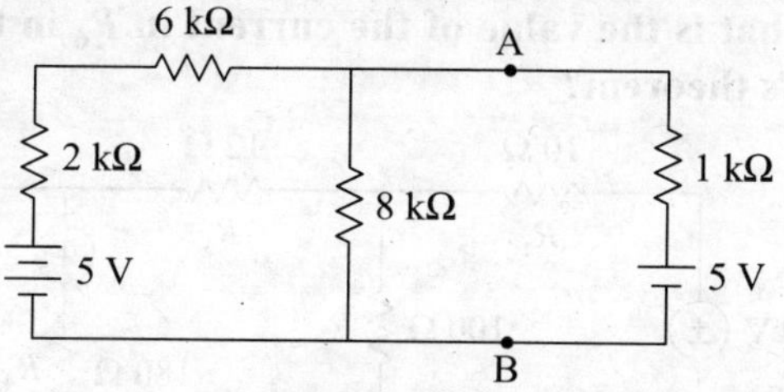

Figure 2.52

Solution Temporarily remove the load resistance and draw the circuit diagram as shown in Figure 2.53(a) to find the open-circuit voltage, i.e., V_{Th}.

By applying the voltage division law

$$V_{Th} = V \cdot \frac{R_3}{R_1 + R_2 + R_3} = 5 \times \frac{8}{2+6+8} = 2.5 \text{ V}$$

Thevenin's resistance can be determined by using Figure 2.53(b).

R_1 and R_2 are connected in series, $R_1 + R_2 = 2 + 6 = 8$ kΩ

$$R_A = 8 \text{ k}\Omega$$

R_A is in parallel with R_3

$$R_{Th} = \frac{R_3 \times R_A}{R_3 + R_A} = \frac{8 \times 8}{8+8} = 4 \text{ k}\Omega$$

Thevenin's equivalent circuit is drawn as shown in Figure 2.53(c).

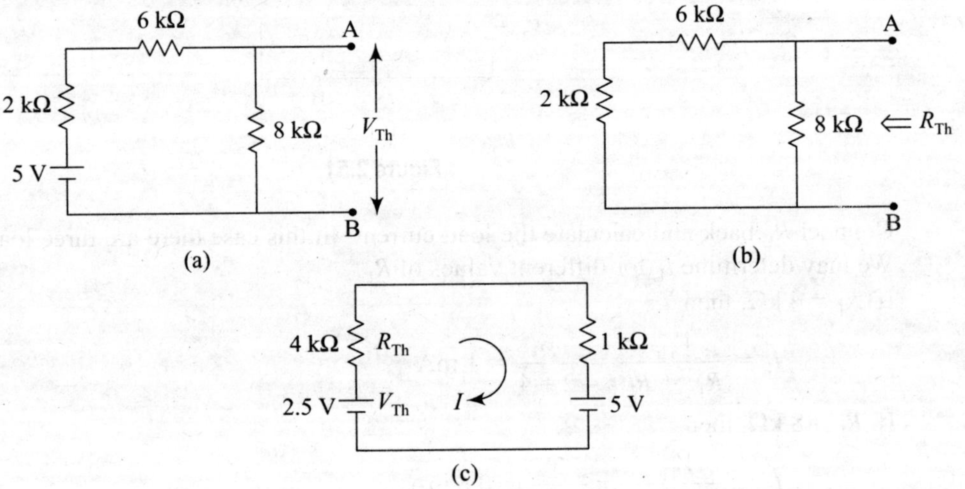

Figure 2.53

By applying the KVL, we get

$$V_{Th} = IR_{Th} + IR_4 + E_2$$
$$2.5 = I(4 + 1) + 5$$
$$I = 0.5 \text{ mA}$$

Hence, the load current is 0.5 mA.

EXAMPLE 2.25 **What is the value of the current in R_6 in the circuit shown in Figure 2.54 by using Thevenin's theorem?**

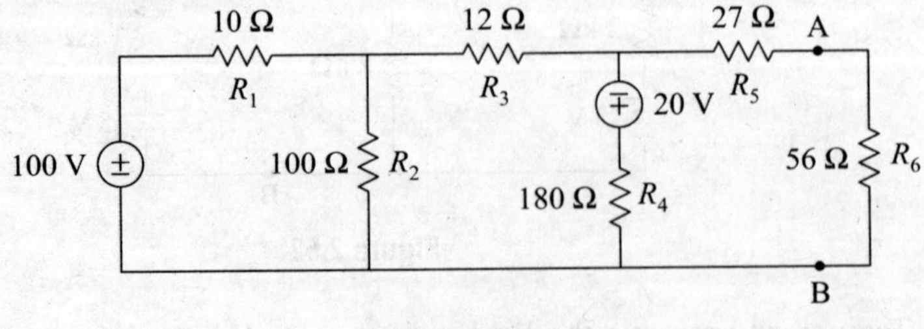

Figure 2.54

Solution Temporarily remove the resistance R_6 and draw the circuit diagram as shown in Figure 2.55(a) and find the open-circuit voltage, i.e., V_{Th}.

The voltage drop across 27 Ω is negligible because no current flows through R_5. Hence, the voltage drop across R_4 and E_2 will give V_{Th}.

To find the current flowing through R_4 we can apply the superposition theorem, as shown in Figure 2.55(b).

Case 1: A 100-V source is acting alone and the E_2 source is short circuited.

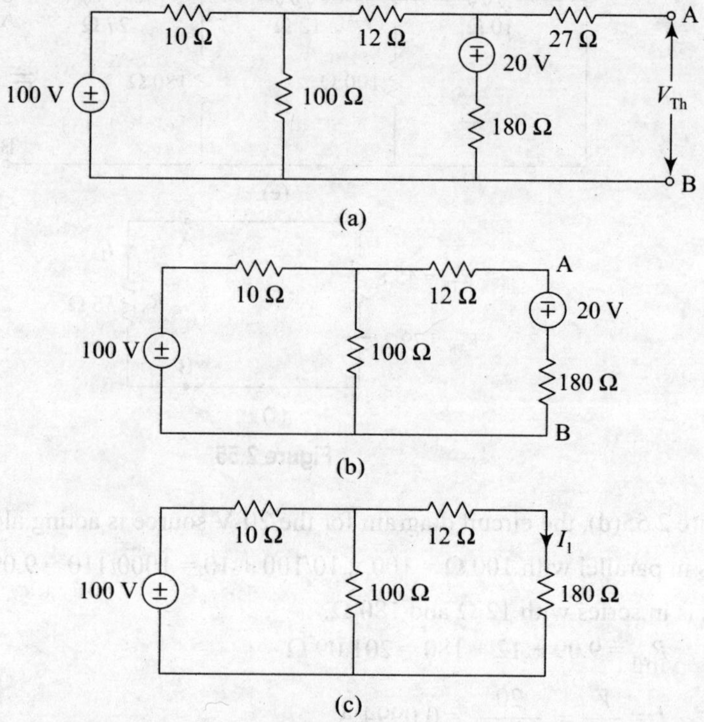

Figure 2.55

180 Ω and 12 Ω are connected in series, 180 + 12 = 192 Ω

192 Ω is parallel to 100 Ω, 192 × 100/192 + 100 = 65.75 Ω

65.75 Ω is in series with 10 Ω

$$R_{eq} = 65.75 + 10 = 75.75\ \Omega$$

$$I = \frac{V}{R_{eq}} = \frac{100}{75.75} = 1.32\ \text{A}$$

To find I_1, apply the current division rule

$$I_1 = I \times \frac{100}{100+192} = 1.32 \times \frac{100}{100+192} = \frac{132}{292} = 0.452\ \text{A}$$

Case 2: A 20-V source is acting alone and E_1 source is short circuited.

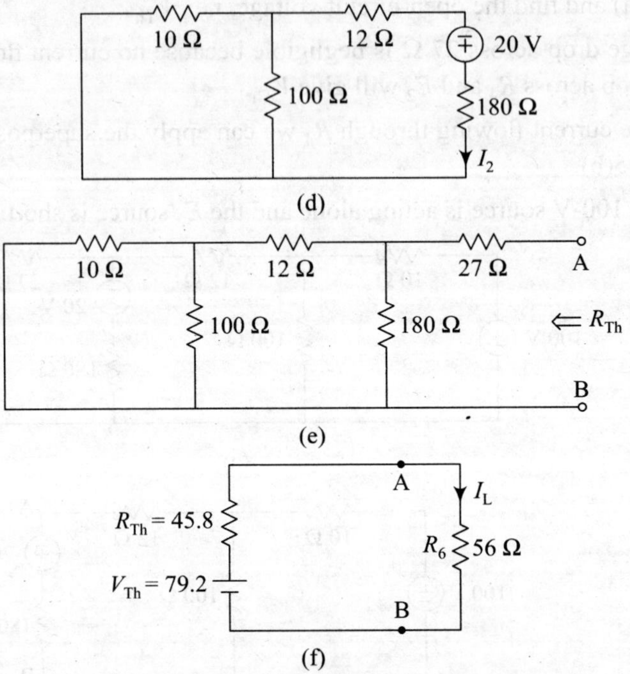

Figure 2.55

In Figure 2.55(d), the circuit diagram for the 20-V source is acting alone.

10 Ω is in parallel with 100 Ω = 100 × 10/100 + 10 = 1000/110 = 9.09 Ω

9.09 Ω is in series with 12 Ω and 180 Ω.

$$R_{eq} = 9.09 + 12 + 180 = 201.09\ \Omega$$

$$I = \frac{V}{R_{eq}} = \frac{20}{201.09} = 0.0994\ \text{A}$$

Current flowing through 180 Ω is 0.0994 A.

$$I_2 = 0.0994\ \text{A}$$

Hence, the algebraic sum of the current flowing through 180 Ω is

$$I_T = 0.452 + 0.0994 = 0.5514\ \text{A}$$

The voltage drop across the 180-Ω resistance is $I_T = 0.55145 \times 180$

$= 99.261$ V

By considering the circuit diagram of Figure 2.55(b), the voltage drop across "AB" is V_{Th}.

$$V_{Th} = 99.262 - 20 = 79.262\ \text{V}$$

The R_{Th} can be determined by using Figure 2.55(e).

As shown in Figure 2.55(e), it is seen that 10 Ω || 100 Ω = 9.09 + 12 = 21.09 Ω || 180 Ω = 18.87 + 27 = 45.88 Ω.

$R_{Th} = 45.88\ \Omega$

Thevenin's equivalent circuit is drawn as shown in Figure 2.55(f).

$$I_L = \frac{V_{Th}}{R_{Th} + R_L} = \frac{79.262}{45.88 + 56} = 0.7781\ \text{A}$$

2.9.3 Norton's Theorem

Norton's theorem may be stated as: "Any two terminals of an active resistance network may be represented by a parallel combination of current source and a resistor. The value of the current source is the short-circuit current at the terminals. The value of the equivalent resistance is equal to the resistance seen looking back into the network with all energy sources replaced by their respective internal resistances."

Using Norton's theorem we are able to mathematically analyze any network to obtain an equivalent circuit which contains a current source I_N and in parallel with internal resistance R_{Th} or R_N, i.e., Norton's resistance.

Note: Thevenin's equivalent resistance and Norton's equivalent resistance are equal.

Norton's theorem is dual of Thevenin's theorem, whereas Thevenin's theorem reduces a two-terminal active network of linear resistances to an equivalent voltage source V_{Th} and series Thevenin's resistance R_{Th}. Norton's theorem replaces the network by an equivalent short circuit current source and a parallel Norton's resistance.

Steps to obtain Norton's equivalent circuit

(i) Temporarily remove the load resistance (if any) across the two given terminals and put a short circuit across them.

(ii) Calculate the short-circuit current I_{sc} or I_N.

(iii) Calculate the resistance of the whole network as looked into these two terminals, after all voltage sources are replaced by a short circuit and all current sources are replaced by an open circuit leaving internal resistance (if any). It is called Thevenin's resistance R_{Th} or Norton's resistance R_N.

(iv) Replace the entire network by a single Norton's current source whose short-circuit current is I_{sc} or I_N and parallel with Norton's resistance R_N or R_{Th} as shown in Figure 2.56.

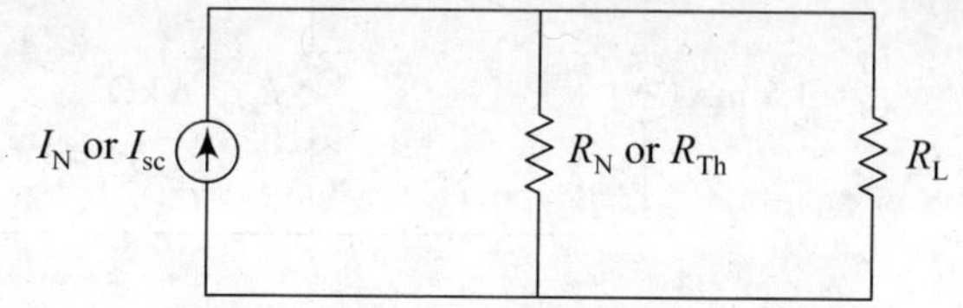

Figure 2.56 Norton's equivalent circuit

(v) Connect the load resistance (R_L) back to its terminals from where it was previously removed.

(vi) Finally, calculate the current flowing through R_L by using the current division rule.

EXAMPLE 2.26 **For the circuit of Figure 2.57, obtain Norton's equivalent circuit between terminals "AB" and find the load current I_L.**

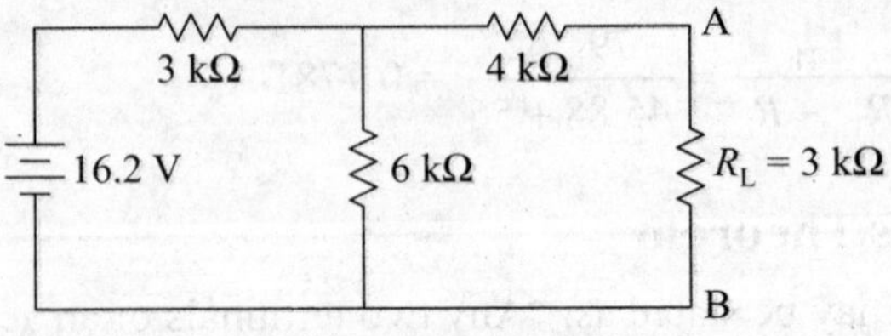

Figure 2.57

Solution To find the short-circuit current, i.e., I_{sc} or I_N. Remove the load resistance and short circuit from the terminals as shown in Figure 2.58(a).

$$R_2 \parallel R_3 = \frac{R_2 R_3}{R_2 + R_3} = \frac{6 \times 4}{4 + 6} = 2.4 \text{ k}\Omega$$

Equivalent resistance, $R_T = 3 + 2.4 = 5.4$ kΩ

Then

$$I_T = \frac{16.2}{5.4} = 3 \text{ mA}$$

Using the current division rule

$$I_N \text{ or } I_{SC} = I_T \cdot \frac{R_2}{R_2 + R_3} = 3 \text{ mA} \times \frac{6}{6+4} = 1.8 \text{ mA}$$

The R_{Th} or R_N can be determined by using Figure 2.58(b).

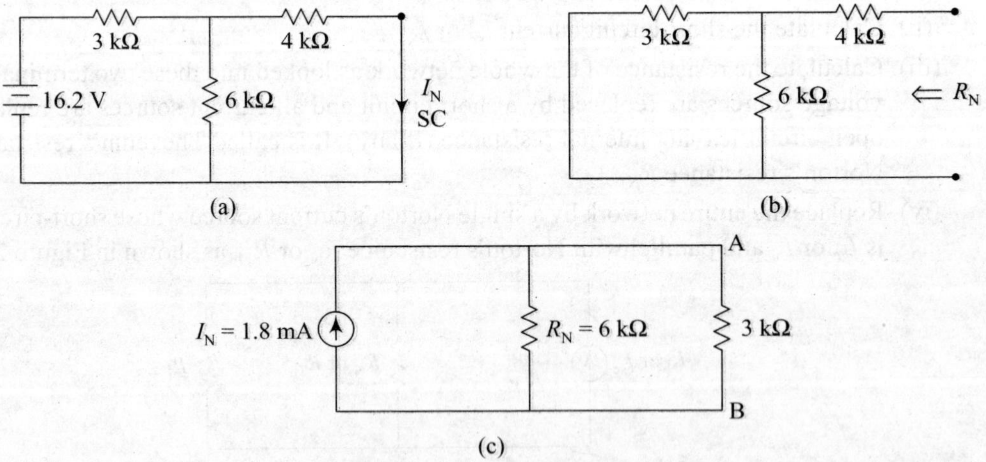

Figure 2.58 Norton's equivalent circuit

$$[R_1 \parallel R_2] + R_3 = \frac{3 \times 6}{3 + 6} + 4 = 6 \text{ k}\Omega$$

$R_{Th} = 6\ k\Omega$

To find I_L, apply the current division rule

$$I_L = I_N \cdot \frac{R_{Th}}{R_{Th} + R_L} = 1.8\ \text{mA} \times \frac{6}{6+3} = 1.2\ \text{mA}$$

EXAMPLE 2.27 **For the circuit shown in Figure 2.59, calculate the current flowing through the 15-Ω load resistance in the circuit.**

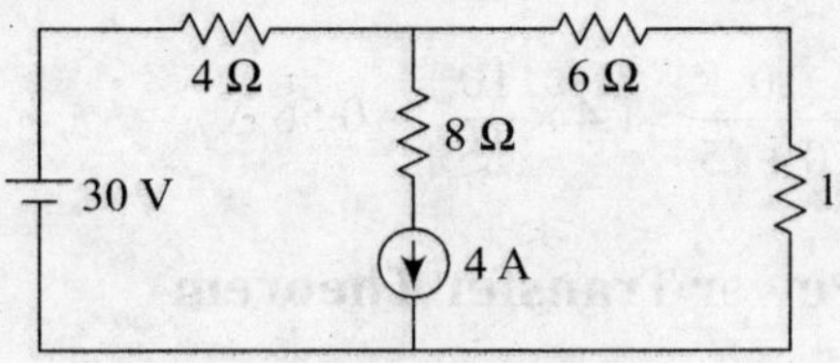

Figure 2.59

Solution To find the short-circuit current, i.e., I_N, we can use the superposition theorem because it contains 30-V, 4-A sources.

Case 1: A 30-V source is acting alone by replacing 4 A with an open circuit as shown in Figure 2.60(a); the 8-Ω resistance is negligible because the branch is open circuited and no current flows through it.

$$I_1 = \frac{30}{4+6} = 3\ \text{A}$$

Case 2: A 4-A source is acting alone by replacing 30 V with a short circuit as shown in Figure 2.60(b).

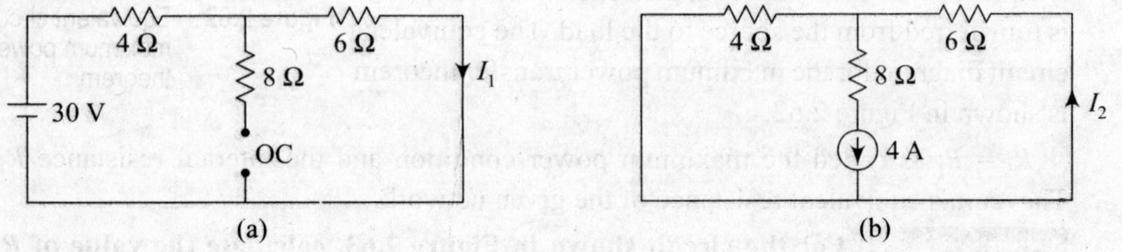

Figure 2.60

To find I_2, apply the current division rule

$$I_2 = I \times \frac{4}{4+6} = 4 \times \frac{4}{4+6} = 1.6\ \text{A}$$

The short-circuit current, $I_N = I_1 - I_2$

$$= 3 - 1.6 = 1.4\ \text{A}$$

R_N can be determined by using Figure 2.61(a).

$$R_N = 4 + 6 = 10\ \Omega$$

Norton's equivalent circuit is drawn as shown in Figure 2.61(b).

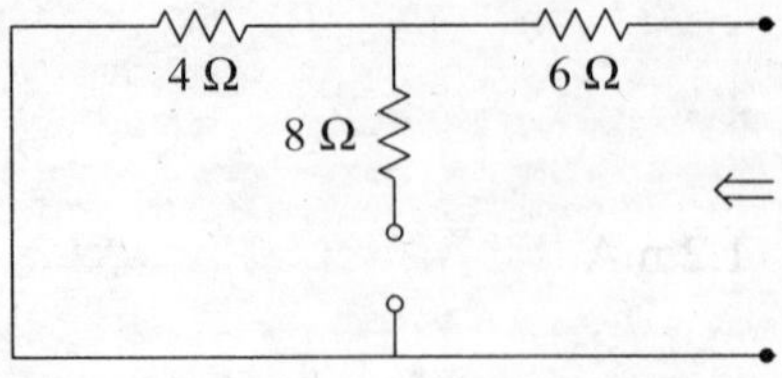

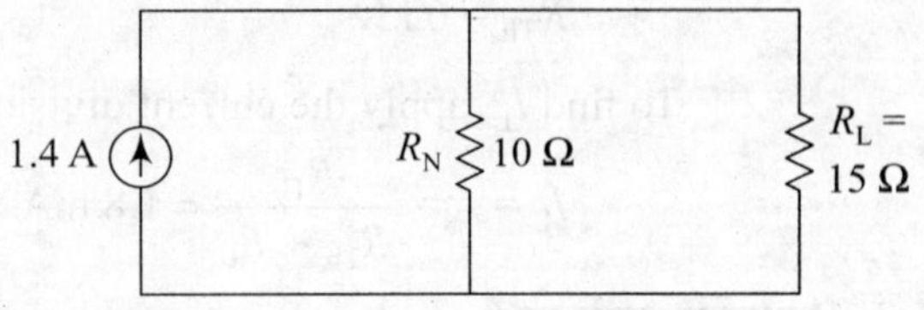

Figure 2.61

$$I_L = I_N \times \frac{10}{10+15} = 1.4 \times \frac{10}{25} = 0.56 \text{ A}$$

2.9.4 Maximum Power Transfer Theorem

This theorem deals with the transfer of maximum power from a source to load and may be stated as: "In a network, the maximum power is transferred from a source to load when the load resistance is made equal to the internal resistance of the source as viewed from the load terminals with the load removed and all emf sources replaced by their internal resistances."

This theorem is particularly useful for analyzing communication networks. The overall efficiency of a network supplying maximum power to any branch is 50%. In the case of electronic and communication networks, the goal is very often to either receive or transmit maximum power specially when the power involved is only a few milli watts (or) micro watts.

When $R_I = R_L$, i.e., the internal resistance of the circuit is equal to the load resistance, then the maximum power is transferred from the source to the load. The equivalent circuit diagram for the maximum power transfer theorem is shown in Figure 2.62.

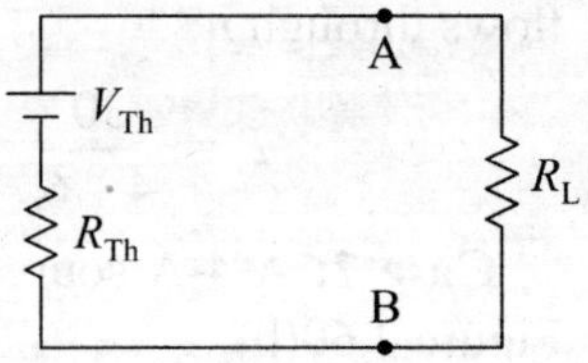

Figure 2.62 Equivalent circuit diagram for maximum power transfer theorem

$R_I = R_L$ is called the maximum power condition and the internal resistance R_I is same as Thevenin's equivalent resistance of the given network.

EXAMPLE 2.28 For the circuit shown in Figure 2.63, calculate the value of R which will absorb maximum power from the circuit. Also, compute the value of maximum power.

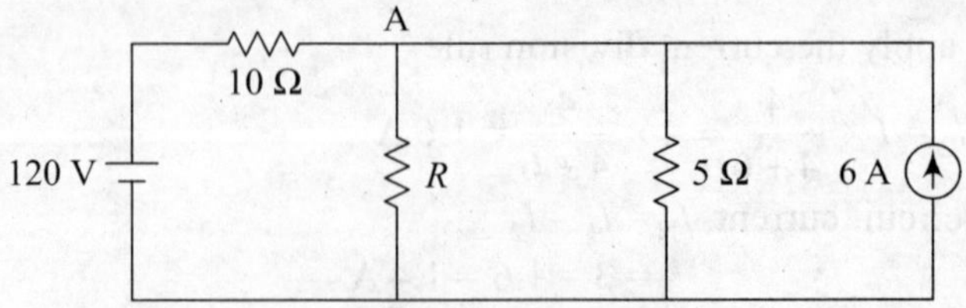

Figure 2.63

Solution Remove "*R*" to find V_{Th} across AB, convert 120 V, and 10 Ω can be converted into equivalent current source and resistance by using source transformation.

By applying the nodal analysis, we can find V_{Th}. To find the R_{Th}, all current sources are replaced by an open circuit.

$$10 \parallel 5 = \frac{10 \times 5}{10 + 5} = \frac{50}{15} = 3.33\ \Omega$$

$$R_I \text{ (or) } R_{Th} = 3.33\ \Omega$$

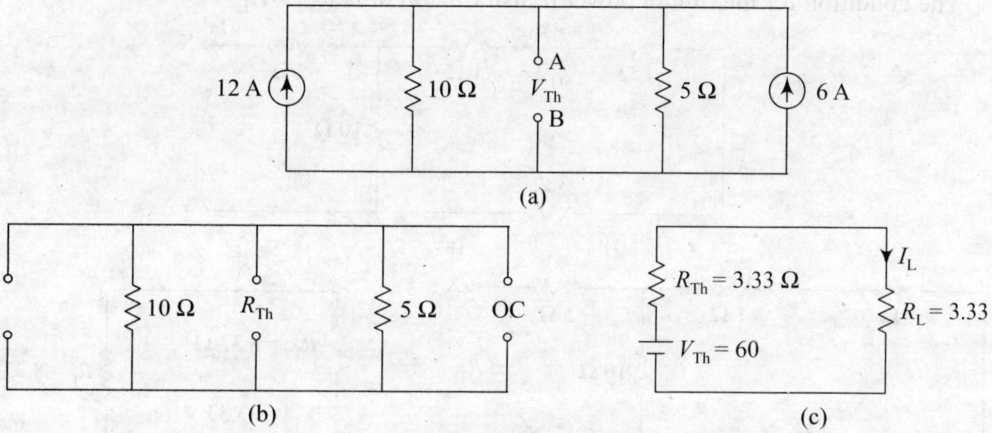

Figure 2.64

The condition for maximum power is $R_I = R_L$.

Hence, the value of *R* is 3.33 Ω.

Maximum power, $P = I^2 R_L = 9^2 \times 3.33 = 270$ W

EXAMPLE 2.29 **For the circuit shown in Figure 2.65, calculate the value of *R* which will absorb maximum power from the source. Find the value of this amount of power.**

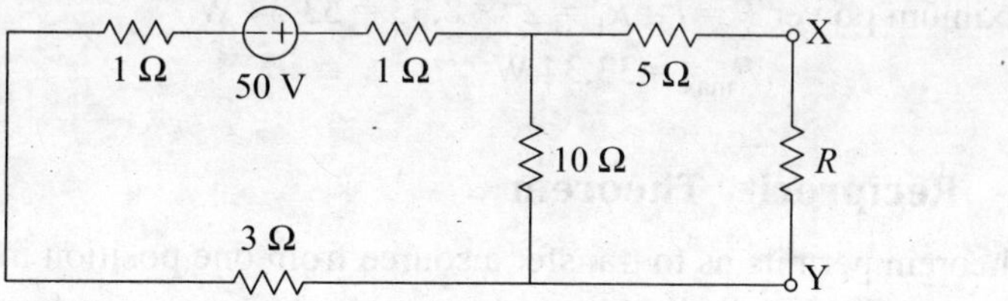

Figure 2.65

Solution Remove *R* to find V_{Th} across XY as shown in Figure 2.65, and the circuit is redrawn as shown in Figure 2.66(a).

The voltage drop across 10 Ω is V_{Th} by neglecting the 5-Ω resistance because no current flows through it.

$$I = \frac{50}{15} = 3.33\ \Omega$$

$$V_{Th} = I \times 10 = 3.33 \times 10 = 33.33\ \text{V}$$

To find R_{Th}, replace the voltage source by short circuit as shown in Figure 2.66(b).

$$1 + 1 + 3 = 5 \parallel 10 = \frac{5 \times 10}{5 + 10} + 5 = 8.33\ \Omega$$

$$R_{Th} = 8.33\ \Omega$$

The condition for maximum power transfer theorem is $R_{Th} = R_L$.

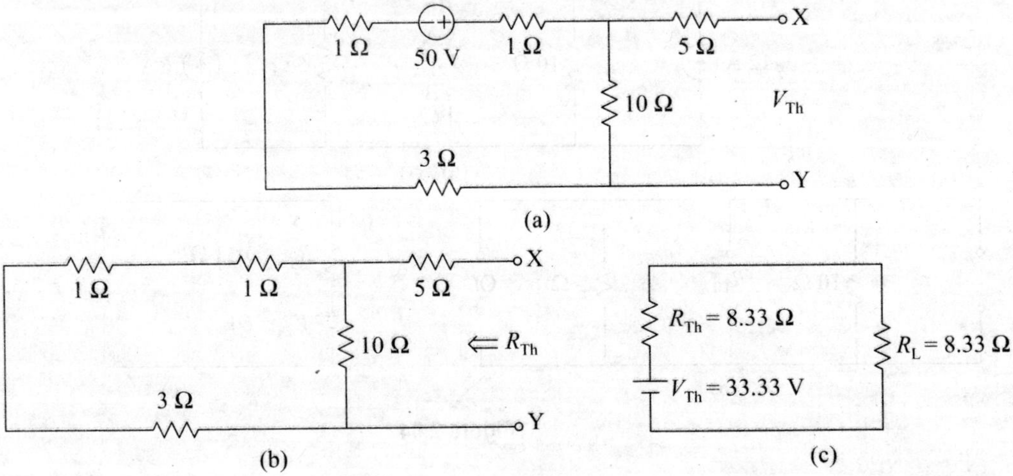

Figure 2.66

Hence, the value of $R = 8.33\ \Omega$

$$I_L = \frac{V_{Th}}{R_{Th} + R_L} = \frac{33.33}{8.33 + 8.33} = 2\ \text{A}$$

Maximum power $P = I^2;\ R_L = 2^2 \times 8.33 = 33.34$ W

$$P_{max} = 33.34\ \text{W}$$

2.9.5 Reciprocity Theorem

This theorem permits us to transfer a source from one position in the circuit to another and may be stated as: "In any linear bilateral network, the ratio of voltage V introduced in one mesh to the current I in any second mesh is the same as the ratio obtained if the positions of V and I are interchanged, other emf being removed."

Consider the network shown in Figures 2.67(a) and (b). As shown in Figure 2.67(a), it is observed that the source V is in series with R_1, and current I_2 is flowing in the resistance R_2. Now let the source V is shifted to the branch containing resistance R_2 and the current I_1 flows through resistance R_1 as shown in Figure 2.67(b).

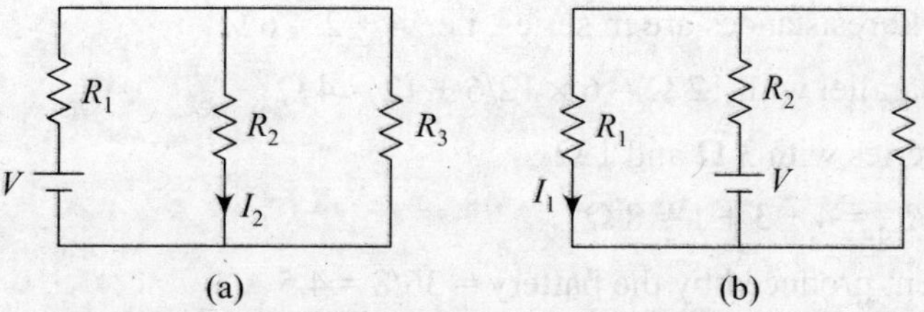

Figure 2.67 Illustration of reciprocity theorem

According to the reciprocity theorem,

$$\frac{V}{I_2} = \frac{V}{I_1} \quad \text{or} \quad I_2 = I_1$$

EXAMPLE 2.30 **Verify the reciprocity theorem for the network shown in Figure 2.68.**

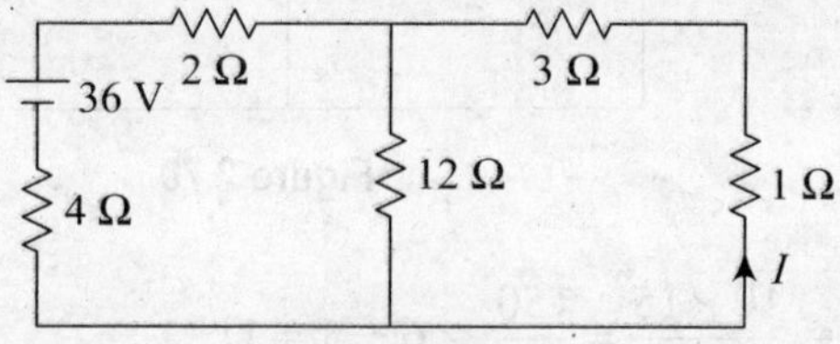

Figure 2.68

Solution The resistances 3 Ω and 1 Ω are connected in series.

$$3 + 1 = 4\ \Omega$$

4 Ω and 12 Ω are connected in parallel

$$\frac{4 \times 12}{4 + 12} = \frac{48}{16} = 3\ \Omega$$

Total equivalent resistance $R_{eq} = 3 + 4 + 2 = 9\ \Omega$

$$\text{Total current produced by battery} = \frac{36}{9} = 4\ \text{A}$$

$$\text{Current, } I = 4 \times \frac{12}{12 + 4} = 3\ \text{A}$$

Thus, the current *I* is 3 A.

By interchanging *V* and *I*, Figure 2.68 is redrawn as shown in Figure 2.69.

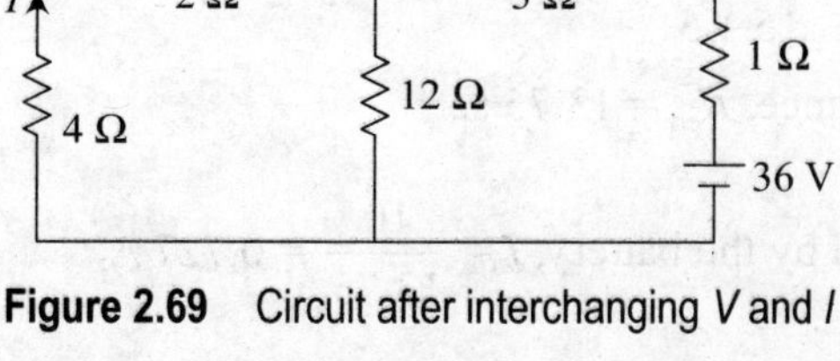

Figure 2.69 Circuit after interchanging *V* and *I*

4 Ω and 2 Ω resistances are in series, i.e., 4 + 2 = 6 Ω

6 Ω is in parallel with 12 Ω = 6 × 12/6 + 12 = 4 Ω

4 Ω is in series with 3 Ω and 1 Ω

$$R_{eq} = 4 + 3 + 1 = 8\ \Omega$$

Total current produced by the battery = 36/8 = 4.5 A

Thus, current $I = 4.5 \times 12/12 + 6 = 3$ A. Hence, in both cases the current is same. The reciprocity theorem is verified.

EXAMPLE 2.31 **Verify the reciprocity theorem for the network given in Figure 2.70.**

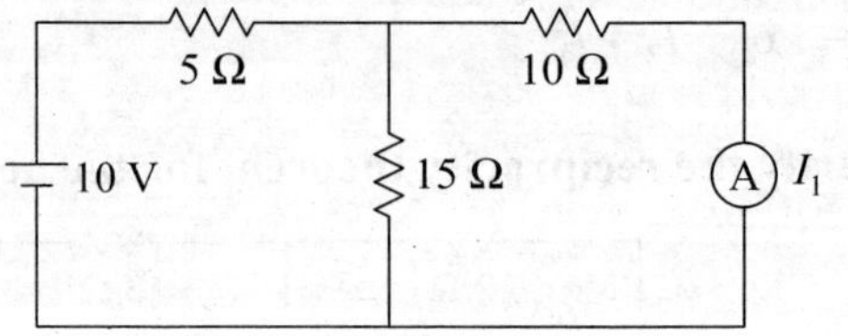

Figure 2.70

Solution $$10 \parallel 15 = \frac{10 \times 15}{10 + 15} = \frac{150}{25} = 6 + 5 = 11\ \Omega$$

Equivalent resistance, $R_{eq} = 11\ \Omega$

Current produced by the battery, $I = \dfrac{10}{11} = 0.909$ A

$$I_1 = I \times \frac{15}{15 + 10} = 0.909 \times \frac{15}{25} = 0.545 \text{ A}$$

By interchanging V and I, Figure 2.70 is redrawn as shown in Figure 2.71.

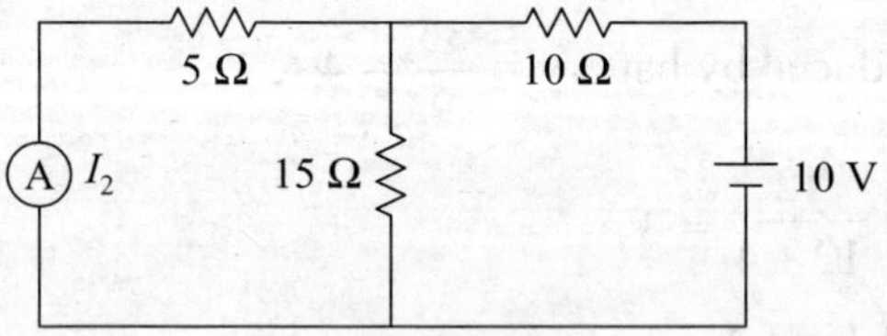

Figure 2.71

$$5 \parallel 15 = \frac{15 \times 5}{15 + 5} = 3.75 + 10 = 13.75\ \Omega$$

Equivalent resistance, $R_{eq} = 13.75\ \Omega$

Current produced by the battery, $I = \dfrac{10}{13.75} = 0.727$ A

$$I_2 = 0.727 \times \frac{15}{20} = 0.545 \text{ A}$$

Both currents are equal. Hence, the reciprocity is verified.

EXAMPLE 2.32 **Using the source transformation technique, find the current through the 6-Ω resistance in the circuit shown in Figure 2.72.**

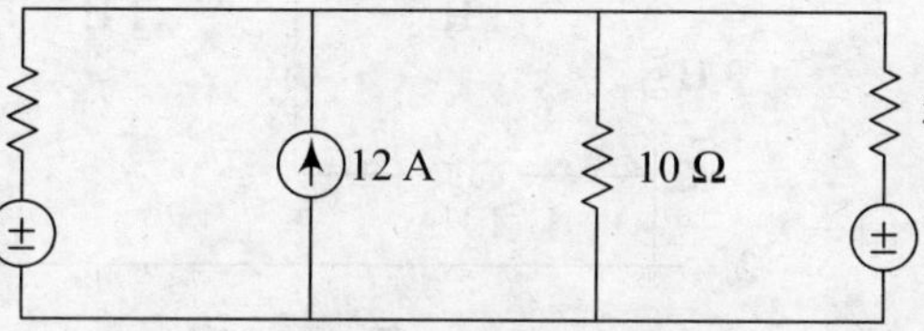

Figure 2.72

Solution By using the source transformation, 60-V source, 6 Ω can be converted into current source in parallel with 6 Ω as shown in Figure 2.73.

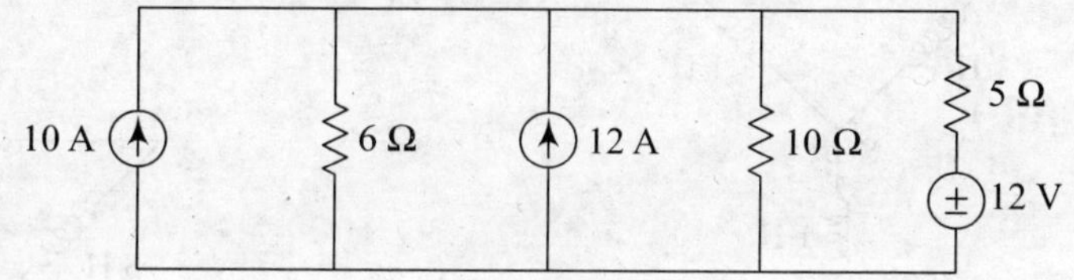

Figure 2.73

$$I = \frac{60}{6} = 10 \text{ A}$$

10-A source parallel with 6 Ω

12-V source, 5 Ω can be converted as $I = 12/5 = 2.4$ A

2.4-A source parallel with 5 Ω

Because of the parallel connection, all current sources are additive.

10 A + 12 A + 2.4 A (because all current source directions are towards upward direction)

10 Ω and 5 Ω are parallel.

$$\therefore \qquad \frac{10 \times 5}{10 + 5} = \frac{50}{15} = 3.33\ \Omega$$

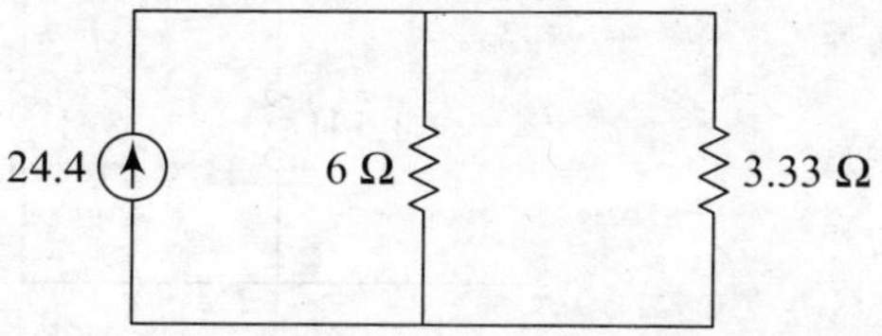

Current through the 6-Ω resistor can be found out by using the current division formula.

$$I_6 = 24.4 \times \frac{3.33}{3.33 + 6} = 8.70 \text{ A}$$

Therefore, current flowing through the 6-Ω resistor is 8.70 A.

EXAMPLE 2.33 Obtain the equivalent inductance at terminals 1 and 2 for the circuit shown in Figure 2.74.

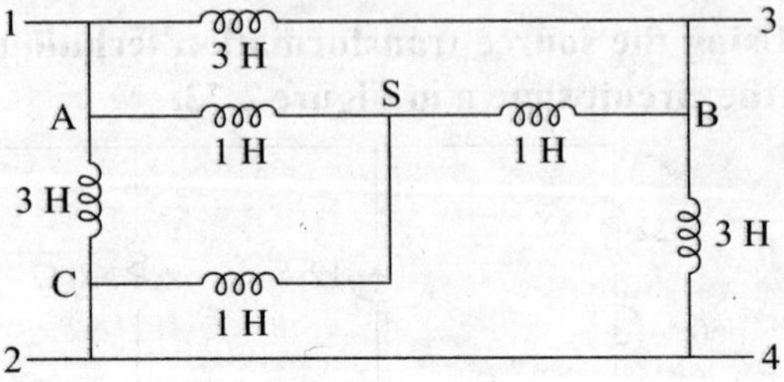

Figure 2.74

Solution Consider "S" is a star point and convert it into delta.

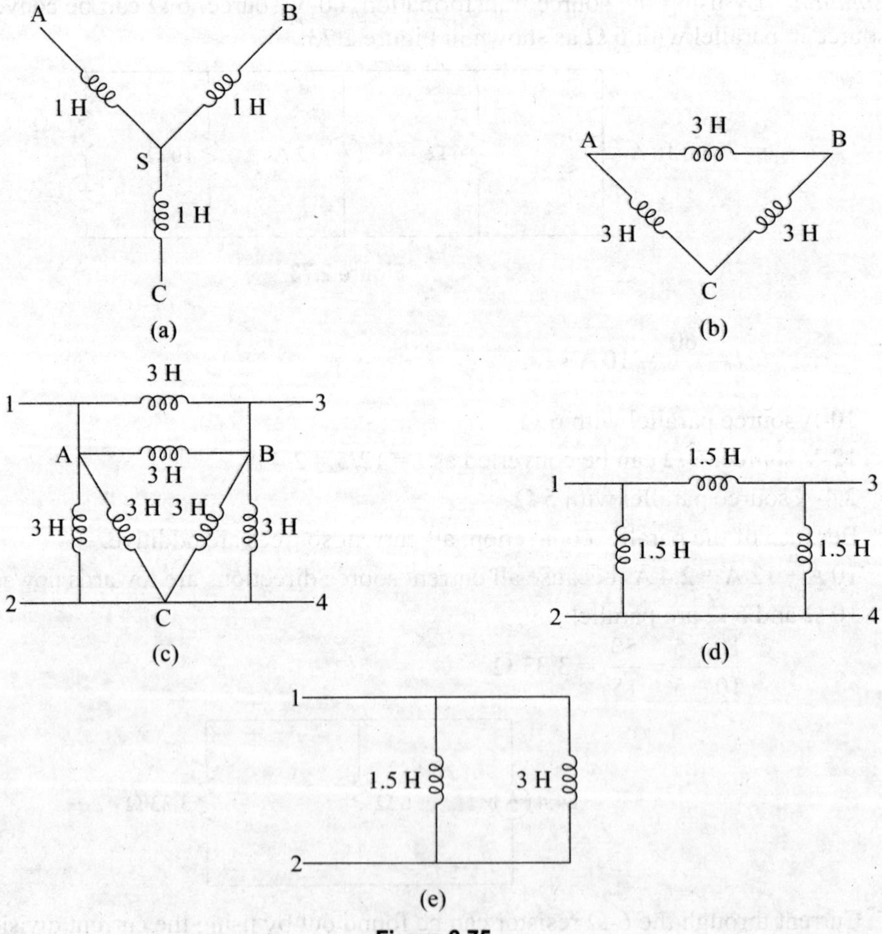

Figure 2.75

The equivalent inductance at terminals 1 and 2, 1–3 inductance at 1.5 H and 3–4 inductance at 1.5 H are in series, so 1.5 + 1.5 = 3 H.

The equivalent inductance between terminals 1 and 2 is 1 H.

EXAMPLE 2.34 **Three capacitors A, B, and C have capacitance 10, 50, and 25 μF, respectively. Calculate**

(i) The charge on each when connected in parallel to a 250-V supply. What is the total capacitance and charge?

(ii) The charge of the circuit when in series and the total capacitance.

Solution

(i) When connected in parallel, the charge is different.

$$C = \frac{Q}{V} \text{ for different capacitors}$$

$$Q_1 = C_1 V$$

$$= 10 \times 250 = 2500\ \mu\text{C}$$

$$Q_2 = 50 \times 250 = 12500\ \mu\text{C}$$

$$Q_3 = 25 \times 250 = 6250\ \mu\text{C}$$

Total charge $Q = Q_1 + Q_2 + Q_3 = 2500 + 12500 + 6250 = 21250\ \mu\text{C}$

$$C = C_1 + C_2 + C_3 = 10 + 50 + 25 = 85\ \mu\text{F}$$

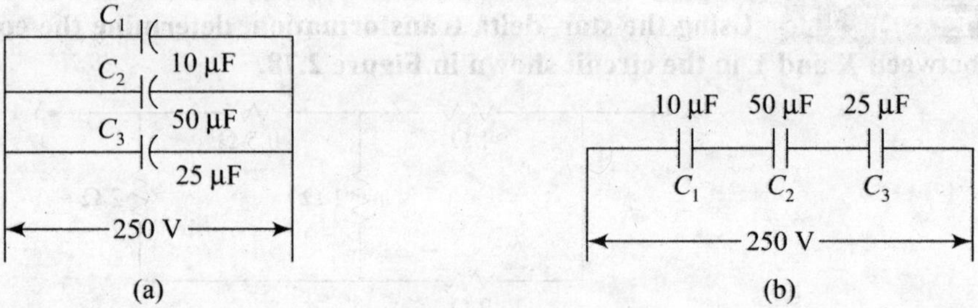

Figure 2.76

(ii) When connected in series, the charge is same.

$$\frac{1}{C_{eq}} = \frac{1}{C_1} + \frac{1}{C_2} + \frac{1}{C_3} = \frac{1}{10} + \frac{1}{50} + \frac{1}{25}$$

$$C_{eq} = 6.25\ \mu\text{F}$$

$$Q = CV = 6.25 \times 250 = 1562.5\ \mu\text{C}$$

$$V_1 = \frac{Q}{C_1} = \frac{1562.5}{10} = 156.25\ \text{V}$$

$$V_2 = \frac{Q}{C_2} = \frac{1562.5}{50} = 31.25\ \text{V}$$

$$V_3 = \frac{Q}{C_3} = \frac{1562.5}{25} = 62.5\ \text{V}$$

EXAMPLE 2.35 For the circuit shown in Figure 2.77, determine the current through the 6-Ω resistor and the power supplied by the current source.

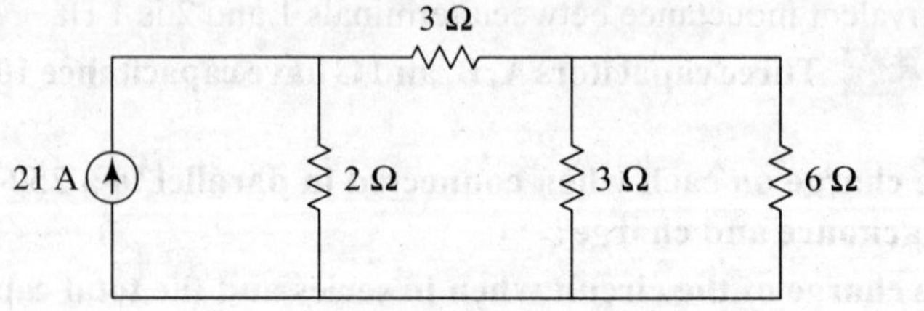

Figure 2.77

Solution Current supplied by the current source, $I = 21$ A

Using the current division rule, the current flowing through the 3-Ω resistance is given by

$$I_3 = \frac{2}{7} = 21 \times \frac{2}{7} = 6 \text{ A}$$

Current through the 6-Ω resistor, $I_6 = 6 \times \dfrac{3}{9} = 2$ A

Power supplied by the source $P = I^2 R_{eq} = (21)^2 R_{eq}$

$$R_{eq} = (3 \parallel 6 + 3) \parallel 2 = 1.428\ \Omega$$

$$P = (21)^2 \times 1.428 = 630 \text{ W}$$

EXAMPLE 2.36 Using the star–delta transformation, determine the equivalent resistance between X and Y in the circuit shown in Figure 2.78.

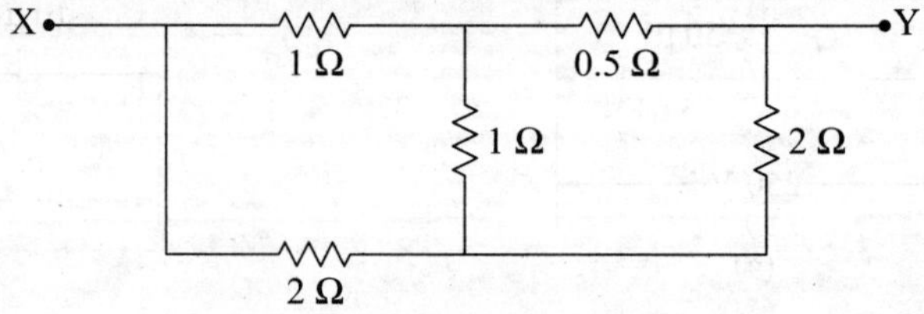

Figure 2.78

Solution From Figure 2.79, it is observed that there are different combinations of star and delta. One of the deltas, i.e., ABCA is converted into star. Their star equivalent resistances are 0.5 Ω, 0.5 Ω, and 0.25 Ω. The circuit is redrawn as shown in Figure 2.79(a).

In Figure 2.79(b), it is seen that $(0.25 + 0.5) \parallel (0.5 + 2) + 0.5$.

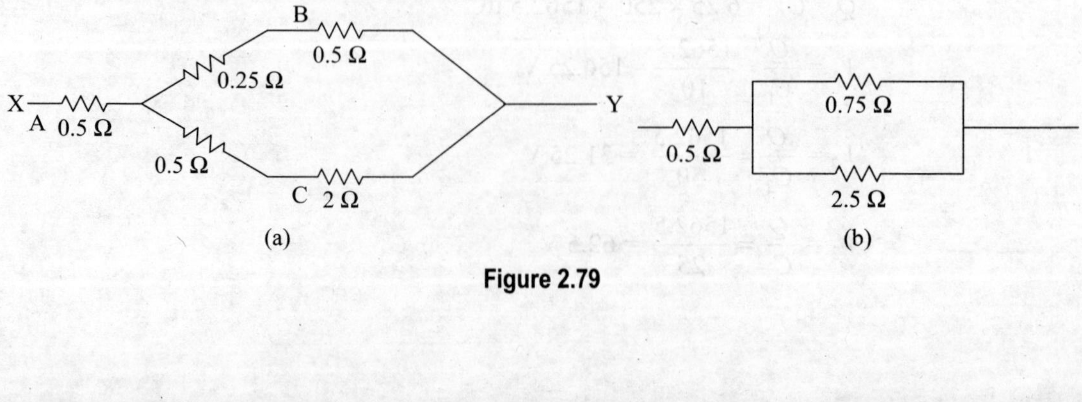

Figure 2.79

$\therefore \quad R_{eq} = 1.0577\ \Omega$

EXAMPLE 2.37 Determine the equivalent circuit seen by the source for the circuit shown in Figure 2.80.

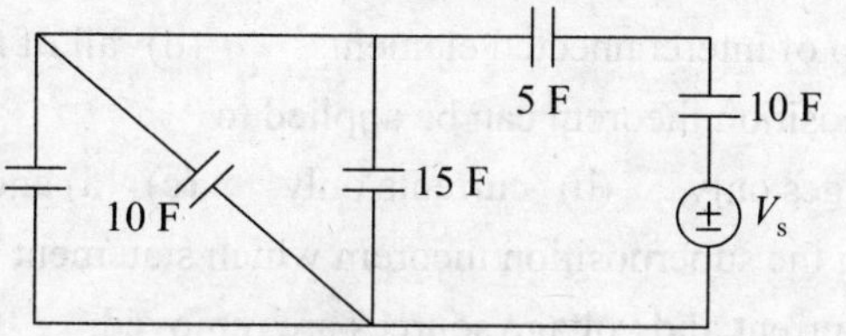

Figure 2.80

Solution Starting from the LHS of the circuit, 5 F and 10 F are in parallel. $C_{1eq} = 5 + 10 = 15$ F. C_{1eq} is in parallel with 15 F. $C_{2eq} = 15 + 15 = 30$ F. Then, 30 F, 5 F, and 10 F are in series and their equivalent is

$$\frac{1}{C_{eq}} = \frac{1}{C_1} + \frac{1}{C_2} + \frac{1}{C_3} = \frac{1}{30} + \frac{1}{5} + \frac{1}{10}$$

$$\frac{1}{C_{eq}} = 0.333 \Rightarrow C_{eq} = 3\ \text{F}$$

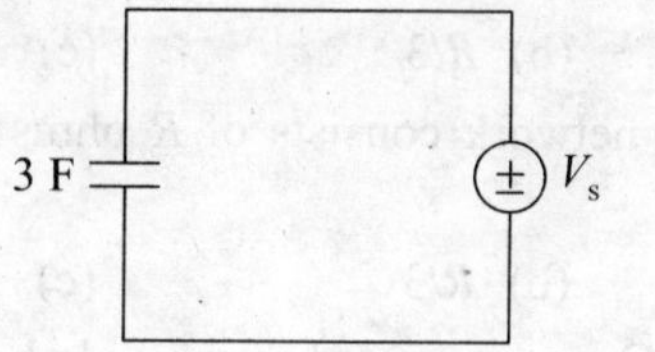

Figure 2.81 Equivalent circuit

CHAPTER AT A GLANCE

- When two resistors R_1 and R_2 are connected in series, the equivalent resistance is $R_1 + R_2$.
- If in a series circuit, $V = V_1 + V_2 + \cdots$, then the current will be same through all the elements.
- If in a parallel circuit, $I = I_1 + I_2 + \cdots$, then the voltage is same across all the elements.
- When two inductors L_1 and L_2 are connected in series, the equivalent inductance, $L = L_1 + L_2$.
- When two resistors are connected in parallel, the equivalent resistance, $1/R = 1/R_1 + 1/R_2$.
- When two capacitors C_1 and C_2 are in series, the equivalent capacitance, $1/C = 1/C_1 + 1/C_2$.
- When two capacitors C_1 and C_2 are in parallel, the equivalent capacitance, $C = C_1 + C_2$.
- For star–delta transformation, $R_{ab} = R_a + R_b + R_a R_b / R_c$.
- For delta–star transformation, $R_a = R_{ab} R_{ac} / R_{ab} + R_{bc} + R_{ca}$.
- For maximum power transfer, the condition is $R_s = R_L$.

MULTIPLE CHOICE QUESTIONS

1. A node in a circuit is defined as a
   (a) junction of two or more elements (b) closed path
   (c) group of interconnected elements (d) all of the above
2. The superposition theorem can be applied to
   (a) voltages only (b) currents only (c) (a) and (b) (d) only (a)
3. In applying the superposition theorem which statement is correct?
   (a) All current and voltage sources are removed.
   (b) Only the current sources are removed.
   (c) Only the voltage sources must be removed.
   (d) Only one source is included at a time.
4. In a network if the number of nodes is n and the number of branches is b, the number of independent mesh equations required to solve the network is
   (a) $b - n + 1$ (b) $b + n + 1$ (c) $b - n - 1$ (d) $b + n - 1$
5. For _____ network, Thevenin's theorem cannot be applied.
   (a) linear (b) active (c) passive (d) nonlinear
6. If a delta-connected network consists of R ohms in each leg of delta, what is its equivalent star?
   (a) R (b) $R/3$ (c) $3R$ (d) $2R$
7. If a star-connected network consists of R ohms in each leg of star, what is its equivalent delta?
   (a) R (b) $R/3$ (c) $3R$ (d) $4R$
8. When 20 Ω and 30 Ω are connected in parallel, the parallel combination is connected in series with another parallel combination of 10 Ω and 15 Ω, and this series and parallel combination is connected across a dc source of 100 V. The current supplied by the source is
   (a) 5.5 A (b) 7 A (c) 8 A (d) 10 A
9. In Question 8, what is the current through the 20-Ω resistance?
   (a) 2.2 A (b) 3.3 A (c) 4.4 A (d) 5.5 A
10. In Question 8, what is the voltage drop across the 30-Ω resistance?
    (a) 33 V (b) 44 V (c) 55 V (d) 66 V
11. Which of the following is an active element in a circuit?
    (a) Current source (b) Resistance
    (c) Inductance (d) Capacitance
12. The circuit having the same properties in either direction is known as
    (a) unilateral (b) reversible (c) bilateral (d) irreversible
13. A network having one or more than one source of emf is known as _____ network.
    (a) passive (b) active (c) linear (d) nonlinear

14. Kirchhoff's laws are valid for

(a) linear circuits only (b) passive time invariant circuits
(c) nonlinear circuits only (d) both linear and nonlinear circuits

15. Kirchhoff's laws are not applicable to circuits with

(a) distributed parameters (b) lumped parameters
(c) passive elements (d) nonlinear resistances

16. The superposition theorem is essentially based on the concept of

(a) reciprocity (b) linearity (c) duality (d) nonlinearity

17. While Thevenizing a circuit between two terminals, V_{Th} is equal to the

(a) short-circuit terminal voltage (b) net voltage available in the circuit
(c) open-circuit terminal voltage (d) emf of the battery nearest to the terminals

18. Thevenin's resistance R_{Th} is determined

(a) by short circuiting the given two terminals
(b) by removing the voltage sources along with their internal resistances
(c) between any two open terminals
(d) between the same open terminals as for V_{Th}

19. While determining R_{Th} of a circuit

(a) voltage and current sources should be left as they are
(b) all independent current and voltage sources are short circuited
(c) all sources should be replaced by their source resistances
(d) none

20. Which of the following theorems is applicable for both linear and nonlinear circuits?

(a) Thevenin's (b) Superposition (c) Norton's (d) None

21. Which of the following statement is correct?

(a) Norton's equivalent resistance is the same as Thevenin's equivalent resistance.
(b) Norton's equivalent is the current equivalent of the network.
(c) The load is connected in parallel to Norton's equivalent resistance and Norton's current source.
(d) (a), (b), and (c)

22. For maximum power transfer, internal resistance of the source should be

(a) equal to load resistance (b) less than that of load
(c) more than that of load (d) zero

23. Three equal resistances are connected in star. If this star is converted into equivalent delta, then

(a) the resistance of the delta network will be smaller than that of the star network
(b) the resistance of both the networks will be equal
(c) the resistance of the delta network will be larger than that of the star network
(d) none

24. The nodal analysis is primarily based on the application of
(a) Ohm's law (b) KCL (c) KVL (d) (a) and (b)

25. Resistance of a 200-W, 250-V lamp will be
(a) 625 Ω (b) 1250 Ω (c) 312.5 Ω (d) 31.25 Ω

26. A 2-Ω resistance having a current of 2 A will dissipate the power of
(a) 2 W (b) 4 W (c) 8 W (d) 10 W

27. An ideal current source has
(a) zero internal conductance (b) zero internal resistance
(c) zero voltage on no-load (d) zero ripple

28. Which of the following statement is incorrect?
(a) Resistance is a passive element (b) Inductor is a passive element
(c) Current source is a passive element (d) Voltage source is an active element

29. According to KVL, the algebraic sum of all *IR* drops and emfs in any closed loop of a network is always
(a) zero (b) negative (c) positive (d) none

30. Which of the following statements is false in case of a series circuit?
(a) The voltage drop across each resistor is the same.
(b) The current flowing through each resistor is the same.
(c) Applied voltage is equal to the sum of voltage drops across individual resistors.
(d) Resistors are additive.

SHORT ANSWER QUESTIONS

1. When two resistors R_1 and R_2 are in series, what is the equivalent resistance?

2. For a given line voltage, four heating coils will produce maximum heat when connected _____.

3. How do you find Thevenin's resistance?

4. How Thevenin's voltage source and Thevenin's resistance are connected?

5. How are Norton's current source and Norton's resistance connected?

6. What is the condition for maximum power transfer to load?

7. What is the efficiency under the condition of maximum power transfer?

8. What is the equivalent inductance of two inductive coils of 15 H and 10 H connected in series?

9. Three resistances of 3 Ω each are connected in delta. What is the value of the resistance in the equivalent star?

10. Three capacitors of 4 μF, 4 μF, and 8 μF are connected in parallel. What is its equivalent capacitance?

REVIEW QUESTIONS

1. Explain the source transformation technique.
2. Explain the star–delta and delta–star transformations.
3. State and explain the superposition theorem.
4. State and explain Thevenin's theorem.
5. State and explain Norton's theorem.
6. State and explain the maximum power transfer theorem.
7. Explain with a suitable example to obtain an equivalent current source from a given voltage source.
8. Explain supernode and supermesh.
9. Explain the mesh analysis with suitable example.
10. Explain the nodal analysis with suitable example.

EXERCISE PROBLEMS

1. Three resistors R_1, R_2, and R_3 are connected in series across a constant voltage V. The voltage across R_1 is 20 V. The power consumed by R_2 is 25 W, $R_3 = 2\ \Omega$. Find the voltage if the current is 5 A?
2. Determine the value of R if the voltage across it is 20 V as shown in Figure 2.82.

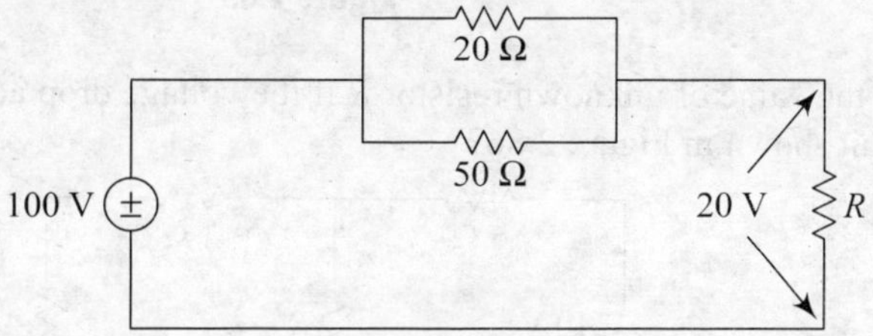

Figure 2.82

3. Determine the currents for the network shown in Figure 2.83.

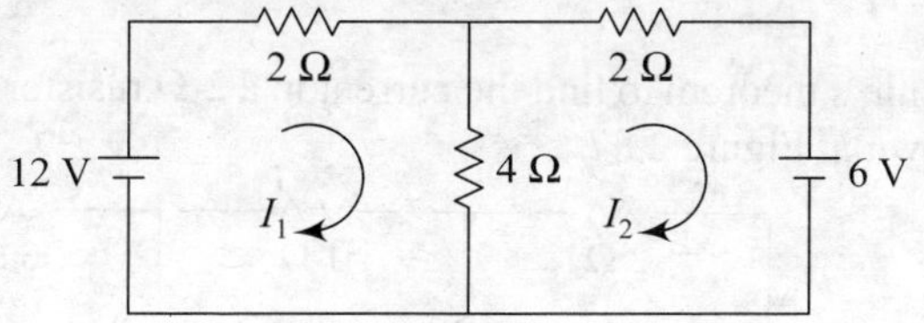

Figure 2.83

4. For the bridge circuit shown in Figure 2.84, determine the currents I_1, I_2, and I_3.

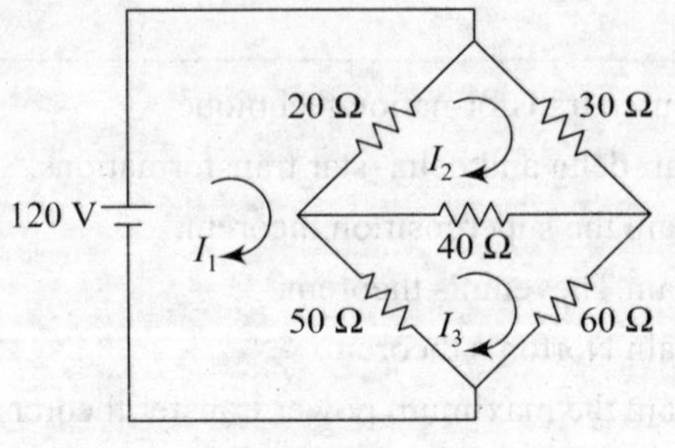

Figure 2.84

5. Calculate the current drawn from 2-V source in the network shown in Figure 2.85.

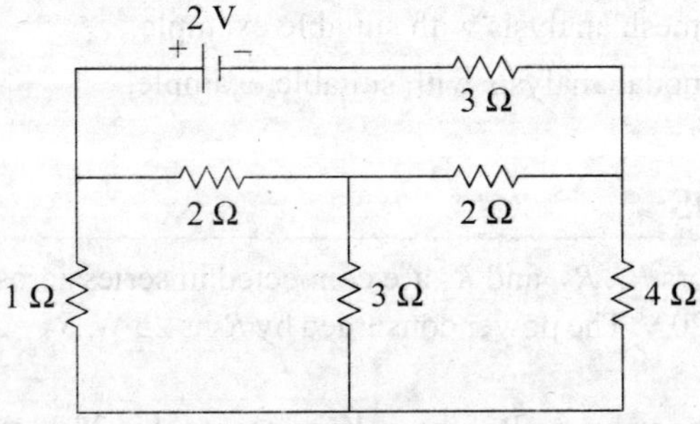

Figure 2.85

6. What is the value of unknown resistor R if the voltage drop across the 4-Ω resistor is 2 V for the circuit shown in Figure 2.86?

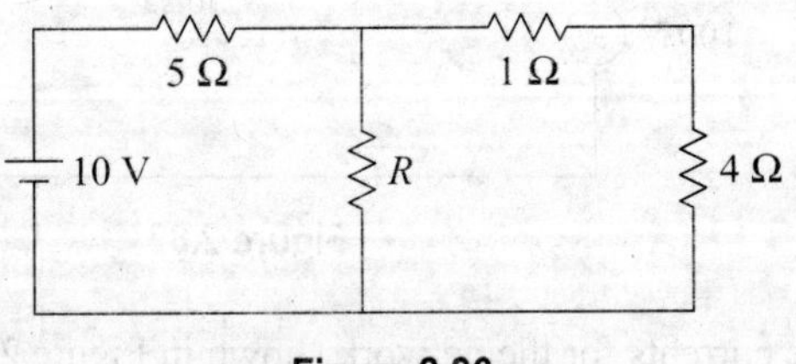

Figure 2.86

7. Use Thevenin's theorem to find the current in a 2-Ω resistor connected between A and B in the circuit shown in Figure 2.87.

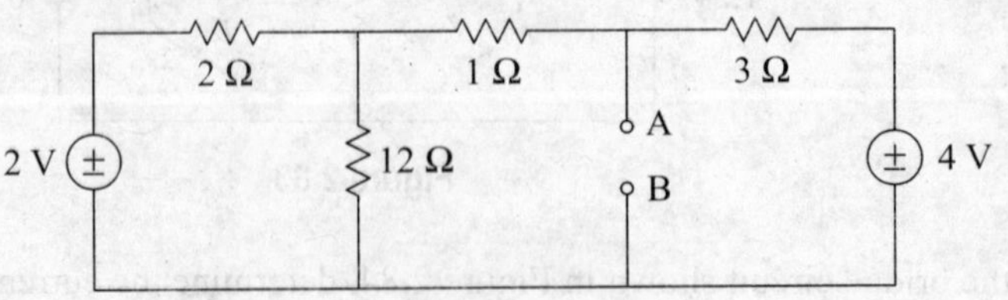

Figure 2.87

8. Use the superposition theorem for Question 7.

9. Find the current in the 10-Ω resistor in the network shown in Figure 2.88 by star–delta transformation.

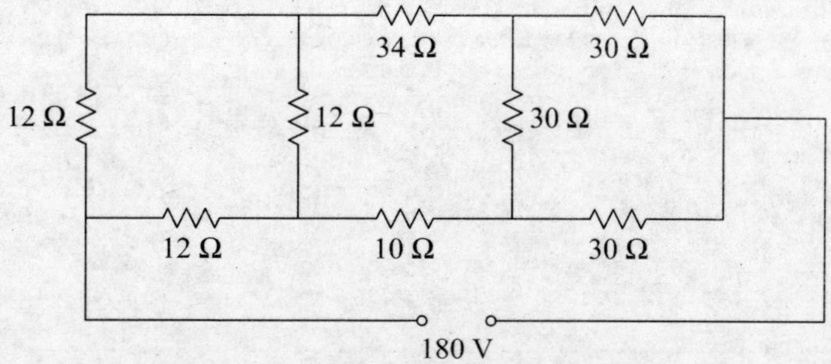

Figure 2.88

10. For Question 9, use Norton's theorem to obtain current.

Chapter 3

Magnetic Circuits

Chapter Objectives

- To explain the concept of electromagnetism
- To explain the magnetic effect of electromagnets, leakage flux, and fringing
- To describe Faraday's laws of electromagnetic induction
- To explain various laws of electromagnetism
- To describe comparison between electric circuits and magnetic circuits

3.1 INTRODUCTION

A body that possesses the property of attracting pieces of iron is known as a *magnet*, and the property of the body by virtue of which this attraction takes place is known as *magnetism*. Magnets are of two types: natural magnets and artificial magnets. Natural magnets are magnets taken out from the earth. Natural magnets are rugged and clumsy and are also comparatively weak; therefore, they do not have much practical utility. Artificial magnets are made either by rubbing a piece of iron or steel with the natural magnet or by passing direct current through a coil wound over a piece of iron or steel. In the first method, a piece of iron or steel is magnetized by repeatedly rubbing a magnet over its entire length in the same direction. The piece converted into magnet has its first terminal the same and the last terminal has the opposite polarity to that of the end of the magnet used for rubbing.

In the second method, direct current is passed through an insulated copper coil wound round a piece of soft iron. Magnets prepared with this method are called *electromagnets*. Magnets may be classified as being permanent or temporary depending on their ability to retain magnetism. The permanent magnets are made from cobalt, nickel alloys, and hardened steel. These substances have high retentivity power and retain their magnetism unaltered for a long time after the removal of the magnetizing force. Permanent magnets are used extensively in telephone receivers, dc motors, electrical measuring instruments, and speedometers. An example of temporary magnets is soft iron. Temporary magnets lose most of their strength when the magnetizing force is removed. These are used in a variety of applications. Generally, these are used in electric generators and motors where it is desirable to control the amount of magnetism present in the magnet. Some important concepts relating to the study of magnetic circuits and electromagnetism are discussed in the following sections.

3.2 CONCEPTS OF MAGNETISM AND ELECTROMAGNETISM

Magnets have two magnetic poles. One of the magnetic poles is known as the north pole and the other as the south pole. It is known from the experiments that similar poles repel each other and opposite poles attract each other.

It is well known that whenever electric current flows through a conductor, a magnetic field immediately comes into existence in the space surrounding the conductor. Hence, it can be said that when electrons are in motion, they produce a magnetic field. The converse is also true, i.e., when a magnetic field embracing a conductor moves relative to the conductor, it produces a flow of electrons in the conductor. This phenomenon whereby an electromotive force (emf) and hence current (i.e., flow of electrons) is induced in any conductor which is cut across or is cut by a magnetic flux is known as *electromagnetic induction*. The historical background of this phenomenon is that after the discovery (by Oersted) that electric current produces a magnetic field, scientists began to search for the converse phenomenon from about 1821 onwards. The problem they put to themselves was how to "convert" magnetism into electricity. It is recorded that Michael Faraday was in the habit of walking about with magnets in his pockets so as to continuously remind him of the problem. After nine years of continuous research and experimentation, he succeeded in producing electricity by "converting magnetism." In 1831, he formulated basic laws underlying the phenomenon of electromagnetic induction (known after his name), upon which is based the operation of most of the commercial apparatus such as motors, generators, and transformers.

3.3 BASIC DEFINITIONS

3.3.1 Magnetic Flux

In a magnetic field, all the lines of force taken together are given the name flux or magnetic flux. Its usual symbol is ϕ and the unit is weber.

3.3.2 Magnetic Flux Density

It is given by the flux passing per unit area through a plane at right angles to the flux. It is usually expressed by the capital letter B and is measured in Wb/m^2. It is a vector quantity.

If ϕ weber is the total magnetic flux passing normally through an area of A square meter, then

$$B = \frac{\phi}{A} = \frac{1}{4\pi r^2} \text{ weber/m}^2 \text{ or tesla (T)} \tag{3.1}$$

Note: Let us find an expression for the flux density at a point distant "r" meters from a unit N-pole (i.e., a pole of strength 1 Wb). Imagine a sphere of radius "r" meters drawn around the unit pole. The flux of 1 Wb radiated out by the unit pole falls normally on a surface of $4\pi r^2$ square meter. Hence,

$$B = \frac{\phi}{A} = \frac{1}{4\pi r^2} \text{ weber/m}^2$$

3.3.3 Magnetic Circuit

Magnetic circuit may be Defined as the route or path which is followed by magnetic flux. The law of magnetic circuits is quite similar to (but not the same as) those of the electric circuit.

Consider a solenoid or a toroidal iron ring having a magnetic path of l meter, area of cross section A square meter, and coil of N turns I amperes of current would be present any where on it as in Figure 3.1. Thus, the field strength inside the solenoid is

$$H = \frac{NI}{l} \text{ ampere-turn/m} \quad (3.2)$$

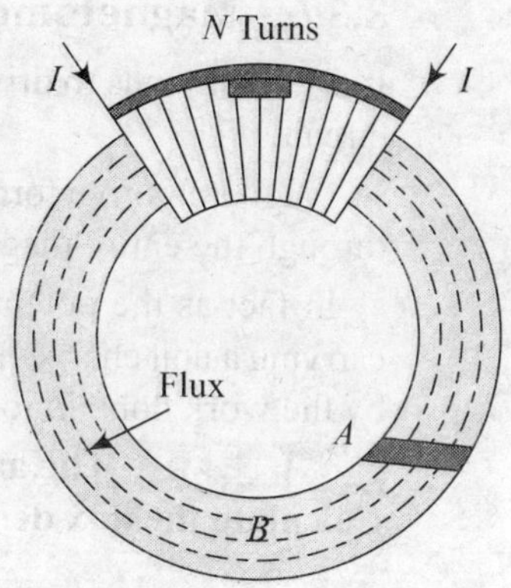

Figure 3.1 Toroidal iron ring

Now,

$$B = \mu_0 \mu_r H = \frac{\mu_0 \mu_r NI}{l} \text{ weber/m}^2 \quad (3.3)$$

$$\text{Total flux produced, } \phi = B \times A = \frac{\mu_0 \mu_r ANI}{l} \text{ weber}$$

$$= \frac{NI/l}{\mu_0 \mu_r A} \quad (3.4)$$

The numerator "NI" which produces magnetization in the magnetic circuit is known as *magnetomotive force* (mmf). Obviously, its unit is ampere-turn (AT). It is analogous to emf in an electric circuit. The denominator $l/\mu_0\mu_r A$ is called the *reluctance* of the circuit and is analogous to resistance in electric circuits. Therefore,

$$\text{Flux} = \frac{\text{emf}}{\text{Resistance}} \quad \text{or} \quad \phi = \frac{F}{S} \quad (3.5)$$

Sometimes, the above equation is called "Ohm's law of magnetic circuit" because it resembles a similar expression in electric circuit, i.e.,

$$\text{Current} = \frac{\text{emf}}{\text{Resistance}} \quad \text{or} \quad I = \frac{V}{R} \quad (3.6)$$

3.3.4 Magnetic Field

The space around the poles of a magnet is called the *magnetic field* and is represented by magnetic lines of forces. For example, the space around the earth and around a permanent magnet is called magnetic field. The total number of lines of forces surrounding a magnet is called the total magnetic flux.

3.3.5 Magnetic Lines of Force

Line of force indicates the existence of magnetic field in the neighborhood of a magnetic pole. These are taken to emanate from N-pole, traverse the surrounding medium, re-enter the S-pole, and pass through the magnet from S-pole to N-pole. Note that the lines of force must form a closed loop.

3.3.6 Magnetomotive Force

It drives or tends to drive flux through a magnetic circuit and corresponds to an emf in an electric circuit.

Magnetomotive force is equal to the work done in joules in carrying a unit magnetic pole once through the entire magnetic circuit. It is measured in ampere-turn.

In fact as the potential difference between any two points is measured by the work done in carrying a unit charge from one point to another, similarly, the mmf between two points is measured by the work done in joules in carrying a unit charge from one point to another.

EXAMPLE 3.1 The area of the face of a pole-shoe is 1.5 cm² and the total flux is 0.18 mWb. Calculate the flux density in air gap.

Solution Flux density, $B = \dfrac{\phi}{A}$

$$= \frac{0.18 \times 10^{-3}}{1.5 \times 10^{-4}}$$

$$= \frac{1.8}{1.5} = 1.2 \text{ Wb/m}^2$$

EXAMPLE 3.2 A coil of 100 turns is wound uniformly over an isolator ring with a mean circumference of 2 m and a uniform sectional area of 0.025 cm². If the coil carries current of 2 A, calculate (i) the mmf of the circuit, (ii) magnetic field intensity, (iii) flux density, and (iv) total flux.

Solution

(i) $\text{mmf} = NI = 100 \times 2 = 200 \text{ AT}$

(ii) $H = \dfrac{\text{mmf}}{I} = \dfrac{200}{2} = 100 \text{ AT/m}$

(iii) $B = \mu_0 H = 4\pi \times 10^{-7} \times 100 = 0.12565 \text{ mWb/m}^2$

(iv) $\phi = B \times A = 1.2565 \times 10^{-3} \times 0.025 \times 10^{-4} = 3.14 \times 10^{-9} \text{ Wb}$

3.3.7 Magnetic Field Intensity

Magnetic field intensity may be defined as the induced pole strength developed per unit area of the bar. Also, it is the magnetic moment developed per unit volume of the bar.

Let m be the pole strength induced in the bar in weber, and A the face or pole area of the bar in square meter. Then,

$$I = \frac{m}{A} \text{ weber/m}^2$$

Hence, it is seen that intensity of magnetization of a substance may be defined as the flux density produced in it due to its own induced magnetism.

If l is the magnetic length of the bar, then the product $(m \times l)$ is known as the *magnetic moment M.*

$$I = \frac{m}{A} = \frac{m \times l}{A \times l} = \frac{M}{v} = \frac{\text{Magnetic moment}}{\text{Volume}} \tag{3.7}$$

3.3.8 Reluctance

Reluctance is the name given to the property of material which opposes the creation of magnetic flux in it. It, in fact, measures the opposition offered to the passage of magnetic flux through a material and is the analogous to the resistance in an electric circuit even in form. Its unit is AT/Wb.

$$\text{Reluctance} = \frac{l}{\mu A} = \frac{l}{\mu_0 \mu_r A}$$

$$\text{Resistance} = \rho \frac{l}{A} = \frac{l}{\sigma A}$$

In other words, the reluctance of magnetic circuit is the number of ampere-turns required per weber of magnetic flux in the circuit. Since 1 AT/Wb = 1 henry, the unit of reluctance is "reciprocal of henry."

3.3.9 Permeance

Permeance is the reciprocal of reluctance and implies the case or readiness with which the magnetic flux is developed. It is analogous to conductance in electric circuit. It is measured in terms of Wb/AT or H.

3.3.10 Permeability

Permeability is the measure of the capability of a material to support the formation of a magnetic field within itself. In other words, it is the level of magnetization that a material obtains in response to an applied magnetic field. Figure 3.2 shows a bar of a magnetic material, say, iron placed in a uniform field of strength "H" newton per weber (N/Wb). Suppose, a flux density of B weber/m^2 is developed in the rod.

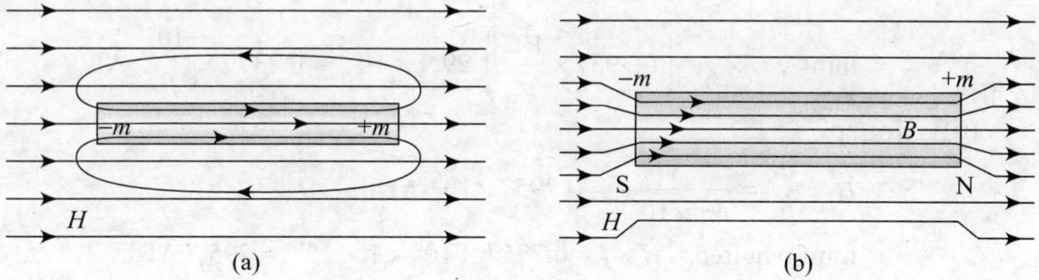

Figure 3.2 A bar of a magnetic material

Absolute permeability and relative permeability: The absolute permeability of the material of the rod is defined as

$$\mu = \frac{B}{H} \text{ henry/m}$$

or

$$B = \mu_0 H = \mu_0 \mu_r H \text{ weber/m}^2 \tag{3.8}$$

When H is established in air (or vacuum), then the corresponding flux density developed in air is

$$B_0 = \mu_0 H$$

Now, when an iron rod is placed in the field, it gets magnetized by induction. If the induced pole strength in the rod is "m" weber, then a flux of "m" weber emanates from its N-pole, re-enters its S-pole, and continues from S- to N-pole within the magnet. If A is the face or pole area of the magnetized iron bar, then the induction flux density in the rod is

$$B_i = \frac{m}{A} \text{ weber/m}^2$$

Hence, the total flux density in the iron rod consists of two parts [Figure 3.2(b)]: (i) B_0—flux density in air even when the rod is not present and (ii) B_i—induction flux density in the rod.

$$B = B_0 + B_i = \mu_0 H + \frac{m}{A}$$

Equation (3.8) may be written as $B = \mu_r \mu_0 H = \mu_r B_0$. Therefore,

$$\mu_r = \frac{B}{B_0} = \frac{B\,(\text{material})}{B_0\,(\text{vacuum})} \quad (\text{for same } H) \tag{3.9}$$

Hence, the relative permeability of a material is equal to the ratio of the flux density produced in that material to the flux density produced in vacuum by the same magnetizing force.

EXAMPLE 3.3 **A mild steel ring of 30 cm mean circumference has a cross-sectional area of 6 cm² and has winding of 500 turns on it. The ring is cut through at a point so as to provide an air gap of 1 mm in the magnetic circuit. It is found that a current of 4 A in the winding produces a flux density of 1 T in the air gap. Find (i) the relative permeability of the mild steel and (ii) the inductance of the winding.**

Solution

(i) Steel ring

$$H = \frac{B}{\mu_0 \mu_r} = \frac{1}{4\pi \times 10^{-7} \times \mu_r} \text{ AT/m} = \frac{0.7957 \times 10^6}{\mu_r} \text{ AT/m}$$

$$\text{mmf} = H \times l = \left(0.7957 \times \frac{10^6}{\mu_r}\right) 29.9 \times 10^{-2} = 0.2379 \times \frac{10^6}{\mu_r} \text{ AT}$$

(ii) Air gap

$$H = \frac{B}{\mu_0} = \frac{1}{4\pi \times 10^{-7}} = 0.7957 \times 10^6 \text{ AT/m}$$

$$\text{mmf required} = H \times l = 0.7957 \times 10^6 \times 1 \times 10^{-3} = 795.7 \text{ AT}$$

$$\text{Total mmf} = \left(0.2379 \times \frac{10^6}{\mu_r}\right) + 795.7$$

$$\text{Total mmf available} = NI = 500 \times 4 = 2000 \text{ AT}$$

Relative permeability of the mild steel,

$$2000 = \left(0.2379 \times \frac{10^6}{\mu_r}\right) + 795.7$$

$$\mu_r = 197.5$$

$$\text{Inductance of the winding} = \frac{N\phi}{I} = \frac{NBA}{I} = \frac{500 \times 1 \times 6 \times 10^{-4}}{4} = 0.075 \text{ H}$$

3.4 CLASSIFICATION OF MAGNETIC MATERIALS

Magnetic materials are classified as (i) paramagnetic materials, (ii) diamagnetic materials, and (iii) ferromagnetic materials.

(i) *Paramagnetic materials:* The materials which are not strongly attracted by a magnet are called paramagnetic materials. Examples are aluminum, platinum, tin, and manganese. These materials slightly get magnetized when placed in a strong magnetic field and act in the direction of the magnetic field.

(ii) *Diamagnetic materials:* The materials which are repelled by a magnet are called diamagnetic materials. Examples are zinc, lead, sulfur, mercury, wood, and bismuth. These materials get magnetized very lightly when they are placed in a strong magnetic field and act in the direction opposite to that of the applied magnetic field.

(iii) *Ferromagnetic/ferrimagnetic materials:* The materials which are strongly attracted by a magnet are called ferromagnetic materials. Examples are iron, steel, and nickel. Their permeability is very high.

For the purpose of electrical engineering analysis, it will be sufficient to classify the materials as nonferromagnetic and ferromagnetic materials. The paramagnetic and diamagnetic materials come under the nonferro category.

Ferromagnetic materials are of two types:

(a) *Soft magnetic materials:* These materials do not retain their magnetism for long time after the removal of magnetizing force. Examples are soft iron, silicon steel, and nickel–iron alloys. These are generally used in transformers, generators and motors, and relays and radars. They have very high relative permeability and low hysteresis and eddy current losses.

(b) *Hard magnetic materials:* These materials retain their magnetism for long time after the removal of magnetizing force. These are used for making permanent magnets. These materials are tungsten alloys, carbon steel, and cobalt steel. They have high residual induction, high magnetic stability, and large energy losses per cycle of magnetization.

3.5 MAGNETIZATION CHARACTERISTICS OF MAGNETIC MATERIALS

The curve plotted between the induction density *B* and the magnetizing force *H* is known as magnetization characteristic.

3.5.1 Magnetic Hysteresis (*B*–*H* Curve)

It may be defined as the lagging of magnetization or induction flux density (*B*) behind the magnetizing force (*H*). Alternatively, it may be defined as that quality of magnetic substance due to which energy is dissipated in it, on the reversal of its magnetism.

Let us take an unmagnetized bar of iron AB and magnetize it by placing it within the field of a solenoid (Figure 3.3). The field $H (= NI/l)$ produced by the solenoid is called the *magnetizing force*. The value of H can be increased or decreased by increasing or decreasing the current through the coil. Let H be increased in steps from zero up to a certain maximum value and the corresponding values of flux density (B) be noted. If we plot a relation between H and B, a curve like OA, as shown in Figure 3.4, is obtained. The material becomes magnetically saturated for H = OM and has at that time a maximum flux density of B_{max} established through it.

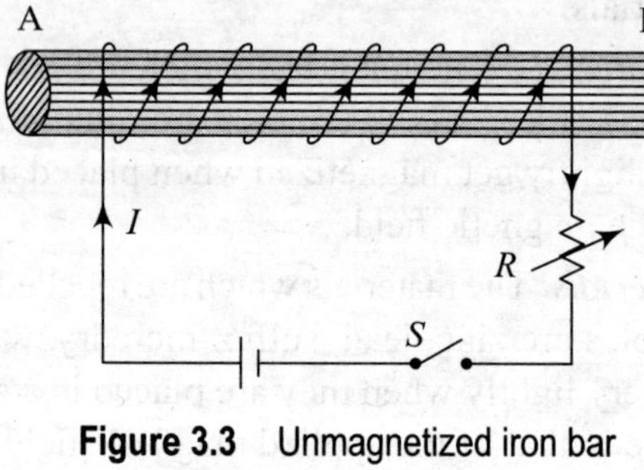

Figure 3.3 Unmagnetized iron bar

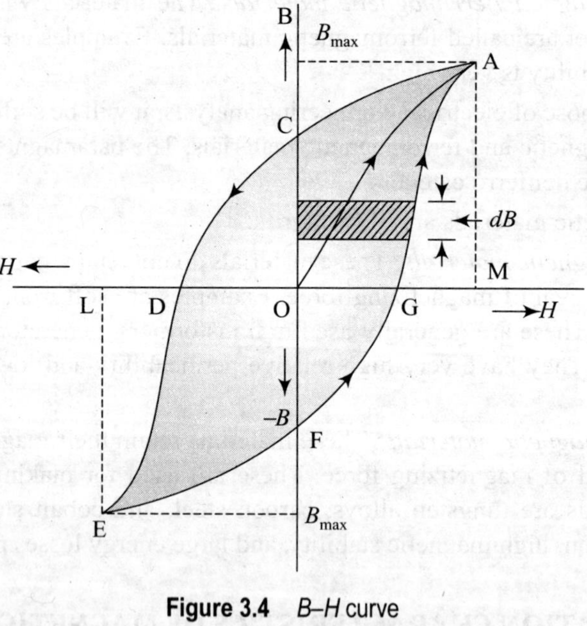

Figure 3.4 *B–H* curve

If H is now decreased gradually (by decreasing solenoid current), the flux density B will not decrease along AO, as might be expected, but will decrease less rapidly along AC. When H is zero, B is not, but has a definite value B_r = OC. It means that on removing the magnetizing force H, the iron bar is not completely demagnetized. This value of B (=OC) measures the *retentivity or remanence* of the material and is called the remanent or residual flux density B_r.

To demagnetize the iron bar, we have to apply the magnetizing force in the reverse direction. When H is reversed (by reversing current through the solenoid), then B is reduced to zero at point D, where H = OD. This value of H required to wipe off the residual magnetism is known as *coercive force* (H_C) and is a measure of the *coercivity* of the material, i.e., its "tenacity" with which it holds on to its magnetism.

If, after the magnetization has been reduced to zero, the value of H is further increased in the "negative," i.e., reversed direction, the iron bar again reaches a state of magnetic saturation, represented by point L. By taking H back its value corresponding to negative saturation (=OL) to its value for positive saturation (=OM), a similar curve EFGA is obtained. If we again start from G, the same curve GACDEFG is obtained once again.

It is seen that B always lag behind H. The two never attain zero value simultaneously. This lagging of B behinds H is given the name "hysteresis" which literally means "to lag behind." The closed loop ACDEFGA which is obtained when the iron bar is taken through one complete cycle of magnetization is known as "hysteresis loop."

One cycle of magnetization of the magnetic material means it is being carried through one reversal of magnetization as shown in Figure 3.5.

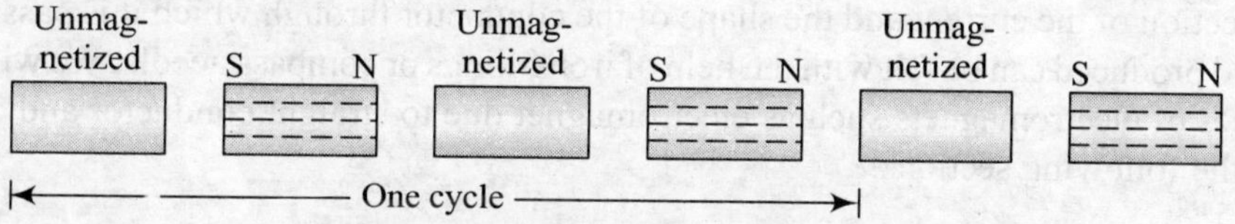

Figure 3.5 One reversal of magnetization

Hysteresis loss

The hysteresis loss is equal to the energy consumed in magnetizing and demagnetizing a magnetic material. It is proportional to (i) the area enclosed in the hysteresis loop, (ii) the frequency of magnetic flux reversal, and (iii) the volume of the magnetic material.

Again the area of the hysteresis loop depends on the flux density at which the material is being operated. The hysteresis loss can be determined by the Steinmetz empirical formula as given below.

$$\text{Hysteresis loss, } P_h = \eta v f B_{max}^{1.6}$$

where η is the constant known as Steinmetz coefficient, which varies with the characteristic of a material; v is the volume of the material in m^3; f is the frequency of magnetic flux reversal; and B_{max} is the maximum value of flux density in the material.

Eddy current loss

The eddy current loss occurs in a magnetic material when it is subjected to a changing magnetic field. Similar to the hysteresis loss, the eddy current loss also results in the rise in temperature of the material. Mathematically, the eddy current loss is given by I^2R, where I is the eddy current magnitude and R is the resistance of the eddy current path. But it is difficult to determine the eddy current magnitude and resistance and hence the eddy current loss. But there is an empirical formula to determine the eddy current loss.

$$P_e = K_e B_{max}^2 f^2 t^2 v$$

where K_e is the eddy current coefficient, f is the number of magnetization cycles per second, B_{max} is the maximum flux density in tesla, t is the thickness of lamination in meters, and v is the volume of the material in m^3.

From the above equation, it is observed that the eddy current loss is proportional to the square of the flux density and the square of the frequency.

3.6 MAGNETIC EFFECT OF AN ELECTROMAGNET

When current passes through a coil or conductor, it produces the magnetic flux around it and the coil works as a magnet. Such a current-carrying coil or conductor is called an *electromagnet*. This is due to the magnetic effect of an electric current. If such a coil is wound around a piece of magnetic material such as steel or iron and carries current, then that material also behaves as an electromagnet. By controlling the current we can control the flux produced and also the flux density.

The direction and the shape of the magnetic field around the coil or conductor depend on the direction of the current and the shape of the conductor through which it is passing. The magnetic field produced can be felt with the help of iron fillings or compass needle. We will discuss different types of electromagnets such as electromagnet due to straight conductor and circular conductor in the following sections.

3.6.1 Magnetic Field Due to Straight Conductor

When a straight conductor carries a current, it produces a magnetic field throughout its length. The lines of magnetic force are in the form of concentric circles in the planes right angled to the conductor.

Figure 3.6 shows a straight conductor carrying current, which is passing through a sheet. The direction of the magnetic flux can be determined by placing compass needle near the conductor. This direction depends on the direction of the current passing through the conductor. For the current direction shown in the figure, i.e., from top to bottom, the direction of flux is clockwise around the conductor.

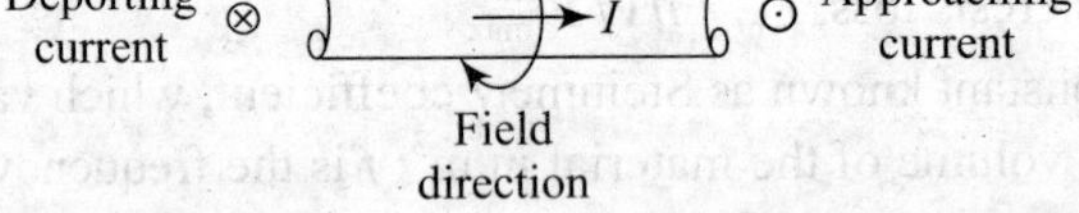

Figure 3.6 Magnetic field due to a straight conductor

Generally, such a current-carrying conductor is represented by a small circle. Then the current through such conductor will either come out of the sheet or go into the plane of the sheet. When the current is going into the plane of the paper, i.e., away from the observer, it is represented by a cross inside the circle indicating the conductors. The cross indicates the rear view of the feathered end of an arrow. The current flowing toward observer, i.e., coming out of the plane of the sheet is represented by a dot inside the circle. The dot indicates front view, i.e., the tip of an arrow. This is shown in Figure 3.7.

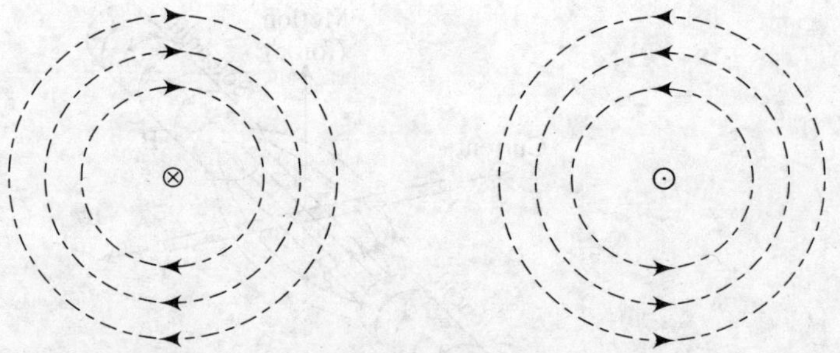

Figure 3.7 Current into the sheet and out of the sheet

3.6.2 Magnetic Field Due to Circular Conductor

A solenoid is a long conductor wound with a number of turns close together to form a coil. Figure 3.8 shows a circular conductor carrying current. The length of the conductor is greater than the diameter of turns. The part or element around which the conductor is wound is called the core of the solenoid. Core may be air or some magnetic material. When current passes through such circular conductor, it produces magnetic field which acts through the coil along its axis and also around the solenoid.

The use of magnetic material for the core produces strong magnet. This is because the current-carrying conductor produces its own flux. In addition to this, the core behaves like a magnet due to magnetic induction, producing its own flux. The direction of the two fluxes is same due to which the resultant magnetic field becomes stronger. The pattern of flux around the solenoid is shown in Figure 3.8.

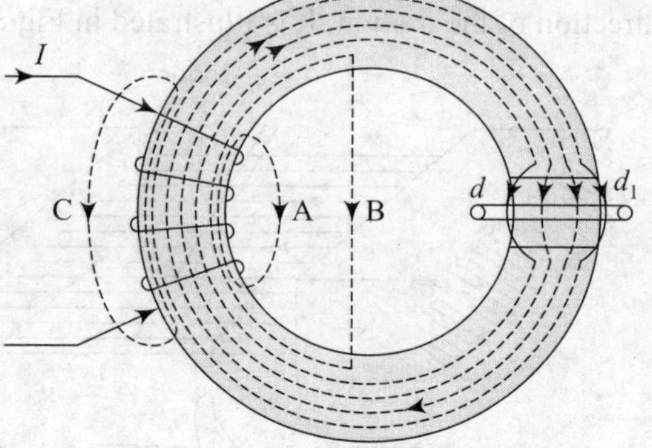

Figure 3.8 Pattern of flux around the solenoid

Force on a current-carrying conductor in a magnetic field

It is found that whenever a current-carrying conductor is placed in a magnetic field, it experiences a force which acts in a direction perpendicular to the direction of both the current and field. Figure 3.9(a) shows a conductor XY lying at right angles to the uniform horizontal field density B Wb/m^2 produced by two solenoids A and B. If l is the length of the conductor lying within this field and I ampere the current carried by it, then the magnitude of the force experienced by it is given by

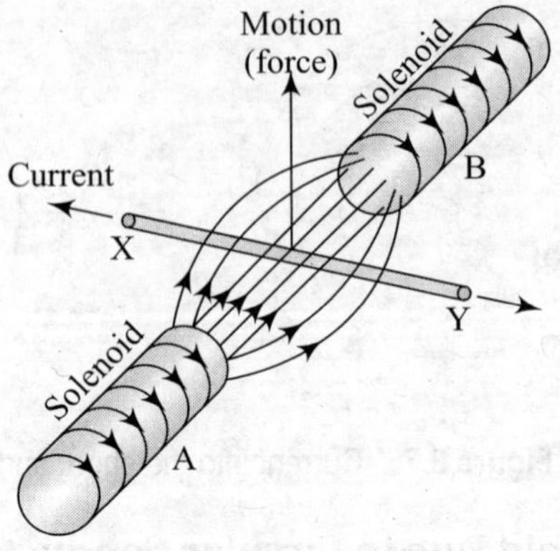

Figure 3.9(a)

$$F = BIl = \mu_0\mu_r HIl \text{ newton} \tag{3.10}$$

Using vector notation,

$$\vec{F} = I\vec{l} \times \vec{B} \text{ and } F = BIl \sin\theta$$

where θ is the angle between $\vec{l}$ and $\vec{B}$, which is 90° in the present case

or

$$F = BIl \sin 90° = BIl \text{ newton} \quad (\because \sin 90° = 1) \tag{3.11}$$

The direction of this force may be easily found by Fleming's left-hand rule.

Hold out your left hand with forefinger, second finger, and thumb at right angles to one another. If the forefinger represents the direction of the field and the second finger that of the current, then thumb gives the direction of the motion. It is illustrated in Figure 3.9(b).

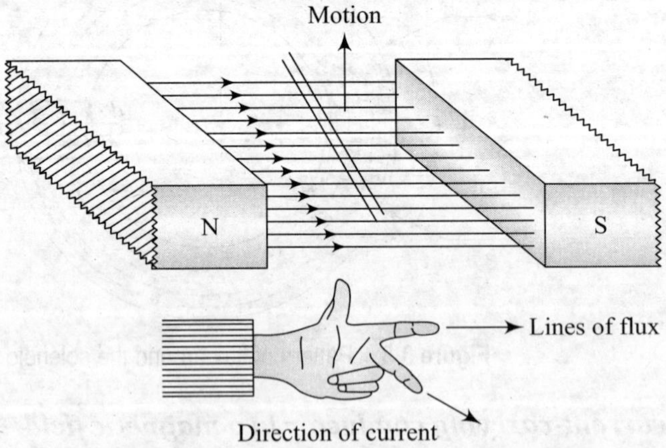

Figure 3.9(b)

Figure 3.9(c) shows another method of finding the direction of force acting on a current-carrying conductor. It is known as flat left-hand rule. If the palm of left hand is so placed that the lines of induction leaving the *N*-pole enter the palm and the four fingers are in the direction of flow of current in the conductor, then the thumb will indicate the direction of force acting on the conductor. The force acts in the direction of the thumb. Obviously, the direction of the motion of the conductor is the same as that of the force.

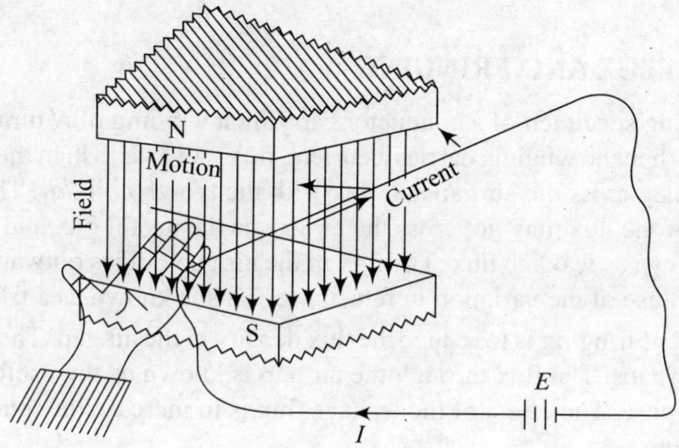

Figure 3.9(c)

It should be noted that no force is exerted on a conductor when it lies parallel to the magnetic field (see Figure 3.10). In general, if the conductor lies at an angle θ with the direction of the field, then B can be resolved into two components, $B \cos \theta$ perpendicular to the conductor. The former produces no effect, whereas the latter is responsible for the motion observed. In that case, $F = BI\,l \sin 90° = BI\,l$ newton, which has been expressed as the cross product of the vector above.

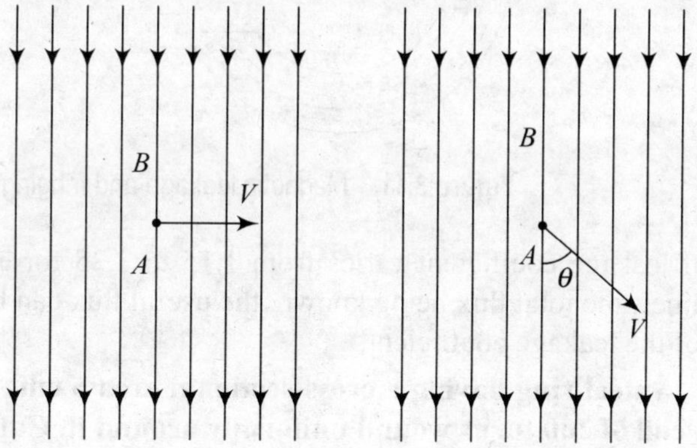

Figure 3.10 Direction of force acting on a current-carrying conductor

EXAMPLE 3.4 Consider a conductor of length 1 m at right angles to a magnetic field of flux density 1 Wb/m² with a velocity of 25 m/s. Calculate the emf induced in the conductor.

Solution $e = Blv = 1 \times 1 \times 25 = 25$ V

EXAMPLE 3.5 A 2-m long cross-bar is kept on the roof rack of a truck traveling at 90 km/h. If the vertical component of the earth magnetic field is 4×10^{-5} T, find the emf induced at the two ends of the cross-bar.

Solution $e = Blv = 4 \times 10^{-5} \times 2 \times \dfrac{90000}{3600}$

$= 2 \times 10^{-3} = 2 \text{ mV}$

3.7 LEAKAGE FLUX AND FRINGING

Consider a ring specimen of a conductor carrying a winding of N turns with air gap as shown in Figure 3.8. When the winding carries a current, flux will establish in the specimen. A small portion of this flux may cross the air gap and link with the search coil dd_1. This flux is called the useful flux. Some of the flux may not cross the air gap as the useful flux and returns by paths A, B, and C, and is known as leakage flux. The flux in the air gap bulges outwards as shown in Figures 3.8 and 3.11 because of the variation in reluctance. This is known as a fringing.

The effect of fringing is to reduce the flux density in the air gap. The effect is more pronounced for larger air gaps. The flux through the air gap is known as the useful flux and can be used for practical purpose. The effect of the leakage flux is to increase the total flux through the exciting winding. Thus,

$$\text{The leakage coefficient} = \frac{\text{Total flux through the exciting winding}}{\text{Useful flux}}$$

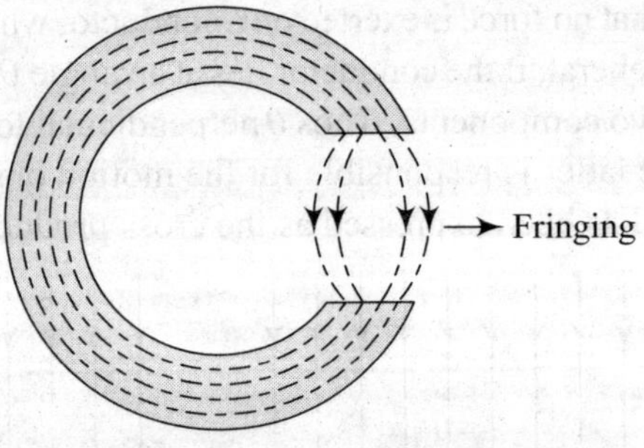

Figure 3.11 Magnetic leakage and fringing

The value of leakage coefficient varies from 1.15 to 1.25 for electrical machines. Thus, in electrical machines, the total flux being known, the useful flux can be determined by assuming a suitable value of the leakage coefficient.

EXAMPLE 3.6 **A steel ring having a cross-sectional area 5 cm² and a mean circumference of 40 cm has a coil of 200 turns wound uniformly around it. Calculate the reluctance of the ring and the current required to produce a flux of 600 μWb in the ring. Assume that the relative permeability of the steel is 400. Neglect the leakage flux.**

Solution Area of cross section, $a = 5 \text{ cm}^2$

Length of the ring, $l = 40$ cm

Relative permeability of steel = 400

Reluctance of the ring, $S = \dfrac{l}{\mu_r \mu_0 a} = \dfrac{0.4}{400 \times 4\pi \times 10^{-7} \times 0.0005} = 159 \times 10^4 \text{ AT/Wb}$

Flux to be produced, $\phi = 600\ \mu\text{Wb}$

Ampere-turns required, $NI = \phi \times S = 600 \times 10^{-6} \times 159 \times 10^4 = 954$

Number of turns of the coil = 200

$$\text{Current required} = \frac{NI}{N} = \frac{954}{200} = 4.77\ \text{A}$$

EXAMPLE 3.7 A ring-shaped electromagnet has an air gap of 6 mm and cross-sectional area 12 cm². The mean length of the core is 60 cm. Calculate the mmf required to produce a flux density of 0.4 Wb/m² in the gap. Take the relative permeability of material as 400.

Solution Area of cross section, $a = 12 \times 10^{-4}\ \text{m}^2$

Flux density, $B = 0.4\ \text{Wb/m}^2$

Relative permeability, $\mu_r = 400$

Flux, $\phi = B \times a = 0.4 \times 12 \times 10^{-4} = 4.8 \times 10^{-4}\ \text{Wb}$

Total reluctance S = Core reluctance + Air gap reluctance

$$= \frac{l_i}{\mu_0 \mu_r a} + \frac{l_g}{\mu_0 a}$$

$$= \left[\frac{60 \times 10^{-2}}{400 \times 12 \times 10^{-4}} + \frac{6 \times 10^{-3}}{12 \times 10^{-4}}\right]\frac{1}{\mu_o} = \frac{6.25}{4\pi \times 10^{-7}} = 497.3 \times 10^4\ \text{AT/Wb}$$

$$\text{mmf} = S \times \phi = 497.3 \times 10^4 \times 4.8 \times 10^{-4} = 2387.04\ \text{AT}$$

3.8 FARADAY'S LAWS OF ELECTROMAGNETIC INDUCTION

Faraday summed up the above facts into two laws known as Faraday's laws of electromagnetic induction.

First law: It states that whenever the magnetic flux linked with a circuit changes, an emf is always induced in it.

or

Whenever a conductor cuts the magnetic flux, an emf is induced in that conductor.

Second law: It states that the magnitude of the induced emf is equal to the rate of change of flux linkages.

Explanation: Suppose a coil has N turns, and the flux through it changes from an initial value of ϕ_1 weber to the final value of ϕ_2 weber in time t seconds. Then, remembering that by flux linkages means the product of the number of turns by the flux linked with the coil, we have

Initial flux linkages = $N\phi_1$; final flux linkages = $N\phi_2$

$$\therefore \quad \text{Induced emf, } e = \frac{N\phi_1 - N\phi_2}{t} \text{ weber/s or volt or } e = N\frac{(\phi_1 - \phi_2)}{t} \text{ volt}$$

Putting the above expression in its differential form, we get

$$e = \frac{d}{dt}(N\phi)$$

$$= N\frac{d\phi}{dt} \text{ volt} \quad (3.12)$$

Usually, a minus sign is given to the right-hand expression to signify the fact that the induced emf sets up current in such a direction that the magnetic effect produced by it opposes the very cause producing it.

$$e = -N\frac{d\phi}{dt} \text{ volt} \tag{3.13}$$

EXAMPLE 3.8 **The field coils of a 6-pole generator, each having 500 turns, are connected in series. When the field is excited, there is a magnetic flux of 0.02 Wb/pole. If the field excited is opened in 0.02 s and the residual magnetism is 0.002 Wb/pole, calculate the average voltage which is induced across the field terminals. In which direction is this voltage directed relative to the direction of the current?**

Solution Total number of turns, $N = 6 \times 500 = 3000$

Total initial flux $= 6 \times 0.02 = 0.12$ Wb

Total residual flux $= 6 \times 0.002 = 0.012$ Wb

Change in flux, $d\phi = 0.12 - 0.012 = 0.108$ Wb

Time of opening the circuit, $dt = 0.02$ s

$$\therefore \quad \text{Induced emf} = N\frac{d\phi}{dt} \text{ volt}$$

$$= 3000 \times \frac{0.108}{0.02}$$

$$= 16200 \text{ V}$$

The direction of this induced emf is the same as the initial direction of the exciting current.

EXAMPLE 3.9 **A coil of 100-Ω resistance is placed in a magnetic field of 1 mWb. The coil has 100 turns and a galvanometer of 400-Ω resistance is connected in series with it. Find the average emf and the current if the coil is moved in 1/10th second from the given field to a field of 0.2 mWb.**

Solution Induced emf $= N\frac{d\phi}{dt}$ volt

$d\phi = 1 - 0.2 = 0.8 \text{ mWb} = 0.8 \times 10^{-3}$ Wb

$dt = 1/10 = 0.1$ s; $N = 100$

$$e = \frac{100 \times 0.8 \times 10^{-3}}{0.1} = 0.8 \text{ V}$$

Total circuit resistance $= 100 + 400 = 500\ \Omega$

Current induced $= 0.8/500 = 1.6 \times 10^{-3}$ A

$= 1.6$ mA

3.8.1 Types of Induced EMFs

The induced emf can be either (i) dynamically induced or (ii) statically induced. In the first case, usually the field is stationary and conductors cut across it (as in dc generators). But in the second case, usually the conductors or the coil remain stationary and the flux linked with it is changed by simply increasing or decreasing the current producing this flux (as in transformers).

Dynamically induced emf

In Figure 3.12, a conductor A is shown in cross section, lying within a uniform magnetic field or flux density B weber/m. The arrow attached to "A" shows its direction of motion. Consider the conditions shown in Figure 3.12 when "A" cuts across at right angles to the flux. Suppose "l" is its length lying within the field and let it move a distance "dx" in time "dt," then the area swept by it is ldx. Hence, flux cut = $l \cdot dx \times B$ weber.

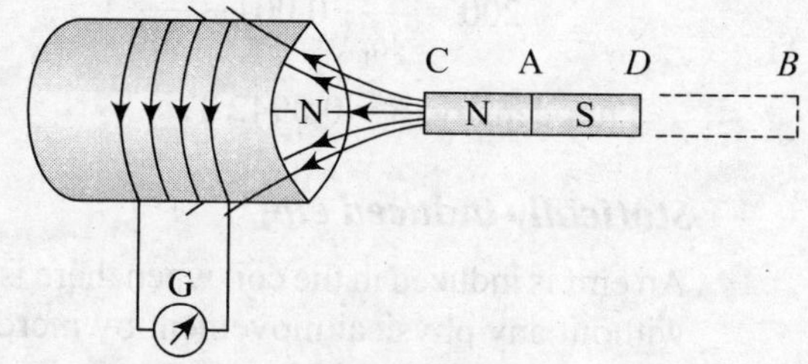

Figure 3.12

Change in flux = $Bl\,dx$ weber

Time taken = dt second

Hence, according to Faraday's laws, the emf induced in it (known as *dynamically induced emf*) is

$$\text{Rate of change of flux linkages} = \frac{Bl\,dx}{dt} = Bl\frac{dx}{dt} = Blv \text{ volt (where } dx/dt = \text{velocity)} \quad (3.14)$$

If conductor A moves at an angle θ with the direction of flux (Figure 3.13), then the induced emf is $e = Blv \sin \theta$ volts = $l\vec{v} \times \vec{B}$ (i.e., as a cross-product vector $\vec{v}$ and $\vec{B}$).

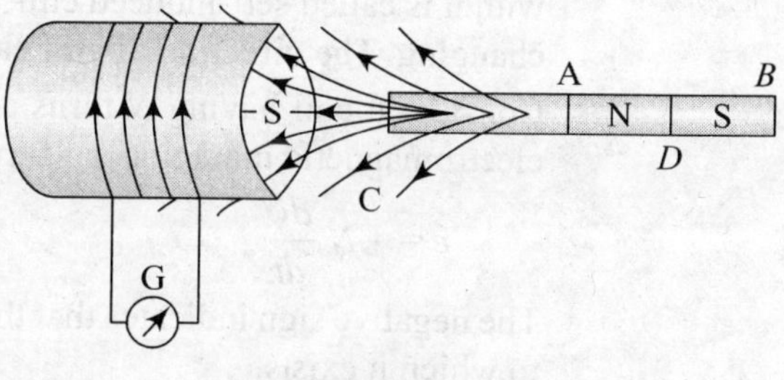

Figure 3.13

The direction of the induced emf is given by Fleming's right-hand rule or flat-hand rule and most easily by vector cross-product given above.

It should be noted that generators work on the production of dynamically induced emf in the conductors housed in a revolving armature lying within a strong magnetic field.

EXAMPLE 3.10 A conductor of length l meter moves at right angles to a uniform magnetic field of flux density 1.5 Wb/m^2 with a velocity of 50 m/s. Calculate the emf induced in it. Also find the value of induced emf when the conductor moves at right angle to the direction of the field.

Solution Here, $B = 1.5$ Wb/m^2, $l = 1$ m, $v = 50$ m/s, and $e = ?$

Now, $e = Blv$

$= 1.5 \times 1 \times 50 = 75$ V

In the second case, $\theta = 90°$

$\therefore \quad \sin 90° = 0.5$

$\therefore \quad e = 75 \times 0.5 = 37.5$ V

EXAMPLE 3.11 An iron ring of mean length 50 cm has an air gap of 1 mm and a winding of 200 turns. If the relative permeability of iron is 300, find the flux density when a current of 1 A flows through the coil?

Solution Let the flux density through the iron as well as through the air gap be B weber/m^2.

Total ampere-turns required = $N \times 1 = 200 \times 1 = 200$

$$\text{Total ampere-turns required} = \frac{B}{\mu_0}\left(l_g + \frac{l_1}{\mu_r}\right)$$

$$200 = \frac{B}{\mu_0}\left(0.001 + \frac{0.5}{300}\right)$$

Flux density, $B = 0.0942$ T

Statically induced emf

An emf is induced in the coil when there is a change in the flux. Such change in flux can be achieved without any physical movement by increasing or decreasing the current producing the flux. The change in flux lines with respect to coil can be achieved without physically moving the coil or the magnet. The emf induced without the physical movement of coil or a magnet is called *statically induced emf*. It can be further subdivided into (i) self-induced emf and (ii) mutually induced emf.

(i) Self-induced emf: The emf induced in a coil due to the change of its own flux linked with it is called self-induced emf. The self-induced emf exists till the current in the coil is changing. The direction of emf can be obtained by Lenz's law.

Consider a coil having N turns and carrying current I amperes. From Faraday's law of electromagnetic induction, self-induced emf can be expressed as

$$e = -N\frac{d\phi}{dt}$$

The negative sign indicates that the direction of this emf is opposing change in current due to which it exists.

(ii) Mutually induced emf: The phenomenon of the generation of induced emf in a circuit by changing the current in a neighboring circuit is called mutual induction. If the flux produced by one coil is getting linked with another coil, and due to change in this flux produced by the first coil there is an induced emf in the second coil, then such an emf is called mutually induced emf. Mutually induced emf is given by

$$e_2 = -M\frac{dI_1}{dt}$$

where e_2 is the induced emf in coil 2, I_1 is the current in coil 1, and M is the mutual inductance.

3.8.2 Lenz's Law

The direction of the induced current may also be found by this law which was formulated by Lenz in 1835. This law states, in effect, that the electromagnetically induced current always flows in such direction that the action of magnetic field set up by it tends to oppose the very cause which produces it.

The statement will be clarified with reference to Figures 3.14 and 3.15. It is found that when N-pole of the bar magnet approaches the coil, the induced current set up by the induced emf flows in the *anticlockwise* direction in the coil as seen from the magnet side. The result is that the face of the coil becomes the N-pole and so tends to oppose the onward approach of the N-pole of the magnet (like poles repel each other). The mechanical energy spent in overcoming this repulsive force is converted into electrical energy which appears in the coil.

When the magnet is withdrawn as in Figure 3.15, the induced current flows in the clockwise direction, thus making the face of the coil (facing the magnet) an S-pole. Therefore, the N-pole of the magnet has to be withdrawn against the attractive force of the S-pole of the coil. Again, the mechanical energy required to overcome this force of attraction is converted into electrical energy.

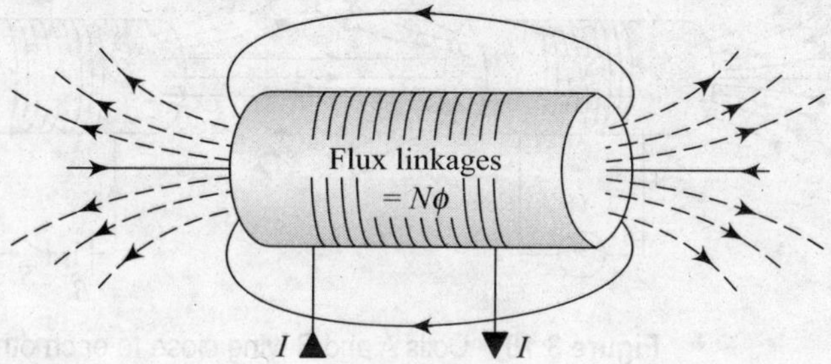

Figure 3.14

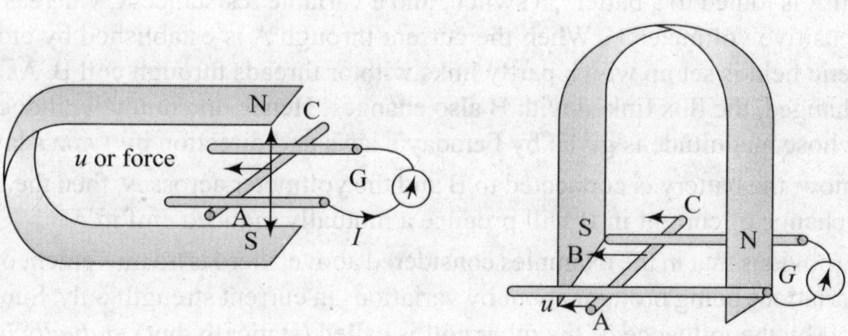

Figure 3.15

It can be shown that Lenz's law is a direct sequence of the law of conservation of energy. Imagine for a moment that when N-pole of the magnet (Figure 3.15) approaches the coil, induced current flows in such a direction as to make the coil face an S-pole. Then, due to inherent attraction between unlike poles, the magnet would be automatically pulled towards the coil without the expenditure of any mechanical energy. It means that we would be able to create electrical energy out of nothing, which is denied by the inviolable law of the conservation of energy. In fact, to maintain the sanctity of this law, it is imperative for the induced current to flow in such a direction that the magnetic effect produced by it tends to oppose the very cause which produces it. In the present case, it is the relative motion of the magnet with respect to the coil which is the cause of the production of the induced current. Hence, the induced current always flows in such a direction to oppose the relative motion, that is, the approach or withdrawal of the magnet.

3.8.3 Self-Induced EMF

Self-induced emf is the emf induced in a coil due to the change in its own flux linked with it. If the current through the coil is changed, then the flux linked with its own turns will also change, which will produce in it the emf called self-induced emf. The direction of this induced emf (as given by

Lenz's law) would be such as to oppose any change of flux which is, in fact, the very cause of its production. Hence, it is also known as the opposing or counter emf of self-induction.

3.8.4 Mutually Induced EMF

Consider two coils A and B lying close to each other (Figure 3.16).

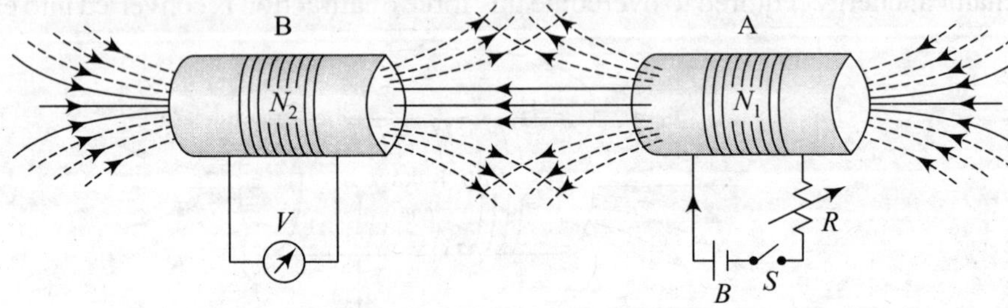

Figure 3.16 Coils A and B lying close to each other

Coil A is joined to a battery, a switch, and a variable resistance R, whereas coil B is connected to a sensitive voltmeter V. When the current through A is established by closing the switch, its magnetic field is set up which partly links with or threads through coil B. As the current through A is changed, the flux linked with B also changes. Hence, the mutually induced emf is produced in B whose magnitude is given by Faraday's laws and direction by Lenz's law.

If, now, the battery is connected to B and the voltmeter across A, then the situation is reversed and a change of current in B will produce a mutually induced emf in A.

It is obvious that in the examples considered above, there is no movement of any conductor, the flux variations being brought about by variations in current strength only. Such an emf induced in one coil by the influence of the other coil is called (statically but) *mutually induced emf.*

3.8.5 Coefficient of Coupling

Consider two magnetically coupled coils A and B having N_1 and N_2 turns, respectively. Their individual coefficients of self-induction are

$$L_1 = \frac{N_1^2}{l/\mu_0\mu_r A} \quad \text{and} \quad \frac{N_2^2}{l/\mu_0\mu_r A}$$

The flux ϕ_1 produced in A due to a current I_1 ampere is

$$\phi_1 = \frac{N_1 I_1}{l/\mu_0\mu_r A}$$

Suppose a fraction k_1 of this flux, i.e., $k_1\phi_1$ is linked with coil B, then

$$M = \frac{k_1\phi_1 \times N_1}{I_1} \quad \text{(where } k \le 1\text{)}$$

Substituting the value of ϕ_1, we have

$$M = k_1 \times \frac{N_1 N_2}{l/\mu_0\mu_r A} \tag{3.15}$$

Similarly, flux ϕ_2 produced due to a current I_2 ampere is

$$\phi_2 = \frac{N_2 I_2}{l/\mu_0 \mu_r A}$$

Suppose a fraction k_2 of this flux, i.e., $k_2\phi_2$ is linked with coil A, then

$$M = \frac{k_2\phi_2 \times N_2}{I_2}$$

$$= k_2 \times \frac{N_1 N_2}{l/\mu_0 \mu_r A} \quad (3.16)$$

Multiplying Eqs. (3.15) and (3.16), we get

$$M^2 = k_1 k_2 \times \frac{N_1^2}{l/\mu_0 \mu_r A} \times \frac{N_2^2}{l/\mu_0 \mu_r A}$$

or

$$M^2 = k_1 k_2 L_1 L_2 \quad (3.17)$$

Substituting $\sqrt{k_1 k_2} = k$, we have

$$M = k\sqrt{L_1 L_2} \text{ or } k = \frac{M}{\sqrt{L_1 L_2}} \quad (3.18)$$

The constant k is called the coefficient of coupling and may be defined as the ratio of the mutual inductance actually present between the two coils to the maximum value. If the flux due to one coil completely links with the other, then the value of k is unity. If the flux of one coil does not at all link with the other, then $k = 0$. In the first case, when $k = 1$, coils are said to be tightly coupled and when $k = 0$, the coils are magnetically isolated from each other.

EXAMPLE 3.12 **Two identical coils A and B, each having 750 turns, lie in parallel planes. A current changing at the rate of 1500 A/s in A induces an emf of 11.25 V in B. Calculate the mutual inductance of the arrangement. If the self-inductance of each coil is 15 mH, calculate the flux produced in coil A per ampere and the percentage of this flux which links the turns of B.**

Solution Now,

$$e_M = M\frac{dI_1}{dt}$$

$$M = \frac{e_M}{dI_1/dt} = \frac{11.25}{1500} = 7.5 \times 10^{-3} \text{ H} = 7.5 \text{ mH}$$

Thus,

$$L_1 = \frac{N_1\phi_1}{I_1}$$

$$\therefore \quad \frac{\phi_1}{I_1} = \frac{L_1}{N_1} = \frac{15 \times 10^{-3}}{750} = 2 \times 10^{-5} \text{ Wb/A}$$

Hence,

$$k = \frac{M}{\sqrt{L_1 L_2}}$$

$$= \frac{7.5\times10^{-3}}{\sqrt{L^2}} = \frac{7.5\times10^{-3}}{15\times10^{-3}} = 0.5 = 50\% \quad (\because L_1 = L_2 = L)$$

EXAMPLE 3.13 **Two coils A and B having 12500 and 16000 turns, respectively, lie in parallel planes so that 60% of the flux produced in coil A links coil B. It is found that a current of 5 A in A produces a flux of 0.6 mWb while the same current in B produces 0.8 mWb. Determine (i) the mutual inductance and (ii) the coupling coefficient.**

Solution

(i) Flux per ampere in A = 0.6/5 = 0.12 mWb

Flux linked with B = 0.12 × 0.6 = 0.072 mWb

$\therefore \quad M = 0.072 \times 10^{-3} \times 16000 = 1.15$ H

Now,

$$L_1 = \frac{12500\times0.6\times10^{-3}}{5} = 1500\times10^{-3}\text{ H}$$

$$L_2 = \frac{16000\times0.8\times10^{-3}}{5} = 2560\times10^{-3}\text{ H}$$

(ii) $$k = \frac{M}{\sqrt{L_1 L_2}}$$

$$= \frac{1.15}{\sqrt{1.5\times2.56}}$$

$$= 0.586$$

3.9 SOME OTHER USEFUL LAWS RELATED TO MAGNETISM

3.9.1 Fleming's Right-Hand Rule

The direction of the emf induced in a moving conductor in a magnetic field can be determined by Fleming's right-hand rule. If the three fingers of the right hand, namely, the forefinger, the middle finger, and the thumb, are held mutually perpendicular such that if the forefinger indicates the direction of the magnetic field and the thumb indicates the direction of motion of the conductor, then the middle finger will indicate the direction of the emf induced. This is shown in Figure 3.17.

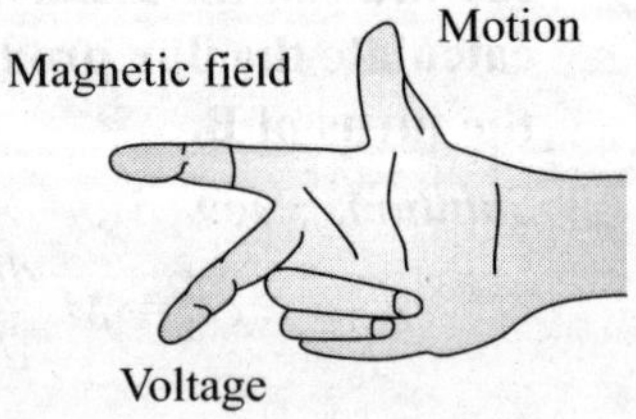

Figure 3.17 Fleming's right-hand rule

3.9.2 Fleming's Left-Hand Rule

The direction of force on a current-carrying conductor in a magnetic field can be determined by Fleming's left-hand rule, in addition to the cross product as given by the equation $F = I(l \times B)$. If the three fingers, namely, the forefinger, the middle finger and the thumb of the left hand, are held mutually perpendicular to each other such that if the forefinger indicates the direction of the magnetic flux and the middle finger indicates the direction

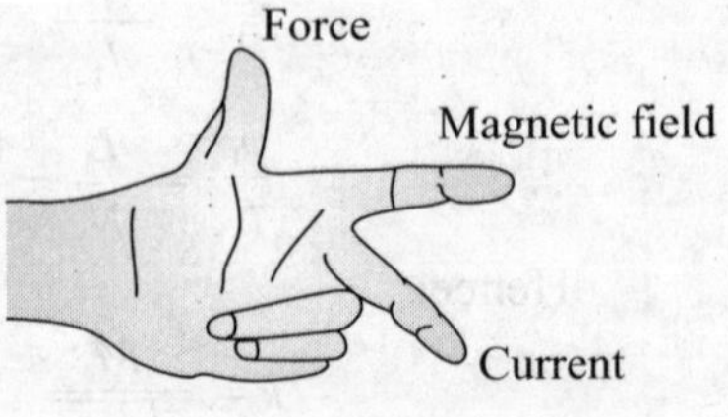

Figure 3.18 Fleming's left-hand rule

of the current, then the thumb indicates the direction of the force experienced by the conductor. This is Fleming's left-hand rule and is shown in Figure 3.18.

EXAMPLE 3.14 A wire 100 m in length carries a current of 10 A and makes an angle of 30° with a uniform magnetic field having a flux density of 1.5 Wb/m². Calculate the magnitude and the direction of the force on the wire?

Solution

Force on the wire $(F) = I\,(l \times B)$

$$|F| = I \cdot l \cdot B \cdot \sin\theta$$

where θ is the angle between I and B.

Given B (magnetic flux density) = 1.5 Wb/m², l = 1 m, I = 10 A, and $\theta = 30°$

Substituting, we get

$$F = 10 \times 1 \times 1.5 \sin 30° = 7.5\ \text{N}$$

3.9.3 Thumb Rule

The right-hand thumb rule states that hold the current-carrying conductor in the right hand such that if the thumb points in the direction of current and is parallel to the conductor, then the curled fingers point in the direction of the magnetic flux around it. Figure 3.19 shows the rule.

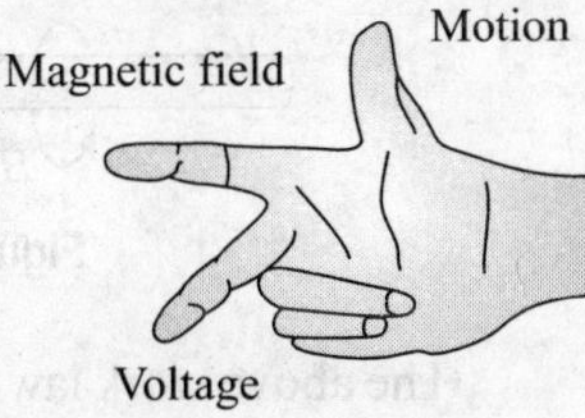

Figure 3.19 Right-hand thumb rule

3.9.4 Cork–Screw Rule

The direction of the magnetic field intensity (H) due to a current-carrying conductor can also be obtained by yet another method called Maxwell's cork–screw rule. Imagine that you are holding a right threaded screw in your hand. Let the screw be rotated such that the screw advances in the direction of current as shown in Figure 3.20. Then the direction of the rotation of the screw gives the direction of the magnetic field as shown in Figure 3.20.

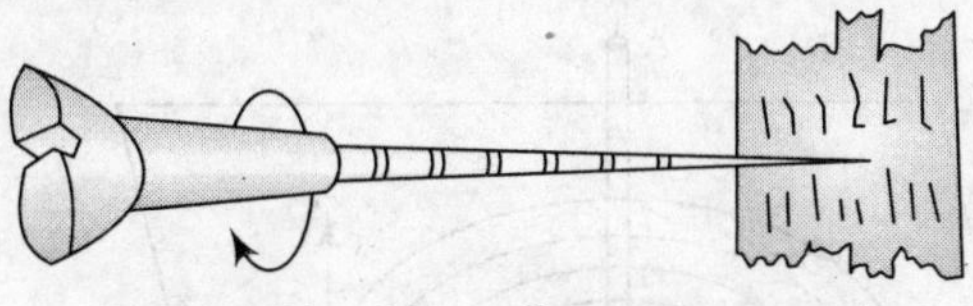

Figure 3.20 Maxwell's cork–screw rule

The magnetic flux lines are expressed in units of webers and are denoted by ϕ (Figure 3.21).

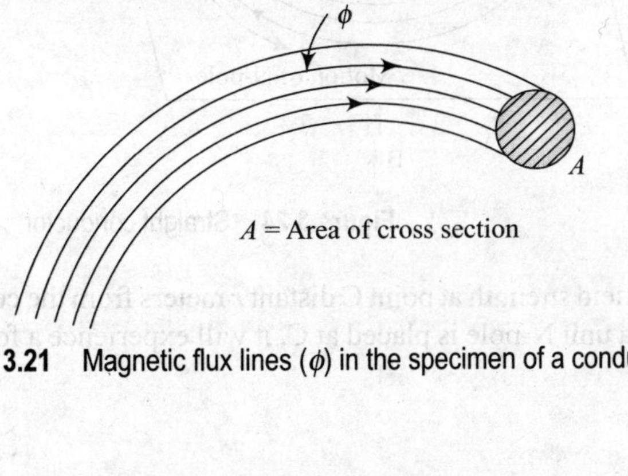

Figure 3.21 Magnetic flux lines (ϕ) in the specimen of a conductor

3.9.5 Ampere Circuit Law

The law states that the mmf (magnetomotive force corresponding to emf, i.e., electromotive force of electric field) around a closed path is equal to the current enclosed by the path. Mathematically, $\oint \vec{H} \cdot d\vec{s} = I$ amperes, where $\vec{H}$ is the vector representing the magnetic field strength in dot product with vector $d\vec{s}$ of the enclosing path S around current I ampere and that is why a line integral ($\oint$) of dot product $\vec{H} \cdot d\vec{s}$ is taken.

The work law is very comprehensive and is applicable to all magnetic fields whatever be the shape of the enclosed path (Figures 3.22 and 3.23). Since path (c) does not enclose the conductor, the mmf around it is zero.

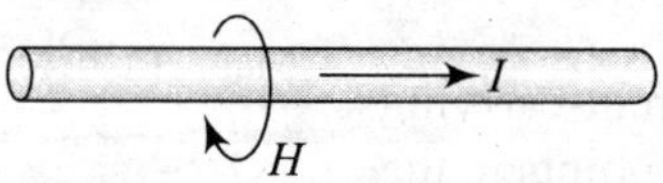

Figure 3.22

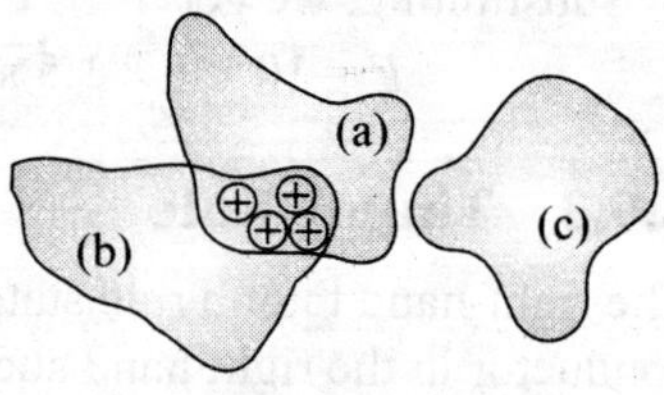

Figure 3.23

The above work law is used for obtaining the value of the mmf around simple idealized circuits such as (i) a long straight current-carrying conductor and (ii) a long solenoid.

Magnetomotive force around a long straight conductor

Figure 3.24 shows a straight conductor which is assumed to extend to infinity in either direction. Let it carry a current of I amperes upward. The magnetic field consists of circular lines of force having their plane perpendicular to the conductor and to the centers of the conductor.

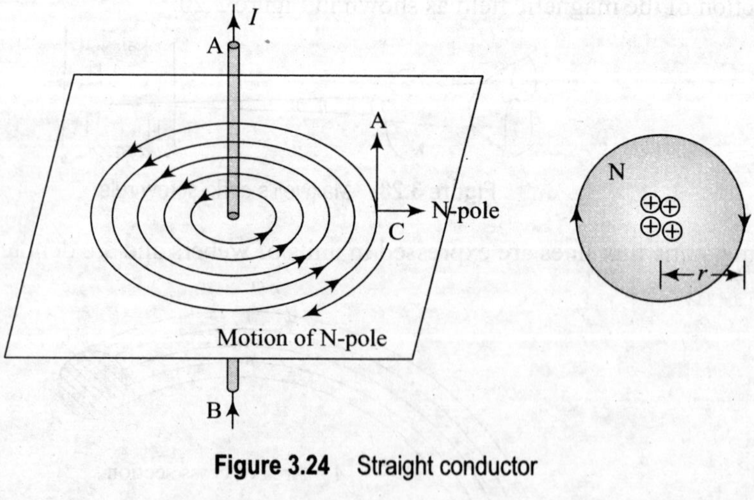

Figure 3.24 Straight conductor

Suppose the field strength at point C distant r meters from the center of the conductor is H. Then, it means that if a unit N-pole is placed at C, it will experience a force H newtons. The direction of this force would be tangential to the circular line of force passing through C. If this unit N-pole is moved once round the conductor against this force, then the work done,

$$\text{mmf} = \text{Force} \times \text{Distance} = I$$

i.e.,

$$I = H \times 2\pi r \text{ joules} = \text{Amperes}$$

or

$$H = \frac{1}{2\pi r} \tag{3.19}$$

$$= \oint \vec{H} \cdot d\vec{s} \text{ joules} = \text{Amperes} = I$$

Obviously, if there are N conductors, then

$$H = \frac{NI}{2\pi r} \text{ ampere/m or oersted} \tag{3.20}$$

and

$$B = \mu_0 \frac{NI}{2\pi r} \text{ weber/m}^2 \text{ or tesla (in air)} \tag{3.21}$$

$$= \frac{\mu_0 \mu_r NI}{2\pi r} \text{ weber/m}^2 \text{ or tesla (in a medium)} \tag{3.22}$$

Magnetic field strength of a long solenoid

Let the magnetic field strength along the axis of a solenoid be H. Let us assume that (i) the value of H remains constants throughout the length l of the solenoid and (ii) the value of H outside the solenoid is negligible.

Suppose a unit N-pole is placed at point A outside the solenoid and is taken once around the completed path (Figure 3.25) in a direction opposite to that of H. Remembering that the force of H newtons acts on the N-pole only over the length l (it being negligible elsewhere), the work done in one round is

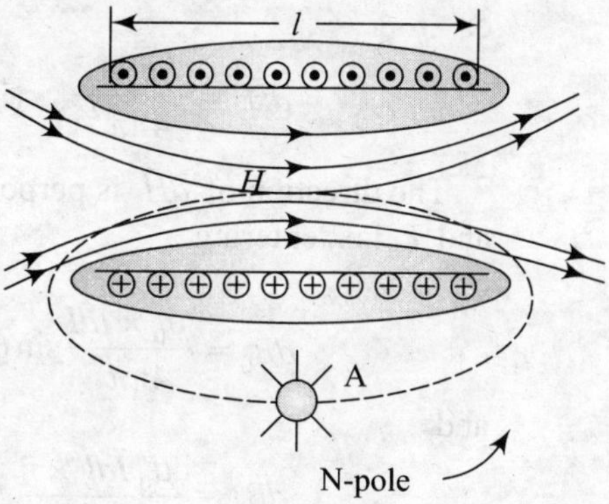

Figure 3.25

Work done = $H \times l$ joules = Amperes

The "ampere-turns" linked with this path are NI, where N is the number of turns of the solenoid and I the current in amperes passing through it. According to the work law,

$$H \times l = NI \text{ or } H = \frac{NI}{l} \text{ ampere/m or oersted} \tag{3.23}$$

Also

$$B = \frac{\mu_0 NI}{l} \text{ weber/m}^2 \text{ or tesla (in air)} \tag{3.24}$$

$$= \frac{\mu_0 \mu_r NI}{l} \text{ weber/m}^2 \text{ or tesla (in a medium)} \tag{3.25}$$

EXAMPLE 3.15 Calculate the magnetizing force and flux density at a distance of 5 cm from a long straight circular conductor having a current of 250 A and placed in air. Draw a curve showing the variation of B from the conductor surface outward if its diameter is 2 mm.

Solution From Eq. (3.19),

$$H = \frac{1}{2\pi r} = \frac{250}{2\pi \times 0.05} = 795.6 \text{ AT/m}$$

$$B = \mu_0 H = 4\pi \times 10^{-7} \times 795.6 = 10^{-3} \text{ Wb/m}^2$$

In general,

$$B = \frac{\mu_0 I}{2\pi r}$$

Now, at the conductor surface, $r = 1$ mm $= 10^{-3}$ m

$$\therefore \quad B = \frac{4\pi \times 10^{-7} \times 250}{2\pi \times 10^{-3}} = 0.05 \text{ Wb/m}^2$$

3.9.6 Biot–Savart Law

Biot–Savart law relates magnetic fields to the currents which are their sources. The expression for the magnetic field strength dH produced at point P by a vanishingly small length dl of a conductor carrying a current of I amperes (Figure 3.26) is given by

$$dH = \frac{Idl \sin\theta}{4\pi\pi^2} \text{ ampere/m}$$

or

$$d\vec{H} = \frac{Id\vec{l} \times \vec{r}}{4\pi\pi^2} \text{ (in vector form)} \qquad (3.26)$$

The direction of $d\vec{H}$ is perpendicular to the plane containing $d\vec{l}$ and $\vec{r}$, i.e., entering.

or

$$dB_0 = \frac{\mu_0 \times Idl}{4\pi\pi^2} \sin\theta \text{ weber/m}^2$$

and

$$d\vec{B}_0 = \frac{\mu_0 Id\vec{l} \times \vec{r}}{4\pi\pi^2} \sin\theta \text{ (in vector form)} \qquad (3.27)$$

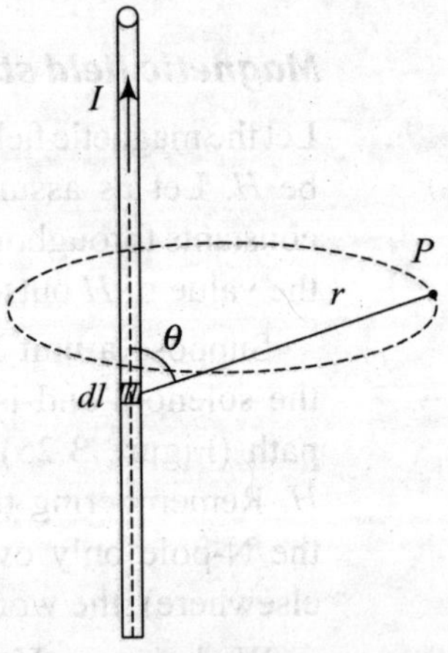

Figure 3.26

EXAMPLE 3.16 A 2.5-m long wire is bent into a square and into a circle. If the current flowing through the wire is 100 A, find the magnetizing force (i) at the center of the square and (ii) at the center of the circle.

Solution

(i) Each side of the square is $2a = 2.5/4 = 0.625$ m

$$\text{Value of } H \text{ at the center of the square} = \frac{\sqrt{2}I}{\pi a}$$

$$= \frac{\sqrt{2} \times 100}{\pi \times 0.3125}$$

$$= 144 \text{ AT/m}$$

(ii) $2\pi r = 2.5$; $r = 0.398$ m

Value of H at the center of the circle = $1/2r = 100/2 \times 0.398 = 125.6$ AT/m

3.10 KIRCHHOFF'S LAWS FOR MAGNETIC CIRCUITS

Similar to Kirchhoff's laws of electric circuits, there are Kirchhoff's laws of magnetic circuits. They are stated as follows:

Kirchhoff's first law: The total magnetic flux directed toward a junction is equal to the sum of the magnetic fluxes directed away from that junction. This law follows from the fact that each line of flux forms a closed path. For example, if an iron core having three limbs L, M, and N is arranged as shown in Figure 3.27 and the coil on the core limb has N turns and carries a current I, then the magnetic flux through the coil gets divided at point P, with a part of the flux passing through the limb M and the remainder passing through the limb N, with both the limbs joining again at point Q. Here, the number of lines of flux going from limb L towards P is exactly the same as the sum of the flux lines going from P toward limbs M and N individually, since the magnetic flux lines are continuous in nature. Therefore,

$$\phi_L = \phi_M + \phi_N \quad \text{or} \quad \phi_L - \phi_M - \phi_N = 0 \tag{3.28}$$

This law is similar to Kirchhoff's current law in an electric circuit. The electrical equivalent of the above is shown in Figure 3.27(b).

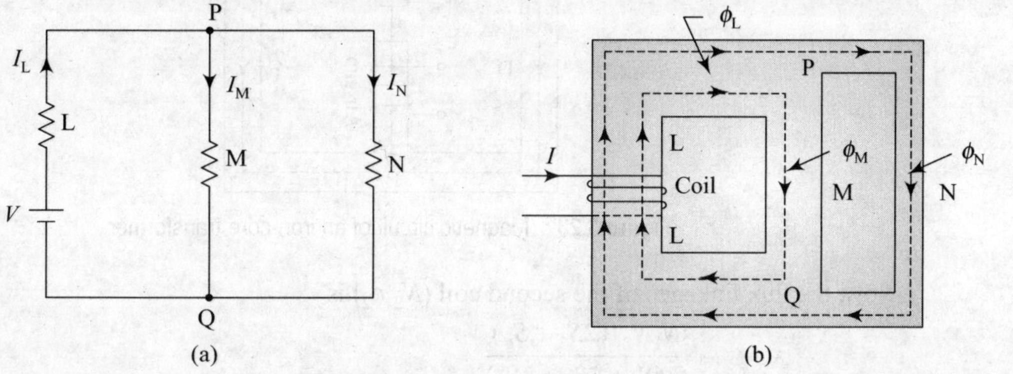

Figure 3.27 Magnetic circuit to illustrate Kirchhoff's law: (a) magnetic circuit and (b) electrical equivalent circuit

Kirchhoff's second law: This law states that in any closed magnetic circuit, the algebraic sum of the product of the magnetizing force and the length of each part of the circuit is equal to the resultant mmf. If H_L is the field strength of limb L and if l_L is the length of the circuit from Q to P via limb L and if H_M, l_M and H_N, l_N are the corresponding values from limbs M and N, then applying the above law to Figures 3.27(a) and (b) for the part QLLP, we get

$$\text{Total mmf} = N_L = H_L\, l_L + H_M\, l_M$$

i.e.,

$$NI = \phi_L R_L + \phi_M R_M \tag{3.29}$$

Similarly, applying the above law to the part LLN, we get

$$NI = \phi_L R_L + \phi_N R_N \tag{3.30}$$

Subtracting Eq. (3.30) from Eq. (3.29), we get

$$0 = \phi_M R_M - \phi_N R_N$$

As an example, consider the following.

Figure 3.28 shows an iron-core transfer of the uniform cross section having a small air gap of thickness t in the central leg. Such air gaps are frequently used in practice in order to increase the reluctance of the magnetic circuits in order to prevent the magnetic saturation of the material. The left leg is wound with N_1 turns of wire and the right leg is wound with N_2 turns. Compute the flux linkages of the coil having N_2 turns, when a current I amperes flows in the other coil. The reluctance of the paths l_1, l_2, and t is given as $S_1 = l_1/A\mu$, $S_2 = l_2/A\mu$, and $S_g = t/A\mu_0$, respectively. Following the usual circuit analysis approach and applying Kirchhoff's laws, the following two equations may be written as:

$$N_1 I = \phi_2 (S_1 + 2S_2 + S_g) - \phi_1 (2S_2 + S_g) \tag{3.31}$$

$$0 = -\phi_1 (2S_2 + S_g) + \phi_2 (S_1 + 2S_2 + S_g)$$

Solving the above equations for ϕ_2, we get

$$\phi_2 = \frac{N_1 I(2S_2 + S_g)}{S_1(S_1 + 4S_2 + 2S_g)} \tag{3.32}$$

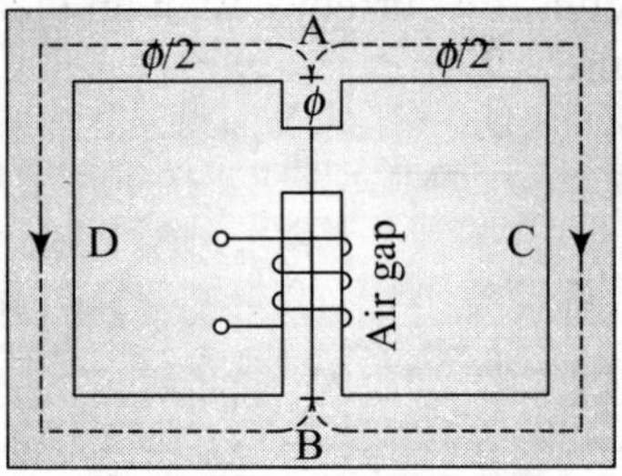

Figure 3.28 Magnetic circuit of an iron-core transformer

Now, the flux linkages of the second coil ($N_2 \phi_2$) is

$$N_2 \phi_2 = \frac{N_1 N_2 I(2S_2 + S_g)}{S_1(S_1 + 4S_2 + 2S_g)} \tag{3.33}$$

3.11 ENERGY STORED IN A MAGNETIC SYSTEM

For establishing a magnetic field, energy must be spent, although no energy is required to maintain it. Take the example of the exciting coils of an electromagnet. The energy supplied to it is spent in two ways: (i) part of it goes to meet I^2R loss and is lost once for all and (ii) part of it goes to create flux and is stored in the magnetic field as potential energy and is similar to the potential energy of a raised weight. When a weight W is raised through a height of h, the potential energy stored in it is W_h. Work is done in raising this weight. But once raised to a certain height, no further expenditure of energy is required to maintain it in that position. This mechanical potential energy can be recovered, so can the electrical energy stored in the magnetic field.

When a current through an inductive coil is gradually changed from zero to a maximum value I, then every change of it is opposed by the self-induced emf produced due to this change. Energy is needed to overcome this opposition. This energy is stored in the magnetic field of the coil and is, later on, recovered when that field collapses. The value of this stored energy may be found in the following two ways:

(i) First method: Let, at any instant,

i = instantaneous value of current

$$e = \text{induced emf at that instant} = L\frac{di}{dt} \tag{3.34}$$

Then, the work done in time dt in overcoming this opposition is

$$dW = ei\,dt$$

$$= L\frac{di}{dt} \times i \times dt$$

$$= Li\,di$$

Then, the work done in establishing the maximum steady current of I is

$$\int_0^W dW = \int_0^1 Li \cdot di = LI^2$$

or

$$W = \frac{1}{2}LI^2 \tag{3.35}$$

This work is stored as the energy of the magnetic field.

$$\therefore \quad E = \frac{1}{2}LI^2 \text{ joules} \tag{3.36}$$

(ii) Second method: If the current increases uniformly from zero value to its maximum steady value I, then the average current is $I/2$. If L is the inductance of the circuit, then the self-induced emf is $e = LI/t$, where "t" is the time for the current change from zero to I. Therefore,

Average power absorbed = Induced emf × Average current

$$= L\frac{I}{t} \times \frac{1}{2}I = \frac{1}{2}\frac{LI^2}{t} \tag{3.37}$$

Total energy absorbed = Power × Time

$$= \frac{1}{2}\frac{LI^2}{t} \times t$$

$$= \frac{1}{2}LI^2$$

$$\therefore \quad \text{Energy stored, } E = \frac{1}{2}LI^2 \text{ joules}$$

It may be noted that in the case of series-aiding coils, energy stored is

$$E = \frac{1}{2}(L_1 + L_2 + M)I^2$$

$$= \frac{1}{2}L_1I^2 + \frac{1}{2}L_2I^2 + \frac{1}{2}MI^2 \tag{3.38}$$

Similarly, for series-opposing coils,

$$E = \frac{1}{2}L_1 I^2 + \frac{1}{2}L_2 I^2 - \frac{1}{2}MI^2 \tag{3.39}$$

3.11.1 Attractive Force of Electromagnetic Lifting Power of Magnet

In Figure 3.29, let P be the pulling force in newtons between two poles and A be the pole area in square meter.

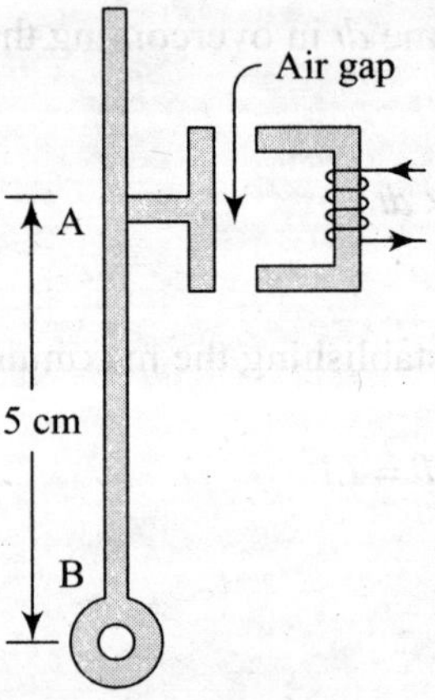

Figure 3.29

If one of the poles (say, upper one) is pulled apart against this attractive force through a distance of dx meters, then

$$\text{Work done} = P \times dx \text{ joules} \tag{3.40}$$

This work goes to provide energy for the additional volume of the magnetic field so created.

Additional volume of the magnetic field created = $A \times dx$ m^3

$$\text{Rate of energy requirement} = \frac{B^2}{2\mu_0} \text{ joule/m}^3$$

$$\therefore \quad \text{Energy required for the new volume} = \frac{B^2 \times Adx}{2\mu_0} \tag{3.41}$$

Equating Eqs. (3.40) and (3.41), we get

$$P \cdot dx = \frac{B^2 \times Adx}{2\mu_0}$$

$$\therefore \quad P = \frac{B^2 A}{2\mu_0} \text{ newton} = 400000 \frac{B^2}{A} \text{ newton}$$

or

$$P = \frac{B^2 A}{2\mu_0} \text{ newton/m}^2 = 40000\, B^2 \text{ newton/m}^2$$

Also,

$$P = \frac{B^2 A}{9.81 \times 2\mu_0} = \frac{B^2 A}{19.62\mu_0} \text{ kg-wt} \tag{3.42}$$

EXAMPLE 3.17 **A horse-shoe magnet is formed out of a bar of wrought iron 45.7 cm long having a cross section of 6.45 cm². Exciting coils of 500 turns are placed on each limb and connected in series. Find the exciting current necessary for the magnet to lift a load of 68 kg assuming that the load has negligible reluctance and makes close contact with the magnet. The relative permeability of iron is 700.**

Solution Force of attraction of each pole = 68/2 = 34 kg

$$= 34 \times 9.81 = 333.5 \text{ N}$$

$$A = 6.45 \text{ cm}^2 = 6.45 \times 10^{-3} \text{ m}^3$$

Since $F = B^2 A/2\mu_0$ newton, therefore

$$333.5 = \frac{B^2 \times 6.45 \times 10^{-4}}{2 \times 4\pi \times 10^{-7}}$$

$$\therefore \quad B = \sqrt{1.3} = 1.14 \text{ Wb/m}^2$$

and

$$H = \frac{B^2}{\mu_0 \times \mu_r} = \frac{1.14^2}{4\pi \times 10^{-7} \times 700} = 1477.4 \text{ AT/m}$$

Length of the plate = 45.7 cm = 0.457 m

$\therefore$ AT required = 1296 × 0.457 = 592.6

Number of turns = 500 × 2 = 1000

$$\therefore \quad \text{Current required} = \frac{592.6}{1000} = 0.593 \text{ A}$$

EXAMPLE 3.18 **The pole face area of an electromagnet is 0.5 m²/pole. It has to lift an iron ingot weighing 1000 kg. If the pole faces are parallel to the surface of the ingot at a distance of 1 mm, determine the coil mmf required. Assume the permeability of iron rod to be infinity and the permeability of free space is $4\pi \times 10^{-7}$ H/m.**

Solution Since iron has permeability of infinity, it offers zero reluctance to the magnet flux.

$$\text{Force at two poles} = 2 \times \frac{B^2 A}{2\mu_0} = \frac{B^2 A}{\mu_0}$$

$$\therefore \quad \frac{B^2 \times 0.5}{4\pi \times 10^{-7}} = 1000 \times 9.8$$

$$B = 0.157 \text{ Wb/m}^2$$

$$\therefore \quad H = \frac{0.157}{4\pi \times 10^{-7}}$$

$$= 125 \times 10^3 \text{ AT/m}$$

$$L = 2 \times 1 = 2 \text{ mm} = 2 \times 10^{-3} \text{ m}$$

$$\therefore \quad \text{AT required} = 125 \times 10^3 \times 2 \times 10^{-3} = 250$$

3.12 SERIES AND PARALLEL MAGNETIC CIRCUIT

Figure 3.30 shows a series and parallel magnetic circuit with two parallel magnetic circuits ACB and ACD connected across the common magnetic path AB which contains an air gap of length l_g. As usual, the flux ϕ in the common core is divided equally at point A between the two parallel paths which have equal reluctance. The reluctance of the path AB consists of (i) the air gap reluctance and (ii) the reluctance of the central core which is comparatively negligible. Hence, the reluctance of the central core AB equals only the air gap reluctance across which are connected two equal parallel reluctances. Hence, the mmf required for the circuit would be the sum of (i) that required for the air gap and (ii) that required for either of the two paths (not both).

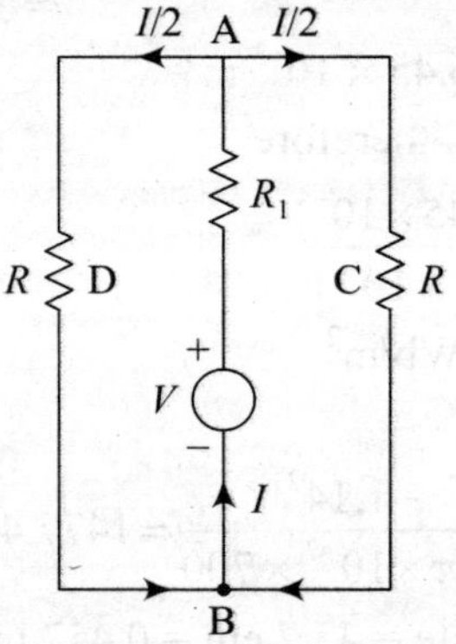

Figure 3.30 Series and parallel magnetic circuit

3.13 COMPARISON BETWEEN MAGNETIC CIRCUITS AND ELECTRIC CIRCUITS

Table 3.1 gives the terms analogous between the magnetic and electric quantities.

Table 3.1 Comparison between magnetic and electric quantities

Magnetic circuits	Symbol	Unit	Electric circuits	Symbol	Unit
Magnetomotive force		AT	Electromotive force	E	volt
Magnetic flux		weber	Current	I	ampere
Magnetic flux density	B	tesla	Current density	J	A/m^2
Reluctance	S	AT/Wb	Resistance	R	ohm
Permeance	P	Wb/AT	Conductance	G	siemens

In an electric circuit, the emf produces the current. Similarly, in a magnetic circuit, the mmf produces the flux. Also, in an electric circuit, the resistance impedes the flow of current, whereas in a magnetic circuit, the reluctance impedes the flow of magnetic flux. The other differences are shown in Table 3.2.

Table 3.2 Difference between magnetic and electric circuits

S. No.	Magnetic circuits	Electric circuits
	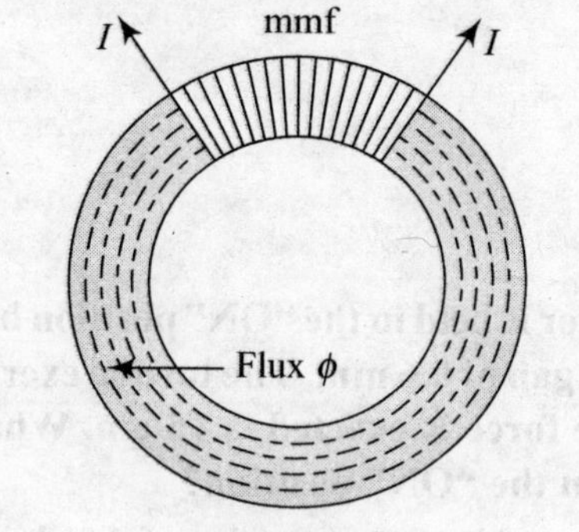 Figure 3.31	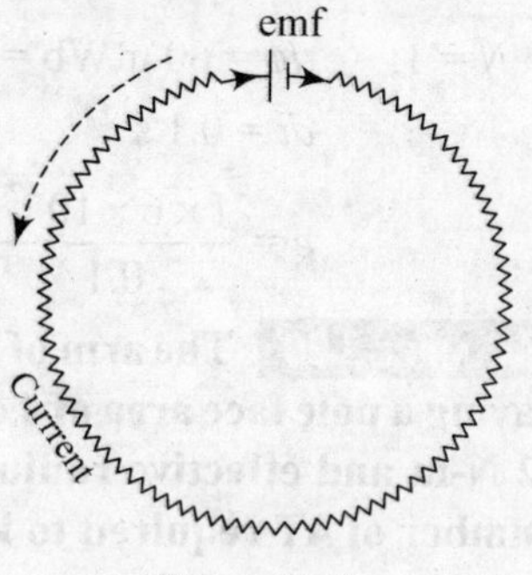 Figure 3.32
1.	$\text{Flux} = \dfrac{\text{mmf}}{\text{Reluctance}}$	$\text{Current} = \dfrac{\text{emf}}{\text{Resistance}}$
2.	mmf (ampere-turns)	emf (volts)
3.	Flux ϕ (webers)	Current I (amperes)
4.	Flux density B (Wb/m²)	Current density (A/m²)
5.	Reluctance $S = \dfrac{l}{\mu A}\left(= \dfrac{l}{\mu_0 \mu_r A}\right)$	Resistance $R = \rho\dfrac{l}{A} = \dfrac{l}{\rho A}$
6.	Permeance (= 1/reluctance)	Conductance (= 1/resistance)
7.	Relucitivity	Resistivity
8.	Permeability (= 1/relucitivity)	Conductivity (= 1/resistivity)
9.	Total mmf $= \phi S_1 + \phi S_2 + \phi S_3 + \cdots$	Total emf $= IR_1 + IR_2 + IR_3 + \cdots$
10.	In a magnetic circuit, there is no real flux but it is only a concept to understand a magnetic circuit.	In an electric circuit, there is a real flow of current when emf is applied.
11.	Once a magnetic circuit is established, the flux can be maintained without an energy source (e.g., permanent magnets).	Energy must be applied to the electric circuits to maintain a constant flow of current.
12.	In a magnetic circuit, it is difficult to find out exactly where the mmf is acting.	In an electric circuit, it is possible to precisely point out where the emf is acting.

EXAMPLE 3.19 **In a 4-pole dynamo, the flux/pole is 15 mWb. Calculate the average emf induced in one of the armature conductors, if armature is driven at 600 rpm.**

Solution It should be noted that each time the conductor passes under a pole (whether N or S) it cuts a flux of 15 mWb.

Hence, the flux cut in one revolution is $15 \times 4 = 60$ mWb

Since conductor its rotating at $600/60 = 10$ rps

Therefore, the time taken for one revolution is 1/10 = 0.1 s

$\therefore$ Average emf generated $= N \dfrac{d\phi}{dt}$ volt

$N = 1$; $d\phi = 60$ mWb $= 6 \times 10^{-2}$ Wb

$dt = 0.1$ s

$$e = \frac{1 \times 6 \times 10^{-2}}{0.1} = 0.6 \text{ V}$$

EXAMPLE 3.20 **The arm of a dc shunt motor is held in the "ON" position by an electromagnet having a pole face area of 4 cm² and an air gap of 0.6 mm. The torque exerted by the spring is 12 N-m and effective radius at which the force is exerted is 15 cm. What is the minimum number of AT required to keep the arm in the "ON" position?**

Solution Let F be the force in newtons exerted by the two poles of the electromagnet.

Torque = Force × Radius

$\therefore$ $12 = F \times 0.15$

$F = 80$ N

$$\text{Force per pole} = \frac{80}{2} = 40 \text{ N}$$

Now,

$$F = \frac{B^2 A}{2\mu_0} \text{ newton}$$

$$\therefore \quad 40 = \frac{B^2 \times 4 \times 10^{-4}}{2 \times 4\pi \times 10^{-7}}$$

$\therefore$ $B = 0.5$ Wb/m²

$\therefore$ $H = 0.5/4\pi \times 10^{-7}$ AT/m

Total air gap $= 2 \times 0.6 \times 10^{-3} = 1.2 \times 10^{-3}$ m

$$\therefore \quad \text{AT required} = Hl = \frac{0.5 \times 1.2 \times 10^{-3}}{4\pi \times 10^{-7}}$$

$$= 477$$

EXAMPLE 3.21 **An iron ring has a X-section of 3 cm² and a mean diameter of 25 cm. An air gap of 0.4 mm has bent cut section of the ring. The ring is wound with a coil of 200 turns through which a current of 2 A is passed. If the total magnetic flux is 0.24 mWb, find the relative permeability of iron, assuming no magnetic leakage.**

Solution $\phi = 0.24$ mWb; $A = 3$ cm² $= 3 \times 10^{-4}$ m²

$$B = \phi/A = \frac{0.24 \times 10^{-3}}{3 \times 10^{-4}} = 0.8 \text{ Wb/m}^2$$

$$\text{AT for iron ring} = H \times l = \frac{B}{\mu_0 \mu_r} \times l = \frac{0.8}{4\pi \times 10^{-7} \times \mu_r} \times 0.25 = 1.59 \times \frac{10^5}{\mu_r}$$

$$\text{AT for air gap} = H \times l = \frac{B}{\mu_0} \times l = \frac{0.8}{4\pi \times 10^{-7}} \times 0.4 \times 10^{-3} = 255$$

$$\text{AT required} = \left(\frac{1.59 \times 10^{-5}}{\mu_r}\right) + 255$$

$$\text{Total AT provided} = 200 \times 2 = 400$$

$$\therefore \quad \left(\frac{1.59 \times 10^{5}}{\mu_r}\right) + 255 = 400 \quad \text{or} \quad \mu_r = 1096$$

CHAPTER AT A GLANCE

- Like poles repel each other and unlike poles attract each other.
- The power of retaining magnetism after the magnetizing force is removed is called retentivity.
- One weber = 10^8 lines of force.
- The opposition offered by the material to the flow of lines of force is called reluctance.
- Current-carrying conductor when placed in the magnetic field experiences a mechanical force due to interaction of the two fluxes. This is the basic principle on which dc electric motor works.
- The direction of the force experienced by the current-carrying conductor placed in the magnetic field can be determined by Fleming's left-hand rule.
- The permeability is defined as the ability or ease with which the magnetic material forces the magnetic flux through a given medium.
- The ratio of magnetic flux density B in a particular medium to the magnetic field strength H producing that flux density is called absolute permeability of that medium, $\mu = B/H$.
- The units of permeability is henry/meter (H/m).
- The permeability of free space = $\mu_0 = 4\pi \times 10^{-7}$ H/m.
- The relative permeability for free space, vacuum, or air, $\mu_r = 1$.
- The force behind the flow of flux in a magnetic circuit is called mmf. Thus, mmf = NI.
- Permeance = 1/Reluctance.
- Current is analogous to flux, emf is analogous to mmf, and resistance is analogous to reluctance.
- The graph between the flux density B and the magnetic field strength H for the magnetic material is called as its magnetization curve or B–H curve.
- According to Faraday's first law, whenever the number of magnetic lines of force (flux) linking with a coil or circuit changes, an emf gets induced in that coil or circuit.
- According to Faraday's second law, the magnitude of induced emf is directly proportional to the rate of change of flux linkages.
- The property of coil which enables to induce an emf in it whenever the current changes is called self-induction.

- The phenomenon of the generation of induced emf in a circuit by changing the current in a neighboring circuit is called mutual induction.
- Self-induced emf = $-L\ (di/dt)$.
- Mutual-induced emf = $M\ (di_1/dt)$.
- The coefficient of coupling $K = (M/\sqrt{L_1L_2}\ )$.

MULTIPLE CHOICE QUESTIONS

1. Flux in a magnetic circuit can be compared in an electric circuit to
   (a) voltage (b) current (c) resistance (d) inductance
2. The magnetic field around a current-carrying conductor
   (a) moves in the direction of current flow
   (b) is parallel to the direction of current flow
   (c) exists at all points along the length of the conductor
   (d) exists only at the start of current flow
3. Current flow in a straight conductor produces
   (a) permeability (b) hysteresis
   (c) a north pole (d) magnetic lines of force
4. The term "permeability" means
   (a) the number of turns on an air core
   (b) the amount of reluctance of 1 m^3 of air
   (c) the mmf required to produce one unit of magnetic flux
   (d) the ability of substance to conduct the magnetic lines of force
5. The ability of a magnetic substance to retain its magnetism after the magnetizing force is removed is called
   (a) hysteresis (b) permeability
   (c) residual magnetism (d) retentivity
6. The residual magnetism is highest in
   (a) wood (b) copper (c) soft iron (d) sheet stell
7. The equivalent t_0 Ohm's law for magnetic circuits in
   (a) permeability = flux density/magnetic intensity
   (b) reluctance = length of flux lines/permeability
   (c) flux = mmf/reluctance
   (d) permeance = 1/reluctance
8. Magnetic mono-pole do not exist and therefore
   (a) magnetic flux lines do not exist (b) magnetic flux lines cross one another
   (c) magnetic flux lines are continuous (d) none of the above

9. The magnetic field intensity (in A/m) due to a straight conductor of radius a and infinite length carrying a current of I amperes at a point ($d > a$) is

(a) $\frac{1}{\pi d}$ (b) $\frac{1}{2\pi a}$ (c) $\frac{1}{4\pi d}$ (d) $\frac{1}{2\pi d}$

10. Relative permeability of vacuum is

(a) $4\pi \times 10^{-7}$ (b) 1 (c) 1 (d) $\frac{1}{4\pi}$

11. The unit of magnetic flux is

(a) weber (b) ampere-turn (c) tesla (d) coulomb

12. The magnetizing force (H) and magnetic flux density (B) are connected by the relation

(a) $B = \mu H$ (b) $B = \mu_0 H/\mu_r$ (c) $B = \mu_0 H/\mu_r$ (d) $B = \mu_0 H/\mu_0$

13. The force experienced by a current-carrying conductor lying parallel to a magnetic field is

(a) BIl (b) $BIl \sin\theta$ (c) HIl (d) zero

14. The unit of magnetomotive force is

(a) weber (b) ampere-turn
(c) henry (d) ampere-turn/weber

15. The permeability in a magnetic circuit corresponds to ________ in an electric circuit.

(a) conductivity (b) resistivity (c) conductance (d) resistance

16. Point out the WRONG statement. Magnetic leakage is undesirable in electrical machines because it

(a) leads to their increased weight (b) increases their cost of manufactures
(c) produces fringing (d) lowers their power efficiency

17. Permanent magnets are normally made of

(a) aluminum (b) wrought iron (c) cast iron (d) alnico alloys

18. Those magnetic materials are best suited for making armature and transformer cores which have ________ permeability and ________ hysteresis loss.

(a) low, high (b) high, low (c) low, low (d) high, high

19. Those materials are well-suited for making permanent magnets which have ________ rentivity and ________ coercivity.

(a) high, high (b) high, low (c) low, low (d) low, high

20. In a magnetic material, hysteresis loss takes place primarily due to

(a) flux density lagging behind magnetizing force
(b) molecular friction
(c) high retentivity
(d) rapid reversals of its magnetization

21. The energy stored by a coil is doubled when its current is increased by

(a) 100% (b) 41.4% (c) 50% (d) 25%

22. The initial rate of rise of current through a coil of $L = 5$ H when suddenly connected to a dc supply of 100 V is ________ A/s.

(a) 500 (b) 0.05 (c) 20 (d) 50

23. The lifting power of an electromagnet depends on its

(a) pole area (b) magnetic flux density

SHORT ANSWER QUESTIONS

1. What is the unit of magnetic flux density in the SI system?
2. Define leakage factor.
3. What is the absolute permeability of vacuum?
4. What do you mean by flux linkage?
5. How the direction of dynamically induced emf in a conductor is determined?
6. How eddy current losses are reduced?
7. What is the energy stored in the magnetic field of a solenoid 30 cm long and 3 cm diameter wound with 1000 turns of wire carrying current of 10 A?
8. What is Faraday's law of electromagnetic induction?
9. What are the factors affecting the mutual inductance between two magnetically coupled coils?
10. Define magnetic flux strength.
11. What is the application of the principle of statically induced emf?
12. What is residual magnetism?

REVIEW QUESTIONS

1. Explain the classification of electromagnetic materials.
2. Define (i) magnetic flux density, (ii) mmf, and (iii) reluctance.
3. Discuss the analogy between electric and magnetic circuits.
4. State and explain Biot–Savart law.
5. State and explain Faraday's laws of electromagnetic induction.
6. Explain about statically induced emf and dynamically induced emf.
7. Draw hysteresis loop and explain briefly the various parts on it.
8. Describe the hysteresis and eddy current losses.
9. Explain (i) residual magnetism and (ii) coercive force of magnetic substance.
10. State and explain Fleming's right-hand rule.

EXERCISE PROBLEMS

1. A magnetic circuit consists of an iron ring of mean circumference 80 cm with cross-sectional area 12 cm^2 throughout. A current of 2 A in the magnetizing coil of 200 turns produces a total flux of 1.2 mWb in the iron. Calculate (a) the flux density in the iron, (b) the absolute and relative permeability of iron, and (c) the reluctance of the circuit.
2. A coil of 500 turns and resistance 20 Ω is wound uniformly on an iron ring of mean circumference 50 cm and cross-sectional area 4 cm^2. It is connected to a 24-V dc supply. Under these conditions, the relative permeability of iron is 800. Calculate the value of (a) the magnetomotive force of the coil, (b) the magnetizing force, (c) the total flux in the iron, and (d) the reluctance of the ring.
3. An iron ring of mean length 100 cm has an air gap of 2 mm and a winding of 500 turns. If the relative permeability of iron is 600 when a current of 3 A flows through the coil, calculate the flux density neglecting fringing.

4. A conductor of active length 30 cm carries a current of 100 A and lies at right angles to a magnetic field of strength 0.4 Wb/m^2. (a) Calculate the force in newtons exerted on it. If the force causes the conductor to move at a velocity of 10 m/s, calculate (b) the emf induced in it and (c) the power in watts developed by it.
5. A straight horizontal wire carries a steady current of 150 A and is situated in a uniform magnetic field of 0.6 Wb/m^2 acting vertically downwards. Determine the magnitude of the force, in kg/m of conductor and the direction in which it works.
6. A conductor of 10 cm length moves with a uniform velocity of 2 m/s at right angles to itself and to a uniform magnetic field having a flux density of 1 Wb/m^2. Calculate the induced emf between the ends of the conductor.
7. An air-cored solenoid has length of 50 m and a diameter of 2 m. Calculate its inductance if it has 1000 turns and also find the energy stored in it if the current rises from zero to 5 A.
8. An air-cored solenoid 1 m in length and 10 cm in diameter has 5000 turns. Calculate (a) self-inductance and (b) the energy stored in the magnetic field when a current of 2 A flows in the solenoid.
9. Determine the force required to separate two magnetic surfaces with contact area of 100 cm^2 if the magnetic flux density across the surface is 0.1 Wb/m^2. Derive formula used, if any.
10. A coil is wound uniformly with 300 turns over a steal string of relative permeability of 900 having a mean circumference of 400 mm and cross-sectional area of 500 mm^2. If a current of 2.5 A is passed through the coil, find (a) the mmf, (b) reluctance of the ring, and (c) the flux.
11. A iron ring of mean length 50 cm has an air gap of 1 mm and a winding of 100 turns. If the relative permeability of iron is 300 and a current of 1 A flows through the coil, calculate the flux density, neglecting fringing.

Chapter

4

Alternating Quantities

Chapter Objectives

- To explain the generation of alternating-current voltages and the expression of electromotive force
- To explain various terms related to alternating quantities
- To explain root mean square and average values
- To describe phasor analysis

4.1 INTRODUCTION

In this chapter, we will discuss the alternating quantities. The circuits which are excited by constant or time-invariant sources are called *direct-current* (*dc* or *DC*) *circuits*. The circuits which are excited by time-varying sources are called *alternating-current* (*ac* or *AC*) *circuits*. Because of the economic advantages ac is widely used throughout the world as compared to dc. We begin our discussion with alternating voltage and current. The alternating voltage is defined as the voltage that varies with respect to time period during a definite cycle both in magnitude and voltage. Similarly, ac is defined as the movement (or flow) of electric charge periodically in the reverse direction or the current that varies with respect to time period during a definite cycle both in magnitude and direction. The usual waveform of an ac power circuit is a sine wave; however, in certain applications, different waveforms are used, such as triangular or square waves. The ac voltage may be increased or decreased by using a transformer. At higher voltages, power can be efficiently transmitted and power losses can also be reduced.

The direction of alternating quantity changes continuously with respect to time. There are many types of alternating waveforms such as cosine wave, saw-tooth wave, and triangular wave. The following are some advantages of sine wave over the other alternating waves:

(i) The nature itself is sinusoidal in characteristic, for example, motion of pendulum and vibration of a spring.

(ii) Any functional waveform can be written in terms of sinusoidal function using the fourier analysis.

(iii) The derivative and integral of sine wave is also sinusoidal and it is easy to handle mathematically.

(iv) The sinusoidal function is easy to generate, transmit, and analyze.

4.2 GENERATION OF ALTERNATING VOLTAGE

The machines which are used to generate alternating quantities are called alternators. It works on the basis of the principle of Faraday's law of electromagnetic induction. The main components of an alternator are stator and rotor. Stator is a stationary element and rotor is a rotating element. Alternating voltage can be generated in a stationary armature with rotating field structure or in a rotating armature with a stationary field as shown in Figure 4.1.

Consider an electromagnetic coil AA′ as shown in Figure 4.1. Let this coil be rotating in anticlockwise direction with constant angular velocity "ω" in a uniform magnetic field between N and S poles.

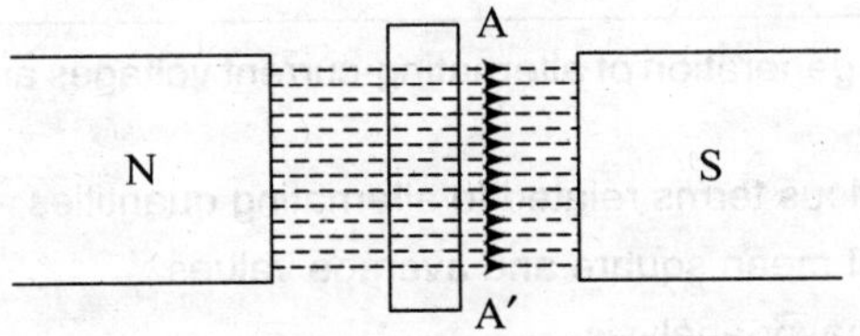

Figure 4.1 Rotating armature with stationary field

The electromotive force (emf) generated in a coil depends upon the position of coil with respect to the uniform magnetic field. When AA′ is perpendicular to the NS axis, the emf generated will be maximum in the coil and if it rotates 90° in the anticlockwise direction, the generated emf will be zero. Similarly, the emf generated is zero at 180°. This completes one-half cycle of the generated emf wave. In the next half cycle the generated emf is negative, maximum at 270° from the initial position, and zero at 360°. This completes another half-cycle of the emf wave.

4.3 EXPRESSION FOR INDUCED EMF

Let it be assumed that the coil has been rotating with angular velocity ω rad/s. Then, in "t" seconds, the coil rotates through angle $\omega t = \theta$ rad from the vertical position, the position of zero emf shown in Figure 4.2(a), and the coil sides A and A′ have a tangential velocity of u m/s. The horizontal component of the velocity tangential, H, can be found out from Figure 4.2(b) as

$$H = u \sin\theta = u \sin\omega t \tag{4.1}$$

The voltage induced in each of the conductor sides A and A′ due to the component of the velocity is perpendicular to the magnetic field. The emf induced in one side of the coil is

$$E = Blu \sin\theta \text{ volts} \tag{4.2}$$

and the total emf induced in the coil is

$$E = 2\,Blu \sin\theta \text{ volts} \tag{4.3}$$

From Eq. (4.3), it may be stated that the generated emf is proportional to sin θ. It follows that when $\theta = 90°$, the generated emf is at its maximum value.

$$E_m = 2\,Blu \text{ volts} \tag{4.4}$$

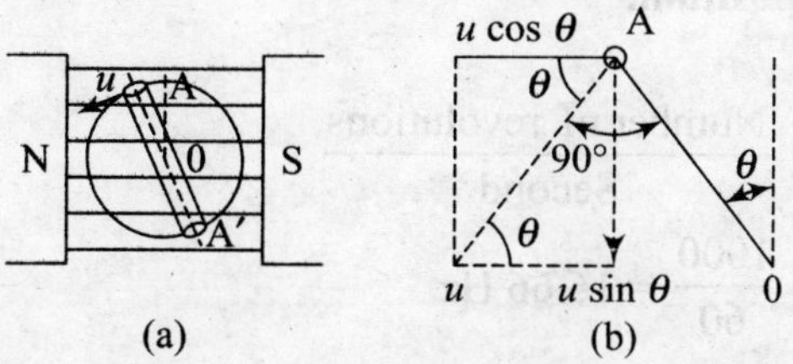

Figure 4.2 Computation of the magnitude of the generated emf

Thus, Eq. (4.3) may be written as

$$E = E_m \sin\omega t \text{ (where } E_m = 2Blu) \tag{4.5}$$

If n is the speed of rotation of the coil in revolutions per second, then

$$u = \pi bn \text{ meter/s} \tag{4.6}$$

and

$$E_m = 2Bl \times \pi bn \text{ volts}$$
$$= 2\pi BAn \text{ volts (where } A = l \times b \text{ is the area of the coil)} \tag{4.7}$$

For a coil having T number of turns in series, each turn having an area of A m^2, the maximum value-generated emf is given by

$$E_m = 2\pi BAnT \text{ volts} \tag{4.8}$$

and the instantaneous value of the generated value can be represented as a function of "θ" as shown in Figure 4.3, where $e(t)$ represents the value of voltage when the coil has rotated through an angle θ from the position of zero emf, and E_m is the maximum value of the induced voltage.

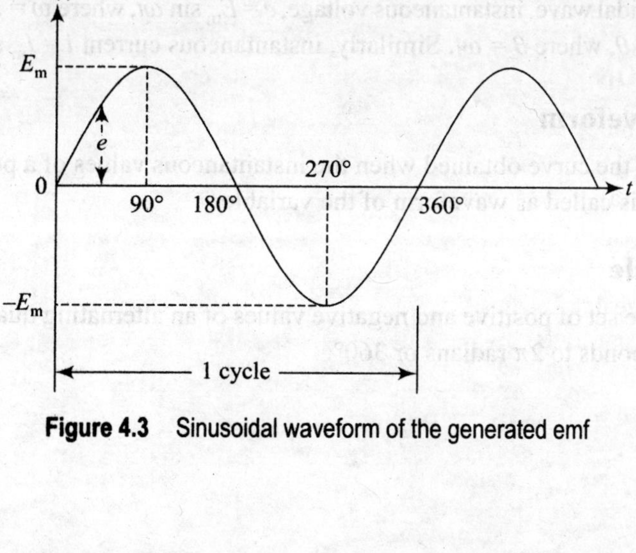

Figure 4.3 Sinusoidal waveform of the generated emf

EXAMPLE 4.1 **A coil of 100 turns is rotated at 1000 rpm in a uniform field having a density of 4000 T, the axis of rotation being at right angles to the direction of the field. The mean area per turn is 200 mm^2. Calculate (i) the frequency, (ii) the period, (iii) the maximum value of the generated emf, and (iv) the value of the generated emf when the coil is rotated by 30° from the zero emf position.**

Solution

(i) $\text{Frequency} = \dfrac{\text{Number of revolutions}}{\text{Second}}$

$= \dfrac{1000}{60} = 16.66 \text{ Hz}$

(ii) $\text{Period} = \dfrac{1}{f} = \dfrac{1}{16.66} = 0.06 \text{ s}$

(iii) The maximum value of induced emf, $E_m = 2\pi BAnT$

$= 2 \times \pi \times 4000 \times 200 \times 16.66 \times 100 \times 10^{-6}$

$= 8374.22 \text{ V}$

(iv) Generated emf at an angle 30°, $e = E_m \sin\theta$

$= 8374.22 \sin 30$

$= 4187.11 \text{ V}$

4.4 TERMINOLOGY RELATED TO ALTERNATING QUANTITIES

The terms generally used in alternating quantity analysis is described in this section.

4.4.1 Instantaneous Value

The value of voltage or current obtained at any instant of time is called instantaneous value. Generally, it is denoted by lower case letters *e*, *i*, etc.

For sinusoidal wave, instantaneous voltage, $e = E_m \sin \omega t$, where ω = angular velocity = $2\pi f$ rad/s, or $e = E_m \sin\theta$, where $\theta = \omega t$. Similarly, instantaneous current $i = I_m \sin \omega t$ or $I_m \sin\theta$.

4.4.2 Waveform

The shape of the curve obtained when the instantaneous values of a periodic variable are plotted against time is called as waveform of the variable.

4.4.3 Cycle

One complete set of positive and negative values of an alternating quantity is called a cycle. One cycle corresponds to 2π radians or 360°.

4.4.4 Time Period and Frequency

Time period

The time taken by an alternating quantity to complete one cycle is called its time period or periodic time and is denoted by T. The cycle repeats after every T seconds. The number of cycles completed per second is f.

$$\text{Periodic time, } T = \text{Time taken in completing one cycle} = \frac{1}{f} \tag{4.9}$$

Frequency

The frequency of an alternating quantity is the number of cycles per second. It is denoted by f and expressed in Hertz or cycle per second.

$$f = \frac{1}{T} \tag{4.10}$$

In India, the standard frequency for power supply is 50 Hz and in some other parts of the world including United States it is 60 Hz.

4.4.5 Amplitude/Maximum Value

The maximum value of voltage or current obtained during positive or negative half-cycle is called amplitude or peak value, denoted by upper case letters E_m, I_m, etc. The maximum value of the emf or current is the maximum value of the instantaneous emf or current reached once in every half-cycle. The ratio of the instantaneous value to the maximum value is

$$\frac{e}{E_m} = \frac{i}{I_m} = \sin\theta \tag{4.11}$$

4.4.6 Angular Frequency/Angular Velocity

A sinusoidal quantity repeats at an interval of 2π rad, the duration T for one complete cycle of a periodic variable is assumed to correspond to a time angle of 2π electrical radians. The angular velocity of tracing the time angle is therefore

$$\omega = \frac{\text{Variation in angle in one cycle}}{\text{Time taken to complete one cycle}} = \frac{2\pi}{T} = 2\pi f \text{ electrical rad/s} \tag{4.12}$$

EXAMPLE 4.2 **Find the amplitude, phase, period, and frequency of the sinusoid $v(t) = 15 \sin(50t + 20°)$.**

Solution Amplitude $V_m = 15$ V

Phase = 20°, $\omega = 50$ rad/s

$$\text{Period} = \frac{2\pi}{\omega} = \frac{2\pi}{50} = 0.125 \text{ s}$$

Frequency, $f = 1/T = 7.96$ Hz

EXAMPLE 4.3 **An alternating current is represented by 10 sin 314t. Find the value of current after 2 s.**

Solution Current after 2 s = 10 sin 314 × 2 = –3.13 A

EXAMPLE 4.4 **A sinusoidal alternating voltage of frequency 50 Hz has a maximum value of 200 V. How much time it will take for the voltage to attain value of 100 V?**

Solution The instant value of voltage is given by

$$v = V_m \sin \omega t \quad \text{or} \quad v = V_m \sin 2\pi f t$$

$$v = 200 \sin 2 \times \pi \times 50 \times t$$

$$= 200 \sin 100\pi t$$

$$100 = 200 \sin 314t$$

$$\sin 100\pi t = 0.5$$

$$100\pi t = \sin^{-1}(0.5) = 30°$$

Time taken to reach 100 V,

$$t = \frac{30°}{100 \times 180°} = 1.66 \times 10^{-3} \text{ s}$$

EXAMPLE 4.5 **A 50-Hz current has a maximum value of 50 A. Find the rate of change in current at time *t* when *t* = 0.01 s after *i* = 0 and is increasing.**

Solution Let $i = 50 \sin \omega t$

$$\omega = 2\pi f = 2\pi \times 50 = 100\pi \text{ rad/s}$$

Rate of change in current

$$\frac{di}{dt} = \frac{d(50 \sin 100\pi t)}{dt}$$

$$= 50 \times 100\pi \cos 100\pi t$$

At $t = 0.01$ s, $di/dt = 50 \times 314 \cos 100\pi \times 0.01 = -15700$ A/s

4.5 RELATION BETWEEN FREQUENCY, SPEED, AND NUMBER OF POLES

In Section 4.2, it is seen that the waveform of the emf developed in an ac generator goes through one complete cycle of variation when the conductors move past one north pole and south pole. The shape of the waveform over the negative half-cycle is exactly similar to that over the positive half-cycle. In a two-pole machine, as the generator rotates through one revolution the voltage generated also completes one cycle of variation. Thus, for p poles, when the ac generator is driven at a speed of N rpm, $P/2$ cycles of variation of emf are produced in one revolution of the generator. The frequency f of the induced emf is given by

$$f = \text{Number of cycles/s}$$

$$= \text{Number of cycles/revolution} \times \text{Number of revolution/s}$$

$$= \frac{P}{2} \times \frac{N}{60}$$

$$\therefore \quad f = \frac{PN}{120} \text{ cycles/s or Hz} \tag{4.13}$$

Hence, for a two-pole machine, the speed at which the machine must be driven to generate an emf of 50 Hz frequency is given by Eq. (4.13) as

$$N = \frac{120 \times f}{P}$$

$$= \frac{120 \times 50}{2}$$
$$= 3000 \text{ rpm}$$

As the minimum number of poles that machine can have is 2, the highest speed at which a 50-Hz ac generator can be operated is 3000 rpm. In practice, this is the operating speed of steam-turbine-driven generating station. For hydroelectric generators, speeds are lower and the number of poles may be 4, 8, 10, 12, or 16.

EXAMPLE 4.6 **If a hydroelectric generator is operating with 12 number of poles to give 50-Hz supply frequency, what is its speed?**

Solution The speed of the alternator is given by

$$N = \frac{120 \times f}{P}$$
$$= \frac{120 \times 50}{12}$$
$$= 500 \text{ rpm}$$

4.6 AVERAGE VALUE AND RMS VALUE OF ALTERNATING QUANTITIES

(i) Average value: The average value of a sine wave over one complete cycle is always zero. Hence, the average of a sine wave is defined over a half-cycle only, i.e., the average value is the arithmetic mean of values at equal interval over a half-cycle. It can be written as

$$V_{avg} = \frac{1}{\pi}\int_0^{\pi} V_m \sin \omega t \, d\omega t \tag{4.14}$$

$$= \frac{V_m}{\pi}(-\cos \omega t)\Big|_0^{\pi}$$

$$= \frac{2V_m}{\pi} = 0.637\, V_m \tag{4.15}$$

In general, it is defined as the value which is obtained by averaging all the instantaneous values over a period of half-cycle.

(ii) Root mean square (rms) value or effective value: The idea of effective value arises from the need to measure the effectiveness of a voltage or current source in delivering power to a resistive load. The rms value of an alternating wave is defined as that steady value (dc) of quantity (voltage or current) which when connected to a circuit for a given time produces the same heating effect produced by the alternating quantity flowing through the same circuit for the same time.

Consider the circuit shown in Figures 4.4(a) and (b). The circuit in Figure 4.4(a) is ac while that in Figure 4.4(b) is dc.

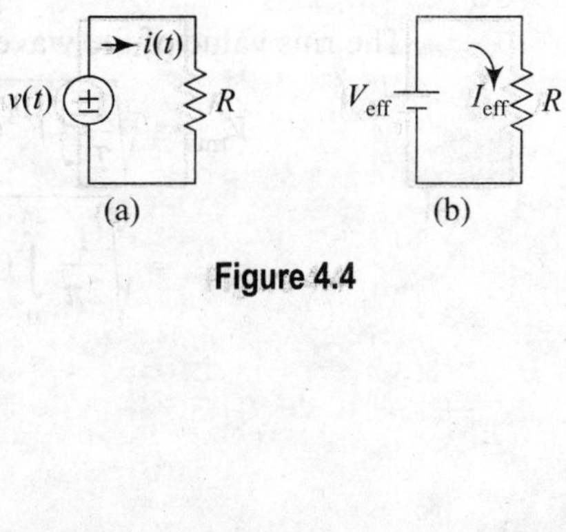

Figure 4.4

The average power absorbed by the resistor in the ac circuit is

$$P = \frac{1}{T}\int_0^T i^2 R\, dt \tag{4.16}$$

$$= \frac{R}{T}\int_0^T i^2\, dt \tag{4.17}$$

The power absorbed in dc circuit is

$$P = I_{\rm rms}^2 R \tag{4.18}$$

Equating the two powers, i.e., Eqs. (4.17) and (4.18), we get

$$I_{\rm rms}^2\, R = \frac{R}{T}\int_0^T i^2\, dt \tag{4.19}$$

Simplifying Eq. (4.19), we get

$$I_{\rm rms}^2 = \frac{1}{T}\int_0^T i^2\, dt$$

$$I_{\rm rms} = \sqrt{\frac{1}{T}\int_0^T i^2\, dt} \tag{4.20}$$

This indicates that the rms value is the square root of the mean of the square of the periodic signal. Thus, the effective value is often known as *rms value*. Similarly, the rms value of a periodic voltage wave $V(t)$ of a time period T can also be expressed as

$$V_{\rm rms} = \sqrt{\frac{1}{T}\int V^2(t)\, dt}$$

So, for a sinusoidal wave shown in Figure 4.5, the rms value can be calculated as follows.

As seen from Figure 4.5, the time period T is equal to 2π rad. So, the wave equation can be written as $V(t) = V_{\rm m} \sin \omega t$.

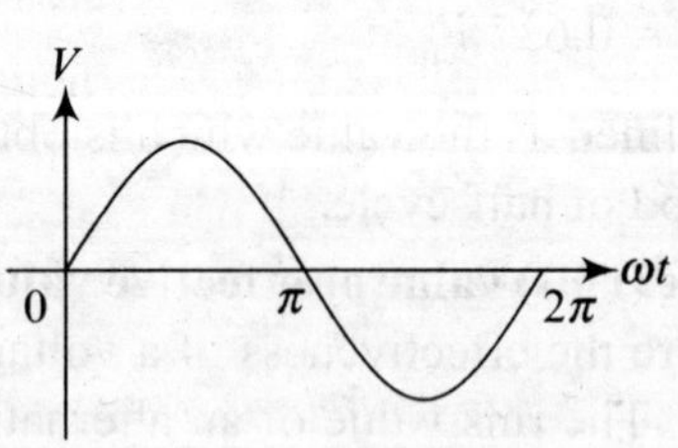

Figure 4.5 Sinusoidal wave

The rms value of the wave is

$$V_{\rm rms} = \sqrt{\frac{1}{T}\int V^2(t)\, d\,\omega t} \tag{4.21}$$

$$= \sqrt{\frac{1}{2\pi}\int_0^{2\pi} (V_{\rm m} \sin \omega t)^2\, d\,\omega t}$$

$$= \sqrt{\frac{1}{2\pi}\int_0^{2\pi} V_m^2 \sin^2 \omega t \, d\,\omega t}$$

$$= \sqrt{\frac{V_m^2}{2\pi}\int_0^{2\pi} \sin^2 \omega t \, d\,\omega t}$$

$$= \sqrt{\frac{V_m^2}{2\pi}\int_0^{2\pi} \frac{(1-\cos \omega t)}{2} \, d\,\omega t}$$

$$= \sqrt{\frac{V_m^2}{2\pi}(2\pi - 0)}$$

$$= \frac{V_m}{\sqrt{2}} \tag{4.22}$$

Similarly, for $i(t) = I_m \sin \omega t$, the rms value $= I_m/\sqrt{2}$.

4.6.1 Form Factor

The form factor is defined as the ratio of an rms value to the average value of an alternating quantity.

$$\text{Form factor} = \frac{\text{rms value}}{\text{Average value}} = \frac{V_{rms}}{V_{avg}} \tag{4.23}$$

For a sine wave,

$$\text{Form factor} = \frac{\left(V_m/\sqrt{2}\right)}{\left(2V_m/\pi\right)}$$

$$= \frac{\pi}{2\sqrt{2}} = 1.11 \tag{4.24}$$

4.6.2 Peak Factor or Crest Factor

It is defined as the ratio of the maximum value to the rms value of an alternating quantity. Therefore,

$$\text{Peak factor} = \frac{\text{Maximum value}}{\text{rms value}} = \frac{V_m}{V_{rms}} \tag{4.25}$$

For a sine wave,

$$\text{Peak factor} = \frac{V_m}{V_m/\sqrt{2}}$$

$$= \sqrt{2} = 1.414 \tag{4.26}$$

4.7 RMS AND AVERAGE VALUE FOR SQUARE WAVE AND TRIANGULAR WAVE

The calculation of rms and average values for triangular and square waves are illustrated with the following illustrative examples.

Consider the square wave shown in Figure 4.6. Here, it can be seen that

For the time interval $0 < t < 0.1$, $V = 20$ V

For the time interval $0.1 < t < 0.3$, $V = 0$ V

The total time period can be obtained as 0.3 s.

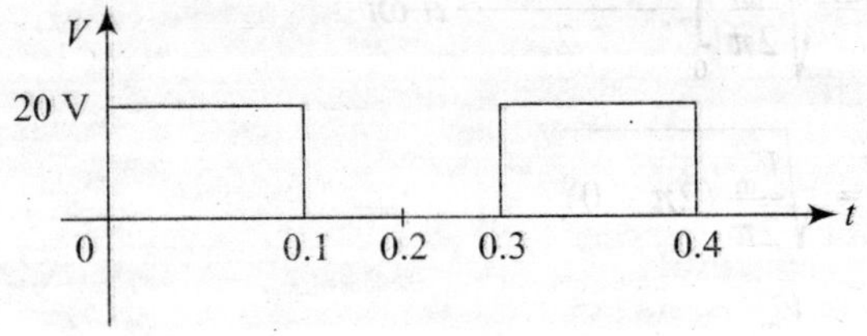

Figure 4.6 Square wave

$$\text{rms value, } V_{\text{rms}} = \sqrt{\frac{1}{T}\int V^2(t)\, dt}$$

$$= \sqrt{\frac{1}{0.3}\int_0^{0.1} 20^2\, dt}$$

$$= \sqrt{\frac{400}{0.3}\, t\Big|_0^{0.1}}$$

$$= \sqrt{\frac{400}{0.3}}$$

$$= 11.5 \text{ V}$$

$$\text{Average value, } V_{\text{avg}} = \frac{1}{T}\int_0^T v(t)\, dt$$

$$= \frac{1}{0.3}\int_0^{0.1} 20\, dt$$

$$= \frac{1}{0.3} \times 20(t)\Big|_0^{0.1}$$

$$= \frac{1}{0.3} \times 20 \times 0.1 = 6.67 \text{ V}$$

Consider the triangular wave shown in Figure 4.7. Here, it is observed that

For the time interval $0 < t < 5$, $i = t$ A

The total time period can be obtained as 5 s.

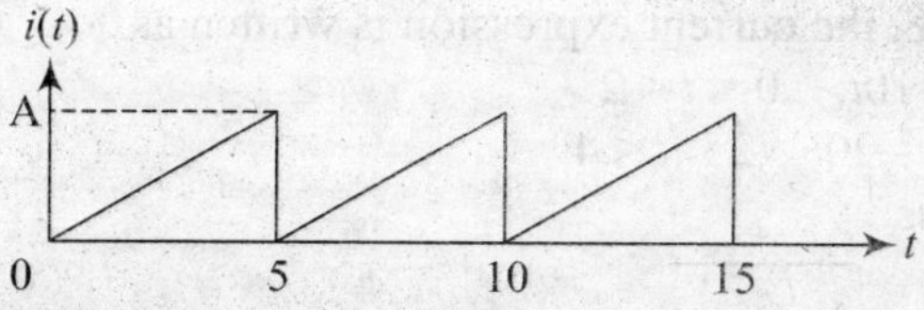

Figure 4.7 Triangular wave

$$\text{rms value, } I_{\text{rms}} = \sqrt{\frac{1}{T}\int i^2(t)\,dt}$$

$$= \sqrt{\frac{1}{5}\int_0^5 t^2\,dt}$$

$$= \sqrt{\frac{1}{5}\frac{t^3}{3}\Bigg|_0^5}$$

$$= \sqrt{\frac{t^3}{15}\Bigg|_0^5}$$

$$= 2.88 \text{ V}$$

$$\text{Average value, } I_{\text{avg}} = \frac{1}{T}\int_0^T i(t)\,dt$$

$$= \frac{1}{5}\int_0^5 t\,dt$$

$$= \frac{1}{5}\frac{t^2}{2}\Bigg|_0^5$$

$$= 2.5 \text{ A}$$

EXAMPLE 4.7 **Calculate the effective value of the current waveform shown in Figure 4.8, if the current is passed through a 4-Ω resistor; find the average power absorbed by the resistor.**

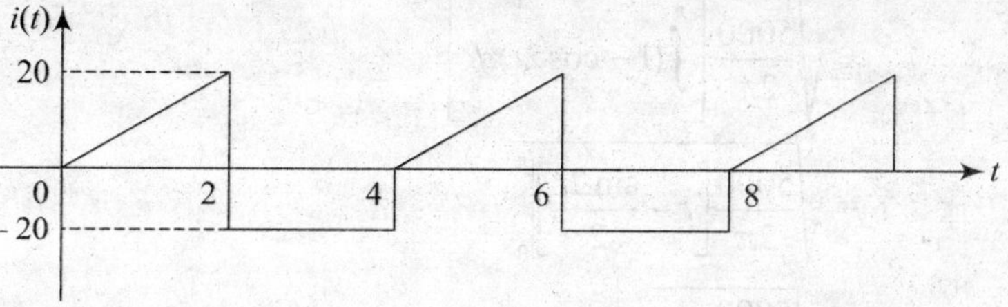

Figure 4.8

Solution Period of the waveform is $T = 4$

From Figure 4.8, the current expression is written as

$$i(t) = 10t; \quad 0 < t < 2$$
$$= -20; \quad 2 < t < 4$$

The rms value is

$$I_{\text{rms}} = \sqrt{\frac{1}{T}\int_0^T i^2\, dt}$$

$$= \sqrt{\frac{1}{4}\int_0^2 (10t)^2\, dt + \int_2^4 (-20)^2\, dt}$$

$$= \sqrt{\frac{100}{4}\left.\frac{t^3}{3}\right|_0^2 + 400\,t|_2^4}$$

$$= 29.44 \text{ A}$$

The power absorbed by the 4-Ω resistor, $P = I_{\text{rms}}^2 R = 29.44^2 \times 4 = 3466.85$ W.

EXAMPLE 4.8 **Find the rms value of half-wave rectified sine wave shown in Figure 4.9.**

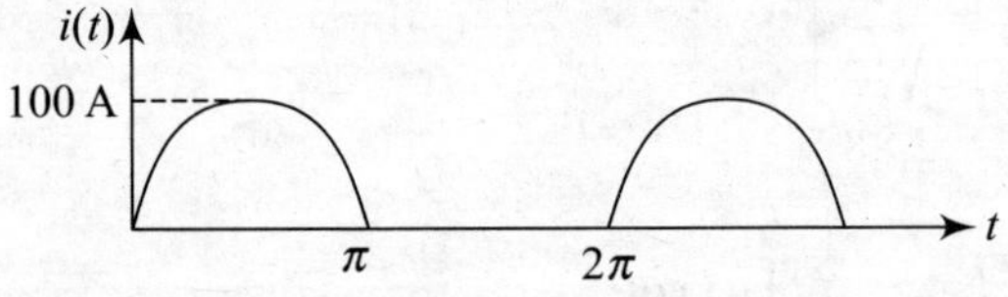

Figure 4.9

Solution The period of the current waveform, $T = 2\pi$. Current is defined as

$$i(t) = 100 \sin t \quad 0 < t < \pi$$
$$= 0 \quad \pi < t < 2\pi$$

$$I_{\text{rms}} = \sqrt{\frac{1}{T}\int_0^T i^2\, dt}$$

$$= \sqrt{\frac{1}{2\pi}\left[\int_0^{\pi} (100 \sin t)^2\, dt + \int_{\pi}^{2\pi} (0)^2\, dt\right]}$$

$$= \sqrt{\frac{5000}{2\pi}\left[\int_0^{\pi} (1 - \cos 2t)dt\right]}$$

$$= \sqrt{\frac{5000}{2\pi}\left[t - \frac{\sin 2t}{2}\right]_0^{\pi}}$$

$$= \sqrt{\frac{5000}{2\pi} \times \pi}$$

$$= 50 \text{ A}$$

EXAMPLE 4.9 **Calculate the form factor and peak factor of the full-wave rectified sine wave.**

Solution

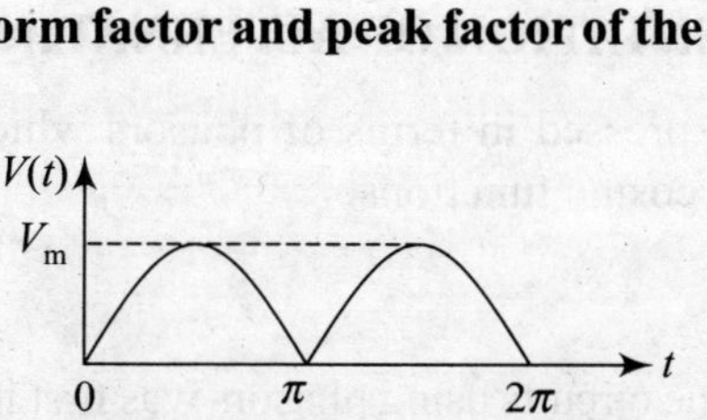

Figure 4.10 Full-wave rectified sine wave

$$\text{Average value, } V_{\text{avg}} = \frac{1}{\pi}\int_0^{\pi} V_{\text{m}} \sin\theta \, d\theta$$

$$= \frac{V_{\text{m}}}{\pi}(-\cos\theta)\Big|_0^{\pi}$$

$$= \frac{2V_{\text{m}}}{\pi}$$

$$V_{\text{rms}} = \sqrt{\frac{1}{T}\int_0^T v^2 \, dt}$$

$$= \sqrt{\frac{1}{\pi}\left[\int_0^{\pi} (V_{\text{m}} \sin\theta)^2 \, d\theta\right]}$$

$$= \sqrt{\frac{V_{\text{m}}^2}{2\pi}\left[\int_0^{\pi} (1 - \cos 2\theta) \, d\theta\right]}$$

$$= \sqrt{\frac{V_{\text{m}}^2}{2\pi}\left[\theta - \frac{\sin 2\theta}{2}\right]_0^{\pi}}$$

$$= \sqrt{\frac{V_{\text{m}}^2}{2\pi} \times \pi}$$

$$= \frac{V_{\text{m}}}{\sqrt{2}}$$

$$\text{Form factor} = \frac{V_{\text{rms}}}{V_{\text{avg}}} = \frac{\frac{V_{\text{m}}}{\sqrt{2}}}{\frac{2V_{\text{m}}}{\pi}} = 1.11$$

$$\text{Peak factor} = \frac{V_{\text{max}}}{V_{\text{rms}}} = \frac{V_{\text{m}}}{\frac{V_{\text{m}}}{\sqrt{2}}} = \sqrt{2} = 1.414$$

4.8 PHASOR REPRESENTATION OF ALTERNATING QUANTITIES

Sinusoids are easily expressed in terms of phasors which are easier to work with than that of trigonometric sine and cosine functions.

4.8.1 Phasor

The method of solving ac circuits using phasors was first introduced by Charles Proteus Steinmetz in 1893. A phasor is a complex number that represents the amplitude and phase of a sinusoid. We can represent the time-domain voltage into frequency domain by the phasor notation. Thus, the source voltage in time domain is given by $v(t) = V_m \sin \omega t$. The voltage in the frequency domain is given by $V = V_m \angle 0°$. This representation is called phasor. It is easy to work with phasors rather than time-varying quantities.

4.8.2 Vector

It is the quantity having both magnitude and direction. Almost all physical quantities are vectors. For example, force is represented in magnitude as well as in direction, i.e., the direction in which it is acting upon.

In electrical engineering, a phasor has magnitude and phase (direction) and it behaves as a vector.

4.8.3 Phase

The phase of sine wave can be defined as the angular measurement of the wave that specifies the position of sine wave relative to a zero reference point as shown in Figure 4.11.

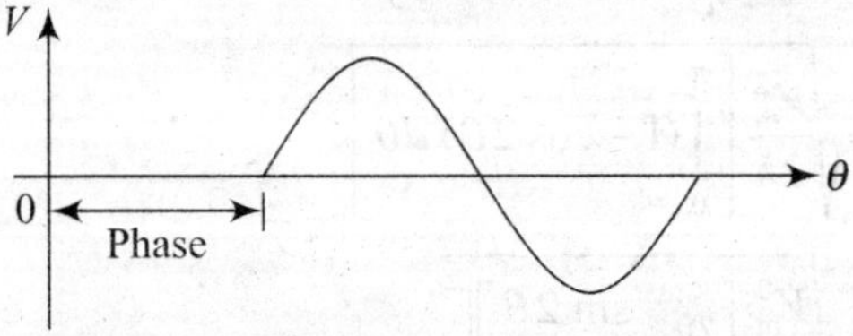

Figure 4.11 Characteristics of sine wave

4.8.4 Phase Difference

Phase difference is defined as the difference between the phases of two alternating quantities as shown in Figure 4.12. It is nothing but the angle difference between the two phasors. If the phase difference is 0°, we say both the quantities are "in phase," otherwise we say that it is "out of phase."

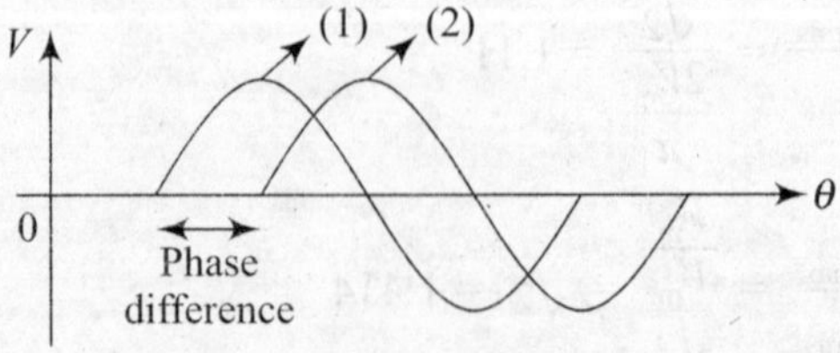

Figure 4.12 Phase difference of sine wave

4.8.5 Phase Angle

In Figure 4.13, the phase of current at point P is $T/4$ s, where T is the time period expressed in terms of angle 90° or $\pi/2$.

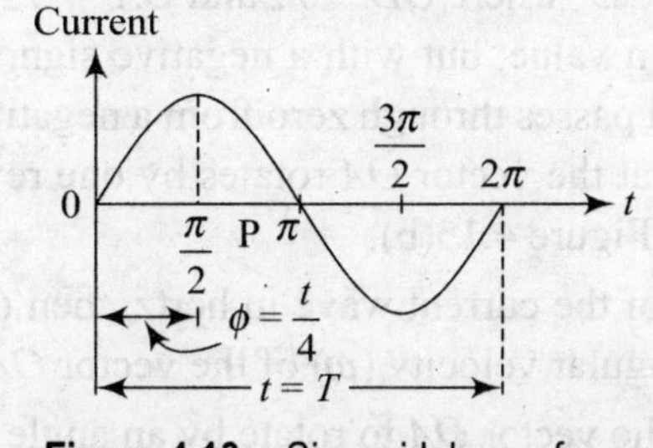

Figure 4.13 Sinusoidal waveform

4.8.6 Phasor Diagram

A graphical representation of phasors is known as the phasor diagram. Figure 4.14 shows the phasor representation. When phasors are out of phase, from the phasor diagram we can say whether the quantity is leading or lagging.

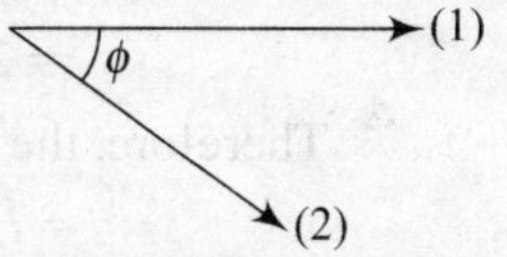

Figure 4.14 Phasor diagram

4.9 PHASOR REPRESENTATION OF ALTERNATING QUANTITY BY ROTATING VECTOR

The rotating vector representation of an alternating current is shown in Figure 4.15. It is observed that as time increases, the sinor rotates on a circle of radius I_m at an angular velocity ω in the counter clockwise direction. Let OA represents to scale, the maximum value of the current, i.e., $OA = I_m$. Figure 4.15(a) shows OA when it is rotated by an angle θ from its original position when the current was passing through its zero value. If AB and AC are its rectangular components, then

$$
\begin{aligned}
OC = AB &= OA \sin\theta \\
&= I_m \sin\theta \qquad (4.27) \\
&= i \text{ the instantaneous value of current at that instant}
\end{aligned}
$$

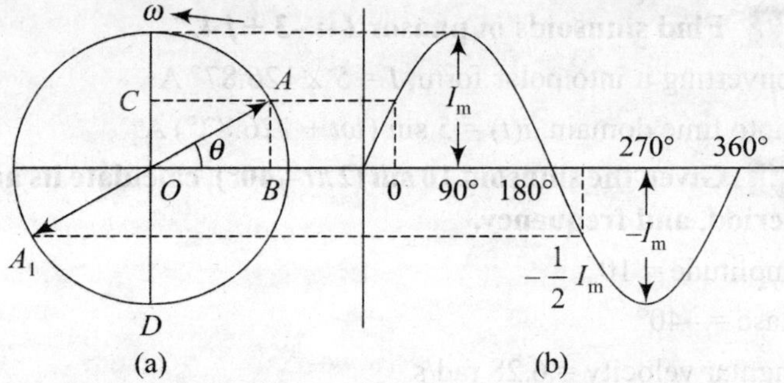

Figure 4.15 Phasor representation of an alternating quantity

Thus, the vertical component of *OA* on the vertical axis (*y*-axis) represents to scale, the instantaneous value of the current. From this it follows that when $\theta = 90°$, $OC = OA = I_m$. When $\theta = 180°$, current = 0. When $\theta = 210°$, the rotating vector is in position OA_1 corresponding to its vertical components *OD*, where $OD = ½$ and $OA = ½\, I_m$. When θ reaches 270°, the current again attains a maximum value, but with a negative sign, i.e., $-I_m$. Finally, when $\theta = 360°$, the current is again zero and passes through zero from a negative to a positive value. From the above discussion, it follows that the vector *OA* rotates by one revolution of 2π rad in one cycle of the current wave, shown in Figure 4.15(b).

If *f* is the frequency of the current wave in hertz, then *OA* rotates through *f* revolution or $2\pi f$ rad in 1 s. Hence, the angular velocity (ω) of the vector *OA* is $2\pi f$ rad/s, i.e., $\omega = 2\pi f$ rad/s.

If the time taken by the vector *OA* to rotate by an angle θ rad is "*t*" seconds, then

$$\theta = \text{Angular velocity} \times \text{Time}$$
$$= \omega \times t$$
$$= 2\pi f t \text{ rad}$$

Therefore, the instantaneous value of the current given by Eq. (4.13) can be written as

$$i = I_m \sin \theta$$
$$= I_m \sin \omega t$$
$$= I_m \sin 2\pi f t \qquad (4.28)$$

EXAMPLE 4.10 **Transform the given time domain voltage into phasor domain.**

$v(t) = 50 \sin (100t - 40°)$ V

Solution The voltage in phasor domain is given by $V = 50 \angle -40°$ volts

EXAMPLE 4.11 **Convert the given voltage phasor into time domain voltage.**

$V = 170 \angle 30°$ V

Solution The time domain voltage is given by $v(t) = 170 \sin (\omega t + 30°)$ volts

EXAMPLE 4.12 **Express these sinusoids as phasors.**

(i) $v = 10 \sin (4t + 50°)$ V; (ii) $i = 7 \sin (10t - 20°)$ A

Solution (i) $v = 10 \angle 50°$; (ii) $i = 7 \angle -20°$

EXAMPLE 4.13 **Find sinusoids of phasor $I = -3 + j\,4$.**

Solution Converting it into polar form, $I = 5 \angle 126.87°$ A

Writing it into time domain, $i(t) = 5 \sin (\omega t + 126.87°)$ A

EXAMPLE 4.14 **Given the sinusoid $10 \sin (2\pi t - 40°)$, calculate its amplitude, phase, angular frequency, period, and frequency.**

Solution Amplitude = 10

Phase = –40°

Angular velocity = 6.28 rad/s

Period = 1 s

Frequency = 1 Hz

4.10 ADDITION AND SUBTRACTION OF TWO PHASORS

Figure 4.16 shows two phasors A and B that are to be added. In the rectangular form the phasors A and B can be written as

$$A = a_1 + ja_2 \quad \text{and} \quad B = b_1 + jb_2$$

The addition of the phasors A and B gives the diagonal C as shown in Figure 4.16(a). If c_1 and c_2 are the real and imaginary components, respectively, of the phasor C, then $C = c_1 + jc_2$.

From Figure 4.16(a)

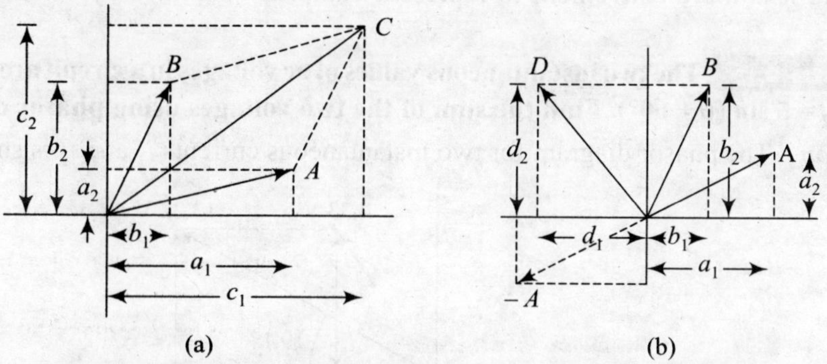

Figure 4.16 (a) Addition of phasors and (b) subtraction of phasors

$$c_1 = (a_1 + b_1) \quad \text{and} \quad c_2 = (a_2 + b_2)$$
$$C = (a_1 + b_1) + j(a_2 + b_2)$$
$$= (a_1 + ja_2) + (b_1 + jb_2)$$
$$= A + B$$

The phasor difference of the phasors A and B is given by the resultant phasor D, as shown in Figure 4.16(b). The subtraction of the phasors A and B is given by the diagonal D of the parallelogram drawn the phasors A and B as its sides, as shown in Figure 4.16(b). If d_1 and d_2 are the real and imaginary components, respectively, of the phasor D, then $D = d_1 + jd_2$. Now, from Figure 4.16(b), if $b_1 < a_1$ is negative, then

$$D = -d_1 + jd_2$$
$$= -a_1 + b_1 + j(b_2 - a_2)$$
$$= (b_1 + jb_2) - (a_1 + ja_2)$$
$$= B - A$$

4.11 MULTIPLICATION AND DIVISION OF TWO PHASORS

Let the two phasors be $A = A\,e^{j\theta}$ and $B = B\,e^{j\phi}$. Then the product of the two phasors can be written as

$$A \cdot B = A\,e^{j\theta} \times B\,e^{j\phi}$$
$$= AB\,e^{j(\theta + \phi)}$$

Similarly, the division of the two phasors can be written as

$$\frac{A}{B} = \frac{A\,e^{j\theta}}{B\,e^{j\phi}}$$

$$= \frac{A}{B}\,e^{j(\theta-\varphi)}$$

From the above equation it is clear that for adding and subtracting phasors it is expeditious to represent complex quantities by the rectangular form, while for multiplication and division of phasors it is more convenient to represent complex quantities by the trigonometric or polar form.

EXAMPLE 4.15 The two instantaneous values of ac voltages in a circuit are given as $v_1 = 10 \sin\theta$ and $v_2 = 5 \sin(\theta + 60°)$. Find the sum of the two voltages using phasor diagram.

Solution The phasor diagram for two instantaneous currents i_1 and i_2 is shown in Figure 4.17.

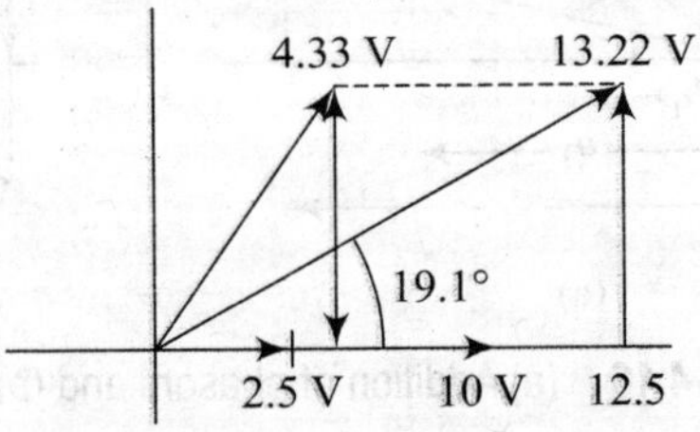

Figure 4.17 Phasor diagram of two instantaneous currents

Resolving the phasors into their horizontal and vertical components, we get

Horizontal component = 10 + 5 cos 60° = 12.5 V

Vertical component = 0 + 5 sin 60° = 4.33 V

$$\text{Resultant voltage, } V = \sqrt{(\text{Horizontal component})^2 + (\text{Vertical component})^2}$$

$$= \sqrt{(12.5)^2 + (4.33)^2}\ \text{V}$$

$$= 13.22\ \text{V} \quad \text{and} \quad \text{angle } \Phi = \tan^{-1}\left(\frac{4.33}{12.5}\right) = 19.1°$$

The resultant voltage equation can be written as $v = 13.22 \sin(\theta + 19.1°)$ V.

EXAMPLE 4.16 If $i_1 = 30 \sin(\omega t + 10°)$ A and $i_2 = 40 \sin(\omega t + 50°)$ A, find $i = i_1 + i_2$. Also find the total power consumed by the circuit if it has a 10-Ω resistance.

Solution The phasor diagram of the currents is shown in Figure 4.18. Resolving the phasors into their horizontal and vertical components, we get

Horizontal component = 30 cos 10° + 40 cos 50° = 55.25 A

Vertical component = 30 sin 10° + 40 sin 50° = 35.85 A

Resultant current,

$$i = \sqrt{(\text{Horizontal component})^2 + (\text{Vertical component})^2}$$

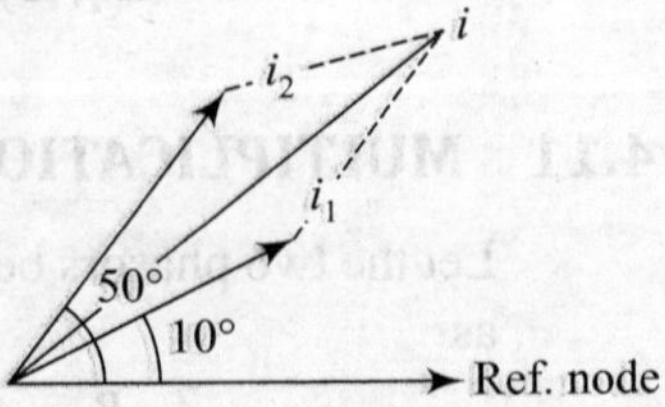

Figure 4.18 Phasor diagram of the currents

$$= \sqrt{(55.25)^2 + (35.85)^2} \text{ V}$$

$$= 65.86 \text{ A} \quad \text{and} \quad \text{angle } \Phi = \tan^{-1}\left(\frac{35.85}{55.25}\right) = 32.97°$$

The resultant current equation can be written as $i = 65.86 \sin(\omega t + 32.97°)$ A.

$$\text{Power loss in the circuit} = I_{\text{rms}}^2 R = \left(\frac{65.86}{\sqrt{2}}\right)^2 \times 10 = 21.687 \text{ kW}$$

EXAMPLE 4.17 Find the difference of two currents given, $i_1 = 25 \sin \omega t$, and $i_2 = 20 \sin(\omega t - 30°)$.

Solution The phasor diagram is shown in Figure 4.19.

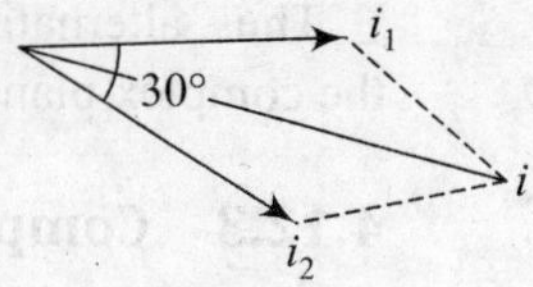

Figure 4.19

$i_1 = 25 \sin \omega t$

$i_2 = 20 \sin(\omega t - 30°)$

The resultant, $i = i_1 - i_2 = 25 \sin \omega t - 20 \sin(\omega t - 30°)$ (1)

Equation (1) can be rewritten as $i = i_1 + (-i_2) = 25 \sin \omega t + (-20 \sin(\omega t - 30°))$. We know that $\sin(\omega t + 180°) = -\sin \omega t$

$i = 25 \sin \omega t + 20 \sin(\omega t + 180° - 30°)$

$i = 25 \sin \omega t + 20 \sin(\omega t + 150°)$

Now, resolving into horizontal and vertical components

Horizontal component = 25 + 20 cos 150° = 7.679 A

Vertical component = 0 + 20 sin 150° = 10 A

$$\text{Resultant voltage } V = \sqrt{(\text{Horizontal component})^2 + (\text{Vertical component})^2}$$

$$= \sqrt{(7.679)^2 + (10)^2} \text{ V}$$

$$= 12.61 \text{ A} \quad \text{and} \quad \text{angle } \Phi = \tan^{-1}\left(\frac{10}{7.679}\right) = 52.47°$$

The resultant current equation can be written as $i = 12.61 \sin(\omega t + 52.47°)$ A.

4.12 VECTOR ANALYSIS

4.12.1 Mathematical Representation of Vectors

Various methods are available to represent vector quantities and to perform operations such as addition and subtraction of vectors analytically rather than doing them graphically.

4.12.2 Significance of "*j*" Operator

The multiplication of any phasors by the operator *j* causes the phasors to rotate through 90° in the counterclockwise direction without affecting its magnitude. Thus, if a phasor *A* is multiplied by the operator *j*, and then *jA* represents a phasor which is at 90° to *A*. When phasor *jA* is again multiplied by the operator *j*, it causes displacement of 180° from the original position of the phasor *A*. The result can be written as

$$j(jA) = j^2 A = -A$$

Hence,

$$j^2 = -1 \quad \text{or} \quad j = \sqrt{-1}$$

Similarly,

$$j^3 A = j(j^2 A) = -jA$$
$$j^4 A = j(j^3 A) = A$$

Figure 4.20 shows the effect of the operator j when applied consequently to phasor A.

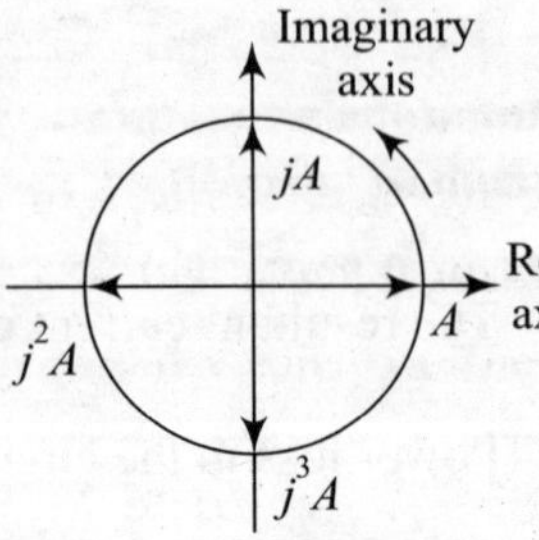

Figure 4.20 Application of the j operator

Thus, alternating voltage and current phasors can be resolved along the real axis and j-axis in the complex plane with the help of the j operator.

4.12.3 Complex Number

A complex number is a number which has one real term and the other imaginary term. For example, $A = 3 + j4$ is a complex number having 3 as the real term and $j4$ as the imaginary term. This complex number can be represented graphically as shown in Figure 4.21 in which the real term forms the horizontal or x-axis component and the imaginary term forms the vertical or y-axis component of the resultant triangle. Nowadays, the real and imaginary terms are also called the positive and negative terms.

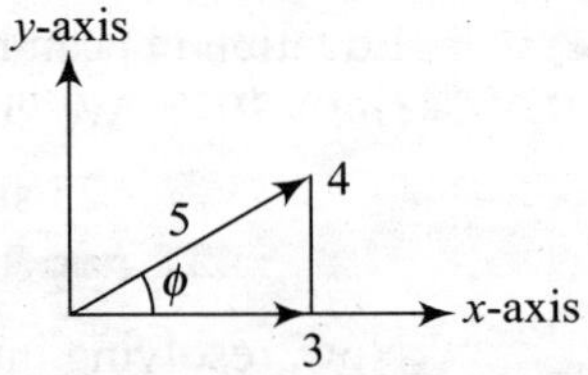

Figure 4.21 Representation of complex numbers

Complex numbers are algebraically represented in the following ways:

$A = a + jb$ (in rectangular or Cartesian coordinates)

$A = A(\cos\phi + j\sin\phi)$ (in trigonometric notation)

$A = |A| \angle \phi$ (in polar coordinates)

4.12.4 Addition and Subtraction of Vectors

Figures 4.22(a) and (b) show two vectors A_1 and A_2 when added and when subtracted, respectively. The vectors are given as

$$A_1 = a_1 + jb_1 \quad \text{and} \quad A_2 = a_2 + jb_2$$

In Section 4.10, it is explained that the resultant of two vectors can be obtained by drawing a parallelogram. The same procedure is adopted to draw Figures 4.22(a) and (b).

Addition

From Figure 4.22(a), it can be seen that a_1 and a_2 are the real (x-axis) components and b_1, b_2 are the imaginary (y-axis) components. From the parallelogram, it is seen that A is the resultant of A_1 and A_2 when added. Therefore,

$$\begin{aligned} A &= a + jb \\ &= (a_1 + a_2) + j(b_1 + b_2) \\ &= (a_1 + jb_1) + (a_2 + jb_2), \quad \text{i.e.,} \quad A = A_1 + A_2 \end{aligned}$$

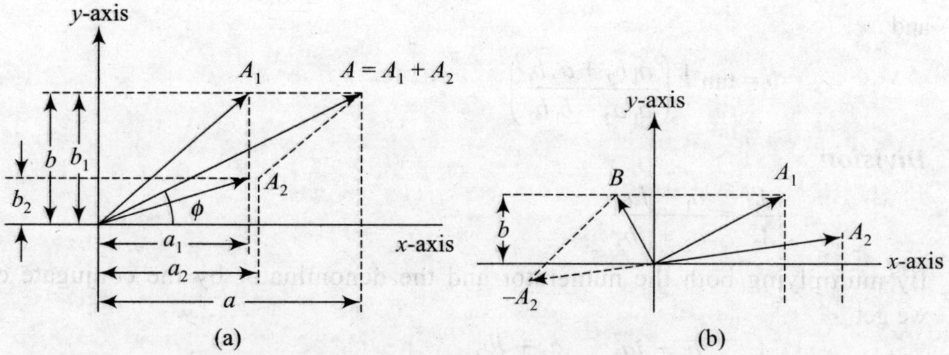

Figure 4.22 (a) Vector addition and (b) subtraction

Here, the real (x-axis) component of $A = a_1 + a_2$, and the imaginary (y-axis) component of $A = b_1 + b_2$. Therefore, the magnitude of

$$A = \sqrt{(a_1 + a_2)^2 + (b_1 + b_2)^2} \quad \text{and} \quad \phi = \tan^{-1}\left(\frac{b_1 + b_2}{a_1 + a_2}\right)$$

Subtraction

From Figure 4.22(b), it can be seen that the difference between the two vectors is given by the resultant vector B, such that

$$A_1 - A_2 = A_1 + (-A_2) = B = -a + jb$$
$$= (a_1 - a_2) + j\,(b_1 - b_2)$$
$$= (a_1 + j\,b_1) - (a_2 + jb_2)$$

From Figure 4.22(b), it is also seen that $(a_1 - a_2)$ is negative and hence a is negative. Therefore, the magnitude of

$$B = A_1 - A_2 = \sqrt{(a_1 - a_2)^2 + (b_1 - b_2)^2}$$

and

$$\phi = \tan^{-1}\left(\frac{b_1 - b_2}{a_1 - a_2}\right)$$

4.12.5 Multiplication and Division of Vectors

Multiplication

Let the two vectors $A_1 = a_1 + jb_1$ and $A_2 = a_2 + jb_2$. Their product

$$A_1 \times A_2 = (a_1 + jb_1) \times (a_2 + jb_2)$$
$$= a_1a_2 + jb_1a_2 + ja_1b_2 + j^2b_1b_2$$
$$|A_1 \times A_2| = (a_1a_2 - b_1b_2)^2 + (a_1b_2 + b_1a_2)^2 \tag{4.29}$$

and

$$\Phi = \tan^{-1}\left(\frac{a_1 b_2 + a_2 b_1}{a_1 a_2 - b_1 b_2}\right) \tag{4.30}$$

Division

$$\frac{A_1}{A_2} = \frac{a_1 + ja_2}{a_2 + jb_2}$$

By multiplying both the numerator and the denominator by the conjugate of $(a_2 + jb_2)$, we get

$$\frac{A_1}{A_2} = \frac{a_1 + ja_2}{a_2 + jb_2} \times \frac{a_2 + jb_2}{a_2 - jb_2}$$

$$= \frac{(a_1a_2 + b_1b_2) + j(b_1a_2 + a_1b_2)}{(a_2^2 + b_2^2)}$$

$$= \frac{(a_1a_2 + b_1b_2)}{(a_2^2 + b_2^2)} + j\frac{(b_1a_2 - a_1b_2)}{(a_2^2 + b_2^2)}$$

Therefore, the magnitude of

$$\frac{A_1}{A_2} = \frac{(a_1a_2 + b_1b_2)}{(a_2^2 + b_2^2)} + j\frac{(b_1a_2 - a_1b_2)}{(a_2^2 + b_2^2)} \tag{4.31}$$

and

$$\Phi = \tan^{-1}\frac{(b_1a_2 - a_1b_2)}{(a_1a_2 + b_1b_2)} \tag{4.32}$$

Equations (4.31) and (4.32) give the result of multiplication and division of two vectors A_1 and A_2 in rectangular coordinates. These calculations are lengthy and time-consuming. However, they become simple and easy if the vectors are represented in polar coordinates.

EXAMPLE 4.18 Find the sum of two currents, $i_1(t) = 5 \sin(\omega t + 20°)$ and $i_2(t) = 6 \sin(\omega t - 10°)$.

Solution First, we convert the given sinusoids into phasors. Since they have the same frequency, we can add both.

Current $i_1(t)$ is expressed as $I_1 = 5 \angle 20°$ A

Current $i_2(t)$ is expressed as $I_2 = 6 \angle{-10°}$ A

Expressing these phasors in rectangular form $I_1 = 4.69 + j1.71$; $I_2 = 5.91 - j1.04$

$$I_1 + I_2 = 4.69 + j1.71 + 5.91 - j1.04 = 10.6 + j0.67 = 10.6 \angle 3.62° \text{A}$$

Writing it as sinusoid $i(t) = 10.6 \sin(\omega t + 3.62°)$ A.

EXAMPLE 4.19 If $v_1 = 20 \sin(20t + 30°)$ V and $v_2 = 30 \sin(20t - 60°)$ V, find $v = v_1 - v_2$.

Solution $V_1 = 20 \angle 30°$ V

$V_2 = 30 \angle{-60°}$ V

$V = V_1 - V_2 = 20 \angle 30° - 30 \angle{-60°}$

$= 17.32 + j10 - 15 + j25.98$

$= 2.32 + j35.98$

$= 36.05 \angle 86.31°$ volts

Writing it into time domain $v(t) = 36.05 \sin (20t + 86.31°)$ volts.

EXAMPLE 4.20 **Evaluate the complex numbers given below.**

(i) $[30 \angle 50° + 20 \angle -20°]^{0.5}$

(ii) $\dfrac{10\angle -40° + (3 + j6)}{(5 + j7)(4 - j8)^*}$

Solution

(i) Converting the polar form into rectangular form

$$30 \angle 50° = 19.28 + j22.98$$

$$20 \angle -20° = 18.79 - j6.84$$

$$30 \angle 50° + 20 \angle -20° = 19.28 + j22.98 + 18.79 - j6.84$$

$$= 38.07 + j16.14$$

$$= 41.3520 \angle 22.97°$$

$$[30 \angle 50° + 20 \angle -20°]^{0.5} = \sqrt{41.35} \angle \frac{22.97}{2}$$

$$= 6.43 \angle 11.48$$

(ii) $$\frac{10 \angle -40° + (3 + j6)}{(5 + j7)(4 - j8)^*} = \frac{7.66 - j6.42 + 3 + j6}{(5 + j7)(4 + j8)}$$

$$= \frac{7.66 - j6.42 + 3 + j6}{(8.6 \angle 54.46°)(8.9 \angle 63.43°)}$$

$$= \frac{10.66 - j0.42}{(76.54 \angle 117.89°)}$$

$$= \frac{10.66 \angle -2.25°}{(76.54 \angle 117.89°)}$$

$$= 0.139 \angle -120.14°$$

EXAMPLE 4.21 **Calculate the phase angle between $i_1 = -10 \sin (\omega t + 60°)$ and $i_2 = 20 \sin (\omega t - 20°)$. State which sinusoid is leading?**

Solution Let us calculate the phase angle in two ways. The first method uses trigonometric identities while the second method uses the graphical approach.

Method 1: In order to compare i_1 and i_2, we must express them in the same form.

$$i_1 = -10 \sin (\omega t + 60°) = 10 \sin (\omega t + 60° + 180°) = 10 \sin (\omega t + 240°)$$

$$i_2 = 20 \sin (\omega t - 20°) = 20 \sin (\omega t + 340°)$$

Comparing the two equations, the phase angle between i_1 and i_2 is 100°. Therefore, i_1 lags i_2 by 100° or in other words i_2 leads i_1 by 100°.

Method 2: The phasor representation of the two currents is shown in Figure 4.23.

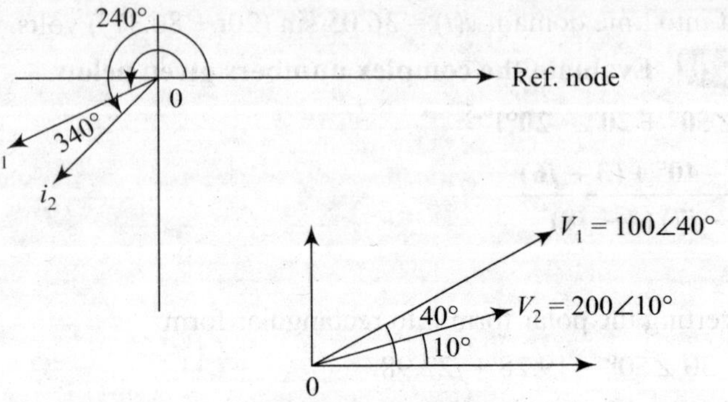

Figure 4.23

From the phasor diagram it is observed that i_2 leads i_1 by 100°.

EXAMPLE 4.22 **Find the sinusoids corresponding to these phasors.**

(i) $I = -20\angle 60°$

(ii) $V = j(4 - j10)$

Solution

(i) $I = -20\angle 60°$

$$i = -20 \sin(\omega t + 60°)$$
$$= 20 \sin(\omega t + 60° + 180°)$$
$$= 20 \sin(\omega t + 240°)$$

(ii) $V = j(4 - j10)$

$$= 10 + j4 = 10.77\angle 21.8° \text{ volts}$$

Converting it into time domain $v(t) = 10.77 \sin(\omega t + 21.8°)$ volts.

EXAMPLE 4.23 **The instantaneous values of three currents are $i_1 = 10 \sin \omega t$, $i_2 = 20 \sin(\omega t + 30°)$, and $i_3 = 30 \sin(\omega t - 30°)$. Determine the resultant current $i = i_1 + i_2 - i_3$.**

Solution Given data

$$i_1 = 10 \sin \omega t, \quad i_2 = 20 \sin(\omega t + 30°), \text{ and } i_3 = 30 \sin(\omega t - 30°)$$

Resolving these currents into horizontal and vertical components

Horizontal component = 10 + 20 cos (30°) – 30 cos (–30°) = 1.34

Vertical component = 0 + 20 sin (30) – 20 sin (–30) = 20

The resultant current, $I = \sqrt{1.34^2 + 20^2} = 20.04$ A

Angle $\Phi = \tan^{-1}\left(\dfrac{20}{1.34}\right) = 86.16°$

The resultant current is given by $i = 20.04 \sin(\omega t + 86.16°)$ A.

EXAMPLE 4.24 **Calculate the maximum value of emf generated in a coil which is rotating at a speed of 100 rps in a uniform magnetic field of 0.6 Wb/m². The coil is wound on a square frame having sides of 5 cm length and wound with 400 turns.**

Solution The maximum value of emf generated is given by $E_m = 2\pi BAnT$ V.

Given

$$A = 5 \times 5 = 25 \text{ cm}^2 = 25 \times 10^{-4} \text{ m}^2$$

$$B = 0.6 \text{ Wb/m}^2$$

$$T = 400$$

$$n = 100 \text{ rps}$$

Therefore, $E_m = 2 \times \pi \times 0.6 \times 25 \times 10^{-4} \times 100 \times 400 = 376.99$ V.

EXAMPLE 4.25 **Represent the following voltages by phasors $v_1 = 100 \sin(\omega t + 40°)$, and $v_2 = 200 \sin(\omega t + 10°)$.**

Solution The given voltages are represented in phasor form as

$$V_1 = 100 \angle 40° \quad \text{and} \quad V_2 = 200 \angle 10°$$

These phasors are also represented through the phasor diagram as shown in Figure 4.24.

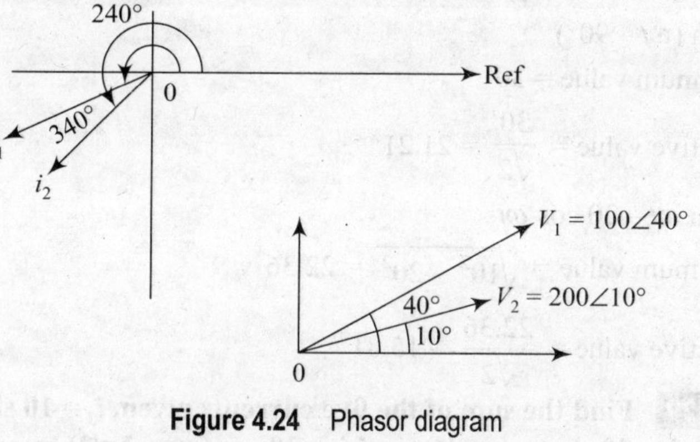

Figure 4.24 Phasor diagram

EXAMPLE 4.26 **Two alternating voltages v_1 and v_2 of frequency 50 Hz have maximum values of 30 V and 40 V, respectively, and v_1 leads v_2 by 30°. Write the equations for v_1 and v_2 and show the voltages graphically and vectorially.**

Solution The equations for v_1 and v_2 are $v_1 = 30 \sin(314t)$ and $v_2 = 40 \sin(314t - 30°)$. The graphical and vector representations of the voltages are shown in Figures 4.25(a) and (b).

EXAMPLE 4.27 **Find the rms value of the current $i(t) = 20 + 20 \sin \omega t$.**

Solution $$I_{rms} = \sqrt{20^2 + \left(\frac{20}{\sqrt{2}}\right)^2} = 24.49 \text{ A}$$

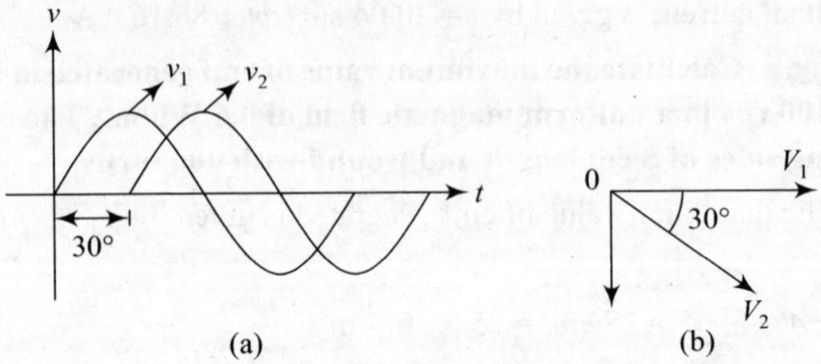

Figure 4.25 (a) Graphical and (b) vector representation

EXAMPLE 4.28 **Calculate the maximum value and the effective value for the following quantities:**

(i) 20 sin ωt, (ii) 30 sin (ωt – 90°), and (iii) 10 sin ωt – 20 cos ωt

Solution

(i) $20 \sin \omega t$

$$\text{Maximum value} = 20$$

$$\text{Effective value} = \frac{20}{\sqrt{2}} = 14.14$$

(ii) $30 \sin (\omega t - 90°)$

$$\text{Maximum value} = 30$$

$$\text{Effective value} = \frac{30}{\sqrt{2}} = 21.21$$

(iii) $10 \sin \omega t - 20 \cos \omega t$

$$\text{Maximum value} = \sqrt{10^2 + 20^2} = 22.36$$

$$\text{Effective value} = \frac{22.36}{\sqrt{2}} = 15.81$$

EXAMPLE 4.29 **Find the sum of the five currents given, $i_1 = 10 \sin \omega t$, $i_2 = 20 \sin (\omega t + \pi/6)$, $i_3 = 5 \cos \omega t$, $i_4 = 20 \sin (\omega t - \pi/4)$, and $i_5 = 30 \cos (\omega t + 2\pi/3)$ by expressing them vectorially. Express the resultant in the form $i = A \sin (\omega t + \phi)$.**

Solution Converting the given sinusoids into phasors

$$i_1 = 10 \sin \omega t = 10 \angle 0°$$

$$i_2 = 20 \sin (\omega t + \frac{\pi}{4}) = 20 \angle 30°$$

$$i_3 = 5 \cos \omega t = 5 \sin (\omega t + 90°) = 5 \angle 90°$$

$$i_4 = 20 \sin (\omega t - \frac{\pi}{4}) = 20 \angle -45°$$

$$i_5 = 30 \cos (\omega t + \frac{2\pi}{3}) = 30 \angle 90° + 120° = 30 \angle 210°$$

Converting phasors into rectangular form and adding

$$i = i_1 + i_2 + i_3 + i_4 + i_5 = 10 + j0 + 17.32 + j10 + j5 + 14.14 - j14.14 - 25.98 - j15$$
$$= 15.48 - j14.14$$
$$= 20.96 \angle -42.41° \text{ A}$$

Current $i = 20.96 \sin(\omega t - 42.41°)$ A.

EXAMPLE 4.30 **Compute the average and rms values of the voltage waveform shown in Figure 4.26.**

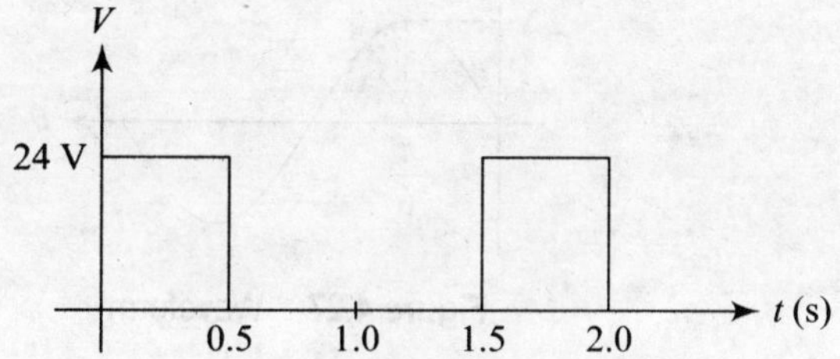

Figure 4.26

Solution For $0 < t < 0.5$, $v = 24$ V and for $0.5 < t < 1.5$, $v = 0$ V, and time period $T = 1.5$ s.

Average value

$$V_{avg} = \frac{1}{1.5}\int_0^{0.5} 24\, dt$$
$$= \frac{1}{1.5} \times 24\, t\Big|_0^{0.5}$$
$$= 8 \text{ V}$$

rms value

$$V_{rms} = \sqrt{\frac{1}{1.5}\int_0^{0.5} 24^2\, dt}$$
$$= \sqrt{\frac{1}{1.5} \times 24^2\, t\Big|_0^{0.5}}$$
$$= \sqrt{192}$$
$$= 13.85 \text{ V}$$

EXAMPLE 4.31 **An alternating current of frequency 50 Hz has a maximum value of 100 A. Calculate its value 1/600 s after the instant of current zero and its value decreasing there afterwards.**

Solution $f = 50$ Hz, $T = \dfrac{1}{50}$, and $\dfrac{T}{2} = \dfrac{1}{100}$

$$\text{Total time} = \frac{1}{100} + \frac{1}{600} = 0.01167$$

$$i = 100 \sin 100 \times 180° \times 0.01167$$
$$= 100 \sin 210°$$
$$= -50 \text{ A}$$

EXAMPLE 4.32 **For the given waveform in Figure 4.27, obtain the equation.**

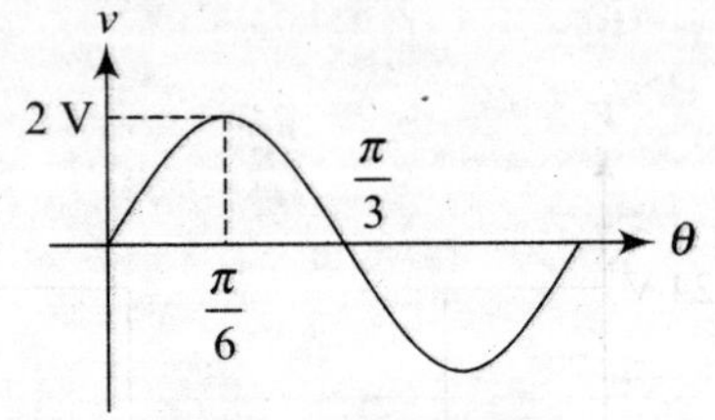

Figure 4.27 Waveform

Solution $v = V_m \sin \omega t = V_m \sin \theta$

When $x = \pi/3$, function is zero.

$$v = 0 = V_m \sin \pi$$

Therefore, $\theta = \pi$. But $x = \pi/3$, $v = 0$. Therefore, $\theta = \pi = 3x$ and $V_m = 2$. The equation is $v = 2 \sin 3x$.

EXAMPLE 4.33 **For the square wave shown in Figure 4.28, find the rms value.**

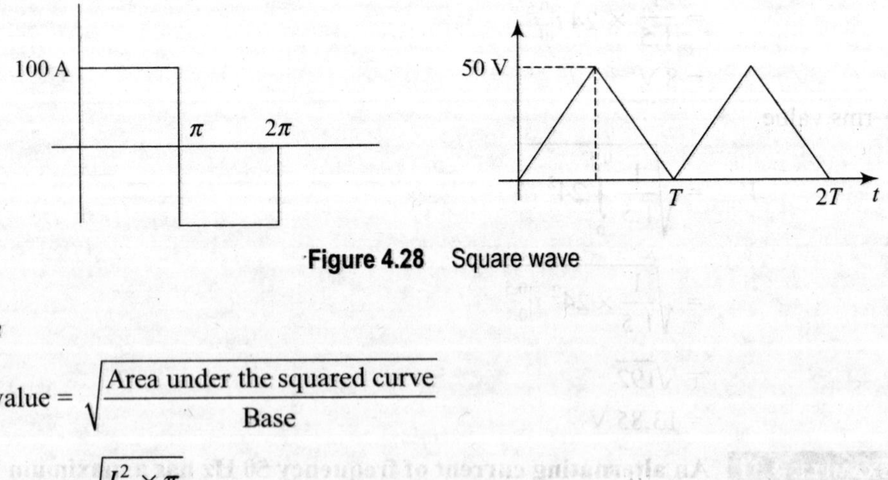

Figure 4.28 Square wave

Solution

$$\text{rms value} = \sqrt{\frac{\text{Area under the squared curve}}{\text{Base}}}$$

$$= \sqrt{\frac{I_m^2 \times \pi}{\pi}}$$

$$= I_m = 100 \text{ A}$$

EXAMPLE 4.34 **Find the average value of the waveform given in Figure 4.29.**

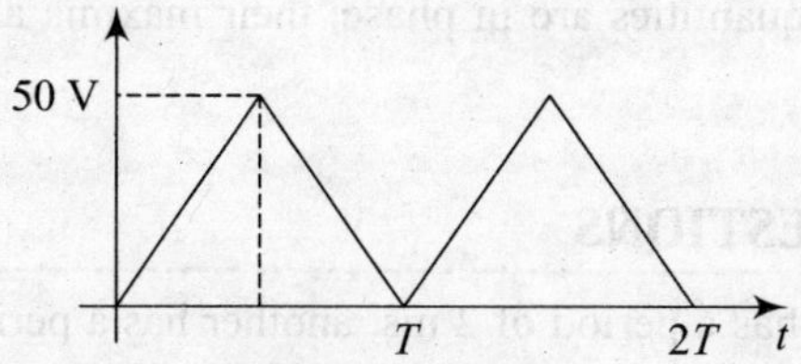

Figure 4.29

Solution $\text{Average value} = \dfrac{\text{Area}}{\text{Base}}$

$$\text{Area} = \frac{1}{2}bh$$

$$= \frac{1}{2} \times T \times 50 = 25\,T$$

$$\text{Base} = T$$

$$\text{Average value} = \frac{25\,T}{T} = 25\text{ V}$$

CHAPTER AT A GLANCE

- The total emf induced in the coil is $E = 2Blu \sin \theta$ volts.
- The maximum value-generated emf is given by $E_m = 2\pi BAnT$ volts.
- Instantaneous value is the value of voltage or current obtained at any instant of time.
- Cycle is one complete set of positive and negative values of an alternating quantity.
- Time period is the time taken by an alternating quantity to complete one cycle.
- Amplitude is the maximum value of voltage or current whether positive or negative.
- Angular velocity, $\omega = \dfrac{\text{Variation in angle in one cycle}}{\text{Time taken to complete a cycle}}$.
- Relation between frequency, speed, and number of poles $f = PN/120$.
- Average value of sinusoidal voltage $V_{avg} = 2V_m/\pi$.
- The rms value of sinusoidal voltage $V_{rms} = V_m/\sqrt{2}$.
- Form factor of sinusoidal wave is 1.11.
- Peak factor of sinusoidal wave is 1.414.
- A phasor is a complex number that represents the amplitude and phase of a sinusoid.
- A complex number can be written either in polar form or in rectangular form.
- Addition and subtraction of complex numbers are better performed in rectangular form and multiplication and division are better done in polar form.
- The graphical representation of phasors is known as a phasor diagram.

- Phasor analysis applies only when frequency is constant.
- When the two quantities are in phase, their maxima and minima reach at exactly the same time.

MULTIPLE CHOICE QUESTIONS

1. One sine wave has a period of 2 ms, another has a period of 5 ms, and other has a period of 10 ms, which sine wave is changing at a faster rate?
   (a) Sine wave with period 2 ms (b) Sine wave with period of 5 ms
   (c) All are at the same rate (d) Sine wave with period of 10 ms
2. How many cycles does a sine wave go through in 10 s when its frequency is 60 Hz?
   (a) 10 cycles (b) 60 cycles (c) 600 cycles (d) 6 cycles
3. If the peak value of a certain sine wave voltage is 5 V, what is the peak-to-peak value?
   (a) 20 V (b) 10 V (c) 5 V (d) 7.07 V
4. An ac current is given as $i = 10 + 10 \sin 314\, t$. The rms value of the current is
   (a) 17.07 A (b) 12.25 A (c) 16.36 A (d) 14.14 A
5. What is the average value of sine wave over a full cycle?
   (a) V_m (b) $V_m/\sqrt{2}$ (c) Zero (d) $\sqrt{2}\, V_m$
6. A sinusoidal current has peak value of 12 A. What is its average value?
   (a) 7.64 A (b) 24 A (c) 8.84 A (d) 12 A
7. Sine wave A has a positive-going zero crossing at 30°. Sine wave B has a positive-going zero crossing at 45°. What is the phase angle between two signals?
   (a) 30° (b) 45° (c) 75° (d) 15°
8. A sine wave A has a positive-going zero crossing at 0° and an rms value of 20 V. What is its instantaneous value at 145°?
   (a) 7.32 V (b) 16.22 V (c) 26.57 V (d) 21.66 V
9. A sine wave voltage is applied across a capacitor; when the frequency of the voltage is increased, the current
   (a) increases (b) decreases
   (c) remains the same (d) becomes zero
10. In SI system the frequency is expressed as
    (a) Hz (b) Hz/s (c) cycles (d) cycles/s
11. The standard supply frequency in India is
    (a) 25 Hz (b) 50 Hz (c) 60 Hz (d) 100 Hz
12. A current is said to be alternating when it changes in
    (a) magnitude only (b) direction only
    (c) both magnitude and direction (d) none
13. An alternating current of frequency 50 Hz and maximum value of 100 A is given as
    (a) $i = 100 \sin 639t$ (b) $i = 141.4 \sin 157t$
    (c) $i = 141.4 \sin 314t$ (d) $i = 100 \sin 314t$

14. A current is said to be dc current when its
(a) magnitude remains constant with time
(b) direction changes with time
(c) magnitude changes with time
(d) magnitude and direction change with time

15. A sinusoidal voltage varies from zero to maximum of 200 V. The voltage at the instant of 30 degrees will be
(a) 200 V (b) 100 V (c) 141.42 V (d) 150 V

16. An ac generator running at 1500 rpm produces emf of 50 Hz. The number of poles on the generator is
(a) 6 (b) 8 (c) 2 (d) 4

17. The time period of an alternating quantity is 0.04 s. Its frequency will be
(a) 25 Hz (b) 50 Hz (c) 60 Hz (d) 0.04 Hz

18. An ac voltage is given as $v = 10 + 10 \sin 314\, t$. The average and rms values of the voltage are
(a) 16.36 A, 17.07 A (b) 10 A, 12.25 A
(c) 10 A, 17.07 A (d) 10 A, 16.36 A

19. An electric heater is rated as 230 V, 500 W ac. The value of 230 V refers to
(a) rms value (b) peak value (c) average value (d) peak-to-peak value

20. The phase difference between the two waveforms can be compared only when they
(a) have the same peak value (b) have the same rms value
(c) have the same frequency (d) are sinusoidal

SHORT ANSWER QUESTIONS

1. Define form factor?
2. Define peak factor?
3. When do you say two quantities are in phase with each other?
4. A constant current of 3 A is flowing through a resistor. What is the rms value of the current?
5. What is the generated emf frequency for p-pole machine running at a speed of N rpm.
6. Define time period.
7. What is the form factor for a sine wave?
8. What is the form factor of dc voltage?
9. A 60-Hz power line voltage of 120 V is applied across a resistance of 10 Ω. What is the frequency of current and rms value of current?
10. For Question 9, what is the phase angle between voltage and current?
11. What is the numerical value of j?
12. What is the value of j^2?
13. What is the conjugate of $-a - jb$?
14. What is the product of $a + jb$ and $c - jd$?

15. What is the product of $a \angle \alpha$ and $b \angle \beta$?
16. What is the division of $a \angle \alpha$ and $b \angle \beta$?
17. What is the importance of j?
18. What is the equivalent of $ae^{j\theta}$?
19. If phasor $c = a + jb$ in rectangular form, what is its equivalent in phasor form?
20. Two sinusoidal currents are given as $i_1 = 100 \sin(\omega t + \pi/4)$ and $i_2 = 50 \sin(\omega t - \pi/6)$. What is the phase difference between two quantities?

REVIEW QUESTIONS

1. Explain the advantages of ac over dc.
2. Explain the generation of an alternating voltage.
3. Define (i) instantaneous value, (ii) time period, (iii) frequency, (iv) cycle, and (v) amplitude.
4. Explain the concept of average value and rms value.
5. Define (i) form factor and (ii) peak factor.
6. Explain how an alternating quantity is represented by a phasor.

EXERCISE PROBLEMS

1. Calculate the frequency of the following values of period.
(a) 0.2 s (b) 50 ms (c) 500 μs
2. Calculate the period for each of the values of frequency.
(a) 60 Hz (b) 500 MHz (c) 1 kHz
3. A certain sine wave has a positive-going zero crossing at 0° and an rms value of 20 V. Calculate its instantaneous value at each of the following angles.
(a) 33° (b) 110° (c) 145° (d) 325°
4. For a particular 0° reference sinusoidal current, the peak value is 200 mA; determine the instantaneous value at each of the following.
(a) 35° (b) 190° (c) 200° (d) 360°
5. Sine wave A lags sine wave B by 30°. Both have peak value of 15 V. Sine wave A is the reference with a positive-going crossing 0°. Determine the instantaneous value of sine wave B at 30°, 90°, 45°, 180°, and 300°.
6. Find the average values of the voltage across R_1 and R_2. In Figure 4.30, values shown are rms.

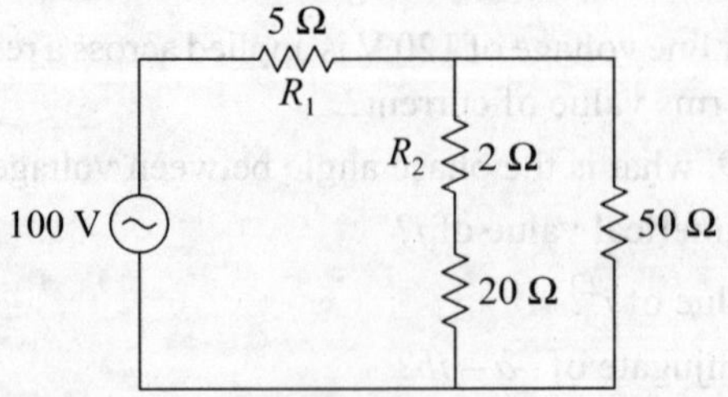

Figure 4.30

7. A sinusoidal voltage is applied to the circuit shown in Figure 4.31, determine rms current, average current, peak current, and peak-to-peak current.

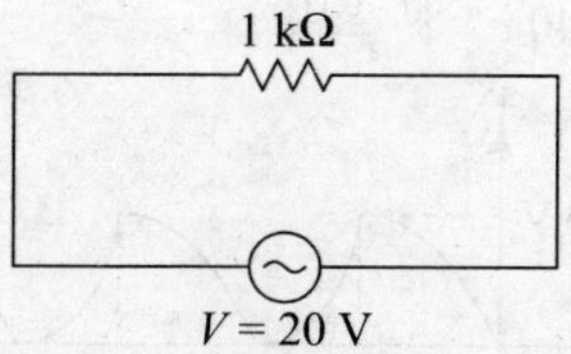

Figure 4.31

8. A sinusoidal voltage of $v(t) = 50 \sin 500t$ applied to a capacitive circuit. Determine the capacitive reactance and the current in the circuit.

9. A sinusoidal voltage source in series with a dc source is shown in Figure 4.32.

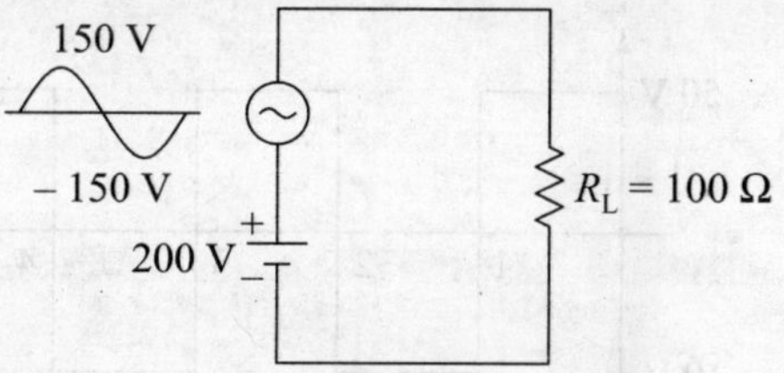

Figure 4.32

Sketch the voltage across R_L. Determine the maximum current through R_L and the average voltage across R_L.

10. Find the effective value of the resultant current in a wire which carries a direct current of 10 A and a sinusoidal current with a peak value of 15 A.

11. An alternating current varying sinusoidally, with a frequency of 50 Hz, has an rms value of 20 A. Write down the equation for the instantaneous value and find this value at (a) 0.0025 s and (b) 0.0125 s after passing through a positive maximum value. (c) At what time, measured from a positive maximum value, will the instantaneous current be 14.14 A?

12. Determine the rms value of the voltage defined by $V = 5 + \sin(314t + \pi/6)$.

13. Find the effective value of the function $V = 100 + 50 \sin \omega t$.

14. A full-wave rectified sine wave is clipped at 0.707 and its maximum value is shown in Figure 4.33. Find the average and effective values of the function.

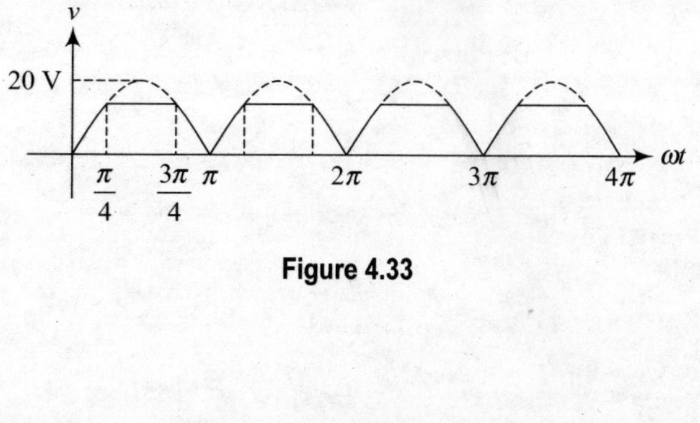

Figure 4.33

15. Find the rms value of the function shown in Figure 4.34 and described as follows:

$0 < t < 0.1 \; V = 40\,(1 - e^{-100t})$

$0 < t < 0.1 \; V = 40\, e^{-50(t-0.1)}$

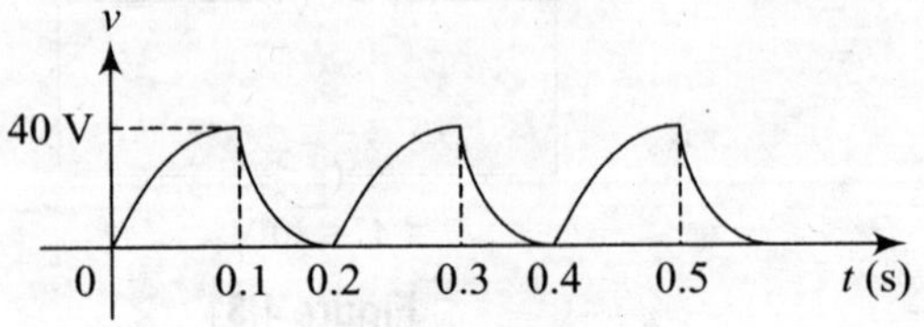

Figure 4.34

16. Calculate the average and effective values of the waveform shown in Figure 4.35. Hence, find the form factor.

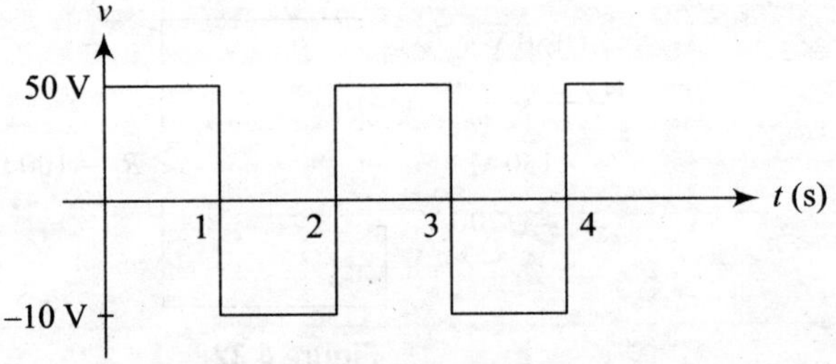

Figure 4.35

17. A full-wave rectified sine wave is clipped such that the effective value is 0.5 V_m as shown in Figure 4.36. Determine the angle at which the wave form is clipped.

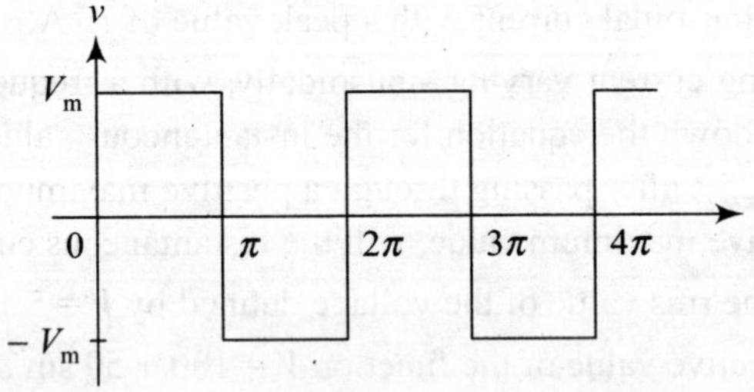

Figure 4.36

Chapter

5

AC Circuit Analysis

Chapter Objectives

- To discuss ac circuits
- To explain the power calculations in ac circuits
- To discuss series and parallel ac circuits
- To explain the concepts of series and parallel resonance
- To discuss the network theorems

5.1 INTRODUCTION

For the sake of easy understanding we have limited our discussion of circuit analysis to dc sources only till now. We now begin the analysis of circuit with time-varying source voltage and current in particular sinusoidally time-varying excitation. In an electric circuit, the direct current (abbreviated as dc) flows continuously in one direction only. But an alternating current (abbreviated as ac) reverses at regular time intervals and has alternatively positive and negative values. The circuits driven by sinusoidal voltage or current sources are called *ac circuits*. Figure 5.1 shows the different waveforms for commercial supplies.

At present a large percentage of the electrical energy being used for residential, industrial, and commercial purpose is ac.

The following are the advantages of ac when compared to dc:

(i) Alternating currents can be generated effectively and distributed over long distances with small power losses.

(ii) The demand of ac is increased after the invention of transformer and induction motor. An ac motor requires less attention in operation and maintenance.

(iii) Many circuits in radios and TVs require the use of ac for their operation which is generated internally by circuits called *oscillators*.

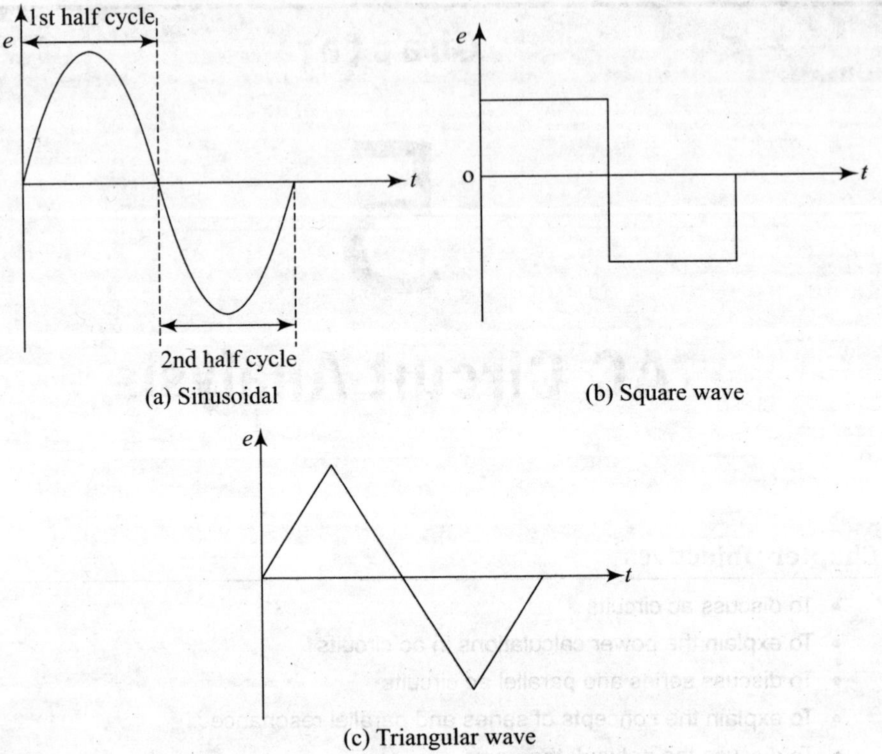

(a) Sinusoidal (b) Square wave

(c) Triangular wave

Figure 5.1 Different alternating waveforms

(iv) Alternating circuits are easily converted to dc, where necessary. But dc cannot be converted to ac so easily.

5.2 SINGLE-PHASE AC CIRCUITS

The single-phase ac circuits containing a pure resistor, pure inductor, and pure capacitor are discussed in the following subsections.

5.2.1 AC Through Pure Resistance

Consider the circuit shown in Figure 5.2(a) and the voltage applied is $v(t) = V_m \sin \omega t$. Let the current be $i(t)$. According to Ohm's law, we can write

$$v(t) = R\, i(t) \tag{5.1}$$

This implies $V_m \sin \omega t = R\, I(t)$. As a result,

$$I(t) = \frac{v(t)}{R} = \frac{V_m \sin \omega t}{R}$$

$$= I_m \sin \omega t \quad \left(\text{where } I_m = \frac{V_m}{R} \right)$$

The time-varying current is given by

$$i(t) = I_m \sin \omega t \tag{5.2}$$

The voltage and current waveforms across a resistor are shown in Figure 5.2(b). From Figure 5.2(b), it is observed that for pure resistor the voltage and current will be in phase with each other. Figure 5.2(c) shows the phasor representation of voltage and current in a resistor. Since both of them are in phase with each other the phasor representation can be seen as a straight line.

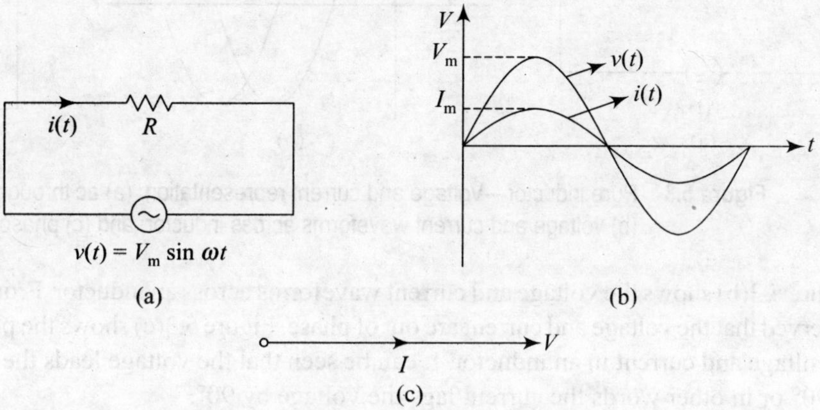

Figure 5.2 Pure resistor—Voltage and current representation: (a) ac through pure resistor; (b) voltage and current waveforms; and (c) phasor diagram

5.2.2 AC Through Pure Inductance

Consider the circuit shown in Figure 5.3(a). Let the current $i(t) = I_m \sin \omega t$ is flowing through the inductor. We know the relationship between voltage and current of an inductor and is given by

$$v(t) = L\frac{di(t)}{dt} \tag{5.3}$$

$$= L\frac{d(I_m \sin \omega t)}{dt}$$

$$= LI_m \cos \omega t$$

$$= wI_m L \cos \omega t$$

$$= wLI_m \sin\left(\omega t + \frac{\pi}{2}\right)$$

$$= X_L I_m \sin\left(\omega t + \frac{\pi}{2}\right) \quad (\text{where } X_L = \omega L)$$

The voltage across the inductor is given by

$$v(t) = X_L I_m \sin\left(\omega t + \frac{\pi}{2}\right) \tag{5.4}$$

Converting the time-domain equation into phasor-domain equation

$$V = X_L I \angle 90° \tag{5.5}$$

In terms of maximum values, $V_m = X_L I_m$. From Eq. (5.5), it is observed that the voltage across the inductor leads the current by +90°, or in other words, the current lags the voltage by 90° in the inductor.

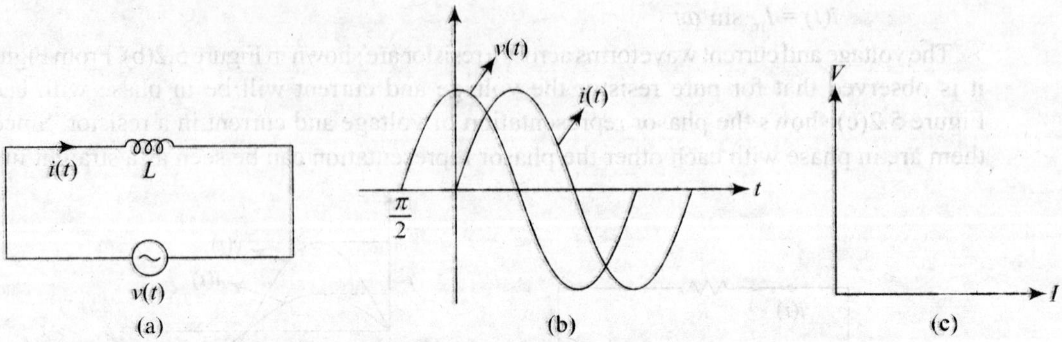

Figure 5.3 Pure inductor—Voltage and current representation: (a) ac through pure inductor; (b) voltage and current waveforms across inductor; and (c) phasor diagram

Figure 5.3(b) shows the voltage and current waveforms across an inductor. From Figure 5.3(b), it is observed that the voltage and current are out of phase. Figure 5.3(c) shows the phasor representation of voltage and current in an inductor. It can be seen that the voltage leads the current by an angle of 90° or in other words the current lags the voltage by 90°.

5.2.3 AC Through Pure Capacitance

Consider the circuit shown in Figure 5.4(a). Let the current $i(t) = I_m \sin \omega t$ is flowing through the capacitor. The voltage across the capacitor is given by

$$v(t) = \frac{1}{C}\int i(t)\,dt \tag{5.6}$$

$$= \frac{1}{C}\int I_m \sin \omega t\,dt$$

$$= \frac{I_m}{C}\,\frac{-\cos \omega t}{\omega}$$

$$= -\frac{I_m}{\omega C}\cos \omega t$$

$$= \frac{I_m}{\omega C}\sin\left(\omega t - \frac{\pi}{2}\right)$$

$$= X_C I_m \sin\left(\omega t - \frac{\pi}{2}\right) \quad (\text{where } X_C = 1/\omega C)$$

The voltage across the capacitor in time domain is given by

$$v(t) = X_C I_m \sin\left(\omega t - \frac{\pi}{2}\right) \tag{5.7}$$

Converting the time-domain quantities into phasor domain, we get

$$V = X_C I \angle -90° \tag{5.8}$$

In terms of maximum values, $V_m = X_C I_m$

From Eq. (5.8), it is observed that the voltage lags the current by 90° or in other words the current leads the voltage by 90°.

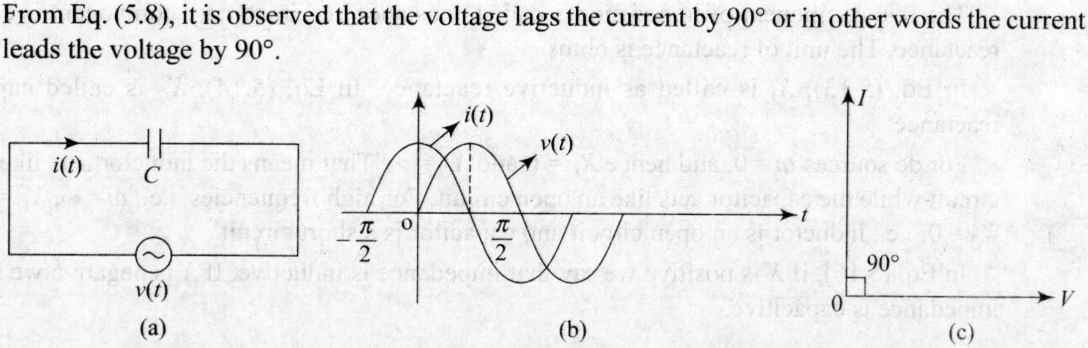

Figure 5.4 Pure capacitor—Voltage and current representation: (a) ac through pure capacitor; (b) voltage and current waveform; and (c) phasor diagram

Figure 5.4(b) shows the voltage and current waveforms across a capacitor. From Figure 5.4(b), it is observed that for capacitor also the voltage and current are out of phase. Figure 5.4(c) shows the phasor representation of voltage and current in a capacitor. It can be seen that the voltage lags the current by an angle of 90°. In other words, the current leads the voltage by an angle of 90°.

5.2.4 Impedance

The current and voltage relationships in phasor domain for all the three passive elements are given by

$$V = RI \tag{5.9}$$

$$V = j\omega LI \tag{5.10}$$

$$V = \frac{I}{j\omega C} \tag{5.11}$$

If these equations are written in phasor voltage and phasor current ratios, then

$$\frac{V}{I} = R \tag{5.12}$$

$$\frac{V}{I} = j\omega L = X_{\text{L}} \tag{5.13}$$

$$\frac{V}{I} = \frac{1}{j\omega C} = X_{\text{C}} \tag{5.14}$$

Impedance is defined as the ratio of the phasor voltage to the phasor current and indicated by symbol Z. In other words, the impedance is the opposition offered to the flow of sinusoidal current in the circuit. The unit of impedance is ohms. Impedance is not a phasor and cannot be transformed to the time domain. Impedances may be combined in series and parallel by the same rules that we have seen for resistors.

The impedance Z may be expressed either in polar form or in rectangular form. In general, the impedance in polar form is expressed as

$$Z = |Z| \angle\theta \tag{5.15}$$

and in rectangular form

$$Z = R + jX \tag{5.16}$$

The real component of impedance is called resistance and imaginary component is called as reactance. The unit of reactance is ohms.

In Eq. (5.13), X_L is called as inductive reactance. In Eq. (5.14), X_C is called capacitive reactance.

For dc sources $\omega = 0$, and hence $X_L = 0$ and $X_C = \infty$. That means the inductor acts like a short circuit while the capacitor acts like an open circuit. For high frequencies, i.e., $\omega = \infty$, $X_L = \infty$, and $X_C = 0$, i.e., inductor is an open circuit and capacitor is a short circuit.

In Eq. (5.16), if X is positive we say that impedance is inductive. If X is negative we say that impedance is capacitive.

5.2.5 Admittance

Admittance is defined as the ratio of phasor current to phasor voltage. It is denoted by Y. Its unit is given by siemens.

$$Y = \frac{1}{Z} \tag{5.17}$$

$$Y = G + jB \tag{5.18}$$

The real part of admittance is called *conductance* (G) and the imaginary part of admittance is called *susceptance* (B).

EXAMPLE 5.1 The voltage $v(t) = 10 \sin(50t + 30°)$ V is applied to a 0.2-H inductor. Find the current through the inductor.

Solution For inductor $V = X_L I$

Representing the voltage in phasor form, $V = 10\angle 30°$ V

$$\text{Current, } I = \frac{V}{X_L} = \frac{V}{j\omega L} = \frac{10\angle 30°}{j \times 50 \times 0.2} = \frac{10\angle 30°}{j10} = \frac{10\angle 30°}{10\angle 90°} = 1\angle{-60°}\text{ A}$$

EXAMPLE 5.2 If voltage $v(t) = 5 \sin(60t - 60°)$ V is applied to a 20-μF capacitor, calculate the current through the capacitor.

Solution Representing the voltage in phasor $V = 5\angle{-60°}$ V

Current through the capacitor is given by $I = j\omega CV$

$$= j \times 60 \times 20 \times 10^{-6} \times 5\angle{-60°}$$

$$= 6\angle 30°\text{ mA}$$

EXAMPLE 5.3 Find the equivalent impedance of the circuit shown in Figure 5.5 and assume that the circuit operates at $\omega = 50$ rad/s.

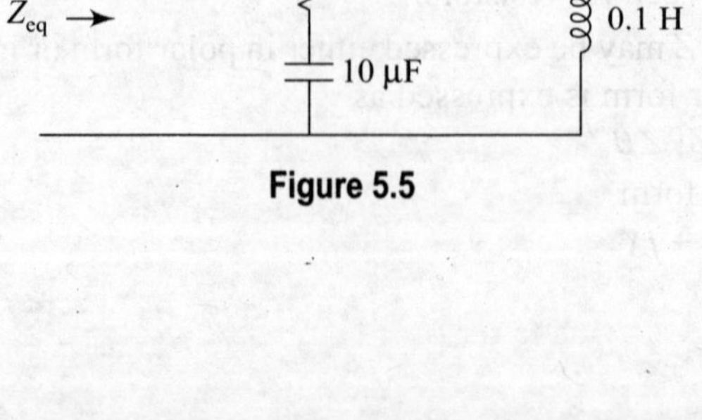

Figure 5.5

Solution Converting all the inductance and capacitance values into inductive and capacitive reactances, we get

$$Z_1 = 3 + jX_L = 3 + j \times 50 \times 0.1 = 3 + j5 = 5.83\ \angle 59.03°\ \Omega$$

$$Z_2 = 2 - jX_C = 2 - \frac{j}{50 \times 10 \times 10^{-6}} = 2 - j2000 = 2000.0\ \angle{-89.94}\ \Omega$$

Let Z_3 be a parallel combination of Z_1 and Z_2.

$$\therefore \quad Z_3 = \frac{Z_1 Z_2}{Z_1 + Z_2} = \frac{5.83\ \angle 59.03° \times 2000\ \angle{-89.94}}{3 + j5 + 2 - j2000} = 5.8\ \angle 58.9°\ \Omega$$

Z_3 is in series with the 5-Ω resistor. The equivalent impedance $Z_{eq} = 5 + 5.8\ \angle 58.9° = 9.4\ \angle 31.8°\ \Omega$.

5.3 POWER IN AC CIRCUITS

Generally, all the energy supplied by the source will be dissipated in the form of heat in a pure resistor. However, in a purely inductive and capacitive circuit, the energy will be stored in the form of magnetic and electric field, respectively. During the first half cycle of the voltage waveform energy will be stored and the energy from inductor or capacitor is returned back to the source during the other half of the waveform. Hence, when a circuit consists of elements such as resistor, inductor, and capacitor, some part of energy is dissipated in the form of heat and other part of energy will be stored and returned alternatively by the reactive part of the circuit. However, this depends on the values of elements considered.

5.3.1 Average Power (or Active Power)

Active power is the average power in watts delivered to a load. It is the only useful power and is the actual power dissipated by the load. Instantaneous power $p(t)$ is the product of the instantaneous values of voltage and current. Let the voltage $v(t) = V_m \sin \omega t\ V$ and current $i(t) = I_m \sin(\omega t + \theta)$ A. The expression for power, $p(t)$, can then be given by

$$p(t) = V_m \sin \omega t \times I_m \sin(\omega t + \theta) \tag{5.19}$$

$$= \frac{1}{2} V_m I_m (\cos\theta - \cos(2\omega t + \theta))$$

$$= \frac{1}{2} V_m I_m \cos\theta - \frac{1}{2} V_m I_m \cos(2\omega t + \theta) \tag{5.20}$$

Equation (5.20) consists of two components: one is $1/2\ V_m I_m \cos\theta$ and is constant irrespective of time. The other is $1/2 V_m I_m \cos(2\omega t + \theta)$ which varies with time.

Hence, the average power is given by

$$P = \frac{1}{T}\int_0^T p(t)\, dt \tag{5.21}$$

$$= \frac{1}{T}\int_0^T \frac{1}{2}(V_m I_m \cos\theta - V_m I_m \cos(2\omega t + \theta)\, dt$$

The integral of the second term is zero and hence the average power is given by $1/2V_mI_m \cos\theta$. This can also be called as active power or real power or true power.

$$\text{Average power or active power, } P = \frac{1}{2} V_m I_m \cos\theta \tag{5.22}$$

Rewriting Eq. (5.22) in terms of rms values

$$P = \left(\frac{V_m}{\sqrt{2}}\right) \cos\theta$$

$$= V_{rms}\, I_{rms} \cos\theta$$

$$= V_{eff}\, I_{eff} \cos\theta$$

$$= VI \cos\theta \tag{5.23}$$

If angle $\theta = 0°$, the voltage and current will be in phase. This implies a purely resistive circuit or resistive load. When $\theta = \pm 90°$, the voltage and current will be out of phase which implies a purely reactive circuit. The power will be zero showing that the reactive element either L or C absorbs no average power. From Eq. (5.23), $P_{avg} = VI \cos\theta$.

It is a standard practice to represent Z, R, and X in the form of a triangle, known as *impedance triangle*. From the impedance triangle, shown in Figure 5.6,

$$\cos\theta = \frac{R}{Z} \tag{5.24}$$

and we know that

$$V = I \times Z \tag{5.25}$$

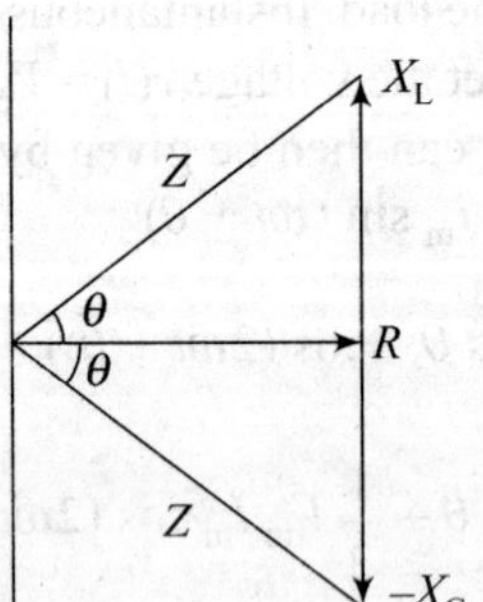

Figure 5.6 Impedance triangle

Substituting Eqs. (5.24) and (5.25) in Eq. (5.23), we obtain

$$P_{avg} = I \times Z \times I \times \left(\frac{R}{Z}\right)$$

$$= I^2R \tag{5.26}$$

Equation (5.26) is the expression for power dissipated in a resistor.

EXAMPLE 5.4 **Find the real power supplied by the voltage source (Figure 5.7).**

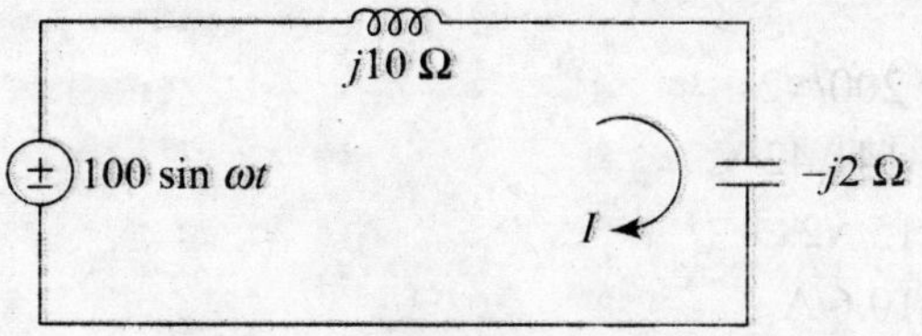

Figure 5.7

Solution Let the circuit current be I.

$$I = \frac{100 \angle 0^\circ}{j10 - j2}$$

$$= \frac{100 \angle 0^\circ}{j8}$$

$$= 12.5 \angle -90^\circ \text{ A}$$

However,

$$I_{rms} = 12.5/\sqrt{2} = 8.83 \angle -90^\circ \text{ A}$$

$$V_{rms} = 100/\sqrt{2} = 70.71 \angle 0^\circ$$

Real power supplied = 70.71 × 8.83 × cos 90° = 0 W. From this we can conclude that the inductor and capacitor cannot dissipate the energy.

5.3.2 Reactive Power

Reactive power is the measure of the energy exchange between the source and the reactive part of the load. The unit of reactive power is volt-ampere reactive (VAR).

We know that the energy storage elements neither dissipate nor supply power, but exchange power back and forth with the rest of the network. In the same way, the reactive power is being transferred back and forth between the source and the load. It represents the lossless interchange between the load and the source.

In an ac circuit, the current can be resolved into active and reactive components. The reactive component can be written as $I \sin \theta$. The power drawn by the circuit due to the reactive component of the current is called *reactive power*. In other words, it is the product of the applied voltage and the reactive component of the current. It is designated by Q. Therefore,

$$Q = VI \sin \theta \tag{5.27}$$

From Figure 5.6, i.e., impedance triangle

$$\sin \theta = \frac{X}{Z} \tag{5.28}$$

Substituting Eqs. (5.25) and (5.28) into Eq. (5.27), we get $Q = I^2X$ volt-ampere reactive.

EXAMPLE 5.5 **In an ac circuit, $v = 200 \sin(\omega t + 30°)$ V, $i = 15 \sin(\omega t - 30°)$ A. Find the reactive power.**

Solution

$$V_{rms} = 200/\sqrt{2}$$
$$= 141.42 \text{ V}$$
$$I_{rms} = 15/\sqrt{2}$$
$$= 10.6 \text{ A}$$

Reactive power, $Q = VI \sin \theta$

But

$$\theta = 30° - (-30°) = 60°$$
$$Q = 141.42 \times 10.6 \times \sin 60°$$
$$= 1298.2 \text{ VAR}$$

5.3.3 Apparent Power

In the case of sinusoidal source applied to a circuit, the product of rms value of voltage and current is called *apparent power*. It is expressed in VA (volt-ampere). The apparent power is so called because it seems apparent that the power should be the voltage–current product, by analogy with dc resistive circuits. It is measured in volt-ampere or VA to distinguish it from real power which is measured in watts.

$$S = VI \tag{5.29}$$

EXAMPLE 5.6 **In an ac circuit, $v = 150 \sin(\omega t + 30°)$ V, $i = 3 \sin(\omega t - 60°)$ A. Find the apparent power.**

Solution

$$V_{rms} = 150/\sqrt{2}$$
$$= 106.06 \text{ V}$$
$$I_{rms} = 3/\sqrt{2}$$
$$= 2.12 \text{ A}$$

Apparent power $= V_{rms} \times I_{rms} = (106.06 \times 2.12) = 224.84$ VA

5.3.4 Power Factor

The power factor is defined as the ratio of average power to apparent power. In other words, it defines cosine of the phase angle between the voltage and current. In terms of resistance and impedance it is defined as the ratio of resistance to impedance.

$$\text{Power factor} = \cos \theta = \frac{R}{Z} = \frac{P}{S} \tag{5.30}$$

The power factor is dimensionless. Angle θ is called as power factor angle. The power factor angle is equal to the angle of the load impedance Z if V is the voltage across the load and I is the current through it. The value of the power factor angle ranges between 0 and 1.

For a purely resistive load, voltage and current will be in phase and $\theta = 0$ and hence power factor = 1. For a purely reactive load, $\theta = \pm 90°$ and power factor = 0. In between these two extreme cases the power factor is said to be leading or lagging.

Leading power factor means that the current leads the voltage, which implies a capacitive load. Lagging power factor means that the current lags the voltage implying an inductive load. For resistive loads reactive power will be zero and power factor is unity. For capacitive loads reactive power is negative and power factor is leading. For inductive loads reactive power will be positive and power factor will be lagging.

EXAMPLE 5.7 **What is the power factor of the circuit shown in Figure 5.8. Assume $\omega = 1$ rad/s.**

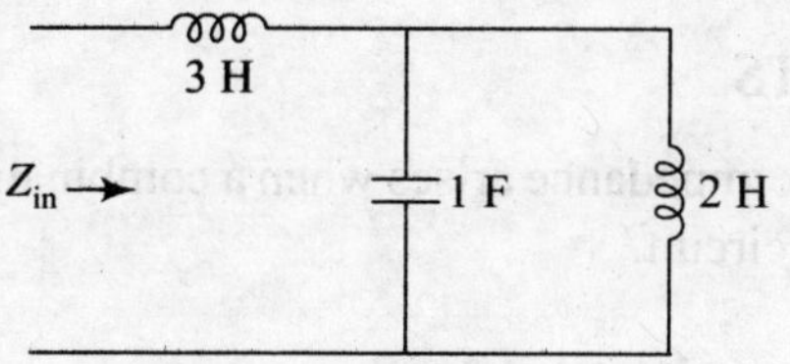

Figure 5.8

Solution In Figure 5.8,

$$Z_{in} = j3\,\omega + \frac{\frac{1}{j\omega} \times j\omega 2}{\frac{1}{j\omega} + j\omega 2}$$

$$= j3\omega + \frac{2}{\frac{1}{j\omega} + 2\,j\omega}$$

Since

$$\omega = 1 \text{ rad/s}$$

$$= j3 + \frac{2}{-j + 2j}$$

$$= j3 + \frac{2}{j} = j3 - 2j = j1 = 1\,\angle 90°$$

Therefore, power factor of the circuit = cos 90° = 0.

EXAMPLE 5.8 **The voltage across the load is $v(t) = 200 \sin(\omega t - 10°)$ V and the current through the element is $i(t) = 2 \sin(\omega t + 50°)$ A. Find the (i) apparent power, (ii) real and reactive powers, and (iii) power factor and load impedance.**

Solution

(i) $V_{rms} = \dfrac{200}{\sqrt{2}}\angle{-10°} = 141.42\,\angle{-10°}$ V

$I_{rms} = \dfrac{2}{\sqrt{2}}\angle 50° = 1.414\,\angle 50°$ A

Apparent power, $S = V_{rms} I_{rms} = 141.42 \times 1.414 = 199.99$ VA

(ii) Real power, $P = V_{rms} I_{rms} \cos\theta = 199.99 \cos(-10° - 50°) = 99.99$ W

Reactive power, $Q = V_{rms} I_{rms} \sin\theta = 199.99 \sin(-10° - 50°) = -173.19$ VAR

(iii) Power factor, = cos θ = cos (−10° − 50°) = 0.5 leading. It is leading because the reactive power is negative.

Load impedance

$$Z = \frac{V_m}{I_m} = \frac{200\angle-10°}{2\angle50°} = 100\angle-60°\ \Omega$$

which is capacitive impedance.

5.4 AC SERIES CIRCUITS

The concept of complex impedance arises when a combination of resistor, inductor, and capacitor is present in series in a circuit.

5.4.1 Series *RL* Circuits

Figure 5.9 shows the circuit consisting of a resistor and an inductor in series with each other. Let a voltage source V be connected to the circuit I flow through the circuit.

Let the voltage be considered as $V = V_m(\cos \omega t + j \sin \omega t)$

$$= V_m e^{j\omega t} \tag{5.31}$$

As a result, the current can also be expressed as

$$I = I_m e^{j\omega t} \tag{5.32}$$

For the *RL* circuit, on the application of the Kirchhoff's voltage law (KVL), we obtain

$$V = V_R + V_L \tag{5.33}$$

$$V = RI_S + L\frac{dI}{dt} \tag{5.34a}$$

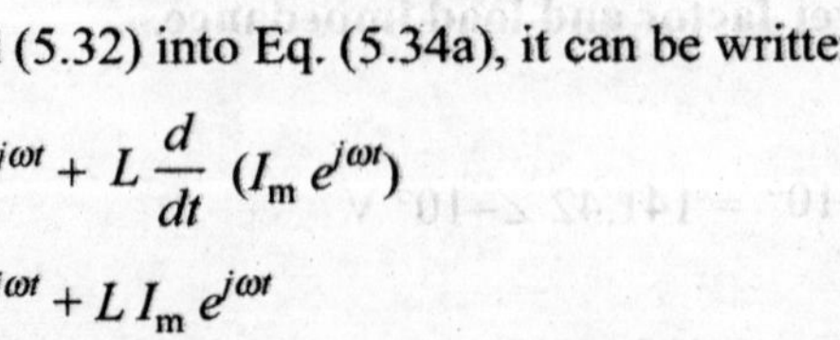

Figure 5.9 Series *RL* circuit

Substituting Eqs. (5.31) and (5.32) into Eq. (5.34a), it can be written as

$$V_m e^{j\omega t} = RI_m e^{j\omega t} + L\frac{d}{dt}(I_m e^{j\omega t})$$

$$= RI_m e^{j\omega t} + L I_m e^{j\omega t}$$

$$= (R + j\omega L) I_m e^{j\omega t}$$

$$V = (R + j\omega L) I$$

$$= ZI \quad (\text{where } Z = R + j\omega L) \tag{5.34b}$$

Figure 5.10(a) shows the impedance diagram for the above case, where $Z = \sqrt{R^2 + X^2}$ and $\theta = \tan^{-1}(X_L/R)$. Figure 5.10(b) shows the voltage diagram relating to V_L and V_R, where $V = \sqrt{V_R^2 + V_L^2}$ and $\theta = \tan^{-1} V_L/V_R$.

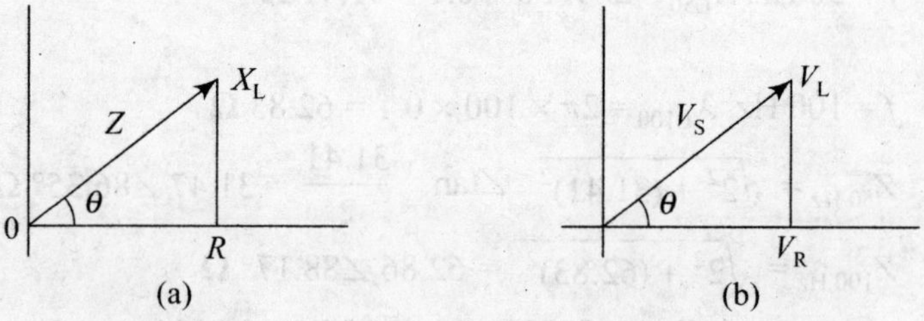

Figure 5.10 (a) Impedance diagram and (b) voltage diagram

EXAMPLE 5.9 **In a series *RL* circuit, *R* = 10 Ω while *L* = 30 mH. The input current lags the supply voltage by 60°. Obtain the value of applied frequency and find the real power supplied if the voltage is *v* = 120 sin *ωt*.**

Solution $R = 10\ \Omega$, $L = 30$ mH, power factor angle $\theta = 60°$

$$\tan\theta = \frac{\omega L}{R}$$

$$\tan 60° = \frac{\omega \times 30 \times 10^{-3}}{10}$$

$$\omega = \frac{10 \times 1.732}{30 \times 10^{-3}} = 577.35 \text{ rad/s}$$

This gives frequency

$$f = \frac{\omega}{2\pi} = \frac{577.35}{2\pi} = 91.88 \text{ Hz}$$

The current is

$$I = \frac{V}{Z} = \frac{120\angle 0°}{10 + j577.35 \times 30 \times 10^{-3}}$$

$$= \frac{120\angle 0°}{10 + j17.32}$$

$$= \frac{120\angle 0°}{19.99\angle 59.99}$$

$$= 6\angle -60° \text{ A}$$

Real power supplied, $P = V_{rms} I_{rms} \cos\theta = 120/\sqrt{2} \times 6/\sqrt{2} \times \cos(60°) = 180$ W.

EXAMPLE 5.10 **In a series *RL* circuit, $R = 2\ \Omega$ and $L = 0.1$ H. Find the impedance at 50 Hz and 100 Hz and the corresponding power factors.**

Solution $Z = R + j X_L$, while $X_L = 2\pi fL$

For

$$f = 50 \text{ Hz}, X_{L50} = 2\pi \times 50 \times 0.1 = 31.41\ \Omega$$

For

$$f = 100 \text{ Hz}, X_{L100} = 2\pi \times 100 \times 0.1 = 62.83\ \Omega$$

$$Z_{50\text{ Hz}} = \sqrt{2^2 + (31.41)^2}\ \angle \tan^{-1}\frac{31.41}{2} = 31.47\ \angle 86.35°\ \Omega$$

$$Z_{100\text{ Hz}} = \sqrt{2^2 + (62.83)^2} = 62.86\ \angle 88.17°\ \Omega$$

Power factor corresponding to 50 Hz = cos (86.35)° = 0.063 lag, and power factor corresponding to 100 Hz = cos (88.17)° = 0.032 lag.

5.4.2 Series *RC* Circuits

Figure 5.11 shows the circuit consisting of a resistor and a capacitor connected in series to each other. Let a voltage source V be connected to the circuit and let a current I flow through the circuit.

Applying the KVL to the circuit, we obtain

$$V = V_R + V_C \tag{5.35}$$

$$= IR + \frac{1}{C}\int I\,dt \tag{5.36}$$

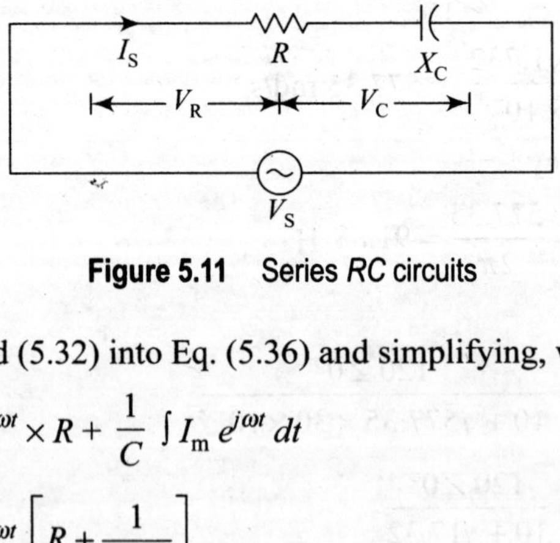

Figure 5.11 Series *RC* circuits

Substituting Eqs. (5.31) and (5.32) into Eq. (5.36) and simplifying, we get

$$V_m e^{j\omega t} = I_m e^{j\omega t} \times R + \frac{1}{C}\int I_m e^{j\omega t}\,dt$$

$$= I_m e^{j\omega t}\left[R + \frac{1}{j\omega C}\right]$$

$$= I_m e^{j\omega t}\left[R - \frac{j}{\omega C}\right]$$

We can write

$$V_m e^{j\omega t} = I_m e^{j\omega t}(R - jX_C) \quad \left(\text{where } X_C = \frac{1}{\omega C}\right) \tag{5.37}$$

$$V = I\,(R - jX_C) = IZ \quad (\text{where } Z = R - jX_C) \tag{5.38}$$

Figure 5.12(a) shows the impedance diagram for the above case, where

$$Z = \sqrt{R^2 + X_C^2} \quad \text{and} \quad \Theta = \tan^{-1}\left(\frac{X_C}{R}\right) \tag{5.39}$$

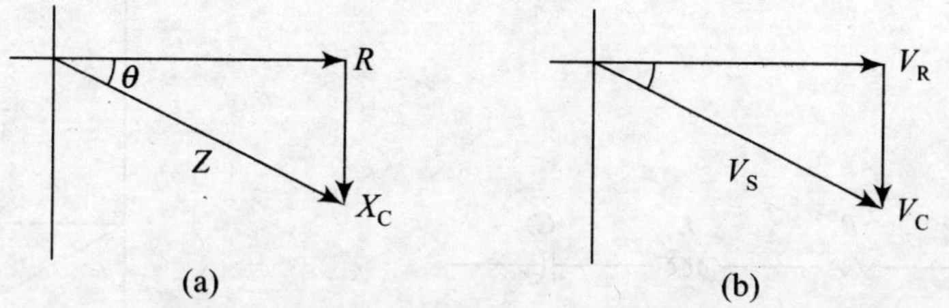

Figure 5.12 Diagrams for a series *RC* circuit

Figure 5.12(b) shows the voltage diagram relating to V_C and V_R, where

$$V = \sqrt{V_R^2 + V_C^2} \quad \text{and} \quad \Theta = \tan^{-1}\left(\frac{V_C}{V_R}\right) \tag{5.40}$$

EXAMPLE 5.11 **In a series *RC* circuit, *R* = 100 Ω and *C* = 25 μF. A sinusoidal voltage of 50 MHz is applied and the maximum voltage across the capacitance is 2.5 V. Find the maximum voltage across the series combination and also determine the apparent power.**

Solution $\omega = 2\pi f = 2 \times \pi \times 50 \times 10^6$ rad/s

$= 314 \times 10^6$ rad/s

The current through the capacitor being given by

$$I = \omega CV$$

$$I_{max} = \omega C V_{max}$$

$$= 314 \times 10^6 \times 25 \times 10^{-9} \times 2.5$$

$$= 19.625 \text{ A}$$

Again,

$$V_{max} = I_{max} Z$$

$$= 19.625 \times \sqrt{100^2 + (314 \times 10^6 \times 25 \times 10^{-9})^2}$$

Maximum voltage across the series circuit = 1968.53 V

The apparent power, $S = VI = 1968.53/\sqrt{2} \times 19.625/\sqrt{2} = 19316.27$ VA

5.4.3 Series *RLC* Circuits

Figure 5.13(a) shows a circuit consisting of a resistor, inductor, and capacitor in series to each other. Let a voltage source *V* be connected to the circuit and let a current *I* flowing through the circuit.

Applying the KVL to the circuit, we obtain

$$V = V_R + V_L + V_C \tag{5.41}$$

$$= I \times R + L\frac{dI}{dt} + \frac{1}{C}\int I\, dt \tag{5.42}$$

Substituting Eqs. (5.31) and (5.32) into Eq. (5.42) and simplifying, we get

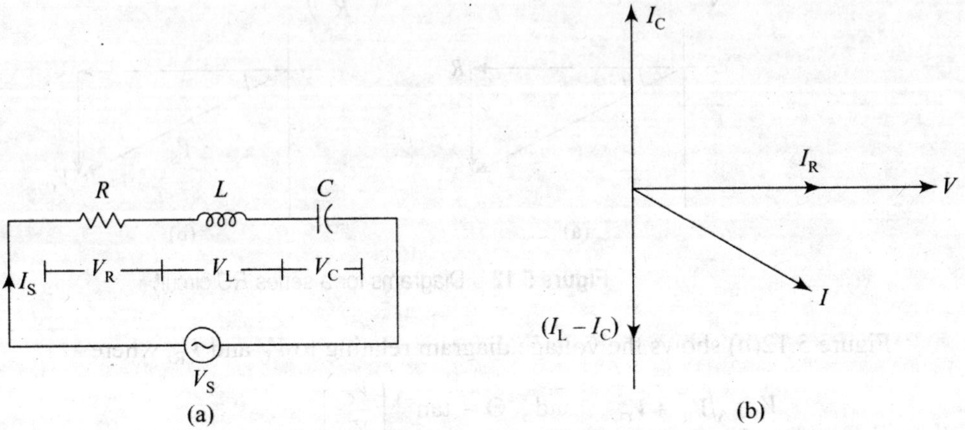

Figure 5.13 (a) Series *RLC* circuit and (b) phasor diagram

$$V_m\, e^{j\omega t} = I_m\, e^{j\omega t} \times R + L\frac{d}{dt}(I_m\, e^{j\omega t}) + \frac{1}{C}\int I_m\, e^{j\omega t}\, dt$$

$$= I_m\, e^{j\omega t}\left[R + j\omega L + \frac{1}{j\omega C}\right]$$

$$= I_m\, e^{j\omega t}\left[R + j\omega L - \frac{1}{j\omega}\right]$$

$$V_m\, e^{j\omega t} = I_m\, e^{j\omega t}\,(R + jX_L - jX_C)$$

$$V = I\,[R + j(X_L - X_C)] \quad \text{(where } X_L = \omega L \text{ and } X_C = 1/\omega C) \tag{5.43}$$

Rewriting Eq. (5.43),

$$V = Z\,I \quad \text{(where } Z = [R + j\,(X_L - X_C)]) \tag{5.44}$$

Voltage drop across resistance $V_R = IR$ in phase with I; voltage drop across inductance $V_L = IX_L$ leading I by 90°; and voltage drop across capacitance $V_C = IX_C$ lagging behind I by 90°.

V_L and V_C are 180° out of phase with each other (or reversed in phase). The phasor diagram of all these drops is shown in Figure 5.13(b).

The applied voltage V being equal to the phasor sum of V_R, V_L, and V_C and is given in magnitude by

$$V = \sqrt{V_R^2 + (V_L - V_C)^2} = \sqrt{(IR)^2 + (IX_L - IX_C)^2} = I\sqrt{(R)^2 + (X_L - X_C)^2}$$

The phase angle between voltage and current is given by

$$\phi = \tan^{-1}\left(\frac{V_L - V_C}{V_R}\right)$$

In terms of R, L, and C,

$$\phi = \tan^{-1}\left(\frac{X_L - X_C}{R}\right)$$

If $X_L > X_C$, the power factor angle ϕ is positive and hence the applied voltage will lead the current. If $X_L < X_C$, power factor angle ϕ is negative and hence applied voltage will lag the current. Power factor of the circuit is given by

$$\cos\phi = \frac{R}{Z} = \frac{R}{\sqrt{R^2 + (X_L - X_C)^2}}$$

EXAMPLE 5.12 **A 100-V, 50-Hz supply is applied across the series circuit consisting of R = 10 Ω, L = 1 mH, and C = 20 μF. Find the input current and voltage across each element.**

Solution $\omega = 2\pi f = 2\pi \times 50 = 314$ rad/s

$$X_L = \omega L = 314 \times 1 \times 10^{-3} = 0.314\ \Omega$$

$$X_C = \frac{1}{\omega C} = \frac{1}{314 \times 20 \times 10^{-6}} = \frac{10^6}{6280} = 159.24\ \Omega$$

$$Z = R + j(X_L - X_C) = 10 + j(0.314 - 159.24)$$

$$= (10 - j\,158.93)\ \Omega = 159.24\ \Omega \angle{-86.4^\circ}$$

$$I = \frac{V}{Z} = \frac{100\angle 0^\circ}{159.24\angle 86.4^\circ} = 0.62\angle 86.3^\circ \text{ A}$$

Also,

$$V_R = IR = 0.62\angle 86^\circ \times 10 = 6.2\angle 86^\circ \text{ V}$$

$$V_L = jX_L I = 0.314\angle 90^\circ \times 0.62\angle 86^\circ = 0.195\angle 176^\circ \text{ V}$$

$$V_C = -jX_C I = 159.24\angle{-90^\circ} \times 0.62\angle 86^\circ$$

$$= 99.05\angle{-4^\circ} \text{ V}$$

5.5 EFFECT OF FREQUENCY VARIATION IN *RLC* SERIES CIRCUIT

The voltage drop across resistance, inductance, and capacitance in a series *RLC* circuit is given by I_R, IX_L, and IX_C. The voltage drop across resistance is independent of frequency. But the voltage drop across the inductor and capacitor is dependent on frequency.

Since $X_L = \omega L = 2\pi fL$, as the frequency increases the inductive reactance increases and hence voltage drop across the inductor increases. $X_C = 1/\omega C = 1/2\pi fC$, as frequency increases, the capacitive reactance decreases and hence voltage drop across the capacitor decreases. That means as the frequency increases the net reactance $X_L - X_C$ is always positive and hence the power factor angle is positive and thus the voltage applied always leads the current or vice versa if the frequency decreases.

5.6 AC PARALLEL CIRCUIT

The analysis of parallel circuits with different methods is explained in this section.

5.6.1 Phasor Method

A parallel circuit with two impedance branches is shown in Figure 5.14(a). The voltage across the two branches is the same (V), but the currents are different (I_1 and I_2). The total current is I.

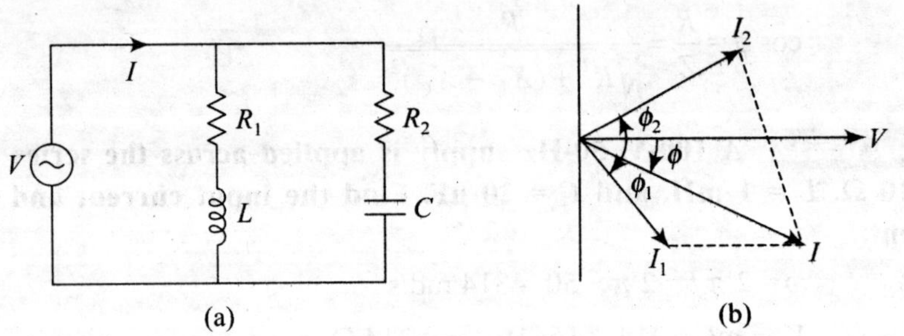

Figure 5.14 Parallel circuit: (a) circuit and (b) phasor diagram

For branch 1:

$$Z_1 = \sqrt{R_1^2 + X_L^2}$$

and

$$I_1 = \frac{V}{Z_1} \quad \text{and} \quad \phi_1 = \tan^{-1}\frac{X_L}{R_1}$$

Current I_1 lags behind the applied voltage (V) by ϕ_1

For branch 2:

$$Z_2 = \sqrt{R_2^2 + X_C^2}$$

and

$$I_2 = \frac{V}{Z_2} \quad \text{and} \quad \phi_2 = \tan^{-1}\frac{X_C}{R_2}$$

Current I_2 leads the applied voltage (V) by ϕ_2.

The resultant current I is the vector sum of the currents I_1 and I_2 as shown in Figure 5.14(b). There are two ways of obtaining the result.

(i) By using the parallelogram law of vectors, in which the resultant is obtained graphically. The graph gives both the magnitude of the resultant current I and the phase angle difference ϕ between I and the applied voltage V. Knowing ϕ, the power factor of circuit can be calculated. This method is tedious and time consuming.

(ii) Alternatively, by resolving currents I_1 and I_2 into their horizontal and vertical components and then obtaining the resultant current by combining these components, as shown in Figure 5.15. From the figure, it can be written that horizontal (x-axis) component is given by

$$I \cos \phi = I_1 \cos \phi_1 + I_2 \cos \phi_2 \tag{5.45}$$

and the vertical (y-axis) component is

$$I \sin \phi = I_1 \sin \phi_1 - I_2 \sin \phi_2 \tag{5.46}$$

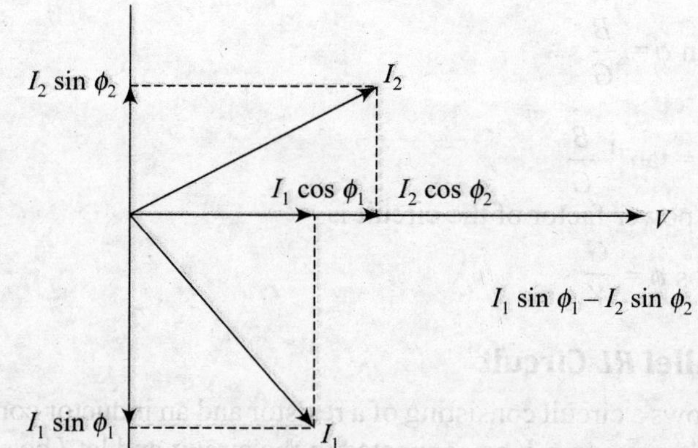

Figure 5.15 Estimation of resultant current I from the horizontal and vertical components of the branch currents I_1 and I_2 of a parallel circuit containing impedances

From Eqs. (5.45) and (5.46), we have

$$I = \sqrt{(I_1 \cos \phi_1 + I_2 \cos \phi_2)^2 + (I_1 \sin \phi_1 - I_2 \sin \phi_2)^2} \quad (5.47)$$

and

$$\phi = \tan^{-1} \frac{(I_1 \sin \phi_1 - I_2 \sin \phi_2)}{(I_1 \cos \phi_1 + I_2 \cos \phi_2)} \quad (5.48)$$

Power factor = $\cos \phi$

If "ϕ" is positive, then I leads V. If "ϕ" is negative, then I lags behind V. In parallel circuits, the applied voltage V is always taken as the reference vector while drawing the phasor diagram.

5.6.2 Admittance Method

It is convenient while solving parallel circuits to use admittance of the circuit instead of its impedance. Admittance of a circuit is the reciprocal of its impedance, i.e., impedance is given as

$$Z = \sqrt{R^2 + X^2}$$

Similarly, admittance

$$Y = \sqrt{g^2 + b^2} \quad \left(\text{where } g = \frac{1}{R} = \text{conductance and } b = \frac{1}{X} = \text{susceptance} \right)$$

The unit of Y, g, and b is mho (1/ohm).

If a circuit consists of two or more parallel branches, then the total conductance $G = g_1 + g_2 + g_3 + \cdots$ and the total susceptance $B = b_1 + b_2 + b_3 + \cdots$ (algebraic sum).

Then, the total admittance

$$Y = \sqrt{G^2 + B^2} \quad (5.49)$$

and the total current

$$I = VY \text{ amperes} \quad (5.50)$$

The phase angle difference (ϕ) between I and V is obtained from

$$\tan \phi = \frac{B}{G}$$

or

$$\phi = \tan^{-1} \frac{B}{G} \tag{5.51}$$

Therefore, the power factor of the circuit is

$$\cos \phi = \frac{G}{Y} \tag{5.52}$$

5.6.3 Parallel *RL* Circuit

Figure 5.16 shows a circuit consisting of a resistor and an inductor connected in parallel with each other. Let a voltage source V be connected to the circuit and let I be source current flowing at the input terminals through the circuit. Let I_1 an I_2 be the current flowing through R and L, respectively.

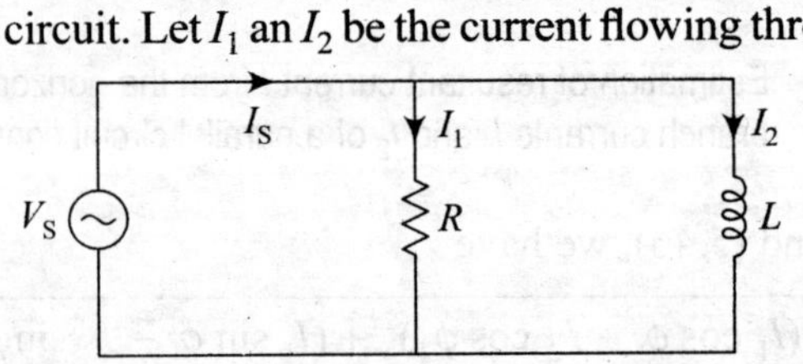

Figure 5.16 Parallel *RL* circuit

The impedance of branch 1 can be written as $Z_1 = R$ and the impedance of branch 2 can be written as $Z_2 = jX_L = j2\pi fL$. The total impedance of the circuit can be written as

$$Z = Z_1 \| Z_2$$

$$Z = \frac{Z_1 Z_2}{Z_1 + Z_2} = \frac{R\,(jX_L)}{R + jX_L} \tag{5.53}$$

The current flowing through the resistor,

$$I_1 = \frac{V_S}{Z_1} = \frac{V_S}{R} \tag{5.54}$$

The current flowing through the inductor,

$$I_2 = \frac{V_S}{Z_2} = \frac{V_S}{jX_L} \tag{5.55}$$

The total current I can be seen as the vector sum of both I_1 and I_2, and its magnitude impedance can be written as $\sqrt{I_1^2 + I_2^2}$. The phasor diagram of Figure 5.16 can be redrawn as shown in Figure 5.17.

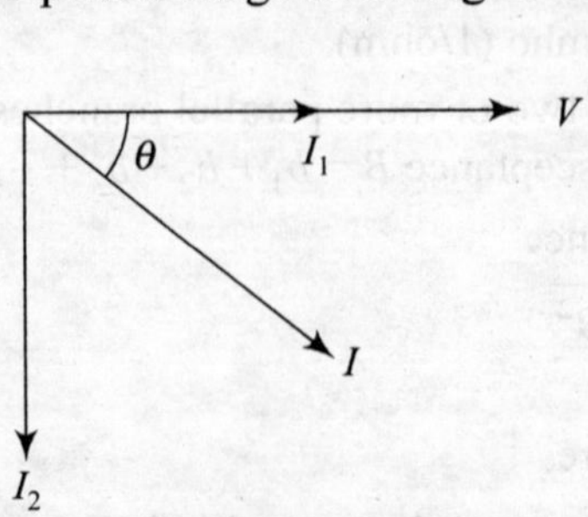

Figure 5.17 Phasor diagram of parallel *RL* circuit

From Figure 5.17, it can be seen that power factor $\cos\phi = I_1/I$.

EXAMPLE 5.13 **In a parallel *RL* circuit, $R = 2\ \Omega$ while $X_L = 4\ \Omega$. A current source of $20\angle 0°$ A feeds the combination. Find the branch currents.**

Solution Source current $I = 20\angle 0° = 20 + j0$ A

Let the current flowing through the resistance is I_R and the inductance is I_L. By the current division rule,

$$I_R = I\frac{jX_L}{R + jX_L} = 20\angle 0° \times \frac{4\angle 90°}{2 + j4}$$

or

$$I_R = \frac{80\angle 90°}{4.47\angle 63.44°} = 17.89\angle -26.56° \text{ A}$$

and

$$I_L = I\frac{R}{R + jX_L} = 20\angle 0°\frac{2\angle 0°}{2 + j4}$$
$$= 8.95\angle -63.44° \text{ A}$$

5.6.4 Parallel *RC* Circuit

Figure 5.18 shows the circuit consisting of a resistor and a capacitor connected in parallel with each other. Let a voltage source V be connected to the circuit and I be the source current flowing at the input terminals through the circuit. Let I_1 and I_2 be the current flowing through R and C, respectively.

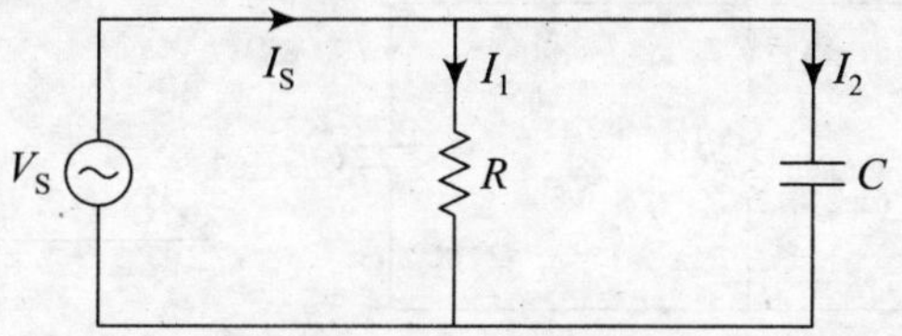

Figure 5.18 Parallel *RC* circuit

The impedance of branch 1 can be written as $Z_1 = R$ and the impedance of branch 2 can be written as $Z_2 = -jX_C = -j/2\pi fC$. The total impedance of the circuit can be written as

$$Z = Z_1 \| Z_2$$

$$Z = \frac{Z_1 Z_2}{Z_1 + Z_2} = \frac{R(-jX_C)}{R - jX_C} \tag{5.56}$$

The current flowing through the resistor,

$$I_1 = \frac{V_S}{Z_1} = \frac{V_S}{R} \tag{5.57}$$

The current flowing through the capacitor,

$$I_2 = \frac{V_S}{Z_2} = \frac{V_S}{-jX_C} \tag{5.58}$$

The total current I can be seen as the vector sum of both I_1 and I_2 and its magnitude impedance can be written as $\sqrt{I_1^2 + I_2^2}$. The phasor diagram of Figure 5.18 can be redrawn as shown in Figure 5.19.

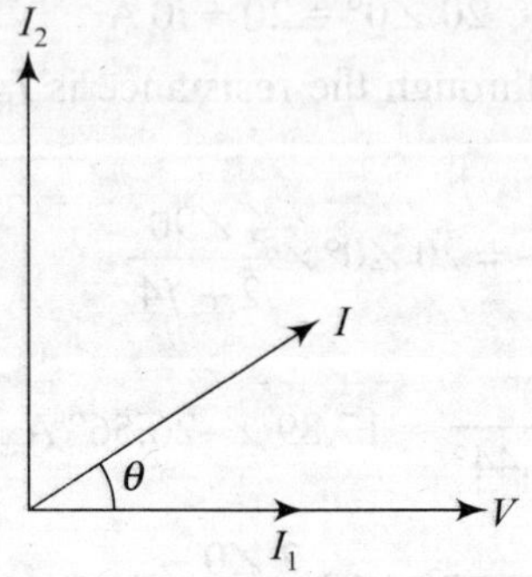

Figure 5.19 Phasor diagram of parallel *RL* circuit

From Figure 5.19, it can be seen that the power factor cos $\phi = I_1/I$.

EXAMPLE 5.14 **Find the impedance, current in each branch, total current, and power factor of the parallel *RC* circuit. The applied voltage is 200 V at 50 Hz. $R = 10\ \Omega$, $C = 120\ \mu F$. Draw the circuit diagram and the phasor diagram of the voltages and currents.**

Solution

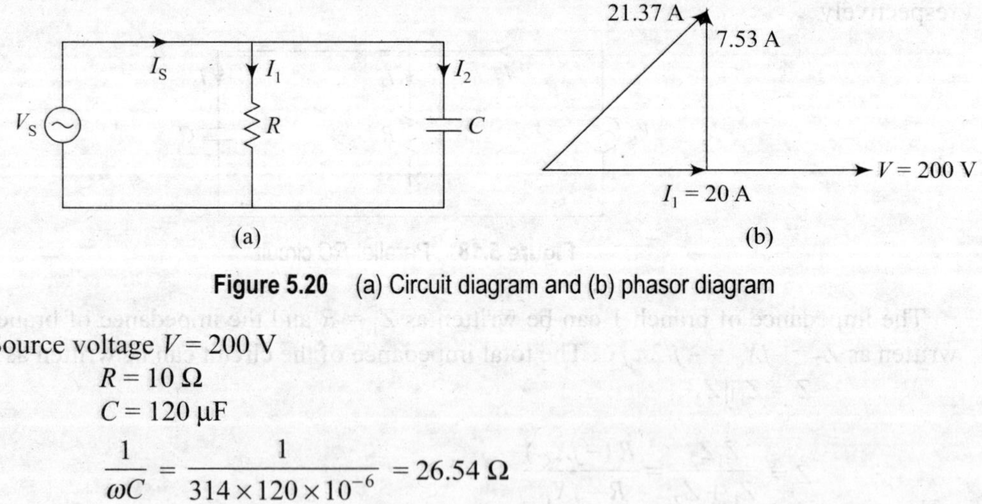

Figure 5.20 (a) Circuit diagram and (b) phasor diagram

Source voltage $V = 200$ V

$R = 10\ \Omega$

$C = 120\ \mu F$

$$\frac{1}{\omega C} = \frac{1}{314 \times 120 \times 10^{-6}} = 26.54\ \Omega$$

$$\text{Current, } I_1 = \frac{V}{R} = \frac{200}{10} = 20\ \angle 0^\circ \text{A}$$

$$\text{Current, } I_2 = \frac{V}{-jX_C} = \frac{200}{26.54\ \angle -90^\circ} = 7.53\ \angle 90^\circ \text{ A}$$

Total current, $I = I_1 + I_2 = 20 + j7.53 = 21.37\ \angle 20.63^\circ$ A

Power factor = cos $(20.63)^\circ = 0.934$ leading

$$\text{Impedance, } Z = \frac{V}{I} = \frac{200}{21.37} = 9.359\ \Omega$$

5.6.5 Parallel *RLC* Circuit

Figure 5.21 shows the circuit consisting of a resistor, an inductor, and a capacitor connected in parallel with each other. Let a voltage source V be connected to the circuit and I be the source current flowing at the input terminals through the circuit. Let I_1, I_2, and I_3 be the current flowing through R, L, and C, respectively.

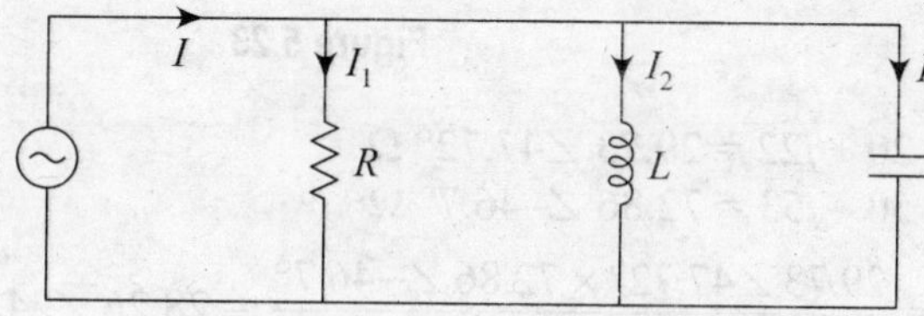

Figure 5.21 Parallel *RLC* circuit

The impedance of branch 1 can be written as $Z_1 = R$. The impedance of branch 2 can be written as $Z_2 = jX_L = j2\pi fL$ and the impedance of branch 3 can be written as $Z_2 = -jX_C = -j/2\pi f_C$. The current flowing in various elements can be given as follows:

$$I_R = \frac{V}{R}, \quad I_L = \frac{V}{jX_L}, \quad I_C = \frac{V}{-jX_C} \tag{5.59}$$

The net current I can be expressed as the vector sum of I_R and $(I_L - I_C)$. It can be shown that

$$I = \sqrt{I_R^2 + (I_L - I_C)^2}$$

The phasor diagram of Figure 5.21 can be redrawn as shown in Figure 5.22.

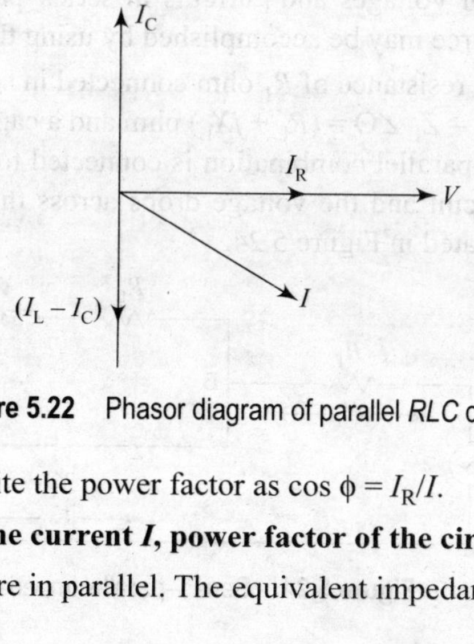

Figure 5.22 Phasor diagram of parallel *RLC* circuit

From Figure 5.22, we can write the power factor as $\cos\phi = I_R/I$.

EXAMPLE 5.15 **Calculate the current *I*, power factor of the circuit shown in Figure 5.23.**

Solution Two impedances are in parallel. The equivalent impedance is given by

$$Z_{eq} = \frac{Z_1 Z_2}{Z_1 + Z_2}$$

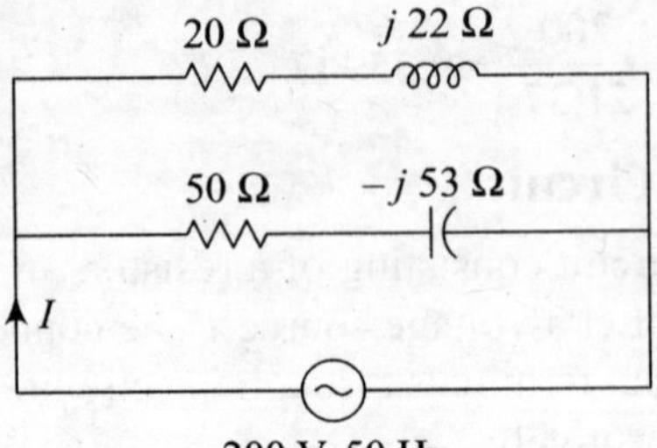

Figure 5.23

$$Z_1 = 20 + j22 = 29.73 \angle 47.72° \ \Omega$$
$$Z_2 = 50 - j53 = 72.86 \angle -46.7° \ \Omega$$
$$Z_{eq} = \frac{29.73 \angle 47.72° \times 72.86 \angle -46.7°}{20 + j22 + 50 - j53} = 28.26 \angle 24.9° \ \Omega$$

Current, $I = \frac{200}{28.26 \angle 24.9°} = 7.07 \angle -24.9° \text{ A}$

Power factor = cos (24.9°) = 0.907 lagging

5.7 SERIES–PARALLEL COMBINATION OF AC CIRCUIT

In ac electrical circuits several impedances are connected in series–parallel combinations, and computations of voltages and currents become complicated. The calculations can be simplified by determining the equivalent impedances of the series–parallel combinations of impedances.

The computations of voltages and currents in series–parallel combinations of impedances connected to the ac source may be accomplished by using the phasor diagram.

Figure 5.24 shows a resistance of R_1 ohm connected in series with an inductance coil having complex impedance $Z_L = Z_L \angle \Theta = (R_L + jX_L)$ ohm and a capacitive reactance X_C ohm connected in parallel. The series–parallel combination is connected to an ac supply. The assumed current distributions in the circuit and the voltage drops across the series impedance and the parallel combinations are indicated in Figure 5.24.

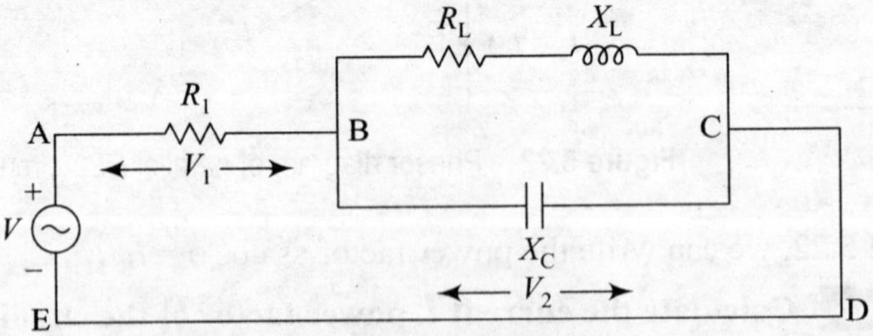

Figure 5.24 Series–parallel impedance circuit

By applying the KVL around the network ABCDEA and the Kirchhoff's current law (KCL) at the junction B, we get

$$V = V_1 + V_2 \quad (5.60)$$

$$I = I_L + I_C \quad (5.61)$$

In the present case, V_2 is chosen as the reference phasor. Thus, $V_2 = V_2 \angle 0°$. Then the current I_C will lead V_2 by 90°, i.e., $I_C = jI_C$. The currents I_L lag V_2 by the impedance angle Θ, where $\Theta = \tan^{-1}(X_l/R)$. The current I_L may be resolved along the reference phasor V_2 and along a direction perpendicular to it. Thus,

$$I_L = I_L(\cos\Theta - j\sin\Theta) \quad (5.62)$$

The phasor sum of I_L and I_C gives the current I. Then

$$V_1 = I_R \quad (5.63)$$

$$V_2 = I_C(jX_C) = I_L Z_L \quad (5.64)$$

Equations (5.62) and (5.64) can be employed to obtain the various currents and voltage drops.

Figure 5.25 shows phasor diagram for the circuit shown in Figure 5.24. The voltage drop across the parallel branch V_2 is the reference phasor and is represented by OV_2. The current through the capacitance I_C leads V_2 by 90° and is OI_C in the phasor diagram. The current through the inductance I_L that lags V_2 by angle Θ is shown as OI_L. The phasor sum of I_L and I_C is I and is marked as OI. The voltage drop V_R across R is in phase with the current I and is shown as OV_1. The phasor sum of V_1 and V_2 is the applied voltage and is drawn as OV.

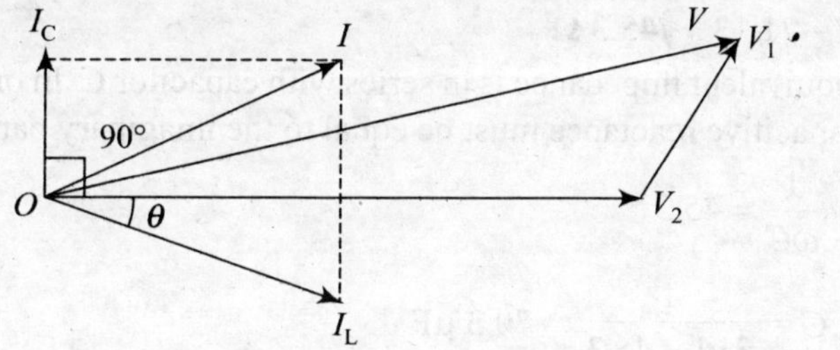

Figure 5.25 Phasor diagram for series–parallel circuit of Figure 5.24

EXAMPLE 5.16 **For the circuit shown in Figure 5.26, calculate the impedance between A and B and then phase angle between voltage and current.**

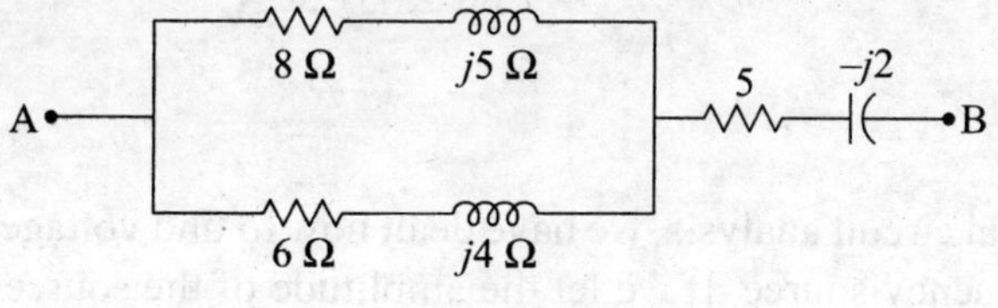

Figure 5.26

Solution $$Z_{AB} = \frac{(8 + j5)(6 + j4)}{(8 + j5 + 6 + j4)} + (5 - j2)$$

$$= \frac{(9.4\angle 32°)(7.21\angle 33.69°)}{(16.6\angle 32.73°)} + (5 - j2)$$

$$= \frac{(67.77 \angle 65.7°)}{(16.6 \angle 32.73°)} + (5 - j2)$$
$$= 4.08 \angle 32.97° + (5 - j2)$$
$$= 3.42 + j2.22 + (5 - j2)$$
$$= 8.42 + j0.22 = 8.42 \angle 1.5°\ \Omega$$

Impedance $Z = 8.42 \angle 1.5°\ \Omega$, the phase angle between voltage and current is 1.5°.

EXAMPLE 5.17 A 100-Ω resistor is shunted by a 0.5-H inductor in series with a capacitor *C*. A voltage of 220 V, 50 Hz is applied to the circuit. Determine (i) the value of *C* to give unity power factor, (ii) total current, and (iii) the current in the inductive branch.

Solution

(i) $X_L = \omega L = 2\pi f L = 314 \times 0.5 = 157\ \Omega$

Resistor and inductor are in parallel. The equivalent impedance is

$$= \frac{100 \times 157 \angle 90°}{100 + j157}$$
$$= \frac{15700 \angle 90°}{186.14 \angle 57.5°}$$
$$= 84.34 \angle 32.5°\ \Omega$$
$$= 71.13 + j45.3\ \Omega$$

The equivalent impedance is in series with capacitor *C*. In order to give unity power factor, the capacitive reactance must be equal to the imaginary part of the equivalent impedance.

$$\frac{1}{\omega C} = 45.3$$

$$C = \frac{1}{314 \times 45.3} = 70.3\ \mu\text{F}$$

(ii) Total current, $I = 220/71.13 = 3.09$ A (since the inductive reactance is cancelled by the capacitive reactance, the equivalent impedance is a pure resistor)

(iii) Current through the inductor $= 3.09 \times 100/(100 + j157)$
$= 1.66 \angle -57.5°$ A

5.8 RESONANCE

In our sinusoidal circuit analysis, we have dealt how to find voltages and currents in a circuit with a constant frequency source. If we let the amplitude of the source remain constant and vary the frequency, we obtain the frequency response. The frequency response of a circuit is the variation in its behavior with change in signal frequency. The frequency response of a circuit is of great importance in many applications especially in communications and control systems.

In many electrical circuits, the resonance is a very important phenomenon. The study of resonance is very useful, particularly in the area of communications. The concept of resonance applies in several areas of science and engineering.

The resonance of a circuit is defined as a condition in an *RLC* circuit in which the capacitive and inductive reactances are equal in magnitude, thereby resulting in purely resistive impedance.

5.8.1 Series Resonance

Consider the series *RLC* circuit shown in Figure 5.27.

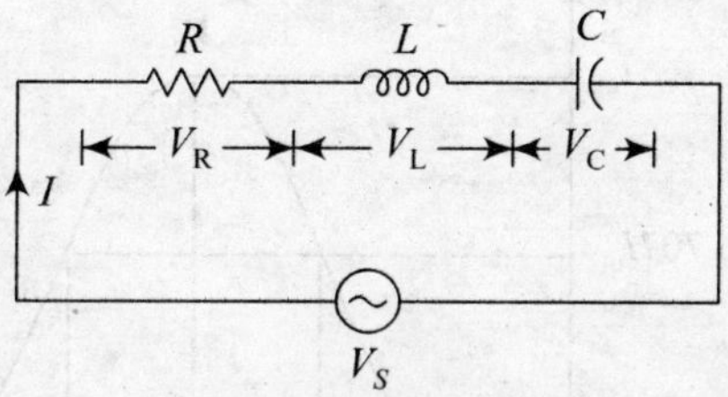

Figure 5.27

The total impedance for the series *RLC* circuit is

$$Z = R + j\,(X_L - X_C)$$

$$= R + j\left(\omega L - \frac{1}{\omega C}\right) \quad (5.65)$$

The circuit is said to be in resonance if the current is in phase with the applied voltage. In a series *RLC* circuit, series resonance occurs when $X_L = X_C$. The frequency at which the resonance occurs is called the *resonant frequency*.

Resonant frequency

When the condition of series resonance is considered, i.e., $X_L = X_C$, the impedance in a series *RLC* circuit becomes purely resistive. At the resonant frequency, f_r, the voltage across the capacitance and inductance is equal in magnitude. Since they are 180° out of phase with each other, they cancel each other and hence zero voltage appears across the *LC* combination.

At resonance

$$X_L = X_C, \text{ i.e., } \omega L = \frac{1}{\omega C} \quad (5.66)$$

Solving for resonant frequency, we get

$$2\pi f_r L = \frac{1}{2\pi f_r C}$$

$$f_r^2 = \frac{1}{4\pi^2 LC}$$

$$\therefore \quad f_r = \frac{1}{2\pi\sqrt{LC}} \quad (5.67)$$

or

$$\omega_r = \frac{1}{\sqrt{LC}} \quad (5.68)$$

Bandwidth, sensitivity, and Q-factor

The bandwidth of any system is the range of frequencies for which the current or output voltage is equal to 70.7% of its value at the resonant frequency. The unit of bandwidth is Hz.

The frequency response of circuit's current magnitude is shown in Figure 5.28.

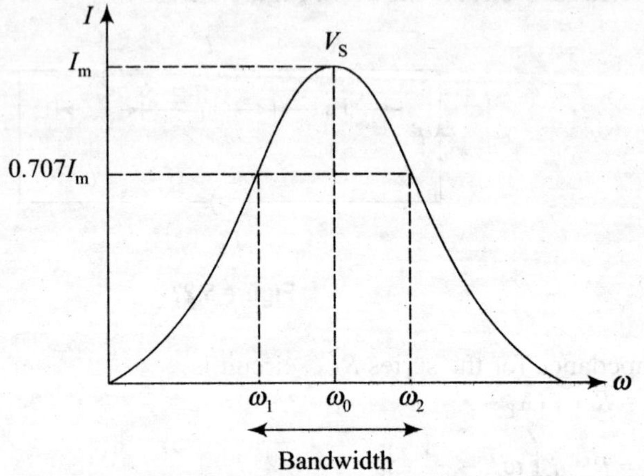

Figure 5.28 Frequency response

If I_0 is the current at resonance, the power delivered to circuit at resonance will be $I_0^2 R$. When the current falls to $0.707I_0$, the power delivered would be

$$\left(\frac{I_0}{\sqrt{2}}\right)^2 R = \frac{1}{2} I_0^2 R$$

i.e., half of the power delivered at resonance. From Figure 5.28, it is observed that

$$\text{Bandwidth} = f_2 - f_1 \text{ Hz} \tag{5.69}$$

where f_1 is called the lower half-power frequency and f_2 is called the upper half-power frequency.

(i) Expression for bandwidth. From Figure 5.28, it is seen that the current at P_1 is $0.707I_{max}$, the impedance of the circuit at this point is $Z = R \pm jR$, i.e., $\sqrt{2}R$ in magnitude.

The half power frequencies are obtained by setting $Z = \sqrt{2}R$ and writing

$$\sqrt{R^2 + \left(\omega L - \frac{1}{\omega C}\right)^2} = \sqrt{2}R \tag{5.70}$$

and solving for ω, squaring Eq. (5.70), we get

$$R^2 + \left(\omega L - \frac{1}{\omega C}\right)^2 = 2R^2 \tag{5.71}$$

$$\left(\omega L - \frac{1}{\omega C}\right)^2 = R^2$$

$$\left(\omega L - \frac{1}{\omega C}\right) = R \tag{5.72}$$

$$\omega^2LC - \omega RC - 1 = 0 \tag{5.73}$$

On solving Eq. (5.73), we get

$$\omega_1 = \frac{-R}{2L} + \sqrt{\left(\frac{R}{2L}\right)^2 + \frac{1}{LC}}\ ; \quad \omega_2 = \frac{R}{2L} + \sqrt{\left(\frac{R}{2L}\right)^2 + \frac{1}{LC}} \tag{5.74}$$

$$\text{Bandwidth} = \omega_1 - \omega_2 \tag{5.75}$$

$$= \left(\frac{R}{2L} + \sqrt{\left(\frac{R}{2L}\right)^2 + \frac{1}{LC}}\right) - \left(\frac{-R}{2L} + \sqrt{\left(\frac{R}{2L}\right)^2 + \frac{1}{LC}}\right)$$

$$\omega_2 - \omega_1 = \frac{R}{L} \tag{5.76}$$

or

$$\text{Bandwidth } f_2 - f_1 = \frac{R}{2\pi L} \tag{5.77}$$

where f_1 is called the lower cutoff frequency and f_2 is called the upper cutoff frequency.

(ii) Resonant frequency in terms of half-power frequencies. The resonant frequency of a circuit may also be expressed as the geometric mean of the half-power frequencies such that $f_r = \sqrt{f_1 f_2}$. Thus, the reactances at these two half-power frequencies must be equal to $\pm R$.

We know that at half-power point P_1, if ω_1 is the angular frequency, then

$$\omega_1 L - \frac{1}{\omega_1 C} = -R \tag{5.78}$$

At half-power point P_2, if ω_2 is the angular frequency, then

$$\omega_2 L - \frac{1}{\omega_2 C} = R \tag{5.79}$$

Adding Eqs. (5.78) and (5.79), we get

$$\omega_1 L - \frac{1}{\omega_1 C} + \omega_2 L - \frac{1}{\omega_2 C} = R - R$$

$$(\omega_1 + \omega_2)L - \frac{1}{C}\left(\frac{1}{\omega_1} + \frac{1}{\omega_2}\right) = 0$$

Further simplifying

$$\omega_1 \omega_2 = \frac{1}{LC} = \omega_r^2$$

$$\omega_r = \sqrt{\omega_1 \omega_2} \tag{5.80}$$

or in other words

$$2\pi f_r = 2\pi\sqrt{f_1 f_2}$$

$$f_r = \sqrt{f_1 f_2} \tag{5.81}$$

where ω_r is called the corner frequency or resonant frequency and ω_1 and ω_2 are called the half-power frequencies.

(iii) Sensitivity. Sensitivity of a series *RLC* circuit may be defined as the ratio of bandwidth to resonant frequency. It is also defined as the reciprocal of quality factor. Therefore,

$$\text{Sensitivity} = \frac{\text{Bandwidth (BW)}}{\text{Resonant frequency } (f_r)} = \frac{f_2 - f_1}{f_r} \tag{5.82}$$

(iv) *Q*-factor. The sharpness of the resonance in a resonant circuit is measured quantitatively by the quality factor Q. It is defined as the ratio of resonant frequency to bandwidth of the circuit. Therefore,

$$Q\text{-factor} = \frac{\text{Resonant frequency } (f_r)}{\text{Bandwidth (BW)}} = \frac{f_r}{f_2 - f_1} \tag{5.83}$$

From the above expression, it is noted that if the bandwidth is narrower, Q would be high. In a series *RLC* circuit, the narrow bandwidth can be obtained with low circuit resistance. The Q-factor can be expressed in different forms as given below.

(i) Q-factor is equal to voltage magnification of the circuit at resonance. It is defined as the ratio of voltage developed across L or C to the applied voltage. Then

$$Q = \frac{IX_L \text{ or } IX_C}{V} \tag{5.84}$$

or in other words

$$Q = \frac{\omega_r L}{R} \quad \text{or} \quad \frac{1}{\omega_r CR}$$

(ii) The Q-factor may also be expressed in terms of energy, i.e.,

$$Q = 2\pi \left[\frac{\text{Maximum energy stored}}{\text{Energy dissipated per cycle}}\right] \tag{5.85}$$

At resonance, I_0 will be maximum and hence the maximum energy stored in the inductor can be written as

$$\text{Maximum energy} = \frac{1}{2} LI_0^2 \quad \text{or} \quad \frac{1}{2} CV^2 \tag{5.86}$$

$$\text{Energy dissipated per cycle} = \left(\frac{I}{\sqrt{2}}\right)^2 \times R \times T$$

$$= \frac{1}{2} I^2 \times R \times T \tag{5.87}$$

$$\therefore \quad Q = 2\pi \frac{\frac{1}{2} LI^2}{\frac{1}{2} RI^2 T} \tag{5.88}$$

$$= 2\pi f_r \frac{L}{R}$$

$$= \frac{\omega_r L}{R} \tag{5.89}$$

Similarly, in terms of capacitance,

$$\therefore \quad Q = 2\pi \frac{\frac{1}{2}CV^2}{\frac{1}{2}RI^2T} \tag{5.90}$$

$$= 2\pi \frac{\frac{1}{2}C\left(\frac{V}{\omega C}\right)^2}{\frac{1}{2}RI^2T}$$

$$= \frac{1}{\omega_r CR}$$

$$= \frac{\omega_r L}{R} = \frac{1}{\omega_r CR} \tag{5.91}$$

The other expression for Q-factor is given by

$$Q = \frac{\omega_r L}{R} = \frac{2\pi f_r L}{R} \tag{5.92}$$

Substituting Eq. (5.67) into Eq. (5.92), we get

$$Q = \frac{2\pi \frac{1}{2\pi\sqrt{LC}} L}{R} = \frac{1}{R}\sqrt{\frac{L}{C}} \tag{5.93}$$

Voltages and currents in a series resonant circuit

At resonant frequency, the inductive reactance is equal to capacitive reactance and hence the impedance is equal to resistance. The current flowing through the circuit is also maximum because of minimum impedance. The voltage drop across the resistance, inductance, and capacitance varies with frequency.

When frequency $f = 0$, $X_L = 0$, and $X_C = \infty$, the inductor acts as a short circuit and the capacitor acts as open circuit. As f is increased from zero, X_L increases from 0, but X_C decreases from ∞ and hence the net reactance $X_L - X_C$ is capacitive and current increases. This continues till $f < f_r$. At $f = f_r$, current reaches its maximum value. When $f > f_r$, $X_L > X_C$, then the net reactance $X_L - X_C$ is inductive and the current starts decreasing, as current decreases the voltage drop V_R also decreases. The drop across resistance is maximum at resonant condition. The variation of current and impedance with frequency is shown Figure 5.29.

The voltage drop across the inductor is given by

$$V_L = IX_L = \frac{V}{Z}X_L = \frac{V\omega L}{\sqrt{R^2 + \left(\omega L - \frac{1}{\omega C}\right)^2}} \tag{5.94}$$

The voltage drop across the capacitor is given by

$$V_C = IX_C = \frac{V}{Z}X_C = \frac{V}{\omega C\sqrt{R^2 + \left(\omega L - \frac{1}{\omega C}\right)^2}} \tag{5.95}$$

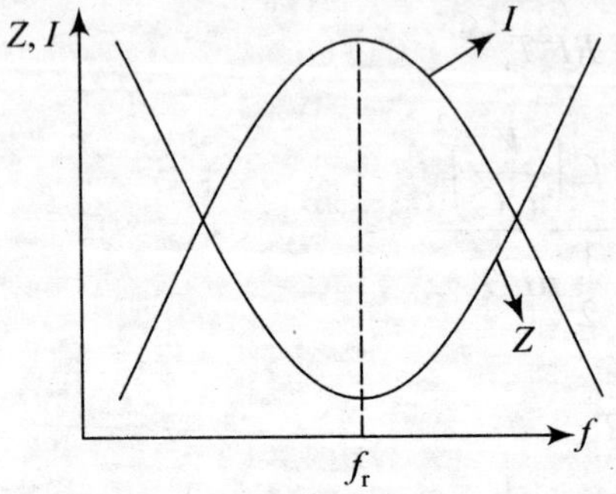

Figure 5.29 Variation of current and impedance with frequency

Derivation of frequency at maximum voltage drop across inductor and capacitor. Equation (5.94) gives the voltage drop across the inductor and is given by

$$V_L = \frac{V\omega L}{\sqrt{R^2 + \left(\omega L - \frac{1}{\omega C}\right)^2}}$$

Differentiating the above equation with respect to ω and equating it to zero gives the condition for maximum voltage drop across the inductor.

$$\frac{dV_L}{d\omega} = 0 \tag{5.96}$$

$$\frac{d}{d\omega}\left[\frac{V\omega L}{\sqrt{R^2 + \left(\omega L - \frac{1}{\omega C}\right)^2}}\right] = 0 \tag{5.97}$$

$$\omega_L = \frac{1}{\sqrt{LC}}\left[\frac{-2}{2 - \frac{R^2 C}{L}}\right] \tag{5.98}$$

Substituting $\omega_L = 2\pi f_L$

$$f_L = \frac{1}{2\pi\sqrt{LC}}\left[\sqrt{\frac{1}{1 - \frac{R^2 C}{2L}}}\right] \tag{5.99}$$

Similarly, the voltage across the capacitor is given by

$$V_C = \frac{V}{\omega C\sqrt{R^2 + \left(\omega L - \frac{1}{\omega C}\right)^2}}$$

Differentiating the above equation with respect to ω and equating it to zero gives the condition for maximum voltage drop across the capacitor.

$$\frac{dV_C}{d\omega} = 0 \tag{5.100}$$

$$\frac{d}{d\omega}\left[\frac{V}{\omega C\sqrt{R^2 + \left(\omega L - \frac{1}{\omega C}\right)^2}}\right] = 0 \tag{5.101}$$

$$\omega_C = \sqrt{\frac{1}{LC} - \frac{R^2}{2L}} \tag{5.102}$$

Substituting $\omega_C = 2\pi f_C$

$$f_C = \frac{1}{2\pi}\left[\sqrt{\frac{1}{LC} - \frac{R^2}{2L}}\right] \tag{5.103}$$

Voltage magnification

Voltage magnification is defined as the ratio of voltage developed across L or C to applied voltage. The voltage across the inductor is given by IX_L. Therefore, the voltage magnification across the inductor is given by

$$\frac{IX_L}{V} = \frac{V\omega_r L}{VR} = \frac{\omega_r L}{R} = Q \tag{5.104}$$

Similarly, the voltage across the capacitor is given by IX_C. Therefore, the voltage magnification across the capacitor is given by

$$\frac{IX_C}{V} = \frac{V}{\omega CRV} = \frac{1}{\omega_r CR} = Q \tag{5.105}$$

Therefore, the voltage magnification across the inductor or capacitor is given by Q. From the above discussion we can conclude that at resonance

(i) The impedance is purely resistive, thus $Z = R$. In other words, the LC series combination acts like a short circuit and the entire voltage is across R.

(ii) The voltage V and the current I are in phase, so that the power factor is unity.

(iii) The equivalent impedance of the circuit has no j-component. (i.e., X_L and X_C components are equal at a particular frequency).

(iv) The inductor voltage and capacitor voltage can be much more than the source voltage.

EXAMPLE 5.18 In a series *RLC* circuit, instantaneous voltage and currents are given as $v = 200 \sin(314t + 30°)$ V and $i = 5 \sin(314t + 30°)$ A. Find the values of *R* and *C* provided $L = 0.5$ H.

Solution $v = 200 \sin(314t + 30°)$ V

$i = 5 \sin(314t + 30°)$ A

From the voltage and current equations it is observed that the instantaneous voltage and currents are in phase, the net circuit impedance being *R*. This indicates that the circuit is at resonance, the net circuit impedance being *R*.

$$R = \frac{V_m}{I_m} = \frac{200}{5} = 40\ \Omega$$

At resonance,

$$X_L = X_C \text{ or } w_0 L = \frac{1}{w_0 C}; \quad C = \frac{1}{w_0^2 L} = 20.2 \times 10^{-6}\ F = 20.2\ \mu\text{F}$$

EXAMPLE 5.19 A series *RLC* circuit consists of $R = 2\ \Omega$, $L = 1$ mH, and $C = 0.2\ \mu$F. Find the resonant frequency.

Solution Resonant frequency,
$$f = \frac{1}{2\pi\sqrt{LC}}$$
$$= \frac{1}{2\pi\sqrt{1 \times 10^{-3} \times 0.2 \times 10^{-6}}}$$
$$= 11.25 \text{ kHz}$$

5.8.2 Parallel Resonance

The parallel circuit shown in Figure 5.30 is dual of the series *RLC* circuit.

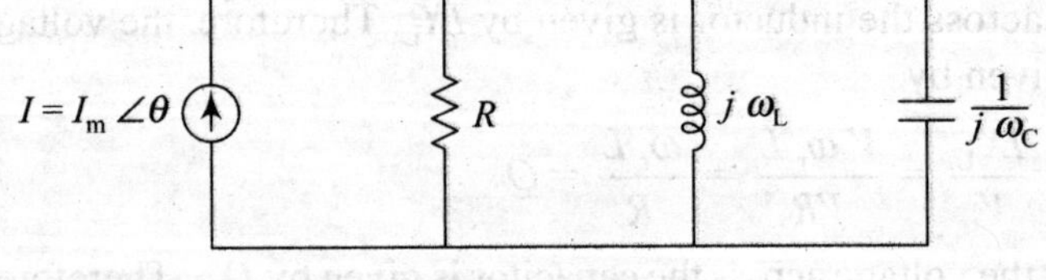

Figure 5.30

Resonant frequency

From Figure 5.30, the total admittance

$$Y = \frac{1}{R} + j\omega C + \frac{1}{j\omega L} \tag{5.106}$$
$$= \frac{1}{R} + j\left(\omega C - \frac{1}{\omega L}\right)$$

Resonance occurs when imaginary part of *Y* is zero, i.e.,

$$\omega C - \frac{1}{\omega L} = 0 \tag{5.107}$$

Simplifying Eq. (5.107), we get

$$\omega^2 = \frac{1}{LC}$$

The resonant frequency is given by

$$\omega_r = \frac{1}{\sqrt{LC}} \text{ rad/s}$$

or

$$f_r = \frac{1}{2\pi\sqrt{LC}} \text{ Hz}$$

which is same as the series resonant circuit. (5.108)

Bandwidth and Q-factor of parallel circuit

The expression for a bandwidth for parallel circuit is same as series circuit and is given as

$$f_2 - f_1 \quad \text{or} \quad \omega_2 - \omega_1 \tag{5.109}$$

In order to obtain the lower half-power frequency in a parallel circuit, we write

$$\omega_1 C - \frac{1}{\omega_1 L} = \frac{-1}{R} \tag{5.110}$$

Solving Eq. (5.110), we have

$$\omega_1^2 + \frac{\omega_1}{RC} - \frac{1}{LC} = 0$$

Further simplifying

$$f_1 = \frac{1}{2\pi}\left(\frac{-1}{2RC} + \sqrt{\left(\frac{1}{2RC}\right)^2 + \frac{1}{LC}}\right) \tag{5.111}$$

Similarly, the upper half-power frequency can be obtained as

$$\omega_2 C - \frac{1}{\omega_2 L} = \frac{1}{R} \tag{5.112}$$

Solving Eq. (5.112) for ω_2

$$\omega_2 = \frac{1}{2RC} + \sqrt{\left(\frac{1}{2RC}\right)^2 + \frac{1}{LC}} \tag{5.113}$$

Writing in terms of f_2

$$f_2 = \frac{1}{2\pi}\left(\frac{1}{2RC} + \sqrt{\left(\frac{1}{2RC}\right)^2 + \frac{1}{LC}}\right) \tag{5.114}$$

$$\text{Bandwidth} = f_2 - f_1 = \frac{1}{2\pi RC}$$

or in other words

$$\omega_2 - \omega_1 = \frac{1}{RC} \tag{5.115}$$

***Q*-factor.** The Q-factor definition is same as series circuit, i.e., it is the ratio of resonant frequency to bandwidth.

$$Q = \frac{\omega_r}{\omega_2 - \omega_1} \tag{5.116}$$

Substituting Eq. (5.115) into Eq. (5.116), we get

$$Q = \frac{\omega_r}{1/RC} = \omega_r RC \tag{5.117}$$

In other words, Q-factor can also be expressed as

$$2\pi \left[\frac{\text{Maximum energy stored}}{\text{Energy dissipated per cycle}} \right]$$

For inductor. In the case of inductor, the maximum energy stored is given by

$$\text{Maximum energy} = \frac{1}{2} L I_0^2 \text{ or } \frac{1}{2} C V^2 \tag{5.118}$$

$$\text{Energy dissipated per cycle} = \left(\frac{I}{\sqrt{2}} \right)^2 \times R \times T$$

$$= \frac{1}{2} I^2 \times R \times T \tag{5.119}$$

$$\therefore \quad Q = 2\pi \frac{\frac{1}{2} L I^2}{\frac{1}{2} R I^2 T} \tag{5.120}$$

$$= \frac{2\pi f L R}{\omega^2 L^2} = \frac{R}{\omega_r L} \tag{5.121}$$

For capacitor. Similarly, in terms of capacitance,

$$\therefore \quad Q = 2\pi \frac{\frac{1}{2} C V^2}{\frac{1}{2} R I^2 T} \tag{5.122}$$

$$= \omega_r CR \tag{5.123}$$

Derivation of resonant frequency for different parallel circuits

Case 1: Consider the parallel circuit shown in Figure 5.31.

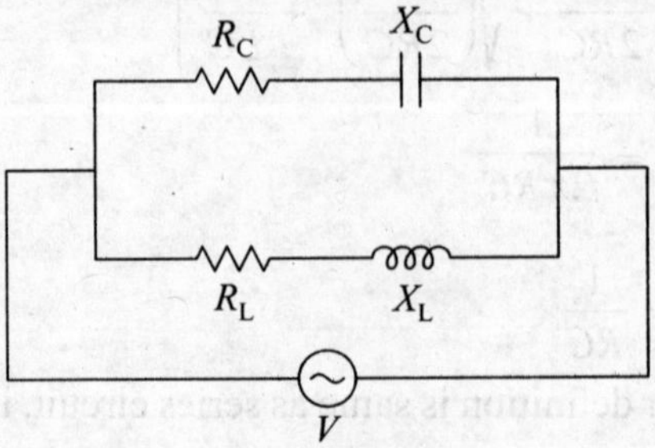

Figure 5.31

The circuit consists of two parallel branches with impedances $Z_1 = R_1 + j\omega L$ and $Z_2 = R_2 - j\omega C$ are connected to a voltage source. To find the resonant frequency we need to determine the equivalent admittance.

$$Y_1 = \frac{1}{Z_1} = \frac{1}{R_1 + j\omega L} = \frac{R_1}{R_1^2 + \omega^2 L^2} - j\frac{\omega L}{R_1^2 + \omega^2 L^2} \tag{5.124}$$

Similarly,

$$Y_2 = \frac{1}{Z_2} = \frac{1}{R_2 - j\omega C} = \frac{R_2}{R_2^2 + \frac{1}{\omega^2 C^2}} + j\frac{\frac{1}{\omega C}}{R_2^2 + \frac{1}{\omega^2 C^2}} \tag{5.125}$$

Total equivalent admittance,

$$Y = Y_1 + Y_2 \tag{5.126}$$

Substituting Eqs. (5.124) and (5.125), we get

$$Y = \left(\frac{R_1}{R_1^2 + \omega^2 L^2} + \frac{R_2}{R_2^2 + \frac{1}{\omega^2 C^2}}\right) + j\left(\frac{\frac{1}{\omega C}}{R_2^2 + \frac{1}{\omega^2 C^2}} - \frac{\omega L}{R_1^2 + \omega^2 L^2}\right) \tag{5.127}$$

The condition for resonance in a parallel circuit, susceptance, is zero, i.e., the imaginary part of admittance Y is equal to zero.

$$\frac{\frac{1}{\omega C}}{R_2^2 + \frac{1}{\omega^2 C^2}} - \frac{\omega L}{R_1^2 + \omega^2 L^2} = 0 \tag{5.128}$$

Solving Eq. (5.128), we get

$$\omega_r = \sqrt{\frac{1}{LC}\left(\frac{L - CR_1^2}{L - CR_2^2}\right)} \tag{5.129}$$

In terms of f_r

$$f_r = \frac{1}{2\pi}\sqrt{\frac{1}{LC}\left(\frac{L - CR_1^2}{L - CR_2^2}\right)} \tag{5.130}$$

If $R_1 = R_2$, then

$$f_r = \frac{1}{2\pi\sqrt{LC}} \tag{5.131}$$

Case 2: Consider the parallel circuit shown in Figure 5.32. It consists of two parallel branches with impedances $Z_1 = R + j\omega L$ and $Z_2 = -j\omega C$ are connected across the voltage source V_S.

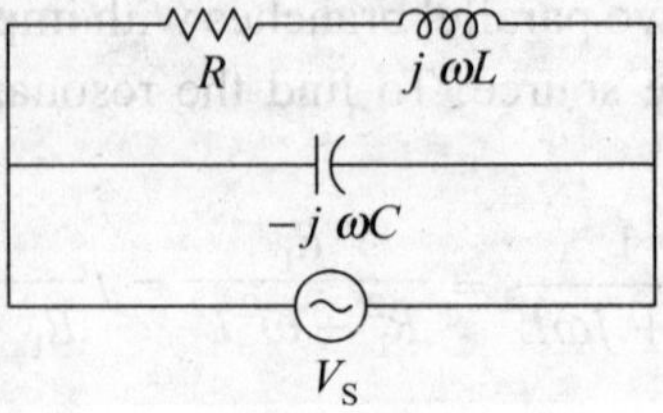

Figure 5.32

The total admittance of the circuit is given by

$$Y = \frac{1}{R + j\omega L} + \frac{1}{-j\omega C} \tag{5.132}$$

$$= \frac{R - jX_L}{R^2 + X_L^2} + \frac{j}{X_C}$$

$$= \frac{R}{R^2 + X_L^2} + j\left(\frac{1}{X_C} - \frac{X_L}{R^2 + X_L^2}\right) \tag{5.133}$$

According to the condition of resonance, imaginary part of Eq. (5.133) must be equal to zero.

$$\left(\frac{1}{X_C} - \frac{X_L}{R^2 + X_L^2}\right) = 0 \tag{5.134}$$

$$\omega C = \frac{\omega L}{R^2 + \omega^2 L^2} \tag{5.135}$$

Solving Eq. (5.135), we get

$$R^2 + \omega^2 L^2 = \frac{L}{C}$$

$$\omega_r = \sqrt{\frac{1}{LC} - \frac{R^2}{L^2}} \tag{5.136}$$

In terms of f_r, Eq. (5.136) can be expressed as

$$f_r = \frac{1}{2\pi}\sqrt{\frac{1}{LC} - \frac{R^2}{L^2}} \tag{5.137}$$

EXAMPLE 5.20 **Determine the resonant frequency of the parallel circuit shown in Figure 5.33.**

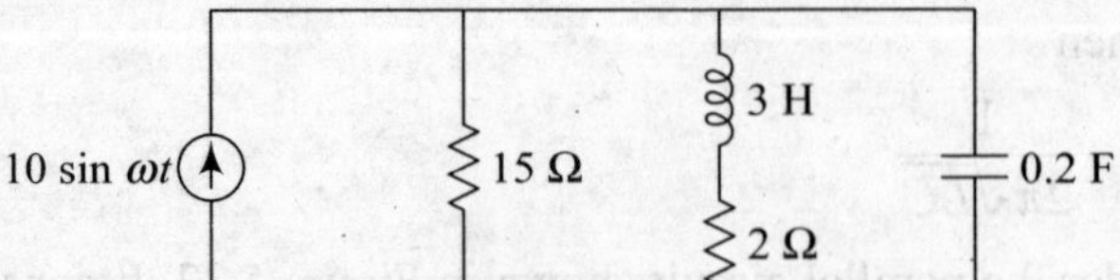

Figure 5.33

Solution The input admittance is

$$Y = \frac{1}{15} + \frac{1}{2 + jw3} + jw0.2$$

$$= 0.066 + j0.2\,\omega + \frac{2 - jw3}{4 + 9w^2}$$

$$= 0.066 + \frac{2}{4 + 9w^2} + j\left(w0.2 - \frac{w3}{4 + 9w^2}\right)$$

At resonance, the imaginary part of admittance is zero.

$$\therefore \quad 0.2\,\omega - \frac{w3}{4 + 9w^2} = 0$$

Solving the above equation for ω, we get $\omega_0 = 1.1$ rad/s.

EXAMPLE 5.21 **Find the resonant frequency for the parallel circuit shown in Figure 5.34.**

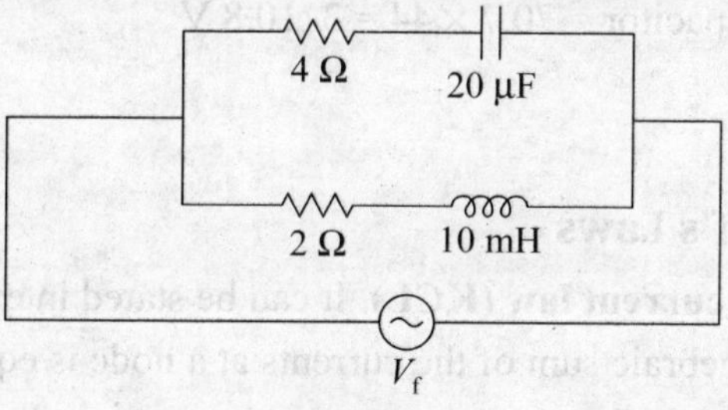

Figure 5.34

Solution Given that $R_1 = 4\ \Omega$, $C_1 = 20\ \mu F$, $R_2 = 2\ \Omega$, $L_2 = 10$ mH.

$$\omega_0 = \frac{1}{\sqrt{L_2 C_1}} \sqrt{\frac{R_1^2 - L_2/C_1}{R_2^2 - L_2/C_1}}$$

$$= \frac{1}{\sqrt{10 \times 10^{-3} \times 20 \times 10^{-6}}} \left[\frac{4^2 - (10 \times 10^{-3})/(20 \times 10^{-6})}{2^2 - (10 \times 10^{-3})/(20 \times 10^{-6})}\right]$$

$$= 2237 \left[\frac{16 - 0.5 \times 10^3}{4 - 0.5 \times 10^3}\right]$$

$$= 2182 \text{ rad/s}$$

$$f_0 = \frac{2182}{2\pi} = 347.2 \text{ Hz}$$

EXAMPLE 5.22 **A coil having a resistance of 5 Ω and an inductance of 0.2 H is connected in series with a 40-μF capacitor. The voltage applied to the circuit is 220 V. Determine the maximum current and corresponding frequency. Also, find the voltage across the capacitor and coil for this frequency.**

Solution Current will be maximum at the resonant frequency and is given by

$$f_r = \frac{1}{2\pi\sqrt{LC}}$$

$$= \frac{1}{2\pi\sqrt{0.2 \times 40 \times 10^{-6}}}$$

$$= 56.26 \text{ Hz}$$

$$\text{Current at this frequency} = \frac{V}{R} = \frac{220}{5} = 44 \text{ A}$$

$$\omega L = \frac{1}{wC} = 353.55 \times 0.2 = 70.7\ \Omega$$

$$\text{Impedance of the coil } Z = \sqrt{5^2 + 70.7^2} = 70.87\ \Omega$$

Voltage across the coil = $IZ = 44 \times 70.87 = 3118.28$ V

Voltage across capacitor = $70.7 \times 44 = 3110.8$ V

5.9 AC NETWORK

5.9.1 Kirchhoff's Laws

(i) Kirchhoff's current law (KCL). It can be stated in either of the following forms:

(a) The algebraic sum of the currents at a node is equal to zero.

(b) The sum of the currents entering into any node is equal to the sum of the currents leaving that node.

In Figure 5.35, the currents I_1, I_2, and I_3 are entering the node and considered as positive; then obviously, the currents I_4 and I_5 leaving the node "p" should be considered negative. The first form on the KCL results in

$$I_1 + I_2 + I_3 + (-I_4) + (-I_5) = 0$$

or

$$I_1 + I_2 + I_3 - I_4 - I_5 = 0 \tag{5.138}$$

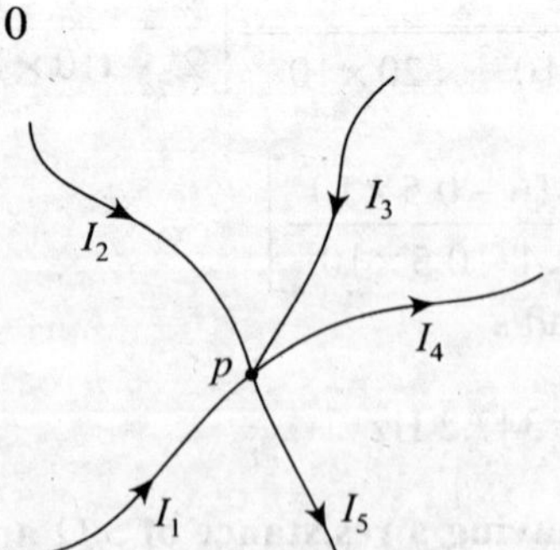

Figure 5.35

or by arranging the terms in Eq. (5.138), we have

$$I_1 + I_2 + I_3 = I_4 + I_5 \tag{5.139}$$

which is actually the second form of the KCL.

In general, the algebraic sum of the currents at any node is expressed as

$$\sum_{i=1}^{n} I_i = 0$$

EXAMPLE 5.23 **Find i_1 and i_2 in Figure 5.36.**

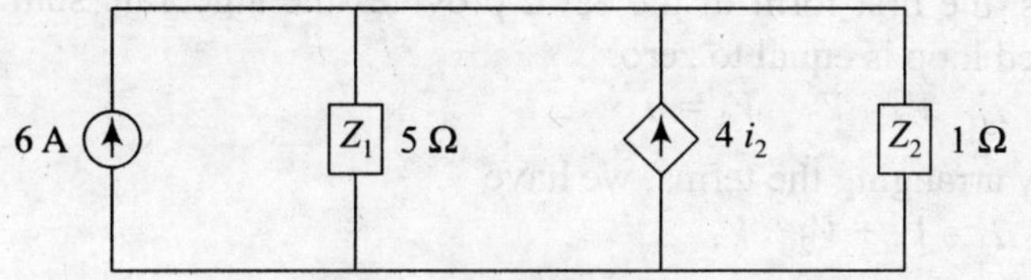

Figure 5.36

Solution Let us assume that the voltage across the 5-Ω or 1-Ω impedance is V volts. Applying the KVL at node 1,

$$i_1 + i_2 = 6 + 4i_2$$

$$i_1 - 3i_2 = 6 \tag{1}$$

However,

$$i_1 = \frac{V}{1} A \quad \text{and} \quad i_2 = \frac{V}{5} \text{ A}$$

Substituting i_1 and i_2 in Eq. (1), we get

$$\frac{V}{1} - \frac{3V}{5} = 6$$

Therefore, $V = 15$ V. Thus, $i_1 = 15$ A and $i_2 = 3$ A.

(ii) Kirchhoff's voltage law (KVL). It can be stated in either of the following forms:

(a) The algebraic sum of the potential rises and drops around any closed loop in an electric circuit is zero. Here, the voltage rise may be taken with a positive sign, while a voltage drop would be taken with a negative sign.

(b) In any closed loop, the sum of voltage rises = the sum of voltage drops.

Let us consider a circuit shown in Figure 5.37. A plus sign is assigned to a potential rise (– to +) and a minus sign is assigned to a potential drop (+ to –).

If we assume the current flow, I, as shown in Figure 5.37, from point "a," a potential drop V_1 (+ to –) will appear across Z_1, V_2 across Z_2, and V_3 across Z_3. Continuing through the voltage source, we have a potential rise E (– to +) before returning to point "a."

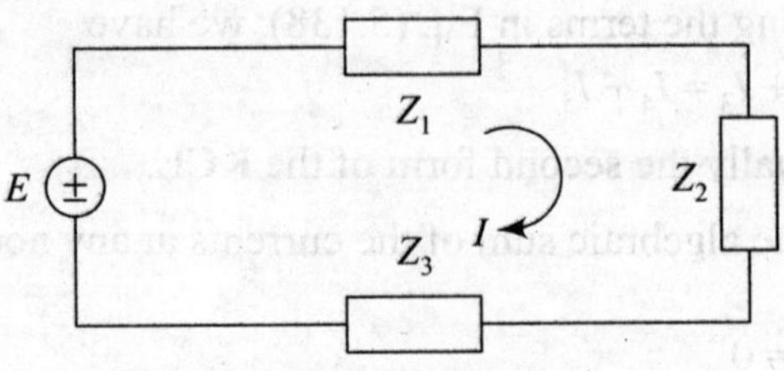

Figure 5.37

Thus, the first form of the KVL provides the algebraic sum of the voltage around any closed loop is equal to zero.

$$E - V_1 - V_2 - V_3 = 0 \quad (5.140)$$

or by arranging the terms, we have

$$E = V_1 + V_2 + V_3 \quad (5.141)$$

which is actually the second form of the KVL.

EXAMPLE 5.24 **Find the current *I* in Figure 5.38.**

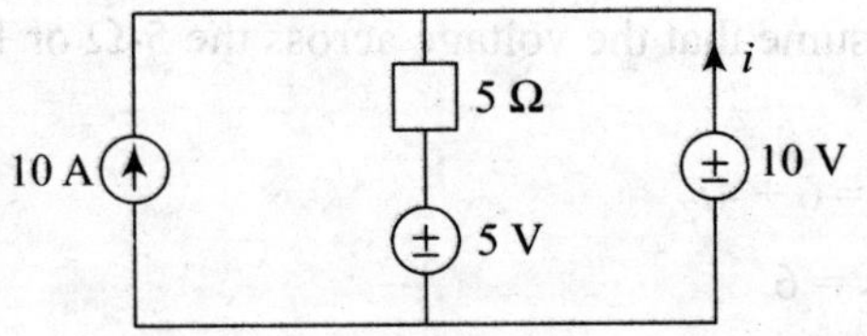

Figure 5.38

Solution The circuit is redrawn as shown in Figure 5.39.

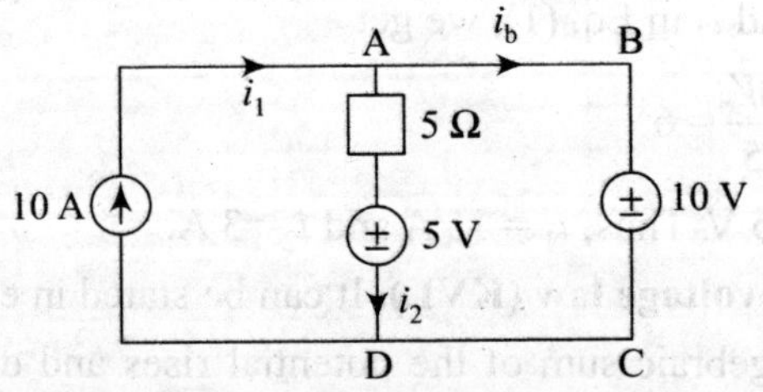

Figure 5.39

Writing the KCL equation at node 1

$$i_2 = i_1 + i = 10 + i \quad (1)$$

Writing the KVL for loop ABCD

$$5i_2 + 5 = 10 \text{ or } i_2 = 1 \text{ A} \quad (2)$$

Using Eqs. (1) and (2), $10 + i = 1$ or $i = -9$ A. Here, the negative sign means that the actual current is flowing opposite to the assumed direction.

EXAMPLE 5.25 **Find the current and power dissipated in the 5-Ω impedance in the circuit shown in Figure 5.40.**

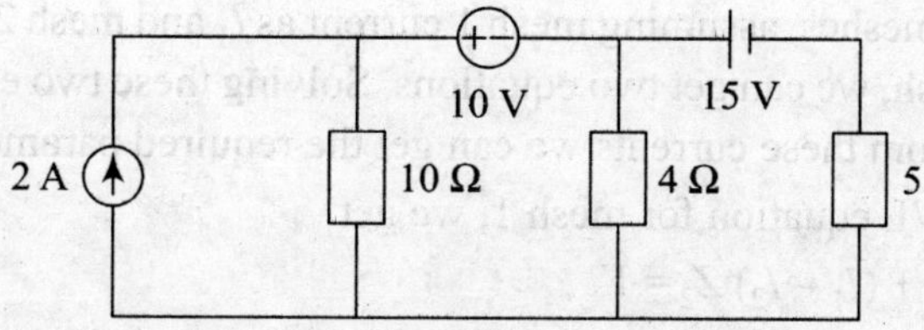

Figure 5.40

Solution The 2-A current source is transformed to an equivalent voltage source and the circuit is redrawn as shown in Figure 5.41.

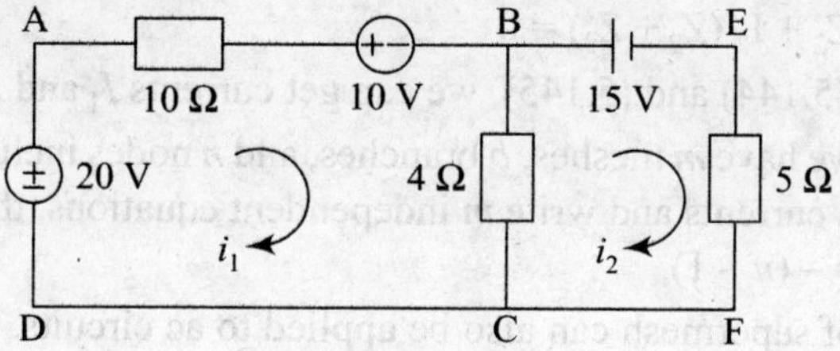

Figure 5.41

Writing the KVL equation for loop ABCD

$$-20 + 10i_1 + 10 + (i_1 - i_2)\,4 = 0$$

$$10i_1 + 4i_1 - 4i_2 = 10$$

or

$$14i_1 - 4i_2 = 10 \tag{1}$$

Writing the KVL equation for loop BEFC

$$(i_2 - i_1)\,4 + 15 + 5i_2 = 0$$

$$-4i_1 - 9i_2 = -15$$

$$4i_1 + 9i_2 = 15 \tag{2}$$

Solving Eqs. (1) and (2), $i_1 = 1.05$ A and $i_2 = 1.197$ A. The current through the 5-Ω resistor is 1.197 A. The power dissipated in the 5-Ω impedance = $(1.197)^2 \times 5 = 7.16$ W.

5.9.2 Mesh Analysis

Consider the circuit, shown in Figure 5.42, containing a voltage source and three impedances.

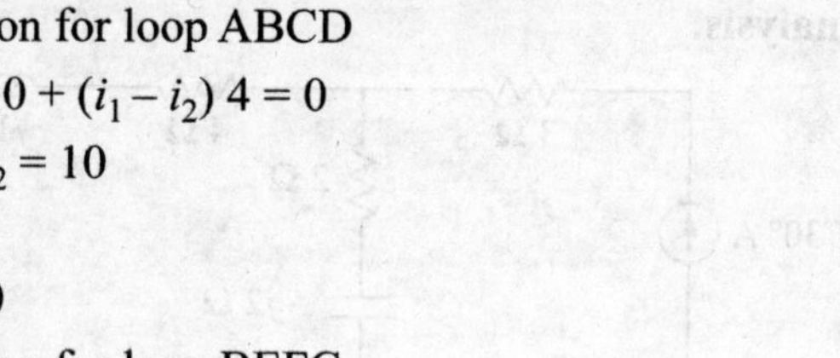

Figure 5.42

There are two meshes, assuming mesh 1 current as I_1 and mesh 2 current as I_2. By applying the KVL for each mesh, we can get two equations. Solving these two equations we get two unknown mesh currents. From these currents we can get the required parameters.

Writing the KVL equation for mesh 1, we get

$$I_1 Z_1 + (I_1 - I_2) Z_2 = V \tag{5.142}$$

Writing the KVL equation for mesh 2, we get

$$Z_2 (I_2 - I_1) + Z_3 I_2 = 0 \tag{5.143}$$

By rewriting Eqs. (5.141) and (5.142), we get

$$I_1 (Z_1 + Z_2) - I_2 Z_2 = V \tag{5.144}$$

$$-I_1 Z_2 + I_2 (Z_2 + Z_3) = 0 \tag{5.145}$$

By solving Eqs. (5.144) and (5.145), we can get currents I_1 and I_2.

In general, if we have *m* meshes, *b* branches, and *n* nodes including the reference node, and we assume *m* branch currents and write *m* independent equations, then the number of mesh currents is given by $m = b - (n - 1)$.

The concept of supermesh can also be applied to ac circuits. We follow the same concepts as discussed in Chapter 2 except that the resistances will be replaced by impedances.

EXAMPLE 5.26 Find the drop voltage across the 2-Ω resistor in the network of Figure 5.43 using the mesh analysis.

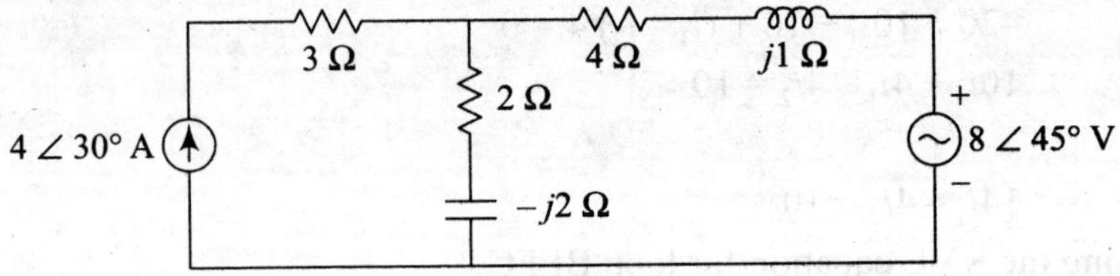

Figure 5.43

Solution Due to the presence of the current source, the mesh current of the first mesh is $I_1 = 4\angle 30°$ A.

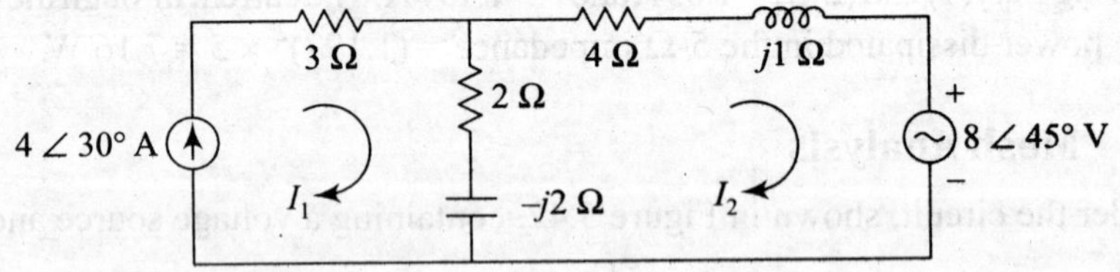

Writing the KVL equation for mesh 2,

$$(4 + j1) I_2 + (2 - 2j) I_2 - 4\angle 30\,(2 - 2j) = -8\angle 45°$$

$$(6 - j1) I_2 - 11.32\angle -15° = -8\angle 45°$$

$$I_2 = \frac{10.04\angle -58°}{6.08\angle -9.46°} = 1.65\angle -48.54° \text{ A}$$

Current flowing through the 2-Ω resistor $= I_1 - I_2$

$= 4\angle 30° - 1.65\angle -48.54°$

$= 2.37 + j3.23$

$= 4.0\angle 537°$ A

Voltage drop across the 2-Ω resistor $= 2(I_1 - I_2)$

$= 2 \times 4\angle 53.7°$ V

EXAMPLE 5.27 Calculate the current I_0 in the circuit shown in Figure 5.44, using the mesh analysis.

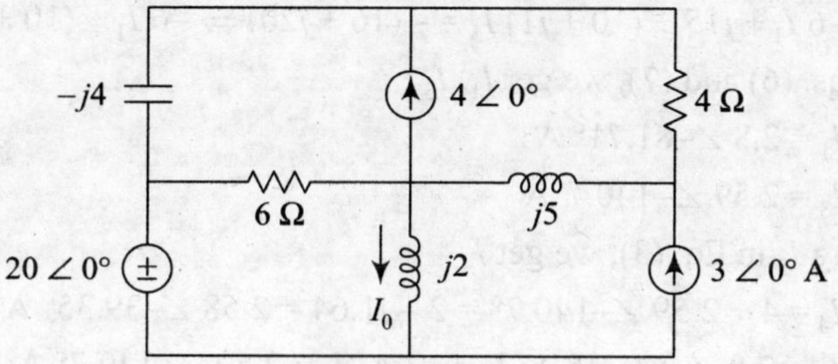

Figure 5.44

Solution There are four meshes for the given circuit. After assigning the mesh currents the circuit is redrawn as shown in Figure 5.45.

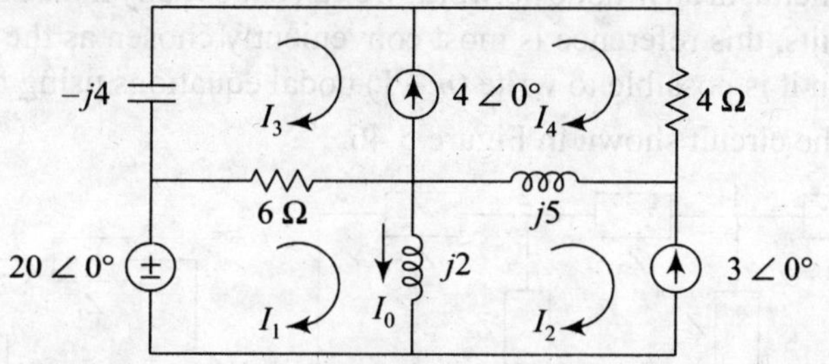

Figure 5.45

Writing the KVL equation for mesh 1,

$$-20 + 6\,(I_1 - I_3) + j2\,(I_1 - I_2) = 0$$

$$(6 + j2)\,I_1 - j2\,I_2 - 6\,I_3 = 20 \qquad (1)$$

Since there is a current source on the periphery of mesh 2, $I_2 = -3$ A.

Meshes 3 and 4 form a supermesh due to the current source between the meshes; hence, writing the supermesh equation,

$$(6 - j4)\,I_3 - 6\,I_1 + (4 + j5)\,I_4 - j5\,I_2 = 0$$

$$-6I_1 - j5\,I_2 + (6 - j4)\,I_3 + (4 + j5)\,I_4 = 0 \qquad (2)$$

Relating the current source and the third and fourth mesh currents

$$I_4 - I_3 = 4 \qquad (3)$$

From Eq. (3), we get

$$I_4 = 4 + I_3 \quad (4)$$

Substituting Eq. (4) into Eq. (2), we get

$$-6I_1 - j5\, I_2 + (6 - j4)\, I_3 + (4 + j5)\,(4 + I_3) = 0$$

$$-6I_1 - j5\, I_2 + (10 + j1)\, I_3 = -(16 + j20) \quad (5)$$

Substituting $I_2 = -3$ A in Eqs. (1) and (5), we get

$$(6 + j2)\, I_1 + j6 - 6\, I_3 = 20 \Rightarrow (6 + j2)\, I_1 - 6\, I_3 = 20 - j6 \quad (6)$$

$$-6\, I_1 + j15 + (10 + j1)\, I_3 = -(16 + j20) \Rightarrow -6\, I_1 + (10 + j1)\, I_3 = -(16 + j35) \quad (7)$$

Solving Eqs. (6) and (7), we get I_1, I_3

$$I_1 = 2.8 \angle{-81.71^\circ}\ \text{A}$$

$$I_3 = 2.59 \angle{-140.7^\circ}\text{A}$$

Substituting I_3 in Eq. (3), we get I_4

$$I_4 = 4 + 2.59 \angle{-140.7^\circ} = 2 - j1.64 = 2.58 \angle{-39.35^\circ}\ \text{A}$$

$$\therefore \quad I_1 = 2.8 \angle{-81.71^\circ}\ \text{A};\ I_2 = -3\ \text{A};\ I_3 = 2.59 \angle{-140.7^\circ}\ \text{A};\ I_4 = 2.58 \angle{-39.35^\circ}\ \text{A}$$

5.9.3 Nodal Analysis

An other method of solving the networks containing complex impedances is the nodal analysis method. In general, in an n-node network, we can choose any node as the reference or datum node. In many circuits, this reference is most conveniently chosen as the common terminal or ground terminal. Then it is possible to write $(n - 1)$ nodal equations using the KCL.

Consider the circuit shown in Figure 5.46.

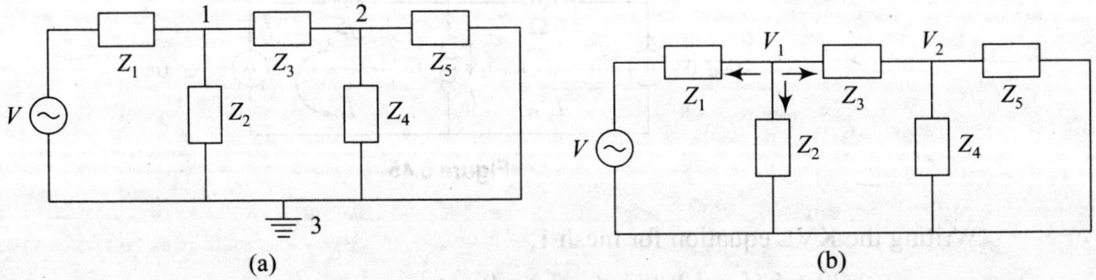

Figure 5.46

The circuit consists of two nodes, and the ground node is taken as the reference node. Let V_1 is the voltage of node 1 with respect to the ground node, V_2 is the voltage of node 2 with respect to the ground node.

In Figure 5.46(b), node 1 is redrawn with all its branches, assuming that all currents are leaving node 1.

Writing the KCL equation for node 1, we get

$$\frac{V_1 - V}{Z_1} + \frac{V_1}{Z_2} + \frac{V_1 - V_2}{Z_3} = 0 \quad (5.146)$$

Similarly, writing the KCL equation for node 2, we get

$$\frac{V_2 - V_1}{Z_3} + \frac{V_2}{Z_4} + \frac{V_2}{Z_5} = 0 \tag{5.147}$$

Rearranging Eqs. (5.146) and (5.147),

$$\left(\frac{1}{Z_1} + \frac{1}{Z_2} + \frac{1}{Z_3}\right)V_1 - \left(\frac{1}{Z_3}\right)V_2 = \frac{V}{Z_1} \tag{5.148}$$

$$\left(\frac{1}{Z_3} + \frac{1}{Z_4} + \frac{1}{Z_5}\right)V_2 - \left(\frac{1}{Z_3}\right)V_1 = 0 \tag{5.149}$$

By solving Eqs. (5.148) and (5.149), we can find the unknown voltages V_1 and V_2.

EXAMPLE 5.28 **In the network shown in Figure 5.47, determine voltages V_1 and V_2.**

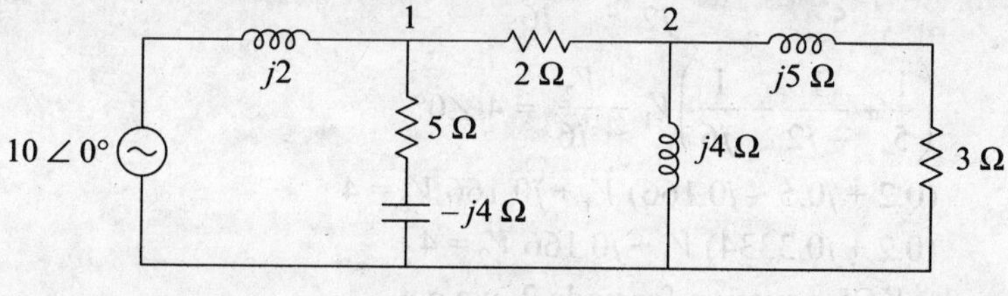

Figure 5.47

Solution Let V_1 and V_2 be the node voltages of nodes 1 and 2, respectively. Writing the KCL equation for node 1, we get

$$\frac{V_1 - 10\angle 0^\circ}{j2} + \frac{V_1}{5 - j4} + \frac{V_1 - V_2}{2} = 0$$

$$\left(\frac{1}{j2} + \frac{1}{5 - j4} + \frac{1}{2}\right)V_1 - \frac{1}{2}V_2 = \frac{10\angle 0^\circ}{j2}$$

$$(0.5 - j0.5 + 0.122 + j0.097)\,V_1 - 0.5\,V_2 = 5\angle -90^\circ$$

$$(0.622 - j0.403)\,V_1 - 0.5\,V_2 = 5\angle -90^\circ \tag{1}$$

Writing KCL equation for node 2, we get

$$\frac{V_2 - V_1}{2} + \frac{V_2}{j4} + \frac{V_2}{3 + j5} = 0$$

$$\left(\frac{1}{2} + \frac{1}{j4} + \frac{1}{3 + j5}\right)V_2 - \frac{1}{2}V_1 = 0$$

$$(0.5 - j0.25 + 0.088 - j0.147)\,V_2 - 0.5\,V_1 = 0$$

$$-0.5\,V_1 + (0.588 - j0.397)\,V_2 = 0 \tag{2}$$

Solving Eqs. (1) and (2), we can get node voltages V_1 and V_2.

$$V_1 = 7.31\angle 219.32^\circ \text{ V}; \; V_2 = 5.15\angle 253.38^\circ \text{ V}$$

EXAMPLE 5.29 **Find I_a in the circuit shown in Figure 5.48 using the nodal analysis.**

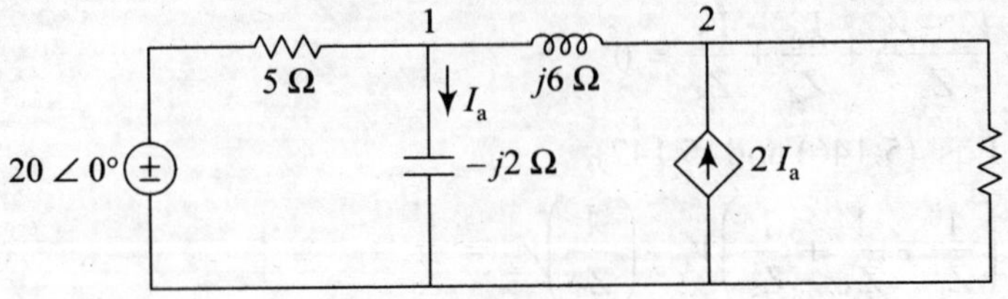

Figure 5.48

Solution Let V_1 and V_2 be the node voltages of nodes 1 and 2, respectively.

Writing the KCL equation for node 1, we get

$$\frac{V_1 - 20\angle 0°}{5} + \frac{V_1}{-j2} + \frac{V_1 - V_2}{j6} = 0$$

$$\left(\frac{1}{5} + \frac{1}{-j2} + \frac{1}{j6}\right)V_1 - \frac{V_2}{j6} = 4\angle 0°$$

$$(0.2 + j0.5 - j0.166)\ V_1 + j0.166\ V_2 = 4$$

$$(0.2 + j0.3334)\ V_1 + j0.166\ V_2 = 4 \qquad (1)$$

Writing the KCL equation for node 2, we get

$$2I_a = \frac{V_2 - V_1}{j6} + \frac{V_2}{3}$$

$$2I_a = \frac{-1}{j6}V_1 + \left(\frac{1}{3} + \frac{1}{j6}\right)V_2 \qquad (2)$$

But

$$I_a = \frac{V_1}{-j2} \qquad (3)$$

Substituting Eq. (3) into Eq. (2), we get

$$2\left(\frac{-V_1}{j2}\right) = \frac{-1}{j6}V_1 + \left(\frac{1}{3} + \frac{1}{j6}\right)V_2$$

$$\left(\frac{1}{j1} + \frac{-1}{j6}\right)V_1 + \left(\frac{1}{3} + \frac{1}{j6}\right)V_2 = 0$$

$$(-j1 + j0.1666)\ V_1 + (0.333 - j0.1666)\ V_2 = 0$$

$$(-j0.8334)\ V_1 + (0.333 - j0.1666)\ V_2 = 0 \qquad (4)$$

Solving Eqs. (1) and (4), we can get V_1 and V_2.

$$V_1 = -19.06\angle 50.98°\ \text{V};\ V_2 = 42.69\angle -12.45°\ \text{V}$$

5.10 AC NETWORK THEOREMS

All the network theorems discussed for dc circuits can also be applied to ac circuits. The only difference is that in ac circuits we have to deal with impedances in place of resistors.

5.10.1 Thevenin's Theorem

The given complex impedance circuit can be solved, using Thevenin's theorem. Thevenin's equivalent circuit consists of an equivalent voltage source called Thevenin's voltage, V_{Th}, and Thevenin's equivalent impedance, Z_{Th}, in series with load impedance as shown in Figure 5.49. The values of Thevenin's equivalent voltage and impedance are determined as per the procedure explained in Chapter 2 for dc circuits.

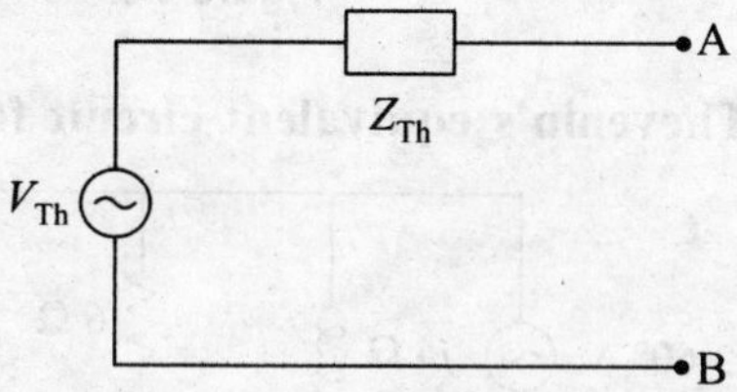

Figure 5.49 Thevenin's equivalent circuit

Thevenin's voltage is equal to the open-circuit voltage across the load terminals, and the impedance is equal to the equivalent impedance seen into the network from the load terminals.

Consider the circuit shown in Figure 5.50. Thevenin's equivalent for the circuit shown in Figure 5.48 between points A and B is found as follows.

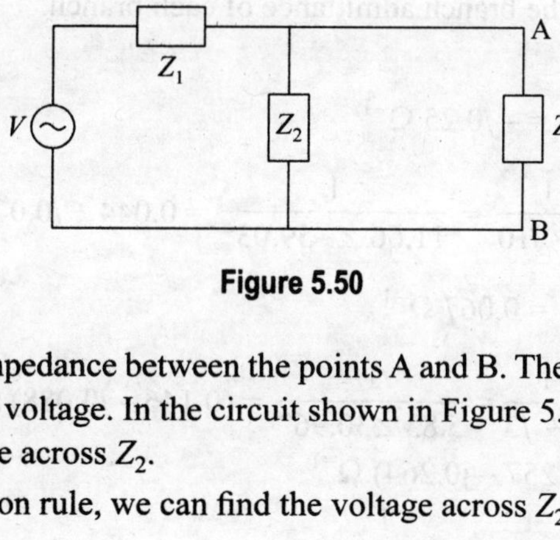

Figure 5.50

First, open circuit the impedance between the points A and B. The voltage across points A and B is Thevenin's equivalent voltage. In the circuit shown in Figure 5.50, the voltage across A and B is the same as the voltage across Z_2.

Using the voltage division rule, we can find the voltage across Z_2.

$$\therefore \quad V_{Th} = V\left(\frac{Z_2}{Z_1 + Z_2}\right) \tag{5.150}$$

To find Thevenin's equivalent impedance, first we should short circuit the voltage source, and looking from points A to B, Z_1 and Z_2 will be in parallel.

$$\therefore \quad Z_{Th} = \left(\frac{Z_1 Z_2}{Z_1 + Z_2}\right) \tag{5.151}$$

Then, Thevenin's voltage source, Thevenin's impedance, and load impedance are connected in series as shown in Figure 5.51.

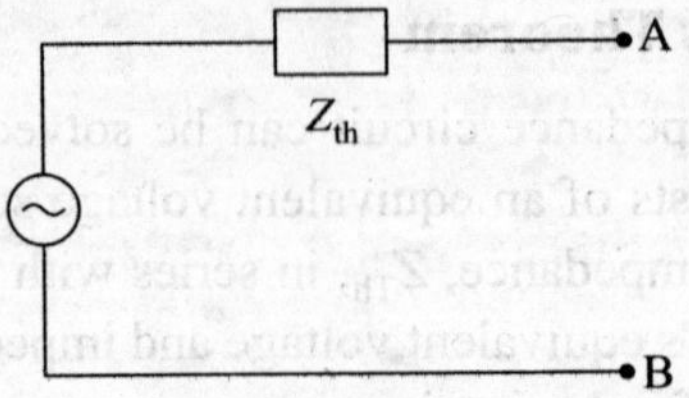

Figure 5.51

EXAMPLE 5.30 **Find Thevenin's equivalent circuit for the circuit shown in Figure 5.52.**

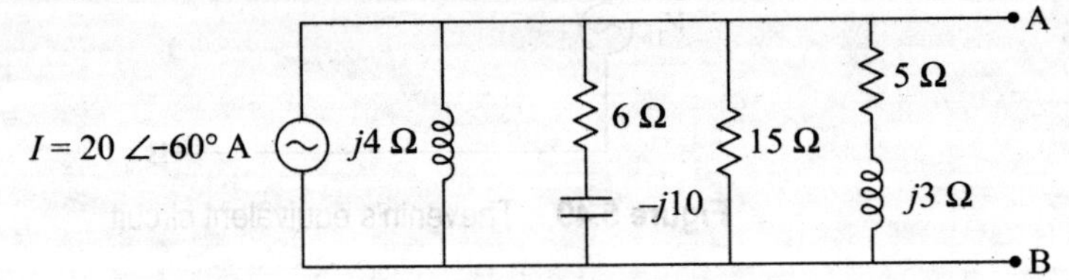

Figure 5.52

Solution Let us open circuit the current source to find Thevenin's equivalent impedance. However, assuming the equivalent admittance to be Y_{eq}, we find that $Y_{eq} = Y_1 + Y_2 + Y_3 + Y_4$, where $Y_1 \cdots Y_4$ is the branch admittance of each branch.

Obviously,

$$Y_1 = \frac{1}{j4} = -j0.25\ \Omega^{-1}$$

$$Y_2 = \frac{1}{6 - j10} = \frac{1}{11.66\angle -59.03^\circ} = 0.044 + j0.074\ \Omega^{-1}$$

$$Y_3 = \frac{1}{15} = 0.067\ \Omega^{-1}$$

$$Y_4 = \frac{1}{5 + j3} = \frac{1}{5.83\angle 30.96^\circ} = (0.146 - j0.088)\ \Omega^{-1}$$

$$\therefore \quad Y_{eq} = (0.257 - j0.264)\ \Omega^{-1}$$

$$Z_{Th} = \frac{1}{Y_{eq}} = \frac{1}{0.257 - j0.264}$$

$$= \frac{1}{0.368\angle -45.7^\circ}$$

$$= 2.72\angle 45.7^\circ\ \Omega$$

$$V_{Th} = V_{AB} = \frac{I}{Y_{eq}} = \frac{20\angle -60^\circ}{0.368\angle -45.7^\circ}$$

$$= 54.34\angle -14.3^\circ\ \text{V}$$

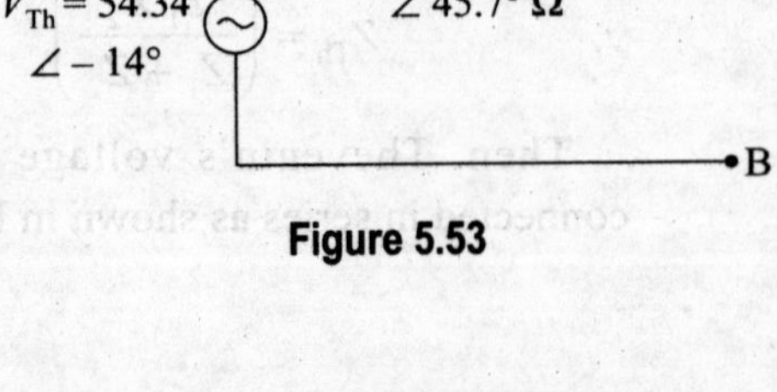

Figure 5.53

Figure 5.53 shows Thevenin's equivalent circuit, where $Z_{Th} = 2.72\angle 45.7^\circ\ \Omega$ and $V_{Th} = 54.34\angle -14.3^\circ$ V.

EXAMPLE 5.31 **Obtain Thevenin's equivalent at terminals A–B of the bridge circuit shown in Figure 5.54.**

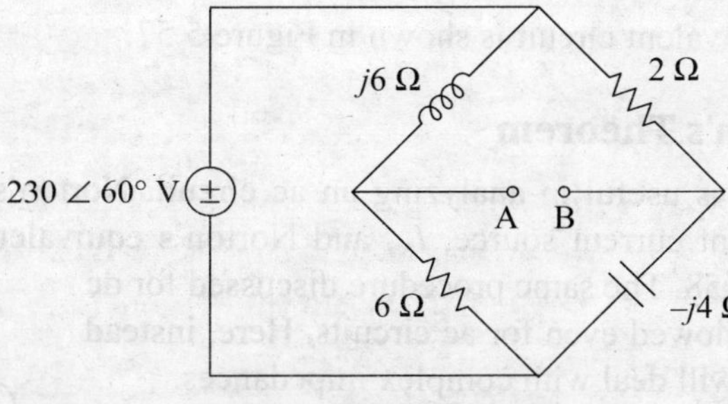

Figure 5.54

Solution To find Z_{Th} by short circuiting the voltage source, circuit is redrawn as shown in Figure 5.55.

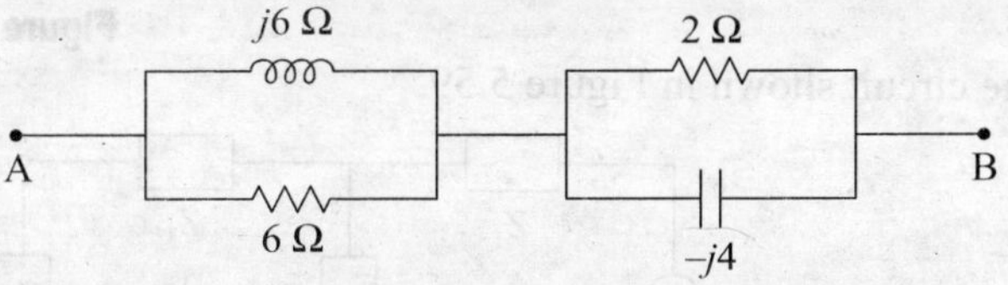

Figure 5.55

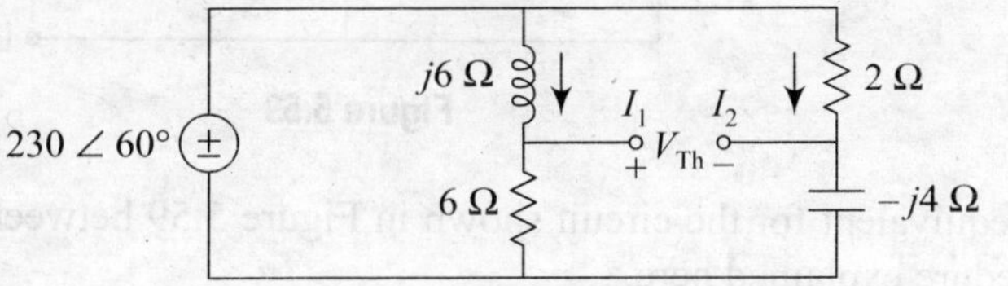

Figure 5.56

$$Z_1 = 6 \parallel j6 = \frac{6 \times j6}{6 + j6} = \frac{36 \angle 90^\circ}{8.48 \angle 45^\circ} = 4.2 \angle 45^\circ = 2.97 + j2.97$$

$$Z_2 = 2 \parallel -j4 = \frac{2 \times (-j4)}{2 - j4} = \frac{8 \angle -90^\circ}{4.47 \angle -63.43^\circ} = 1.78 \angle -26.57^\circ = 1.59 - j0.796$$

$$Z_{Th} = Z_1 + Z_2 = 2.97 + j2.97 + 1.59 - j0.796 = 4.56 + j2.174 = 5.05 \angle 25.49^\circ$$

To find V_{Th} the circuit is redrawn as shown in Figure 5.56.

$$I_1 = \frac{230 \angle 60^\circ}{6 + j6} = 27.12 \angle 15^\circ \text{ A}$$

$$I_2 = \frac{230 \angle 60^\circ}{2 - j4} = 51.45 \angle 123.43^\circ \text{ A}$$

$$V_{Th} = 2I_2 - j6\,(I_1)$$

$$= 2 \times 51.45 \angle 123.43^\circ - j6 \times 27.12 \angle 15^\circ$$

$$= -56.68 + j85.87 + 42.11 - j157.17$$

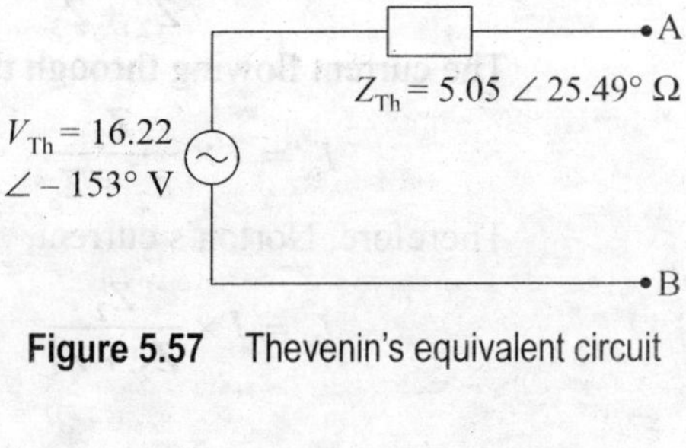

Figure 5.57 Thevenin's equivalent circuit

$= -14.57 - j71.3$

$= 16.22 \angle -153.92° \text{ V}$

Thevenin's equivalent circuit is shown in Figure 5.57.

5.10.2 Norton's Theorem

Norton's theorem is useful in analyzing an ac circuit. Norton's equivalent circuit consists of Norton's equivalent current source, I_N, and Norton's equivalent impedance, Z_N, arranged as shown in Figure 5.58. The same procedure discussed for dc circuits will be followed even for ac circuits. Here, instead of resistances we will deal with complex impedances.

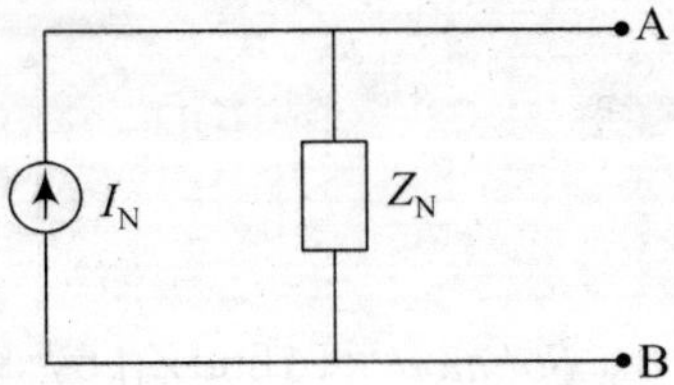

Figure 5.58 Norton's equivalent circuit

Norton's current is equal to the current passing through the short-circuited load terminals, and the values of impedance are equal to the impedance seen into the network across the load terminals.

Consider the circuit shown in Figure 5.59.

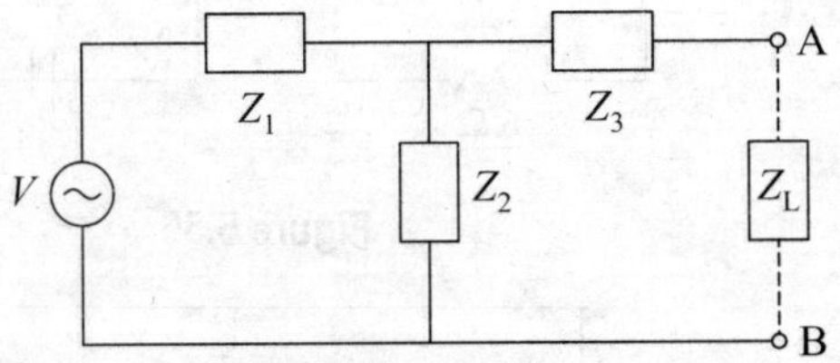

Figure 5.59

Norton's equivalent for the circuit shown in Figure 5.59 between points A and B is found as per the procedure explained here.

First, short circuit the load impedance Z_L between points A and B and find the short-circuit current. This is called as Norton's equivalent current source.

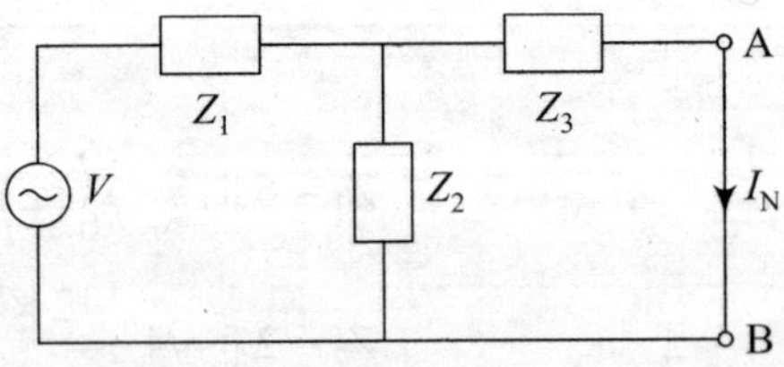

Figure 5.60

Figure 5.60 shows the circuit when load impedance is short circuited. From Figure 5.60, it is observed that Z_2 and Z_3 are in parallel and the resultant is in series with Z_1. Therefore, the current supplied by voltage source is

$$I = \frac{V}{Z_1 + \dfrac{Z_2 Z_3}{Z_2 + Z_3}} \tag{5.152}$$

The current flowing through the short-circuited terminals A and B

$$I_{sc} = I \times \frac{Z_2}{Z_2 + Z_3}$$

Therefore, Norton's current,

$$I_N = I \times \frac{Z_2}{Z_2 + Z_3} \tag{5.153}$$

To find Norton's impedance, open circuit the load impedance between points A and B and then short circuit the voltage source (open circuit if it is the current source). Looking from open-circuited terminals, find the equivalent impedance which is called as Norton's equivalent impedance.

From Figure 5.61, it is observed that Z_1 and Z_2 are in parallel and the resultant is in series with Z_3.

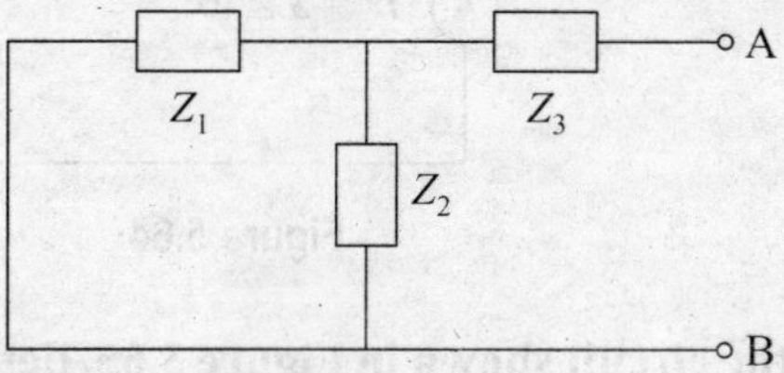

Figure 5.61

Therefore,

$$Z_N = \left(\frac{Z_1 Z_2}{Z_1 + Z_2}\right) + Z_3 \tag{5.154}$$

Norton's equivalent circuit is shown in Figure 5.62.

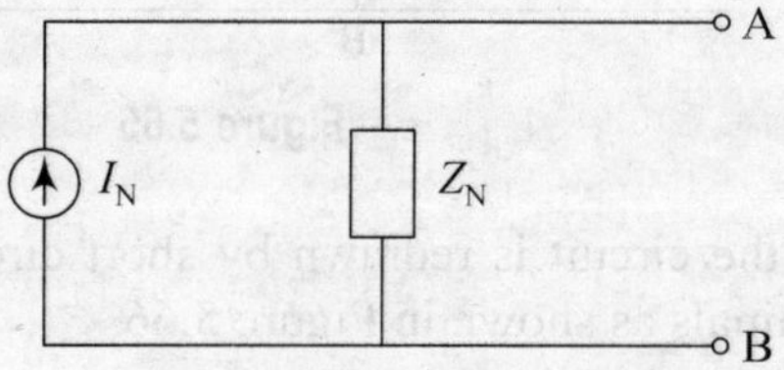

Figure 5.62

EXAMPLE 5.32 Thevenin's equivalent circuit is shown in Figure 5.63. Determine Norton's equivalent circuit.

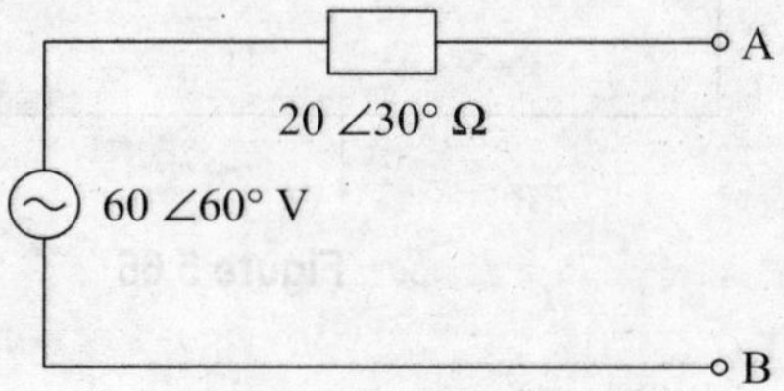

Figure 5.63

Solution Thevenin's voltage = $V_{Th} = 60\ \angle 60°$ V

Thevenin's impedance = $Z_{Th} = 20\ \angle 30°\ \Omega$

$$\therefore \quad I_N = \frac{V_{Th}}{Z_{Th}} = \frac{60\ \angle 60°}{20\ \angle 30°} = 3\ \angle 30°\ \text{A}$$

Obviously, Thevenin's impedance will be equal to Norton's impedance, hence

$$Z_N = Z_{Th} = 20\ \angle 30°\ \Omega$$

Norton's equivalent is shown in Figure 5.64.

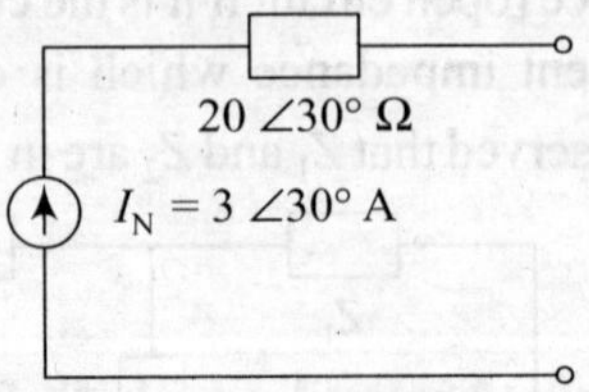

Figure 5.64

EXAMPLE 5.33 **For the circuit shown in Figure 5.65, determine the load current I_L by using Norton's theorem.**

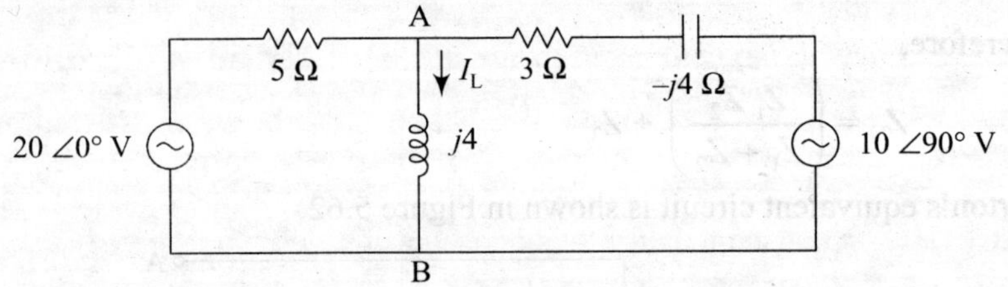

Figure 5.65

Solution To find Z_N, the circuit is redrawn by short circuiting the voltage sources and open circuiting the load terminals as shown in Figure 5.66.

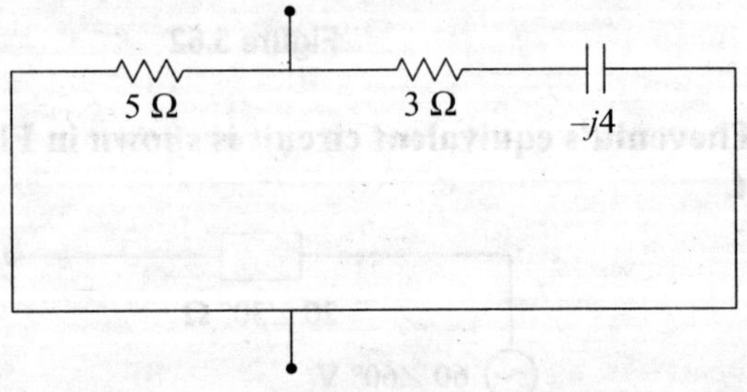

Figure 5.66

$$Z_N = Z_{AB} = 5 \parallel (3 - j4)$$

$$= \frac{5 \times (3 - j4)}{(5 + 3 - j4)}$$

$$= \frac{25 \angle -53.13^\circ}{8.94 \angle -26.56^\circ}$$

$$= 2.79 \angle -26.57^\circ \ \Omega$$

To find I_N, by short circuiting the load terminals, the circuit is redrawn as shown in Figure 5.67.

$$I_N = I_{AB} = \frac{20 \angle 0^\circ}{5} + \frac{10 \angle 90^\circ}{3 - j4}$$

$$= 4 + j0 - 1.59 + j1.2$$
$$= 2.41 + j1.2 = 2.69\ \angle 26.46°\ \text{A}$$

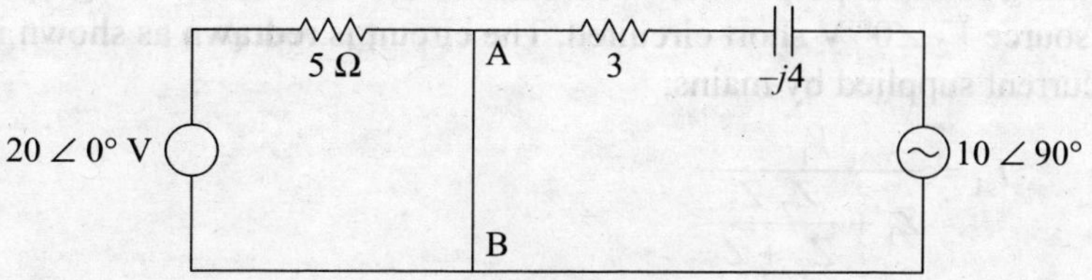

Figure 5.67

Norton's equivalent circuit is drawn as shown in Figure 5.68.

The current through the load terminals is determined using the current division rule.

$$I_L = I_N \times \frac{2.79\ \angle -26.57°}{2.79\ \angle -26.57° + 4\ \angle 90°}$$

$$= 2.69\ \angle 26.46° \times \frac{2.79\ \angle -26.57°}{2.49 - j1.24 + j4}$$

$$= 2.69\ \angle 26.46° \times 0.75\ \angle -74.47°$$

$$= 2.01\ \angle -47.84°\ \text{A}$$

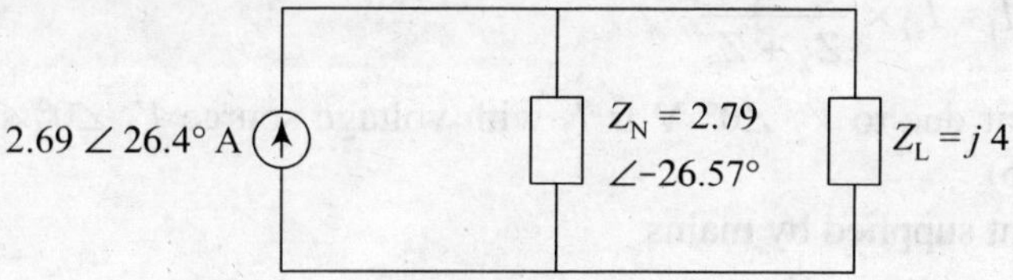

Figure 5.68

5.10.3 Superposition Theorem

Similar to dc circuits, the superposition theorem can also be used to analyze ac circuits containing more than one source. This theorem states that the response in any element in a circuit is the phasor sum of the responses that can be expected to flow if each source acts independently of the other sources. While considering each source, all the other sources are replaced by their internal impedances. If no internal impedance is present, then the voltage source is short circuited and current source is open circuited. This theorem is valid only for linear systems.

Consider a circuit as shown in Figure 5.69.

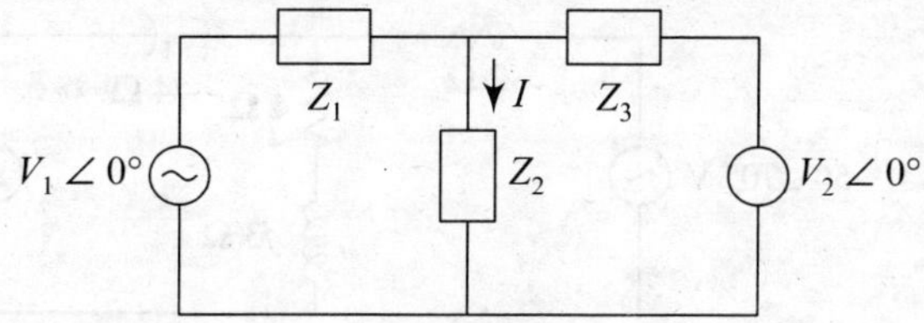

Figure 5.69

Now, let us find the current I passing through the impedance Z_2.

According to the superposition theorem, the current due to voltage source $V_1 \angle 0°$ is I_1 with voltage source $V_2 \angle 0°$ V short circuited. The circuit is redrawn as shown in Figure 5.70(a).

The current supplied by mains,

$$I_{s1} = \frac{V_1}{Z_1 + \dfrac{Z_2 Z_3}{Z_2 + Z_3}} \tag{5.155}$$

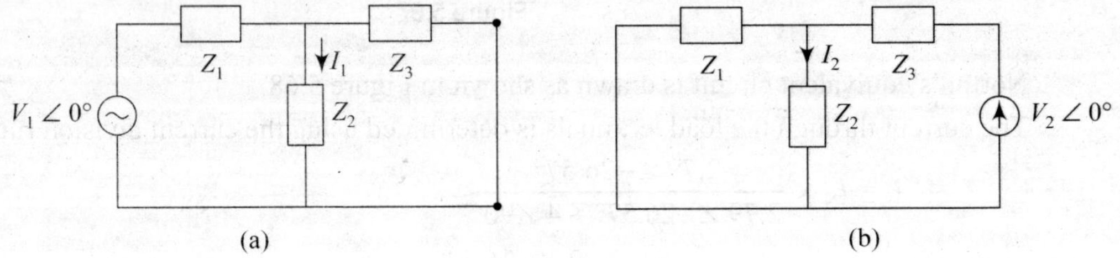

Figure 5.70

The current flowing through Z_2, i.e.,

$$I_1 = I_{s1} \times \frac{Z_3}{Z_2 + Z_3} \tag{5.156}$$

The current due to $V_2 \angle 0°$ V is I_2 with voltage source $V_1 \angle 0°$ short circuited as shown in Figure 5.70(b).

The current supplied by mains

$$I_{s2} = \frac{V_2}{Z_3 + \dfrac{Z_1 Z_2}{Z_1 + Z_2}} \tag{5.157}$$

The current flowing through Z_2, i.e.,

$$I_2 = I_{s2} \times \frac{Z_1}{Z_1 + Z_2} \tag{5.158}$$

The total current passing through the impedance Z_2 is

$$I = I_1 + I_2 \tag{5.159}$$

EXAMPLE 5.34 For the circuit shown in Figure 5.71, determine the current in (4 + *j*3) Ω by using the superposition theorem.

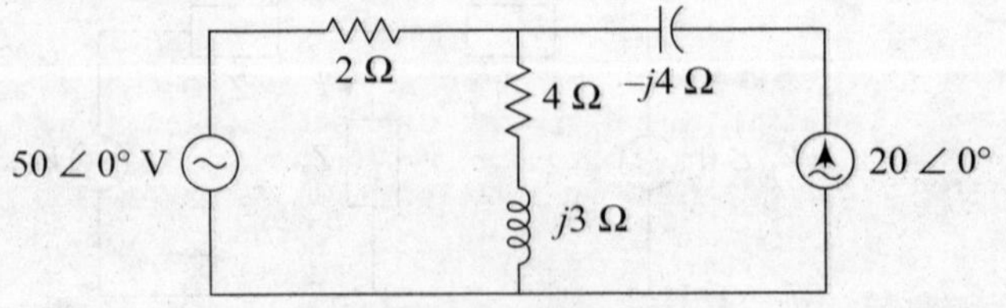

Figure 5.71

Solution The current in (4 + *j*3) Ω, when the voltage source 50 ∠0° acting alone is

$$I_1 = \frac{50\angle 0^\circ}{(6+j3)}$$

$$= \frac{50\angle 0^\circ}{6.7\angle 26.56^\circ}$$

$$I_1 = 7.46\angle -26.56^\circ$$

Current in $(4 + j3)\ \Omega$, when the current source $20\angle 0^\circ$ A acting alone is

$$I_2 = 20\angle 0^\circ \times \frac{2}{(6+j3)}$$

$$= \frac{40\angle 0^\circ}{6.7\angle 26.56^\circ}$$

$$= 5.97\angle -26.56^\circ \text{ A}$$

Total current in $(4 + j3)\ \Omega$ due to both the sources is

$$I = I_1 + I_2$$

$$= 7.46\angle -26.56^\circ + 5.97\angle -26.56^\circ$$

$$= 6.67 - j3.33 + 5.34 - j2.66$$

$$= 12.64 - j5.99$$

$$= 13.43\angle -26.34^\circ \text{ A}$$

Total current in $(4 + j3)\ \Omega$ is $I = 13.43\angle -26.34^\circ$ A.

EXAMPLE 5.35 **Use the superposition theorem to find I_0 in the circuit shown in Figure 5.72.**

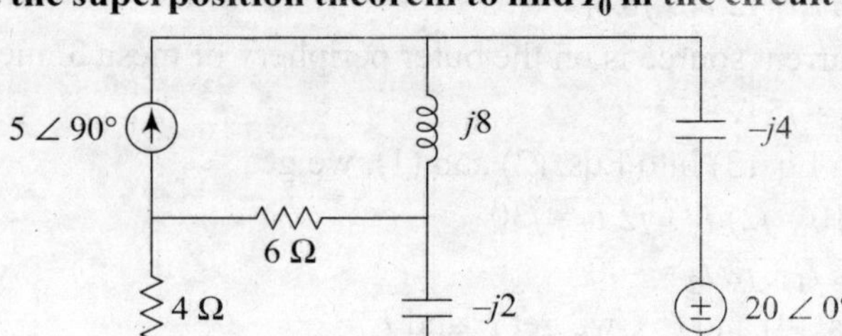

Figure 5.72

Solution Let I_1 and I_2 are currents due to the voltage and current source, respectively. When the voltage source $20\angle 0^\circ$ is acting alone, the current source will be open circuited and the circuit is redrawn as shown in Figure 5.73(a).

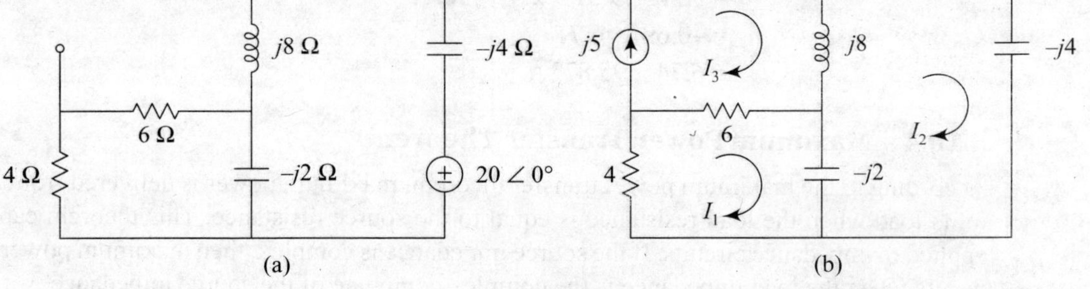

Figure 5.73

The equivalent impedance,

$$Z_1 = (10 \parallel (-j2) + j8 - j4$$
$$= \frac{10 \times -j2}{10 - j2} + j4$$
$$= \frac{20 \angle -90^\circ}{10.19 \angle -11.31^\circ} + j4$$
$$= 0.38 - j1.92 + j4$$
$$= 0.38 + j2.08$$
$$= 2.11 \angle 79.6^\circ \ \Omega$$

Current, $I_1 = \dfrac{20 \angle 0^\circ}{2.11 \angle 79.6^\circ} = 9.47 \angle -79.6^\circ$ A

When the current source 5 ∠90° is acting alone, the voltage source will be short circuited and the circuit is redrawn as shown in Figure 5.73(b).

To get I_2 we use the mesh analysis. Writing the KVL equation for mesh 1, we get

$$4I_1 + 6\,(I_1 - I_3) - j2\,(I_1 - I_2) = 0$$
$$(10 - j2)\,I_1 + j2\,I_2 - 6\,I_3 = 0 \qquad (1)$$

Writing the KVL equation for mesh 2, we get

$$-j4\,(I_2) - j2\,(I_2 - I_1) + j8\,(I_2 - I_3) = 0$$
$$j2\,I_1 + j2\,I_2 - j8\,I_3 = 0 \qquad (2)$$

Since the current source is on the outer periphery of mesh 3, mesh 3 current is

$$I_3 = j5 \text{ A} \qquad (3)$$

Substituting Eq. (3) into Eqs. (2) and (1), we get

$$(10 - j2)\,I_1 + j2\,I_2 = j30 \qquad (4)$$
$$j2\,I_1 + j2\,I_2 = -40 \qquad (5)$$

Solving Eqs. (4) and (5), we get I_1 and I_2.

$$I_1 = \frac{40 + j30}{10 - j4} = \frac{50 \angle 36.87^\circ}{10.77 \angle -21.8^\circ} = 4.64 \angle 858.67^\circ \text{ A}$$

$$I_2 = \frac{32.1 + j4.82}{-2j} = \frac{32.45 \angle 8.54^\circ}{2 \angle -90^\circ} = 16.2 \angle 98.54^\circ \text{ A}$$

The current I_0 is given by

$$I_0 = I_1 + I_2 = 9.47 \angle -79.6^\circ + 16.2 \angle 98.54^\circ$$
$$= 1.71 - j9.31 - 2.4 + j16.02$$
$$= -0.69 + j6.71$$
$$= 6.74 \angle 95.87^\circ \text{ A}$$

5.10.4 Maximum Power Transfer Theorem

According to the maximum power transfer theorem, maximum power is delivered from a source to its load when the load resistance is equal to the source resistance. This theorem can also be applied to impedance circuits. If the source impedance is complex, then maximum power transfer occurs when the load impedance is the complex conjugate of the source impedance.

Consider the circuit shown in Figure 5.74.

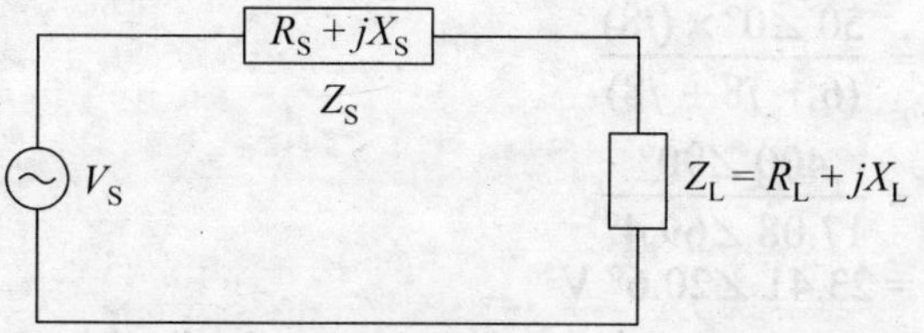

Figure 5.74

Current passing through the circuit is given by

$$I = \frac{V_S}{(R_S + jX_S) + (R_L + jX_L)} \tag{5.160}$$

$$\text{Magnitude of current, } I = |I| = \frac{V_S}{\sqrt{(R_S + R_L)^2 + (X_S + X_L)^2}} \tag{5.161}$$

Power delivered to the circuit is

$$P = I^2 R_L = \frac{V_S^2 R_L}{(R_S + R_L)^2 + (X_S + X_L)^2} \tag{5.162}$$

In Eq. (5.162), if R_L is fixed, the value of P is maximum when $X_S = -X_L$. Then the power

$$P = \frac{V_S^2 R_L}{(R_S + R_L)^2} \tag{5.163}$$

If R_L is variable, maximum power is transferred when the load resistance is equal to the source resistance. If $R_L = R_S$ and $X_L = -X_S$, then $Z_L = Z_S^*$. This means that maximum power transfer occurs when the load impedance is equal to the complex conjugate of the source impedance Z_S.

EXAMPLE 5.36 **For the circuit shown in Figure 5.75(a), find the load impedance for which the source delivers maximum power to the load. Also, find the amount of maximum power.**

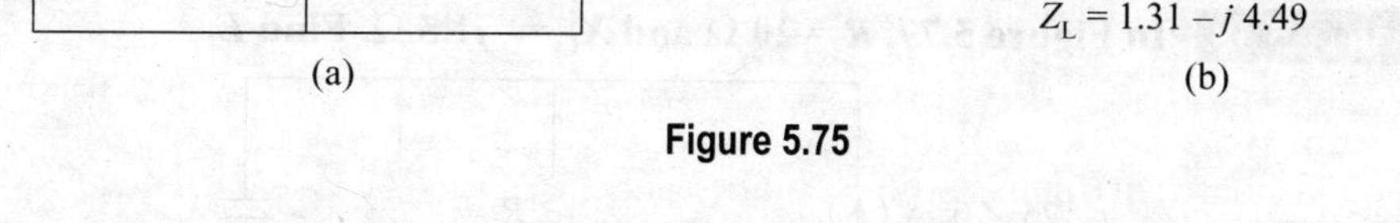

Figure 5.75

Solution First, let us remove the load impedance Z_L. Looking back into the network through the open-circuited load terminals, we find Thevenin's equivalent circuit.

$$Z_{Th} = \frac{j8(6 + j8)}{6 + j8 + j8} = \frac{-64 + j48}{6 + j16} = 4.68 \angle 73.7° \ \Omega = 1.31 + j4.49 \ \Omega$$

Using the voltage division rule

$$V_{Th} = \frac{50 \angle 0° \times (j8)}{(6 + j8 + j8)}$$

$$= \frac{400 \angle 90°}{17.08 \angle 69.4°}$$

$$= 23.41 \angle 20.6° \text{ V}$$

As per the maximum power transfer theorem, the load impedance (Z_L) must be equal to the complex conjugate of the source impedance (i.e., Z_{Th}), i.e., $Z_L = Z_{Th}^* = (1.31 - j\,4.49)\ \Omega$. The circuit is redrawn by connecting the load impedance in series with Thevenin's equivalent circuit as shown in Figure 5.75(b).

Maximum power is given by

$$P_{max} = I^2 R_L$$

$$= \left(\frac{23.41}{2.62}\right)^2 \times 1.31$$

$$= 104.58 \text{ W}$$

EXAMPLE 5.37 **Find the value of R_L in Figure 5.76 that will absorb maximum power and find that maximum power.**

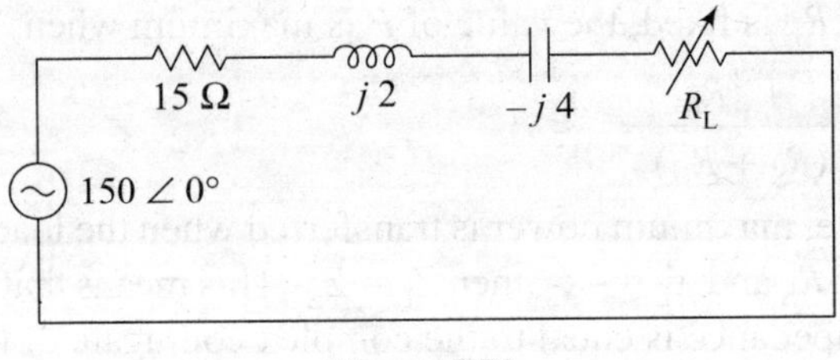

Figure 5.76

Solution Let us remove R_L. Thevenin's impedance is $Z_{Th} = 15 + j2 - j4 = (15 - j2)\ \Omega$. Thevenin's voltage across the open circuit of R_L is given by $V_{Th} = 150 \angle 0°$ V. As per the maximum power transfer theorem, the load resistance must be equal to the source resistance.

$$R_L = 15\ \Omega$$

Maximum power transfer is

$$P_{max} = \frac{V_{Th}^2}{4R_L} = \frac{150^2}{4 \times 15} = 375 \text{ W}$$

EXAMPLE 5.38 **In Figure 5.77, $R = 20\ \Omega$ and $X_C = -j1.5\ \Omega$. Find E.**

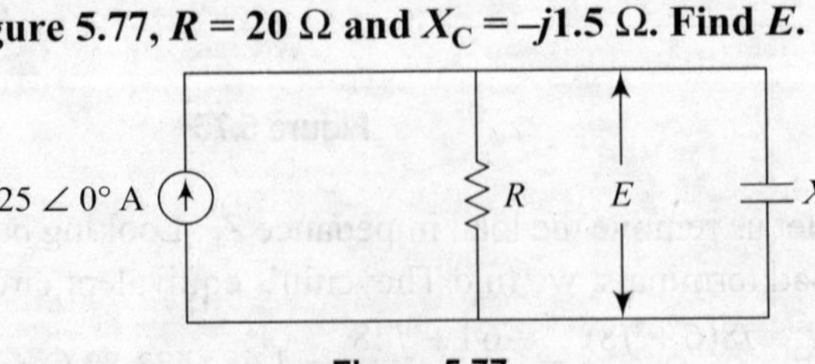

Figure 5.77

Solution Current through the resistance

$$R = 25\angle 0° \times \frac{-j1.5}{20 - j1.5}$$

$$= \frac{37.5\angle{-90°}}{20.05\angle{-4.29°}}$$

$$= 1.87\angle{-85.7°}\text{ A}$$

$$E = IR = 1.87\angle{-85.7°} \times 20 = 37.4\angle{-85.7°}\text{ V}$$

EXAMPLE 5.39 **In Figure 5.78, find *I*.**

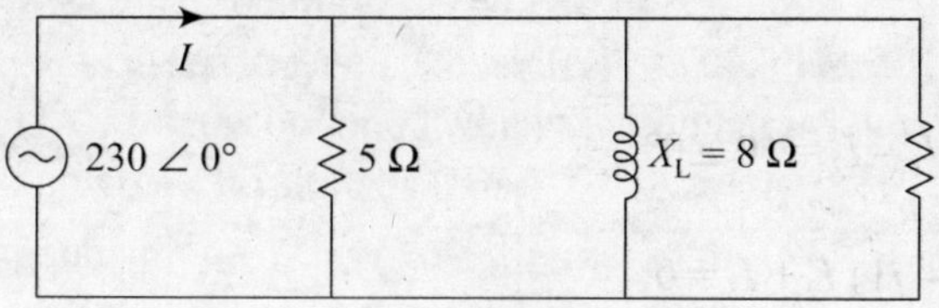

Figure 5.78

Solution $Y_R = G = \frac{1}{R}\angle 0° = 0.2\angle 0° = (0.2 + 0)\ \Omega^{-1}$

$$Y_{X_L} = B_L\angle{-90°} = \frac{1}{X_L}\angle{-90°} = 0.125\angle{-90°}\ \Omega^{-1}$$

$$Y_{X_C} = B_C\angle 90° = \frac{1}{X_C}\angle 90° = 0.05\angle 90°\ \Omega^{-1}$$

$$Y_{eq} = Y_R + Y_{X_L} + Y_{X_C}$$

$$= 0.2\angle 0° + 0.125\angle{-90°} + 0.05\angle 90°$$

$$= (0.2 + j0) + (0 - j0.125) + (0 + j0.05)$$

$$= (0.2 + j0.075)\ \Omega^{-1}$$

$$= 0.214\angle{-20.6°}$$

$$Z_{eq} = \frac{1}{Y_{eq}} = \frac{1}{0.214\angle{-20.6°}}$$

$$= 4.67\angle 20.6°\ \Omega$$

Thus,

$$I = \frac{V}{Z} = \frac{230\angle 0°}{4.67\angle 20.6°} = 49.12\angle{-20.6°}\text{ A}$$

EXAMPLE 5.40 **For the circuit shown in Figure 5.79, what should be the value of V_2 such that its current will be zero?**

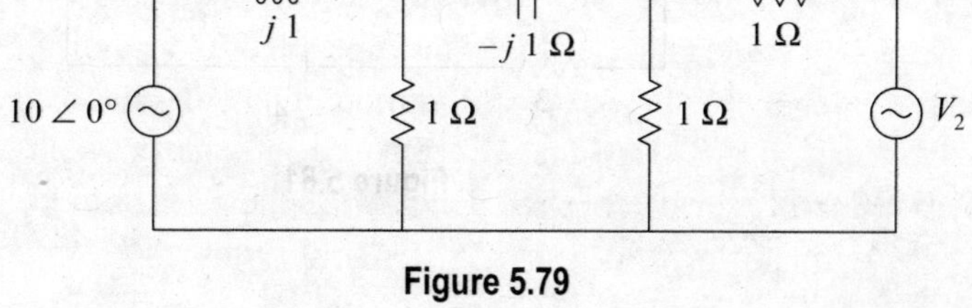

Figure 5.79

Solution After assigning the mesh currents the circuit is redrawn as shown in Figure 5.80. Let I_1, I_2, and I_3 are the mesh currents of meshes 1, 2, and 3, respectively.

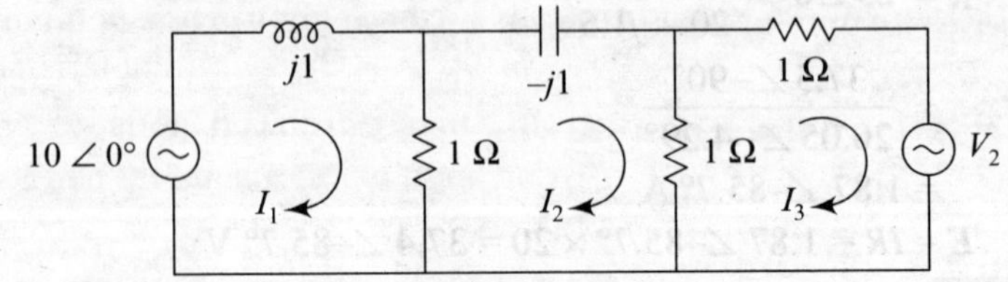

Figure 5.80

For mesh 1,

$$(1 + j1)\, I_1 - I_2 = 10\,\angle 0°$$

For mesh 2,

$$-I_1 + (2 - j1)\, I_2 + I_3 = 0$$

For mesh 3,

$$I_2 + 2I_3 = V_2$$

Writing these equations in matrix form

$$\begin{bmatrix} 1+j1 & -1 & 0 \\ -1 & 2-j1 & 1 \\ 0 & 1 & 2 \end{bmatrix} \begin{bmatrix} I_1 \\ I_2 \\ I_3 \end{bmatrix} = \begin{bmatrix} 10\,\angle 0° \\ 0 \\ V_2 \end{bmatrix}$$

For $I_3 = 0$

$$I_3 = \frac{\begin{vmatrix} 1+j1 & -1 & 10\,\angle 0° \\ -1 & 2-j1 & 0 \\ 0 & 1 & V_2 \end{vmatrix}}{\begin{vmatrix} 1+j1 & -1 & 0 \\ -1 & 2-j1 & 1 \\ 0 & 1 & 2 \end{vmatrix}} = \frac{V_2\,(2 + j1) - 10}{3 + j1} = 0$$

$$\text{Therefore, } V_2 = \frac{10\,\angle 0°}{2 + j1} = 4.48\,\angle{-26.56°}\text{ V}$$

EXAMPLE 5.41 For the circuit shown in Figure 5.81, the voltmeter connected across the 5-Ω resistor reads 50 V. Find current *I*.

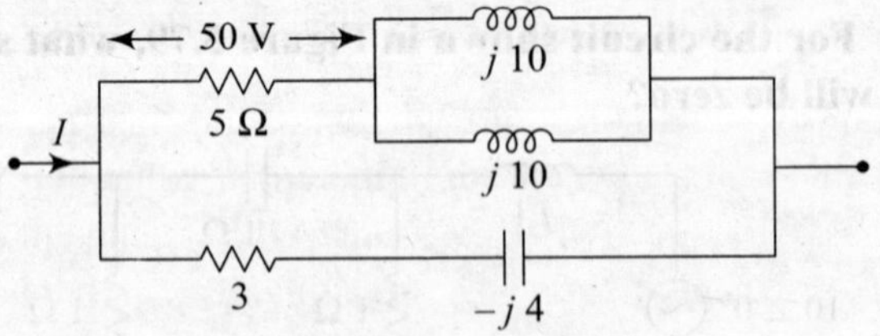

Figure 5.81

Solution The voltage across the 5-Ω resistor is 50 V. Therefore, the current flowing through the 5-Ω resistor is 50/5 = 10 A.

For the first parallel branch, equivalent impedance $= j10 \parallel j10 = j5\ \Omega$
Impedance, $Z_1 = (5 + j5)\ \Omega$
Voltage across impedance, $Z_1 = 10 \times (5 + j5) = 50 + j50$

$$\text{Current through } Z_1 = I_1 = \frac{50 + j50}{5 + j5} = 10\text{ A}$$

$$\text{Current through } Z_2 = I_2 = \frac{50 + j50}{3 - j4} = 14.14\ \angle 98.13°\text{ A} = -2 + j14\text{ A}$$

Current, $I = I_1 + I_2 = 10 - 2 + j14 = 8 + j14 = 16.12\ \angle 60.2°$ A

EXAMPLE 5.42 **For the circuit shown in Figure 5.82, find the supply voltage.**

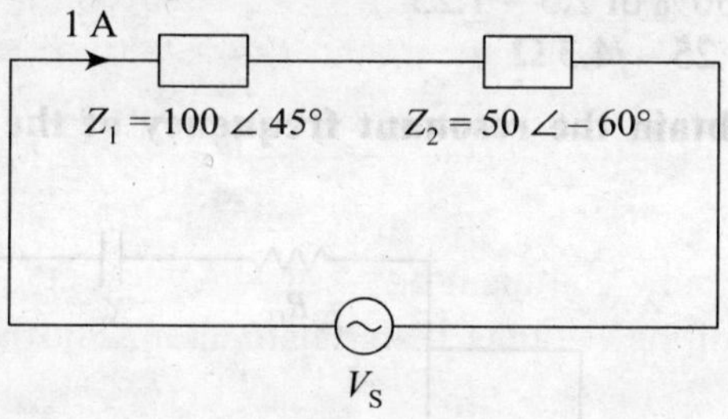

Figure 5.82

Solution The supply voltage

$$\begin{aligned} V_S &= I\,(Z_1 + Z_2) \\ &= 1(100\ \angle 45° + 50\ \angle -60°) \\ &= 95.7 + j27.4\text{ V} \end{aligned}$$

EXAMPLE 5.43 **Two elements are connected in series in a circuit. The voltage and current in the circuit are $v = 1000 \sin (1.732t + 30°)$, $i = 10 \sin (1.732t - 30°)$. What are the two elements and their values?**

Solution It is noted from the voltage and current equations that current i lags voltage v by 60°. Hence, the elements are resistance R and inductance L.

$$Z = \frac{V_{max}}{I_{max}} = \frac{1000}{10} = 100\ \Omega$$

Power factor $= \cos\theta = \cos 60° = 0.5$

But

$$\cos\theta = \frac{R}{Z}$$

Therefore,

$$\frac{R}{Z} = 0.5$$

$$R = 100 \times 0.5 = 50\ \Omega$$

$$X_L = \sqrt{Z^2 - R^2} = \sqrt{100^2 - 50^2} = 86.60\ \Omega$$

For $\omega = 1.732$, the inductance $L = 50$ H.

EXAMPLE 5.44 **In a series circuit the voltage across the circuit $V = 250\angle 0°$ and $I = 50\angle -60°$. If the resistance is reduced to 50%, what is the new impedance?**

Solution The old impedance, $Z_{old} = \dfrac{250}{50} = 5\ \Omega$

Power factor angle, $\phi = 60°$

Power factor $= \cos 60° = 0.5$

$$\therefore \quad \frac{R}{Z_{old}} = 0.5$$

$$R = 2.5\ \Omega$$

$$X_C = \sqrt{Z^2 - R^2} = \sqrt{5^2 - 2.5^2} = 4.3\ \Omega$$

As

$$R_{new} = 50\% \text{ of } 2.5 = 1.25$$

$$Z_{new} = 1.25 - j4.3\ \Omega$$

EXAMPLE 5.45 **Obtain the resonant frequency of the given parallel circuit shown in Figure 5.83.**

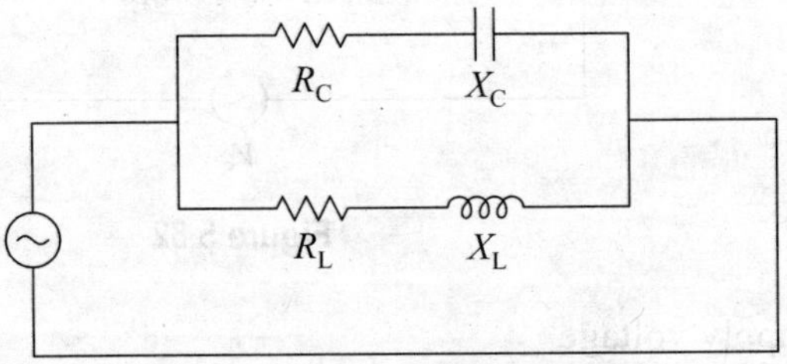

Figure 5.83 Parallel circuit

Solution The admittance of the given circuit is as follows:

$$Y = \frac{1}{R_L + j\omega L} + \frac{1}{R_C - j/\omega C}$$

$$= \frac{R_L - j\omega L}{R_L^2 + \omega^2 L^2} + \frac{R_C + j/\omega C}{R_C^2 + (1/\omega C)^2}$$

$$= \frac{R_L}{R_L^2 + \omega^2 L^2} + \frac{R_C}{R_C^2 + (1/\omega C)^2} + j\left[\frac{1j/\omega C}{R_C^2 + (1/\omega C)^2} - \frac{\omega L}{R_L^2 + \omega^2 L^2}\right]$$

At resonance the susceptance part becomes zero.

$$\frac{\omega_r L}{R_L^2 + \omega_r^2 L^2} = \frac{1/\omega_r C}{R_C^2 + (1/\omega_r C)^2}$$

$$\omega_r L\left[R_C^2 + \frac{1}{\omega_r^2 C^2}\right] = \frac{1}{\omega_r C}\left[R_L^2 + \omega_r^2 L^2\right]$$

$$\omega_r^2\left[R_C^2 + \frac{1}{\omega_r^2 C^2}\right] = \frac{1}{LC}\left[R_L^2 + \omega_r^2 L^2\right]$$

$$\omega_r^2 R_C^2 - \frac{\omega_r^2 L}{C} = \frac{1}{LC} R_L^2 - \frac{1}{C^2}$$

$$\omega_r^2\left[R_C^2 - \frac{L}{C}\right] = \frac{1}{LC}\left[R_L^2 + \frac{L}{C}\right]$$

$$\omega_r = \frac{1}{\sqrt{LC}}\sqrt{\frac{R_L^2 - L/C}{R_C^2 - L/C}} \qquad (1)$$

The condition for resonant frequency is given by Eq. (1). As a special case, if $R_L = R_C$, then Eq. (1) becomes $\omega_r = 1/\sqrt{LC}$. Therefore, $f_r = 1/2\pi\sqrt{LC}$.

EXAMPLE 5.46 **A sinusoidal voltage $v = 100 \sin \omega t$ is applied to a series R_L circuit. The current in the circuit is given by $i = 50 \sin (\omega t - 30°)$. Determine the (i) apparent power, (ii) power factor, and (iii) real power.**

Solution

(i) Apparent power, $S = V_{rms} \times I_{rms} = \frac{100}{\sqrt{2}} \times \frac{50}{\sqrt{2}} = 2500$ VA

(ii) Power factor = cos 30° = 0.866

(iii) Real power, $P = S \times$ Power factor $= 2500 \times 0.866 = 2165$ W

EXAMPLE 5.47 **Determine the voltage across each element of the circuit as shown in Figure 5.84.**

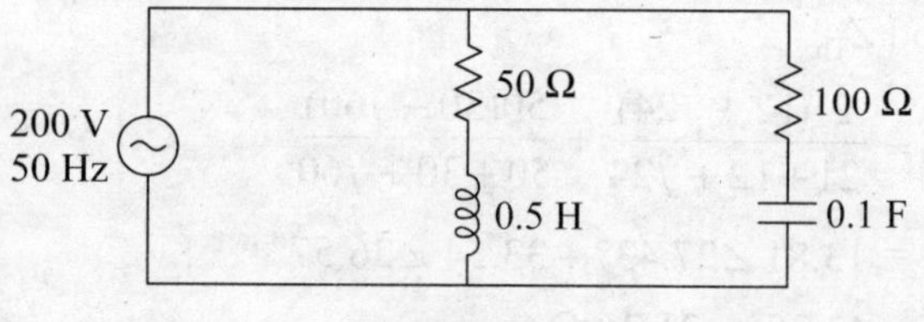

Figure 5.84

Solution $X_L = 2\pi f L = 314 \times 0.5 = 157.08\ \Omega$

$$X_C = \frac{1}{2\pi f C} = \frac{1}{314 \times 0.1} = 0.032\ \Omega$$

$$Z_1 = \sqrt{R^2 + X_L^2} = \sqrt{50^2 + 157.08^2} = 164.84\ \Omega$$

$$Z_2 = \sqrt{R^2 + X_C^2} = \sqrt{100^2 + 0.032^2} = 100\ \Omega$$

$$I_1 = \frac{V}{Z_1} = \frac{200}{164.84} = 1.21\text{ A}$$

$$I_2 = \frac{V}{Z_2} = \frac{200}{100} = 2\text{ A}$$

$$V_{R_1} = I_1 R_1 = 1.21 \times 50 = 60.5\text{ V}$$

$$V_{R_2} = I_2 R_2 = 2 \times 100 = 200\text{ V}$$

$$V_L = I_1 X_L = 1.21 \times 157.08 = 190.06\text{ V}$$

$$V_C = I_2 X_C = 2 \times 0.032 = 0.064\text{ V}$$

$$\theta_1 = \tan^{-1}\left(\frac{X_L}{R_1}\right) = \tan^{-1}\left(\frac{157.08}{50}\right) = 72.34°$$

$$\theta_2 = \tan^{-1}\left(\frac{-X_C}{R_2}\right) = \tan^{-1}\left(\frac{-0.032}{100}\right) = -0.02°$$

EXAMPLE 5.48 Obtain Thevenin's equivalent for the bridge circuit shown in Figure 5.85.

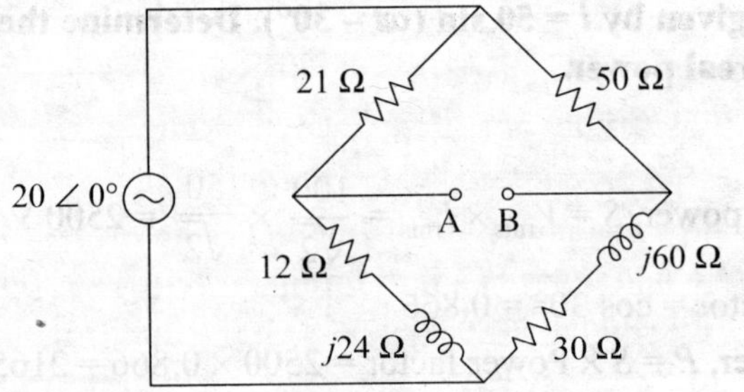

Figure 5.85

Solution To find Z_{Th},

$$Z_{Th} = \frac{21(12 + j24)}{21 + 12 + j24} + \frac{50(30 + j60)}{50 + 30 + j60}$$

$$= 13.81 \angle 27.43° + 33.54 \angle 26.57°$$

$$= 42.25 + j21.36\ \Omega$$

$$= 47.34 \angle 26.81°$$

To find V_{Th},

$$V_{Th} = V_{AB} = V_A - V_B$$

$$= \frac{20 \angle 0°\ (12 + j24)}{(33 + j24)} - \frac{20 \angle 0°\ (30 + j60)}{(80 + j60)}$$

$$= 13.15 \angle 27.43° - 13.416 \angle 26.37°$$

$$= 0.362 \angle 164°\text{ V}$$

Thevenin's equivalent circuit is drawn as shown in Figure 5.86.

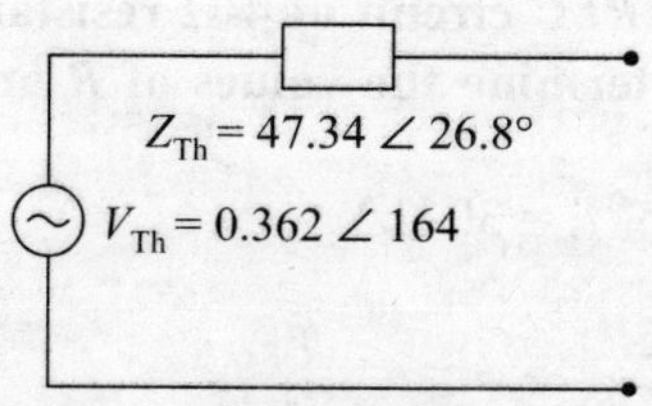

Figure 5.86

EXAMPLE 5.49 **In the circuit shown in Figure 5.87, determine the voltage applied across AB which will lead to a current of 10 A to flow in the capacitor.**

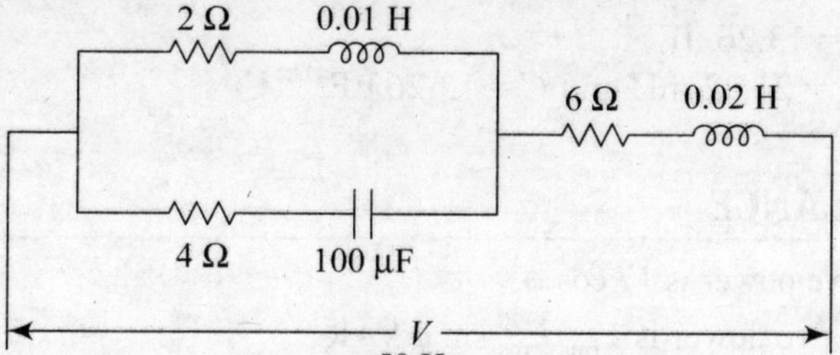

Figure 5.87

Solution Z_1 and Z_2 are in parallel

$$Z_1 = 2 + j314 \times 0.01 = 2 + j3.14\ \Omega$$

$$Z_2 = 4 - j\frac{1}{314 \times 100 \times 10^{-6}} = 4 - j31.84\ \Omega$$

$$Z_3 = 6 + j314 \times 0.02 = 6 + j6.28 = 8.68 \angle 46.3°\ \Omega$$

Current through Z_2 is 10 A.

The voltage across parallel branch $= Z_2 \times 10$

$$= (4 - j31.84) \times 10$$

$$= 40 - j318.4 \text{ V}$$

$$\text{Current through } Z_1 = \frac{40 - j318.4}{2 + j3.14} = \frac{320 \angle -82.84°}{3.72 \angle 57.5°}$$

$$= 86.02 \angle -139.5° \text{ A}$$

Total current $I = 10 + 86.02 \angle -139.5° = 10 - 65.41 - j55.86$

$$= -55.41 - j55.86 = 78.68 \angle -134.76° \text{ A}$$

Voltage across $Z_3 = 8.68 \angle 46.3° \times 78.68 \angle -134.76°$

$$= 682 \angle -88.460 = 18.32 - j681.75 \text{ V}$$

Voltage across AB $= 40 - j318.4 + 18.32 - j681.75$ V

$$= 58.32 - j1000.15 = 1001.84 \angle -86.66° \text{ V}$$

EXAMPLE 5.50 **A series *RLC* circuit of 4-Ω resistance should be designed to have a pass bandwidth of 30 Hz. Determine the values of *R* and *C* at which the circuit resonates at 300 Hz.**

Solution Bandwidth $= f_2 - f_1 = 30$ Hz

$$f_2 - f_1 = \frac{R}{2\pi L}$$

$$30 = \frac{4}{2\pi L} \Rightarrow L = 21.22 \text{ mH}$$

$$f_r = \frac{1}{2\pi\sqrt{LC}} \Rightarrow C = \frac{1}{4\pi^2 L f_r^2} = \frac{1}{4\pi^2 \times 21.22 \times 10^{-3} \times 300^2}$$

$$C = 13.26\ \mu\text{F}$$

Therefore, $L = 21.22$ mH and $C = 13.26$ μF.

CHAPTER AT A GLANCE

- The active power is $VI \cos \phi$.
- The reactive power is $V_{rms} I_{rms} \sin \phi$ VAR.
- The apparent power is VI.
- Power factor is the cosine of the phase angle between voltage and current (or) power factor = $\cos \phi$.
- The resonance of a circuit is defined as "the condition that occurs when a circuit's inductive and capacitive reactances are exactly equal."
- Resonant frequency, $f_r = 1/2\pi\sqrt{LC}$.
- Kirchhoff's current law (KCL) states that the algebraic sum of the currents at a node is equal to zero.
- Kirchhoff's voltage law (KVL) states that the algebraic sum of the potential rises and drops around any closed loop in an electric circuit is zero.
- Thevenin's voltage is open-circuit voltage across the load terminals.
- Thevenin's impedance is the impedance seen from load terminals keeping sources off.
- The maximum power is transferred when the load resistance is equal to the source resistance.

MULTIPLE CHOICE QUESTIONS

1. In pure resistor, the voltage and currents are
   (a) out of phase (b) in phase
   (c) 90° out of phase (d) 45° out of phase
2. In pure capacitor, the voltage
   (a) is in phase with the current (b) is out of phase with the current
   (c) lags behind the current by 90° (d) leads the current by 90°

3. The current in a pure inductor

(a) lags behind the voltage by 90° (b) leads the voltage by 90°

(c) is in phase with the voltage (d) lags behind the voltage by 45°

4. The phasor combination of restive power and reactive power is called

(a) true power (b) apparent power

(c) reactive power (d) average power

5. Apparent power is expressed in

(a) volt-ampere (b) watts

(c) volt-ampere or watts (d) VAR

6. A power factor of 0 indicates

(a) purely resistive element (b) purely reactive element

(c) both (a) and (b) (d) none of these

7. For a certain load, the true power is 100 W and the reactive power is 100 VAR. What is the apparent power?

(a) 200 VA (b) 100 VA (c) 141.4 VA (d) 120 VA

8. If a load is purely resistive and the true power is 5 W, what is the apparent power?

(a) 10 VA (b) 5 VA (c) 25 VA (d) √50 VA

9. True power is defined as

(a) $VI \cos \phi$ (b) VI (c) $VI \sin \phi$ (d) none of these

10. In a certain *RL* circuit, $V_R = 2$V and $V_L = 3$ V. What is the magnitude of the total voltage?

(a) 2 V (b) 3 V (c) 5 V (d) 3.61 V

11. When the frequency of applied voltage in a series *RL* circuit increases, what happens to the inductive reactance?

(a) Decreases (b) Remains the same

(c) Increases (d) Becomes zero

12. In a parallel *RL* circuit, $R = 50\ \Omega$, $X_L = 75\ \Omega$. What is the admittance?

(a) 0.024 S (b) 75 S (c) 50 S (d) 1.5 S

13. What is the phase angle between the inductor current and the applied voltage?

(a) 0° (b) 45° (c) 90° (d) 30°

14. In a certain series *RC* circuit, $V_R = 4$ V and $V_C = 6$ V. What is the magnitude of the total voltage?

(a) 7.2 V (b) 4 V (c) 6 V (d) 52V

15. When the frequency of applied voltage in a series *RC* circuit increases, what happens to the capacitive reactance?

(a) Decreases (b) Remains the same

(c) Increases (d) Becomes zero

16. In a certain parallel *RC* circuit, $R = 50\ \Omega$, $X_C = 75\ \Omega$. What is the admittance?

(a) 0.033 S (b) 0.02 S (c) 50 S (d) 75 S

17. What is the phase angle between the capacitor current and the applied voltage?

(a) 0° (b) 45° (c) 90° (d) 180°

18. In a certain series *RLC* circuit, $V_R = 24$ V, $V_L = 15$ V, and $V_C = 45$ V. What is the source voltage?
    (a) 38.42 V (b) 45 V (c) 15 V (d) 24 V
19. The superposition theorem is valid
    (a) only for ac circuits (b) only for dc circuits
    (c) for both ac and dc (d) neither of the above
20. When applying the superposition theorem to any circuit
    (a) the voltage source is shorted, the current source is opened
    (b) the voltage source is opened, the current source is shorted
    (c) both are opened
    (d) both are shorted
21. When applying Thevenin's theorem, Thevenin's voltage is equal to the
    (a) short-circuit voltage at the terminals (b) open-circuit voltage at the terminals
    (c) voltage of the source (d) total voltage available in the circuit
22. Thevenin's impedance Z_{Th} is found
    (a) by short circuiting the two given terminals
    (b) between any two open terminals
    (c) by removing the voltage source along with the terminal resistance
    (d) between the same open terminals as for V_{Th}
23. The maximum power transfer theorem can be applied
    (a) only to dc circuits (b) only to ac circuits
    (c) to both dc and ac circuits (d) neither of the above

SHORT ANSWER QUESTIONS

1. What is the relationship between current and voltage in a pure inductor?
2. If the supply frequency of a purely inductive circuit is doubled. Then what will be the circuit current?
3. What is the unit of capacitive reactance?
4. What is the power factor for a purely resistor circuit?
5. If an alternating triangular voltage is applied to a resistor, what will be the shape of current waveform?
6. What capacitance will give the same reactance as an inductor of 1 H at 50 Hz?
7. What are the limits for power factor?
8. What is the meaning of leading power factor?
9. What is the expression to calculate the reactive power?
10. What is the importance of symbol *j*?
11. What is the numerical value of operator j^2?
12. What is the unit of susceptance?
13. What are the characteristic features of series resonant circuit?

14. In a series RLC circuit, how is the magnitude of resonant frequency varied?
15. Define bandwidth?
16. What is the line current at resonance in an RLC parallel circuit?
17. What is the condition for maximum power transfer in an ac circuit?
18. What is the expression for Ohm's law in an ac circuit?
19. A circuit contains $(1+ j2)$ V source and draws a current $(2 + j3)$ A. What is the power in watts?
20. A circuit contains $(1 + j2)$ V source and draws a current $(2 + j3)$ A. What is the power factor?

REVIEW QUESTIONS

1. Define and explain real power and reactive power in an ac circuit.
2. Explain the superposition theorem for ac circuits.
3. State and explain Thevenin's theorem.
4. State and explain Norton's theorem.
5. Explain the maximum power transfer theorem for ac circuits.
6. Define power factor.
7. Explain what is impedance and power triangle.
8. Draw the vector diagram for a series RL circuit.
9. Define bandwidth.
10. Draw the phasor diagram for series RC circuit.

EXERCISE PROBLEMS

1. A sinusoidal voltage $v = 50 \sin \omega t$ is applied to a series RL circuit. The current in the circuit is given by $i = 25 \sin (\omega t - 53°)$. Determine the (a) apparent power, (b) average power, and (c) power factor.
2. Determine the (a) power factor, (b) true power, (c) reactive power, and (d) apparent power in the circuit shown in Figure 5.88.

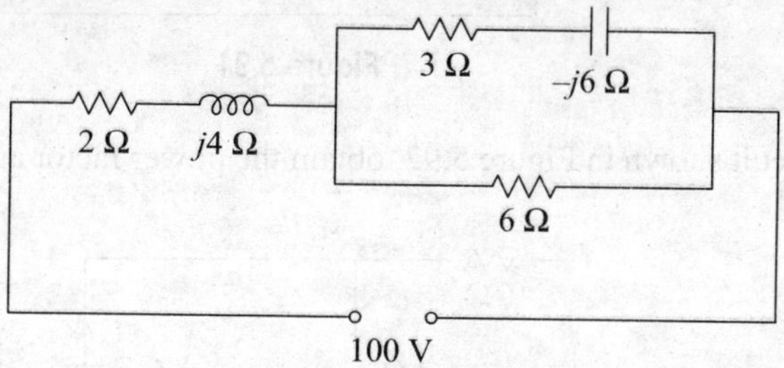

Figure 5.88

3. In a certain RC circuit, the true power is 300 W and the reactive power is 100 W. What is the apparent power?

4. For the total current in the circuit in Figure 5.89, determine the voltage across resistor V_R, and across capacitor V_C.

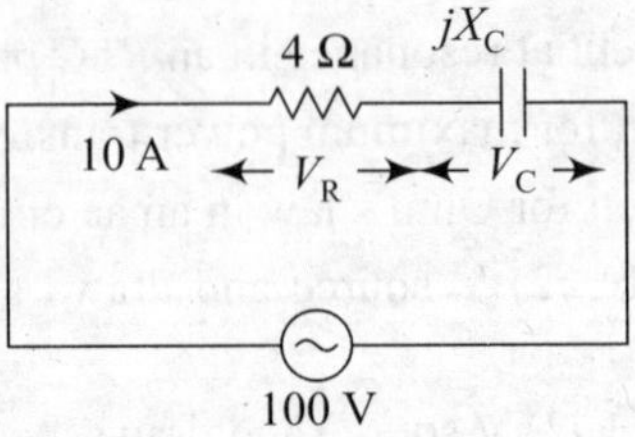

Figure 5.89

5. In a parallel circuit having resistance $R = 5\ \Omega$ and $L = 0.02$ H, the applied voltage is $v = 100 \sin(1000\,t + 50°)$ V. Find the total current and the angle by which the currents lag the applied voltage.
6. A coil has a resistance of 20 Ω and inductance of 80 mH and is connected in series with a 100-μF capacitor. Determine the circuit impedance at resonance and also find the resonant frequency.
7. Determine the resonant frequency for the circuit shown in Figure 5.90.

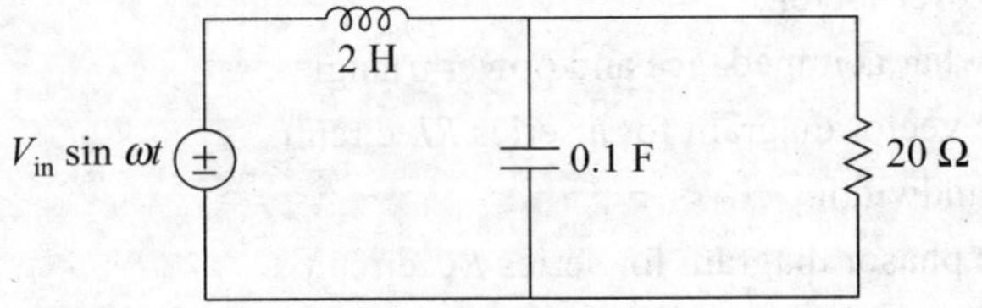

Figure 5.90

8. Find the resonant frequency in the ideal parallel LC circuit shown in Figure 5.91.

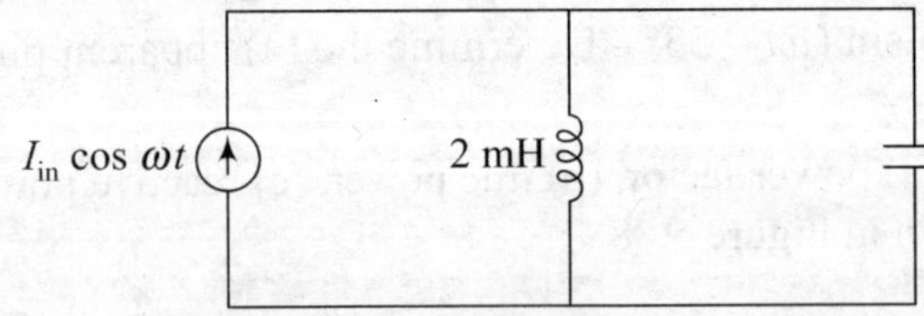

Figure 5.91

9. For the circuit shown in Figure 5.92, obtain the power factor and state whether it is leading or lagging?

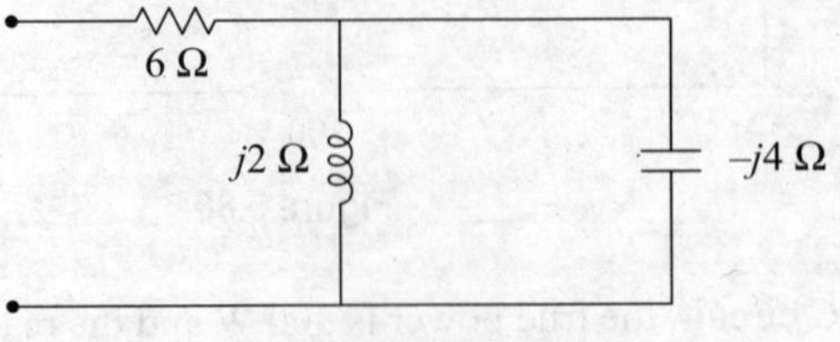

Figure 5.92

10. Determine the currents I_1 and I_2 in the circuit shown in Figure 5.93.

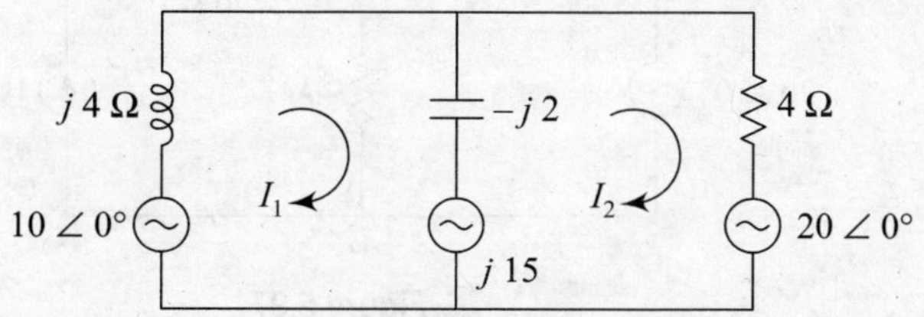

Figure 5.93

11. For the circuit shown in Figure 5.94, find the voltage across XY terminals if 10 A current is flowing through the capacitor.

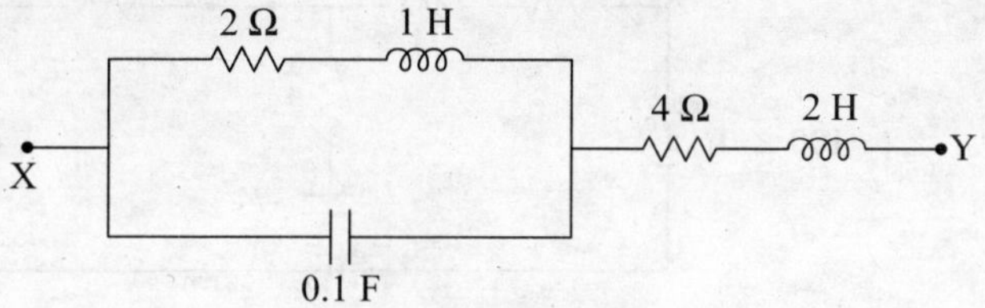

Figure 5.94

12. Two impedances are connected in parallel as shown in Figure 5.95. If the total current is 25 A, find the complex power taken by each impedance and draw its phasor diagram.

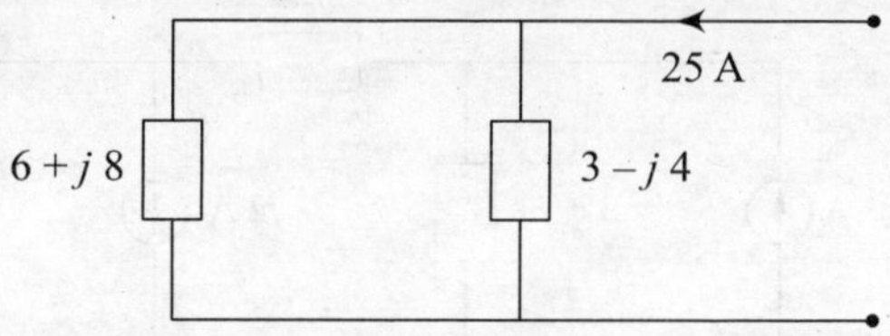

Figure 5.95

13. For the circuit shown in Fig 5.96, determine Thevenin's equivalent circuit between the output terminals.

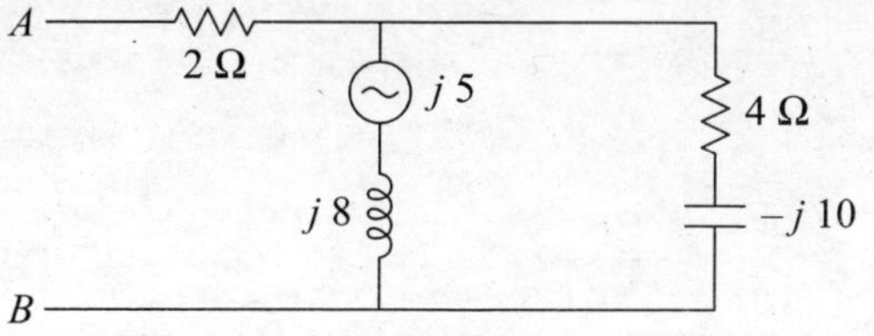

Figure 5.96

14. For the circuit shown in Figure 5.97, determine Norton's equivalent circuit between the output terminals, AB.

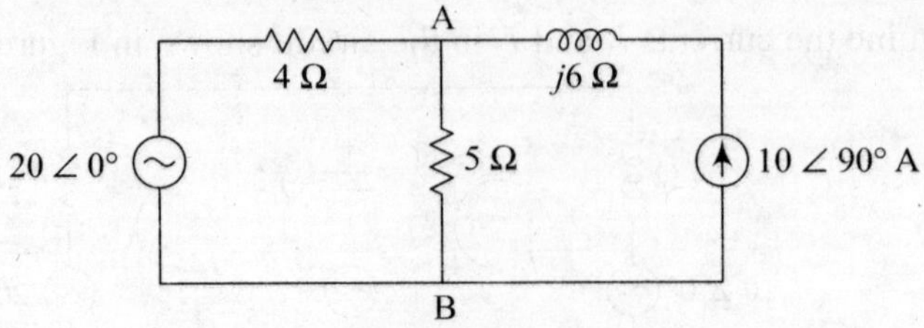

Figure 5.97

15. For the circuit shown in Figure 5.98, find the value of load impedance for which the source delivers maximum power. Calculate the value of the maximum power.

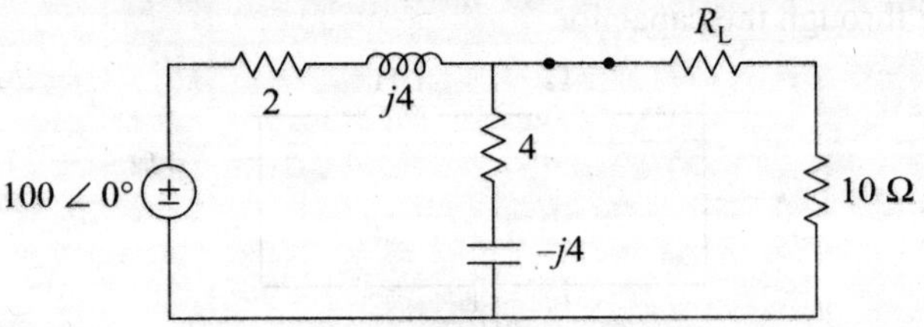

Figure 5.98

16. Determine the voltage across (2 + *j*5) Ω impedance shown in Figure 5.99 by using the superposition theorem.

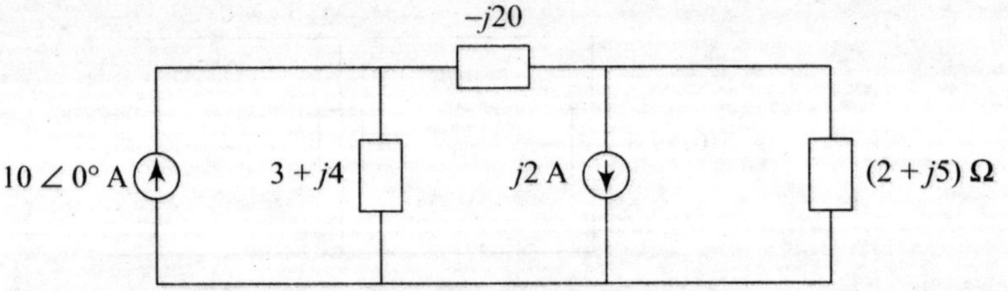

Figure 5.99

Chapter 6

Three-Phase Systems

Chapter Objectives

- To give the comparison of single-phase and three-phase systems
- To explain how three-phase voltages are produced
- To discuss how three-phase supply and loads are connected
- To explain the different configurations of star and delta
- To discuss the method to measure the three-phase power

6.1 INTRODUCTION

So far we have dealt with single-phase circuits. A single-phase alternating system consists of a generator connected through a pair of lines to a load. The three-phase electric power generation, transmission, and distribution are a common method used worldwide. It is a type of poly-phase system, and is the most common method used to distribute power. It is also used to power large motors and other large loads.

In a three-phase system, three circuit conductors carry three alternating currents, of the same frequency, which reach their instantaneous peak values at different times. Taking one conductor as the reference, the other two conductors are delayed in time by one-third and two-thirds of one cycle of the electrical current. This delay between phases has the effect of giving constant power transfer over each cycle of the current, and also makes it possible to produce a rotating magnetic field in an electric motor. In a balanced three-phase system, there are three sinusoidal voltage sources of equal amplitude and equal frequency of variation but out of phase with each other by $2\pi/3$ or 120° electrical.

Three-phase systems may or may not have a neutral wire. A neutral wire allows the three-phase system to use a higher voltage while still supporting lower voltage single-phase appliances. In high-voltage distribution situations, it is common not to have a neutral wire as the loads can simply be connected between phases (phase–phase connection).

A three-phase system has properties that make it very desirable in electric power systems:

(i) The phase currents tend to cancel out one another, summing to zero in the case of a linear balanced load. This makes it possible to eliminate or reduce the size of the neutral conductor; all the phase conductors carry the same current and so can be the same size for a balanced load.

(ii) Power transfer into a linear balanced load is constant, which helps to reduce generator and motor vibrations.

(iii) Three-phase systems can produce a magnetic field that rotates in a specified direction which simplifies the design of electric motors.

The third-order phase is the lowest phase to exhibit all of these properties. Most household loads are single phase. In India and in some other countries, the three-phase power generally does not enter homes. Even in the areas where it does, it is typically split out at the main distribution board and the individual loads are fed from a single phase. The three phases are typically indicated by colors which vary from country to country.

In this chapter, the principles of the generation of three-phase voltages and how the windings are connected are explained. The star and delta configurations of windings are also explained. The relationships between phase and line currents and voltages for both types of connection are developed. The methods of connection for measuring the power in three-phase systems are also discussed.

6.2 COMPARISON OF SINGLE-PHASE AND THREE-PHASE SYSTEMS

In the earlier days, the alternating current was used for the applications such as resistive loads, e.g., electric lamps and electric heaters, and worked satisfactorily for these loads. However, when ac motors were developed, it was found that the single-phase ac supply did not work satisfactorily as it was unable to produce the starting torque. An important advantage of the three-phase power supply is that the three-phase motors have a non-zero starting torque, unlike their single-phase counterpart. For a given frame size, the output of a three-phase machine is greater than that of a single-phase machine.

Nearly all electric power is generated and distributed in three phases at the operating frequency of 50 Hz in India and 60 Hz in some other parts of the world. When one- or two-phase inputs are required, they are taken from the three-phase system rather than generated independently.

The instantaneous power in a three-phase system is constant. This causes uniform power transmission and less vibration of three-phase machines.

The generation, transmission, and distribution of power are usually accomplished by means of balanced three-phase circuits. For a fixed voltage between conductors, a three-phase system requires 75% of the weight of the conductor as compared to a single-phase system. Thus, a three-phase system is more economical than a single-phase system.

Although theoretically any number of phases might be present, such as 2 phases, 6 phases, or 12 phases, it is a common practice to employ 3-phase circuits. However, in some applications, such as aluminium industry where 48 phases are required for melting purposes can be obtained by manipulating the three phases provided.

6.3 THREE-PHASE VOLTAGE SUPPLY

The generation of three-phase voltage supply and analysis are discussed in this section.

6.3.1 Generation of Three-Phase Voltages

Three-phase voltages are often produced with a three-phase ac generator (or alternator) whose cross-sectional view is shown in Figure 6.1. It consists of a rotating magnet called *rotor* surrounded by a stationary winding called *stator*. It consists of three identical stationary coils aa_1, bb_1, and cc_1, with their axes 120° apart from each other, are placed over the periphery of the stator frame. Here, "a," "b," and "c" are the finish terminals and a_1, b_1, and c_1 are the start terminals of the three coils.

Each of these coils constitutes a phase of the machine. As the rotor rotates its magnetic field cuts the flux from the three coils and induces voltages in the coils. Because the coils are placed 120° apart, the induced voltages in the coils are equal in magnitude but out of phase by 120°.

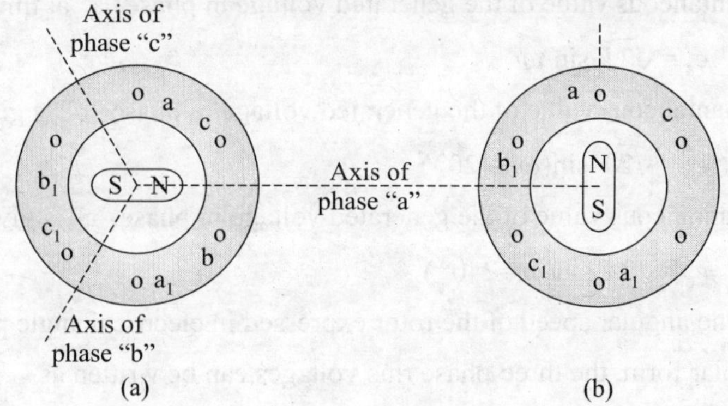

Figure 6.1 Generation of three-phase voltages

For the position of coil aa_1 with respect to the poles as shown in Figure 6.1(a), the voltage generated or the emf induced in the coil is zero. When the poles move in an anticlockwise direction through 90°, as shown in Figure 6.1(b), the emf induced in the coil is maximum and its direction around the loop is from the start a_1 to the finish "a." Let this direction be regarded as positive. The emf induced in loop aa_1 for one complete rotation of the poles is represented by the curve e_a in Figure 6.2(a).

Since the poles rotate in the anticlockwise direction, as is evident from Figure 6.1, the emf induced in loop bb_1 has exactly the same amplitude as that generated in coil aa_1, but lags by 120°, and its direction around the loop is from the start b_1 to the finish "b." Similarly, the emf induced in loop cc_1 has the same amplitude as that generated in coil bb_1 but lags the phase in coil "b" by 120°, and its direction around the loop is from the start c_1 to the finish "c." Hence, the emfs generated in loops aa_1, bb_1, and cc_1 are represented by the three equally spaced curves e_a, e_b, and e_c, respectively, in Figure 6.2.(a). The induced emfs are assumed to be positive when their directions around the loops are from the start terminal to the finish terminal of their respective loops. Thus, the three coils, generating voltages equal in magnitude but differing in phase by 120° from each other, can be viewed as three independent voltage sources.

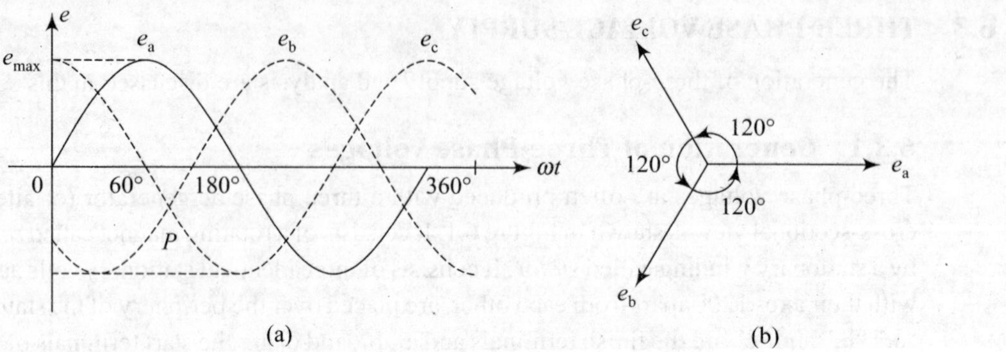

Figure 6.2 (a) Positive sequence waveforms of three-phase voltages and (b) phasor diagram of positive sequence voltages

The instantaneous value of the generated voltage in phase "a" at time t is given by

$$e_a = \sqrt{2}\, V \sin \omega t \tag{6.1}$$

then the instantaneous value of the generated voltage in phase "b" is given by

$$e_b = \sqrt{2}\, V \sin(\omega t - 120°) \tag{6.2}$$

and the instantaneous value of the generated voltage in phase "c" is given by

$$e_c = \sqrt{2}\, V \sin(\omega t - 240°) \tag{6.3}$$

where ω is the angular speed of the rotor expressed in electrical angle per second.

In the polar form, the three-phase rms voltages can be written as

$$V_a = V\angle 0° \tag{6.4a}$$

$$V_b = V\angle -120° \tag{6.4b}$$

$$V_c = V\angle -240° \tag{6.4c}$$

If the voltage sources have the same amplitude and frequency w and are out of phase with each other by 120 degrees, the voltages are said to be balanced. This implies that

$$|V_a| = |V_b| = |V_c|;\ V_a + V_b + V_c = 0 \tag{6.5}$$

Thus, balanced phase voltages are equal in magnitude and are out of phase with each other by 120°.

6.3.2 Phase Sequence

The phase sequence is important in a three-phase power distribution. It determines the direction of the rotation of a motor connected to the power source.

The phase sequence is defined as the time order in which the phases pass through their respective maximum values. The phase sequence is determined by the order in which the phasors pass through a fixed point in the phase diagram. In Figure 6.3(a), as the phasors rotate in the anticlockwise direction with angular frequency, w, they pass through the horizontal axis in a sequence abcabca....

This sequence is called as "abc" sequence or positive sequence. It can also be represented by RYB, where R stands for red, Y stands for yellow, and B stands for blue. This sequence of voltages is produced when the rotor in Figure 6.1 rotates in the anticlockwise direction. In the positive phase sequence, the voltage of phase "a" leads that of phase "b" which in turn leads phase "c." Similarly, for the phasors in Figure 6.3(b), as they rotate in the anticlockwise direction, they pass through the horizontal axis in a sequence acbacba.... This is called as "acb" sequence or negative sequence. In the negative sequence, V_a leads V_c which in turn leads V_b. This sequence of voltages is produced when the rotor rotates in the clockwise direction.

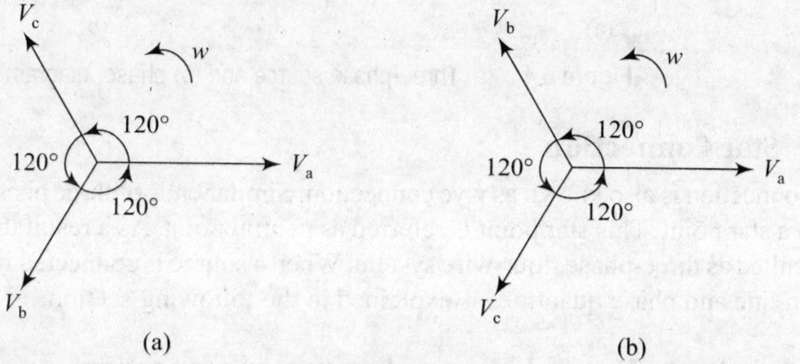

Figure 6.3 (a) abc or positive sequence and (b) acb or negative sequence

EXAMPLE 6.1 **Determine the phase sequence of the voltages given below.**

$$V_a = 100 \sin(\omega t + 20°)$$

$$V_b = 100 \sin(\omega t - 220°)$$

$$V_c = 100 \sin(\omega t - 100°)$$

Solution The voltages can be expressed in polar form as

$$V_a = 100 \angle 20°$$

$$V_b = 100 \angle -220°$$

$$V_c = 100 \angle -100°$$

V_a leads V_c by 120° and V_c leads V_b by 120°. Hence, the given sequence is acb sequence.

6.4 METHODS OF CONNECTION OF THREE-PHASE SUPPLY

A balanced three-phase supply available to the users of electric energy may be closely approximated by three sinusoidal ideal voltage sources, which are equal in magnitude and 120° out of phase with each other. The three voltage sources forming a three-phase source are V_a, V_b, V_c, with $V_a = V\angle 0°$, $V_b = V\angle -120°$, $V_c = V\angle -240°$ [see Figure 6.4(a)]. The phasor diagram for these voltages is shown in Figure 6.4(b).

The three single-phase voltages may be arranged in a star or delta configuration to form a three-phase supply system.

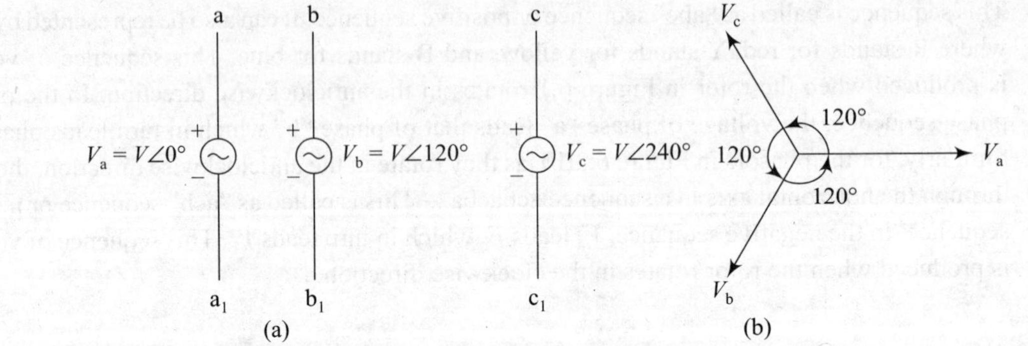

Figure 6.4 (a) Three-phase source and (b) phasor diagram

6.4.1 Star Connection

A star connection is also known as wye connection. Similar ends of three phases are joined together to form a star point. This star point is referred as neutral point. As a result this type of connection is also called as three-phase, four-wire system. When a source is connected in star, the relationship between line and phase quantities is explained in the following section.

Relation between line and phase values in star connection

In this section, we will discuss the relation between line and phase quantities of a star connection. In a star connection, the negative polarity terminals a_1, b_1, and c_1 of the voltage sources [see Figure 6.4(a)] are joined together to form the neutral terminal "n," and "a," "b," and "c" the positive polarity terminals, are brought out to form a three-phase star-connected supply as shown in Figure 6.5.

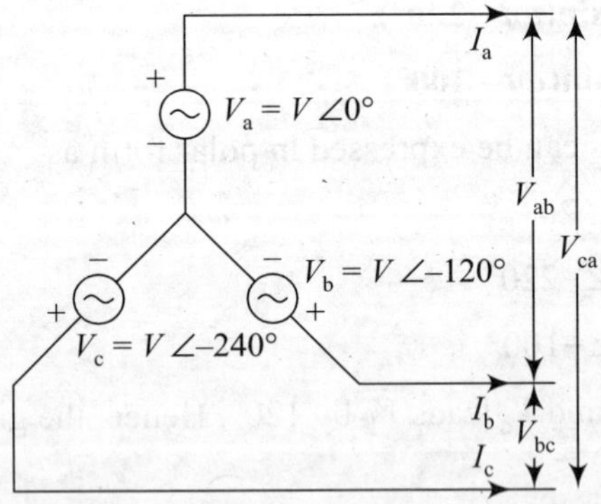

Figure 6.5 Three-phase star-connected supply

Relation between voltages. Now, let us consider the star-connected voltages as shown in Figure 6.5. The terminals "a," "b," and "c" are called line terminals. The voltage between any line and neutral point is called the phase voltage, V_{Ph}. The voltage between any two lines is called line voltage, V_L. The voltages V_a, V_b, and V_c are, respectively, between lines "a," "b," and "c," and the neutral point "n." These voltages are called phase voltages. The voltage phasors of the three-phase supply are then given as

$$V_a = V\angle 0°; \quad V_b = V\angle -120°; \quad V_c = V\angle -240°$$

The three-phase star-connected supply has three terminals "a," "b," and "c," called the line terminals, and may or may not have the fourth terminal "n," called the neutral terminal. The voltages across the terminals V_{ab}, V_{bc}, and V_{ca} are called line voltages. The relationships between the line and phase quantities for a star-connected system may be derived from the circuit diagram and the phasor diagram for the same supply system shown in Figure 6.6.

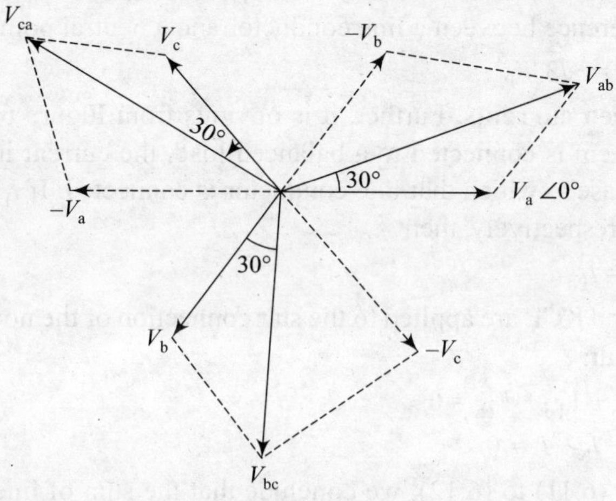

Figure 6.6 Phasor diagram for a star-connected supply

The value of voltage acting between the terminals "a" and "b" is the phase difference of V_a and V_b. Hence, $-V_b$ is drawn equal and opposite to V_b and added to V_a giving V_{ab} as the voltage acting from "b" to "a" via "n." It may be noted that the positive direction of the voltage V_{ab} is from "b" to "a." Thus, in terms of phasor voltages

$$\begin{aligned} V_{ab} = V_a - V_b &= V\angle 0° - V\angle -120° \\ &= V(1 + j0) - V(-0.5 - j0.866) \\ &= V(1.5 + j0.866) \\ &= V\sqrt{(1.5^2 + 0.866^2)}\ \tan^{-1}(0.866/1.5) \\ &= \sqrt{3}\ V\angle 30° \end{aligned} \tag{6.6}$$

Similarly,

$$\begin{aligned} V_{bc} = V_b - V_c &= V\angle -120° - V\angle -240° \\ &= V(-0.5 - j0.866) - V(-0.5 + j0.866) \\ &= -j1.732 \\ &= \sqrt{3}\ V\angle -90° \end{aligned} \tag{6.7}$$

Similarly,

$$\begin{aligned} V_{ca} = V_c - V_a &= V\angle -240° - V\angle 0° \\ &= V(-0.5 + j0.866) - V(1 + j0) \\ &= V(-1.5 + j0.866) \\ &= V\sqrt{((-1.5^2) + 0.866^2)}\ \tan^{-1}(0.866/-1.5) \\ &= \sqrt{3}\ V\angle 150° \end{aligned} \tag{6.8}$$

From Eqs. (6.6) to (6.8), it is observed that the magnitude of line voltages is equal, i.e., $|V_{ab}| = |V_{bc}| = |V_{ca}|$ and out of phase of 120°. Thus, it forms a balanced three-phase voltage system. The magnitude of the line voltages is $\sqrt{3}$ times the phase voltage, and the line voltages lead their corresponding phase voltages by 30°. Thus, in general, if V_L represents the line voltage, i.e., the voltage difference between two line conductors and V_{Ph} represents the phase voltage, or the voltage difference between a line conductor and a neutral point, then

$$V_L = \sqrt{3}\, V_{Ph} \tag{6.9}$$

Relation between currents. Further, it is obvious from Figure 6.5 that when a star-connected three-phase system is connected to a balanced load, the current in a line conductor is identical to that of the phase to which that line conductor is connected. If I_L and I_{Ph} represent the line and phase currents, respectively, then

$$I_L = I_{Ph} \tag{6.10}$$

If the KVL and KCL are applied to the star connection or the node in Figure 6.5, the following expressions result:

$$V_{ab} + V_{bc} + V_{ca} = 0 \tag{6.11}$$

$$I_a + I_b + I_c = 0 \tag{6.12}$$

Thus from Eqs. (6.11) to (6.12), we conclude that the sum of line voltages and line currents is equal to zero, respectively.

From the above analysis, we can conclude that for a balanced star-connected supply,

(i) $V_L = \sqrt{3}\, V_{Ph}$
(ii) All line voltages are equal in magnitude and are displaced by 120° from each other
(iii) All phase voltages are equal in magnitude and are displaced by 120° from each other
(iv) All line voltages are 30° ahead of their respective voltages
(v) $I_L = I_{Ph}$

EXAMPLE 6.2 **Calculate the line currents when a star-connected balanced load with an impedance of $(15 + j6)\ \Omega$ is connected to a 120-V, three-phase balanced supply in positive sequence.**

Solution Impedance per phase = $(15 + j6)\ \Omega = 16.15\ \angle 21.8°$

Since load is connected to balanced supply in positive sequence, $V_a = 120\angle 0°$; $V_b = 120\ \angle{-120°}$; $V_c = 120\angle{-240°}$.

Line current

$$I_a = \frac{120\ \angle 0°}{16.15\ \angle 21.8°} = 7.43\ \angle{-21.8°}\text{ A}$$

According to the positive sequence, I_b and I_c will be

$$I_b = I_a\angle{-120°} = 7.43\ \angle{-141.8°}\text{ A}$$

$$I_c = I_a\angle{-240°} = 7.43\ \angle{-261.8°}\text{ A}$$

EXAMPLE 6.3 **The phase voltage of a three-phase star-connected load is 120 V. Find the line voltages of the load (assume positive sequence).**

Solution Since it is connected in positive sequence, the phase voltages are given by

$$V_a = 120\ \angle 0°\text{ V}$$

$V_b = 120\angle{-120°}$ V

$V_c = 120\angle{-240°}$ V

Let V_{ab}, V_{bc}, and V_{ca} be the line voltages. We know the relationship between the line voltages and phase voltages. The line voltages are 1.732 times the phase voltages and lead by 30°.

Hence, the line voltages are written as

$$V_{ab} = \sqrt{3} \times 120\angle 0° + 30° = 207.84\angle 30° \text{ V}$$

$$V_{bc} = \sqrt{3} \times 120\angle -120° + 30° = 207.84\angle -90° \text{ V}$$

$$V_{ab} = \sqrt{3} \times 120\angle -240° + 30° = 207.84\angle -210° \text{ V}$$

6.4.2 Delta Connection

A delta connection is also known as mesh connection. In this connection, the dissimilar ends are joined together. The three line conductors are taken from the three junctions of the mesh or delta connection to feed the three-phase load. This constitutes a three-phase, three-wire system. There is no common terminal in this connection. When source is connected in delta, the relationship between the phase and line quantities is explained in the following section.

Relation between line and phase values in delta connection

The three voltage sources of Figure 6.4 can be delta connected by joining "a_1" to "b," "b_1" to "c," and "c_1" to "a" together so as to form a closed circuit as shown in Figure 6.7. Three lines can be taken out from the nodes thus forming a delta-connected supply.

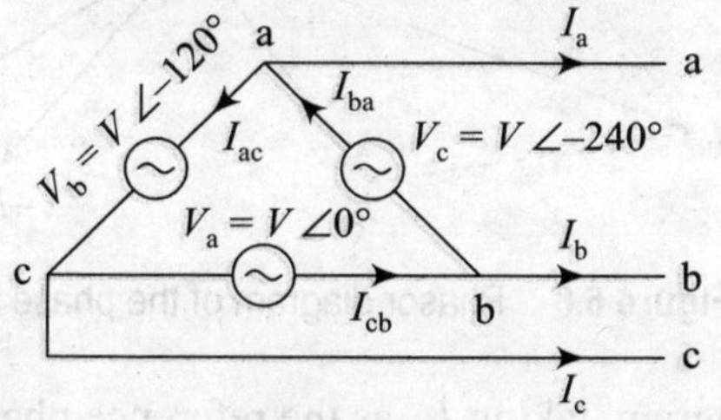

Figure 6.7 Three-phase delta-connected supply

Relation between voltages. The voltage phasor of the three phases is given by

$$V_a = V\angle 0°;\; V_b = V\angle{-120°};\; V_c = V\angle{-240°}$$

Voltages across the two line terminals are V_{ab}, V_{bc}, and V_{ca}. These are called as line voltages. Since each phase is connected across two line terminals, the line voltages are also referred as phase voltages in the delta-connected system. It is obvious from Figure 6.7 that the line voltages in a delta-connected supply are equal to the phase voltage and are given by

$$\left.\begin{aligned} V_{ab} &= V_a = V\angle 0° \\ V_{bc} &= V_b = V\angle{-120°} \\ V_{ca} &= V_c = V\angle{-240°} \end{aligned}\right\} \quad (6.13)$$

Thus, in general, if V_L represents the line voltage and V_{Ph} represents the phase voltage, then

$$V_L = V_{Ph} \tag{6.14}$$

Relation between currents. The currents flowing in the phases of a delta-connected supply are denoted by I_{ba}, I_{ac}, and I_{cb}.

The line currents are denoted by I_a, I_b, and I_c. The value of line currents may be found out by applying the KCL at the nodes "a," "b," and "c."

$$I_a = I_{ba} - I_{ac} \tag{6.15}$$

$$I_b = I_{cb} - I_{ba} \tag{6.16}$$

$$I_c = I_{ac} - I_{cb} \tag{6.17}$$

The phasor diagram of the balanced phase currents is shown in Figure 6.8. When the phase currents are balanced, the line currents can be obtained from the phasor diagram as well.

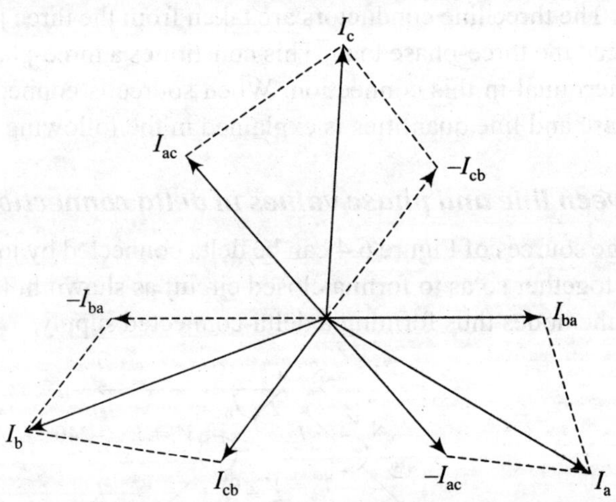

Figure 6.8 Phasor diagram of the phase and line currents

From the phasor diagram, taking I_{ba} as the reference phasor, i.e., $I_{ba} = I_{ba} \angle 0° = I_{Ph} \angle 0°$,

$$\left.\begin{aligned} I_{ba} &= I_{ba} \angle 0° \\ I_{cb} &= I_{ba} \angle -120° \text{ A} \\ I_{ac} &= I_{ba} \angle -240° \text{ A} \end{aligned}\right\} \tag{6.18}$$

Using Eqs. (6.15)–(6.17), the line currents obtained as

$$\begin{aligned} I_a &= I_{ba} - I_{ac} \\ &= I_{ba} - I_{ba} \angle -240° \\ &= I_{ba} (1 - 1 \angle -240°) \\ &= I_{ba} (1 + 0.5 - j0.866) \\ &= \sqrt{3}\, I_{ba} \angle -30° \text{ A} \end{aligned} \tag{6.19}$$

Thus, the above equation shows that the magnitude of line current is $\sqrt{3}$ times the magnitude of phase current I_{Ph} and lags behind by 30°. Similarly, the remaining line currents can also be determined. Thus, the line currents are given by

$$\left.\begin{aligned} I_a &= \sqrt{3}\, I_{Ph} \angle{-30°} \\ I_b &= \sqrt{3}\, I_{Ph} \angle{-150°} \\ I_c &= \sqrt{3}\, I_{Ph} \angle 90° \end{aligned}\right\} \qquad (6.20)$$

where I_{Ph} is the magnitude of the phase current. If I_{Ph} and I_L represent the magnitudes of the phase and line current, respectively, in a balanced delta-connected system, then

$$I_L = \sqrt{3}\ I_{Ph} \qquad (6.21)$$

where $I_L = |I_a| = |I_b| = |I_c|$

and

$$I_{Ph} = |I_{ba}| = |I_{cb}| = |I_{ac}|$$

From the above analysis, we can conclude that for the delta-connected system

(i) $I_L = \sqrt{3}\, I_{Ph}$
(ii) All line currents are equal in magnitude and are displaced by 120°
(iii) The line currents are 30° behind their respective phase currents
(iv) $V_L = V_{Ph}$

EXAMPLE 6.4 **Find the line currents of the given delta-connected system whose phase current is 10∠0°A (assume positive sequence).**

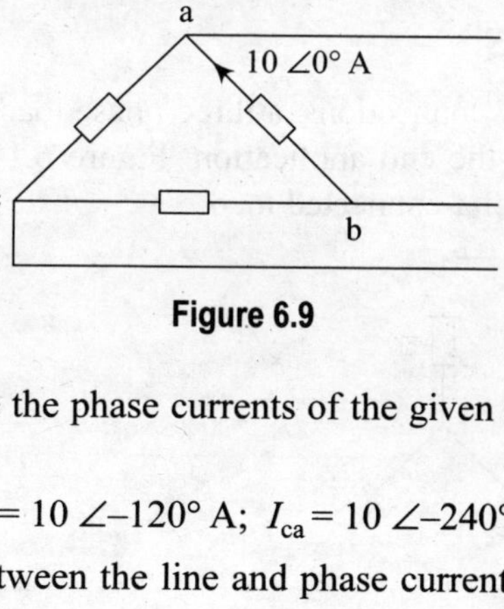

Figure 6.9

Solution Let I_{ab}, I_{bc}, and I_{ca} be the phase currents of the given system and I_a, I_b, and I_c be the line currents of the system.

$$I_{ab} = 10 \angle 0°\text{ A};\ I_{bc} = 10 \angle{-120°}\text{ A};\ I_{ca} = 10 \angle{-240°}\text{ A}$$

We know the relationship between the line and phase currents of the delta connection. The line currents are $\sqrt{3}$ times the phase currents and lag by 30°. Therefore, the line currents are given by

$$I_a = \sqrt{3} \times 10 \angle 0°{-30°} = 17.32 \angle{-30°}$$

$$I_b = 17.32 \angle{-30-120} = 17.32 \angle{-150°}$$

$$I_c = 17.32 \angle{-30-240} = 17.32 \angle{-270°}$$

EXAMPLE 6.5 A balanced delta-connected load having an impedance of $(20 + j15)\ \Omega$ is connected to the positive sequence generator having voltage of $230\ V\angle 0°$. Determine the phase and line currents of the load.

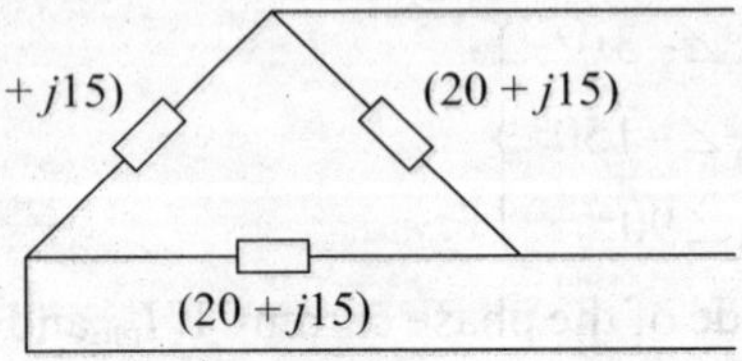

Figure 6.10

Solution The load is connected in delta. Phase voltages are equal to line voltages. The line current is equal to $\sqrt{3}$ times the phase current.

$$\text{Impedance per phase} = 20 + j15 = 25\ \angle 36.87°$$

$$\text{Phase current, } I_{ab} = \frac{230\ \angle 0°}{25\ \angle 36.87°} = 9.2\ \angle -36.87°\ \text{A}$$

$$I_{bc} = 9.2\ \angle -36.87° - 120° = 9.2\ \angle -156.87°\ \text{A}$$

$$I_{ca} = 9.2\ \angle -36.87° - 240° = 9.2\ \angle -276.87°\ \text{A}$$

$$\text{Line current, } I_a = \sqrt{3}\ Ip\ \angle -30° = 15.93\ \angle -66.87°\ \text{A}$$

$$I_b = 15.93\ \angle -186.87°\text{A}$$

$$I_c = 15.93\ \angle -306.87°\text{A}$$

6.5 LOAD CONNECTIONS

Like the generator supply connections, a three-phase load can be either star-connected or delta-connected depending on the end application. Figure 6.11(a) shows a star-connected load and Figure 6.11(b) shows a delta-connected load.

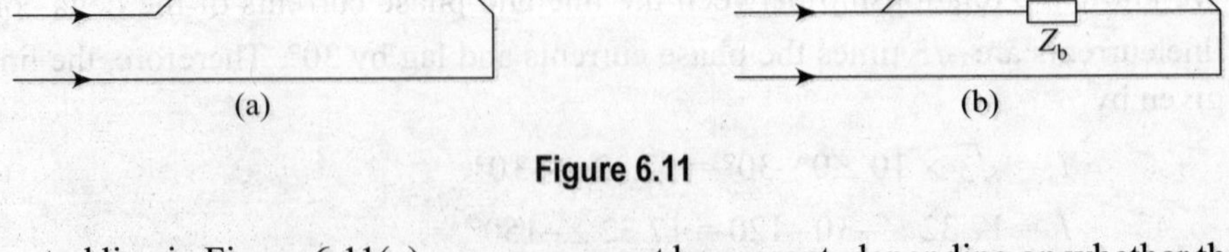

Figure 6.11

The neutral line in Figure 6.11(a) may or may not be present, depending on whether the system is four- or three-wire. And of course, a neutral connection is topologically impossible for a delta connection. A star- or delta-connected load is said to be balanced if the phase impedances are equal in magnitude and in phase. Load is said to be unbalanced if the phase impedances are not equal in magnitude or phase. For balanced star-connected load,

$$Z_1 = Z_2 = Z_3 = Z_Y \tag{6.22}$$

where Z_Y is the star impedance per phase. For balanced delta-connected load,

$$Z_a = Z_b = Z_c = Z_\Delta \tag{6.23}$$

where Z_Δ is the load impedance per phase. We recall the transformation,

$$Z_\Delta = 3Z_Y \text{ or } Z_Y = \frac{1}{3} Z_\Delta \tag{6.24}$$

Hence, we can transform a star-connected load into a delta-connected load or vice versa using Eq. (6.24). When loads are connected in star, the relations between line and phase quantities derived earlier in Section 6.4.1 hold good for loads also. When loads are connected in delta, the relations between line and phase quantities derived in Section 6.4.2 hold good for loads also.

6.5.1 Balanced Star-Connected Load

Figure 6.12 shows a three-phase, four-wire balanced star-connected load. In the three-phase, three-wire balanced star-connected load, neutral point is absent.

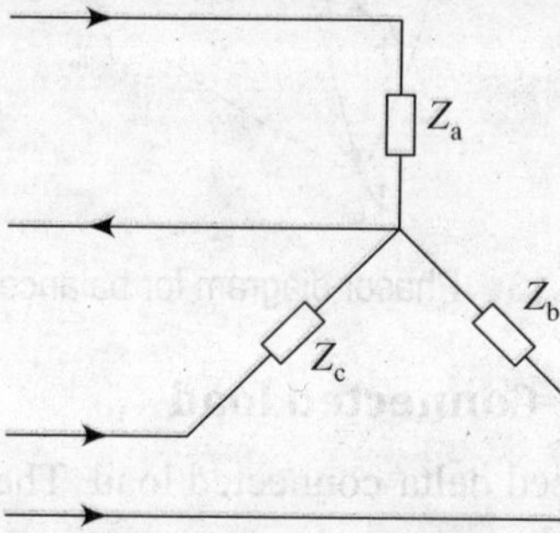

Figure 6.12 Three-phase, four-wire star-connected load

In a balanced star-connected load, load impedance is equal and is given by $Z_Y = Z \angle\phi$. Applying the KCL at the point "N," we get

$$I_n = I_a + I_b + I_c \tag{6.25}$$

$$I_a = \frac{V_{an}}{Z\angle\phi};\ I_b = \frac{V_{bn}}{Z\angle\phi};\ I_c = \frac{V_{cn}}{Z\angle\phi} \tag{6.26}$$

Assuming the voltages are also balanced $V_{an} = V_{Ph}\angle 0°$; $V_{bn} = V_{Ph}\angle -120°$; $V_{cn} = V_{Ph}\angle -240°$, Eq. (6.26) can be rewritten as

$$I_a = \frac{V_{Ph}\angle 0°}{Z\angle\phi} = \frac{V_{Ph}}{Z}\angle -\phi° \Bigg\}$$

$$\left.\begin{aligned} I_b &= \frac{V_{Ph}\angle -120^\circ}{Z\angle\phi} = \frac{V_{Ph}}{Z}\angle -120^\circ - \phi^\circ \\ I_c &= \frac{V_{Ph}\angle -240^\circ}{Z\angle\phi} = \frac{V_{Ph}}{Z}\angle -240^\circ - \phi^\circ \end{aligned}\right\} \tag{6.27}$$

Substituting Eq. (6.27) into Eq. (6.25), we get

$$I_n = \frac{V_{Ph}}{Z}\angle -\phi^\circ + \frac{V_{Ph}}{Z}\angle -120^\circ - \phi^\circ + \frac{V_{Ph}}{Z}\angle -240^\circ - \phi^\circ = 0 \tag{6.28}$$

Since the load is balanced the current flowing through the neutral wire is zero. From Eq. (6.27), it is observed that line currents I_a, I_b, and I_c are equal in magnitude and displaced by an angle of 120° from each other.

The phasor diagram for balanced star-connected load is shown in Figure 6.13.

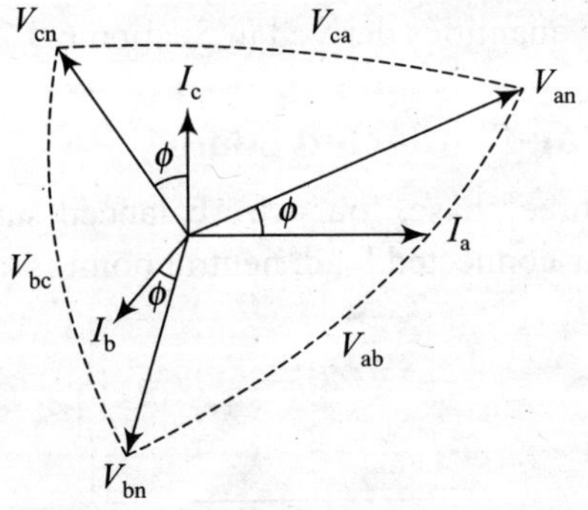

Figure 6.13 Phasor diagram for balanced star-connected load

6.5.2 Balanced Delta-Connected load

Figure 6.14 shows a balanced delta-connected load. The load impedance is equal in magnitude and is given by $Z_\Delta = Z\angle\phi$ ohms.

For the delta connection, the phase voltages and line voltages are equal in magnitude and displaced by 120°.

$$\therefore \quad V_{ab} = V_{Ph}\angle 0^\circ;\ V_{bc} = V_{Ph}\angle -120^\circ;\ V_{ca} = V_{Ph}\angle -240^\circ \tag{6.29}$$

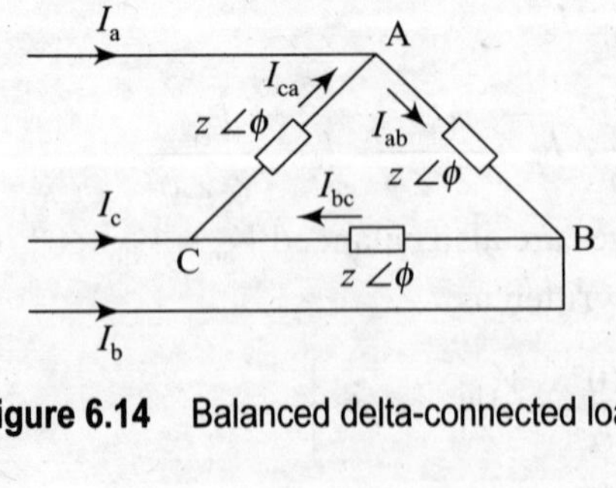

Figure 6.14 Balanced delta-connected load

The phase currents are represented by I_{ab}, I_{bc}, and I_{ca} and are given by

$$I_{ab} = \frac{V_{ab}\angle 0°}{Z\angle\phi°} = \frac{V_{Ph}}{Z}\angle -\phi° \qquad (6.30)$$

$$I_{bc} = \frac{V_{bc}\angle -120°}{Z\angle\phi°} = \frac{V_{Ph}}{Z}\angle -120° - \phi° \qquad (6.31)$$

$$I_{ca} = \frac{V_{ca}\angle -240°}{Z\angle\phi°} = \frac{V_{Ph}}{Z}\angle -240° - \phi° \qquad (6.32)$$

The line currents are denoted by I_a, I_b, and I_c and are calculated using the KCL. Applying the KCL at point A

$$I_a = I_{ab} - I_{ca} \qquad (6.33)$$

Substituting Eqs. (6.30) and (6.32) into Eq. (6.33), we get

$$I_a = \frac{V_{Ph}}{Z}\angle -\phi° - \frac{V_{Ph}}{Z}\angle -240° - \phi° \qquad (6.34)$$

Applying the KCL at point B, we get

$$I_b = I_{bc} - I_{ab} \qquad (6.35)$$

Substituting Eqs. (6.30) and (6.31) into Eq. (6.35), we get

$$I_b = \frac{V_{Ph}}{Z}\angle -120° - \phi° - \frac{V_{Ph}}{Z}\angle -\phi° \qquad (6.36)$$

Applying the KCL at point C, we get

$$I_c = I_{ca} - I_{bc} \qquad (6.37)$$

Substituting Eqs. (6.31) and (6.32) into Eq. (6.37), we get

$$I_c = \frac{V_{Ph}}{Z}\angle -240° - \phi° - \frac{V_{Ph}}{Z}\angle -120° - \phi° \qquad (6.38)$$

By using Eqs. (6.34), (6.36), and (6.38), we can calculate the line currents in the balanced delta-connected load.

EXAMPLE 6.6 **A balanced star-connected load is supplied from a symmetrical three-phase, 440-V system. The current in each phase is 20 A and lags 30° behind the phase voltage. Calculate the impedance in each phase.**

Solution Line voltage, $V_L = 440$ V

Phase voltage, $V_{Ph} = \frac{440}{\sqrt{3}}\angle 0° = 254.03\angle 0°$ V

Phase current, $I_{Ph} = 20\angle -30°$ A

In star connection, phase current is equal to line current.

Impedance

$$Z = \frac{V_{Ph}}{I_{Ph}} = \frac{254.03\angle 0°}{20\angle -30°} = 12.7\angle 30°\ \Omega$$

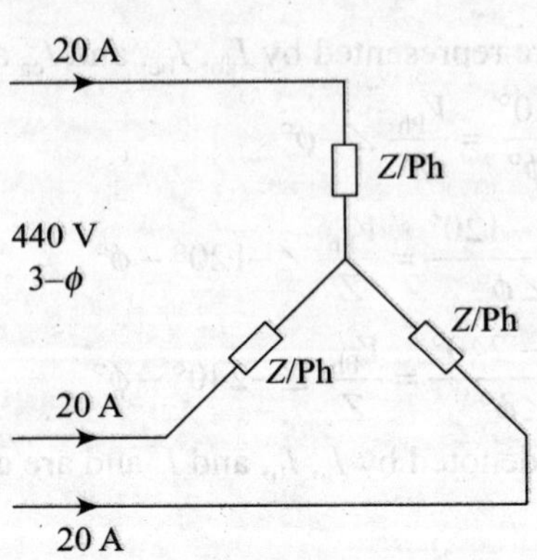

Figure 6.15

EXAMPLE 6.7 Three identical impedances of 20 ∠45° are connected in a delta to three-phase, three-wire, 400-V system. Determine the line currents.

Solution

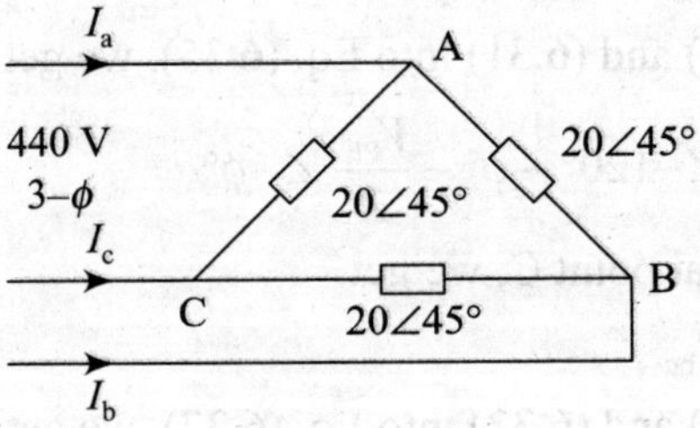

Figure 6.16

With the given data, the delta connection can be drawn as shown in Figure 6.16.

Impedance $Z = 20\angle 45° = 14.14 + j14.14$

Line voltage V_L = phase voltage V_{Ph} = 400 V

Phase currents are calculated as

$$I_{ab} = \frac{V_{Ph}}{Z}\angle -\phi° = \frac{400}{20}\angle -45° = 20\angle -45° = 14.14 - j14.14$$

$$I_{bc} = \frac{V_{Ph}}{Z}\angle -120° - \phi° = \frac{400}{20}\angle -120° - 45° = 20\angle -165° = -19.32 - j5.17$$

$$I_{ca} = \frac{V_{Ph}}{Z}\angle -240° - \phi° = \frac{400}{20}\angle -240° - 45° = 20\angle -285° = 5.17 + j19.32$$

The line currents are calculated as

$$I_a = I_{ab} - I_{ca} = (14.14 - j14.14) - (5.17 + j19.32) = 8.97 - j33.46 = 34.64\angle -74.9° \text{ A}$$

$I_b = I_{bc} - I_{ab} = (-19.32 - j5.17) - (14.14 - j14.14) = -33.46 + j8.97 = 34.64\angle 164.99°$ A

$I_c = I_{ca} - I_{bc} = (5.17 + j19.32) - (-19.32 - j5.17) = 24.49 + j24.49 = 34.63\angle 45°$ A

6.6 ANALYSIS OF THREE-PHASE CIRCUITS

The main aim of the analysis of a three-phase circuit is to determine the currents and the voltages in all the phases in the various elements/components of a given network. Since both the three-phase source, and the three-phase load can be either star- or delta-connected or vice versa, we have four possible combinations, i.e., (i) star–star, (ii) star–delta, (iii) delta–delta, and (iv) delta–star. In the subsequent section, we will consider two of these possible configurations, i.e., (i) and (ii). In practice, source is rarely connected in delta. Hence, the analyses of (iii) and (iv) configurations are not covered in this book.

6.6.1 Star-Connected Supply and Star-Connected Balanced Load

Figure 6.17 shows a balanced star–star system. It is a three-phase system with a balanced star-connected source and a balanced star-connected load. We assume balanced load so that the load impedances are equal. The per phase load impedance is represented by $Z = Z\angle\theta°$. Assuming the positive sequence, the phase voltages are represented by V_a, V_b, V_c and the line voltages are represented by V_{ab}, V_{bc}, V_{ca}. The three-phase, four-wire star–star system is shown in Figure 6.17. We have already discussed the relationships between the line and phase voltages of a star connection.

This type of supply is normally employed in high- and medium-voltage distribution systems. For this connection, the currents from the phases of the source and the load are the same.

Applying the KVL to each phase in Figure 6.17, we obtain the line currents as

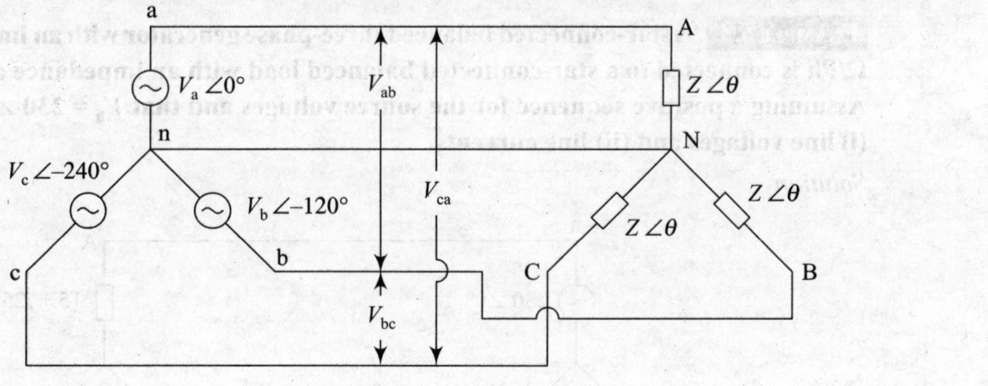

Figure 6.17 Star-connected supply connected to star-connected load

$$I_a = \frac{V_a}{Z} \tag{6.39}$$

$$I_b = \frac{V_b}{Z} = \frac{V_a\angle -120°}{Z} = I_a\angle -120° \tag{6.40}$$

$$I_c = \frac{V_c}{Z} = \frac{V_a\angle -240°}{Z} = I_a\angle -240° \tag{6.41}$$

Therefore, $I_a + I_b + I_c = 0$. So that the neutral current is

$$I_n = -(I_a + I_b + I_c) = 0 \tag{6.42}$$

The voltage between point nN is

$$V_{nN} = Z_n I_n = 0 \tag{6.43}$$

i.e., the voltage across the neutral wire is zero. Thus, the neutral line can be removed without affecting the system.

In fact, in a long distance power transmission, earth itself acts as the neutral conductor. The line current is the current in each line, the phase current is the current in each phase of the source or load. In star–star system, the line current is the same as the phase current. Figure 6.18 gives the phasor diagram showing all the currents and the voltages. It may be noted that joining the tips of phase voltage phasors in the phasor diagram gives the line voltages.

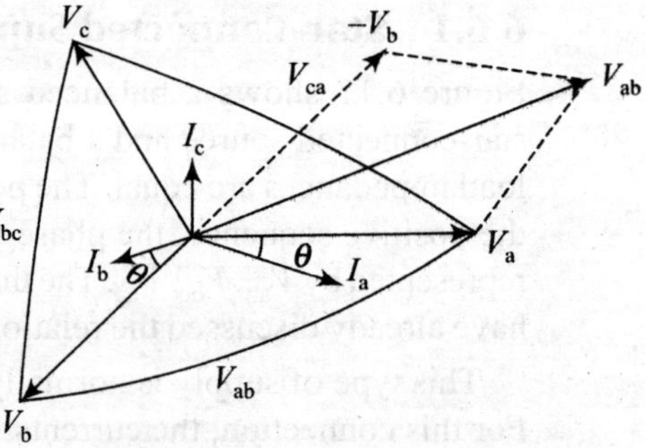

Figure 6.18 Phasor diagram

For a three-phase, four-wire star-connected supply system, the three-phase voltages and the neutral conductor are available at the supply terminals. The neutral conductor connects the neutral point of the source with the neutral point of the star-connected load. This connection is generally employed for the low-voltage distribution system. If the three-phase load is unbalanced, the neutral conductor carries current I_n.

EXAMPLE 6.8 **A star-connected balanced three-phase generator with an impedance of (1 + *j*2) Ω/Ph is connected to a star-connected balanced load with an impedance of (15 + *j*25) Ω/Ph. Assuming a positive sequence for the source voltages and that $V_a = 230\angle 0°$, determine the (i) line voltages and (ii) line currents.**

Solution

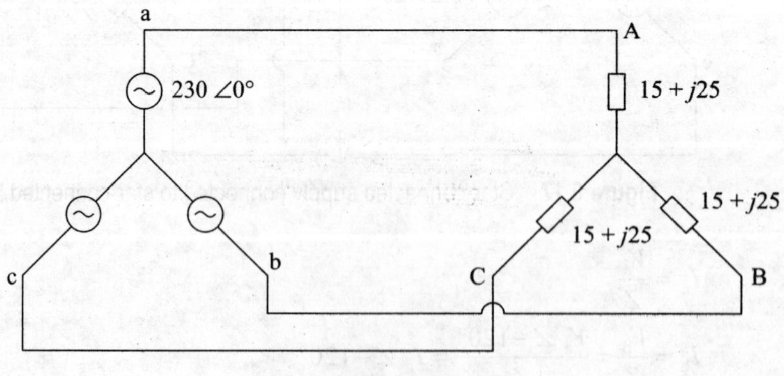

Figure 6.19

Total impedance per phase $Z = (1 + j2) + (15 + j25)$

$$= (16 + j27)\ \Omega$$

$$= 31.38\ \angle 59.34°\ \Omega$$

Both generator and load are star-connected. The line voltages are $\sqrt{3}$ times the phase voltages and lead by 30°.

(i) Line voltages, $V_{ab} = \sqrt{3}\ V_a \angle 30°$

$$= \sqrt{3} \times 230\ \angle 0° + 30°$$

$$= 398.37\ \angle 30°\ \text{V}$$

$$V_{bc} = 398.39\ \angle -9\,0°$$

$$V_{ca} = 398.39\ \angle -210°$$

(ii) Line currents, $I_a = \dfrac{V_a}{Z/\text{Ph}}$

$$= \frac{230\ \angle 0°}{31.38\ \angle 59.34°}$$

$$= 7.33\ \angle -59.34°\ \text{A}$$

$$I_b = I_a \angle -120°$$

$$= 7.33\ \angle -120° - 59.34°$$

$$= 7.33\ \angle 179.34°\ \text{A}$$

$$I_c = I_a \angle -240°$$

$$= 7.33\ \angle -240° - \angle 59.34°$$

$$= 7.33\ \angle 299.34°\ \text{A}$$

6.6.2 Star-Connected Supply and Delta-Connected Balanced Load

Figure 6.20 shows a balanced star–delta system, where the source is star-connected and the load is delta-connected. No neutral connection from the source to the load is seen. The per phase load impedance is represented by $Z = Z\angle\theta°$. Assuming the positive sequence, the phase voltages are given by

$$\left.\begin{aligned} V_a &= V_{Ph} \angle 0° \\ V_b &= V_{Ph} \angle -120° \\ V_c &= V_{Ph} \angle -240° \end{aligned}\right\} \qquad (6.44)$$

and the balanced line voltages are given by

$$\left.\begin{aligned} V_{ab} &= \sqrt{3}\ V_{Ph} \angle 30° = V_{AB} \\ V_{bc} &= \sqrt{3}\ V_{Ph} \angle -90° = V_{BC} \\ V_{ca} &= \sqrt{3}\ V_{Ph} \angle -210° = V_{CA} \end{aligned}\right\} \qquad (6.45)$$

Equation (6.45) shows that the line voltages are equal to the voltages across the load impedances for this configuration. From these voltages, we can obtain the phase currents as

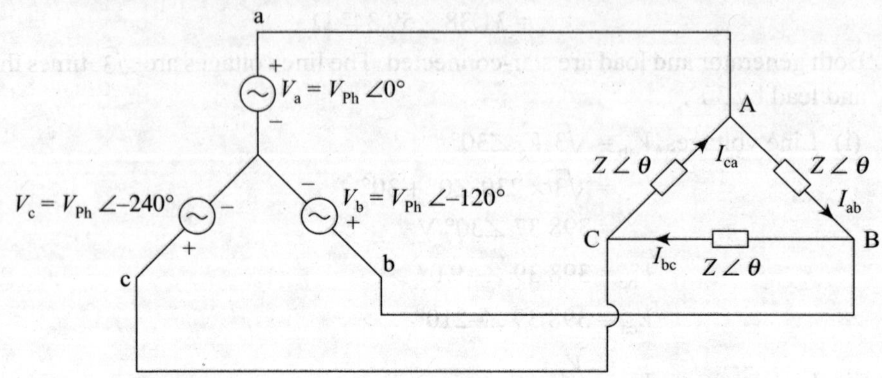

Figure 6.20 Delta-connected load across star-connected supply

$$\left.\begin{aligned} I_{AB} &= V_{AB}/Z = (\sqrt{3}\ V_{Ph} \angle 30°)/Z \angle\theta = (\sqrt{3}\ V_{Ph}/Z) \angle(30 - \theta)° \\ I_{BC} &= V_{BC}/Z = (\sqrt{3}\ V_{Ph} \angle -90°)/Z \angle\theta = (\sqrt{3}\ V_{Ph}/Z) \angle(-90 - \theta)° \\ I_{CA} &= V_{CA}/Z = (\sqrt{3}\ V_{Ph} \angle -210°)/Z \angle\theta = (\sqrt{3}\ V_{Ph}/Z) \angle(-210 - \theta)° \end{aligned}\right\} \tag{6.46}$$

The line currents are obtained from the phase currents by applying the KCL at nodes A, B, and C as shown in Figure 6.20.

$$\left.\begin{aligned} I_a &= I_{AB} - I_{CA} \\ I_B &= I_{BC} - I_{AB} \\ I_C &= I_{CA} - I_{BC} \end{aligned}\right\} \tag{6.47}$$

$$\begin{aligned} I_a &= I_{AB} - I_{CA} \\ &= I_{AB} - I_{AB} \angle -240° \\ &= I_{AB} (1 - 1 \angle -240°) \\ &= I_{AB} (1 + 0.5 - j0.866) \\ I_a &= \sqrt{3}\ I_{AB} \angle -30° \text{ A} \\ I_b &= I_{BC} - I_{AB} \\ &= I_{AB} \angle -120° - I_{AB} \\ &= I_{AB} (1 \angle -120° - 1) \\ &= I_{AB} (-0.5 - j0.866 - 1) \\ I_b &= \sqrt{3}\ I_{AB} \angle -150° \text{ A} \\ I_c &= I_{CA} - I_{BC} \\ &= I_{AB} \angle -240° - I_{AB} \angle -120° \\ &= I_{AB} (1 \angle -240° - 1 \angle -120°) \end{aligned}$$

$$= I_{AB}(-0.5 + j0.866 + 0.5 + j0.866)$$

$$I_c = \sqrt{3}\ I_{AB} \angle -90° \text{ A}$$

Thus, the line currents are given by

$$\left.\begin{aligned} I_a &= \sqrt{3}\ I_{AB} \angle -30° \text{ A} \\ I_b &= \sqrt{3}\ I_{AB} \angle -150° \text{ A} \\ I_c &= \sqrt{3}\ I_{AB} \angle -90° \text{ A} \end{aligned}\right\} \qquad (6.48)$$

It may be seen from Eq. (6.48) that the phase currents in the load are equal in magnitude and are displaced from each other by 120°. Thus, the magnitude of the phase current, $I_{Ph} = \sqrt{3}\ V_{ph}/Z$, similarly, the line currents are also a balanced set of currents. Figure 6.21 gives the phasor diagram of voltages and currents of a delta-connected load across a star-connected supply.

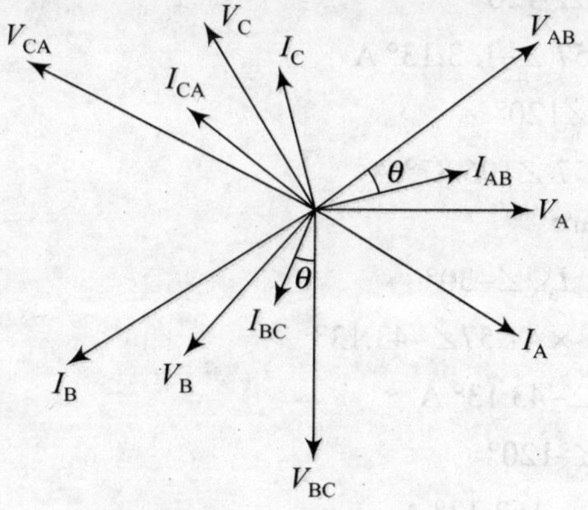

Figure 6.21 Phasor diagram

EXAMPLE 6.9 **A balanced star-connected source with V_a = 120 ∠10° V is connected to a delta-connected balanced load of (3 + *j*4) Ω/Ph. Calculate the phase and line currents. Assume "abc" sequence.**

Solution

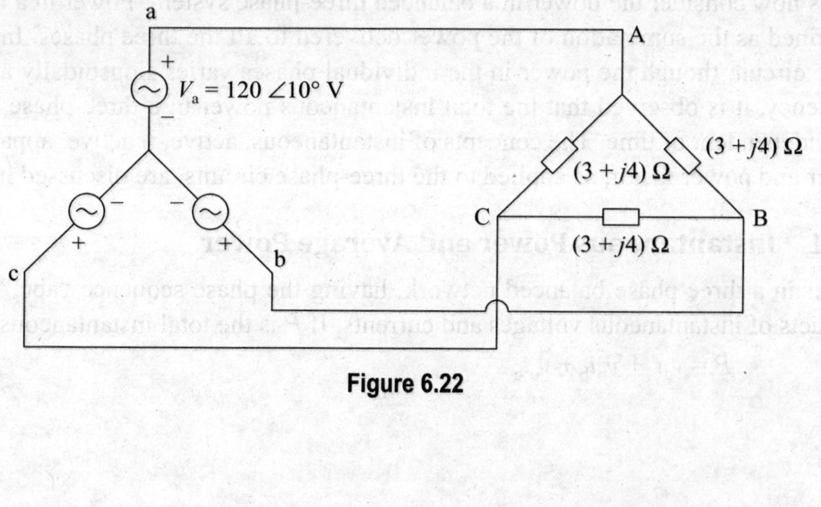

Figure 6.22

Load impedance per phase, $Z = 3 + j4\ \Omega$

$$= 5\ \angle 53.13^\circ\ \Omega$$

Phase voltage, $V_a = 120\ \angle 10^\circ$

Since the source is star-connected, its line voltages are given by

$$V_{ab} = \sqrt{3}\ V_a\ \angle 30^\circ$$
$$= \sqrt{3} \times 120\ \angle 10^\circ + 30^\circ$$
$$= 207.85\ \angle 40^\circ$$

The same V_{ab} will appear across the delta-connected load. Hence, the phase currents are

$$I_{ab} = \frac{V_{ab}}{Z} = \frac{207.85\ \angle 40^\circ}{5\ \angle 53.13^\circ} = 41.57\ \angle -13.13^\circ\ \text{A}$$

$$I_{bc} = I_{ab}\ \angle -120^\circ$$
$$= 41.57\ \angle -133.13^\circ\ \text{A}$$

$$I_{ca} = I_{ab}\ \angle 120^\circ$$
$$= 41.57\ \angle 106.87^\circ\ \text{A}$$

The line currents are

$$I_a = \sqrt{3}\ I_{ab}\ \angle -30^\circ$$
$$= \sqrt{3} \times 41.57 \angle -43.13^\circ$$
$$= 72\ \angle -43.13^\circ\ \text{A}$$

$$I_b = I_a\ \angle -120^\circ$$
$$= 72\ \angle -163.13^\circ\ \text{A}$$

$$I_c = I_a \angle -240^\circ$$
$$= 72 \angle 76.87^\circ\ \text{A}$$

6.7 POWER IN THREE-PHASE AC SYSTEMS WITH BALANCED LOAD

Let us now consider the power in a balanced three-phase system. Power in a three-phase circuit is defined as the summation of the power delivered to all the three phases. In a balanced three-phase circuit, though the power in the individual phases varies sinusoidally at twice the supply frequency, it is observed that the total instantaneous power in a three-phase circuit is constant and independent of time. The concepts of instantaneous, active, reactive, apparent, and complex power and power factor, as applied to the three-phase circuits, are discussed in this section.

6.7.1 Instantaneous Power and Average Power

Power in a three-phase balanced network, having the phase sequence "abc," is the sum of the products of instantaneous voltages and currents. If P is the total instantaneous power, then

$$P = v_a i_a + v_b i_b + v_c i_c \tag{6.49}$$

The instantaneous values of phase voltages v_a, v_b, and v_c and phase currents i_a, i_b, and i_c can be written as

$$v_a = \sqrt{2}\, V_{Ph} \sin \omega t$$
$$v_b = \sqrt{2}\, V_{Ph} \sin (\omega t - 120°)$$
$$v_c = \sqrt{2}\, V_{Ph} \sin (\omega t - 240°)$$
$$i_a = \sqrt{2}\, I_{Ph} \sin (\omega t - \phi)$$
$$i_b = \sqrt{2}\, I_{Ph} \sin (\omega t - \phi - 120°)$$
$$i_c = \sqrt{2}\, I_{Ph} \sin (\omega t - \phi - 240°)$$

where V_{Ph} is the rms value of the phase voltage; I_{Ph} is the rms value of the phase current; and ϕ is the angle between the phase voltage and the corresponding phase current.

The total instantaneous power in the three phases then becomes

$$\begin{aligned} P &= v_a i_a + v_b i_b + v_c i_c \\ &= \sqrt{2}\, V_{Ph} \sin \omega t \ \sqrt{2}\, I_{Ph} \sin (\omega t - \phi) \\ &\quad + \sqrt{2}\, V_{Ph} \sin (\omega t - 120°)\ \sqrt{2}\, I_{Ph} \sin (\omega t - \phi - 120°) \\ &\quad + \sqrt{2}\, V_{Ph} \sin (\omega t - 240°)\, \sqrt{2}\, I_{Ph} \sin (\omega - \phi - 240°) \qquad (6.50) \\ &= 2V_{Ph} I_{Ph} [\sin \omega t \sin (\omega t - \phi) + \sin (\omega t - 120°) \sin (\omega t - \phi - 120°) \\ &\quad + \sin (\omega t + 120°) \sin (\omega t - \phi + 120°)] \\ &= 3 V_{Ph} I_{Ph} \cos \phi \end{aligned}$$

Thus, the instantaneous power becomes

$$P = 3 V_{Ph} I_{Ph} \cos \phi \qquad (6.51)$$

The total instantaneous power in a balanced three-phase system is constant. It does not change with time as the instantaneous power of each phase does. This result is true whether the load is star- or delta-connected. This is one important reason for using the three-phase system to generate and transmit the power.

The average power per phase for either delta-connected or star-connected load is $P/3$ or

$$P_{Ph} = V_{Ph} I_{Ph} \cos \phi \qquad (6.52)$$

In a star-connected system, the line voltage $V_L = \sqrt{3}\, V_{Ph}$ and $I_L = I_{Ph}$ substituting for V_{Ph}, I_{Ph} in Eq. (6.51), the power in a star-connected network becomes

$$P = \sqrt{3}\, V_L I_L \cos \phi$$

Similarly, for a delta-connected system, $V_L = V_{Ph}$ and $I_L = \sqrt{3}\, I_{Ph}$ substituting for V_{Ph} and I_{Ph}, the power in a delta-connected network becomes

$$P = \sqrt{3}\, V_L I_L \cos \phi$$

Hence, it can be concluded that in a balanced three-phase network whether connected in star or delta,

Average power (or active power), $P = \sqrt{3}\, V_L I_L \cos \phi$ (6.53)

It may be noted that unlike in single-phase systems, where the instantaneous power is pulsating, the total instantaneous power for a three-phase system is constant. Thus, a three-phase supply system has a distinct advantage over a single-phase system, especially in the running of large industrial drives.

6.7.2 Reactive Power and Apparent Power

The reactive power in a three-phase system is equal to the sum of the reactive powers of the individual phases. For a balanced three-phase system, the reactive power

$$Q = 3V_{Ph} I_{Ph} \sin \phi \tag{6.54}$$
$$= \sqrt{3}\, V_L I_L \sin \phi$$

The apparent power or the volt-ampere for a three-phase circuit is the sum of the volt-ampere of the three individual phases. For a balanced system, the apparent power

$$S = 3V_{Ph} I_{Ph} \tag{6.55}$$
$$= \sqrt{3}\, V_L I_L$$

It may be observed that in a balanced three-phase system,

$$P = 3V_{Ph} I_{Ph} \cos \phi = S \cos \phi$$
$$Q = 3V_{Ph} I_{Ph} \sin \phi = S \sin \phi$$
$$S = \sqrt{P^2 + Q^2} \tag{6.56}$$

6.7.3 Complex Power

For convenience in calculations, the concept of complex power is useful. It is a complex number; the real part of this number is the total active power and the imaginary part is the total reactive power in a circuit. Thus,

$$S = P + jQ = S(\cos \phi + j \sin \phi) = S \angle \phi$$

Complex power in terms of V_{Ph} and I_{Ph} can be expressed as

$$S = 3V_{Ph} I^*_{Ph} \tag{6.57}$$

Let $V_{Ph} = Ve^{j0}$ and $I_{Ph} = Ie^{j(0-\phi)}$, where I_{Ph} lags V_{Ph} by an angle ϕ. Then, for balanced load

$$S = 3V_{Ph} I^*_{Ph} \tag{6.58}$$
$$= 3Ve^{j0}\, Ie^{j(0+\phi)}$$
$$= 3VIe^{j\phi}$$
$$= 3VI(\cos \phi + j \sin \phi)$$
$$= P + jQ = S(\cos \phi + j \sin \phi) \tag{6.59}$$

Remember that V_{Ph}, I_{Ph}, V_L, and I_L are all rms values and that ϕ is the angle between the phase voltage and phase current.

EXAMPLE 6.10 Determine the active, reactive, and apparent powers of the balanced delta-connected load of (10 + *j*8) Ω when it is connected to a 440-V, three-phase supply in the "ABC" sequence.

Solution

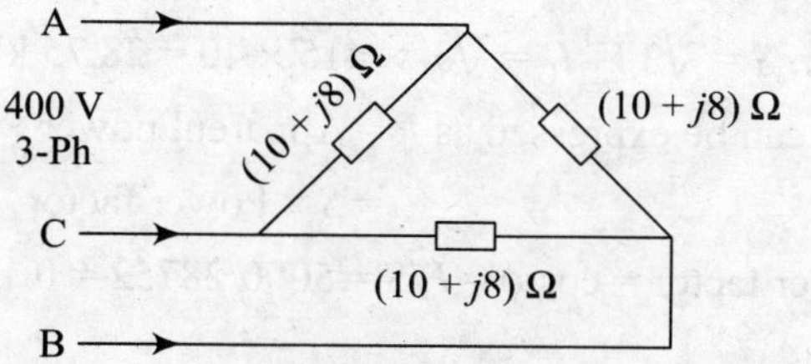

Figure 6.23

Since it is a delta connection, line voltage = phase voltage = 440 V

Impedance per phase = $10 + j8\ \Omega = 12.81\angle 38.66°\ \Omega$

Phase current, $I_p = \dfrac{440}{12.81\angle 38.66°} = 34.35\angle -38.66°$ A

Line current, $I_L = \sqrt{3} \times 34.35\angle -30° - 38.66° = 59.5\angle -68.66°$ A

Active power, $P = \sqrt{3}\, V_L I_L \cos\phi$

$= \sqrt{3} \times 440 \times 59.5 \times \cos(-68.66°)$

$= 16.50$ kW

Reactive power, $Q = \sqrt{3}\, V_L I_L \sin\phi$

$= \sqrt{3} \times 440 \times 59.5 \times \sin(68.66°)$

$= 42.216$ kVAr

Apparent power, $S = \sqrt{3}\, V_L I_L$

$= \sqrt{3} \times 440 \times 59.5$

$= 45.345$ kVA

EXAMPLE 6.11 **A three-phase motor is treated as a balanced star-connected load. It draws 5 kW, when the line voltage is 415 V and the line current is 40 A. Determine the power factor of the motor.**

Solution

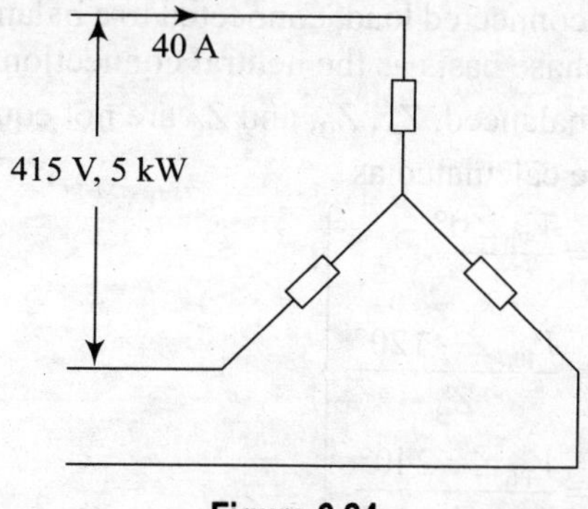

Figure 6.24

Real power of the motor, $P = 5000$ W

Apparent power, $S = \sqrt{3}\, V_L I_L = \sqrt{3} \times 415 \times 40 = 28.75$ kVA

The real power can be expressed as $P =$ Apparent power × Power factor

$= S \times$ Power factor

Therefore, power factor $= \cos\phi = P/S = 5000/28752 = 0.174$ lag.

6.8 UNBALANCED THREE-PHASE CIRCUITS

An unbalanced three-phase system is one in which the phases are not balanced. This may happen due to unequal magnitudes of the phase quantities, unequal phase differences between the phase quantities, or both. This may also be caused by unbalanced loading, unbalanced supply voltage, or both. Unbalanced three-phase systems are solved by the direct application of the mesh analysis or nodal analysis.

6.8.1 Unbalanced Star Connection

Figure 6.25 shows a three-phase, four-wire unbalanced star-connected load.

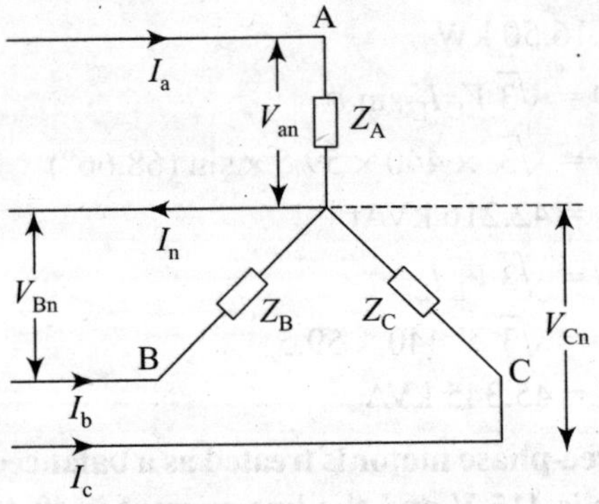

Figure 6.25 Unbalanced three-phase, four-wire star-connected load

Some simplifications are possible under specific conditions. For instance, the simplification for an unbalanced star-connected load, connected to a balanced three-phase four-wire supply, can be carried out on per-phase basis as the neutral connection is available.

Since the load is unbalanced, Z_A, Z_B, and Z_C are not equal.

The line currents are calculated as

$$\left.\begin{aligned} I_A &= \frac{V_A}{Z_A} = \frac{V_{Ph}\angle 0^\circ}{Z_A} \\ I_B &= \frac{V_B}{Z_B} = \frac{V_{Ph}\angle -120^\circ}{Z_B} \\ I_C &= \frac{V_C}{Z_C} = \frac{V_{Ph}\angle -240^\circ}{Z_C} \end{aligned}\right\} \tag{6.60}$$

This set of unbalanced line currents produces current in the neutral line, which is not zero as in the case of balanced system. Applying the KCL at node N gives the current in the neutral wire

$$I_N = -(I_A + I_B + I_C) \tag{6.61}$$

In a three-wire system where the neutral is absent, we can still find the line currents using the mesh analysis. The power calculations are also done as per phase in an unbalanced system. For example,

Active power, $P_{Ph} = V_{Ph} I_{Ph} \cos\phi$

Reactive power, $Q_{Ph} = V_{Ph} I_{Ph} \sin\phi$

Apparent power, $S_{Ph} = V_{Ph} I_{Ph}$

In a three-phase three-wire system where the neutral is absent we can still find the line currents I_A, I_B, and I_C using the mesh analysis or transformation technique. In the transformation technique, the given star-connected unbalanced load is converted into delta-connected unbalanced load. Then for this delta connection, line currents are found. The line currents of this unbalanced delta are exactly equal to the line currents of the original star connection.

6.8.2 Unbalanced Delta Connection

If the loads connected in delta are not equal either in magnitude or in angle, then we say that the load is an unbalanced delta connection. Figure 6.26 shows such an unbalanced delta connection.

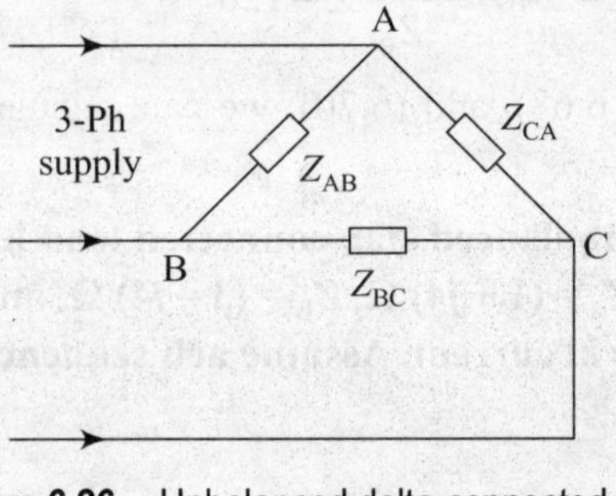

Figure 6.26 Unbalanced delta-connected load

Assuming the supply connected is of balanced type, we know that the phase voltage is equal to the line voltage and displaced each other by 120°.

Let I_A, I_B, and I_C are the line currents and I_{AB}, I_{BC}, and I_{CA} are the phase currents.

The phase currents are given by

$$I_{AB} = \frac{V_{AB}\angle 0°}{Z_A} = \frac{V_{Ph}\angle 0°}{Z_A} \tag{6.62}$$

$$I_{BC} = \frac{V_{BC}\angle -120°}{Z_B} = \frac{V_{Ph}\angle -120°}{Z_B} \tag{6.63}$$

$$I_{CA} = \frac{V_{CA}\angle -240^\circ}{Z_C} = \frac{V_{Ph}\angle -240^\circ}{Z_C} \tag{6.64}$$

The line currents are calculated using the KCL, i.e., applying the KCL at points A, B, and C, we can determine the line currents similar to the balanced case. Applying the KCL at point A

$$I_A = I_{AB} - I_{CA} \tag{6.65}$$

Substituting Eqs. (6.62) and (6.64) into Eq. (6.65), we get

$$I_A = \frac{V_{Ph}}{Z_A}\angle 0^\circ - \frac{V_{Ph}}{Z_C}\angle -240^\circ \tag{6.66}$$

Applying KCL at point B, we get

$$I_B = I_{BC} - I_{AB} \tag{6.67}$$

Substituting Eqs. (6.62) and (6.63) into Eq. (6.67), we get

$$I_B = \frac{V_{Ph}}{Z_B}\angle -120^\circ - \frac{V_{Ph}}{Z_A}\angle 0^\circ \tag{6.68}$$

Applying KCL at point C, we get

$$I_C = I_{CA} - I_{BC} \tag{6.69}$$

Substituting Eqs. (6.63) and (6.64) into Eq. (6.69), we get

$$I_C = \frac{V_{Ph}}{Z_C}\angle -240^\circ - \frac{V_{Ph}}{Z_B}\angle -120^\circ \tag{6.70}$$

By using Eqs. (6.66), (6.68), and (6.70), we can calculate the line currents in the unbalanced delta-connected load.

EXAMPLE 6.12 **An unbalanced star-connected load has balance voltages of 200 V and the load impedances are $Z_a = (1 + j4)\ \Omega$, $Z_b = (3 - j4)\ \Omega$, and $Z_C = (6 + j10)\ \Omega$. Calculate the line currents and the neutral current. Assume acb sequence.**

Solution

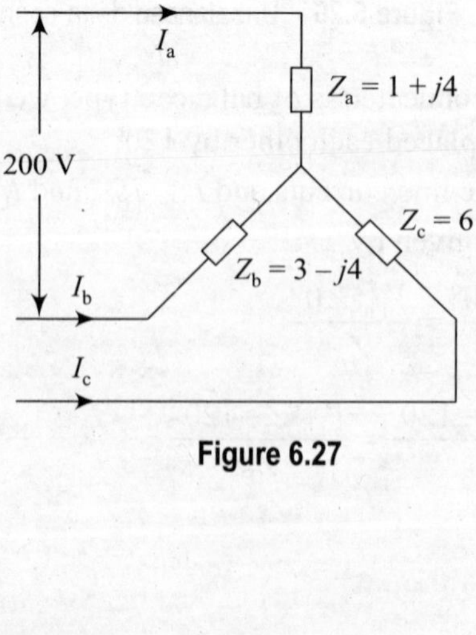

Figure 6.27

The line currents are

$$I_A = \frac{200\angle 0°}{1+j4} = \frac{200\angle 0°}{4.12\angle 75.96°}$$

$$= 48.54\angle -75.96° \text{ A}$$

$$I_B = \frac{V_B}{Z_B} = \frac{200\angle -240°}{3-j4}$$

$$= \frac{200\angle -240°}{5\angle -53.13°}$$

$$= 40\angle -186.87° \text{ A}$$

$$I_C = \frac{V_C}{Z_C} = \frac{200\angle -120°}{6+j10}$$

$$= \frac{200\angle -120°}{11.66\angle 59°}$$

$$= 17.15\angle -179° \text{ A}$$

The current in the neutral wire is

$$I_n = -(I_A + I_B + I_C)$$

$$= -(48.54\angle -75.96° + 40\angle -186.87° + 17.15\angle -179°)$$

$$= -(11.77 - j47.09 - 39.71 + j4.78 - 17.147 - j0.299)$$

$$= -(-45.087 - j42.61)$$

$$= 45.087 + j\,42.61$$

$$= 62.03\angle 43.4° \text{ A}$$

EXAMPLE 6.13 **An unbalanced delta-connected load with impedances of $Z_{ab} = 10\angle -30°$; $Z_{bc} = 30\angle 60°$; and $Z_{ca} = 50\angle 100°$ is connected to a balanced line-to-line voltage of 440 V in the positive sequence. Find the line currents and the total power absorbed.**

Solution

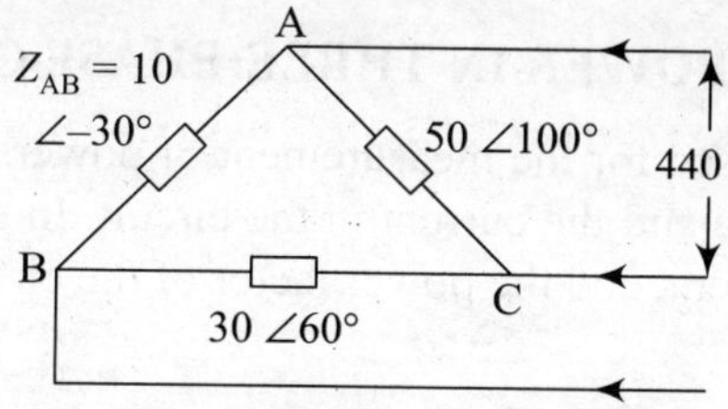

Figure 6.28

The load impedances are

$$Z_{ab} = 10\angle -30° = 8.66 - j5;\ Z_{bc} = 30\angle 60° = 15 + j25.9;\ Z_{ca} = 50\angle 100° = -8.68 + j49.24$$

The phase currents are calculated as

$$I_{ab} = \frac{V_{ab}}{Z_{ab}} = \frac{440\angle 0^\circ}{10\angle -30^\circ} = 44\angle 30^\circ \text{ A}$$

$$I_{bc} = \frac{V_{bc}}{Z_{bc}} = \frac{440\angle -120^\circ}{30\angle 60^\circ} = 14.66\angle -180^\circ \text{ A}$$

$$I_{ca} = \frac{V_{ca}}{Z_{ca}} = \frac{440\angle -240^\circ}{50\angle 100^\circ} = 8.8\angle -340^\circ \text{ A}$$

The line currents are

$$\begin{aligned} I_a &= I_{ab} - I_{ca} \\ &= (38.1 + j22) - (8.27 + j3) \\ &= 29.83 + j19 \text{ A} \\ &= 35.37\angle 32.5^\circ \text{ A} \end{aligned}$$

$$\begin{aligned} I_b &= I_{bc} - I_{ab} \\ &= -14.66 - (38.1 + j22) \\ &= (-52.76 - j22) \text{ A} \\ &= 57.16\angle -57^\circ \text{ A} \end{aligned}$$

$$\begin{aligned} I_c &= I_{ca} - I_{bc} \\ &= (8.27 + j3) - (-14.66) \\ &= 22.93 + j3 \\ &= 23.12\angle -7.45^\circ \text{ A} \end{aligned}$$

Total power absorbed is

$$\begin{aligned} P &= I_{ab}^2 R_{ab} + I_{bc}^2 R_{bc} + I_{ca}^2 R_{ca} \\ &= 44^2 \times 8.66 + 14.66^2 \times 15 + 8.8^2 \times (-8.68) \\ &= 19.317 \text{ kW} \end{aligned}$$

6.9 MEASUREMENT OF POWER IN THREE-PHASE CIRCUITS

Generally wattmeter is used for the measurement of power. In a dc circuit, power consumed is the product of the voltage and the current in the circuit. In an ac circuit, it is the product of the rms voltage, the rms current, and the power factor of the circuit.

6.9.1 Wattmeter

Figure 6.29 shows the connection diagram of a wattmeter. It has two coils namely current coil and voltage coil or pressure coil. The current coil is connected in series with the line to measure the line current. The voltage coil is connected between any two lines and measures the voltage. In Figure 6.29, ML shows the current coil and CV shows the voltage coil. The current coil will have less number of turns as compared to the voltage coil because the wattmeter should not cause wrong estimation of the current. If it has more turns, the current flowing through the circuit will be reduced and it shows wrong value. The interaction of torque produced by the voltage coil and current coil causes the needle of the wattmeter to deflect.

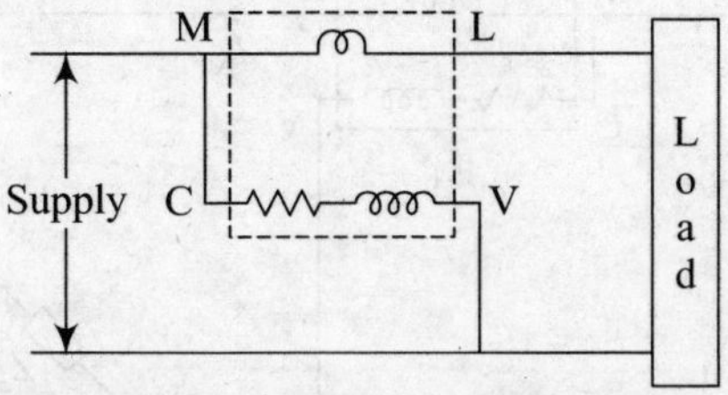

Figure 6.29 Connection diagram of wattmeter

The total active power in a three-phase network is the summation of the active power delivered to all the three phases of the load. In the case of a star-connected balanced load, with the neutral terminal accessible, only one wattmeter is needed to measure the three-phase power; for an unbalanced three-phase load, two wattmeters are needed for the measurement of active power. There are three methods to measure power. All these three methods are discussed in this section.

6.9.2 One-Wattmeter Method

A single wattmeter can measure the average power in a three-phase system that is balanced, so that $W_1 = W_2 = W_3$. The total power is three times the reading of that one wattmeter. In a three-phase star-connected balanced load, with the neutral terminal accessible, power can be measured by connecting the current coil of the wattmeter in one phase and its voltage coil between the same phase and the neutral as shown in Figure 6.30.

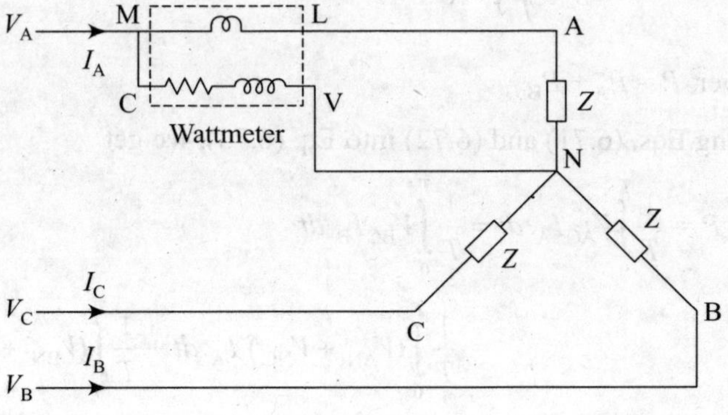

Figure 6.30 Measurement of power in a star-connected balanced load

The reading shown by the wattmeter is the active power per phase. The total active power of the star-connected balanced load is equal to three times the wattmeter reading. However, two or three single-phase wattmeters are necessary to measure power if the system is unbalanced.

6.9.3 Two-Wattmeter Method

It is the most commonly used method for the three-phase power measurement. The connection diagram for this method is shown in Figure 6.31.

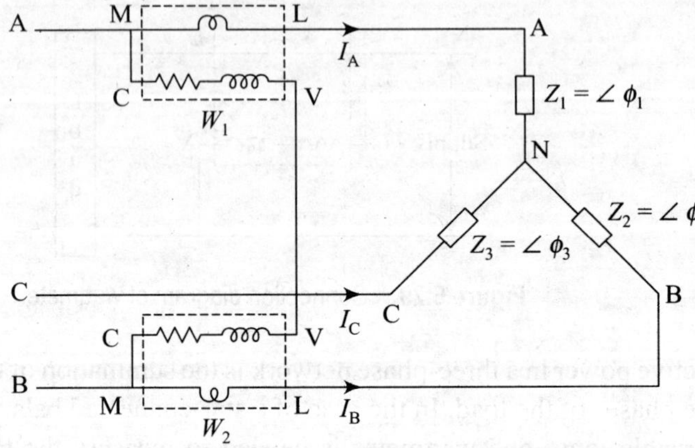

Figure 6.31 Connection diagram for two-wattmeter method

From Figure 6.31, it is observed that there are only two wattmeters used in this method. The current coils are connected in the lines A and B. The voltage coil is connected from the lines A and B to the line C, i.e., third line. Therefore,

$$\text{Power in phase A, } P_A = \frac{1}{T}\int_0^T V_{AC} I_A \, dt \tag{6.71}$$

$$\text{Power in phase B, } P_B = \frac{1}{T}\int_0^T V_{BC} I_B \, dt \tag{6.72}$$

$$\text{Total power, } P = P_A + P_B \tag{6.73}$$

Substituting Eqs. (6.71) and (6.72) into Eq. (6.73), we get

$$P = \frac{1}{T}\int_0^T V_{AC} I_A \, dt = \frac{1}{T}\int_0^T V_{BC} I_B \, dt \tag{6.74}$$

$$= \frac{1}{T}\int_0^T (V_{AN} + V_{NC})\, I_A \, dt + \frac{1}{T}\int_0^T (V_{BN} + V_{NC})\, I_B \, dt$$

$$= \frac{1}{T}\int_0^T ((V_{AN} + V_{NC})\, I_A + (V_{BN} + V_{NC}) I_B) \, dt$$

$$= \frac{1}{T}\int_0^T ((V_{AN} - V_{CN}) I_A + (V_{BN} - V_{CN}) I_B) \, dt$$

$$= \frac{1}{T}\int_0^T (V_{AN} I_A + V_{BN} I_B - V_{CN}\ (I_A + I_B)\ dt$$

$$= \frac{1}{T}\int_0^T (V_{AN} I_A + V_{BN} I_B + V_{CN} I_C)\ dt \quad [\because\ I_A + I_B + I_C = 0] \tag{6.75}$$

Equation (6.75) represents the power in a three-phase system using two-wattmeter method.

Power factor measurement using two-wattmeter method

The phasor diagram for the balanced star-connected load is shown in Figure 6.32. Let P_1 and P_2 be the two wattmeter readings. In the phasor diagram, V_A, V_B, and V_C are the rms values of the phase voltages appearing across the loads and I_A, I_B, and I_C are the corresponding values of the rms phase currents. Phasor V_{AB} is the phase difference between V_A and V_B, phasor V_{BC} is the phase difference between V_B and V_C, and phasor V_{CA} is the phase difference between V_C and V_A. Since the line current I_A flows through the current coil of wattmeter W_1, its voltage coil measures the voltage between phase A and phase C; hence, power P_1 is given by

$$P_1 = V_{AC}\ I_A \cos\ (30° - \phi) \tag{6.76}$$

Similarly, the current coil and voltage coil of wattmeter W_2 measure I_B and V_{BC}. Hence, power P_2 is given by

$$P_2 = V_{BC}\ I_B \cos\ (30° + \phi) \tag{6.77}$$

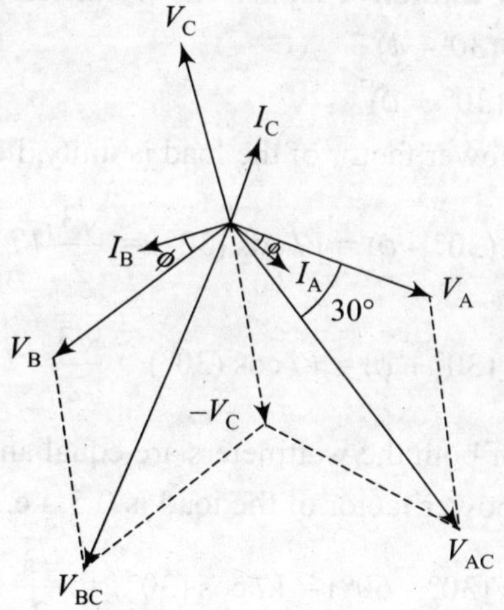

Figure 6.32 Phasor diagram for a balanced star-connected load

The total power is given by

$$P_1 + P_2 = V_{AC}\ I_A \cos\ (30° - \phi) + V_{BC}\ I_B \cos\ (30° + \phi) \tag{6.78}$$

Since the load is balanced, substituting $V_{BC} = V_{AC} = V$ and $I_A = I_B = I$ in Eq. (6.78), we get

$$P_1 + P_2 = VI \cos\ (30° - \phi) + VI \cos\ (30° + \phi)$$

$$= \sqrt{3}\ VI \cos\phi \tag{6.79}$$

Subtracting Eq. (6.76) from Eq. (6.77) and simplifying, we get

$$P_1 - P_2 = VI\,[\cos(30° - \phi) - VI\cos(30° + \phi)]$$

$$= VI \sin\phi \tag{6.80}$$

Dividing Eq. (6.80) by Eq. (6.79) gives

$$\frac{P_1 - P_2}{P_1 + P_2} = \frac{VI \sin\phi}{\sqrt{3}\,VI\cos\phi}$$

$$\tan\phi = \sqrt{3}\left(\frac{P_1 - P_2}{P_1 + P_2}\right) \tag{6.81}$$

$$\phi = \tan^{-1}\left[\sqrt{3}\left(\frac{P_1 - P_2}{P_1 + P_2}\right)\right] \tag{6.82}$$

Equation (6.82) provides an expression for the power factor angle ϕ of the load by the two-wattmeter method.

Effect of load power factor on wattmeter readings

This method can be used for different power factor angles without any problem.

Wattmeter readings for inductive loads. The wattmeter readings are given by

$$P_1 = VI\cos(30° - \phi)$$

$$P_2 = VI\cos(30° + \phi)$$

Case (i): When the power factor of the load is unity, i.e., $\phi = 0°$.

$$P_1 = VI\cos(30° - \phi) = VI\cos(30°) = \frac{\sqrt{3}}{2}VI \tag{6.83}$$

$$P_2 = VI\cos(30° + \phi) = VI\cos(30°) = \frac{\sqrt{3}}{2}VI \tag{6.84}$$

Readings of both the wattmeters are equal and positive.

Case (ii): When the power factor of the load is 0.5, i.e., $\phi = 60°$.

$$P_1 = VI\cos(30° - 60°) = VI\cos(30°) = \frac{\sqrt{3}}{2}VI \tag{6.85}$$

$$P_2 = VI\cos(30° + 60°) = VI\cos(90°) = 0 \tag{6.86}$$

Reading of wattmeter W_1 will give the total power of the load while the second wattmeter reads zero.

Case (iii): When the power factor of the load is zero, i.e., $\phi = 90°$.

$$P_1 = VI\cos(30° - 90°) = VI\cos(60°) = \frac{1}{2}VI \tag{6.87}$$

$$P_2 = VI\cos(30° + 90°) = VI\cos(120°) = -\frac{1}{2}VI \tag{6.88}$$

Readings of the two wattmeters are equal but of opposite sign and the first meter reading is negative.

Case (iv): Let power factor angle increases from 0° to 60°, i.e., power factor decreases from 1 to 0.5. In this case, both the wattmeter readings are positive.

Case (v): Let power factor angle increases from 60° to 90°, i.e., power factor decreases from 0.5 to 0. In this case, W_1 is positive and W_2 is negative. It must also be noted that if one of the wattmeters gives negative reading then it must be corrected by reversing the connections of one of the terminals of its pressure coil. Under these circumstances its reading is taken as negative, and the total power is then the difference of the readings. Thus, in general, the total power is given by the algebraic sum of the two readings.

Wattmeter readings for capacitive loads. The two wattmeter readings for capacitive loads are given by

$$P_1 = VI\cos(30° - (-\phi)) = VI\cos(30° + \phi) \tag{6.89}$$

$$P_2 = VI\cos(30° + (-\phi)) = VI\cos(30° - \phi) \tag{6.90}$$

The readings of the wattmeters are interchanged for leading power loads, i.e., capacitive loads.

The above results are tabulated as given in Table 6.1.

Table 6.1 Summary of wattmeter readings

Power factor ($\cos\phi$)	Wattmeter reading 1 (P_1)	Wattmeter reading 2 (P_2)	Remark
1	Positive	Positive	Equal
0.5 lag	Positive	0	One reads zero
0	Negative	Positive	Equal
Leading power factor	$VI\cos(30° + \phi)$	$VI\cos(30° - \phi)$	Readings are interchanged

Although these results are derived from a balanced star-connected load, they are equally valid for a balanced delta-connected load. However, the two-wattmeter methods cannot be used for power measurement in a three-phase four-wire system unless the current through the neutral line is zero. We use the three-wattmeter method to measure the real power in a three-phase four-wire system.

6.9.4 Three-Wattmeter Method

The connection diagram for a three-wattmeter method is shown in Figure 6.33.

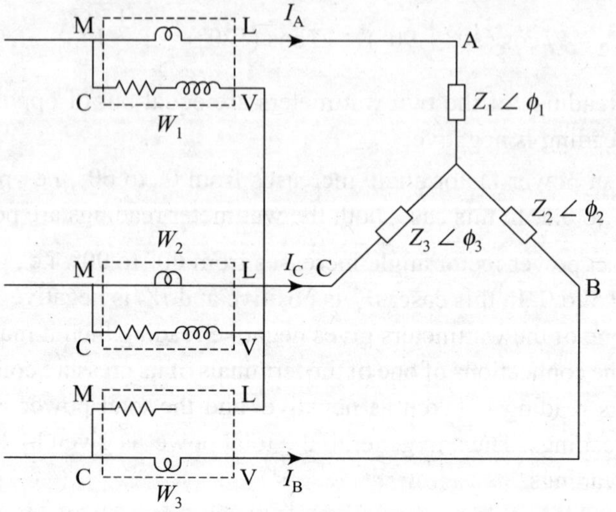

Figure 6.33 Connection diagram for a three-wattmeter method

From Figure 6.33, it is observed that three wattmeters are connected in all the three phases. The three-wattmeter method of power measurement will work regardless of whether the load is balanced or unbalanced, star- or delta-connected. The three-wattmeter method is well suited for power measurement in a three-phase system where the power factor is constantly changing. The total average power is the algebraic sum of the three-wattmeter readings.

$$P_T = P_1 + P_2 + P_3 \tag{6.91}$$

where P_1, P_2, and P_3 are the readings of three wattmeters, respectively.

$$P_T = \frac{1}{T}\int_0^T (V_{AN}\, I_A + V_{BN} I_B + V_{CN} I_C)\, dt$$

If the load is balanced, $P_1 = P_2 = P_3$; if the load is unbalanced, $P_1 \neq P_2 \neq P_3$; total power $= P_1 + P_2 + P_3$ (for unbalanced load) $= 3P_1$ or $3P_2$ or $3P_3$ (for balanced load).

Note: Thus, practically, single wattmeter as well as three-wattmeter methods are rarely used for the industrial loads, due to their limitations.

EXAMPLE 6.14 **The two-wattmeter method shows the wattmeter readings $W_1 = 2000$ W; $W_2 = 1000$ W when connected to a star-connected load. The line voltage is 440 V. Determine the (i) average power per phase, (ii) reactive power per phase, and (iii) power factor.**

Solution The wattmeter readings are $W_1 = 2000$ W; $W_2 = 1000$ W

(i) Total average power, $P = W_1 + W_2 = 2000 + 1000 = 3000$ W

Average power per phase, $P/\text{Ph} = 3000/3 = 1000$ W

(ii) Total reactive power, $Q = \sqrt{3}\,(W_2 - W_1) = \sqrt{3}\,(1000 - 2000)$

$= -1732.05$ VAR

Reactive power per phase, $Q/\text{Ph} = Q/3 = -1732.05/3 = -577.35$ VAR

(iii) Power factor angle $= \tan^{-1}\left(\dfrac{Q}{P}\right)$

$$= \tan^{-1}\left(\frac{-1732.05}{3000}\right) = -30°$$

Power factor = cos (30°) = 0.866 lead

EXAMPLE 6.15 **A three-phase balanced star-connected load has impedance per phase $Z = 3 + j6\ \Omega$. If the load is connected to the 440-V lines, determine the readings of the wattmeters W_1 and W_2 and also find the total real and reactive power.**

Solution Per phase impedance, $Z/\text{Ph} = 3 + j6 = 6.7\angle 63.43°\ \Omega$. Therefore, the power factor angle is 63.43°.

Line voltage, $V_L = 440$ V

Line current, $I_L = \dfrac{440}{\sqrt{3} \times 6.7} = 37.91$ A

Wattmeter readings are given by

$$W_1 = V_L I_L \cos(\theta + 30°)$$

$$= 440 \times 37.91 \times \cos(63.43° + 30°) = -997.97 \text{ W}$$

$$W_2 = V_L I_L \cos(\theta - 30°)$$

$$= 440 \times 37.91 \times \cos(63.43° - 30°) = 13920.78 \text{ W}$$

Total active power, $P = W_1 + W_2 = -997.97 + 13920.78 = 12922.81$ W

Total reactive power, $Q = \sqrt{3}\,(W_2 - W_1) = \sqrt{3}\,(13920.78 - (-997.97)) = 25.84$ kVAR

EXAMPLE 6.16 **In a three-phase four-wire system, loads of 14 kW, 12 kW, and 10 kW are connected between the three-line conductors and the neutral point as shown in Figure 6.34. The line voltage is 440 V. Calculate the (i) line currents and (ii) current in the neutral wire.**

Solution The phase voltages are given by

$$V_a = \frac{440\angle 0°}{\sqrt{3}} \text{ V}$$

$$V_b = 254.03\angle -120°$$

$$V_c = 254.03\angle -240°$$

(i) The line currents are given by

$$I_a = \frac{14000}{254.03\angle 0°} = 55.11\angle 0° \text{ A}$$

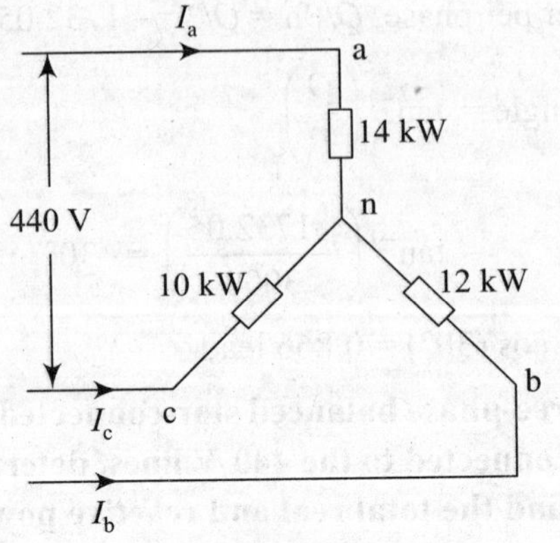

Figure 6.34

$$I_b = \frac{12000}{254.03\ \angle -120°} = 47.28\ \angle 120°\ \text{A}$$

$$I_c = \frac{10000}{254.03\ \angle -240°} = 39.36\ \angle 240°\ \text{A}$$

(ii) The current in the neutral wire is

$$\begin{aligned} I_n &= I_a + I_b + I_c \\ &= 55.11\ \angle 0° + 47.28\ \angle 120° + 39.36\ \angle 240° \\ &= 55.11 - 23.64 + j40.94 - 19.68 - j34.08 \\ &= 11.79 + j6.86 \\ &= 13.64\ \angle 30.2°\ \text{A} \end{aligned}$$

EXAMPLE 6.17 **A delta-connected load as shown in Figure 6.35 is connected to a 415-V, three-phase, 50-Hz supply. Calculate the (i) phase currents and (ii) line currents.**

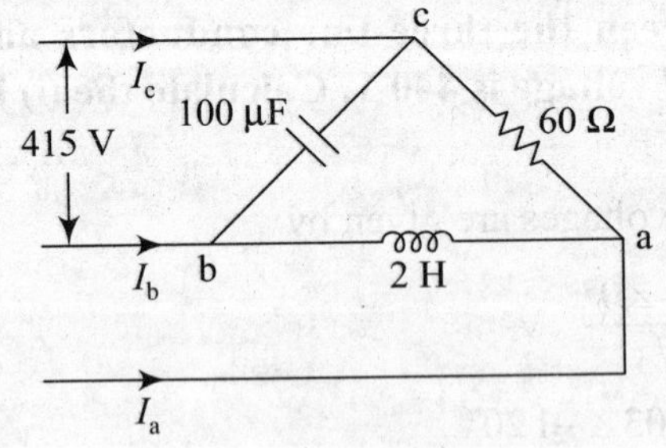

Figure 6.35

Solution Given sequence is positive sequence, hence

$$V_{ab} = 415\ \angle 0°\ \text{V}$$

$$V_{bc} = 415\ \angle -120°\ \text{V}$$

$$V_{ca} = 415\ \angle -240°\ \text{V}$$

In some legs of the delta the reactance value is not given in ohms. We need to convert those into ohms. The reactance between "a" and "b" = $X_{ab} = j\omega L = j2\pi fL = j2 \times \pi \times 50 \times 2 = j628\ \Omega$

The reactance between "b" and "c" = $X_{bc} = \dfrac{1}{j\omega c} = \dfrac{-j}{\omega c} = \dfrac{-j}{2 \times \pi \times 50 \times 100 \times 10^{-6}}$

$$= -j31.8\ \Omega$$

(i) The phase currents are calculated as

$$I_{ab} = \frac{415 \angle 0°}{628 \angle 90°} = 0.66 \angle -90°\ \text{A} = -j0.66\ \text{A}$$

$$I_{bc} = \frac{415 \angle -120°}{31.8 \angle -90°} = 13.05 \angle -30°\ \text{A} = 11.30 - j6.525\ \text{A}$$

$$I_{ca} = \frac{415 \angle -240°}{60} = 6.92 \angle -240°\ \text{A}$$

(ii) The line currents are calculated as

$$I_a = I_{ab} - I_{ca}$$
$$= -j0.66 - (-3.46 + j5.99)$$
$$= 6.35 \angle -57°\text{A}$$

$$I_b = I_{bc} - I_{ab}$$
$$= (11.30 - j6.525) + j0.66$$
$$= 11.30 - j5.865\ \text{A} = 12.73 \angle -27.43°\ \text{A}$$

$$I_c = I_{ca} - I_{bc}$$
$$= -3.46 + j5.99 - (11.30 - j6.525)$$
$$= -7.84 - j12.515 = 14.76 \angle -122°\ \text{A}$$

EXAMPLE 6.18 **A balanced star-connected load of (3 + j4) Ω per phase is connected to a 400-V, three-phase, 50-Hz supply. Find the (i) line current, (ii) power factor, (iii) real power, (iv) reactive power, and (v) apparent power.**

Solution

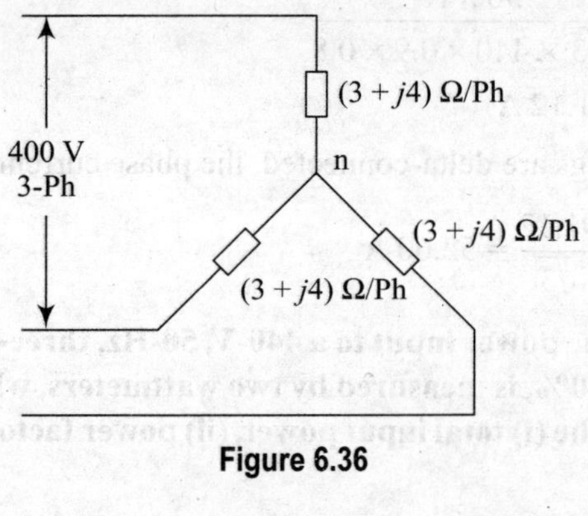

Figure 6.36

(i) Load impedance, $Z_L = 3 + j4 = 5\angle 53.13°$

The line current, $I_a = \dfrac{V_a}{Z_L} = \dfrac{400}{\sqrt{3} \times 5\angle 53.13°} = 46.18\angle -53.13°$ A

As the load is balanced, the phase currents are also balanced. They are equal in magnitude but are displaced from each other by 120°.

$\therefore \quad I_b = 46.18\angle -53.13° - 120°$ A $= 46.18\angle -173.13°$

$I_c = 46.18\angle -53.13° - 240°$ A $= 46.18\angle -293.13°$

(ii) The power factor, $\cos\phi = \cos(53.13°) = 0.6$ lagging

(iii) Real power, $P = \sqrt{3}\, V_L I_L \cos\phi$

$= \sqrt{3} \times 400 \times 46.18 \times 0.6$

$= 19.19$ kW

(iv) Reactive power, $Q = \sqrt{3}\, V_L I_L \sin\phi$

$= \sqrt{3} \times 400 \times 46.18 \times 0.8$

$= 25.59$ kVAR

(v) Apparent power, $S = \sqrt{3}\, V_L I_L = \sqrt{3} \times 400 \times 46.18$

$= 31.99$ kVA

EXAMPLE 6.19 A three-phase motor operating on a 440-V, 50-Hz supply system is developing 50 kW at an efficiency of 90% and a power factor of 0.8. Calculate the (i) line current and (ii) phase current if the windings are delta-connected.

Solution

(i) Efficiency, $\eta = \dfrac{\text{Output power}}{\text{Input power}}$

$$0.9 = \frac{50 \times 10^3}{\sqrt{3} \times 440 \times I_L \times 0.8}$$

$$I_L = \frac{50 \times 10^3}{\sqrt{3} \times 440 \times 0.9 \times 0.8}$$

$= 91.12$ A

(ii) If the windings are delta-connected, the phase current is given by

$$I_{Ph} = \frac{91.12}{\sqrt{3}} = 52.60 \text{ A}$$

EXAMPLE 6.20 The power input to a 440-V, 50-Hz, three-phase motor, running on full load at an efficiency of 80%, is measured by two wattmeters, which indicate 600 kW and 300 kW, respectively. Find the (i) total input power, (ii) power factor, (iii) line currents, and (iv) horse power output.

Solution

(i) Total input power = $W_1 + W_2 = 600 + 300 = 900$ kW

(ii) Power factor angle, $\phi = \tan^{-1}\left(\sqrt{3}\,\dfrac{W_1 - W_2}{W_1 + W_2}\right)$

$$= \tan^{-1}\left(\sqrt{3}\,\frac{600-300}{600+300}\right)$$

$$= 30°$$

Power factor, $\cos\phi = \cos 30° = 0.866$ lag

(iii) The line current can be determined from

$$P = \sqrt{3}\ V_L I_L \cos\phi$$

$$I_L = \frac{900 \times 10^3}{\sqrt{3} \times 440 \times 0.866} = 1363.7 \text{ A}$$

(iv) Efficiency, $\eta = \dfrac{\text{Output power}}{\text{Input power}}$

Output power = $\eta \times$ Input power = $0.8 \times 900 = 720$ kW (we know that 1 hp = 746 W)

$$\text{Horse power output} = \frac{720 \times 10^3}{746} = 965.14 \text{ hp}$$

EXAMPLE 6.21 **Two wattmeters are used to measure power in a three-phase balanced load. The wattmeter readings are 8.2 kW and 7.5 kW. Calculate the power factor of the load.**

Solution The two-wattmeter readings are $W_1 = 8.2$ kW; $W_2 = 7.5$ kW.

$$\text{Power factor angle } \phi = \tan^{-1}\left(\sqrt{3}\,\frac{W_1 - W_2}{W_1 + W_2}\right)$$

$$= \tan^{-1}\left(\sqrt{3}\,\frac{8200-7500}{8200+7500}\right)$$

$$= 4.415°$$

Power factor, $\cos\phi = \cos 4.415° = 0.99$ lag

EXAMPLE 6.22 **Two wattmeters are used to measure the power in a three-phase balanced circuit. What is the power factor of the load when (i) both the meters read equal, (ii) both the meters read equal but one is negative, and (iii) one reads twice the other.**

Solution

(i) Both meters read equal, i.e., $W_1 = W_2$.

$$\text{Power factor angle, } \phi = \tan^{-1}\left(\sqrt{3}\,\frac{W_1 - W_2}{W_1 + W_2}\right)$$

$$= \tan^{-1}(0)$$

$$= 0°$$

Power factor, $\cos\phi = \cos 0° = 1$

(ii) Both the meters read equal but one is negative, i.e., $W_1 = -W_2$.

Power factor angle, $\phi = \tan^{-1}\left(\sqrt{3}\dfrac{W_1 - W_2}{W_1 + W_2}\right)$

$= \tan^{-1}(\infty)$

$= 90°$

Power factor, $\cos\phi = \cos 90° = 0$

(iii) One reads twice the other, i.e., $W_1 = 2W_2$.

Power factor angle, $\phi = \tan^{-1}\left(\sqrt{3}\dfrac{2W_2 - W_2}{2W_2 + W_2}\right) = \tan^{-1}\left(\dfrac{1}{\sqrt{3}}\right) = 30°$

Power factor, $\cos\phi = \cos 30° = 0.866$

EXAMPLE 6.23 **A three-phase 500-V motor operates at a power factor of 0.4 and takes an input power of 30 kW. Two wattmeters are employed to measure the input power. Find the readings of each instrument.**

Solution Input power, $P = 30$ kW

$$P = \sqrt{3}\, V_L I_L \cos\phi$$

$$30000 = \sqrt{3} \times 500 \times I_L \times 0.4$$

$$I_L = \frac{30000}{\sqrt{3} \times 500 \times 0.4} = 86.6 \text{ A}$$

Power factor angle, $\phi = \cos^{-1}(0.4) = 66.42°$

Reading of wattmeter 1, $W_1 = V_L I_L \cos(30° - \phi)$

$= 500 \times 86.6 \times \cos(30° - 66.42°)$

$= 34.84$ kW

Reading of wattmeter 2, $W_2 = V_L I_L \cos(30° + \phi)$

$= 500 \times 86.6 \times \cos(30° + 66.42°)$

$= -4.84$ kW

EXAMPLE 6.24 **Three similar resistors are connected in star across a 450-V, three-phase circuit. The line current is 10 A if the same resistors are connected in delta across the same supply. Calculate the (i) phase current, (ii) line current, and (iii) power consumed.**

Solution Load is balanced and star-connected.

$$I_L = I_{Ph} = 10 \text{ A}$$

$$V_L = 450 \text{ V}$$

$$V_{Ph} = \frac{450}{\sqrt{3}} = 259.8 \text{ V}$$

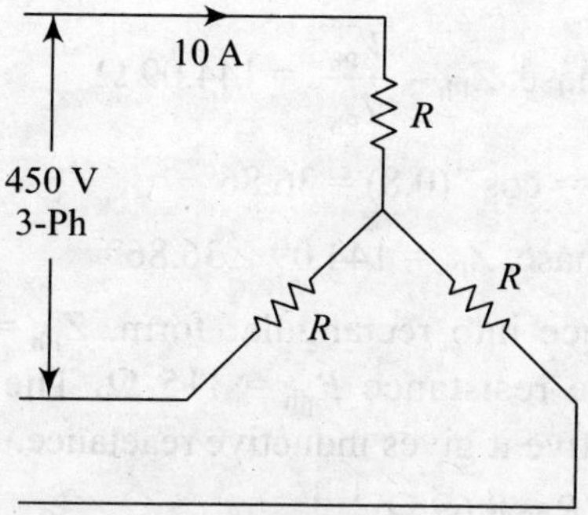

Figure 6.37

Resistance $R = \frac{V_{Ph}}{I_{Ph}} = \frac{259.8}{10} = 25.98\ \Omega$

If the same resistors are connected in delta,

(i) Phase current, $I_{Ph} = \frac{450}{25.98} = 17.32$ A

(ii) Line current, $I_L = \sqrt{3}\, I_{Ph} = 30$ A

(iii) Power consumed, $P = \sqrt{3}\, V_L I_L \cos\phi$

$= \sqrt{3} \times 450 \times 30 \times 1$

$= 23.38$ kW

EXAMPLE 6.25 **Three identical coils are connected in delta to a three-phase, three-wire, 50-Hz, 415-V supply and takes a line current of 5 A at 0.8 power factor (lag). Calculate the resistance and conductance of coils.**

Solution

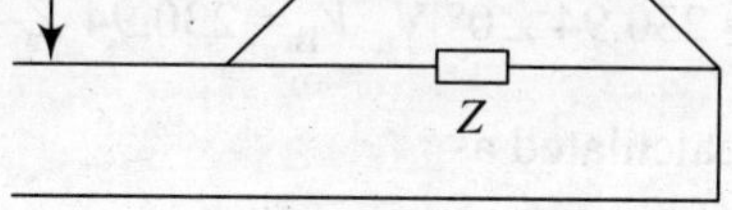

Figure 6.38

Since the coils are connected in delta, the line voltage is equal to the phase voltage,

$\therefore$ Phase voltage, $V_{Ph} = 415$ V

Line current, $I_L = 5$ A

Phase current, $I_{Ph} = \frac{I_L}{\sqrt{3}} = 2.88$ A

The impedance per phase, $Z_{Ph} = \dfrac{V_{Ph}}{I_{Ph}} = 144.09\ \Omega$

Power factor angle, $\phi = \cos^{-1}(0.8) = 36.86°$

The impedance per phase, $Z_{Ph} = 144.09 \angle 36.86°$

Writing the impedance into rectangular form, $Z_{Ph} = (115.28 + j86.43)\ \Omega$. The real part of impedance gives the resistance $R_{ph} = 115\ \Omega$. The imaginary part gives the reactance 86.43 Ω. Since it is positive it gives inductive reactance.

The conductance = $1/R = 8.69\ \Omega^{-1}$.

EXAMPLE 6.26 **An unbalanced star-connected load of $Z_A = 1\angle 100°$, $Z_B = 5\angle -90°$, and $Z_C = 4\angle 90°$ is supplied from a three-phase, 400-V, three-wire supply. Calculate the line currents. Take phase sequence ABC.**

Solution

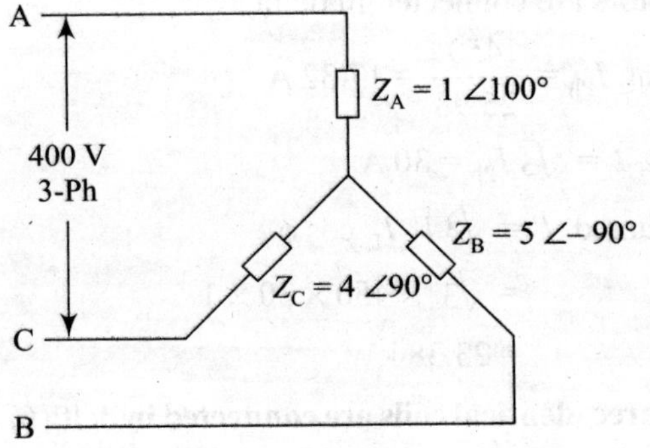

Figure 6.39

The impedances given are

$$Z_A = 1\angle 100°,\ Z_B = 5\angle -90°,\ Z_C = 4\angle 90°$$

The phase voltages are

$$V_A = \frac{400}{\sqrt{3}} = 230.94\angle 0°\ \text{V},\ V_B = 230.94\angle -120°\ \text{V},\ V_C = 230.94\angle -240°\ \text{V}$$

The line currents are calculated as

$$I_A = \frac{V_A}{Z_A} = \frac{230.94\angle 0°}{1\angle 100°} = 230.94\angle -100°\ \text{A}$$

$$I_B = \frac{V_B}{Z_B} = \frac{230.94\angle -120°}{5\angle -90°} = 46.18\angle -30°\ \text{A}$$

$$I_C = \frac{V_C}{Z_C} = \frac{230.94\angle -240°}{4\angle 90°} = 57.73\angle -330°\ \text{A}$$

EXAMPLE 6.27 **A balanced three-phase star-connected load of 120 kW takes a leading current of 150 A with a line voltage of 11 kV, 50 Hz. Find the circuit constants of the load per phase.**

Solution Phase voltage, $V_{Ph} = \dfrac{11000}{\sqrt{3}} = 6350.85 \text{ V}$

Since it is star-connected, the line current is equal to the phase current, $I_{Ph} = 150$ A

Phase impedance, $Z_{Ph} = \dfrac{V_{Ph}}{I_{Ph}} = \dfrac{6350.85}{150} = 42.33\ \Omega$

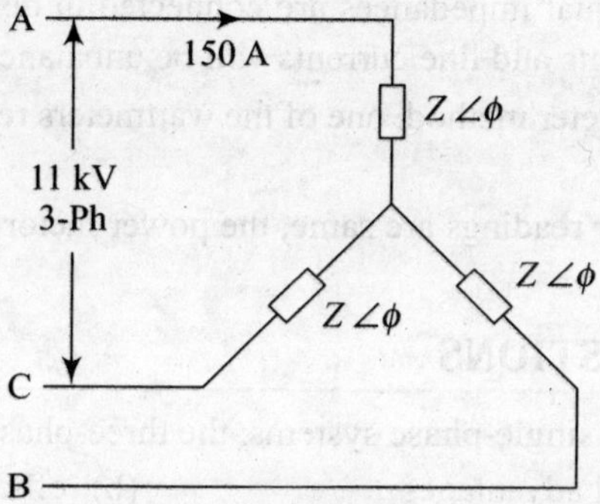

Figure 6.40

Power per phase, $P_{Ph} = \dfrac{120}{3} = 40 \text{ kW}$

Power in terms of resistance, $P_{Ph} = I^2R = 40000$

$\therefore$ Resistance, $R = \dfrac{40000}{150^2} = 1.77\ \Omega$

Reactance, $X = \sqrt{Z^2 - R^2} = \sqrt{42.33^2 - 1.77^2} = 42.29\ \Omega$

Because of the leading current this must be capacitive

$$\frac{1}{wC} = 42.29$$

$$C = \frac{1}{2 \times \pi \times 50 \times 42.29} = 0.75\ \mu\text{F}$$

CHAPTER AT A GLANCE

- For a star-connected system: $V_L = 1.732\ V_{Ph}$ and $I_L = I_{Ph}$.
- For a delta-connected system: $V_L = V_{Ph}$ and $I_L = \sqrt{3}\ I_{Ph}$.
- Active power in a three-phase network, $P = 3V_{Ph} I_{Ph} \cos\phi = \sqrt{3}\ V_L I_L \cos\phi$.
- Reactive power in a three-phase network, $Q = 3V_{Ph} I_{Ph} \sin\phi = \sqrt{3}\ V_L I_L \sin\phi$.

- Apparent power in a three-phase network, $S = 3V_{Ph} I_{Ph} = \sqrt{3}\, V_L I_L$.
- The total power in a balanced three-phase network with the neutral available is three times the reading shown by the wattmeter.
- The total power measured in a three-phase network (balanced or unbalanced), by the two-wattmeter method, is $P_1 + P_2$, where P_1 and P_2 are the readings on the wattmeters. Power factor angle, $\phi = \tan^{-1}\left[\sqrt{3}(P_1 - P_2)\right]/(P_1 + P_2)$.
- In three-phase systems, the emfs are 120° apart.
- When three unequal impedances are connected in delta to a three-phase, three-wire system, both phase currents and line currents will be unbalanced.
- In the two-wattmeter method, one of the wattmeters reads zero reading when the power factor is 0.5 lag.
- If both wattmeter readings are same, the power factor is unity.

MULTIPLE CHOICE QUESTIONS

1. Compared to the single-phase systems, the three-phase power systems are
   (a) operational advantages (b) efficient
   (c) economical (d) all of the above
2. In a two-phase system, phase voltages differ by
   (a) 60° (b) 120° (c) 90° (d) 180°
3. Which of the following are the necessary conditions for an entire three-phase system to be balanced?
   (a) Line voltages are equal in magnitude
   (b) Impedance in each phase must be identical
   (c) Phase difference between successive line voltages are equal
   (d) All of the above
4. Which of the following is true related with a three-phase delta-connected circuit?
   (a) Line currents are equal to phase currents
   (b) Line voltage is equal to phase voltage
   (c) Line currents are 90 degrees apart
   (d) Line voltage is 1.732 times the phase voltage
5. Which of the following is not true related to a three-phase power system?
   (a) The instantaneous sum of power delivered is always constant
   (b) A non-zero starting torque for three-phase motors
   (c) The output of a three-phase machine is less than that of a single-phase machine for the same frame size
   (d) For the same voltage the weight of a conductor is 25% less than in a single-phase
6. In a three-phase system, the phase sequence is used to indicate the
   (a) frequency of the voltages
   (b) magnitude of the voltages

(c) angle between the voltages
(d) order in which the phase voltages attain their peak values

7. What is the meaning of positive phase sequence?
(a) P'ase voltage *a* leads the phase voltage *b* by 120°
(b) Phase voltage *c* leads the phase voltage *a* by 120°
(c) Phase voltage *b* leads the phase voltage *c* by 120°
(d) All of the above

8. Which of the following represents a negative phase sequence?
(a) c–b–a (b) a–c–b (c) b–a–c (d) All of the above

9. How delta connection is formed?
(a) The starting end of one coil is connected to the finishing end of the other
(b) All the starting ends of the coils are connected together
(c) The starting and the finishing end points of each coil are left unconnected
(d) None of the above

10. The sum of the instantaneous voltages around a delta-connected supply is equal to
(a) the maximum line voltage
(b) twice the maximum line voltage
(c) $\sqrt{2}$ times the rms value of the phase voltage
(d) zero

11. What is the expression for real power in a three-phase system?
(a) $V_{Ph} I_{Ph} \cos \phi$ (b) $3V_{Ph} I_{Ph} \cos \phi$ (c) $\sqrt{3}\, V_L I_L \cos \phi$ (d) (b) and (c)

12. A 415-V, three-phase, star-connected system is connected to a delta-connected balanced load of 100 Ω each. How much is the line current?
(a) 2.866 A (b) 4.15 A (c) 7.18 A (d) 2.39 A

13. What is the expression for the neutral current of a three-phase four-wire system?
(a) $I_n = I_a + I_b + I_c$ (b) $I_n = 0$ (c) $I_n = I_a + I_b$ (d) $I_n = I_a + I_c$

14. What is the expression for the reactive power in a balanced three-phase system?
(a) $\sqrt{3}\, V_L I_L \cos \phi$ (b) $\sqrt{3}\, V_L I_L \sin \phi$ (c) $\sqrt{3}\, V_L I_L$ (d) $3V_L I_L$

15. For what type of load, the one-wattmeter method will be used to measure power in a three-phase circuit?
(a) Balanced
(b) Balanced star-connected load
(c) Balanced star-connected, with the neutral accessible load
(d) Balanced delta-connected load

16. To which of the following circuits, the two-wattmeter method is not suitable to measure power in a three-phase circuit?
(a) Unbalanced star-connected loads with the neutral inaccessible
(b) Balanced star- or delta-connected loads
(c) Unbalanced delta-connected loads
(d) None of these

17. The power in a three-phase system is given by $\sqrt{3}\, V_L I_L \cos \phi$, where ϕ is the phase angle between
    (a) line voltage and line current (b) line voltage and phase current
    (c) phase voltage and phase current (d) phase voltage and line current
18. Which of the following systems is a four-wire system?
    (a) Star (b) Delta (c) (a) and (b) (d) Star with neutral
19. In a balanced three-phase star-connected system, E_{RY} lags behind E_B by how many electrical degrees?
    (a) 90 (b) 60 (c) 30 (d) 120
20. In a three-phase balanced star-connected load, the neutral current is equal to
    (a) I_L (b) I_P (c) 1 (d) 0
21. The relationship between the line and phase voltage of a delta-connected circuit is given by
    (a) $V_L = V_P$ (b) $V_L = V_P/3$ (c) $V_L = \sqrt{3}\, V_P$ (d) $V_L = 2V_P$
22. The minimum number of wattmeters to measure power in a three-phase unbalanced star-connected load is
    (a) 3 (b) 1 (c) 4 (d) 2
23. A three-phase star-connected symmetrical load consumes P watts of power from a balanced supply. If the same load is connected in delta to the same supply, the power consumption will be
    (a) $3P$ (b) P (c) $\sqrt{3}P$ (d) $2P$
24. W_1 and W_2 are the readings of two wattmeters used to measure power of a three-phase balanced load. The reactive power drawn by the load is
    (a) $W_1 + W_2$ (b) $W_1 - W_2$ (c) $\sqrt{3}\,(W_1 - W_2)$ (d) $\sqrt{3}\,(W_1 + W_2)$
25. In the measurement of the three-phase power by the two-wattmeter method, if the two-wattmeter readings are equal, the power factor of the load is
    (a) 1 (b) 0 (c) 0.8 lagging (d) 0.8 leading

SHORT ANSWER QUESTIONS

1. What is the phase angle between the emfs in a three-phase system?
2. What will happen if the phase sequence at the three-phase load reverses?
3. What are the conditions for the entire system to be balanced?
4. What is the magnitude of neutral current for three-phase, four-wire balanced star-connected load?
5. A circuit draws 2 A when it is connected to a 50-V supply. The wattmeter connected in the circuit reads 80 W. What is the power factor of the circuit?
6. A three-phase 500-V motor load has a power factor of 0.707. Two wattmeters are connected to measure the power that shows an input of 30 W. What is the reactive power?
7. What is the relationship between the line voltage and phase voltage in a delta-connected system?
8. A three-phase star-connected symmetrical load consumes P watts of power from a balanced supply. If the same load is connected in delta to the same supply, what will be the power consumption?

9. In a delta-connected load, if one of the resistors is removed what will be the power?
10. In the two-wattmeter method, if the readings of two wattmeters are equal. What is the type of load connected?
11. In a two-wattmeter measurement of the three-phase power, one of the wattmeters is showing zero reading. Find the power factor of the circuit.

REVIEW QUESTIONS

1. Explain the advantages of three-phase systems.
2. Explain the comparison between single-phase and three-phase systems.
3. Explain the principle of generation of three-phase voltages.
4. Derive the relation between line and phase quantities of voltages and currents for a star-connected system.
5. Derive the relation between line and phase quantities of voltages and currents for a delta-connected system.
6. Derive the expression for three-phase real power, reactive power, and apparent power.
7. Explain the various methods of power measurement.
8. Explain how power is measured using two wattmeters.
9. Explain the measurement of power using the three-wattmeter method.
10. Explain how neutral current is calculated in a three-phase four-wire system.

EXERCISE PROBLEMS

1. Three balanced loads are connected in delta and take a power of 30 kW at 0.8 power factor from a three-phase, 415-V supply. Calculate the line current. If the above loads are reconnected in star to the same supply, what power would be consumed?
2. Three identical noninductive resistances are connected in star across a 400-V, three-phase, 50-Hz supply. Calculate (a) the value of each resistor if the line current is 5 A and (b) the total power consumed. If the resistors are now connected in delta across the same supply, calculate (c) the line current and (d) the power consumed.
3. A three-phase delta-connected load, each phase of which has an inductive reactance of 40 Ω and resistance 25 Ω, is fed from the secondary of a three-phase star-connected transformer, which has a phase voltage of 240 V. Draw the circuit diagram of the system and calculate (a) the current in each phase of the load; (b) the potential difference across each phase of the load; (c) the current in the transformer secondary windings; and (d) the total power taken from the supply and its power factor?
4. A three-phase star-connected alternator supplies a delta-connected load with a 415-V, 50-Hz supply. Each phase of the load has a resistance of 15 Ω and reactance of 25 Ω. Calculate (a) the current in each phase of the load; (b) the alternator phase voltage and current; (c) the power factor of the load; and (d) the total power taken by the load. Sketch the circuit diagram and show on it the phase and line voltages and currents.

5. Three similar impedances are connected in delta across a three-phase, 400-V, 50-Hz supply. The line current is 12 A at a power factor of 0.8 lagging. Calculate the inductance and resistance of each impedance?
6. Three chokes, resistance 40 Ω, and reactance 30 Ω are connected (a) in star and (b) in delta, to a 400-V, three-phase balanced supply. What is the line current in each case and the total power dissipated?
7. A balanced star-connected load of 2.5 resistance and 0.03 H inductance in series in each phase is connected across a 415-V, three-phase, 50-Hz supply. Calculate the line current, power input, kVA, and kVAR taken by the load.
8. A 415-V, three-phase star-connected alternator supplies a delta-connected induction motor. The motor has a full load efficiency of 0.87 pu and power factor 0.8. Calculate the current in each phase of the motor and the alternator if the output of the motor is 14.92 kW.
9. A 400-V, three-phase, star-connected motor takes 37.3 kW at power factor 0.91. Calculate the phase and line currents. If the motor were reconnected in delta, calculate the correct voltage of a three-phase supply suitable for this load.
10. A three-phase delta-connected motor operating on a 400-V supply is developing 20 hp at an efficiency of 0.86 and a power factor of 0.85. Determine the line and phase currents.
11. Each phase of a balanced star-connected load consists of $R = 10\ \Omega$ and $C = 10\ \mu F$. Calculate the line currents when a symmetrical 400-V, 50-Hz, three-phase supply is applied to it.
12. Three impedances $Z_1 = 10\angle 0°$, $Z_2 = 10\ \angle 26.8°$, and $Z_3 = 10\ \angle{-26.8°}\ \Omega$ are connected to a three-phase, four-wire star connection and the supply is 440 V. Find the line currents, neutral currents, and total power.
13. An unbalanced delta-connected load with impedances of $Z_{ab} = 10 - j5$; $Z_{bc} = 16$; $Z_{ca} = 8 + j6$ is connected to a balanced line with line voltage of 200 V in the positive sequence. Find the line currents.
14. A delta-connected load with impedance $Z_1 = 10\ \angle 30°$, $Z_2 = 25\ \angle 0°$, $Z_3 = 20\ \angle{-30°}\ \Omega$ and the supply is 500 V. Assuming the sequence be abc, find the line currents and total power.
15. A balanced three-phase load consisting of three coils, each with a resistance of 3 Ω and inductance of 0.04 H is connected to a three-phase, 415-V, 50-Hz supply. Determine the total active power when the coils are (a) star-connected and (b) delta-connected.
16. A balanced three-phase, star-connected load of 6 kW takes a leading current of 10 A with a line voltage of 415 V, 50 Hz. Find the circuit constants of the load per phase.
17. A balanced star-connected load is supplied from a symmetrical three-phase, 440-V system. The current in each phase is 20 A and lags 30° behind the phase voltage. Find the (a) phase voltage and (b) total power. Also, draw the phasor diagram showing the currents and voltages.
18. A balanced mesh-connected load of $(12 + j9)\ \Omega$ per phase is connected to a three-phase, 400-V supply. Find the (a) line current, (b) power factor, (c) power, (d) reactive volt-amperes, and (e) total volt-amperes.
19. For a three-phase balanced supply of 10 kV, three impedances of 15 Ω, $(10 + j5)\ \Omega$, and $(6 - j8)\ \Omega$ are connected each one in each phase. Find the line currents and neutral current.
20. Three identical coils, each having a resistance of 20 Ω and a reactance of 20 Ω, are connected in star across a 440-V, three-phase line. Determine the line current, phase current, and reading on each of the two wattmeters connected to measure power.

Chapter 7

Transformers

Chapter Objectives

- To explain the importance of transformers
- To deal with various types of transformers
- To discuss how to develop an equivalent circuit of a transformer
- To discuss how to find the performance of a transformer
- To explain about special transformers

7.1 INTRODUCTION

The electric energy is seldom used at the point where it is generated. It has to be transmitted over long distances before it is used. Generally the generation of ac voltage at a generating station is 11–25 kV. But for economic reasons power will be transmitted at higher voltages such as 132, 220, 400, 765 kV, etc. At the same time, the voltage should be reduced to the safe level for the utilization of power at 400 V/230 V (commercial and residential loads). Hence, a device is needed to increase or decrease the voltage levels. Such a device is called *transformer*.

The transformer is a static device which transfers the electrical energy from one circuit to another circuit without changing the frequency. It does not involve any rotating parts; the efficiency of a transformer will be higher than that of a rotating equipment of the same rating. This is because there are no rotational losses in a transformer. For low capacity below 5 kVA, the efficiency is nearly 90–96% and for high capacity above 15 kVA to several hundred kVA, the efficiency varies from 96% to 99%. Transformers are extensively used in the day-to-day life. The ratings of transformers can be as small as a few volt-amperes or may be as high as several thousand volt-amperes.

7.2 BASIC PRINCIPLE OF TRANSFORMER

The principle of the operation of transformer is based on Faraday's laws of electromagnetic induction discovered by Michael Faraday (1791–1867). Figure 7.1 shows a simple transformer. It consists of two windings and is placed on the single magnetic core made up of the laminated and rewetted sheet of steel plate insulated and staggered together to form a magnetic core. The winding which is connected to ac supply is known as *primary* and the winding which is connected to the load is known as *secondary.* If the primary winding is excited with ac supply it draws the exciting current in primary and produces the required flux which flows through the magnetic core. The primary and secondary windings cut this flux and induce emf in the windings without changing frequency.

For example, if the primary winding is excited with an ac supply of $V_1 = V_m \sin \omega t$, the primary winding will draw the exciting current I_0 and produces the flux $\phi = \phi_m \sin \omega t$ and this flux circulates in the magnetic core as shown in Figure 7.1. The flux ϕ cuts both the windings, so there is self-induced emf e_1 in the primary winding and the mutual-induced emf e_2 in the secondary winding by the principle of magnetic induction.

$$\text{Induced emf, } e = -N\frac{d\phi}{dt} \tag{7.1}$$

$$\text{Induced emf on primary winding, } e_1 = -N_1\frac{d\phi}{dt} \tag{7.2}$$

$$\text{Induced emf on secondary winding, } e_2 = -N_2\frac{d\phi}{dt} \tag{7.3}$$

The magnitude of the induced emf depends on the number of turns of the windings.

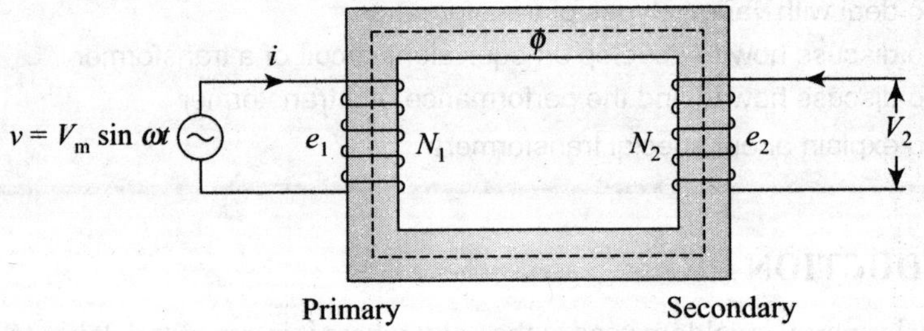

Figure 7.1 Simple transformer

Note: Why the transformers are not operated on dc supply?

The transformers cannot be connected to dc supply because the flux produced in the transformer core will not vary and remain constant in magnitude. Hence, the emf will not be induced in the windings. Thus, the transformer is not capable of raising or lowering the dc voltage. There is also possibility of getting the winding burnt because the winding resistance is quite low and high current starts flowing. That is why dc is never applied to a transformer.

7.3 CONSTRUCTION OF TRANSFORMERS

The constructional details of the transformer are shown in Figure 7.2. The construction has to ensure efficient removal of heat from the two sheets of heat generation core and winding. In all types of transformers, the core is constructed of transformer steel sheet laminations assembled to provide a continuous magnetic path with a minimum of airgap included. The steel used is of high silicon content. Sometimes heat is treated to produce a high permeability and a low hysteresis loss at the usual operating flux densities. The eddy current loss is minimized by laminating the core, the laminations being insulated from each other by a light coat of the core plate of varnish or by an oxide layer on the surface. The thickness of laminations varies from 0.35 mm for a frequency of 50 Hz to 0.5 mm for a frequency of 25 Hz. It is seen that the joints in the alternate layer are staggered in order to avoid the presence of narrow gaps right through the cross section of the core. Such staggered joints are said to be imbricated.

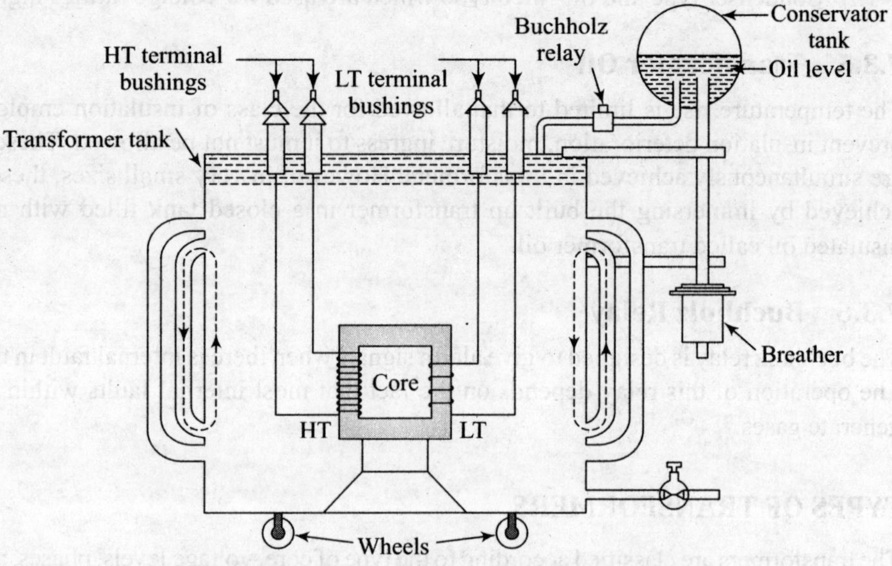

Figure 7.2 Constructional view of a transformer

7.3.1 Tank

To facilitate natural oil circulation and to increase the cooling surface exposed to the ambient, tubes or fans are provided on the outside of the tank walls. In large-size transformers, tubes may be cooled by air. For still large installations the best cooling system appears to be that in which the oil is circulated by a pump from the top of the transformer tank to a cooling plant returning when cold to the bottom of the tank.

In small sizes the transformers are directly placed in a protective housing or are enclosed in hard rubber molding and are air-cooled.

7.3.2 Conservator

The conservator is a smaller sized tank placed on the top of the main tank. This arrangement ensures that the surface area of transformer oil exposed to the atmosphere is limited so as to prevent the first oxidization and then the consequent deterioration of insulating properties of the oil.

7.3.3 Breather

Power transformers are provided with a conservator through which the transformer breathes into the atmosphere. Transformer oil exposed directly to the atmosphere may absorb moisture and dust from the environment and may lose its electrical properties in a very short time. To avoid this from happening, a breather is provided.

7.3.4 Bushings

The purpose of bushings is to provide proper insulation for the output leads to be taken out from the transformer tank. Bushings used are generally of two types.

(i) Porcelain type which are used for voltage ratings of up to 33 kV.

(ii) Condenser type and oil-filled type which are used for voltage ratings higher than 33 kV.

7.3.5 Transformer Oil

The temperature rise is limited to that allowed for the class of insulation employed. Further to prevent insulation deterioration, moisture ingress to it must not be allowed. These two objectives are simultaneously achieved in transformers. However, in very small sizes, these objectives are achieved by immersing the built up transformer in a closed tank filled with noninflammable insulated oil called transformer oil.

7.3.6 Buchholz Relay

The buchholz relay is designed to give alarm signals when there is internal fault in the transformers. The operation of this relay depends on the fact that most internal faults within the transformer generate gases.

7.4 TYPES OF TRANSFORMERS

The transformers are classified according to the type of core, voltage levels, phases, and applications.

7.4.1 Core

According to the core, the transformers are classified as follows:

Core-type transformer

In this case coils are wound around the limbs. Each limb carries one-half of the primary and secondary windings in order to reduce the leakage reactance of the windings. Here, the low-voltage (LV) winding is wound near to the core and the high-voltage (HV) winding is over to the LV winding reference to the core so as to reduce the insulation material.

Small transformers have rectangular or square cross-section with rectangular or circular coils. But it is wasteful in case of large capacity transformers. In case of large-sized transformers, a cruciform core with circular cylindrical coils is employed. From the economic point of view, this cruciform core is larger while in circular coils the winding is easy and provides more mechanical strength. The most important advantages of the cruciform structures are the high space factor and the turns of mean length which is reduced, finally reducing the copper losses. A core-type transformer is shown in Figure 7.3.

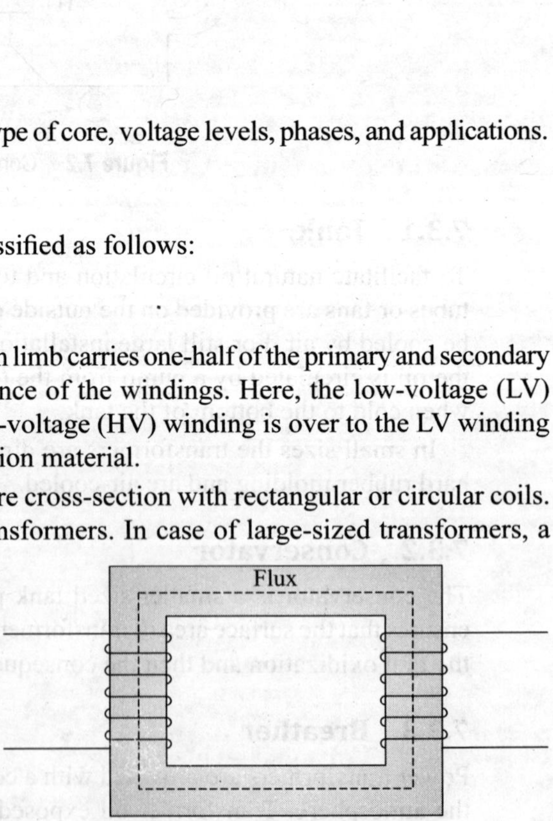

Figure 7.3 Core-type transformer

Shell-type transformer

In this case, a core is wound around the coils. Here, the total flux flows through the central limb and decides equally to the outer limbs. Ultimately, the cross section of the central limb is twice that of the outer limb. Generally, the sandwich-type winding is preferred for this type of transformer. Here, most of the core is supported by mechanical injuries to the windings. The shell-type construction is commonly used for small transformers. To have economic benefits it is advised to go for a square or rectangular core. Generally, for high-voltage transformers, a shell-type construction is preferred. Here, the disadvantage is that natural cooling does not exist. Because the windings are surrounded by the core and if one wants to remove any winding for maintenance, large number of laminations are required to be removed. Figure 7.4 shows the shell-type transformers.

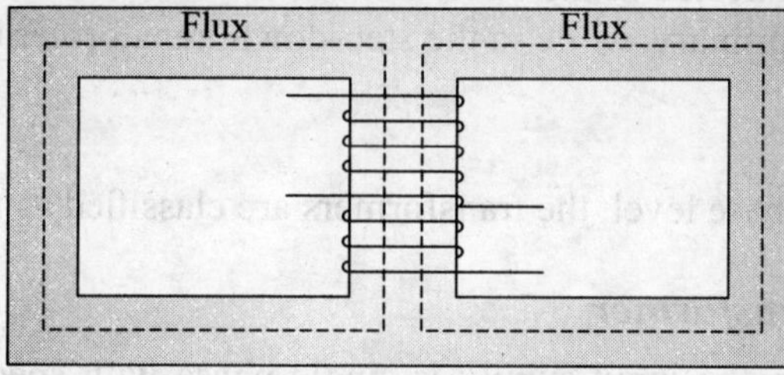

Figure 7.4 Shell-type transformer

Spiral-type transformer

In this case, magnetic circuit is distributed and this magnetic circuit is independent of having more than two circuits. In this core construction looks like spokes of the wheel. Here, the spokes are independent magnetic circuit. In this case, first the LV winding is wound and then the HV winding is wound. This is called as *Berry-type transformer*. Figure 7.5 shows Berry-type transformer.

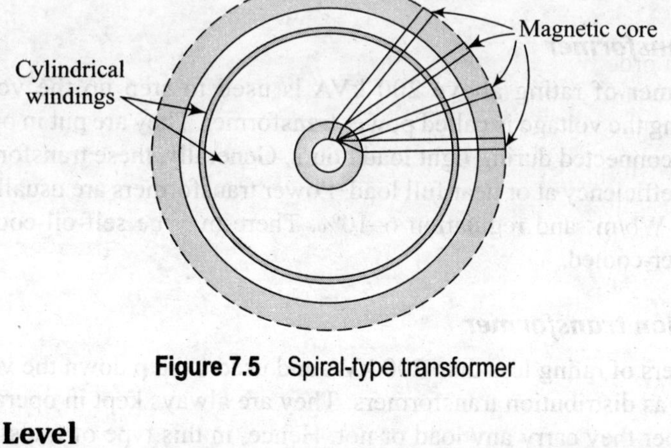

Figure 7.5 Spiral-type transformer

7.4.2 Voltage Level

According to the voltage level, the transformers are classified as follows:

Step-down transformer

A transformer in which the output voltage is less than its input (primary) voltage is called *step-down transformer.* It also depends on the number of turns. If $N_2 < N_1$, then the transformer is called step-down transformer. But in practice, there is small difference between the terminal voltage and the induced voltage. The terminal voltage is the voltage available at the load.

Step-up transformer

A transformer in which the output voltage is greater than its input (primary) voltage is called *step-up transformer.* In terms of number of turns, if $N_2 > N_1$, then the transformer is called step-up transformer.

The same transformer can be used as step-up transformer or step-down transformer depending on the way it is connected in the circuit. When the transformer is used as a step-up transformer the LV winding is primary, while in the step-down transformer the HV winding is primary.

7.4.3 Phase

According to the phase level, the transformers are classified as follows:

Single-phase transformer

In this transformer, the input supply is single phase with specified frequency. So, it is called single-phase transformer.

Three-phase transformer

In this type of transformer, the input supply is three phase with specified frequency. Here, the phase difference between any two phases is 120° because 360/3 = 120°.

7.4.4 Application

According to the application, the transformers are classified as follows:

Power transformer

A transformer of rating above 200 kVA is used to step up the voltage for generating and transforming the voltage is called power transformer. They are put in operation during load hours and are disconnected during light load hours. Generally, these transformers are designed to have maximum efficiency at or near full load. Power transformers are usually made to use flux density of 1.5–1.7 Wb/m^2 and regulation 6–10%. There may be self-oil-cooled, forced oil-cooled, or forced water-cooled.

Distribution transformer

Transformers of rating less than 200 kVA and used to step down the voltage at distribution level are known as distribution transformers. They are always kept in operation in the entire day, i.e., 24 h whether they carry any load or not. Hence, in this type of transformer iron losses occur at all the time. But copper losses occur only during the load time. They are of the self-cooling type and are almost invariably oil-immersed. The regulation of this type of transformers is 4–8% only. They make use of cold rolled steel with flux density of 1.7 Wb/m^2.

Instrument transformer

Instrument transformers are divided into two types. These are current transformers and potential transformers. Current transformers are used to measure high current ratings of kilo-amperes and mega-amperes. In current transformers, the primary winding has few turns of heavy wire whereas the secondary winding has many turns of very fine wire.

Potential transformers are used to measure high voltages in the order of kilo volts. The primary winding has many turns and is connected across the high-voltage line. The secondary winding has few turns and is connected to a voltmeter.

7.5 EMF EQUATION OF TRANSFORMER

The transformer's primary winding is excited with ac supply V_1. So, the exciting current establishes the flux ϕ in the core. Consequently the primary winding has a flux linkage of λ_1 which induces the emf in primary winding with N_1 turns.

Primary winding flux linkage, $\lambda_1 = N_1\phi$ (7.4)

Instantaneous value of primary winding induced emf, $e_1 = \dfrac{d\lambda_1}{dt}$ (7.5)

$$\therefore \quad e_1 = N_1 \frac{d\phi}{dt} \tag{7.6}$$

Let the flux in the core be written as $\phi = \phi_m \sin \omega t$

where ϕ_m is the maximum value of the flux; ϕ is the instantaneous value of the flux, Wb; ω is the frequency of the supply, $2\pi f$ rad/s; and N_1 is the number of turns in primary winding.

$$\therefore \quad e_1 = N_1 \frac{d(\phi_m \sin \omega t)}{dt} \tag{7.7}$$

$$= \omega N_1 \phi_m \cos \omega t$$

$$= 2\pi f N_1 \phi_m \sin\left(\omega t + \frac{\pi}{2}\right) \tag{7.8}$$

$$= E_m \sin\left(\omega t + \frac{\pi}{2}\right) \tag{7.9}$$

where

E_m = Maximum value of induced emf in primary winding = $2\pi f N_1 \phi_m$ (7.10)

But the rms value of the induced emf = $(1/\sqrt{2}) \times$ maximum value of induced emf. The rms value of the induced emf in primary winding,

$$E_1 = \frac{1}{\sqrt{2}} [2\pi f N_1 \phi_m] \tag{7.11}$$

Therefore, the induced emf in primary winding,

$$E_1 = 4.44\phi N_1 \phi_m \text{ volts} \tag{7.12}$$

Similarly, the secondary winding induced emf,

$$E_2 = 4.44\phi\, N_2\phi_m \text{ volts} \tag{7.13}$$

where N_2 is the number of turns in secondary winding.

7.5.1 Turns Ratio (Transformation Ratio)

The ratio of secondary to primary induced emf is the same as the ratio of secondary to primary turns, i.e., $N_2/N_1 = E_2/E_1$ is known as the *transformation ratio* and is denoted by the letter k. Based on the transformation ratio, the transformers are divided as

(i) If $k > 1$ or $N_2 > N_1$, the secondary induced emf or terminal voltage is greater than the primary induced emf or supply voltage. Such a transformer is known as the *step-up transformer.*

(ii) If $k < 1$ or $N_2 < N_1$, the secondary induced emf is less than the primary induced emf. Such a transformer is known as the *step-down transformer*.

(iii) If $k = 1$ or $N_1 = N_2$, the primary induced emf is equal to secondary induced emf. Such a transformer is known as the one-to-one transformer and is used generally where the isolation is required between the two circuits.

7.6 IDEAL TRANSFORMERS

7.6.1 Ideal Transformer on No Load

The ideal transformer has the following characteristics:

(i) It has no losses.

(ii) Its primary and secondary winding resistances are negligible.

(iii) Its leakage flux and leakage inductances are zero.

(iv) Its permeability of core is so high that the negligible current is required to establish the flux in it.

Consider an ideal transformer, as shown in Figure 7.6, when V_1 is applied to the transformer at no load, i.e., $I_2 = 0$. Here, I_1 chooses a closed path of primary winding. The magnetic circuit is magnetized due to I_1. Therefore, I_1 is called *magnetizing current* (I_m). The transformer is ideal, when the winding resistance is negligible and is purely inductive in nature. In a single inductive ac circuit, the current lags voltage by 90°. Therefore, I_m lags V_1 by 90° but I_m and ϕ are in phase because I_m is due to ϕ only.

The flux links both the windings producing induced emfs E_1 and E_2 in both primary and secondary windings, respectively. But according to Lenz's law the induced emf opposes the cause of producing it which is V_1; therefore, E_1 and V_1 are in antiphase with each other but equal in magnitude. The induced emf E_2 also opposes V_1. Therefore, E_2 is also in antiphase with V_1. Hence, E_1 and E_2 are in phase for ideal transformer on no load as shown in Figure 7.7.

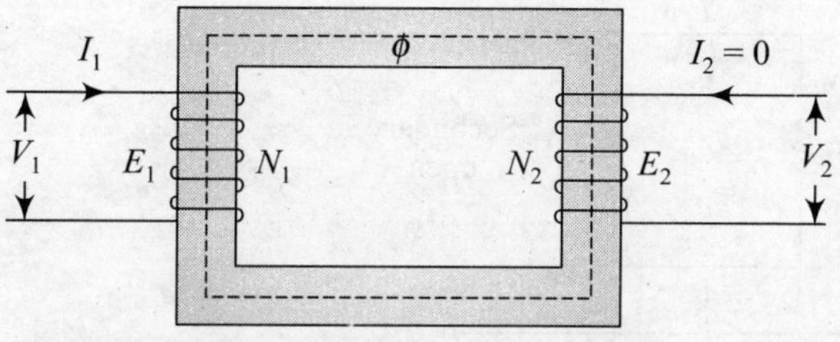

Figure 7.6 Ideal transformer

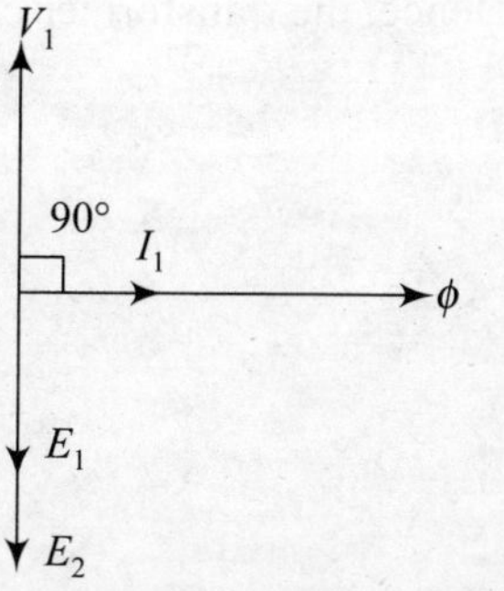

Figure 7.7 Phasor diagram of ideal transformer

From the phasor diagram as shown in Figure 7.7, it is observed that ϕ is taken as reference because it is common for both windings and I_m is due to ϕ only. Therefore, ϕ and I_m are in phase. Here, V_1 leads ϕ or I_m by 90°; in other words, I_m or ϕ lags V_1 by 90° because it is purely an inductive circuit.

E_1 and E_2 are in phase and both oppose supply voltage V_1. The power input to the transformer is $V_1 I_m \cos\phi$. Since $\phi = 90°$, the power input to transformer is zero. This is because on no-load output power is zero and for ideal transformer there are no losses. Hence, the input power is also zero.

7.6.2 Ideal Transformer on Load

If the secondary is loaded, the current I_2 flows through the load. The secondary mmf $N_2 I_2$ tends to oppose the core flux according to Lenz's law, but the excitation of the primary does not allow any change in the core flux. The two mmfs are equal (see Figure 7.8).

$$N_1 I_1 = N_2 I_2$$

$$\therefore \quad \frac{I_1}{I_2} = \frac{N_2}{N_1} = k \tag{7.14}$$

From the above equation, we can say that the current transforms in an ideal transformer is the inverse ratio of its turns. The phasor diagram for an ideal transformer under the load condition is shown in Figure 7.9 for power factor $\cos\theta$.

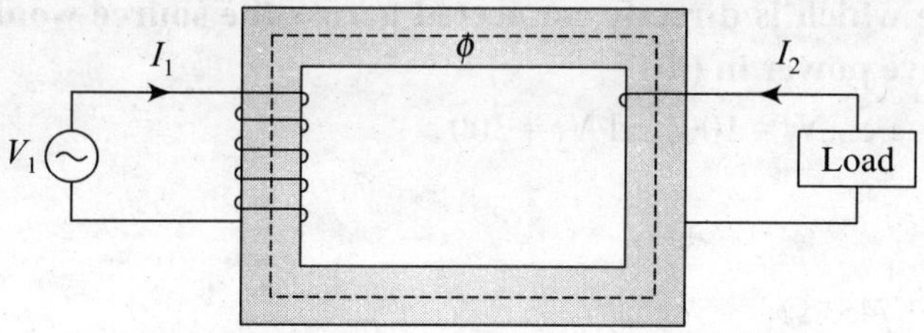

Figure 7.8 Ideal transformer on load

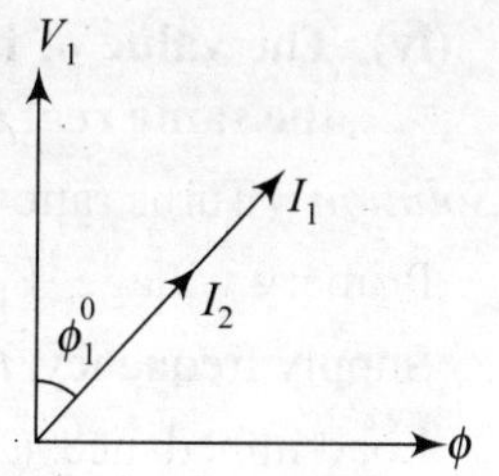

Figure 7.9 Phasor diagram on load

7.6.3 Equivalent Circuit of Ideal Transformer on No Load

When the secondary is open circuited the current flowing through the secondary is zero. There will be only very small amount of no-load primary current. The losses are now only iron losses.

Hence, the transformer can be represented on no load as shown in Figure 7.10.

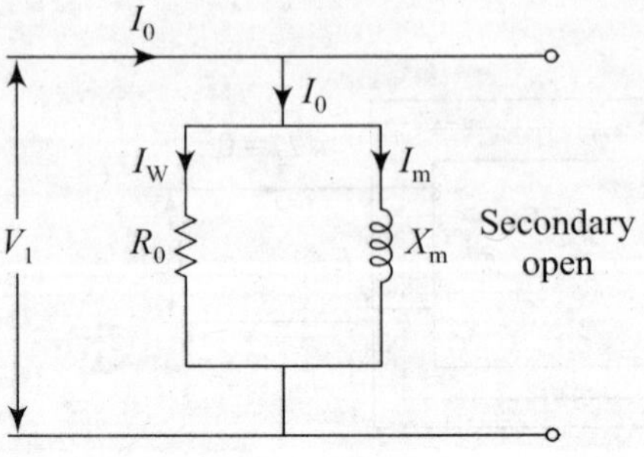

Figure 7.10 Equivalent circuit of a transformer on no load

$$R_0 = \frac{V_1}{I_w}$$

$$X_m = \frac{V_1}{I_m}$$

The primary no-load current and the transformer core loss are functions of the source voltage and hence the shunt branch consisting of R_0 and X_0 is connected across the supply source.

Note: Why rating of the transformer is always expressed in kVA?

Generally, the available output power depends on secondary power factor. As the secondary power factor can change depending on the load, the rating of the transformer is not specified in watts or kilowatts but indicated in as the product of voltage and current called kVA rating.

EXAMPLE 7.1 **The turns ratio of a transformer is 100/300, the primary winding is connected to a source of 3.3 kV, 50 Hz supply. The load impedance of (110 + *j*45) Ω is connected across the secondary terminals. Calculate the following:**

(i) Value of maximum flux in the core

(ii) Primary and secondary currents

(iii) The real and reactive power supplied by the source to the primary transformer

(iv) The value of impedance which is directly connected across the source would draw the same real and reactive power in (iii)

Solution Turns ratio = 100/300, i.e., $N_1 = 100$ and $N_2 = 300$

Primary voltage, $V_1 = 3.3$ kV

Supply frequency, $f = 50$ Hz

Load impedance, $(Z_2) = (110 + j45)\ \Omega$

(i) The transformer induced emf equation is

$$E_1 = \sqrt{2}\,\pi f \phi_m N_1$$

$$\phi_m = \frac{V_1}{\sqrt{2}\pi\, f\, N_1}$$

Therefore, the value of maximum flux in the core,

$$\phi_m = \frac{3.3\times10^3}{\sqrt{2}\pi\times50\times100} = 0.1486 \text{ Wb}$$

(ii) Transformation ratio, $K = N_2/N_1$ (or) E_2/E_1 (or) I_1/I_2

$$K = \frac{N_2}{N_1} = \frac{300}{100} = 3$$

$$K = \frac{E_2}{E_1} \Rightarrow E_2 = KE_1$$

Secondary voltage, $E_2 = 3 \times 3.3 \times 10^3 = 9.9$ kV

$$I_2 = \text{Secondary current} = \frac{E_2}{Z_L} = \frac{9.9\times10^3}{(110+j45)} = 83.3\angle-22.25 \text{ A}$$

$$K = \frac{I_1}{I_2} \Rightarrow I_1 = KI_2$$

Primary current, $I_1 = 3 \times 83.3 \angle{-22.25}$

$$= 249.89 \angle-22.25 \text{ A}$$

(iii) The real and reactive power supplied by the source to the transformer is

$$S = V_1 I_1^* = V_2 I_2^*$$

$$= 3.3\times10^3\times(249.89\angle22.25)$$

$$= 763.24 + j312.25$$

Real power, $P_1 = 763.24$ W

Reactive power, $Q_1 = 312.25$ VAr

(iv) The value of impedance which is connected directly across the source would draw the same real and reactive power, i.e., Z_L is transferred to primary.

$$Z_L' = Z_{L01} = \frac{Z_L}{(K)^2} = \frac{(110+j45)}{(3)^2} = (12.22+j5)\ \Omega$$

EXAMPLE 7.2 **A single-phase 50-Hz transformer has 100 turns on primary winding and the secondary winding turns are 500. The cross-sectional area of the core is 220 cm². A 240-V, 50-Hz voltage source is connected to the primary winding of the transformer. Calculate**

(i) The emf induced in the secondary winding.

(ii) The maximum value flux density in the core (B_m).

Solution Given data

Primary voltage, $V_1 = 240$ V, frequency of supply, $f = 50$ Hz

Primary turns, $N_1 = 100$, secondary turns, $N_2 = 500$

Area of the core, $A = 220$ cm^2

(i) The transformation ratio of the transformer is

$$K = \frac{N_2}{N_1} \text{ or } \frac{E_2}{E_1} \text{ or } \frac{I_1}{I_2}$$

$$K = \frac{500}{100} = 5$$

$$E_2 = KE_1 = 5 \times 240 = 1200 \text{ V}$$

(ii) The transformer-induced emf, $E_1 = \sqrt{2}\pi f \phi_m N_1$ volts

But maximum flux, $\phi_m = B_m \times A$,

where B_m = Maximum flux density

A = Area of the core

$$B_m = \frac{E_1}{\sqrt{2}\pi \times f \times A \times N_1} = \frac{240}{\sqrt{2}\pi \times 50 \times 220 \times 10^{-4} \times 100} = 0.4911 \text{ Wb/m}^2$$

EXAMPLE 7.3 **A 50-Hz single-phase transformer has 6600 V/400 V. Having emf per turn is 10 V and the maximum flux density in the core is 1.6 T. Find the**

(i) Suitable number of primary and secondary turns

(ii) Cross-sectional area of the core

Solution Given data

Primary voltage, V_1 = 6600 V, secondary voltage, V_2 = 400 V

The emf/turn in the transformer = 10 V

Maximum flux density in the core = 1.6 T

(i) Number of primary turns, $N_1 = \frac{\text{Primary voltage}}{\text{emf/turn}} = \frac{6600}{10} = 660$

Number of secondary turns, $N_2 = K \times N_1 = 40$

(ii) The emf equation of transformer is

$$E_1 = \sqrt{2}\pi f B_m A N_1 \text{ volts}$$

$$\text{Area of the core, } A = \frac{6600}{\sqrt{2} \times \pi \times 50 \times 1.6 \times 660} = 281.3 \text{ cm}^2$$

EXAMPLE 7.4 **A single-phase core-type 50-Hz transformer is in the square shape having 25 cm side, the maximum flux density in the core 1.2 Wb/m². Calculate the number of turns per limb on the high-voltage (HV) side and low-voltage (LV) side for a 3400 V/240 V ratio.**

Solution Given data

Maximum flux density in the core, B_m = 1.2 Wb/m^2

Voltage ratio = 3400 V/240 V

Primary voltage, V_1 = 3400 V, secondary voltage, V_2 = 240 V

Supply frequency, f = 50 Hz

In this problem, HV is primary and LV is secondary.

Secondary voltage, $V_2 = \sqrt{2}\pi fB_m A N_1$ volts

$$240 = \sqrt{2}\pi \times 50 \times 1.2 \times 25 \times 10^{-2} \times 25 \times 10^{-2} N_2$$

$$= 16.66\, N_2$$

Secondary or LV turns, $N_2 = \dfrac{240}{16.66} = 14.41 \approx 15$

But in transformer the number of turns takes even numbers so the number of secondary turns or LV winding turns, $N_2 \approx 16$.

The transformation ratio, $K = V_2/V_1 = 240/3400 = 0.0706$. But $K = N_2/N_1$. Primary or HV turns, $N_1 = N_2/K = 16/0.0706 = 226.62 \approx 228$ turns.

Note: In a transformer, to reduce the leakage reactance, half of the turns should be on primary and half of the turns on secondary.

So the secondary turns on each limb $= N_2/2 = 16/2 = 8$ and the primary turns on each limb $= N_1/2 = 228/2 = 114$.

7.7 ACTUAL TRANSFORMER OR REAL TRANSFORMER

As in the ideal case of the transformer, the losses in core and winding are negligible. But when an actual transformer is loaded, there is always core loss or iron loss in the core and copper losses in the windings, which cannot be neglected. The operation of the transformer can be considered in two cases

(i) Under no load

(ii) With load

7.7.1 Transformer on No Load

Under no-load condition, the transformer draws the no-load current, I_0. It meets the no-load losses such as core losses. Figure 7.11 shows the transformer under no-load condition and assuming there is no leakage flux.

When the transformer is excited, it draws the no-load current of I_0 in the primary winding. This I_0 will set up the required flux in the core that will be used to induce the emf in secondary by the mutual induction principle. The induced emfs in primary and secondary are E_1 and E_2 and they are dependent on the primary and secondary turns N_1 and N_2, respectively.

$$E_1 = -N_1 \frac{d\phi}{dt}; \quad E_2 = -N_2 \frac{d\phi}{dt} \text{ volts}$$

The drawn current I_0 also meets the no-load losses in the transformer, i.e., iron losses in the core (hysteresis plus eddy current loss) and a small amount of copper loss in the primary winding, but the losses in the secondary are zero because of no load. The no-load current I_0 lags V_1 by an angle ϕ_0; it is always less than 90°.

The corresponding loss is given by $W_0 = V_1 I_0 \cos \phi_0$, where $\cos \phi_0$ = no-load power factor.

From the phasor diagram shown in Figure 7.12, the no-load current has two components, i.e., magnetizing component, $I_m = I_0 \sin \phi_0$, and working component, $I_w = \cos \phi_0$, and I_w is always very small as compared to I_m.

The no-load current is given by

$$I_0 = \sqrt{I_w^2 + I_m^2} \tag{7.15}$$

where I_m is the magnetizing component of the no-load current and I_w is the working component of the no-load current.

I_0 is very small as compared to the full-load current. Usually, it is about 5–10% of the full-load current. Therefore, the copper losses of the primary winding under no-load conditions are negligible. So the total power input under no-load conditions is equal to the iron loss in the transformer. These losses are constant throughout the operation of transformer at any load and power factor.

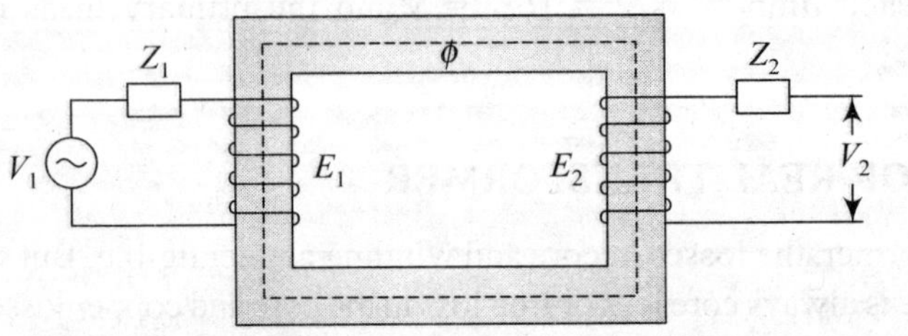

Figure 7.11 Transformer on no load

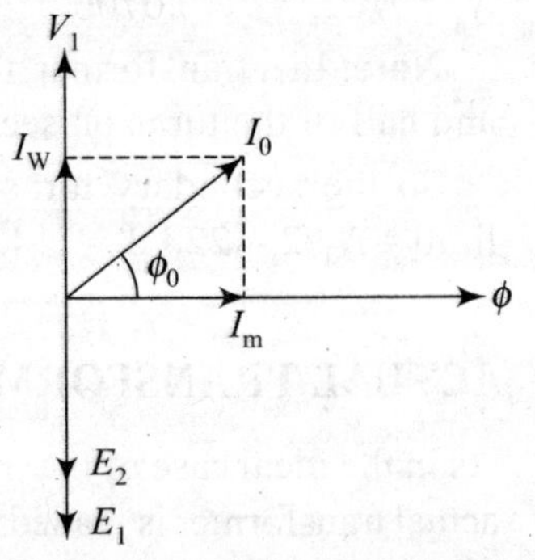

Figure 7.12 Phasor diagram

7.7.2 Transformer on Load

If the secondary of the transformer is connected to the load as shown in Figure 7.13, it supplies a load current or secondary current of I_2 amperes, at a terminal voltage of V_2 volts. The magnitude of the load current depends on the nature of the load connected. As per Faraday's law, the secondary current I_2 sets up its own flux ϕ_2 by its own mmf N_2I_2. This flux ϕ_2 is opposite to the main flux ϕ which is set up by the primary mmf N_1I_0. Hence, the net flux in the core is going to reduce momentarily which in turn reduces the primary induced emf E_1. Therefore, the primary winding draws the extra current of I_2', i.e., unequal in magnitude of I_2. This extra current I_2' produces the extra flux of ϕ_2' in the primary winding by the mmf of $I_2'N_1$ which is opposite to the secondary flux of ϕ_2. So, the magnetic effect produced by the secondary load current is neutralized by the additional primary current I_2', and the net flux available in the core always remains constant. So, the flux under no load and on load is always same.

$$\phi_1 = \phi_2$$

$$I_2'N_1 = N_2 I_2 \tag{7.16}$$

$$I_2' = \left(\frac{N_2}{N_1}\right) I_2$$

$$= KI_2 \tag{7.17}$$

The transformer under no-load draws only the no-load current in the primary. When it is loaded, it draws both the no-load current (I_0) and load component in the primary, I_2'. Therefore, the net primary current is $\overline{I_1} = \overline{I_0} + \overline{I_2'}$ (phasor sum). The phasor diagram when the transformer is on load for resistive load and inductive load is shown in Figures 7.13(a) and (b).

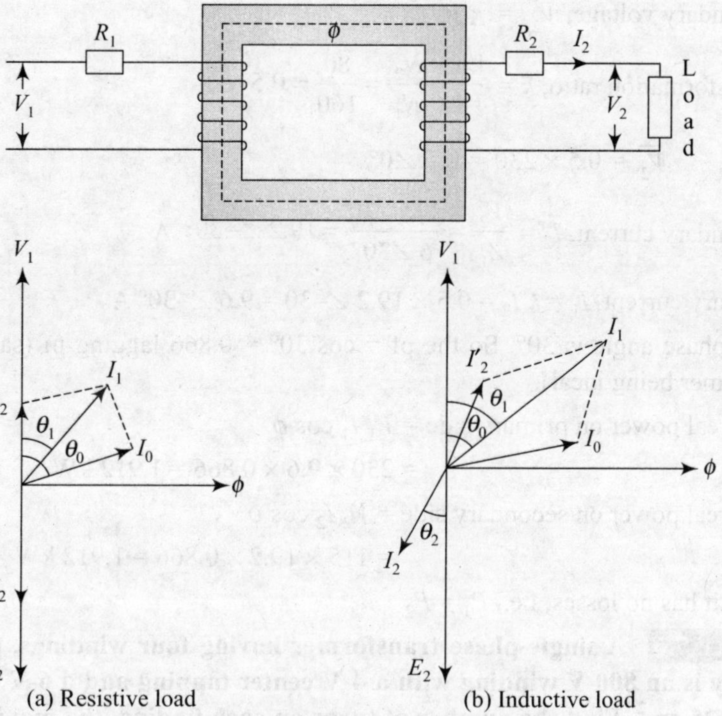

(a) Resistive load (b) Inductive load

Figure 7.13 Transformer on load

EXAMPLE 7.5 **A transformer is having 160 turns on primary and 80 turns on secondary. The primary winding is excited by the 230-V, 50-Hz supply and the secondary side having a load impedance of 6 ∠30° is connected as shown in Figure 7.14. Neglect the losses of the transformer. Calculate (i) the primary and secondary currents, (ii) their power factor, and also (iii) the primary and secondary real powers.**

Solution Given data

Primary turns, N_1 = 160, secondary turns, N_2 = 80. Load impedance, $Z_L = 6\angle 30°$, primary voltage, V_1 = 230 V.

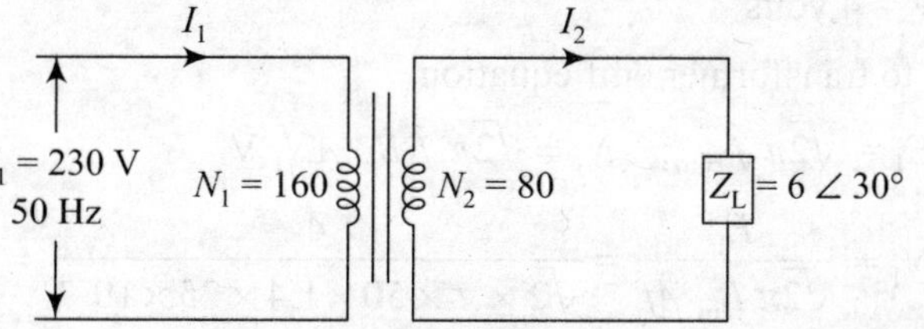

Figure 7.14

Assume the transformer to be ideal because of neglecting loss, take V_1 as the reference

Primary voltage, $\overline{V_1} = 230 \angle 0°$ V

Secondary voltage, $\overline{V_2} = K\overline{V_1}$

Transformation ratio, $k = \dfrac{V_2}{V_1} = \dfrac{N_2}{N_1} = \dfrac{80}{160} = 0.5$

$\therefore \quad \overline{V_2} = 0.5 \times 230 = 115 \angle 0°$

Secondary current, $I_2 = \dfrac{V_2}{Z_L} = \dfrac{115 \angle 0°}{6 \angle 30°} = 19.2 \angle{-30°}$ A

Primary current, $I_1 = k\, I_2 = 0.5 \times 19.2 \angle{-30} = 9.6 \angle{-30°}$ A

The phase angle is 30°. So the pf = cos 30° = 0.866 lagging pf (same as a secondary-side transformer being ideal).

The real power on primary side $= V_1\, I_1 \cos\phi$

$= 230 \times 9.6 \times 0.866 = 1.912$ kW

The real power on secondary side $= V_2\, I_2 \cos\phi$

$= 115 \times 19.2 \times 0.866 = 1.912$ kW

That it has no losses, i.e., $P_1 = P_2$

EXAMPLE 7.6 A single-phase transformer having four windings, primary is 230 V. The primary is an 800-V winding with a 4-V center tapping and a 6-V winding for a net core area of 25 cm². Find the number of turns on each finding, the maximum value of flux not exceed 1.4 T.

Solution Given data

Primary voltage, $V_1 = 230$ V; Area of the core, $A = 25$ cm^2

Maximum value of flux density, $B_m = 1.4$ T

Three secondary winding areas, 800 V, 4 V, 6 V with center tappings

The center tapping winding must have even number of turns and also first find the number of turns for low-voltage winding and from this find the induced voltage/turn, take the low-voltage winding as a 4 V winding.

$V = 4$ volts

According to transformer emf equation

$$E_1 = \sqrt{2}\pi f \phi_m; \quad N_1 = \sqrt{2}\pi f B_m A N_1 \text{ V}$$

$$N_1 = \frac{E_1}{\sqrt{2}\pi B_m A f} = \frac{4}{\sqrt{2} \times \pi \times 50 \times 1.4 \times 25 \times 10^{-4}} = 5.15$$

Take the turns are even numbers, so 5.15 ≈ 6 turns and from these the emf induced/turn = voltage/number of turns = 4/6 = 0.67 V/turn. The induced emf/turn in a transformer is constant.

Number of turns on 230 V winding primary = voltage/emf/turn

$$= \frac{230}{0.67} = 343.28 \approx 344 \text{ turns}$$

Number of turns on the 4-V winding = voltage/emf/turn = $\frac{4}{0.67} = 6$ turns

Number of turns on the 6-V winding = $\frac{6}{0.67} = 8.95 \approx 10$ turns

Number of turns on the 800-V winding = $\frac{800}{0.67} = 1194$ turns

7.7.3 Resistance and Leakage Reactance of the Transformer

In the preceding section, an ideal transformer is discussed. According to our assumptions, it does not contain any resistance and leakage flux. But in practice, it is not possible to get such an ideal transformer.

In real transformer, both primary and secondary windings consist of finite resistances R_1 and R_2. Due to these resistances copper losses and voltage drop occur in the windings. The secondary terminal voltage V_2 is less than the secondary induced emf E_2. Mathematically, it can be written as $V_2 = E_2 - I_2 R_2$.

Similarly, in primary winding the induced emf, $E_1 = V_1 - I_1 R_1$. In both the cases, the effect of magnetic leakage is negligible.

The only flux that exists in an ideal transformer is ϕ_m. Previously it was assumed that all the flux links both the primary and secondary windings. But in actual practice, it is impossible to get such condition. A small amount of flux flows through the airgap and this is usually called the *leakage flux*. The primary leakage flux ϕ_1 linking with primary winding is produced by the primary ampere-turns only; therefore, it is proportional to primary current. On no-load since the primary current is negligible, the leakage flux produced by it is also negligible. But on load conditions, the primary current is high hence the leakage flux also increases. The primary leakage flux ϕ_1 produces self-induced emf E_1.

Its magnitude is given by $E_1 = 2\pi f L_1 I_1$, where L_1 is the self-inductance of the primary winding. Thus, the reactance of the primary winding,

$$X_1 = \frac{E_1}{I_1} = \frac{2\pi f L_1 I_1}{I_1} = 2\pi f L_1 \tag{7.18}$$

Similarly, the secondary leakage flux ϕ_2 is established by secondary ampere-turns and is proportional to the secondary current I_2. No leakage flux exists on no-load since no current flows. But on load, leakage flux ϕ_2 produces induced emf E_2 in the secondary winding. Mathematically, it can be written as $E_2 = 2\pi f L_2 I_2$, where L_2 is the self-inductance of the secondary winding. The secondary reactance is obtained by

$$X_2 = \frac{E_2}{I_2} = \frac{2\pi f L_2 I_2}{I_2} = 2\pi f L_2 \tag{7.19}$$

The effect of magnetic leakage is thus to produce emfs of self-inductances and, therefore, equivalent in effect to the addition of an inductive coil in series with each winding, the reactance is called the leakage reactance. A transformer with magnetic leakage and winding resistance is equivalent to an ideal transformer having inductive and resistive coils connected in series with each winding as shown in Figure 7.15.

7.8 EQUIVALENT RESISTANCE AND REACTANCE OF TRANSFORMER

In an actual transformer the windings have resistance and leakage reactances. Let the values be R_1, R_2, X_1, and X_2, respectively, where subscript 1 represents the primary winding and 2 represents the secondary winding. It is advantageous to transfer the resistances and reactances of the winding to any one winding. The main advantage is that the calculations become simpler. The copper loss in the secondary winding is $I_2^2 R_2$. This loss is supplied by the primary current I_1. Hence, if R_2' is the equivalent resistance in primary which would cause the same copper loss as R_2 in secondary, then

$$I_1^2 R_2' = I_2^2 R_2$$

$$\therefore \quad R_2' = \left(\frac{I_2}{I_1}\right)^2; \quad R_2 = \frac{R_2}{K^2}$$

Similarly, the equivalent resistance of the primary winding referred to the secondary side is $R_1' = R_1 K^2$.

With these transformations we can calculate the total resistance of the transformer winding either referred to primary or secondary and is given by

$$\text{Equivalent resistance referred to primary, } R_{01} = R_1 + R_2' = R_1 + \frac{R_2}{k^2}$$

$$\text{Equivalent resistance referred to secondary, } R_{02} = R_2 + R_1' = R_2 + k^2 R_1$$

Similarly, leakage reactances can also be transferred from one side to another side and are given by

$$\text{Equivalent reactance referred to primary, } X_{01} = X_1 + X_2' = X_1 + \frac{X_2}{k^2}$$

$$\text{Equivalent reactance referred to secondary, } X_{02} = X_2 + X_1' = X_2 + k^2 X_1$$

The total impedance referred to either primary or secondary side is given by

$$Z_{01} = \sqrt{R_{01}^2 + X_{01}^2} \quad \text{and} \quad Z_{02} = \sqrt{R_{02}^2 + X_{02}^2}$$

7.8.1 Equivalent Circuit

The analysis of the transformer can be better understood if the transformer equivalent circuit is known. The no-load primary current has two components I_m and $I_w \cdot I_m$ produces the flux and is assumed to flow through reactance X_0 called no-load reactance, while I_w is the working component representing core losses, hence is assumed to flow through the resistance R_0. This circuit consisting of R_0 and X_0 in parallel is called exciting circuit.

When the load is connected to the transformer then the secondary current I_2 flows. This causes voltage drop across R_2 and X_2. Due to I_2, the primary draws an additional current which is given by $I_2' = I_2/K$. Hence, primary current is the phasor addition of I_0 and I_2'. This causes the voltage drop across the primary resistance R_1 and reactance X_1. Hence, the equivalent circuit can be shown as in Figure 7.15(a).

The equivalent circuit is further simplified by transferring all the values to the primary or secondary. This makes calculations much simpler.

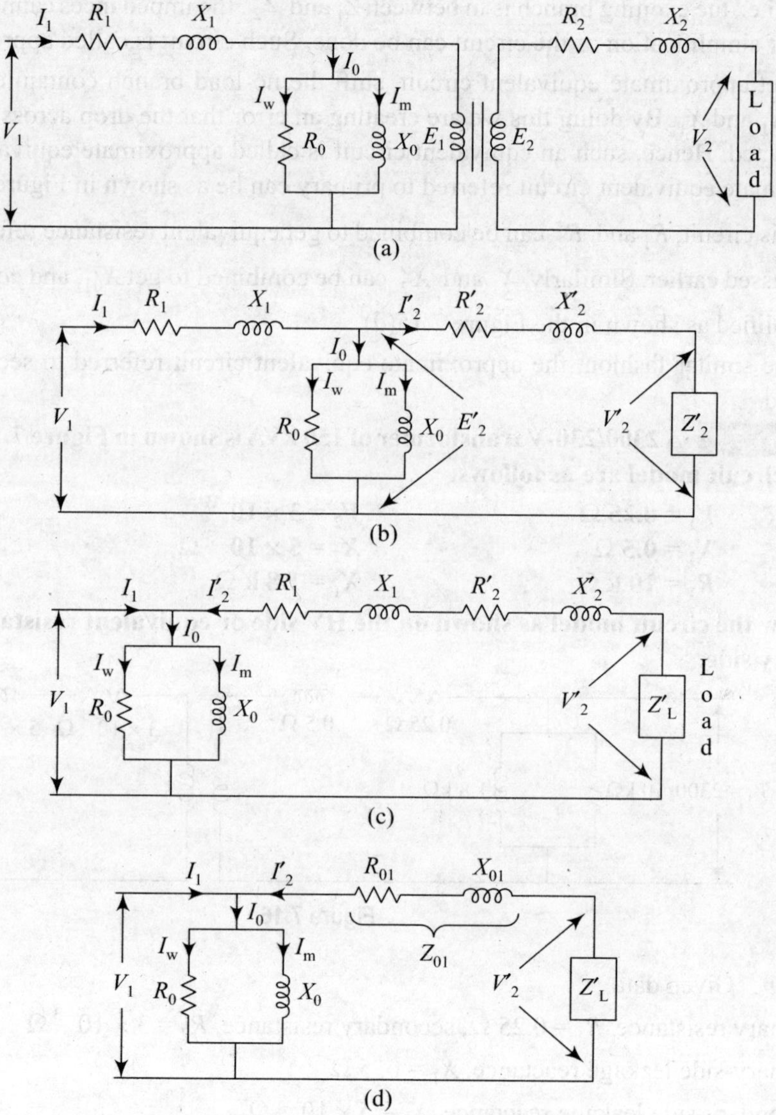

Figure 7.15 Equivalent circuit of transformer

Transferring secondary quantities to the primary side, we get

$$R_2' = \frac{R_2}{K^2}, \quad X_2' = \frac{X_2}{K^2}, \quad Z_2' = \frac{Z_2}{K^2}$$

while

$$E_2' = \frac{E_2}{K},\ I_2' = KI_2$$

Thus the exact equivalent circuit referred to primary can be shown as in Figure 7.15(b). Similarly, we can also develop an exact equivalent circuit referred to secondary. Now as long as the no-load branch, i.e., the exciting branch is in between Z_1 and Z_2', the impedances cannot be combined. So, a further simplification of the circuit can be done. Such circuit is called approximate circuit.

To get approximate equivalent circuit, shift the no-load branch containing R_0 and X_0 to the left of R_1 and X_1. By doing this we are creating an error that the drop across R_1 and X_1 due to I_0 is neglected. Hence, such an equivalent circuit is called approximate equivalent circuit. So, the approximate equivalent circuit referred to primary can be as shown in Figure 7.15(c).

In this circuit, R_1 and R_2' can be combined to get equivalent resistance referred to primary R_{01} as discussed earlier. Similarly, X_1 and X_2' can be combined to get X_{01}, and equivalent circuit can be simplified as shown in the Figure 7.15(d).

In the similar fashion, the approximate equivalent circuit referred to secondary can also be obtained.

EXAMPLE 7.7 **A 2300/230-V transformer of 150 kVA is shown in Figure 7.16. The parameters of the circuit model are as follows:**

$V_1 = 0.25\ \Omega$	$V_2 = 3 \times 10^{-3}$
$X_1 = 0.5\ \Omega$	$X_2 = 5 \times 10^{-3}\ \Omega$
$R_1 = 10\ \text{k}\ \Omega$	$X_1 = 1.8\ \text{k}\ \Omega$

Draw the circuit model as shown on the HV side or equivalent resistance referred to the primary side.

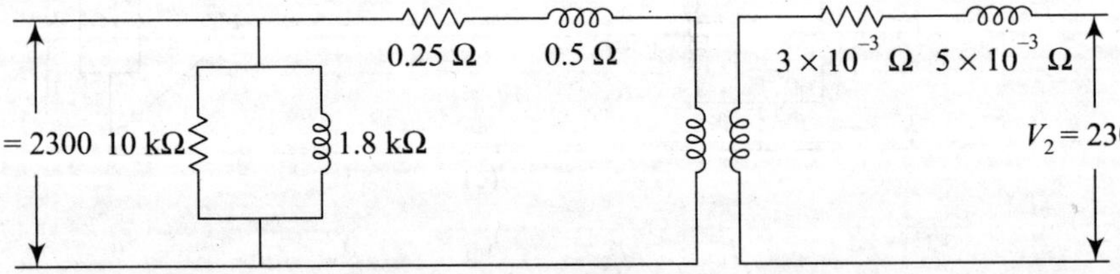

Figure 7.16

Solution Given data

Primary resistance, $R_1 = 0.25\ \Omega$, secondary resistance, $R_2 = 3 \times 10^{-3}\ \Omega$

Primary-side leakage reactance, $X_1 = 0.\ 5\ \Omega$

Secondary-side leakage reactance, $X_2 = 5 \times 10^{-3}\ \Omega$

Magnetizing reactance or iron-cored reactance, $X_i = 1.8\ \text{k}\Omega$

Magnetizing resistance or iron-cored resistance, $R_i = 10\ \text{k}\Omega$

The transformation ratio of transformer, $k = \frac{V_2}{V_1} = \frac{230}{2300} = 0.1$

The total resistance referred to the primary side, $R_{01} = R_1 + \dfrac{R_2}{K^2} = 0.25 + \dfrac{3\times10^{-3}}{(0.1)^2} = 0.55\ \Omega$

The total reactance referred to the primary side, $X_{01} = X_1 + \dfrac{X_2}{(K)^2} = 0.5 + \dfrac{5\times10^{-3}}{(0.1)^2} = 1\ \Omega$

The total impedance referred to the primary side or HV side, $Z_{01} = R_{01} + j\,X_{01}$

$$= (0.55 + j1)\ \Omega$$

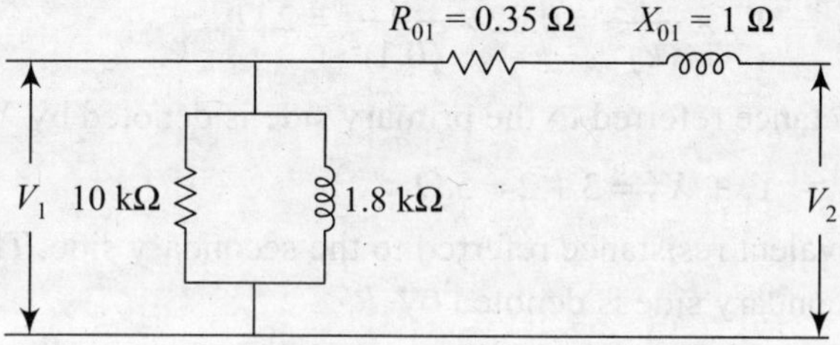

Figure 7.17

The equivalent circuit diagram of the total impedance referred to HV side is shown in Figure 7.17.

EXAMPLE 7.8 **A 2300/230-V, 50-Hz, single-phase transformer rating is 25 kVA. It has the following resistances and leakage reactances**

$$\mathbf{R_1 = 0.6\ \Omega \quad and \quad X_1 = 3.0\ \Omega}$$
$$\mathbf{R_2 = 0.008\ \Omega \quad and \quad X_2 = 0.02\ \Omega}$$

Calculate

(i) The equivalent resistance referred to the primary side

(ii) The equivalent reactance referred to the primary side

(iii) The equivalent resistance referred to the secondary side

(iv) The equivalent reactance referred to the secondary side

Solution Given data

Rating of the transformer = 25 kVA

Primary voltage, $V_1 = 2300$ V

Secondary voltage, $V_2 = 230$ V

Primary resistance, $R_1 = 0.6\ \Omega$

Primary leakage reactance, $X_1 = 3\ \Omega$

Secondary resistance, $R_2 = 0.008\ \Omega$

Secondary leakage reactance, $X_2 = 0.02\ \Omega$

The transformation ratio of transformer, $k = \dfrac{V_2}{V_1} = \dfrac{230}{2300} = 0.1$

(i) The equivalent resistance referred to the primary side. The secondary-side resistance referred to the primary side is denoted by R_2'.

$$\therefore \quad R_2' = R_2 \times \frac{1}{(K)^2} = 0.008 \times \frac{1}{(0.1)^2} = 0.8$$

The equivalent resistance referred to the primary side is denoted by R_{01}.

$$\therefore \quad R_{01} = R_1 + R_2' = 06 + 0.8 = 1.4\ \Omega$$

(ii) The equivalent reactance referred to the the primary side. The secondary reactance referred to the primary side is denoted by X_2'.

$$\therefore \quad X_2' = X_2 \times \frac{1}{(k)^2} = 0.02 \times \frac{1}{(0.1)^2} = 2\ \Omega$$

Total reactance referred to the primary side is denoted by X_{01}.

$$\therefore \quad X_{01} = X_1 + X_2' = 3 + 2 = 5\ \Omega$$

(iii) The equivalent resistance referred to the secondary side. The primary resistance referred to the secondary side is denoted by R_1'.

$$\therefore \quad R_1' = R_1 \times (k)^2 = 0.6 \times (0.1)^2 = 0.006\ \Omega$$

The total resistance referred to the secondary side is denoted by R_{02}.

$$\therefore \quad R_2' = R_1' + R_2 = 0.006 + 0.008 = 0.014\ \Omega$$

(iv) The total reactance referred to the secondary side. The primary reactance refered to the secondary side is denoted by X_1'.

$$\therefore \quad X_1' = X_1 \times (k)^2 = 3 \times (0.1)^2 = 0.03\ \Omega$$

The total reactance referred to the secondary side is denoted by X_{02}.

$$\therefore \quad X_2' = X_1' + X_2 = 0.03 + 0.02 = 0.05\ \Omega$$

7.9 LOSSES IN TRANSFORMER

Since a transformer is a static device, it does not contain any rotational losses. Hence, major losses that occur in a transformer on load can be divided into two groups:

(i) Core losses

(ii) Copper losses

7.9.1 Core Losses

The core flux in a transformer remains practically constant for all loads. Therefore, these losses are constant at any load. It is the sum of two losses, i.e., (i) hysteresis loss and (ii) eddy current loss.

(i) **Hysteresis loss.** These losses are directly proportional to the flux density of core, frequency, and volume of core.

Hysteresis losses, $W_h \propto B_{max}^{\eta} fv$

$$W_h = kB_{max}^{\eta} fv \text{ watts} \tag{7.20}$$

where η is the Steimntz constant, which is 1.6; B_m is the maximum flux density; f is the frequency; and v is the volume of the core. These losses can be reduced by selecting a suitable high silicon steel content material.

(ii) Eddy current loss. These losses are directly proportional to the square of flux density, frequency, and thickness.

Eddy current losses, $W_e \propto B_{max}^2 f^2 t^2$

$$W_e = kB_{max}^2 f^2 t^2 \text{W} \tag{7.21}$$

where k is the proportional constant; t is the thickness; f is the frequency; and B_{max} is the maximum flux density. These losses can be reduced by selecting thin sheets of laminations. If thin sheets are selected, automatically thickness is reduced.

EXAMPLE 7.9 **The transformer core losses are found to be 50 W at 40 Hz and 86 W at 60 Hz measured at same peak flux density. Compute the hysteresis and eddy current loss at 50 Hz.**

Solution Given data

Core losses at 40 Hz = 50 W

Core losses at 60 Hz = 86 W

In both cases, the flux density is same.

Total core losses, $W_i = Af + Bf^2$ (or) $W_i/f = A + Bf$

$$\frac{50}{40} = A + B \times 40$$

or

$$A + 40B = 1.25 \tag{1}$$

$$\frac{86}{60} = A + B \times 60$$

or

$$A + 60B = 1.43 \tag{2}$$

Solving Eqs. (1) and (2), we get

$$A = 0.89 \quad \text{and} \quad B = 0.009$$

The hysteresis and eddy current loss at 50 Hz are

$$\begin{aligned} W_i &= Af + Bf^2 \\ &= 0.89\,(50) + (0.009)\,(50)^2 \\ &= W_h + W_e \\ &= 44.5 + 22.5 = 67 \text{ W} \end{aligned}$$

Hysteresis loss at 50 Hz, $W_h = 44.5$ W

Eddy current loss at 50 Hz, $W_e = 22.5$ W

EXAMPLE 7.10 A 10-kg specimen of steel sheet laminations of the maximum flux density and waveform factor are maintained constant. The test will be conducted on this transformer, and the following results are obtained:

Frequency (Hz):	**20**	**40**	**50**	**60**	**80**
Total losses (W):	**16**	**35**	**48**	**65**	**102**

Calculate the eddy current loss per kg at a frequency of 50 Hz.

Solution Given data

Weight of the specimen = 10 kg

When the flux density and waveform factor remain constant, the expression for iron loss

$$W_i = Af + Bf^2 \text{ (or) } \frac{W_i}{f} = A + Bf$$

The values of W_i/f for different frequencies are as follows:

Frequency (Hz)	20	40	50	60	80
$\frac{W_i}{f}$	0.8	0.875	0.96	1.083	1.275

The graph between f and W_i/f has been plotted as shown in Figure 7.18.

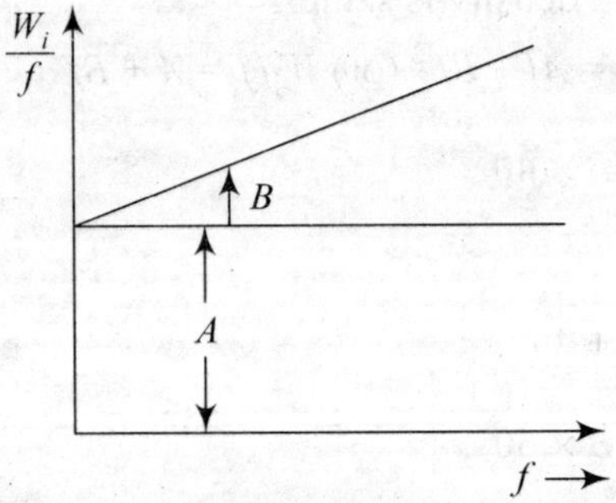

Figure 7.18

At

$$20 \text{ Hz} = \frac{W_i}{f} = A + Bf$$

$\Rightarrow$ $$0.8 = A + 20B$$

or

$$A + 20B = 0.8 \quad (1)$$

at

$$40 \text{ Hz} \Rightarrow 0.875 = A + 40B$$

$$A + 40B = 0.875 \quad (2)$$

Solving Eqs. (1) and (2), we get

$$A = 0.725 \quad \text{and} \quad B = 0.00375$$

Eddy current loss at 50 Hz = $Bf^2 = (0.00375)(50)^2 = 9.375$ W

$$\text{Eddy current loss/kg} = \frac{9.375}{10} = 0.9375 \text{ W}$$

EXAMPLE 7.11 A transformer is connected to an 1100-V, 50-Hz supply, the core loss is 1100 W of which 700 W is the hysteresis loss and 400 W is the eddy current loss. If the applied voltage is raised to 2200 V and frequency to 100 Hz, find the core losses.

Solution Given data

Applied voltage, $V = 1100$ V

Frequency, $f = 50$ Hz

Core losses = 1100 W

Hysteresis losses, $W_h = 700$ W

Eddy current loss, $W_e = 400$ W

If the applied voltage and frequency are doubled, find the new core losses, i.e., $V = 2200$ V and $f = 100$ Hz. Here, voltage and frequency both are doubled, therefore, B_{max} is constant. With 1100 V at 50 Hz, the hysteresis loss, $W_h = Af$.

$$\Rightarrow \quad 700 = A \times 50$$

$$\therefore \quad A = \frac{700}{50} = 14$$

Eddy current loss, $W_e = Bf^2$

$$\Rightarrow \quad 400 = B(50)^2$$

$$\therefore \quad B = \frac{400}{(50)^2} = 0.16$$

With 2200 V at 100 Hz

Hysteresis loss, $W_h = Af = 14 \times 100 = 1400$ W

Eddy current loss, $W_e = Bf^2 = 0.16 \leftrightarrow (100)^2 = 1600$ W

New core losses, $W_i = W_h + W_e = 1400 + 1600 = 3000$ W

7.9.2 Copper Losses

Basically the copper losses are due to ohmic resistances of the transformer windings. If R_1 and R_2 are the resistances of primary and secondary windings, respectively, then the copper losses of primary and secondary are $I_1^2 R_1$ and $I_2^2 R_2$.. Therefore, the total copper losses of transformer are $I_1^2 R_1 + I_2^2 R_2$. Generally, these losses vary as square of that particular current. These losses are also called as variable losses.

7.10 TESTING OF TRANSFORMER

The performance of a transformer can be analyzed by its equivalent circuit parameters shown in Figure 7.15. In order to determine the equivalent circuit parameters, the following two tests are performed on the transformer.

(i) Open-circuit (OC) test
(ii) Short-circuit (SC) test

7.10.1 Open-Circuit Test

This test is done to determine the core parameters such as R_0 and X_0. It is also known as core loss test. The circuit diagram for the OC test is shown in Figure 7.19.

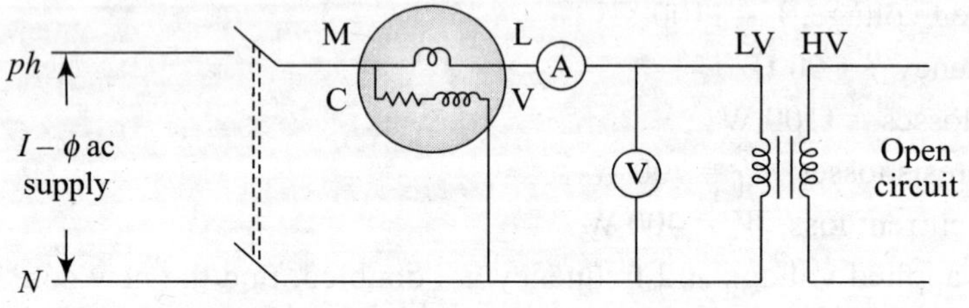

Figure 7.19 Circuit diagram for OC test

The primary is connected to the rated supply of the normal frequency through the wattmeter, ammeter, and voltmeter. The secondary of the transformer is left open. Normally for the OC test, the meters are connected on the LV side and the HV side will be open circuited to avoid the high current at the OC side. The wattmeter used is the low power factor (LPF) meter. Since the secondary is open circuited, there will not be any load current and hence no loss in the secondary side. In the primary, the no-load current is very less, i.e., 5–10% of the full-load current, so that the copper loss in the primary winding can also be neglected. Therefore, the power is utilized only to meet the core losses in this test.

Wattmeter reading, W_0 = Core losses

Voltmeter reading, V_0 = Applied voltage

Ammeter reading, I_0 = No-load current

$$\therefore \quad W_0 = P_i \text{ (core loss)}$$

$$= V_0 I_0 \cos \phi_0$$

No-load power factor, $\cos \phi_0 = \dfrac{W_0}{V_0 I_0}$

Working component of the no-load current, $I_w = I_0 \cos \phi_0$

Magnetizing component of the no-load current, $I_m = I_0 \sin \phi_0$

The shunt branch parameters can be calculated as

$$Y_0 = G_i - jB_m$$

$$= \frac{I_0}{V_0}$$

$$V_0^2 G_i = W_0$$

$$G_i = \frac{W_0}{V_0^2}; \quad P = \frac{V^2}{R}$$

$$B_m = \sqrt{Y_0^2 - G_i^2}$$

The values what we get are LV-side parameters. If we want to refer these values to the HV side, the values can be converted using the transformation ratio.

7.10.2 Short-Circuit Test

This test is conducted to determine the winding parameters such as R_{01}, X_{01} or R_{02}, X_{02}. It is also known as copper loss test. In this test we measure the voltage, current, and power to calculate the resistance and reactance of the winding.

The circuit diagram for SC test is shown in Figure 7.20.

A low voltage is applied to the primary winding such that the rated full-load current passes through the winding. The secondary winding is short circuited by a fine conductor. Normally, in this test, meters will be connected on the HV side and the LV winding will be short circuited. The wattmeter used in this test is unity power factor (upf) wattmeter. Since the applied voltage on the primary is very less, the flux setup is small so the core loss can be neglected.

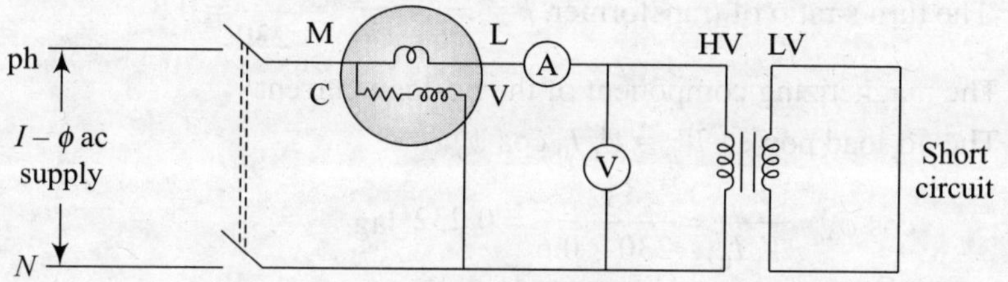

Figure 7.20 Circuit diagram for the SC test

Wattmeter reading, W_{SC} = Copper loss

Voltmeter reading, V_{SC} = Applied voltage

Ammeter reading, I_{SC} = Full-load current

$$\therefore \quad Z_{01} = \frac{V_{SC}}{I_{SC}}$$

and

$$R_{01} = \frac{W_{SC}}{I_{SC}^2}$$

$$X_{01} = \sqrt{Z_{01}^2 - R_{01}^2}$$

The values are referred to the HV side. We can transfer these values to the LV side by using the transformation ratio.

EXAMPLE 7.12 **The no-load test is conducted on a single phase transformer. The following test data are obtained:**

Primary voltage, $V_1 = 230$ V, secondary voltage, $V_2 = 115$ V

Primary current, $I_0 = 0.6$ A, power input, $W_0 = 32$ W

Resistance of the primary winding, $R_1 = 0.5\ \Omega$

Find the following:

(i) Turns ratio

(ii) The magnetizing component of the no-load current

(iii) Its working (or) loss component

(iv) Iron loss

Draw no-load phasor diagram to scale.

Solution Given data

Primary voltage, $V_1 = 230$

Secondary voltage, $V_2 = 115$ V

Primary current, $I_0 = 0.6$ A

Power input, $W_0 = 32$ W

Primary winding resistance, $R_1 = 0.5\ \Omega$

(i) The turn's ratio of transformer, $k = \dfrac{N_2}{N_1} = \dfrac{V_2}{V_1} = \dfrac{115}{230} = 0.5$

(ii) The magnetizing component of the no-load current

The no-load power, $W_0 = V_1 I_0 \cos\phi_0$

$$\cos\phi_0 = \frac{W_0}{V_1 I_0} = \frac{32}{230 \times 0.6} = 0.232 \text{ lag}$$

$$\phi_0 = \cos^{-1}(0.232)\ 76.6$$

$$\sin\phi_0 = 0.973$$

The magnetizing current, $I_m = I_0 \sin\phi_0 = 0.6 \times 0.973 = 0.584$ A

(iii) Working component, $I_w = I_0 \cos\phi_0 = 0.6 \times 0.232 = 0.1392$ A

Primary copper losses $= I_0^2 R_1 = (0.6)^2 \times 0.5 = 0.18$ W

(iv) The core loss (or) iron loss = Input power (W_0) – Copper loss

$= 32 - 0.18 = 31.82$ W

The phasor diagram is shown in Figure 7.21.

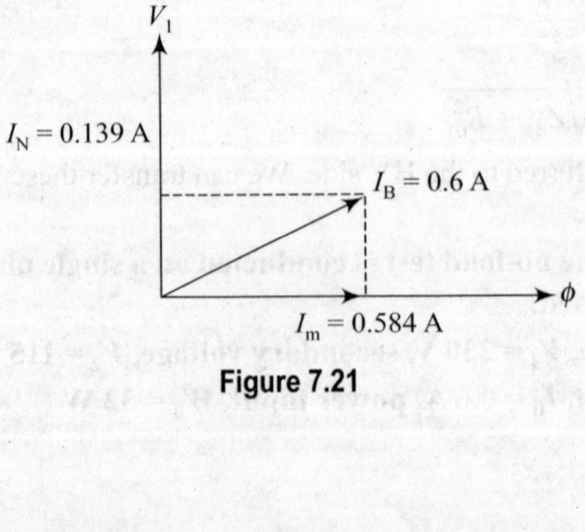

Figure 7.21

EXAMPLE 7.13 **A 25-kVA, 2200/220-V, 50-Hz, single-phase transformer obtained the following test results:**

OC test (LV side) = 220 V, 1.2 A, 100 W

SC test (HV side) = 100 V, 7 A, 310 W

Calculate the parameters of the equivalent circuit of the transformer referred to the LV side, and draw the equivalent circuit.

Solution Given data

Rating of the transformer = 25 kVA

Primary voltage, $V_1 = 2200$ V

Secondary voltage, $V_2 = 220$ V

From the OC test:

No-load power, $W_0 = V_0 I_0 \cos\phi_0$

No-load power factor, $\cos\phi_0 = \dfrac{W_0}{I_0 V_0} = \dfrac{100}{1.2 \times 220} = 0.38$ lag

$\sin\phi_0 = 0.925$

Working component current, $I_w = I_0 \cos\phi_0 = 0.38 \times 1.2 = 0.456$ A

Magnetizing component current, $I_m = I_0 \sin\phi_0 = 0.925 \times 1.2 = 1.11$ A

$$\text{Magnetizing core resistance, } R_0 = \frac{V_0}{I_w} = \frac{220}{0.456} = 482.46\ \Omega$$

$$\text{Magnetizing core reactance, } X_0 = \frac{V_0}{I_m} = \frac{220}{1.11} = 198.2\ \Omega$$

From the SC test:

$$Z_{SC} = \frac{V_{SC}}{I_{SC}} = \frac{100}{7} = 14.3\ \Omega$$

Copper loss $(P_{SC}) = I_{SC}^2 R_{SC}$

$$R_{SC\,HV} = \frac{P_{SC}}{I_{SC}^2} = \frac{310}{7^2} = 6.33\ \Omega$$

$$X_{SC\,HV} = \sqrt{Z_{SC}^2 - (R_{SC\,HV})^2} = \sqrt{(14.3)^2 - (6.33)^2} = 12.82\ \Omega$$

Equivalent parameters referred to the LV side

$R_0 = 482.46\ \Omega$ and $X_0 = 198.2\ \Omega$

$$\text{The transformation ratio of transformer, } K = \frac{V_2}{V_1} = \frac{220}{2200} = 0.1$$

$R_{SC\,LV} = K^2 R_{SC\,HV}\ (0.1)^2 \times 6.33 = 0.0633$

$X_{SC\,LV} = X_{SC\,HV}\ (K)^2 = 12.82\ (0.1)^2 = 0.1282\ \Omega$

The equivalent circuit diagram referred to the LV side is shown in Figure 7.22.

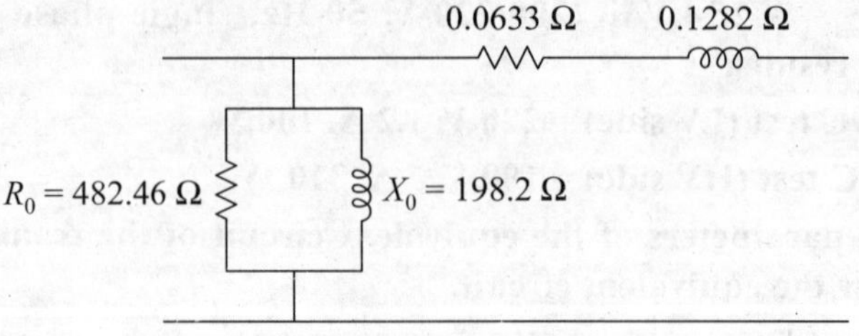

Figure 7.22

7.11 REGULATION OF TRANSFORMER

When the transformer on the secondary side is connected to load with constant excitation voltage on the primary side, the secondary voltage may decrease or increase with the lagging and leading power factors, respectively. This change in terminal voltage from no load to full load when the load is thrown off or removed at a specified power factor is known as the *regulation of a transformer.*

Figure 7.23 shows the appropriate equivalent circuit diagram by neglecting the shunt branch because the magnetizing current will not have any effect while calculating voltages.

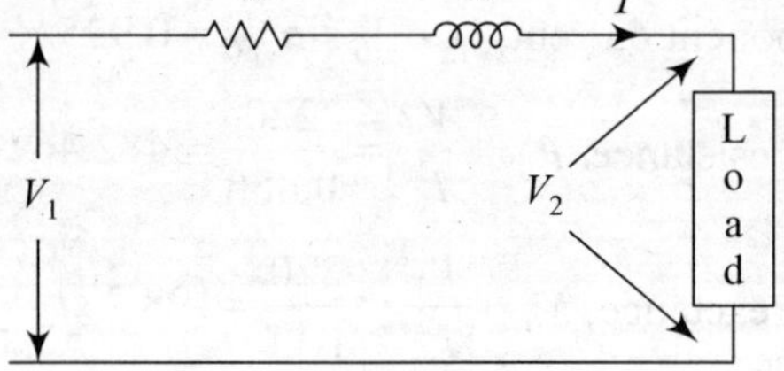

Figure 7.23 Approximate equivalent circuit

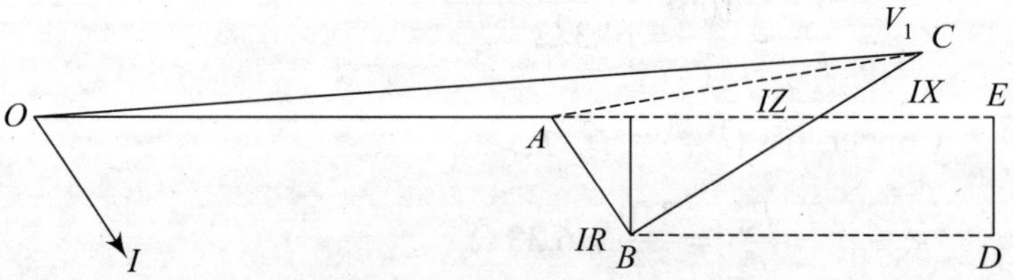

Figure 7.24 Phasor diagram

From Figure 7.24,

$$V_1 \cong OE$$

$$AE = V_1 - V_2 = IR\cos\phi + IX\sin\phi, \text{ for lagging pf}$$

$$AE = V_1 - V_2 = IR\cos\phi - IX\sin\phi, \text{ for leading pf}$$

When the load is removed, ${}_0V_2 = V_1$

$$\therefore \quad V_1 - V_2 = {}_0V_2 - V_2$$

Then

$\% \text{ Regulation} = \dfrac{{}_0V_2 - V_2}{V_2} \times 100$ is known as the percentage of voltage regulation up

$\% \text{ Regulation} = \dfrac{{}_0V_2 - V_2}{{}_0V_2} \times 100$ is known as the percentage of voltage regulation down

$$\% \text{ Regulation up} = \frac{IR\cos\phi + IX\sin\phi}{V} \tag{7.22}$$

$$= \frac{I_2\,(R_{02}\cos\phi + X_{02}\sin\phi)}{V_2} \text{ (referred to secondary)}$$

$$\therefore \quad \frac{I_2R_{02}}{V_2} = R_{\text{pu}} \quad \text{and} \quad \frac{I_2X_{02}}{V_2} = X_{\text{pu}} \tag{7.23}$$

The per unit regulation can be determined by $R_{\text{pu}}\cos\phi \pm X_{\text{pu}}\sin\phi$

7.11.1 Condition for Maximum Regulation

(i) For lagging power factor: The condition for the maximum regulation can be calculated by differentiating the regulation with respect to $d\phi$ and equating it to zero.

$$\frac{d\,(\text{reg})}{d\phi} = 0$$

$$\frac{d(IR\cos\phi + IX\sin\phi)}{d\phi} = 0 \tag{7.24}$$

$$-R\sin\phi + X\cos\phi = 0$$

$$R\sin\phi - X\cos\phi = 0$$

$$\tan\phi = \frac{X}{R} \Rightarrow \phi = \tan^{-1}\left(\frac{X}{R}\right) \tag{7.25}$$

$$\cos\phi = \frac{R}{\sqrt{R^2 + X^2}} \tag{7.26}$$

The voltage regulation will be maximum when the load power factor is equal to the impedance angle of the transformer.

(ii) For leading power factor:

$$\frac{d(\text{reg})}{d\phi} = 0$$

$$\frac{d(IR\cos\phi - IX\sin\phi)}{d\phi} = 0 \tag{7.27}$$

$$-R\sin\phi - X\cos\phi = 0$$

$$\tan\phi = \frac{-X}{R} \Rightarrow \phi = \tan^{-1}\left(\frac{-X}{R}\right) \tag{7.28}$$

The voltage regulation is negative for the leading power factor. Figure 7.25 shows the variation of regulation for both lagging and leading power factors.

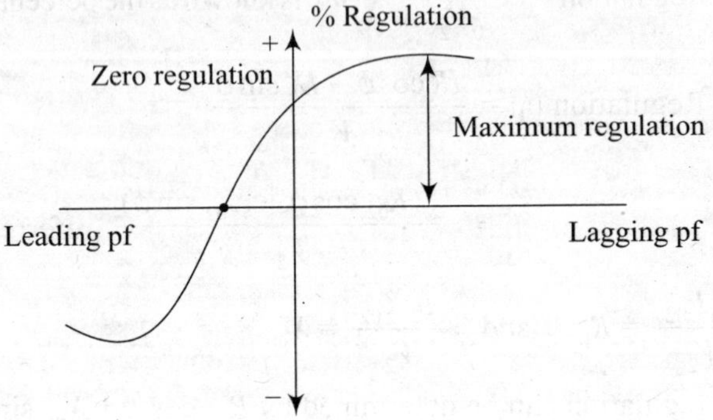

Figure 7.25 Variation of regulation with power factor

7.11.2 Condition for Zero Regulation

The condition for zero regulation is obtained as

$$I(R\cos\phi + X\sin\phi) = 0 \tag{7.29}$$

$$\tan\phi = \frac{-R}{X} \tag{7.30}$$

$$\phi = \tan^{-1}\left(\frac{-R}{X}\right) \tag{7.31}$$

$$\cos\phi = \cos\left[\tan^{-1}\left(\frac{-R}{X}\right)\right]$$

At this $\cos\phi$ the voltage regulation of the transformer becomes zero.

EXAMPLE 7.14 **The OC and SC tests were conducted on a 50-kVA transformer and the following test results were obtained:**

OC test: primary voltage, 3300 V; secondary voltage, 415 V; and primary power (W_0), 440 W.

SC test: primary voltage, 120 V; primary current, 15.15 A; primary power, 440 W; and secondary currents equal to full-load current. Find the voltage regulation for 0.7 power factor (i) lagging and (ii) leading.

Solution Given data

Rating of transformer = 50 kVA

$V_{OC} = 3300$ V, $W_0 = 440$ W, secondary voltage = 415 V

$V_{SC} = 120$ V, $I_{SC} = 15.15$ A, $P_{SC} = 440$ W

Secondary current = Full-load current = Short-circuit current

$$\text{Full-load current, } I_2 = I_{SC} = \frac{\text{Transformer rating}}{\text{Voltage}}$$

$$I_{SC} = I_{f1} = \frac{50 \times 1000}{415} = 120.5 \text{ A}$$

$$R_1' = \frac{P_{SC}}{(I_{SC})^2} = \frac{440}{(15.15)^2} = 1.92\ \Omega$$

$$Z_1' = \frac{V_{SC}}{I_{SC}} = \frac{120}{15.15} = 7.92\ \Omega$$

$$X_1' = \sqrt{(Z_1')^2 - (R_1')^2} = \sqrt{(7.921)^2 - (1.92)^2} = 7.685\ \Omega$$

(i) At 0.7 pf lagging, the transformation ratio of transformer,

$$k = \frac{V_2}{V_1} = \frac{415}{3300} = 0.126$$

$$R_2' = R_1'(k)^2$$
$$= 1.92 \times (0.126)^2 = 0.0304\ \Omega$$

$$X_2' = X_1'(k)^2 = 7.685 \times (0.126)^2 = 0.122\ \Omega$$

$$\cos\phi_0 = 0.7 \text{ lag}, \phi_0 = \cos^{-1} 0.7 = 45.6° \Rightarrow \sin(45.6°) = 0.714$$

$$\%\text{ Regulation} = \frac{I_2 R_2' \cos\phi + I_2 X_2' \sin\phi}{V_2} \times 100$$

$$\%\text{ Regulation} = \frac{120.5 \times 0.0304 \times 0.7 + 120.5 \times 0.122 \times 0.714}{415}$$

$$= 0.0315 \times 100 = 3.15\ \%$$

(ii) $$\%\text{ Regulation at 0.7 pf leading} = \frac{I_2 R_2' \cos\phi - I_2 \times X_2' \sin\phi_2}{V_2} \times 100$$

$$= \frac{120.5 \times 0.0304 \times 0.7 - 120.5 \times 0.122 \times 0.714}{415} \times 11$$

$$= -0.0191 \times 100 = -1.91\%$$

EXAMPLE 7.15 **A single-phase, 50-Hz, 2.5-kVA, 415/240-V transformer has the following parameters referred to the HV side:**

$$\boldsymbol{R_{01} = 3.5\ \Omega \text{ and } X_{01} = 4.5\ \Omega}$$

Determine the regulation of transformer when operating at (i) full load with 0.8 pf lagging, (ii) full load with 0.8 pf leading, and (iii) half load at 0.8 pf leading.

Solution Given data

Rating of transformer = 2.5 kVA

Primary voltage, V_1 = 415 V

Secondary voltage, V_2 = 240 V

Total resistance referred to the primary side, $R_{01} = 3.5\ \Omega$

Total reactance referred to the primary side, $X_{01} = 4.5\ \Omega$

(i) At 0.8 pf lagging

Primary current at full load, $I_1 = I_p = \dfrac{2.5 \times 10^3}{415} = 6.02$ A

$$\cos\phi = 0.8 \Rightarrow \phi_0 = \cos^{-1}(0.8) = 36.86$$

$$\sin\phi = \sin(36.86) = 0.6$$

$$\% \text{ Regulation} = \frac{I_1 R_{01}\cos\phi + I_1 X_{01}\sin\phi}{V_1} \times 100$$

$$= \frac{6.02 \times 3.5 \times 0.8 + 6.02 \times 4.5 \times 0.6}{415} \times 100$$

$$= 0.0798 \times 100$$

$$= 7.98\%$$

(ii) For 0.8 leading pf, % regulation $= \dfrac{I_1 R_{01}\cos\phi - I_1 X_{01}\sin\phi}{V_1} \times 100$

$$= \frac{6.02 \times 3.5 \times 0.8 - 6.02 \times 4.5 \times 0.6}{415} \times 100$$

$$= 0.00145 \times 100$$

$$= 0.145\%$$

(iii) Half load at 0.8 pf leading

Half full-load current $= 6.02 \times \dfrac{1}{2} = 3.01$ A

$$\% \text{ Regulation} = \frac{I_1 R_{01}\cos\phi - I_1 X_{01}\sin\phi}{V_1} \times 100$$

$$= \frac{3.01 \times 3.5 \times 0.8 - 3.01 \times 4.5 \times 0.6}{415} \times 100$$

$$= 7.25 \times 10^{-4} \times 100$$

$$= 0.0725\%$$

EXAMPLE 7.16 A transformer is having the copper loss of 3% of the output and the reactance drop of 6%. Calculate the voltage regulation when the power factor is (i) 0.8 pf lagging and (ii) 0.8 pf leading.

Solution Given data

Copper loss = 3% of the output

Reactance drop = 6%

(i) % Regulation at 0.8 pf lagging

Copper loss $= (I_S)^2 R_{02} = 0.03\,(V_S I_S)$

$$I_S R_S = 0.03 V_S$$

Reactance drop = $I_S X_{02} = 0.06V_S$

$$\% \text{ Regulation for lagging pf} = \frac{I_2 R_{02} \cos\phi + I_2 X_{02} \sin\phi}{V_2} \times 100$$

Here,

$R_S = R_2, I_2 = I_S, V_2 = V_S$ and $X_S = X_2$

$\cos\phi = 0.8 \Rightarrow \phi = \cos^{-1} 0.8 = 36.86°$

$\sin\phi = \sin(36.86) = 0.6$

$$\% \text{ Regulation} = \frac{(0.03V_S) \times 0.8 + (0.06V_S)\,0.6}{V_S} \times 100 = \frac{0.06V_S}{V_S} \times 100 = 6\%$$

(ii) Regulation at 0.8 pf leading

$$\% \text{ Regulation for leading pf} = \frac{I_S R_{02} \cos\phi - I_s X_{02} \sin\phi}{V_S} \times 100$$

$$= \frac{(0.03V_S) \times 0.8 - (0.06V_S) 0.6}{V_S} \times 100$$

$$= \frac{-0.012V_S}{V_S} \times 100 = -1.2\%$$

EXAMPLE 7.17 A single-phase 6600/415-V transformer has an equivalent resistance of 0.015 pu and an equivalent reactance of 0.045 pu. Calculate the full-load voltage regulation at 0.8 pf lag if the primary voltage is 6600 V. Find also the secondary terminal voltage at full load.

Solution Given data

Primary-side voltage, $V_1 = 6600$ V

Secondary-side voltage, $V_2 = 415$ V

Resistance drop = 0.015 pu

Reactance drop = 0.045 pu

(i) % Voltage regulation at full load with 0.8 pf lagging

Per unit voltage regulation = $E_r \cos\phi + E_x \sin\phi$

E_r = Per unit resistance drop = 0.015 pu

E_x = Per unit reactance drop = 0.045 pu

$\cos\phi = 0.8$ and $\phi = \cos^{-1}(0.8) = 36.86$

$\sin\phi = \sin(36.86) = 0.6$

$$\frac{E_2 - V_2}{V_2} = E_r \cos\phi + E_x \sin\phi$$

$$= 0.015 \times 0.8 + 0.045 \times 0.6 = 0.039$$

(ii) For a primary voltage of 6600 V, the secondary no-load voltage (E_2) is 415 V. The change in secondary terminal voltage

$$\frac{E_2 - V_2}{V_2} = 0.039$$

$$\frac{415 - V_2}{V_2} = 0.039$$

The secondary terminal voltage, $V_2 = 399.42$ V

EXAMPLE 7.18 **A single-phase transformer has 1000 turns on the primary and 250 turns on the secondary side. The no-load current is 2.4 A and the no-load power factor is 0.21 lagging. Calculate the current and power factor of the primary circuit when the secondary draws a current 310 A at 0.8 pf lagging.**

Solution Given data

Primary turns, $N_1 = 1000$

Secondary turns, $N_2 = 250$

No-load current, $I_0 = 2.4$ A

No-load pf = 0.217 lagging

Secondary current, $I_2 = 310$ A

The transformation ratio, $k = \frac{N_2}{N_1} = \frac{250}{1000} = 0.25$

The relation from the turns and current

$$k = \frac{N_2}{N_1} = \frac{I_1}{I_2} \Rightarrow I_1 = \frac{N_2}{N_1} \times I_2$$

$$I_1 = \frac{250}{1000} \times 310 = 77.5 \text{ A}$$

$$I_0 = 2.4 \text{ A}$$

$$\cos\phi_0 = 0.217 \Rightarrow \phi_0 = \cos^{-1}(0.217) = 77.47°$$

$$\sin\phi_0 = \sin(77.47°) = 0.976$$

$$I_1\cos\phi_1 = I_2^1\cos\phi_2 + I_0\cos\phi_0$$

$$= 77.5 \times 0.8 + 2.4 \times 0.217 = 62.52 \text{ A}$$

$$I_1\sin\phi_1 = I_2\sin\phi_2 + I_0\sin\phi_0$$

$$= 77.5 \times 0.6 + 2.4 \times 0.976 = 48.84 \text{ A}$$

$$I_1 = \sqrt{(I_1\cos\phi_1)^2 + (I_1\sin\phi_1)^2}$$

$$= \sqrt{(62.52)^2 + (48.84)^2} = 79.34\, A$$

$$\tan\phi_1 = \frac{I_1\sin\phi_1}{I_1\cos\phi_1} = \frac{48.84}{62.52} = 0.7812$$

$$\phi_1 = \tan^{-1}(0.7812) = 38.63°$$

Primary power factor, $\cos\phi_1 = \cos 38.63 = 0.781$ lagging.

7.12 EFFICIENCY OF TRANSFORMER

The efficiency of any device can be defined as the ratio of the output power to the input power. So, the transformer efficiency can be defined as

$$\text{Efficiency, } \eta = \frac{\text{Output power}}{\text{Input power}}$$

But due to losses present in transformer the output power will not be equal to the input power

Input power = output power + total losses

$$\eta = \frac{\text{Output power}}{\text{Output power + Total losses}}$$

$$= \frac{\text{Output power}}{\text{Output power} + I_2^2 R_{o2} + P_i}$$

Here, the output power = $V_2 I_2 \cos\phi_2$

R_{02} = Total resistance of the transformer referred to secondary side

Efficiency at any particular load "x" of transformer is given as

$$\eta_{\text{at particular load}} = \frac{V_2 I_2 \cos\phi_2 \cdot x}{x \cdot V_2 I_2 \cos\phi_2 + x^2 P_{cu} + P_i} \tag{7.32}$$

From previous discussion, we come to know that the net flux is constant irrespective of loading conditions. Hence, the iron losses are constant. But the copper losses vary with the load current. Hence efficiency also varies with variation in load. At light loads the efficiency is very poor and as the load increases the increased efficiency becomes constant after 20–30% of the load and there becomes a slight fall at around 120% of the load. The efficiency versus load curves for different power factors is shown in Figure 7.26.

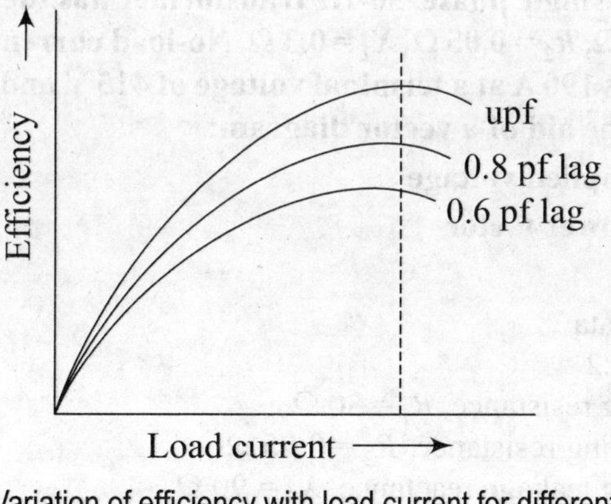

Figure 7.26 Variation of efficiency with load current for different power factors

7.12.1 Condition for Maximum Efficiency

According to the efficiency equation of transformer, the efficiency is function load current I_2. Let us assume that $\cos\phi_2$ is constant. Therefore, the secondary terminal voltage V_2 is also constant.

The condition for the maximum efficiency of transformer is

$$\frac{d\eta}{dI_2} = 0 \tag{7.33}$$

$$\frac{d\eta}{dI_2} = \frac{d}{dI_2}\left[\frac{V_2\, I_2 \cos\phi_2}{V_2\, I_2 \cos\phi_2 + I_2^2\, R_{02} + P_i}\right] = 0 \tag{7.34}$$

Simplifying further

$$P_{\text{i}} + I_2^2\, R_{02} - 2I_2^2\, R_{02} = 0$$

$$P_{\text{i}} - I_2^2 R_{02} = 0$$

$$P_{\text{i}} = I_2^2 R_{02} \tag{7.35}$$

$$P_{\text{i}} = P_{\text{cu}} \tag{7.36}$$

The condition to achieve maximum efficiency of a transformer is that copper losses must be equal to iron losses.

kVA supplied at maximum efficiency

At constant V_2, the kVA supplied is function of load current

$$\text{kVA at } \eta_{\text{max}} = I_2\, V_2$$

$$= V_2\, I_2 \times \sqrt{\frac{P_{\text{i}}}{P_{\text{cu}} \text{ at fl}}} \tag{7.37}$$

$$\text{kVA supplied at } \eta_{\text{max}} = \text{kVA rating at fl} \sqrt{\frac{P_{\text{i}}}{P_{\text{cu}} \text{ at fl}}} \tag{7.38}$$

EXAMPLE 7.19 **A single phase, 50-Hz transformer has the following data: turns ratio 40:2, $R_1 = 30\ \Omega$, $X_1 = 90\ \Omega$, $R_2 = 0.05\ \Omega$, $X_2 = 0.3\ \Omega$. No-load current is 1.4 A, leading flux by 35°. The secondary delivers 190 A at a terminal voltage of 415 V and at a pf of 0.8 lagging. Determine the following by the aid of a vector diagram:**

The primary applied voltage

The primary power factor

The efficiency

Solution Given data

Turns ratio = 40:2

Primary winding resistance, $R_1 = 30\ \Omega$

Secondary winding resistance, $R_2 = 0.05\ \Omega$

Primary winding leakage reactance, $X_1 = 90\ \Omega$

Secondary winding leakage reactance, $X_2 = 0.3\ \Omega$

No-load current, $I_0 = 1.4$ A, leading flux by an angle = 35°

The secondary current, $I_2 = 190$ A and $V_2 = 415$ V at a pf of 0.8 lagging

(i) Take V_2 as reference vector

$$V_2 = 415\ \angle 0° = (415 + j0)\ \text{V}$$

$I_2 = 190(0.8 - j0.6) = (152 - j114)$ A

$Z_2 = (0.05 + j0.3)\ \Omega$

$E_2 = V_2 + I_2 Z_2$

$= (415 + j0) + (152 - j114) \times (0.05 + j0.3)$

$E_2 = (456.8 + j39.9) = 458.54 \angle 5°$ V

Obviously $\beta = 5°$

$$E_1 = \frac{E_2}{k}$$

The transformation ratio of transformer is

$$k = \frac{E_2}{E_1} \Rightarrow E_1 = \frac{E_2}{k} = 20\,(456.8 + j39.9) = (9136 + j798) - E_1 = (-9136 - j798)$$

$$= 9170.79 \angle -175°$$

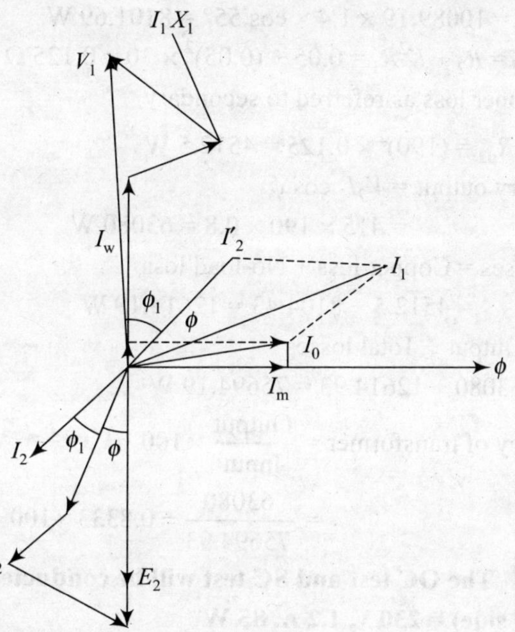

Figure 7.27

Secondary current referred to primary

$I_2' = -kI_2 = -0.05\,(152 - j114) = (-7.6 + j5.7) = 9.5 \angle -143.13°$A

$I_2' = 95 \angle -143.13°$ A

As seen from the vector diagram shown in Figure 7.27, I_0 leads by an angle

$5° + 90° + 35° = 130°$

$I_0 = 1.4 \angle 130° = 1.4\,(\cos 130 + j \sin 130°)$

$= 1.4\,(-0.643 + j\,0.766) = (-0.9 + j\,1.073)$

Primary current, $I_1 = -I_2' + I_0$

$$= (-7.6 + j\,5.7) + (-0.9 + j1.073) = (-8.5 + j\,6.773)$$

$$I_1 = 10.87\angle 141.45^\circ \text{ A}$$

$$V_1 = -E_1 + I_1 Z_1$$

$$= (-9136 - j798) + (-8.5 + j6.733)\,(30 + j90)$$

$$= (-9996.97 - j1361.01) = 10089.19\angle -172.25^\circ \text{ V}$$

(ii) Primary power factor

Phase angle between I_1 and V_1, $\phi_1 = 172.3^\circ - 141.45^\circ = 30.85^\circ$

$\therefore$ Primary power factor = cos (30.85°) = 0.858 lagging

(iii) Efficiency of the transformer

No-load input primary power, $W_0 = V_1 I_0 \cos \phi_0$

$$W_0 = 10089.19 \times 1.4 \times \cos 55^\circ = 8101.69 \text{ W}$$

$$R_{02} = R_2 + K^2 R_1 = 0.05 + (0.05)^2 \times 30 = 0.125\ \Omega$$

Total copper loss as referred to secondary

$$I_2^2 R_{02} = (190)^2 \times 0.125 = 4512.5 \text{ W}$$

Secondary output = $V_2 I_2 \cos \phi$

$$= 415 \times 190 \times 0.8 = 63080 \text{ W}$$

Total losses = Copper loss + No-load loss

$$= 4512.5 + 8104.43 = 12614.19 \text{ W}$$

Input = Output + Total losses

$$= 63080 + 12614.93 = 75694.19 \text{ W}$$

$$\text{Efficiency of transformer} = \frac{\text{Output}}{\text{Input}} \times 100$$

$$= \frac{63080}{75694.93} = 0.8333 \times 100 = 83.33\%$$

EXAMPLE 7.20 **The OC test and SC test will be conducted on a 230/460-V transformer**

OC test (LV side) = 230 V, 1.2 A, 85 W

SC test (HV side) = 30 V, 14 A, 105 W (LV winding short circuited)

Determine

(i) The circuit constant

(ii) The applied voltage and efficiency when the output is 12 A, 415 V

Solution Given data

Primary voltage, $V_1 = 230$ V

Secondary voltage, $V_2 = 460$ V

From OC test:

Open-circuit voltage, $V_{OC} = 230$ V, no-load current, $I_0 = 1.2$ A

No-load power, $W_0 = 85$ W

From SC test:

Short-circuit voltage, $V_{SC} = 30$ V, short-circuit current, $I_{SC} = 14$ A

Short-circuit power, $W_{SC} = 105$ W

(i) The circuit constants

The transformation ratio of transformer, $k = \frac{V_2}{V_1} = \frac{460}{230} = 2$

No-load power, $W_0 = V_0 I_0 \cos \phi_0$

$$\cos \phi_0 = \frac{W_0}{I_0 V_0} = \frac{85}{1.2 \times 230} = 0.308 \text{ lag}$$

Watt full component (or) working component of current, $I_w = I_0 \cos \phi_0$

$$= 1.2 \times 0.308 = 0.37 \text{ A}$$

Magnetizing component of current, $I_m = I_0 \sin \phi_0$

$\cos \phi_0 = 0.308 \Rightarrow \phi_0 = \cos^{-1} (0.308) = 72.1$

$\sin (72.1) = 0.9514$

$I_m = I_0 \sin \phi_0 = 1.2 \times 0.9514 = 1.142$ A

Now resistance representing the core losses

$$R_0 = \frac{V_{OC}}{I_w} = \frac{230}{0.37} = 621.62\ \Omega$$

$$\text{Magnetizing reactance, } X_0 = \frac{V_{OC}}{I_m} = \frac{230}{1.142} = 201.4\ \Omega$$

As the primary is short circuited all values are referred to secondary winding

$$R_{02} = \frac{P_{SC} \text{ (or) } W_{SC}}{(I_{SC})^2} = \frac{105}{(14)^2} = 0.536\ \Omega$$

$$Z_{02} = \frac{V_{SC}}{I_{SC}} = \frac{30}{14} = 2.143\ \Omega$$

$$X_{02} = \sqrt{(Z_{02})^2 - (R_{02})^2} = \sqrt{(2.143)^2 - (0.536)^2} = 2.075\ \Omega$$

As R_{02} and X_{02} are referred to primary, let us transfer these values to primary as follows:

$$R_{01} = \frac{R_{02}}{(K)^2} = \frac{0.536}{(2)^2} = 0.134\ \Omega$$

$$X_{01} = \frac{X_{02}}{(K)^2} = \frac{2.075}{(2)^2} = 0.519\ \Omega$$

$$Z_{01} = \frac{Z_{02}}{(K)^2} = \frac{2.143}{(2)^2} = 0.536\ \Omega$$

The equivalent circuit is shown in Figure 7.28.

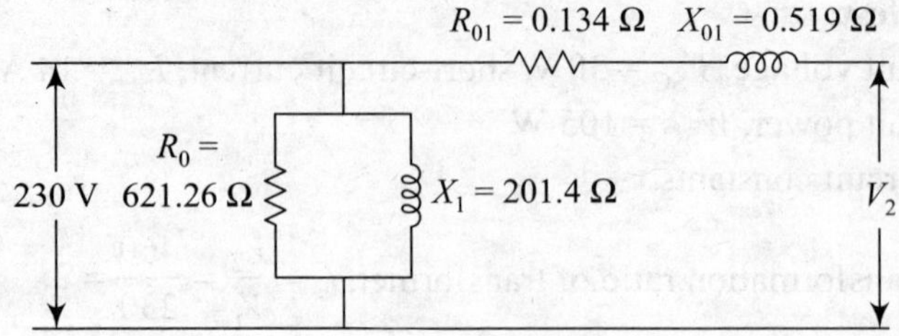

Figure 7.28

(ii) From Figure 7.29, the applied voltage V_1 is the vector sum of V_1 and $I_1 Z_{01}$ output current, $I_2 = 12$ A

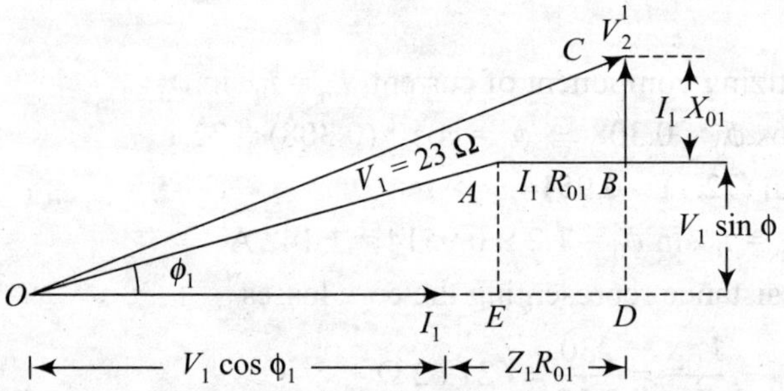

Figure 7.29

Input current, $I_1 = I_2 K = 12 \times 2 = 24$ A

Primary current, $I_1 = 24$ A

Now,

$$I_1 R_{01} = 24 \times 0.134 = 3.216 \text{ V}$$
$$I_1 X_{01} = 24 \times 0.519 = 12.46 \text{ V}$$
$$\cos \phi_1 = 0.8 \Rightarrow \phi_1 = \cos^{-1} 0.8 = 36.86°$$
$$\sin 36.86 = 0.6$$

Refer from the phasor diagram neglecting the angle between V_1 and V_1^1, we have

$$OC^2 = OD^2 + DC^2$$
$$OC = V_1^1 = \sqrt{(OE + ED^2) + (DB + BC)^2}$$
$$= \sqrt{(V_1 \cos \phi_1 + I_1 R_{01})^2 + (V_1 \sin \phi_1 + I_1 X_{01})^2}$$
$$= \sqrt{(230 \times 0.8 + 3.216)^2 + (V_1 \sin \phi_1 + I_1 X_{01})^2} = 240.18 \text{ V}$$

Hence, applied voltage, $V_1 = 240.18$ V

Iron loss (or) no-load losses = 85 W

Copper loss = $W_{SC} = I_2^2 R_{02} = (12)^2 \times 0.536 = 77.184$ W

Output = $V_2 I_2 \cos \phi_2 = 415 \times 12 \times 0.8 = 3984$ W

Input = Output + Losses = 3984 + 85 + 77.184 = 4146.184 W

$$\text{Efficiency, } \eta = \frac{\text{Output}}{\text{Input}} \times 100$$

$$= \frac{3984}{4146.184} \times 100 = 96.08\%$$

EXAMPLE 7.21 **A single-phase, 50-Hz, 50-kVA transformer has a full-load primary current of 260 A and total resistance referred to primary is 0.005 Ω if the iron loss amounts to 210 W. Find the efficiency at full load, and half load at (i) unity power factor and (ii) 0.8 power factor.**

Solution Given data

Transformer rating = 50 kVA

Full-load primary current, $I_1 = 260$ A

Total resistance referred to primary, $R_{01} = 0.005\ \Omega$

Iron losses, $W_i = 210$ W

Full-load copper loss, $W_c = I_1^2 R_{01} = (260)^2\ (0.005) = 338$ W

Total losses at full load = $W_i + W_c = 210 + 338 = 548$ W

(i) Unity power factor

Full load = 50 × 1 = 50 kW

$$\text{Efficiency at full load with upf} = \frac{\text{Output}}{\text{Output + Losses}} = \frac{50 \times 10^3}{50 \times 10^3 + 548} \times 100$$

$$= 98.92\%$$

(ii) η at 0.8 pf

Full load = $50 \times 0.8 \times 10^3 = 40$ kW

$$\text{Efficiency at full load with 0.8 pf} = \frac{40}{40 + 0.548} \times 100 = 98.64\%$$

Efficiency at half load:

$$\text{Half full-load output} = \frac{1}{2} \times 50 = 25 \text{ kVA}$$

Efficiency at 0.8 power factor:

Half full-load output = 25 × 0.8 = 20 kW

$$\text{Copper loss at half load} = \left(\frac{I_1}{2}\right)^2 \times R_{01}$$

$$= \left(\frac{260}{2}\right)^2 \times 0.005 = 84.5 \text{ W}$$

$$\eta \text{ at half load at 0.8 pf} = \frac{20 \times 10^3}{20 \times 10^3 + \text{Losses}}$$

Total losses = $W_c + W_i = 210 + 84.5 = 294.5$ W

$$\eta = \frac{20 \times 10^3}{20 \times 10^3 + 294.5} \times 100 = 98.55\%$$

Efficiency at half-load upf:

Half load at upf = 25 × 1 = 25 kW

$$\eta = \frac{\text{Output}}{\text{Output} + \text{Total losses}} = \frac{25 \times 10^3}{25 \times 10^3 + 294.5} \times 100 = 98.84\%$$

EXAMPLE 7.22 **A transformer 1100/230 V has primary and secondary resistances of 0.4 Ω and 0.03 Ω, respectively, if the iron loss amounts to be 250 W. Determine (i) the secondary current at which maximum efficiency occurs and (ii) the maximum efficiency at 0.8 pf lagging.**

Solution Given data

Primary voltage, $V_1 = 1100$ V

Secondary voltage, $V_2 = 230$ V

Primary resistance, $R_1 = 0.4\ \Omega$

Secondary resistance, $R_2 = 0.03\ \Omega$

Iron losses = 250 W

(i) The transformation ratio of transformer, $K = \frac{V_2}{V_1} = \frac{230}{1100} = 0.2091$

Total resistance referred to secondary, $R_{02} = R_2 + R_1 (K)^2 = 0.03 + 0.4(0.2091)^2 = 0.0475\ \Omega$

Let I_2 be the secondary current at maximum efficiency.

At maximum efficiency:

Copper loss = Iron loss

$$I_2^2 R_{02} = 250$$

$$I_2^2 = \frac{250}{R_{02}} = \frac{250}{0.0475} = 5263.16$$

$$I_2 = 72.55 \text{ A}$$

Hence, the secondary current at maximum efficiency = 72.55 A

(ii) Output at maximum efficiency at 0.8 power factor is

$$V_2 I_2 \cos \phi_2 = 230 \times 72.55 \times 0.8 = 13349.2 \text{ W}$$

At maximum efficiency:

Copper loss = Iron loss

$$P_c = P_i$$

The total losses at maximum efficiency = $P_i + P_i = 2P_i$

$$P_{total} = 250 \times 2 = 500 \text{ W}$$

$$\text{Maximum efficiency} = \frac{\text{Output}}{\text{Output} + \text{Total losses}} \times 100$$

$$= \frac{13349.2}{13349.2 + 500} \times 100 = 96.39\%$$

EXAMPLE 7.23 **The efficiency of a 250-kVA, single-phase, 50-Hz transformer is 95% when delivering full load at 0.8 power factor lagging and 96.2% at delivering half load at upf. Calculate the efficiency at 75% of the full load at 0.8 power factor lagging.**

Solution Given data

Rating of the transformer = 250 kVA

Efficiency at full load with 0.8 pf = 95%

Efficiency at half load with upf = 96.2%

At full load:

Output at 0.8 pf lagging = 250 × 0.8 = 200 kW

$$\text{Input} = \frac{\text{Output}}{\text{Efficiency}} = \frac{200 \times 10^3}{0.95} = 210.53 \text{ kW}$$

Total losses = Input – Output

$$= 210.53 - 200 = 10.53 \text{ kW} \quad (1)$$

At half load:

$$\text{Output at upf} = \frac{250}{2} \times 1 = 125 \text{ kW}$$

$$\text{Input} = \frac{\text{Output}}{\text{Efficiency}} = \frac{125 \times 10^3}{0.962} = 129.94 \text{ kW}$$

Total losses = 129.94 – 125 = 4.94 kW

$$\text{Copper loss at half load} = \left(\frac{1}{2}\right)^2 P_c = \frac{1}{4} P_c$$

$$\text{Total losses} = P_i + \frac{1}{4} \times P_c = 4.94 \text{ kW} \quad (2)$$

Subtracting Eq. (2) from Eq. (1), we get

$$\frac{3}{4} P_c = 10.53 - 4.94 = 5.59 \text{ kW}$$

$$P_c = 5.59 \times \frac{4}{3} = 7.45 \text{ kW}$$

$$P_i = 10.53 - 7.45 = 3.08 \text{ kW}$$

Copper loss at 75% of full load = $(0.75)^2 \times P_c = (0.75)^2 \times 7.45 = 4.191$ kW

Total loss at 75% at full load = $P_i + P_c = 3.08 + 4.191 = 7.27$ kW

Output at 75% of full load at 0.8 pf = 250 × 0.75 × 0.8 = 150 kW

$$\text{Efficiency} = \frac{\text{Output}}{\text{Input}} = \frac{150 \times 10^3}{150 \times 10^3 + 7.27 \times 10^3} \times 100 = 95.38\%$$

EXAMPLE 7.24 **A 10-kVA, 415/230-V, single-phase, 50-Hz transformer has maximum efficiency of 93% at 81% of full load and upf. Determine the efficiency at 0.8 pf lagging.**

Solution Given data

Rating of the transformer = 10 kVA

Primary voltage, $V_1 = 415$ V

Secondary voltage, $V_2 = 230$ V

The maximum efficiency occurs at 81% of full load = 93%, i.e.,

$$\eta_{max} = 93\% \text{ at } 81\% \text{ of full load.}$$

Output at 81% of full load at upf $= 10 \times 0.81 \times 1 = 8.1$ kW

$$\text{Input at 81\% of full load} = \frac{\text{Output}}{\text{Efficiency}} = \frac{8.1 \times 10^3}{0.93} = 8.709 \text{ kW}$$

Total losses = Input – Output = 8.709 – 8.1 = 0.609 kW

At maximum efficiency

Copper loss = Core loss

$$P_i = (0.81)^2 P_c = 0.6561 P_c$$

Total losses at maximum efficiency $= 2P_i = 0.609$ kW

$$P_i = \frac{0.609}{2} = 0.3045 \text{ kW}$$

$$P_c = \frac{P_i}{0.6561} = \frac{0.3045}{0.6561} = 0.4641 \text{ kW}$$

Total full-load losses $= P_i + P_c = 0.3045 + 0.4641 = 0.7686$ kW

Full-load output at 0.8 pf lagging $= 10 \times 0.8 = 8$ kW

Efficiency of full load at 0.8 pf lagging is

$$\frac{\text{Output}}{\text{Output + Total losses}} = \frac{8}{8 + 0.7686} \times 100 = 91.235\%$$

EXAMPLE 7.25 **A 10-kVA, 440/230-V, single-phase transformer gave the following results: short-circuit test (HV side), 50 V, 22.73 A, 160 W, the maximum efficiency occurs at upf and 1.3 times of full-load current. Determine the full-load efficiency and also the maximum efficiency.**

Solution Given data

Rating of the transformer = 10 kVA

Primary voltage, $V_1 = 440$ V

Secondary voltage, $V_2 = 230$ V

From the SC test (HV side)

$$V_{SC} = 50 \text{ V}, I_{SC} = 22.73 \text{ A and } W_{SC} = 160 \text{ W}$$

The maximum efficiency occurs at upf and 1.3 times of full-load current.

$$\text{Full-load current on HV side} = \frac{10 \times 10^3}{440} = 22.73 \text{ A}$$

The short-circuit test is conducted on the HV side only, therefore, the full-load copper loss at upf will be 160 W.

The maximum efficiency occurs at 1.3 times of full-load current.

Hence at upf, corresponding copper loss = $(1.3)^2 \times 160 = 270.4$ W

At maximum efficiency

Copper loss = Core loss at upf

Input power = $1.3 \times 10000 \times 1 = 13000$ W

Hence, the maximum efficiency at upf is

$$\frac{13}{13 + 0.27 + 0.27} \times 100 = 0.9601 \times 100 = 96.01\%$$

Full-load efficiency at 0.8 pf

Output power at full load and 0.8 pf = $10000 \times 0.8 = 8$ kW

Core losses = 270.4 W

Copper loss at full load = 160 W

Total losses = $P_i + P_c = 270.4 + 160 = 430.4$ W

$$\eta = \frac{\text{Output}}{\text{Output + Losses}} \times 100 = \frac{8 \times 10^3}{8 \times 10^3 + 430.4} \times 100 = 94.89\%$$

EXAMPLE 7.26 **A single-phase, 50-Hz, 40-kVA transformer has an iron loss of 450 W and full-load copper loss of 900 W. Find the load at which the maximum efficiency is achieved at upf.**

Solution Given data

Rating of the transformer = 40 kVA

Iron losses = 450 W

Copper loss = 900 W (full-load copper loss)

If

X = fraction of rated load at which maximum efficiency

P_i = Iron losses = 450 W

P_c = full-load copper loss = 900 W

$X^2 P_c = P_i$

Substituting the values of P_i and P_c in the above equation, we get

$$X^2 (900) = 450 \Rightarrow X^2 = \frac{1}{2}$$

$$\therefore \quad X = 0.707$$

Hence, the efficiency is maximum at 70.7% of the rated load.

At these loads, $P_i = P_c$

Corresponding output = $40 \times 0.707 = 28.28$ kW

$$\text{Corresponding efficiency} = \frac{28.28}{28.28 + 0.45 + 0.45} \times 100 = 96.92\%$$

EXAMPLE 7.27 **A transformer has a resistance of 2.0% and a reactance of 6%. At full load, what is the power factor at which the regulation will be**

(i) Zero

(ii) Positive maximum

(iii) If the maximum efficiency of the transformer occurs at full load (at upf) what will be the efficiency under these condition?

Solution Given data

Resistance = 2.0%

Reactance = 6.0%

Approximate percentage regulation = 2.0 cos $\phi \pm$ 6.0 sin ϕ

(i) If regulation is zero, negative sign must be applicable. This happens at leading pf

$$\text{Corresponding pf} = \tan\phi = \frac{2}{6} = 0.333$$

$$\phi = \tan^{-1}(0.333) = 18.44° \text{ leading}$$

(ii) For a maximum positive regulation lagging pf the result can be obtained.

Corresponding $\tan\phi = 6/2 = 3$

$$\phi = \tan^{-1} 3 = 71.57°$$

$$\%\text{ voltage regulation} = 2 \times \cos 71.57 + 6 \times \sin 71.57 = 6.325\%$$

(iii) Maximum efficiency occurs at such load when

Iron losses = Copper losses

The copper loss depends on the resistance of transform

Take iron losses = 2 copper losses

$$\text{Efficiency} = \frac{100}{100 + 2 \times 2} = \frac{100}{104} \times 100 = 96.15\%$$

EXAMPLE 7.28 **If P_1 and P_2 are the iron loss and copper loss of a transformer on full load, find the ratio of P_1 and P_2 such that maximum efficiency occurs at 75% of full load.**

Solution Given data

If P_1 = Iron losses

P_2 = Copper losses at full load

$$\text{Copper loss at 75\% of full load} = P_2(0.75)^2 = \frac{9}{16} P_2$$

At maximum efficiency, $P_{\text{Iron}} = P_{\text{Cu}}$

$$P_1 = \frac{9}{16} P_2 = \frac{P_1}{P_2} = 9/16$$

EXAMPLE 7.29 **The rating of a 2300/230-V, 50-Hz single-phase transformer is 25 kVA. The efficiency at upf is 96% at rated load and also at half rated load. Determine the transformer core losses and ohmic loss.**

Solution Given data

Rating of the transformer = 25 kVA

Primary voltage, $V_1 = 2300$ V

Secondary voltage, $V_2 = 230$ V

The efficiency at upf full load = 96%, and also at half load the efficiency is same.

The transformer core losses and ohmic loss

$$\text{At full load, } \eta = 1 - \frac{\text{Losses}}{\text{Out put + Losses}}$$

$$0.96 = 1 - \frac{P_C + P_{SC}}{25000 \times 1 + P_C + P_{SC}}$$

$$0.04 = \frac{P_C + P_{SC}}{25000 + (P_C + P_{SC})}$$

$$P_C + P_{SC} = 1000 + 0.04\, P_C + 0.04\, P_{SC}$$

$$P_C + P_{SC} = \frac{1000}{0.96} = 1041.67 \text{ W}$$

$$\text{At half load, } \eta = 1 - \frac{P_C + \left(\frac{1}{4}\right) P_{SC}}{\text{Output} + P_C + \left(\frac{1}{4}\right) P_{SC}}$$

$$= 1 - \frac{P_C + \left(\frac{1}{4}\right) P_{SC}}{(25000) \times \frac{1}{2} + P_C + \left(\frac{1}{4}\right) P_{SC}}$$

$$0.96 = 1 - \frac{P_C + 0.25\, P_{SC}}{12500 + P_C + 0.25\, P_{SC}}$$

$$\Rightarrow \quad 0.04 = \frac{P_C + 0.25\, P_{SC}}{12500 + P_C + 0.25\, P_{SC}}$$

$$P_C + 0.25\, P_{SC} = 500 + 0.04\, P_C + 0.01\, P_{SC}$$

$$P_C + \left(\frac{1}{4}\right) P_{SC} = \frac{500}{0.96} = 520.83 \text{ W}$$

$$P_C + P_{SC} = 1041.67 \text{ W}$$

$$P_C + \frac{1}{4} P_{SC} = 520.83 \text{ W}$$

Solving the above equations, we get

$$P_{SC} = P_C = 694.44 \text{ W}$$

EXAMPLE 7.30 **The maximum efficiency of a single phase, 50 Hz, 100 kVA transformer is 96% and it occurs at 82% of the full load at 0.8 power factor. If the leakage impedance is 6%, find the voltage regulation at rated load of 0.8 pf lagging.**

Solution Given data

Rating of the transformer = 100 kVA

Maximum efficiency = 96%

The load at which maximum efficiency occurs = 82% of full load

Leakage impedance = 6%

$$\text{Efficiency } (\eta) = \frac{\text{Output}}{\text{Output + Losses}}$$

$$\frac{1}{\eta} = \frac{\text{Output + Losses}}{\text{Output}} = 1 + \frac{\text{Losses}}{\text{Output}}$$

$$\therefore \quad \text{Losses} = \left(\frac{1}{\eta} - 1\right)\text{Output}$$

Output at maximum efficiency = (100000) (0.82) (0.8) = 65600 W

$$\text{Total transformer losses} = \left[\frac{1}{0.96} - 1\right] \times [65600] = 2733.33 \text{ W}$$

At maximum efficiency, ohmic loss = Core loss

$$P_C = P_{SC} = \frac{2733.33}{2} = 1366.67 \text{ W}$$

This ohmic losses of 1366.67 W occur at 82% of full-load current.

$$\therefore \quad \text{Ohmic loss at full load} = 1366.67 \left[\frac{100}{82}\right] = 1666.671 \text{ W}$$

$$R_{epu} = \frac{1666.67}{100000} = 0.0167 = E_{rpu}$$

It is given that $Z_{epu} = 0.06$

$$X_{epu} = \sqrt{(0.06)^2 - (0.0167)^2} = 0.05763 = E_x \text{ pu}$$

Voltage regulation = $E_r \cos \phi_2 + E_x \sin \phi_2$

$$\cos \phi_2 = 0.8 \Rightarrow \phi_2 = \cos^{-1}(0.8) = 36.86°$$

$$\sin \phi_2 = \sin (36.86°) = 0.6$$

Voltage regulation = $0.0167 \times 0.8 + 0.05763 \times 0.6 = 0.0479$

% Voltage regulation = $0.0479 \times 100 = 4.794\%$

EXAMPLE 7.31 **A 15-kVA, 1100/110-V, 50-Hz, single-phase transformer has the following test results:**

OC test (LV side): 110 V, 0.8 A, 90 W

SC test (HV side): 70 V, 12 A, 100 W

Determine the following:

(i) Core losses of the transformer

(ii) Equivalent resistance and leakage reactance referred to HV side

(iii) Equivalent resistance and leakage reactance referred to LV side

(iv) Regulation of transformer at full load and half load at 0.8 pf lagging

(v) Transformer terminal voltage at full load at 0.8 pf lagging

(vi) Efficiency of the transformer at full load and half load at 0.8 pf lagging

Solution Given data

Rating of the transformer = 15 kVA

Primary voltage, $V_1 = 1100$ V

Secondary voltage, $V_2 = 110$ V and frequency, $f = 50$ Hz

No-load voltage, $V_{0c} = 110$ V

No-load current, $I_0 = 0.8$

No-load power, $W_0 = 90$ W

Short-circuit voltage, $V_{SC} = 70$ V

Short-circuit current, $I_{SC} = 12$ A

Short-circuit power, $W_{SC} = 100$ W

Transformation ratio of transformer, $K = \dfrac{V_2}{V_1} = \dfrac{110}{1100} = 0.1$

(i) From SC test $V_{SC} = 70$ V, $I_{SC} = 12$ A, $P_{SC} = 100$ W

$$Z_{01} = \frac{V_{SC}}{I_{SC}} = \frac{70}{12} = 5.833\ \Omega$$

$$R_{01} = \frac{P_{SC}}{(I_{SC})^2} = \frac{100}{(12)^2} = 0.694\ \Omega$$

$$X_{01} = \sqrt{(Z_{01})^2 - (R_{01})^2} = \sqrt{(5.833)^2 - (0.694)^2} = 5.792\ \Omega$$

(ii) Equivalent resistance and leakage reactance referred to the HV side (or) primary side are 0.694 Ω and 5.792 Ω.

(iii) Equivalent resistance and leakage reactance referred to the LV side (or) secondary side

$$R_{02} = [R_{01}] \times k^2 = 0.694 \times [0.1]^2 = 0.00694\ \Omega$$
$$X_{02} = [X_{01}] \times k^2 = 5.792 \times [01]^2 = 0.05792\ \Omega$$

(iv) % Regulation of transformer at full load and half load at 0.8 pf lagging

(a) Secondary current at full load, $I_2 = \dfrac{15 \times 10^3}{110} = 136.36$ A

$$\cos\phi = 0.8 \Rightarrow \phi = \cos^{-1} 0.8 = 36.86$$
$$\sin\phi = \sin(36.86) = 0.6$$

Approximate voltage drop = $I_2 (R_{02} \cos \phi + X_{02} \sin \phi)$

$$= 136.36 (0.00694 \times 0.8 + 0.05792 \times 0.6) = 5.496 \text{ V}$$

Voltage regulation of the transformer = $\dfrac{5.496}{110} = 0.04996$ pu

% Regulation = $0.04996 \times 100 = 4.996\%$

(b) Secondary current at half load, $I_2 = \dfrac{7.5 \times 10^3}{110} = 68.2$ A

At half load, the approximate voltage drop of transformer = $I_2 (R_{02} \cos \phi + X_{02} \sin \phi)$

Regulation in pu = $\dfrac{2.75}{110} = 0.025$ pu

% Voltage regulation = $0.025 \times 100 = 2.5\%$

At half-load terminal voltage, $V_2 = 110 - 2.75 = 107.5$ V

(v) At full load the transformer terminal voltage at 0.8 pf lagging = $110 - 5.496$ V = 104.504 V

(vi) Efficiency of the transformer

(a) At full load at 0.8 pf lagging

Copper loss at full load, $P_{SC} = I_2^2 R_{02} = (136.36)^2 \times 0.00694 = 129.04$ W

Total losses of the transformer = $P_i + P_{SC} = 90 + 129.04 = 219.04$ W

At full-load output power = $V_2 I_2 \cos \phi_2$

$$= 104.504 \times 136.36 \times 0.8 = 11400.132 \text{ W}$$

$$\text{Efficiency, } \eta = \frac{\text{Output}}{\text{Output} + \text{Losses}} \times 100$$

$$= \frac{11400.132}{11400.132 + 219.04} \times 100 = 98.12\%$$

(b) η at half load at 0.8 pf

$$\text{Copper loss at half load} = \left(\frac{1}{2}\right)^2 \times \text{Full-load copper loss}$$

$$= \frac{1}{4} \times 129.04 = 32.26 \text{ W}$$

Total losses at half load = $P_i + P_c = 90 + 32.26 = 122.26$ W

Output at half load = $V_2 I_2 \cos \phi_2$

$$= 107.5 \times \left(\frac{136.36}{2}\right) \times 0.8 = 5863.48 \text{ W}$$

$$\text{Efficiency, } \eta = \frac{5863.48}{5863.48 + 122.26} \times 100 = 97.96\%$$

7.13 ALL-DAY EFFICIENCY OF TRANSFORMER

The all-day efficiency of a transformer can be defined as the ratio of output energy over 24 h to input energy for the same time

$$\eta_{AD} = \frac{\text{Output in kWh}}{\text{Input in kWh}}$$

Generally, the all-day efficiency of a transformer can be calculated for distribution transformer. Since the distribution transformer does not supply the rated load for the entire day. Therefore, all-day efficiency of transformer will always be less than the commercial or ordinary efficiency of transformer.

EXAMPLE 7.32 **A 400-kVA distributed transformer whose copper and iron losses at full load are 5 kW and 4 kW, respectively. During a day of 24 h it is loaded as under**

Number of hours	Loading in kW	Power factor
6	320	0.8
10	240	0.75
4	80	0.8
4	0	—

Find the all-day efficiency.

Solution Given data

Rating of the transformer = 400 kVA

Full-load copper loss = 5 kW

Iron losses = 4 kW

Load of 320 kW at 0.8 pf $= \frac{320}{0.8} = 400$ kVA

Similarly for 240 kW at 0.75 pf $= \frac{240}{0.75} = 320$ kVA

For 100 kW at 0.8 pf $= \frac{80}{0.8} = 100$ kVA, i.e., one-fourth of full load

Copper loss at full load of 400 kVA = 5 kW

Copper loss at 320 kVA $= 5 \times \left[\frac{320}{400}\right]^2 = 3.2$ kW

Copper loss at 100 kVA $= 5 \times \left[\frac{100}{400}\right]^2 = 0.3125$ kW

Total copper loss in 24 h = $(6 \times 5) + (10 \times 32) + 0.3125 + 4 \times 0$

$\Rightarrow \quad (6 \times 5) + (10 \times 3.2) + 0.3125 \times 4 + 4 \times 0 = 63.25$ kWh

Iron loss for 24 h = $24 \times 4 = 96$ kWh

Total transformer losses for 24 h = $63.25 + 96 = 159.25$ kWh

Transformer output for 24 h = (6 × 320) + (10 × 240) + (4 × 80) + 4 × 0

= 4640 kW

$$\text{All-day efficiency} = \frac{\text{Output for 24 h}}{\text{Output for 24 h + Total losses for 24 h}}$$

$$\eta_{AD} = \frac{4640}{4640 + 159.25} \times 100 = 96.68\%$$

EXAMPLE 7.33 **A 125-kVA transformer is loaded as follows: load is increased from 0 to 83.25 kVA in 3 h from 7:00 a.m. to 10:00 a.m., stays at 83.25 kVA from 10:00 a.m. to 6:00 p.m., and then the transformer is disconnected till next day. Assuming the load to be resistive and the core loss is equal to the copper loss of 1.2 kW, determine the all-day efficiency and ordinary efficiency of transformer.**

Solution Given data

Rating of transformer = 125 kVA

From 7:00 a.m. to 10:00 a.m., the load increased from 0 to 83.25 kVA in 3 h.

$$\text{The average load in 3 h} = \left(\frac{0 + 83.25}{2}\right) = 41.67 \text{ kVA}$$

$$41.63 \text{ kVA} \left(\frac{1}{3} \text{ of full load kVA}\right)$$

$$\text{Load from 10:00 a.m. to 6:00 p.m.} = 83.25 \text{ kVA} \left(\frac{2}{3} \text{ of full load kVA}\right)$$

Copper loss = Iron loss = 1.2 kW

Ordinary efficiency:

In this case, load variations are not relevant

Output = 125 × 1 = 125 kW, Iron loss = Copper loss = 1.2 kW

In this case the load is resistive so take pf as upf.

Total losses = 2.4 kW = $P_i + P_c$

$$\text{Ordinary efficiency} = \frac{\text{Output}}{\text{Output + Losses}} = \frac{125}{125 + 2.4} \times 100 = 98.12\%$$

All-day efficiency:

$$\text{Copper loss from 7 a.m. to 10 a.m.} = 3 \times \left(\frac{1}{3}\right)^2 \times 1.2 = 0.4 \text{ kWh}$$

$$\text{Copper loss from 10 a.m. to 6 p.m.} = 8 \times \left(\frac{2}{3}\right)^2 \times 1.2 = 4.3 \text{ kWh}$$

Total copper loss for 24 h = 0.4 + 4.3 = 4.67 kWh

Total iron loss for 24 h = 24 × 1.2 = 28.8 kWh

Losses for a day of 24 h = 28.8 + 4.67 = 33.47 kWh

Output for 24 h = 3 × (41.67) + 8 × (83.25) = 791.01 kWh

$$\text{All-day efficiency, } \eta = \frac{\text{Output for 24 h}}{\text{Output for 24 h + Total losses for 24 h}} \times 100$$

$$= \frac{791.01}{791.01 + 33.47} \times 100 = 95.94\%$$

EXAMPLE 7.34 **A 10-kVA, 1–ϕ, transformer has a core loss of 50 W and full-load ohmic loss of 110 W. The daily variation of load on the transformer is as follows:**

6:00 a.m. to 1:00 p.m. 3 kW at 0.60 pf

1:00 p.m. to 5:00 p.m. 8 kW at 0.8 pf

5:00 p.m. to 1:00 a.m. full load at upf

1:00 a.m. to 1:00 p.m. no load

Solution Given data

Rating of the transformer = 10 kVA

Core losses = 50 W

Copper losses (full load) = 110 W

Fractional loading (x) and output kWh corresponding to load variations can be worked out in the tabulation form as below.

S. No.	Number of hours	$x = \dfrac{\text{Load kVA}}{\text{Transformer rating}}$	$x^2 P_c$ in kW	Output in kWh	Copper loss in kWh
1.	7	$\dfrac{3/0.6}{10} = 0.5$	$(0.5)^2 \times 0.11 = 0.0275$	$3 \times 7 = 21$	$0.0275 \times 7 = 0.1925$
2.	4	$\dfrac{8/0.8}{10} = 1$	$(1)^2 \times 0.11 = 0.11$	$8 \times 4 = 32$	$0.11 \times 4 = 0.44$
3.	8	$\dfrac{10/1}{10} = 1$	0.11	$10 \times 8 = 80$	$0.11 \times 8 = 0.88$
4.	5	Zero	Zero	Zero	Zero

Output in kWh = 21 + 32 + 80 = 133

Total ohmic loss in kWh = 0.1925 + 0.44 + 0.88 = 1.513 kWh

$$\text{Core losses during 24 h} = \frac{50}{1000} \times 24 = 1.2 \text{ kWh}$$

$$\text{Energy efficiency = All-day efficiency} = \frac{133}{133 + 1.513 + 1.2} \times 100 = 98\%$$

EXAMPLE 7.35 **The all-day efficiency of a 250-kVA transformer is 97% when it is loaded as follows.**

No. of hours	Load (kW)	Power factor
10	150	0.8
9	180	0.9
5	No load	—

If maximum efficiency of a transformer occurs at 82% of full load, find the iron loss and full-load copper loss.

Solution Given data

Rating of the transformer = 250 kVA

All-day efficiency of transformer = 97%

Maximum efficiency occurs at 82% of full load.

For 10 h, the load is 150 kW at 0.8 pf.

For 9 h, the load is 180 kW at 0.9 pf.

For 5 h, the load is zero.

Output/day = 150 × 10 + 180 × 9 + 5 × 0 = 3120 kWh

$$\text{All-day efficiency} = \frac{\text{Output in kWh}}{(\text{Output} + \text{Losses}) \text{ in kWh}}$$

$$0.97 = \frac{3120}{3120 + \text{Losses}}$$

$$0.97\,(3120 + \text{Losses}) = 3120$$

$$\text{Losses} = \frac{3120 - 3026.4}{0.97} = 96.5 \text{ kWh}$$

Let P_i be the iron loss in kW and P_c be the full-load copper loss in kW.

$$\text{Copper loss at 150 kW} = \frac{\left(\frac{150}{0.8}\right)^2}{(250)^2} \times P_c = 0.5625\,P_c$$

$$\text{Copper loss for 10 h at 150 kW} = 0.5625 \times 10 \times P_c$$
$$= 5.625\,P_c \text{ kWh}$$

$$\text{Copper loss at 180 kW} = \frac{(180/0.9)^2}{(250)^2} \times P_c = 0.64\,P_c$$

$$\text{Copper loss at 180 kW for 9 h} = 0.64 \times 9 \times P_c = 5.76\,P_c$$
$$= 5.76\,P_c \text{ kWh}$$

Iron loss for 24 h = $P_i \times 24$ kWh

Total losses/day = $(5.625\,P_c + 5.76\,P_c) + 24\,P_i = 11.385\,P_c + 24\,P_i$ kWh

But total losses/day = 96.5 kWh

$$11.385\,P_c + 24\,P_i = 96.5 \quad (1)$$

Copper loss at 82% of full load = $(0.82)^2 P_c = 0.6724\,P_c$

At maximum efficiency, $P_i = 0.6724\,P_c$

Substituting this P_i value in Eq. (1), we get

$$11.385\,P_c + 24\,(0.6724\,P_c) = 96.5$$

$$11.385\,P_c + 16.14\,P_c = 96.5 \Rightarrow P_c = \frac{96.5}{27.53} = 3.51\text{ kW}$$

Full-load copper loss = 3.51 kW

Iron loss, $P_i = 0.6724\,P_c$

$= 0.6724 \times 3.51 = 2.36$ kW

$$P_c = 3.51\text{ kW}$$

$$P_i = 2.36\text{ kW}$$

7.14 AUTOTRANSFORMER

Figure 7.30 shows an autotransformer. The winding AC with N_1 number of turns is connected across V_1 supply, whose induced emf is E_1. Let the winding BC consists of N_2 turns and terminal voltage V_2, whose induced emf is E_2. But $E_2/E_1 = N_2/N_1$ if the circuit is closed through the load, then E_2 drives a current I_2 through the load. Here the two mmfs $N_1 I_1$ and $N_2 I_2$ will be equal and opposite. If B is a variable point then the output voltage V_2 is varied.

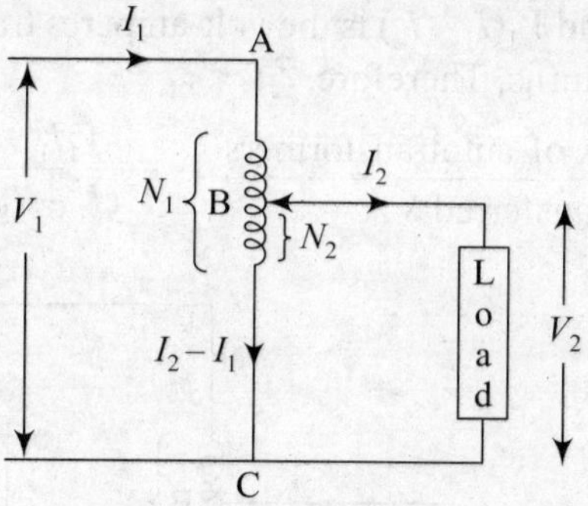

Figure 7.30 Autotransformer

(i) Step-down autotransformer. In Figure 7.30, if V_1 is greater than V_2 it is called as step-down autotransformer. According to ideal condition

$$\frac{N_2}{N_1} = \frac{V_2}{V_1} = \frac{I_1}{I_2} = k \quad (7.39)$$

The currents are distributed through the windings as shown in Figure 7.30. The VA delivered to the load is $V_2 I_2$

$$V_2 I_2 = V_2 I_1 + V_2 (I_2 - I_1) \quad (7.40)$$

where $V_2 I_1$ is the volt-amperes transferred conductively to the load through the AB portion of $(N_2 - N_1)$ turns and $V_2 (I_2 - I_1)$ is the volt-amperes transferred inductively to the load through the BC portion having N_2 turns. Therefore,

$$\frac{\text{Output VA of autotransformers}}{\text{Transferred VA}} = \frac{\text{Output VA of a transformer}}{\text{Output VA of equivalent ordinary transformer}}$$

$$= \frac{V_2 I_2}{V_2(I_2 - I_1)} = \frac{1}{1-k} \tag{7.41}$$

An ordinary transformer connected as a step-down transformer will have a volt-ampere rating 1/(1 – K) times of rating of ordinary transformer. Therefore, this result shows that the ordinary transformer connected as an autotransformer will have greater VA rating than those connected as ordinary transformers.

(ii) Step-up autotransformer. Figure 7.31 shows a step-up autotransformer. If V_1 is less than V_2 or the number of turns in portion AB, i.e., $(N_1 - N_2)$, is less than the number of turns in portion BC, i.e., N_2, it is called step-up autotransformer. The current directions are distributed as shown in Figure 7.31. According to ideal conditions

$$\frac{N_2}{N_1} = \frac{V_2}{V_1} = \frac{I_1}{I_2} = k \tag{7.42}$$

The volt-ampere input $V_1 I_1$ can be written as

$$V_1 I_1 = V_2 I_2 + (I_1 - I_2) \tag{7.43}$$

where $V_2 I_1$ is the volt-amperes transferred conductively from input to the load through the $(N_1 - N_2)$ turns and $V_1(I_1 - I_2)$ is the volt-amperes transferred inductively from input to the load through N_2 turns. Therefore,

$$\frac{\text{Output VA of autotransformers}}{\text{Transferred VA}} = \frac{V_1 I_1}{V_1(I_1 - I_2)} = \frac{K}{K-1} \tag{7.44}$$

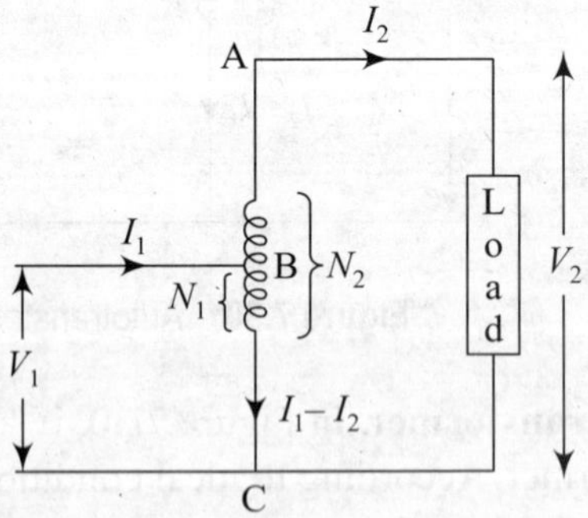

Figure 7.31 Step-up autotransformer

This result shows that a step-up transformer can be obtained from a step-down transformer by interchanging the currents $(I_1 - I_2)$ and I_2 provided that "K" is defined in the same manner in both cases.

7.14.1 Copper Saving in an Autotransformer

For the same capacity and voltage ratio, an autotransformer requires less winding material than a two winding transformer. The cross section of the conductor is proportional to the current carried by the conductor and length of the conductor is proportional to the number of turns. Hence, the weight of the copper is proportional to the product of the number of turns and current. For ordinary transformers,

Weight of copper in primary $\propto N_1 I_1$

Weight of copper in secondary $\propto N_2 I_2$

Total weight of copper $\propto N_1 I_1 + N_2 I_2$

But in an autotransformer (Figure 7.30) the turns $(N_1 - N_2)$ across AB carry a current of I_1. The weight of copper in section AB is proportional to $(N_1 - N_2)I_1$. In the same way, the turns N_2 across BC carry a current of $(I_2 - I_1)$. The weight of copper in section BC is proportional to $N_2(I_2 - I_1)$.

$$\text{Total weight of copper} \propto (N_1 - N_2) I_1 + N_2 (I_2 - I_1) \tag{7.45}$$

$$\propto N_2 I_2 + (N_1 + 2N_2) I_1$$

$$\frac{\text{Weight of copper in autotransformers}}{\text{Weight of copper in ordinary transformer}} = \frac{N_2 I_2 + (N_1 - 2N_2) I_1}{N_1 I_1 + N_2 I_2} \tag{7.46}$$

Divide entire equation by $N_2 I_1$

$$\frac{\dfrac{N_2 I_2}{N_2 I_1} + \dfrac{I_1 N_1 - 2N_2 I_1}{N_2 I_1}}{\dfrac{N_1 I_1}{N_2 I_1} + \dfrac{N_2 I_2}{N_2 I_1}}$$

$$\frac{\dfrac{I_2}{I_1} + \dfrac{N_1}{N_2} - 2}{\dfrac{N_1}{N_2} + \dfrac{I_2}{I_1}}$$

But

$$\frac{N_2}{N_1} = \frac{I_1}{I_2} = K$$

$$= 1 - K$$

These results show that the saving in the copper of autotransformer is $(1 - K)$ times the weight of copper in ordinary transformer. If K is nearer to 1 the saving is very large. Hence, the autotransformer is more useful for transformation ratio near to 1.

EXAMPLE 7.36 A load of 8 kW is supplied by an autotransformer at 140 V and at upf if the primary voltage is 230 V. Determine the (i) transformation ratio, (ii) secondary current, (iii) primary current, (iv) number of turns across the secondary if the total number of turns is 300, (v) power transformed, and (vi) power conducted directly from supply mains to load.

Solution Given data

Load = 8 kW

Primary voltage, $V_1 = 230$ V

Secondary voltage, $V_2 = 140$ V

Total number of turns on primary, $N_1 = 300$

(i) Transformation ratio, $K = \dfrac{V_2}{V_1} = \dfrac{140}{230} = 0.61$

(ii) Secondary current, $I_2 = \dfrac{8 \times 1000}{V_2 \cos \phi} = \dfrac{8000}{140 \times 1} = 57.14$

(iii) Primary current, $I_1 = KI_2 = 0.61 \times 57.14 = 34.86$ A

(iv) Number of turns across secondary, $N_2 = KN_1 = 0.61 \times 300 = 183$

(v) Power transformed = Load × $(1 - K)$ = 8 (1 – 0.61) = 3.12 kW

(vi) Power conducted directly from the supply mains to load

= Total load supplied by the autotransformer – Power transformed

= 8 – 3.12 = 4.88 kW

EXAMPLE 7.37 **The primary and secondary voltage of an autotransformer are 500 V and 400 V, respectively. Show with the aid of a diagram the current distribution in the windings when the secondary current is 200 A. Calculate the economy in copper.**

Solution Given data

Primary voltage, $V_1 = 500$ V

Secondary voltage, $V_2 = 400$ V

Secondary current, $I_2 = 200$ A

Transformation ratio, $K = \dfrac{V_2}{V_1} = \dfrac{400}{500} = 0.8$

Primary current, $I_1 = KI_2 = 0.8 \times 200 = 160$ A

We know that

$$\frac{\text{Weight of copper on autotransformer}}{\text{Weight of copper on ordinary transformer}} = 1 - K$$

Economy in copper

$$= \frac{\text{Weight of copper on ordinary transformer} - \text{Weight of copper on autotransformer}}{\text{Weight of copper on ordinary transformer}}$$

$$= 1 - \frac{\text{Weight of copper on autotransformer}}{\text{Weight of copper on ordinary transformer}}$$

$$= 1 - (1 - K)$$

$$= 1 - (1 - 0.8) = 80\%$$

EXAMPLE 7.38 **A 230/330-V autotransformer draws power from a 230-V source and supplies a 6-kW load with a power factor of 0.8 lagging. A second load of 2 kW is supplied at upf. Neglecting losses, calculate the current drawn by the transformer from the 230-V line and its power factor.**

Solution Given data

$V_2 = 230$ V

$V_1 = 330$ V

6 kW at 0.8 pf lagging

2 kW at upf

The connection diagram is shown in Figure 7.32.

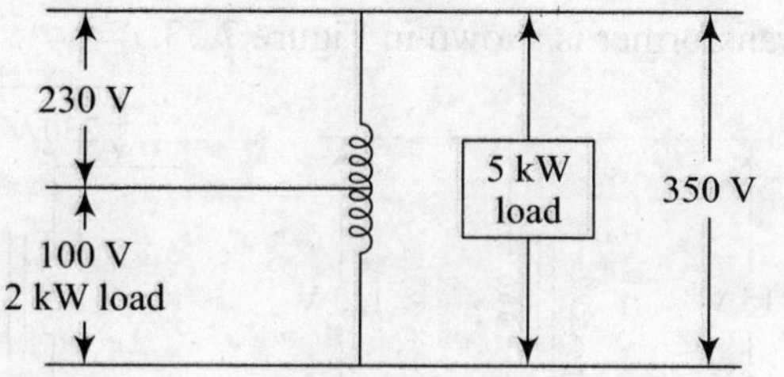

Figure 7.32

Transformation ratio for primary and 6-kW load $= \dfrac{330}{230} = 1.435$

Current drawn by the first load, $I_1 = \dfrac{6 \times 1000}{330 \times 0.8} = 22.73$ A

$\therefore$ Current drawn by the primary to this supply load $= 22.73 \times 1.435$

$= 32.62$ A

$= I_2 \times K = I_1$ (at 0.8 pf lagging)

Current drawn by the secondary load $= \dfrac{2 \times 1000}{100 \times 1} = 20$ A

Transformation ratio for primary and 2-kW load $= \dfrac{100}{230} = 0.435$

$\therefore$ Current drawn by primary to supply this load $= 0.435 \times 20 = 8.7$ A at upf

Hence, the total primary current drawn from the 230-V supply is the vector sum of

(i) 32.62 A at 0.8 pf lagging

(ii) 8.7 A at upf

Resolving these currents into their *X*- and *Y*-components, we get

X-component $= 32.62 \times 0.8 + 8.7 \times 1 = 34.8$ A

Y-component $= 32.62 \times 0.6 = 19.57$ A

$\therefore$ Total primary current, $I_3 = 34.8 + j\,19.57 = 39.93$ A

Power factor $= \dfrac{34.8}{39.93} = 0.872$ lagging

EXAMPLE 7.39 **A 5-kVA, 1 – ϕ, 50-Hz transformer has a full-load efficiency of 96% and iron of 50 W. The transformer is connected now as an autotransformer to 230-V supply if it delivers 5 kW load at upf to a 115-V circuit, calculate the efficiency of the operation and the current drawn by the high voltage side.**

Solution Given data

Rating of the two-winding transformers = 5 kVA

Full-load efficiency, $\eta = 96\%$

Iron loss = 50 W

The transformer is connected as an autotransformer with 230-V supply.

The load is 5 kW at upf 115-V circuit.

Two-winding transformer is shown in Figure 7.33.

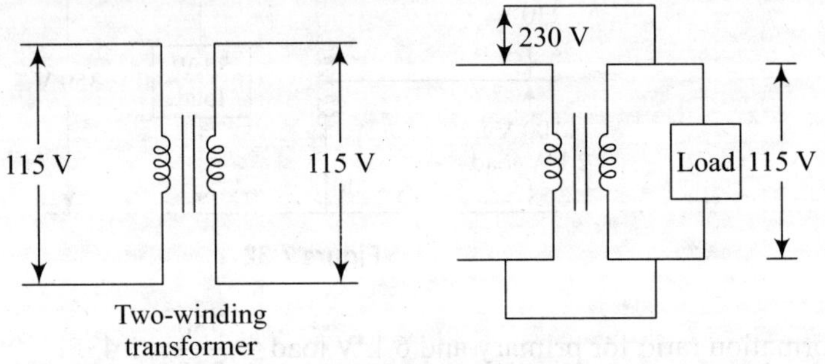

Figure 7.33

The same transformer unit is connected as an autotransformer to a 230 V supply. Since two windings are connected in series hence the voltage across each is 115 V.

In both the connections the iron loss would remain the same since each of the autotransformer windings will carry the currents but half of the currents as the conventional two-winding transformer, the copper loss will be one-fourth of the previous value.

Two-winding transformer:

Efficiency, $\eta = 96\%$

$$0.96 = \frac{\text{Output}}{\text{Output} + \text{Losses}} = 0.96 = \frac{5000}{5000 + \text{Losses}}$$

Losses = (Iron loss + Copper loss)

$$0.96 = \frac{5000}{5000 + (\text{Iron loss} + \text{Copper loss})} = 0.96 = \frac{5000}{5000 + 50 + \text{Copper loss}}$$

$$5000 = 4800 + 48 + \text{Copper loss}$$

Copper loss = 5000 – 4848 = 152 W

Copper loss = 152 W

Autotransformer:

$$\text{Copper loss} = \frac{152}{4} = 38 \text{ W}$$

Iron loss = 50 W

$$\therefore \quad \text{Efficiency} = \frac{5000}{5000 + 38 + 50} \times 100 = 98.27\%$$

EXAMPLE 7.40 A two-winding transformer is rated at 2300/230 V, 50 kVA. It is reconnected as a step-up autotransformer with 2300 V input. Calculate the rating of the autotransformer and the inductively and conductively transferred powers while delivering the rated output at upf.

Solution Given data

Two-winding transformer rating = 50 kVA

$V_1 = 2300$ V and $V_2 = 230$ V

The two-winding transformer is connected to an autotransformer

$V_{1auto} = 2300$ V

The inductively and conductively transferred powers at rated output at upf:

The rated primary current of two-winding transformer $= \frac{50000}{2300} = 21.74$ A

The rated secondary current of two-winding transformer $= \frac{50000}{2300} = 21.74$ A

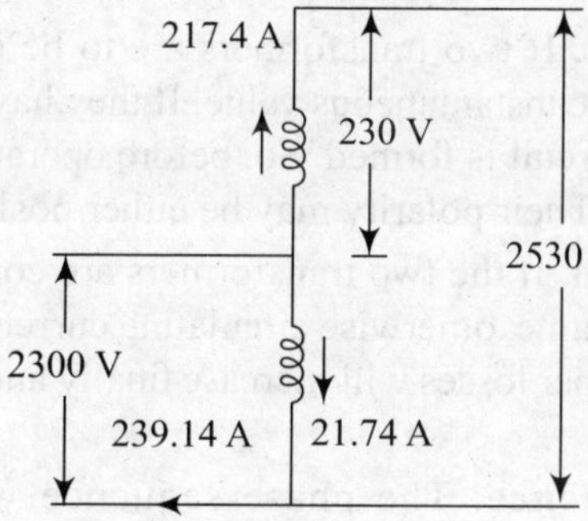

Figure 7.34

With required reconnection the 2300-V winding works as a common winding to input and the output current can be 217.4 A with voltage of 2530 V which means that the output of these autotransformer = 2530 × 217.4 = 550.022 kV

The corresponding input current $= 217.4 \times \frac{2530}{2300} = 239.14$ A

The ratio of turns in these cases is given by $K = \frac{2530}{2300} = 1.1$

With step-up job and $k = 1.1$,

$$\frac{\text{Rating of autotransformer}}{\text{Rating of two-winding transformer}} = \frac{K}{K-1} = \frac{1.1}{0.1} = 11$$

This gives the rating of autotransformer of 550 kVA.

At upf, the rated load = 550 W

Out of this, inductively transferred power = Power handled by the common winding

$= (2300 \times 21.74) \times 10^{-3} = 50.002$ kW

Rated output of two-winding transformer = 50.002 kW

The remaining power of 500 kW is conductively transferred as is clear from the division of current at the input mode out of the total current of 239.14 A from the source of 217.43 A goes straight to the output of the remaining current of 21.74 A is through the common and inductive path as marked in Figure 7.34.

7.15 PARALLEL OPERATION OF TRANSFORMERS

Sometimes the load on substation is more than the rating of the existing transformer. Here, one more transformer is needed to share the load which is connected in parallel so the operation of parallel transformers should have the following conditions:

(i) The same polarity
(ii) The same voltage ratio
(iii) The same phase sequence
(iv) Equal per unit impedances in magnitude and phase angle

(a) **Polarity test.** If two transformers are to be connected in parallel they should not have the same instantaneous value. If they have the same instantaneous value then a dead short circuit is formed. So, before operating the transformers in parallel check the polarity. Their polarity may be either positive or negative.

(b) **Voltage ratio.** If the two transformers are connected in parallel, their voltage ratio must be the same, otherwise circulating currents will exist. Once circulation currents exist the copper losses will increase finally and the efficiency of the transformer will decrease.

(c) **Phase sequence.** The phase sequence will be applicable for three-phase transformers.

(d) **Per unit impedances.** Transformers connected in parallel may not have the same kVA rating. But they will share the load on the ratio of their rating. If the primary and secondary sides of transformers are parallel with their per unit impedances then their voltage regulation will be same, otherwise it will differ.

If X/R ratios of two transformers are not equal, they would not operate at the same power factor.

7.15.1 Load Sharing in Equal kVA

Figure 7.35 shows the equivalent circuit for two transformers A and B which are connected in parallel. Transformer A supplies load current of I_A and transformer B supplies a current of I_B to load. Total load current is $I = I_A + I_B$ of impedance Z. Let the voltage across load is V. The secondary emfs are E_A and E_B with impedances Z_A and Z_B, respectively. If E_A and E_B are equal in magnitude and in phase, then there will not be any circulating current between the transformers.

Then

$$E_A = E_B$$
$$I_A Z_A = I_B Z_B$$

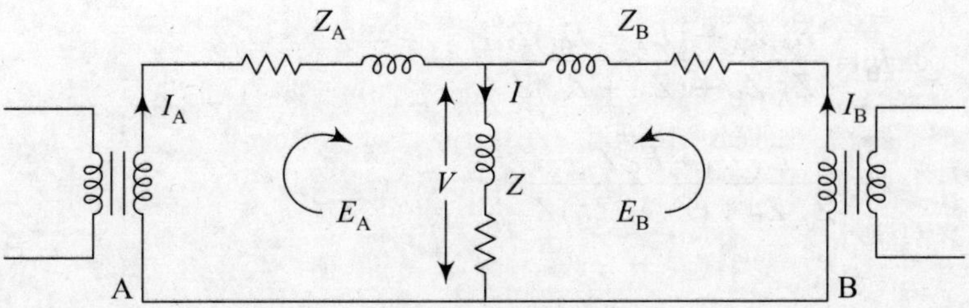

Figure 7.35

But

$$I = I_A + I_B$$

From the current division principle,

$$I_A = I \cdot \frac{Z_B}{Z_A + Z_B} \tag{7.47}$$

Similarly,

$$I_B = I \cdot \frac{Z_A}{Z_A + Z_B} \tag{7.48}$$

If Eqs. (7.47) and (7.48) are multiplied by V, then we get VA delivered by each transformer. VI_A is W_A, i.e., VA delivered by transformer A and VI_B is W_B, i.e., VA delivered by transformer B. Then VI is the total volt-ampere.

$$W_A = W \cdot \frac{Z_B}{Z_A + Z_B}$$

$$W_B = W \cdot \frac{Z_A}{Z_A + Z_B}$$

In the above equations, Z_A and Z_B can be expressed in ohms or in pu. From the above equations, it can be concluded that the transformers will share the load in proportion to their kVA rating. But leakage impedances are inversely proportional to their kVA rating.

7.15.2 Load Sharing in Unequal kVA

Applying the KVL to the left loop of Figure 7.35,

$$E_A = I_A Z_A + IZ$$
$$= I_A Z_A + (I_A + I_B)Z$$

Similarly,

$$E_B = I_B Z_B + IZ$$
$$= I_B Z_B + (I_A + I_B)Z$$

The solution for the above equation

$$I_A = \frac{E_A Z_B + (E_A - E_B)Z}{Z_A Z_B + (Z_A + Z_B)Z} \tag{7.49}$$

$$I_B = \frac{E_A Z_A - (E_A - E_B)Z}{Z_A Z_B + (Z_A + Z_B)Z} \tag{7.50}$$

$$I = \frac{E_A Z_B + E_B Z_A}{Z_A Z_B + (Z_A + Z_B)Z} \tag{7.51}$$

But

$$V = I.Z$$

$$= \frac{E_A Z_B + E_B Z_A}{\frac{Z_A Z_B}{Z} + (Z_A + Z_B)}$$

From the above equation, it can be observed that if E_A and E_B are not equal then the circulating current will exist between, and the current is given by

$$I = \frac{(E_A - E_B) Z}{Z_A Z_B + (Z_A + Z_B)Z} \tag{7.52}$$

7.16 SPECIAL TRANSFORMERS

These are classified as two types based on the application in electronics and telecommunications:

(i) Audio frequency transformers

(ii) Current transformers

7.16.1 Audio Frequency Transformers

These transformers are used in audio frequency ranges from 20 to 20,000 Hz. They are used in electronic and telecommunication (control systems and measurements are also used for boosting the voltage or impedance) circuits. Basically, they are small in size and have core losses.

The most important point in electronic circuit is distortion and it should be as low as possible. The input voltage should be amplified equally and the phase should be zero. For all frequencies, here two characteristics are important. One is amplitude frequency characteristics (ratio of output voltage to input versus frequency) and other is phase characteristics (i.e., phase angle between output voltages and input voltage versus frequency) and these are used for frequency response. Generally, a flat response is desirable over audio frequency response and this frequency response can be determined by exact equivalent circuit diagram.

At low frequencies the leakage reactances are not important but the shifting of magnetizing branch is important. The ratio of V_2/V_1 decreases with decrease in frequency, intermediate frequency range; none of the inductances is effective and the equivalent circuit reduces to purely resistive circuit. The ratio of V_2/V_1 is constant in this range. In high-frequency ranges the leakage reactance becomes more. Due to inductances and capacitances the ratio of V_2/V_1 decreases with increase in frequency, the response is shown in Figure 7.36.

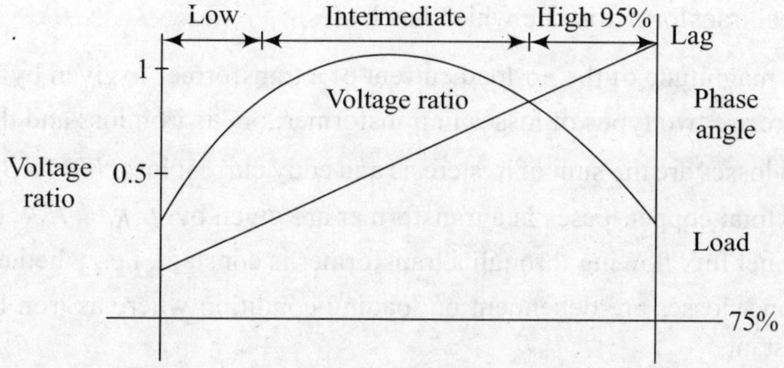

Figure 7.36

7.16.2 Current Transformers

The function of a current transformer is to step-down a high-magnitude alternating current. It can be easily measured by a low-range ammeter. The primary winding of a current transformer will consist of a few turns of thick wire and is connected in series with the load. The secondary winding has a large number of turns of thick wire and is usually in the range of 1–5 A. Therefore, a current transformer is a step-up transformer. Its secondary is always short circuited by a small resistance or relay coil.

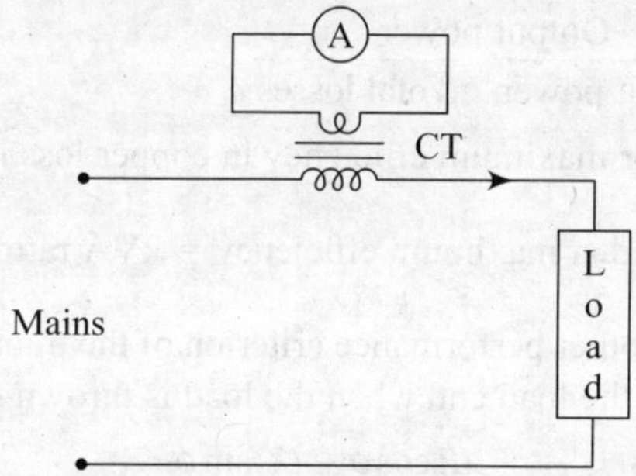

Figure 7.37 Current transformer

CHAPTER AT A GLANCE

- Transformer is a static device used for power transmission from one voltage level to another without changing the supply frequency.
- Transformer basically works on the principle of electromagnetic induction.
- It mainly consists of two windings and one core.
- These can be used at different applications such as power transformers, distribution transformers, and instrument transformers.
- The induced emf in primary winding is given by $E_1 = 4.44 f N_1 \phi_m$ volts.
- The induced emf in secondary winding is given by $E_2 = 4.44 f N_2 \phi_m$ volts.
- The transformation ratio is given by $N_2/N_1 = E_2/E_1 = I_1/I_2 = k$.
- An ideal transformer is one which does not have any losses.

- A real transformer is one which has losses.
- The magnitude of the no-load current of a transformer is given by $I_0 = \sqrt{I_w^2 + I_m^2}$.
- There are two types of losses in transformer: one is iron loss and the other is copper loss.
- Iron losses are the sum of hysteresis and eddy current loss, $W_h = kB_{max}^{\eta} fv + W_e = kB_{max}^2 f^2 t^2$.
- The total copper losses in a transformer are given by $I_1^2 R_1 + I_2^2 R_2$.
- The net flux flowing through a transformer is constant, i.e., whether it is loaded or unloaded.
- Copper losses are dependent on loading condition where as iron losses are independent and constant.
- Mathematically, an analysis can be made if we know the electrical equivalent circuit of the transformer.
- The total resistance referred to a primary is given by $R_{01} = R_1 + R_2' = R_1 + (R_2/K^2)$.
- The total resistance referred to a secondary is given by $R_{02} = R_2 + R_1' = R_2 + R_1 K^2$.
- The total reactance referred to a primary is given by $X_{01} = X_1 + X_2' = X_1 + (X_2/K^2)$.
- The rotal reactance referred to a secondary is given by $X_{02} = X_2 + X_1' = X_2 + X_1 K^2$.
- Without loading a transformer, the efficiency can be predetermined with open-circuit test and short-circuit test.
- The efficiency of a transformer is defined as the ratio of output to input in percent.

$$\eta = \frac{\text{Output power}}{\text{Output power + Total losses}}$$

- The condition for maximum efficiency in copper loss is equal to iron loss.
- The kVA supplied at maximum efficiency = kVA rating at fl $\sqrt{\frac{P_i}{P_{cu} \text{ at fl}}}$.
- Regulation is another performance criterion of the transformer. It is defined as the ratio of rise in the voltage at the load end when the load is thrown off to the full-load voltage.

$$\frac{{}_0V_2 - V_2}{V_2} \times 100 = \frac{IR\cos\phi + IX\sin\phi}{V_2} \times 100$$

- Regulation will be high for lagging power factor and low and sometimes even negative for leading power factors.
- All-day efficiency of a transformer is defined as the ratio of output energy to the input energy for a period of 24 h, η_{AD} = output in kWh/input in kWh.
- In autotransformer, only one winding is used in primary winding and secondary winding.
- The saving in the copper of autotransformer is $(1 - K)$ times the weight of copper in an ordinary transformer.
- When the load is heavy, then two or more transformers can be operated in parallel.
- While transformers are operating in parallel they must satisfy certain conditions: the polarity, voltage ratio, and phase sequence must be same.
- These can also be used as instrument transformers such as current transformers (CT) and potential transformers (PT).

MULTIPLE CHOICE QUESTIONS

1. The primary and secondary of a transformer are ____________ coupled.
   (a) electrically (b) magnetically
   (c) electrically and magnetically (d) none
2. If a transformer core has airgaps, then
   (a) reluctance of magnetic path is decreased
   (b) hysteresis loss is decreased
   (c) magnetizing current is greatly increased
   (d) eddy current is increased
3. A transformer is so designed that primary and secondary have
   (a) high leakage reactance (b) large resistance
   (c) tight magnetic coupling (d) good electric coupling
4. An ideal transformer is one which
   (a) has no losses and leakage reactance
   (b) does not work
   (c) has same number of primary and secondary turns
   (d) none of the above
5. A low-voltage outdoor lighting system uses a transformer that steps 120 V down to 24 V for safety. The equivalent resistance of all low-voltage lamps is 906 Ω.What is the current in the secondary coil. Assume the transformer is ideal and there are no losses.
   (a) 2 A (b) 2.5 A (c) 4.5 A (d) 1.5 A
6. The no-load ratio of a 50-Hz single-phase transformer is 6000/250 V. The maximum flux in the core is 0.06 Wb. What is the number of primary turns?
   (a) 450 (b) 900 (c) 350 (d) 210
7. Calculate the core area required for a 1600-kVA, 6600/250-V, 50-Hz single-phase core-type power transformer. Assume the maximum flux density of 1.2 Wb/m^2 and induced voltage per turn of 30 V.
   (a) 975 cm^2 (b) 1100 cm^2 (c) 1125 cm^2 (d) 1224 cm^2
8. An iron-cored transformer working at a maximum flux density of 0.8 Wb/m^2 is replaced by silicon steel working at a maximum flux density of 1.2 Wb/m^2. If the total flux is to remain unchanged, what is the reduction in volume expressed as a percentage of the original volume? The frequency and voltage per turn are the same.
   (a) 33% (b) 9% (c) 22% (d) 11%
9. A single-phase 50-Hz, core-type transformer has square cores of 20 cm side. The permissible flux density is 1 Wb/m^2. If the stacking factor is 0.9, the voltage induced per turn is
   (a) 6 (b) 11 (c) 8 (d) 14
10. The no-load pf of a transformer is small because
    (a) the iron loss component of I_0 is large
    (b) the magnetizing component of I_0 is large
    (c) the magnetizing component of I_0 is small
    (d) none of the above

11. A transformer takes a current of 0.6 A and absorbs 64 W when the primary is connected to its normal supply of 200 V, 50 Hz, the secondary being on open circuit. The iron loss current is

(a) 0.2 A (b) 0.43 A (c) 1 A (d) 0.32 A

12. The transformer operates poorly at very low frequencies because

(a) permeability of core is increased
(b) magnetizing current is abnormally high
(c) primary reactance is too much increased
(d) none of the above

13. The effect of the leakage flux in a transformer is to

(a) increase copper losses (b) decrease copper losses
(c) cause voltage drop in the windings (d) none of the above

14. Coils charge transformer are built up to nearly circular cross section in order to reduce

(a) leakage reactance (b) iron losses
(c) eddy current loss (d) copper loss

15. The open-circuit test on a transformer is always made on

(a) low-voltage winding (b) high-voltage winding
(c) either low- or high-voltage winding (d) none of the above

16. When the secondary of a transformer is short-circuited, the primary inductance

(a) is decreased (b) is increased
(c) remains unchanged (d) none of the above

17. When load on a transformer is increased, the eddy current

(a) is increased (b) remains unchanged
(c) is decreased (d) none of the above

18. When load on a transformer is decreased ______________ loss is decreased.

(a) eddy current (b) hysteresis (c) copper (d) friction

19. In a practical transformer, copper losses account for about ____________ of the total losses.

(a) 10% (b) 85% (c) 25% (d) 50%

20. The core-type transformer is generally suitable for

(a) high voltage and small output (b) high voltage and high output
(c) low voltage and high output (d) none of the above

21. A 10-Ω resistive load is to be impedance matched by a transformer to a source with 6250 Ω of internal resistance. The ratio of primary to secondary turns of transformer should be

(a) 10 (b) 15 (c) 20 (d) 25

22. A single-phase transformer with a ratio of 6600/600 V has a load impedance $(4 + j3)\ \Omega$ connected across the terminals of low-voltage winding. The power delivered to the load is

(a) 24.2×100 W (b) 57.6×1000 W
(c) 1.2×1000 W (d) 2.4×10000 W

23. A 10-kVA, 2000/400-V, single-phase transformer has a primary resistance and inductive reactance of 5 Ω and 12 Ω, respectively. The secondary values are 0.2 Ω and 0.48 Ω, respectively. The equivalent impedance of the transformer referred to the primary side is

(a) 12 Ω (b) 28 Ω (c) 31 Ω (d) 26 Ω

24. A 230/460-V transformer has a primary resistance of 0.2 Ω and a reactance of 0.5 Ω and the corresponding values for the secondary are 0.75 Ω and 1.8 Ω, respectively. The total voltage drop in the secondary when supplying 10 A at 0.8 pf lagging is

(a) 35.2 (b) 18.6 (c) 13.7 V (d) 24.2 V

25. A 2 kVA transformer has iron loss of 150 W and full load copper loss of 250 W. The maximum efficiency of the transformer would occur when the total loss is

(a) 400 W (b) 300 W (c) 600 W (d) 500 W

26. The primary and secondary winding of an ordinary transformer always have

(a) different number of turns (b) same size of copper wire
(c) a common magnetic circuit (d) separate magnetic circuits

27. Regarding the working component I_w and magnetizing component I_m of the no-load primary current of a transformer, which of the following statement is true?

(a) $I_w < I_m$ (b) $I_w > I_m$ (c) $I_w = I_m$ (d) I_w/I_m = constant

28. In performing a short-circuit test on a transformer, usually low-voltage side is short circuited because it has

(a) lower terminal voltage and higher current rating
(b) more number
(c) low insulation
(d) easy access

29. During short-circuit test, the iron loss of a transformer is negligible because

(a) the entire input is just sufficient to meet copper losses only
(b) the voltage applied across the HV side is a small fraction of the rated voltage and so is the flux
(c) the iron core becomes fully saturated
(d) the supply frequency is held constant

30. When a 400-Hz transformer is operated at 50 Hz its kVA rating is

(a) reduced to 1/8 (b) increased 8 times
(c) unaffected (d) determined by load on secondary

31. Under heavy loads, transformer efficiency is comparatively low because

(a) the voltage drops both in primary and secondary become large
(b) the secondary output is much less as compared to primary input
(c) the copper loss becomes high in proportion to the output
(d) the iron loss is increased considerably

32. For same kVA rating, an autotransformer has ______________ as compared to a conventional two-winding transformer.

(a) higher efficiency (b) lower core loss
(c) smaller size (d) all of the above

33. The iron loss of a transformer at 400 Hz is 10 W. Assuming that eddy current and hysteresis losses vary as the square of flux density, the iron loss of the transformer at rated voltage but at 50 Hz would be
(a) 80 (b) 640 (c) 1.25 (d) 100

34. The voltage applied to the HV side of a transformer during the short-circuit test is 2% of its rated voltage. The core loss will be ______ percent of the rated core loss.
(a) 4 (b) 0.4 (c) 0.25 (d) 0.04

35. Transformers are rated in kVA instead of kW because
(a) the load power factor is often not known
(b) kVA is fixed whereas kW depends on load pf
(c) the total transformer loss depends on volt-amperes
(d) it has become customary

36. An autotransformer having a transformation ratio of 0.8 supplies a load of 3 kW. The power transferred conductively from primary to secondary is ______ kW.
(a) 0.6 (b) 2.4 (c) 1.5 (d) 0.27

37. As compared to delta-todelta bank, the capacity of the voltage-to-voltage bank is ______ %.
(a) 57.7 (b) 2.4 (c) 50 (d) 86.6

38. For supplying a balanced three-phase load of 40 kVA, the rating of each transformer in voltage-to-voltage bank should be ______ kVA.
(a) 20 (b) 23 (c) 34 (d) 25

39. The percentage over the loading of two 20-kVA transformers when supplying a balanced three-phase load of 40 kVA in a voltage-to-voltage bank is ______ %.
(a) 15 (b) 57.7 (c) 73.2 (d) 33.3

40. The main reason why a voltage-to-boltage transformer cannot be paralleled with a delta-tostar transformer is that
(a) whereas one bank has low windings, the other has three windings
(b) there is 30°primary to secondary phases shift in the latter
(c) usually the kVA ratings of the two banks are different
(d) the ratio of line-toline voltages is different in the two banks

41. Leakage fluxes of a transformer may be minimized by
(a) keeping the magnetizing current to the minimum
(b) reducing the reluctance of the iron core to the minimum
(c) avoiding overloads
(d) sectionalizing and interleaving the primary and secondary windings

42. Three units of 1:5 transformers are connected in delta to star to supply a three-phase load from a 400-V, three-phase source. The line voltage on the load side is ______ volt.
(a) 2000 (b) 80 (c) 3464 (d) $80\sqrt{3}$

43. A 100-Ω load has a pf of 0.8 A; 1 kW generator supplies power to the load through a transformer. Best matching will be produced when the transformer turn ratio is approximately
(a) 3:1 (b) 8:1 (c) 10:1 (d) 2:1

44. An ideal transformer is one which
(a) has no losses and leakage reactance
(b) does not work
(c) has same number of primary and secondary turns
(d) none of the above

45. The most suitable material for transformer core is
(a) hot rolled grain-oriented steel (b) cold rolled grain-oriented steel
(c) cast steel (d) aluminum

46. The grain-oriented laminated sheet steel in transformer reduces the
(a) copper loss (b) eddy current loss
(c) hysteresis loss (d) none of the above

47. The ratio of weight of copper on an autotransformer to that on an ordinary two winding transformer is given as where N is turns ratio
(a) N (b) 1–N (c) N^2 (d) 1 + N

48. A step-up transformer increases
(a) voltage (b) current (c) power (d) both (a) and (b)

49. A transformer transforms
(a) frequency (b) voltage (c) current (d) both (b) and (c)

50. The rating of a transformer is usually measured in
(a) volts (b) amperes (c) kW (d) kVA

51. A transformer operates on the principle of
(a) self-inductance (b) mutual inductance
(c) both (d) none of the above

52. The emf equation of a transformer is
(a) 4.44f $N\phi_m$ (b) 4.44f $NB_m A$ (c) both (a) and (b) (d) none of the above

53. A transformer 2000 kVA, 250 Hz is operated at 50 Hz. Its kVA rating should be revised to
(a) 2000 kVA (b) 400 kVA (c) 10,000 kVA (d) none of the above

54. A 400-V/100-V, 10-kVA, two-winding transformer is reconnected as an autotransformer across a suitable voltage source, the maximum rating of such an arrangement could be
(a) 50 kVA (b) 15 kVA (c) 12.5 kVA (d) 8.75 kVA

55. Transformer cores are laminated in order to
(a) decrease iron loss (b) prevent eddy current loss
(c) eliminate magnetic hysteresis (d) none of the above

SHORT ANSWER QUESTIONS

1. How eddy current losses are minimized?
2. What is the main purpose of using core in a transformer?
3. When will the transformer get negative regulation?

4. How the copper loss varies with power factor?
5. What is the condition for getting maximum efficiency of transformer?
6. Does the transformer draw any current when its secondary is open circuited?
7. What will be the magnitude of the no-load primary current?
8. Define the regulation of transformer?
9. On what factors does the all-day efficiency of a transformer depend primarily?
10. What are the essential conditions for the parallel operation of a transformer?
11. What is the function of a transformer oil in a transformer?
12. What are the desirable properties of a transformer core material?
13. What is the function of breather in a transformer?
14. What do you mean by good voltage regulation of transformer?
15. Why transformer efficiency is high compared to other electrical devices?

REVIEW QUESTIONS

1. Explain the principle and operation of a transformer.
2. Explain the construction details of a transformer.
3. Deduce an emf equation of a transformer.
4. Explain different losses in a single-phase transformer.
5. Explain the transformer on no load with phasor diagram.
6. Explain transformer on load with phasor diagram.
7. Explain and draw the equivalent circuit of a transformer.
8. Explain the OC and SC test of a transformer.
9. Explain the operation of autotransformer.
10. Explain the parallel operation of the transformer.

EXERCISE PROBLEMS

1. A 3300/250-V, 50-Hz, single-phase transformer is built on a core having an effective cross-sectional area of 125 cm^2 and 70 turns on the low-voltage winding. Calculate the (a) maximum value of flux density and (b) number of turns on HV winding.
2. The no-load current of a transformer is 5.0 A at 0.3 pf lag when supplied at 230-V, 50-Hz supply. The number of turn on the primary winding is 200. Calculate the (a) maximum value of flux in the core, (b) core loss, and (c) magnetizing current.
3. A 220/110-V, 50-Hz, 1.5-kVA, has primary and secondary winding resistance of 1 Ω and 2 Ω, and reactance of 3 and 5, respectively, find the (a) total extra resistance equilibrium reactance and (b) equilibrium impedance referred by primary and secondary.
4. A 100-kVA, single-phase transformer with ratio of 10000/200 V, 50 Hz, requires 300 V at the HV side to circulate full-load current with LV winding shorted, and the corresponding intake is 1000 W. Calculate the percentage regulation and secondary terminal voltage on full load at 0.8 pf lagging.

5. In a 25-kVA, 2000/200-V transformer, the iron and copper losses at full load are 350 and 450 W, respectively. Calculate the efficiency at upf (a) full load, (b) half full-load, and (c) load for maximum efficiency.
6. Two single-phase transformers P and Q of equal voltage ratio are operated in parallel and they are supplying a load requiring 500 A at 0.8 pf lagging at a terminal voltage of 400 V. The equivalent impedances of the transformers as referred to the secondary windings are $(2 + j3)$ and $(3 + j5)$ Ω, respectively. Calculate the current supplied by each transformer.
7. An autotransformer supplies a load of 3 kW at 115 V at a upf. If the primary voltage is 230 V, determine the power transferred inductively to the load.

Chapter

8

DC Machines

Chapter Objectives

- To discuss the basic principles of dc generator and dc motor
- To explain the significance of back emf
- To discuss mechanical power developed by motor armature
- To explain types of dc motors

8.1 INTRODUCTION

The law of conservation of energy states that "energy can neither be created nor be destroyed, but can be transformed from one form to another form." If the electrical energy is converted into mechanical energy it is known as the *electromechanical energy conversion*. Based on the electromechanical energy conversion, electrical machines are of two types, namely, direct current machines and alternating current machines.

If the magnitude of the current or voltage is constant with respect to time and is unidirectional, it is called as *direct current* (dc), whereas if the magnitude of the current changes in both positive and negative directions with respect to time, it is called as *alternating current* (ac).

A dc machine can operate either as a dc motor or as a dc generator. If the electrical energy is converted into mechanical energy, it is treated as a motor, and if the mechanical energy is converted into electrical energy, it is treated as a generator. The concepts of the generator and motor can be observed from Faraday's laws of electromagnetic induction.

8.2 BASIC PRINCIPLES

8.2.1 Generator

Whenever a conductor cuts the magnetic field, an emf is induced in that conductor and the magnitude of the induced emf depends on the rate of change of flux ($d\phi/dt$). The direction of the induced emf is given by Fleming's right-hand rule.

We can achieve the rate of change of flux by the following two ways:

(i) By keeping the conductor stationary and rotating the magnetic field system; the induced emf thus produced is known as *statically induced emf.*

(ii) By keeping the field system stationary and rotating the conductor; the induced emf thus produced is known as *dynamically induced emf.*

Fleming's right-hand rule

According to Fleming's right-hand rule, if the forefinger, middle finger, and thumb of the right hand are stretched naturally perpendicular to each other as shown in Figure 8.1, the forefinger indicates the direction of the magnetic field, the middle finger indicates the direction of the induced emf, and the thumb indicates the direction of motion of the conductor.

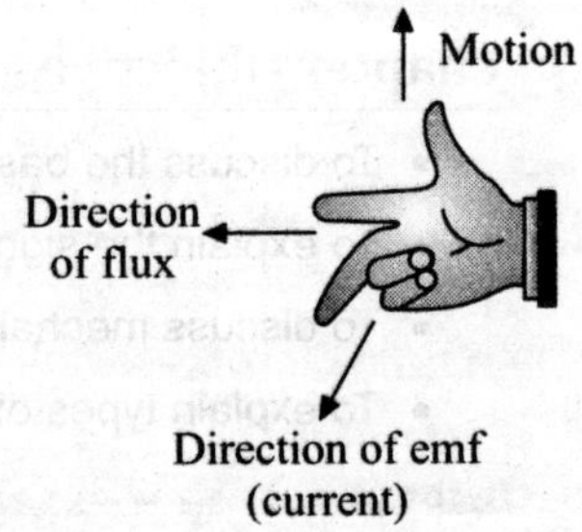

Figure 8.1 Fleming's right-hand rule

8.2.2 Motor

Whenever a current-carrying conductor is placed inside a magnetic field, a force is exerted by that conductor. The direction of the force is shown by Fleming's left-hand rule.

Fleming's left-hand rule

According to Fleming's left-hand rule, if the forefinger, middle finger, and thumb of the left hand are stretched naturally perpendicular to each other as shown in Figure 8.2, the forefinger indicates the direction of the magnetic field, the middle finger indicates the direction of the current, and the thumb indicates the direction of the force developed.

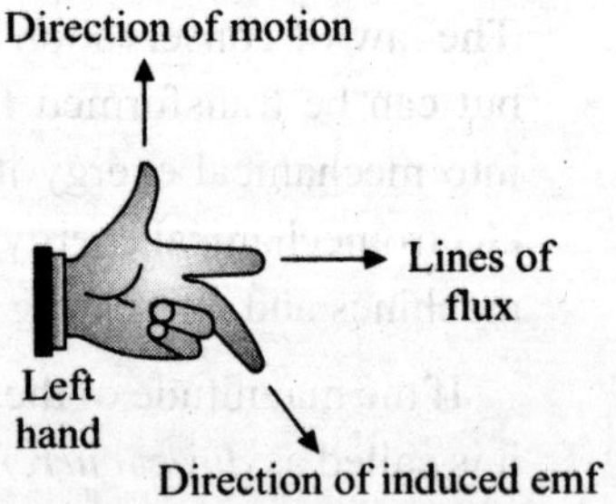

Figure 8.2 Fleming's left-hand rule

8.3 CONSTRUCTIONAL DETAILS OF DC MACHINE

A dc machine basically consists of two parts: the stationary part and the rotating part. The stationary part of the machine is known as stator, and it consists of field poles and outer frame called yoke. The rotating part of the machine is known as rotor, and it consists of armature, commutator, and shaft. The radial distance between these two is known as air gap and is normally very small.

On the basis of constructional view, the dc generator consists of three main parts: (i) magnetic field system, (ii) armature, and (iii) commutator and brushes. Figure 8.3 shows the cross-sectional view of a dc machine.

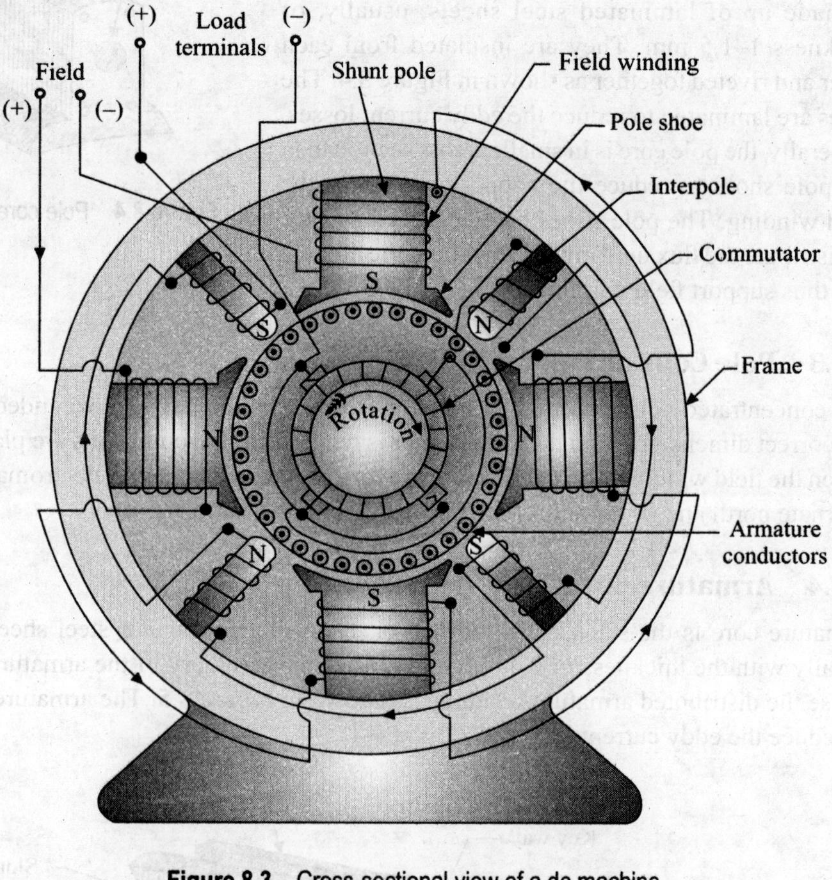

Figure 8.3 Cross-sectional view of a dc machine

8.3.1 Yoke

Yoke is the outer frame of the dc machine. It is a hollow cylinder made up of cast steel or rolled steel, and it is an unlamented ferromagnetic material. Generally, in small generators the yoke is made of cast iron. But for large machines, usually, cast steel or rolled steel is used. The following are the important functions of yoke:

(i) It provides the mechanical support for the poles.

(ii) It provides the return path for magnetic flux produced by the field system.

(iii) It acts as a protecting cover for the whole machine.

8.3.2 Pole Cores and Pole Shoes

The magnetic field system consists of pole cores and pole shoes. Pole cores, basically, carry field windings to produce the magnetic flux (ϕ). An even number of pole cores are bolted to the yoke. The pole core is made up of laminated steel sheets, usually, of thickness 1–1.5 mm. They are insulated from each other and riveted together as shown in Figure 8.4. The poles are laminated to reduce the eddy current losses. Generally, the pole core is in smaller cross section than the pole shoes to reduce the copper content for the field winding. The pole shoes have a curved surface to distribute the flux uniformly throughout the air gap, and thus support field winding.

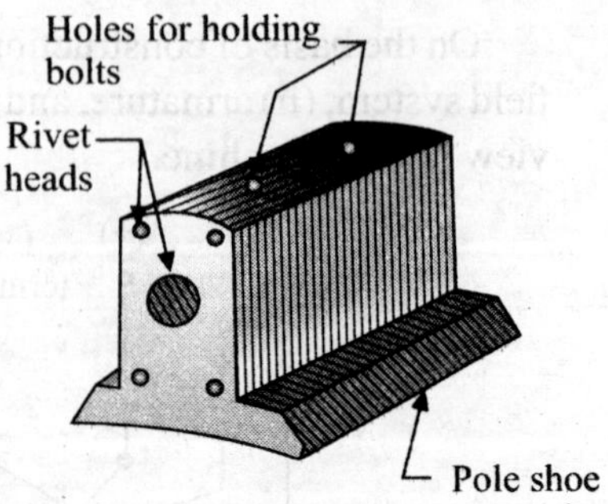

Figure 8.4 Pole core and pole shoes

8.3.3 Pole Coils or Field Windings

The concentrated field windings or pole coils consist of copper wire wounded on a former with the correct dimensions, and then the former is removed and wound coils are placed over the core. When the field windings are excited with dc current, they behave like electromagnets and produce alternate north and south poles (i.e., electro-polar constitution).

8.3.4 Armature Core

Armature core is the stack of cylindrical or drum-shaped circular steel sheet disc laminations usually with the thickness of 0.4–0.6 mm. The outer periphery of the armature core is slotted to house the distributed armature windings as shown in Figure 8.5. The armature core is laminated to reduce the eddy current loss.

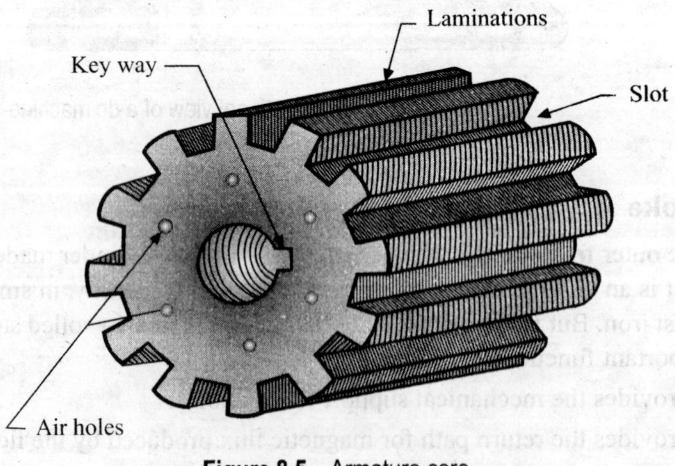

Figure 8.5 Armature core

8.3.5 Commutator

Commutator is a group of wedge-shaped copper segments. The function of the commutator is to facilitate the collection of current from armature coils and convert the alternating current induced in the armature into unidirectional current. It is of cylindrical shape made up of copper segments placed side by side to form a ring. These segments are insulated from each other with thin mica sheets. All the commutator segments are connected to the armature coils by means of copper lugs or strips. The construction of a commutator is shown in Figure 8.6.

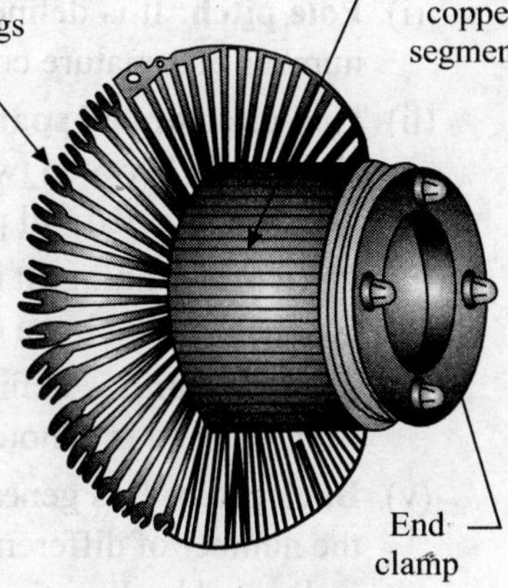

Figure 8.6 Commutator

8.3.6 Brush and Brush Holders

The brushes are made with graphite or composition of carbon and graphite to have good conductivity and low coefficient of friction to reduce the electrical loss and excessive wear, respectively. The brushes are placed in the brush holders with the required pressure to connect the commutator segments. The brushes collect the armature current from the commutator segments in case of a dc generator or feed into the armature winding in case of a dc motor. Figure 8.7 shows the arrangement of brush and brush holders.

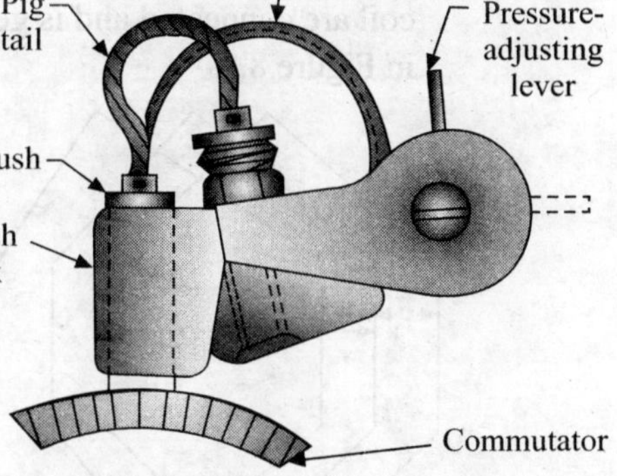

Figure 8.7 Brush and brush holders

8.3.7 Armature Windings

Armature winding is wound around the armature core through the peripheral slots of the armature core. While studying about armature winding we come across conductors, coil, and winding. There is a slight difference among these three terms.

Conductor is the active element which generates the emf, since it experiences magnetic field. A coil is a combination of two conductors, which is in the form of a ring. Similarly, a number of coils are present and are suitably connected which is termed as a winding.

The fashion or style in which these coils are connected is of two types: (i) lap winding and (ii) wave winding.

Another type of winding classification is done by the number of conductors or coils sides placed in the slots, namely, (i) single-layer winding in which only one conductor is placed in the armature slot and (ii) double-layer winding in which two conductors are housed in the armature slot.

To know about windings in detail, following terms used in connection with the windings are defined:

(i) **Pole pitch:** It is defined as the distance between two successive poles or is equal to the number of armature conductors per pole.

(ii) **Coil pitch or coil span:** It is the distance measured in terms of armature slots or armature conductors between two sides of a coil. If the coil pitch is equal to the pole pitch, then the winding is called full pitched.

(iii) **Pitch of a winding:** It is defined as the distance round the armature between two successive conductors which are directly connected together.

(iv) **Front pitch:** It is defined as the number of armature conductors spanned by a coil on the front end and is denoted by Y_F.

(v) **Back pitch:** It is generally measured in terms of the armature conductors and is equal to the number of difference of the conductors connected to a given signal of commutator. It is denoted by Y_B.

(vi) **Resultant pitch:** It is the distance between the beginning of one coil and that of the next coil to which it is connected. It is denoted by Y_R.

(vii) **Commutator pitch:** It is the distance between the segments to which the two ends of the coil are connected and is generally measured in terms of commutator segments as shown in Figure 8.8.

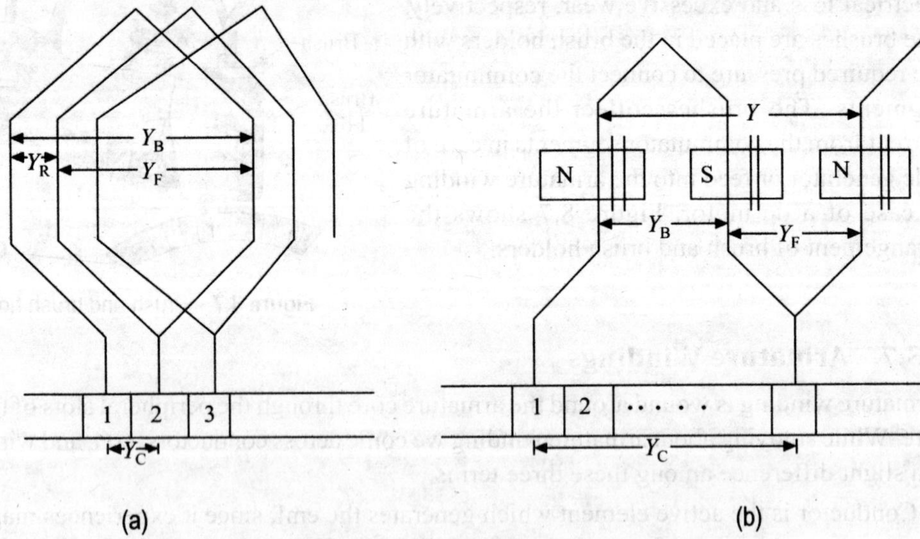

Figure 8.8 (a) Lap winding and (b) wave winding

Simple lap winding

In lap winding, the finishing end of each coil is connected to the starting end of the next coil. The windings may be progressive or retrogressive, that is, it may progress in the same direction as that of the coils are wound or in opposite direction.

The back pitch Y_B is an odd integer for this type of winding and is nearly equal to the coil sides per pole.

$$\text{Back pitch, } Y_B = \frac{Z}{p} \pm 1$$

$$\text{Average pitch, } Y_A = \frac{Y_B + Y_F}{2} = \frac{Z}{p}$$

Y_F is the front pitch; $Y_B > Y_F$ for progressive winding.

Simple wave winding

In a wave winding, the finishing end of a coil under a particular pole pair is connected to the starting of the coil, under the next pole pair.

$$\text{Average pitch, } Y_A = \frac{Z+2}{p}$$

Y_B and Y_F are odd integers and nearly equal to coil sides per pole. So,

$$Y_A = \frac{Y_B + Y_F}{2} = \frac{Z}{p}$$

Table 8.1 shows the differences between lap and wave windings

Table 8.1 Differences between lap and wave windings

S. No.	Description	Lap winding	Wave winding
1.	Coil span	$\frac{\text{No. of slots } (S)}{\text{Pole } (p)}$	$\frac{\text{No. of slots } (S)}{\text{Pole } (p)}$
2.	Parallel paths	No. of poles (p)	2
3.	Back pitch (Y_B)	$Y_B = \frac{2C}{p} + k$	$Y_B = \frac{2C}{p} \pm k$
4.	Winding pitch (Y)	$Y = Y_B - Y_F$	$Y = Y_B + Y_F$
5.	Average pitch (Y_A)	$Y_A = \frac{Y_B + Y_F}{2}$	$Y_A = \frac{Y_B + Y_F}{2}$
6.	Commutator pitch (Y_C)	$Y_C = \pm 1$	$Y_C = Y_A$
7.	Armature current	Conductor current × p	Conductor current × 2
8.	Dummy coils	Not needed	May be needed
9.	Equalizer rings	Needed	Not needed
10.	Applications	Low-voltage high-current machine	High-voltage low-current machine

8.4 OPERATING PRINCIPLE OF DC GENERATOR

For understanding the operation of a dc generator, let us consider a single-turn rectangular copper coil ABCD rotating about its own axis in a magnetic field. The two ends of the coil are joined to the two slip rings "a" and "b" as shown in Figure 8.9. These slip rings are insulated from each other. Two brushes "B_1" and "B_2" are placed such that they press against the slip rings. An external load resistance R_L is connected between the brushes.

Assume that the coil is rotating in a clockwise direction in the magnetic field system. So the magnitude of the flux linked with the coil changes; therefore, according to Faraday's laws of electromagnetic induction, an emf induced in the coil is proportional to the rate of change of flux linkage ($d\phi/dt$). The flux and rate of change of flux linkages with the coil for various positions analyzed for one complete revolution, that is, 0° to 360°, is described with Figure 8.9.

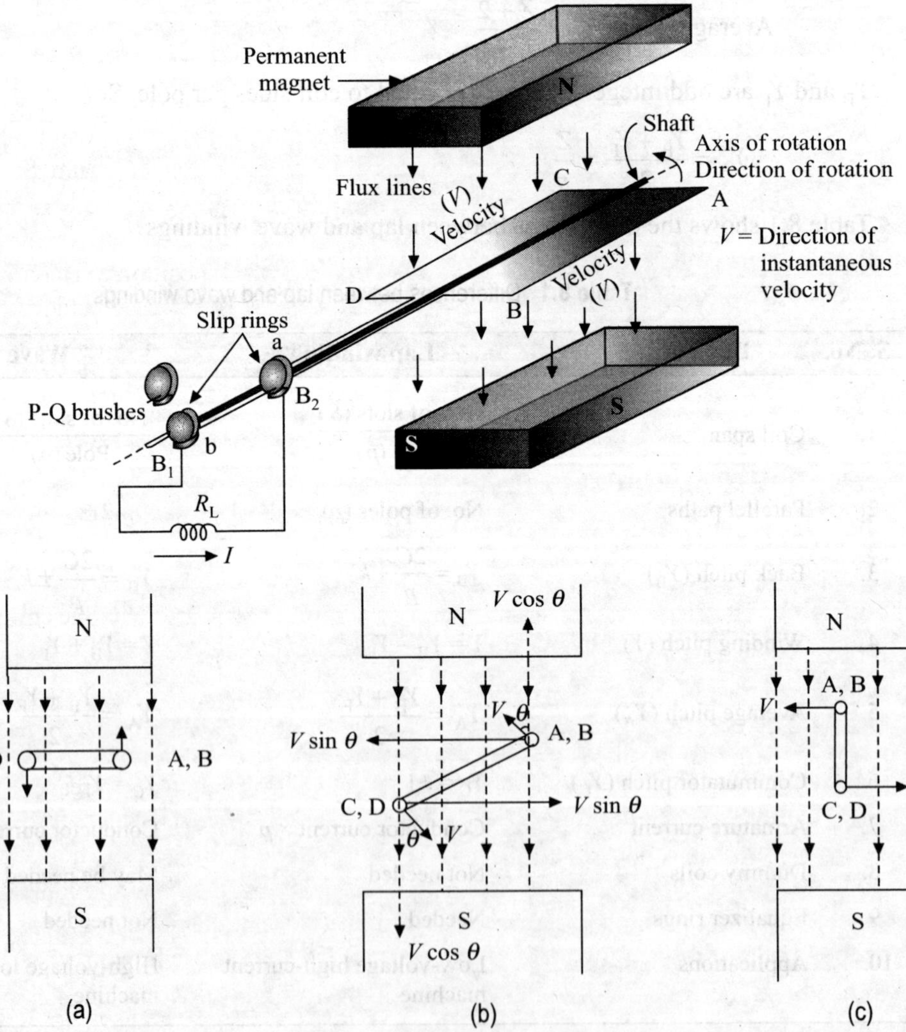

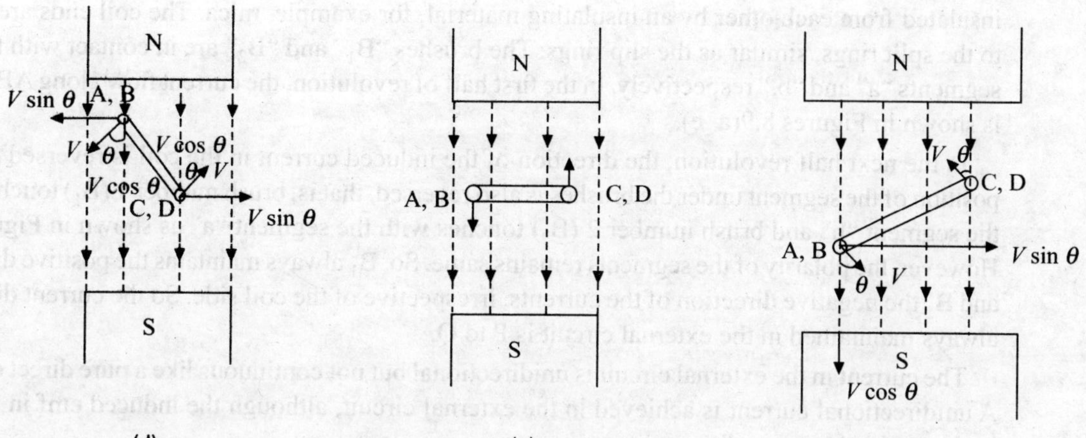

Figure 8.9 Single-turn generator

When the plane of the coil is at right angles to the lines of flux as in position 1, the flux linked with the coil is maximum, but the rate of change of flux linkage is minimum. So the induced emf at this position is zero. This is taken as the starting position. Here the angle of rotation is zero as shown in Figure 8.9(a).

As the coil continues to rotate for further 90°, it reaches to position 2. Here the rate of change of flux linkage increases gradually, so the induced emf also increases gradually. At position 2 the coil plane is horizontal to the lines of flux; the flux linked with the coil is minimum but the rate of change of flux linkages is maximum. The induced current flows from A to B and C to D as shown in Figure 8.9(b).

In the second quarter revolution, that is, from 90° to 180° till it reaches to position 3, the flux linked with the coil decreases gradually, but the rate of change of flux linkages decreases. Hence, the emf induced decreases up to zero as shown in Figure 8.9(c). By applying Fleming's right-hand rule, the direction of the current flow is ABPQCD as shown in Figure 8.9(c).

In the next quarter revolution, that is, from 180° to 270°, the coil plane is horizontal to the lines of flux. The coil sides AB and CD are interchanged. The flux linked with the coil is minimum and the rate of change of flux linkages is maximum. So the induced emf is maximum, and it increases from zero to maximum in the direction as shown in Figure 8.8(d). The direction of the induced current is from D to C and B to A as shown in Figure 8.9(d).

In the last quarter revolution, that is, from angle 270° to 360°, till position 1 is reached, the flux linked with the coil increases while the rate of change of flux linkages decreases to zero gradually. Hence, the emf induced is gradually decreased to zero as shown in Figure 8.9(e). The current flow is along DCQPBA, which is just the reverse of the previous direction of flow.

Hence, we see that the current obtained is reversed after every half revolution. This periodic reversal of current is known as alternating current. This alternating current not only reverses its direction but also does not maintain its magnitude constant while flowing in any one direction.

The flow of current is made unidirectional in the external circuit by replacing the slip rings as split rings. Split rings are conducting cylinders which are cut into two halves or segments and are insulated from each other by an insulating material, for example, mica. The coil ends are joined to the split rings, similar as the slip rings. The brushes "B_1" and "B_2" are in contact with the two segments "a" and "b," respectively. In the first half of revolution, the current flow along ABPQCD is shown in Figures 8.9(a–c).

In the next half revolution, the direction of the induced current in the coil is reversed and the position of the segment under the brushes is also reversed, that is, brush number 1 (B_1) touches with the segment "b" and brush number 2 (B_2) touches with the segment "a" as shown in Figure 8.9. However, the polarity of the segments remains same. So, B_1 always maintains the positive direction and B_2 the negative direction of the currents, irrespective of the coil side. So the current direction always maintained in the external circuit is P to Q.

The current in the external circuit is unidirectional but not continuous like a pure direct current. A unidirectional current is achieved in the external circuit, although the induced emf in the coil is alternating because of the split rings.

In practical machines, the number of coils increases, so, obviously, the number of split rings also increases. Hence, the split rings are replaced by a device called commutator.

8.5 EMF EQUATION OF DC GENERATOR

Let us consider a generator having "Z" number of conductors and the magnetic field system producing "ϕ" flux per pole and running at a speed of "N" revolutions per minute. So

ϕ = Flux per pole in webers
Z = Total number of conductors in armature
p = Number of poles
N = Speed of the armature in rpm
A = Number of parallel paths in armature winding
E_g = Induced emf

$$\text{Average emf generated per conductor} = N\frac{d\phi}{dt} \text{ volts} \tag{8.1}$$

$$= \frac{d\phi}{dt} \quad (\because N = 1) \tag{8.2}$$

Flux cut by conductor in one revolution = ϕp

$$\therefore \quad d\phi = \phi p \tag{8.3}$$

$$\text{Number of revolution per second} = \frac{N}{60}$$

$$\text{Time taken for one revolution} = \frac{60}{N}$$

$$\therefore \quad dt = \frac{60}{N} \text{ seconds}$$

$$E_g = \frac{d\phi}{dt} = \frac{\phi p}{60/N} = \frac{\phi NP}{60}$$

The induced emf or generated emf for "Z" number of conductors is the emf generated in any one of the armature parallel path.

$$\therefore \quad E_g = \frac{\phi ZNp}{60} \text{ volts} \tag{8.4}$$

If the dc generator is having "A" number of parallel paths, then

$$E_g = \frac{\phi ZNp}{60A} \text{ volts} \tag{8.5}$$

Case 1: If the dc generator is having wave-type winding in armature, then the number of parallel paths, $A = 2$.

$$E_g = \frac{\phi ZNp}{120} \text{ volts}$$

Case 2: If the dc generator is having lap-type winding in armature, then the number of parallel paths, $A = p$.

$$E_g = \frac{\phi ZNp}{60 \times p} = \frac{\phi ZN}{60} \text{ volts}$$

EXAMPLE 8.1 **A 4-pole wave-wound dc generator is having 50 slots with 20 conductors per slot and rotating at 1500 rpm. The flux per pole is 0.018 Wb. Calculate the emf generated.**

Solution Given data

Number of poles, $p = 4$
Flux per pole, $\phi = 0.018$ Wb
Speed of the armature, $N = 1500$ rpm
Number of slots = 50
Number of conductor per slot = 20
$\therefore$ Total number of conductors, $Z = 50 \times 20 = 1000$

Generated emf, $E_g = \dfrac{\phi ZNp}{60A}$

Since the machine is wave wound, hence the number of parallel paths $A = 2$.

$$\therefore \quad E_g = \frac{0.018 \times 1000 \times 1500 \times 4}{60 \times 2} = 900 \text{ V}$$

EXAMPLE 8.2 **A dc generator develops an emf of 220 V at 1200 rpm with a flux per pole 0.04 Wb. What will be the flux value to induce 240 V at 1400 rpm?**

Solution Given data

Case 1: Generated emf, $E_{g1} = 220$ V
Speed, $N_1 = 1200$ rpm
Flux per pole, $\phi_1 = 0.04$ Wb

Case 2: Generated emf, $E_{g2} = 240$ V
Speed of the rotor, $N_2 = 1400$ rpm
Flux per pole, $\phi_2 = ?$

We know that generated emf, $E_g = \dfrac{\phi Z N p}{60A}$ (where Z, p, and A are constants).

$\therefore \quad E_g \propto \phi N$

Case 1: $E_{g1} = 220$ V (Given)

$N_1 = 1200$ rpm

$\phi_1 = 0.04$ Wb

Case 2: $E_{g2} = 240$ V (Given)

$N_2 = 1400$ rpm

$$\frac{E_1}{E_2} = \frac{\phi_1 N_1}{\phi_2 N_2}$$

$$\frac{220}{240} = \frac{0.04 \times 1200}{\phi_2 \times 1400}$$

$$\phi_2 \times 1400 = 0.04 \times 1200 \times \frac{240}{220}$$

$$\phi_2 = \frac{52.36}{1400} = 0.037 \text{ Wb}$$

Therefore, the flux required is 0.037 Wb.

EXAMPLE 8.3 A 40-kW, 280-V dc shunt generator has armature and field resistances of 0.04 Ω and 120 Ω, respectively. Calculate the total power developed by the armature when it delivers full output power.

Solution Given data

Power output of the generator, $P = 40$ kW

Terminal voltage of the generator = 280 V

Armature resistance of the generator, $R_a = 0.04\ \Omega$

Shunt field resistance of the generator, $R_{sh} = 120\ \Omega$

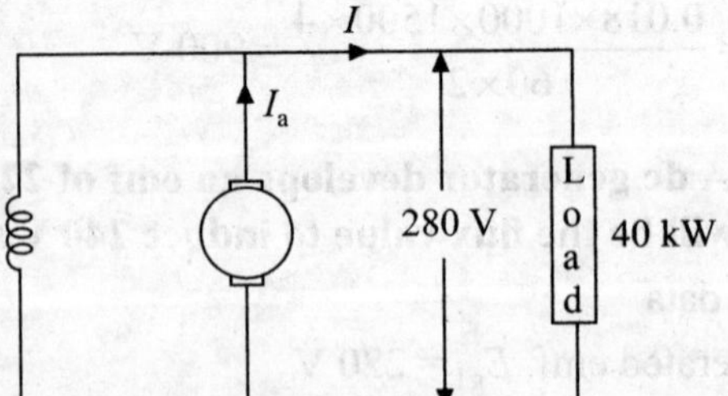

Figure 8.10 Shunt generator

Load current, $I_L = \dfrac{P}{V} = \dfrac{40000}{280} = 142.85$ A

Shunt field current, $I_{sh} = 280/120 = 2.3$ A

From Figure 8.10, $I_a = I_L + I_{sh}$

The voltage drop across armature = $I_a R_a$

$$I_a R_a = 145.15 \times 0.04 = 5.80 \text{ V}$$

Generated voltage, $E_g = V + I_a R_a = 280 + 5.80 = 285.80$ V

Power developed in armature is $E_g I_a$

$$E_g I_a = 285.80 \times \frac{145.15}{1000} = 41.48 \text{ kW}$$

EXAMPLE 8.4 **Calculate the generated emf of a 4-pole, wave-wound armature having 38 slots with 18 conductors per slot when driven at 1000 rpm. The flux per pole is 0.018 Wb.**

Solution Given data

Number of poles, $p = 4$

$A = 2$ (Since it is wave winding)

Number of slots = 38

Number of conductors per slot = 18

Speed of the rotor, $N = 1000$ rpm

Flux per pole, $\phi = 0.018$ Wb

Generated emf = ?

$$\text{Generated emf, } E_g = \frac{\phi ZN}{60} \times \left(\frac{p}{A}\right) \text{ volts}$$

Total number of conductors, Z = Number of slots × Number of conductors per slot

$$= 38 \times 18 = 684$$

The induced emf in the generator, $E_g = 410.4$ V

EXAMPLE 8.5 **A 6-pole lap-wound dc generator has 400 conductors on its armature. The flux per pole is 0.02 Wb. Calculate**

(i) The speed at which the generator must be run to generate 400 V.

(ii) The speed if the generator were wave wound to generate 200 V.

Solution Given data

Number of poles, $p = 6$

$A = p = 6$ (Since it is lap winding)

Number of conductors, $Z = 400$

Flux per pole, $\phi = 0.02$ Wb

(i) $E_g = 400$ V (Given)

$$\text{Generated emf, } E_g = \frac{\phi ZN}{60} \times \left(\frac{p}{A}\right) \text{ volts}$$

$$400 = \frac{0.02 \times 400 \times N}{60} \times \frac{6}{6}$$

$N = 3000$ rpm

(ii) $E_g = 220$ V (Given)

$$\text{Generated emf, } E_g = \phi \frac{ZN}{60} \times \left(\frac{p}{A}\right) \text{ volts}$$

$$200 = \frac{0.02 \times 400 \times N}{60} \times \frac{6}{6}$$

$N = 1500$ rpm

EXAMPLE 8.6 **A dc generator develops an emf of 200 V when driven at 900 rpm with a flux per pole of 0.04 Wb. It is desired that this emf be increased to 250 V at 1000 rpm. What should be the value of flux per pole under new circumstances?**

Solution Given data

Case 1: Generated emf, $E_{g1} = 200$ V

Speed of the generator, $N_1 = 900$ rpm

Flux per pole, $\phi_1 = 0.04$ Wb

Case 2: Generated emf, $E_{g2} = 250$ V

Speed of the generator, $N_2 = 1000$ rpm

Flux per pole, $\phi_2 = ?$

$$\text{Generated emf, } E_g = \frac{\phi ZN}{60}\left(\frac{p}{A}\right) \text{ volts}$$

where, Z, p, and A are constants.

So, $E_g \propto \phi N$

In the first case,

$200 \propto 0.04 \times 900$

In the second case,

$250 \propto \phi \times 1000$

$$\frac{250}{200} = \frac{1000\phi}{900 \times 0.04}$$

$\phi = 0.045$ Wb

The required flux per pole to develop 250 V at 1000 rpm is 0.045 Wb.

EXAMPLE 8.7 **A 4-pole lap-wound dc shunt generator has a useful flux per pole of 0.06 Wb. The armature winding consists of 240 turns each of 0.0035 Ω resistance. Calculate the terminal voltage when running at 920 rpm, if the armature current is 60 A.**

Solution Given data

Number of poles, $p = 4$

$A = p = 4$ (Since the type of winding is lap winding)

Flux per pole, $\phi = 0.06$ Wb

Number of turns = 240

Resistance of each turn = 0.0035 Ω

Speed of the armature, $N = 920$ rpm

Armature current, $I_a = 60$ A

Number of conductors, Z = 240 × 2 = 480 (Since each turn has two conductor sides)

Generated emf, $E_g = \frac{\phi ZN}{60} \times \left(\frac{p}{A}\right)$ volts

$$= \frac{0.06 \times 480 \times 920}{60}\left(\frac{4}{A}\right)$$

$$= 441.6 \text{ V}$$

Total resistance of 240 turns or 480 conductors = 240 × 0.0035 = 0.84 Ω

There are four parallel paths in armature due to the lap winding.

∴ Resistance of each path $= \frac{0.84}{4} = 0.21\ \Omega$

Armature resistance = R_a

$$\frac{1}{R_a} = \frac{1}{0.21} + \frac{1}{0.21} + \frac{1}{0.21} + \frac{1}{0.21}$$

$$R_a = 0.0525\ \Omega$$

Armature voltage drop = $I_a R_a = 60 \times 0.0525 = 3.15$ V

Terminal voltage, $V = E_g - I_a R_a = 441.6 - 3.15 = 438.45$ V

Example 8.8 The resistance of the field circuit of a shunt-excited dc generator is 200 Ω when the output of the generator is 120 kW, the terminal voltage is 525 V, and the generated emf is 550 V. Calculate the (i) armature resistance and (ii) generated emf if the terminal voltage is 540 V at 80 kW of the output.

Solution Given data

Case 1:

Shunt field resistance, $R_{sh} = 200\ \Omega$

Power output of the generator, $P = 120$ kW

Terminal voltage of the generator, $V = 525$ V

Generated emf, $E_{g1} = 550$ V

Case 2:

Power output, $P = 80$ kW

Terminal voltage, $V = 540$ V

Generated emf, $E_{g2} = ?$

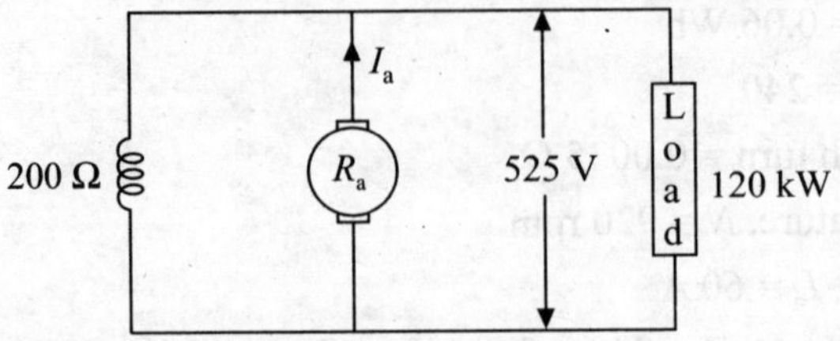

Figure 8.11 A dc shunt generator

(i) Output current, $I_L = P/V = 120000/525 = 228.57$ A

Shunt current, $I_{sh} = V/R_{sh} = 525/200 = 2.625$ A

Armature current, $I_a = I_L + I_{sh} = 231.19$ A

Armature voltage drop due to armature resistance

$$= I_a R_a = 550 - 525 = 25 \text{ V}$$

or

$$231.19 R_a = 25 \text{ V}$$

Armature resistance, $R_a = 0.108\ \Omega$

(ii) Output current = 80000/540

= 148.14 A

$$I_{sh} = 540/200 = 2.7 \text{ A}$$

But $I_a = I_L + I_{sh}$ 148.14 + 2.7 = 150.84 A

$$I_a R_a = 150.84 \times 0.108 = 16.29 \text{ V}$$

∴ Generated emf, $E_g = V + I_a R_a$

$= 540 + 16.29 = 556.29$ V

EXAMPLE 8.9 **The armature of a 4-pole wave-wound dc generator is required to generate an emf of 480 V on open circuit when revolving at a speed of 620 rpm. Calculate the magnetic flux per pole required. The armature has 160 slots with 2 coil sides per slot and each coil consisting of 3 turns. The armature is wave wound.**

Solution Given data

Number of poles, $p = 4$

Generated emf, $E_g = 480$ V

Speed of the generator, $N = 620$ rpm

$A = 2$ (Since it is wave winding)

Number of slots = 160

Number of coil sides per slot = 2

Number of turns per coil side = 3

Number of conductors, $Z = 160 \times 2 \times 3 = 960$

Induced emf = E_g

$$\text{Generated emf, } E_g = \frac{\phi ZN}{60}\left(\frac{p}{A}\right) \text{ volts}$$

$$480 = \frac{\phi \times 960 \times 620}{60} \times \frac{4}{2}$$

$$\phi = 0.024 \text{ Wb}$$

Flux required per pole, ϕ =0.024 Wb

8.6 ARMATURE REACTION AND COMMUTATION

The generated emf in the air gap fluxes is weakened if the direction of the armature flux, due to the armature reaction, is in the same direction as that of the main field flux, it is termed as cross magnetization. In this case the armature of a dc generator causes the current to flow in the distributed armature winding. This then produces the flux in armature. This armature flux disturbs the main field flux. This effect is called *armature reaction.*

If the direction of the armature flux, due to the armature reaction, opposes the main flux, it is termed as *demagnetizing effect*. In this case the air gap fluxes are strengthened.

8.6.1 Armature Reaction

The effect of armature flux on the main field flux is called the armature reaction. Consider that the armature conductors are not carrying any current but the field winding is excited, which sets up main flux, as shown in Figure 8.12. The magnetic neutral axis (MNA) coincides with the geometrical neutral axis (GNA). Now consider that the armature conductors are only carrying current (i.e., field excitation is zero) as shown in Figure 8.13.

For the assumed direction of the rotation, the armature conductors set up flux whose direction is as shown in Figure 8.13 and is shown by a vector OF_A which is parallel to the brush axis.

But under actual load condition, both the main mmf and armature mmf will be acting simultaneously in the generator as shown in Figure 8.14.

The result is that the flux through the armature is not uniform and symmetrical about the pole axis, rather it is distorted. It also appears that the flux is crowded at trailing pole tips and weakened at leading pole tips.

The resultant mmf can be found by considering both the mmfs (i.e., the armature mmf and the main field mmf). The resultant mmf OF is found vectorially equal to OF_A and OF_m. The position of MNA which is always perpendicular to the resultant mmf OF is shown in Figure 8.14. Since the MNA is shifted through an angle θ, the brushes also need to be shifted through the same value, in order to align themselves with the MNA. Due to this brush shift, the armature mmf gets redistributed, that is, some of the conductors, which are under the influence of the south pole previously, come under the influence of the north pole and vice versa.

This regrouping is shown in Figure 8.15. Now the armature mmf, which is represented by OF_A will not be horizontal but is inclined by an angle θ to the left. This can now be resolved into

two rectangular components: OF_d parallel to the polar axis and OF_c perpendicular to the polar axis.

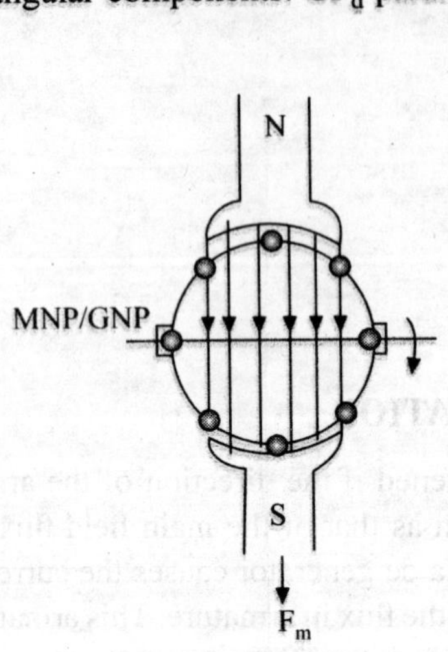

Figure 8.12

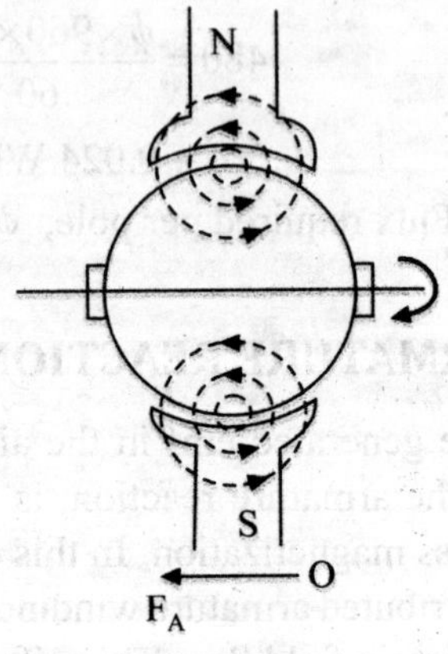

Figure 8.13

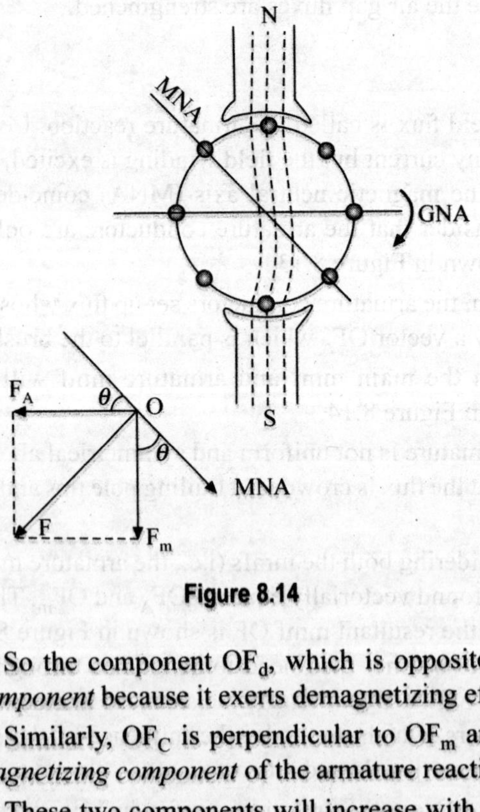

Figure 8.14

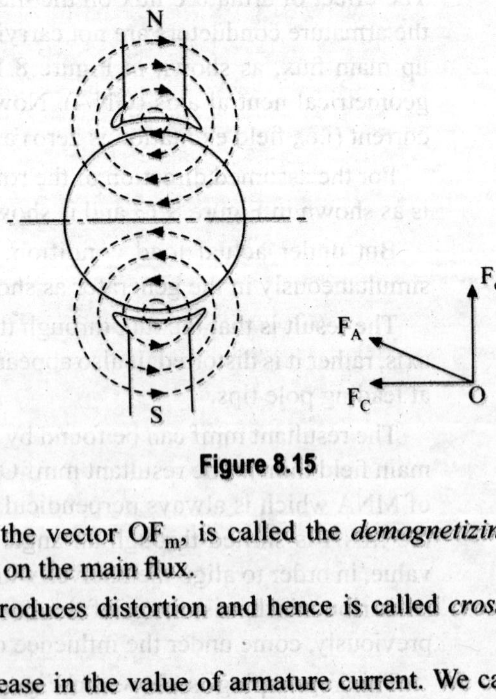

Figure 8.15

So the component OF_d, which is opposite to the vector OF_m, is called the *demagnetizing component* because it exerts demagnetizing effect on the main flux.

Similarly, OF_C is perpendicular to OF_m and produces distortion and hence is called *cross-magnetizing component* of the armature reaction.

These two components will increase with increase in the value of armature current. We can calculate the demagnetizing ampere-turns per pole by the following methods:

The total number of armature conductors in angles AOC and BOD = $\frac{4\theta_m}{360}Z$

As the two conductors constitute one turn,

∴ Total number of turns in these angles = $\frac{2\theta_m}{360}Z$

Demagnetization ampere-turns per pair of poles = $\frac{2\theta_m}{360}ZI$

Demagnetization ampere-turns per pole, $AT_d = \frac{\theta_m}{360}ZI$

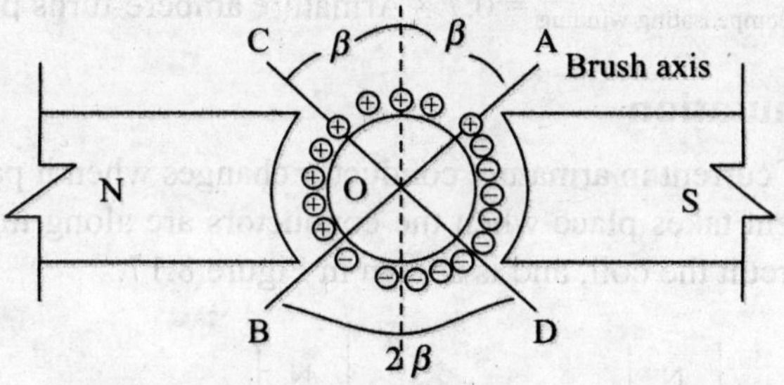

Figure 8.16 Demagnetizing ampere-turns per pole

For the conductors which are responsible for cross-magnetizing, ampere-turns are between BOC and AOD as shown in Figure 8.16. Therefore, the cross-magnetizing ampere-turns per pole can be calculated by subtracting the demagnetizing ampere-turns per pole from the total armature ampere-turns per pole.

Total armature ampere-turns per pole $= \frac{ZI}{2p}$

Total armature conductors per pole = $\frac{Z}{P}$

Demagnetizing ampere-turns per pole, $AT_d = \frac{\theta_m}{360}Z\,I$

Demagnetizing conductors per pole $= \frac{2\theta_m}{360}Z$

Cross-magnetizing conductors per pole = $AT_c = \frac{Z}{P}\frac{2\theta_m}{360}Z$

Cross-magnetizing ampere-turns conductors per pole = $ZI\left[\frac{1}{p} - \frac{2\theta_m}{360}\right]$

Cross-magnetizing ampere-turns per pole = $ZI\left[\frac{1}{2p} - \frac{\theta_m}{360}\right]$

$$\frac{AT_c}{\text{Pole}} = ZI\left[\frac{1}{2p} - \frac{\theta_m}{360}\right]$$

Compensation windings

For large dc machines, subject to large fluctuations, the compensating windings are used. Their function is to neutralize the cross-magnetizing effect of armature reaction. The windings are embedded in the pole shoe slots and are connected in series with the armature winding in such a way that the current in these windings will be opposite to that flowing in armature conductors under the pole shoe. Because of the cost and additional space taken by this winding, it is provided on large direct current machines which are subject to violent fluctuations in loads.

Number of ampere-turns per pole for compensations winding is given by (pole arc per pole pitch) armature ampere-turns per pole

$$AT_{\text{compensating winding}} = 0.7 \times \text{Armature ampere-turns per pole}$$

8.6.2 Commutation

The direction of current in armature conductor changes when it passes under a brush, that is, the reversal of current takes place when the conductors are along the brush axis, that is, when the brushes short circuit the coil, and is shown in Figure 8.17.

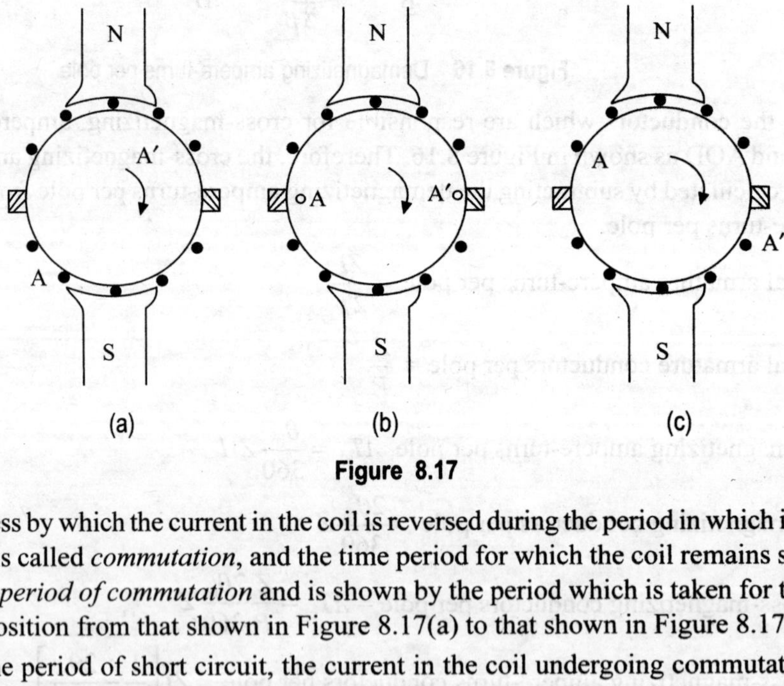

Figure 8.17

The process by which the current in the coil is reversed during the period in which it passes under the brushes is called *commutation*, and the time period for which the coil remains short circuited is called the *period of commutation* and is shown by the period which is taken for the coil *AA′* to change its position from that shown in Figure 8.17(a) to that shown in Figure 8.17(c).

During the period of short circuit, the current in the coil undergoing commutation should be reversed to its full value as shown graphically by line "a" in Figure 8.18 (for a value $+I$ to a value $-I$). If, however, this change does not take place completely, the difference current jumps from the commutator bar to brush, thereby causing sparking.

The failure of the coil to get the current reversed to its full value during commutation is due to the production of the self-induced emf in the coil undergoing commutation. This induced emf is called the "reactance voltage" and opposes the reversal of current. To avoid

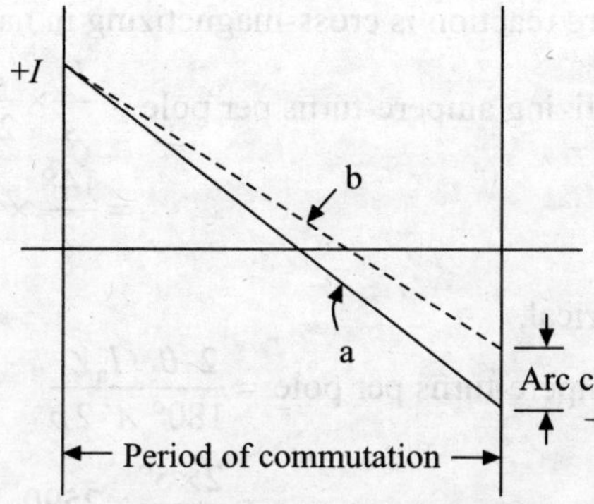

Figure 8.18 Commutation period

the reactance voltage being induced in the coils undergoing commutation, the brushes are to be shifted to the MNA. The amount of brush shift depends on the load delivered by the armature. Since each time shifting the brushes to the MNA under load (which varies) is not practicable, the brushes are fixed at the GNA, and following suitable methods are derived:

(i) Use of high brush contact resistance.

(ii) Shifting of brushes to the new MNA.

(iii) Use of commutating poles (also known as compoles or interpoles).

The polarity of an interpole in a dc generator must be the same as that of the main pole immediately ahead.

(iv) Use of compensating winding.

EXAMPLE 8.10 **A 6-pole, 145-A dc shunt generator has 420 conductors and is wave wound. Its field current is 3 A. Find the demagnetizing and cross-magnetizing ampere-turns per pole at full load, if**

(i) Brushes are on GNA

(ii) Brushes are shifted from GNA by 5° electrical

(iii) Brushes are shifted from GNA by 5° mechanical

Solution Given data

Number of poles, $p = 6$

Line current, $I_L = 145$ A

Total number of conductors, $Z = 420$

Number of parallel paths, $A = 2$ (Since the winding is of wave type)

Field current, $I_{sh} = 3$ A

Armature current, $I_a = I_L + I_{sh} = 145 + 3 = 148$ A

(i) With brushes on GNA, $\theta = 0$; therefore, demagnetizing ampere-turns per pole = 0

The entire armature reaction is cross-magnetizing in nature.

$\therefore$ Cross-magnetizing ampere-turns per pole $= \dfrac{I_a}{A} \times \dfrac{Z}{2p}$

$= \dfrac{148}{2} \times \dfrac{420}{2 \times 6} = 2590$ AT/pole

(ii) Here, $\theta = 5°$ electrical,

Demagnetizing ampere-turns per pole $= \dfrac{2 \cdot \theta}{180°} \dfrac{I_a Z}{A \cdot 2p}$

$= \dfrac{2 \times 5°}{180°} \times 2590$

$= 143.88$ AT/pole

Cross-magnetizing ampere-turns per pole $= 2590 - 143.88$

$= 2446.11$ AT/pole

(iii) Here, $\theta_{\text{elect}} = \dfrac{P}{2} \theta_{\text{mech}}$

$\theta = \dfrac{6}{2} \times 5° = 15°$ electrical

Demagnetizing ampere-turns per pole $= \dfrac{2 \times 15°}{180°} \times 2590$

$= 431.66$ AT/pole

Cross-magnetizing ampere-turns per pole $= 2590 - 431.66 = 2158$ AT/pole

EXAMPLE 8.11 A 225-kW, 400-V, 6-pole dc generator has 700 lap-wound conductors. It is given a brush lead of 2.0 angular degrees (mechanical). From the geometric neutral, calculate the cross-magnetizing and demagnetizing ampere-turns per pole. Neglect the shunt field current.

Solution Given data

Output power, $P = 225$ kW
Terminal voltage, $V = 400$ V
Number of poles, $p = 6$
Total number of conductors, $Z = 700$
Number of parallel paths, $A = p = 6$
Brush are lead by an angle $= \theta_{\text{mech}} = 2°$
AT_d and $AT_c = ?$

Armature current, $I_a = \dfrac{225 \times 10^3}{400} = 562.5$ A

Conductor current, $I_c = \dfrac{562.5}{6} = 93.75$ A

Total armature ampere-turns, $AT_a = \dfrac{1}{2}\left(\dfrac{700 \times 93.75}{6}\right) = 5468$ AT/pole

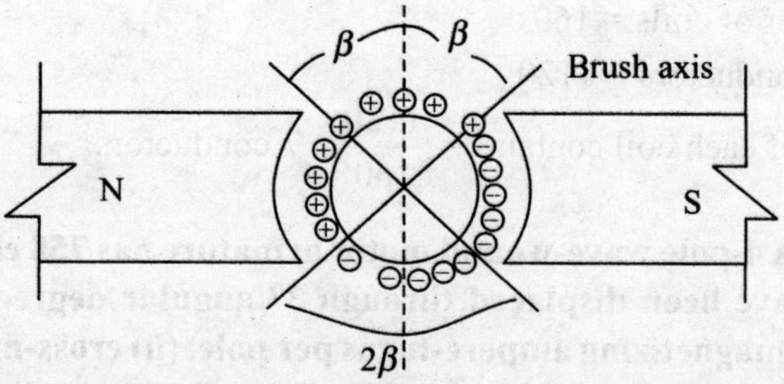

Figure 8.19 Demagnetization ampere-turns and cross-magnetization ampere-turns

It is easily observed that in 180° electrical conductors in the belt of 2β electrical degrees are demagnetizing, where β is the brush shift in electrical degrees.

$$\beta = 2° \left(\frac{6}{2}\right) = 6° \text{ electrical } \beta$$

$$\text{Cross-magnetizing ampere-turns} = \left(1 - \frac{2\times 6°}{180°}\right) = 5103$$

$$\text{Demagnetizing ampere-turns} = 5468 \times \frac{2\times 6°}{180°}$$

$$= 364.5 \text{ AT/pole}$$

EXAMPLE 8.12 **A 6-pole generator has an output of 200 A at 500 V. The lap-connected armature has 1120 conductors and 160 commutator segments. If the brushes are advanced by 3 segments from no-load neutral axis, estimate demagnetizing and cross-magnetizing ampere-turns per pole.**

Solution Given data

Number of poles, $p = 6$
Output current, $I_L = 200$ A
Terminal voltage, $V = 500$ V
Number of conductors, $Z = 1120$
Number of commutator segments = 160
Brushes are advanced by 3 segments.
Current per path, $I = 200/6 = 33.33$

$$\theta_m = 3 \times 360/160 = 6.75°$$

$$\text{Demagnetizing ampere-turns per pole} = ZI\frac{\theta_m}{360°}$$

$$= \frac{1120\times 33.33\times 6.75°}{360°} = 700 \text{ AT/pole}$$

$$\text{Cross-magnetizing ampere-turns per pole} = ZI\left(\frac{1}{2p} - \frac{\theta_m}{360°}\right)$$

$$= 1120\times 33.33\left(\frac{1}{2\times 6} - \frac{6.75°}{360°}\right)$$

$$= 2410 \text{ AT/pole}$$

Note: Number of coils = 160

Number of conductors = 1120

Hence, side of each coil contains = $\frac{1120}{160} = 7$ conductors

EXAMPLE 8.13 A 4-pole wave-wound motor armature has 750 conductors and delivers 100 A. The brushes have been displaced through 3° angular degrees from the geometrical axis. Calculate (i) demagnetizing ampere-turns per pole, (ii) cross-magnetizing ampere-turns per pole, and (iii) the additional shunt field current for neutralizing the demagnetizing if the field winding has 1000 turns/pole.

Solution Given data

Number of poles, $p = 4$

Number of conductors, $Z = 750$

Number of parallel paths, $A = 2$ (Since winding is of wave type)

Armature current, $I_a = 100$ A

Brush displaced by an angle $\theta_m = 3°$

Armature current per path, $I = \frac{100}{2} = 50$ A

(i) Demagnetizing, $AT_d = 750 \times 50 \times \frac{3°}{360°} = 313$ AT

(ii) Cross-magnetizing, $AT_c = 750 \times 50 \left(\frac{1}{2 \times 4} - \frac{3°}{360°} \right) = 4375$ AT

Total armature ampere-turns per pole = 750 × 50/4 = 9375 AT

Therefore, the cross-magnetizing ampere-turns per pole,

$AT_c = 9375 - 313 = 9062$ AT

(iii) An extra shunt field turns = AT_d/I_{sh}

Therefore, an additional field current required for neutralizing the effect of demagnetization

$$I_{sh} = \frac{AT_d}{\text{Shunt field turns}}$$

$$I_{sh} = \frac{313}{1000} = 0.313 \text{ A}$$

EXAMPLE 8.14 A 4-pole, wave-wound 220-V dc shunt generator has a full-load current of 45 A and a shunt field current of 3 A. Find the demagnetizing and cross-magnetizing ampere-turns per pole if the brushes are advanced through 3-communtator segments at full-load. There are 120 segments in the commutator of the machine. Assume the machine has single turn coils.

Solution Given data

Number of poles, $p = 4$

Number of parallel paths, $A = 2$ (Since winding is of wave-type)
Terminal voltage, $V = 220$ V
Load current, $I_L = 45$ A
Field current, $I_f = 3$ A
Number of commutator segments = 120
The machine has single turn coil, therefore,
Brushes are advanced by = 3 commutator segments

$$\theta_m = \frac{360° \times 3°}{120} = 9°$$

Number of conductors = 2 × Number of commutator segments
Since every commutator segments will have two conductors, therefore,

$$Z = 2 \times 120 = 240$$

Armature current, $I_a = I_L + I_f = 45 + 3 = 48$ A

Armature current per path, $I = \frac{I_a}{A}$

$$I = \frac{48}{2} = 24 \text{ A}$$

Number of demagnetization ampere-turns per pole,

$$AT_d/\text{pole} = Z \times I \times \frac{\theta_m}{360°}$$

$$= 240 \times 24 \times \frac{9°}{360°}$$

$$= 144 \text{ AT/pole}$$

Similarly, cross-magnetization ampere-turns per pole, $AT_c = ZI\left(\frac{1}{Zp} - \frac{\theta_m}{360°}\right)$

$$= 240 \times 24\left(\frac{1}{2 \times 4} - \frac{9°}{360°}\right)$$

$$= 576 \text{ AT/pole}$$

EXAMPLE 8.15 Determine the demagnetizing and cross-magnetizing ampere-turns per pole and series turns required to balance the demagnetizing ampere-turns of a dc generator having the following data: 450 conductors, total current 150 A, 4 poles, 2 circuit wave winding, angle of lead = 7°, leakage coefficient = 1.5.

Solution Given data

Number of conductors, $Z = 450$
Armature current, $I_a = 150$ amps
Number of parallel paths, $A = 2$ (Since the winding is of wave type)
Brush are lead by an angle, $\theta_m = 7°$
Armature current per path, $I = 150/2 = 75$ A

(i) Cross-magnetizing ampere-turns per pole

$$AT_c/\text{pole} = ZI\left(\frac{1}{2p} - \frac{\theta_m}{360°}\right)$$

$$= 450 \times 75\left(\frac{1}{2 \times 4} - \frac{7°}{360°}\right)$$

$$= 3562.5 \text{ AT/pole}$$

(ii) Demagnetizing ampere-turns per pole,

$$AT_d/\text{pole} = \frac{Z \times I \times \theta_m}{360}$$

$$AT_d/\text{pole} = 450 \times 75 \times \frac{7°}{360°}$$

$$= 656.25$$

(iii) Series turns required to balance the demagnetizing ampere-turns,

$$= \lambda \frac{AT_d}{i_a}$$

$$= 1.5 \times \frac{656.25}{150}$$

$$= 6.56$$

Series turns required to balance the demagnetizing ampere turns are 7 turns.

EXAMPLE 8.16 **A 350-V, 900-A lap-wound dc machine has 6 poles and 850 armature conductors. Calculate the number of conductors in the pole face to give full compensation if the pole face covers 75% of pole span.**

Solution Given data

Terminal voltage, $V = 350$ V

Armature current, $I_a = 900$ A

Number of poles, $p = 6$

Number of conductors, $Z = 850$

Number of parallel paths, $A = p = 6$ (Since lap winding)

Ampere-turns per pole for compensating winding = AT_{cw}

Current in each armature conductor, $I = 900/6 = 150$ A

$$AT_{cw} = \text{Armature ampere-turns per pole} \times \frac{\text{Pole area}}{\text{Pole pitch}}$$

$$= 0.75 \times \frac{ZI}{2p}$$

Ampere-turns per pole for compensating winding $AT_{cw} = 0.75 \times \frac{850 \times 150}{2 \times 6} = 7968.75$ AT

8.7 TYPES OF GENERATOR

The dc generators are broadly classified into two groups in which the field system is excited:

(i) Separately excited generators
(ii) Self-excited generators

8.7.1 Separately Excited Generators

Separately excited generators are those whose field system is energized from an external independent source as shown in Figure 8.20.

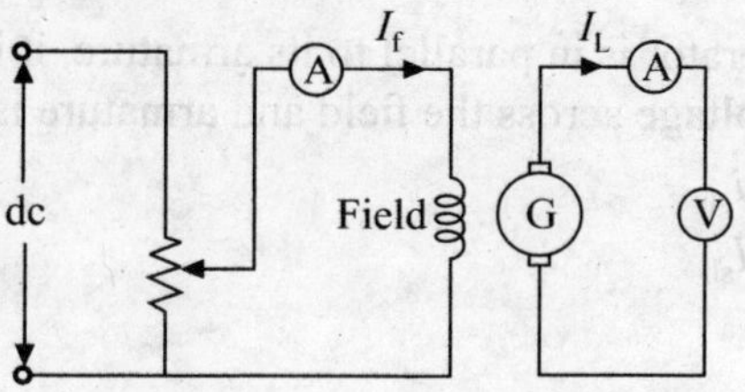

Figure 8.20 Separately excited generator

8.7.2 Self-Excited Generators

Self-excited generators are those whose field system is energized by the current produced by the generator themselves. By the property of magnets, the field system retains the flux even if the magnetizing source is removed away. This small retained flux is known as the *residual flux*. This residual flux causes the induced emf in the armature (nearly 3 V). So there is a small circulating current in the armature, which produces the corresponding flux. This flux is in phase with the main flux, hence the resultant flux will increase. This increase in flux causes an increase in the induced emf in the armature, which will again cause more circulating current. This process continues till the field system gets saturated. At saturation point, the rated voltage is induced. This process is known as *voltage build up* in dc machine.

According to the arrangement of the field with respect to the armature, the self-excited generators are classified into

(i) Series generator
(ii) Shunt generator
(iii) Compound generator
 (a) Long-shunt generator and (b) Short-shunt generator

Series generator

In series generator, the field winding is in series with the armature as shown in Figure 8.21. The performance equations are

$$I_{se} = I_a = I_L \qquad (\because \text{ series}) \tag{8.6}$$

$$E_g = V + I_a (R_a + R_{se}) \tag{8.7}$$

There exists a drop in the brush contact, so the voltage equation becomes

$$E_g = V + I_a (R_a + R_{se}) + V_{bc} \tag{8.8}$$

where V_{bc} is the voltage drop across the brushes.

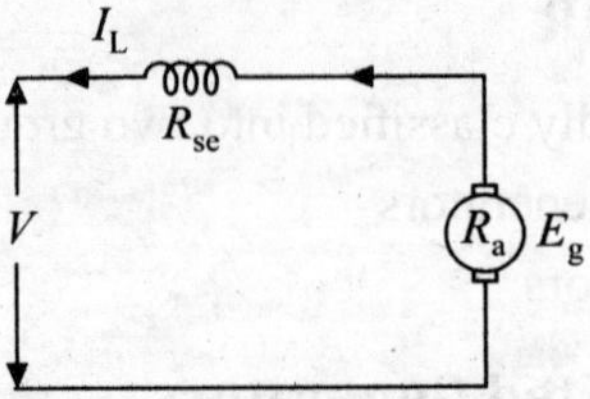

Figure 8.21 Series generator

Shunt generator

If the field of the generator is in parallel to its armature, it is known as shunt generator as shown in Figure 8.22. The voltage across the field and armature is the same due to shunt arrangement.

$$I_a = I_L + I_{sh} \tag{8.9}$$

$$I_L = I_a - I_{sh}$$

$$I_{sh} = \frac{V}{R_{sh}} \tag{8.10}$$

$$E_g = V + I_a R_a \tag{8.11}$$

There exists a drop in the brush contact, so the voltage equation becomes

$$E_g = V + I_a R_a + V_{bc} \tag{8.12}$$

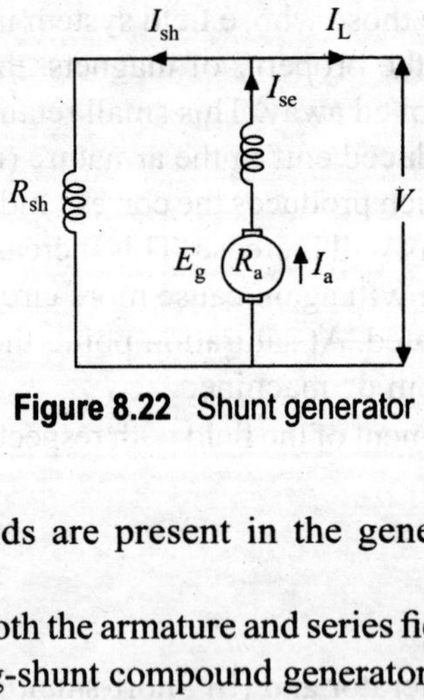

Figure 8.22 Shunt generator

Compound generator

If both the shunt and series fields are present in the generator, it is known as compound generator.

(i) If the shunt field is across both the armature and series field winding, this type of compound generator is termed as long-shunt compound generator as shown in Figure 8.23(a).

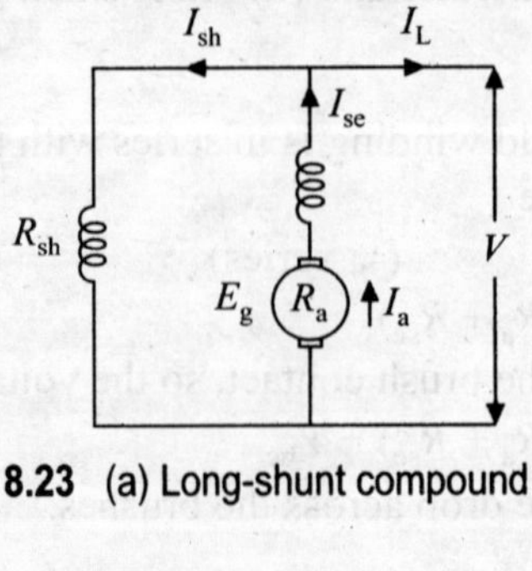

Figure 8.23 (a) Long-shunt compound generator

Performance equations,

$$I_a = I_{se} \tag{8.13}$$

$$I_a = I_{se} = I_{sh} + I_L \tag{8.14}$$

$$I_{sh} = \frac{V}{R_{sh}}$$

$$I_L = I_a - I_{sh} \quad \text{or} \quad I_L = I_{se} - I_{sh}$$

$$V + I_a R_a + I_{se} R_{se} = E_g$$

$$V + I_a (R_a + R_{se}) = E_g \tag{8.15}$$

There exists a drop in the brush contact, so the voltage equation becomes

$$V + I_a (R_a + R_{se}) + V_{bc} = E_g \tag{8.16}$$

(ii) If the shunt field winding is only parallel to the armature, then the compound generator is termed as short-shunt compound generator as shown in Figure 8.23(b).

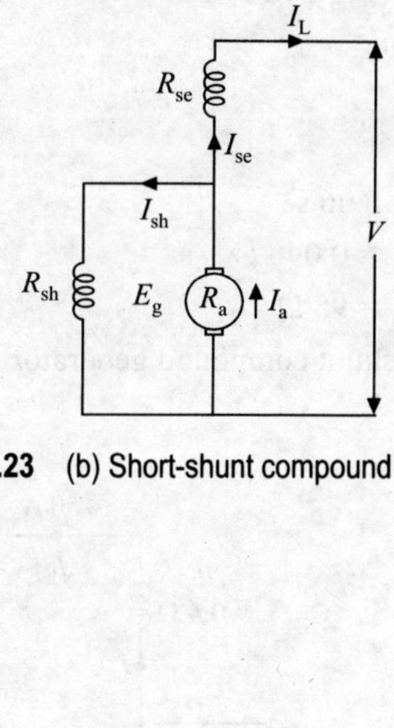

Figure 8.23 (b) Short-shunt compound generator

Performance equations,

$$I_L = I_{se}$$

$$I_a = I_L + I_{sh}$$

$$I_L = I_a - I_{sh}$$

$$I_{sh} = \frac{V - I_{se} R_{se}}{R_{sh}}$$

$$V + I_a R_a + I_{se} R_{se} = E_g \tag{8.17}$$

There exists a drop in the brush contact, so the voltage equation becomes

$$V + I_a R_a + I_{se} R_{se} + V_{bc} = E_g \tag{8.18}$$

In case the shunt field winding aids the series field winding, it is known as *cumulative compound generator*, whereas if the shunt field winding and the series field winding oppose each other, it is known as *differential compound generator*.

8.7.3 Applications of DC Generators

(i) **Separately excited generators:** These are generally used in special applications such as electroplating and electrorefining of materials.

(ii) **Shunt generators:** Commonly used in battery charging and for ordinary lighting purposes.

(iii) **Series generators:** These are commonly used as boosters on dc feeders and as a constant current generators for welding and arc lamps.

(iv) **Cumulative compound generator:** These are used for domestic lighting purposes and to transmit energy over long distance.

(v) **Differential compound generator:** The use of this type of generators is very rare and is used for special application such as electric arc welding.

EXAMPLE 8.17 A short-shunt compound generator delivers 200 V, 50 A to load. The resistance of armature, series field, and shunt field are 0.06 Ω, 0.04 Ω, and 95 Ω, respectively. Calculate the voltage generated allowing a brush of 2 V.

Solution Given data

Terminal voltage = 200 V

Load current, $I_L = 50$ A

Armature resistance, $R_a = 0.06\ \Omega$

Series field resistance, $R_{se} = 0.04\ \Omega$

Shunt field resistance, $R_{sh} = 95\ \Omega$

Type of generator: short-shunt compound generator

Brush drop = 2 V

$E_g = ?$

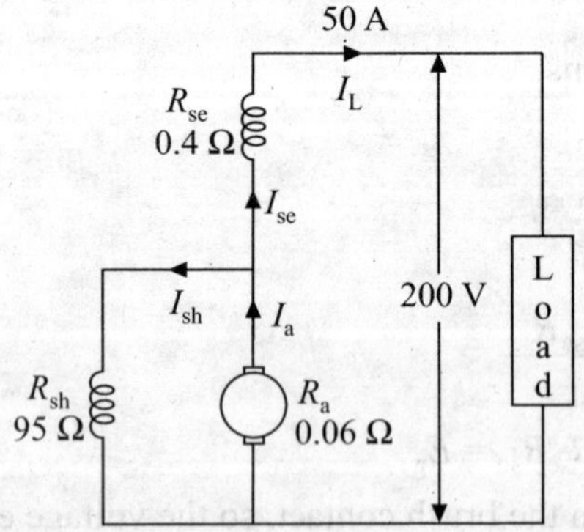

Figure 8.24 Short-shunt compound generator

Voltage across shunt coil = 200 – 2.0 = 198 V

$$\text{Shunt current, } I_{sh} = \frac{198}{95} = 2.08 \text{ A}$$

Armature current, $I_a = I_L + I_{sh} = 50 + 2.08 = 52.08$ A

Armature drop, $R_a I_a = 52.08 \times 0.06 = 3.125$ V

Generated voltage, $E_g = V + R_a I_a + R_{se} I_{se} +$ brush drop

$= 200 + 3.125 + 2 + 2.0 = 207.12$ V

EXAMPLE 8.18 A dc shunt generator supplies a load of 9 kW at 220 V. Calculate the induced emf if the armature resistance is 0.8 Ω and the field resistance is 70 Ω.

Solution: Given data

Output of generator = 9 kW

Terminal voltage, $V = 220$ V

Armature resistance, $R_a = 0.8\ \Omega$

Field resistance, $R_{sh} = 70\ \Omega$

$E_g = ?$

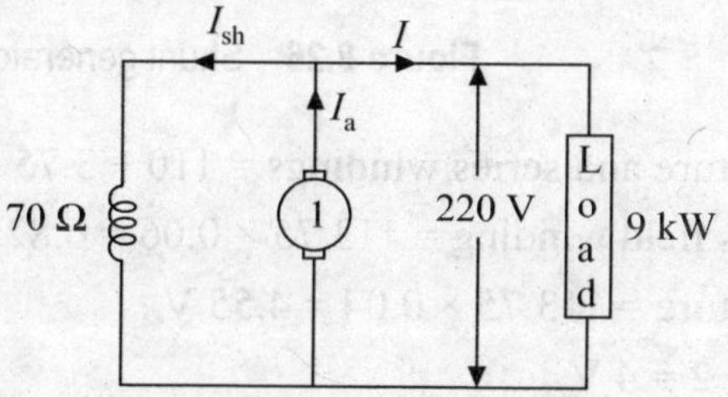

Figure 8.25 Shunt generator

Load current, $I = 9 \times 1000/220$

$= 40.9$ A

Shunt current, $I_{sh} = V/R_{sh} = 220/70 = 3.142$ A

Armature current, $I_a = I + I_{sh} = 40.90 + 3.142 = 44.05$ A

Armature voltage drop, $I_a R_a = 44.4 \times 0.8 = 35.27$ V

Emf generated in armature, E_g = Terminal voltage + Armature drop

$= 220 + 35.232 = 255.24$ V

EXAMPLE 8.19 A 4-pole long-shunt compound generator supplies 110 A at terminal voltage of 450 V. Its armature resistance is 0.04 Ω, the series field resistance is 0.06 Ω, and the shunt field resistance is 120 Ω. Find the generated emf. Take drop per brush as 2 V. Ignore armature reaction.

Solution Given data

Number of poles, $p = 4$

Load current, $I_L = 110$ A

Armature resistance, $R_a = 0.04\ \Omega$

Series field resistance, $R_{se} = 0.06\ \Omega$

Shunt field resistance, $R_{sh} = 120\ \Omega$

Drop per brush = 2 V

Brush drop = 2 × 2 = 4 V

$I_a = I_{se}$

$I_{se} = 110 + 3.75 = 113.75$ A

Shunt current, $I_{sh} = 450/120 = 3.75$ A

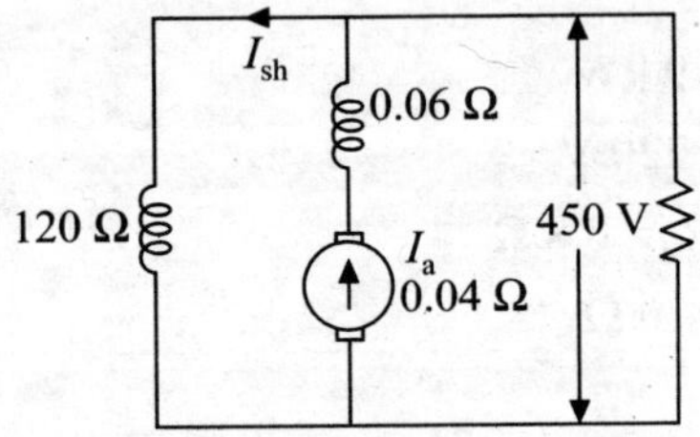

Figure 8.26 Shunt generator

Current through armature and series windings = 110 + 3.75 = 113.75 A

Voltage drop on series field winding = 113.75 × 0.06 = 6.825 V

Voltage drop on armature = 113.75 × 0.04 = 4.55 V

Drop at brushes = 2 × 2 = 4 V

Generated voltage, $E_g = V + I_{se}R_{se} + I_aR_a$ + brush drop

$= 450 + 6.845 + 4.563 + 4 = 465.375$ V

EXAMPLE 8.20 A short-shunt cumulative compound dc generator supplies 9 kW at 230 V. The shunt field, series field, and armature resistances are 110, 0.5, and 0.6 Ω, respectively. Calculate the induced emf and the load resistance.

Solution Given data

Output power, $P = 9$ kW

Terminal voltage, $V = 230$ V

Armature resistance, $R_a = 0.6\ \Omega$

Series field resistance, $R_{se} = 0.5\ \Omega$

Shunt field resistance, $R_{sh} = 110\ \Omega$

$E_g = ?$

Load current, $I = 9 \times 1000/230 = 39.13$ A

Voltage drop in series winding

$= 0.5 \times 39.13 = 19.56$ V

Voltage across shunt winding

$= 230 - 19.56 = 210.44$ V

Shunt current, $I_{sh} = 210.44/110 = 1.913$ A

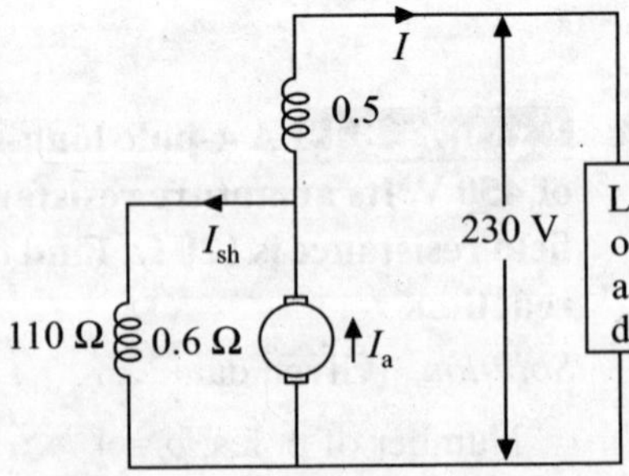

Figure 8.27 Short-shunt generator

Armature current, $I_a = I_{sh} + I = 1.913 + 39.13 = 41.043$ A

Armature drop, $I_aR_a = 41.04 \times 0.6 = 24.62$ V

Induced emf = 230 + 24.62 + 19.56 = 274.18 V

If "R" is load resistance,

We know $P = \dfrac{V^2}{R}$

$$R = \frac{(230)^2}{9\times10^3} = 5.87\ \Omega$$

EXAMPLE 8.21 **A 4-pole dc shunt generator with a wave-wound armature having 420 conductors has to supply a load of 500 lamps each of 75 W at 250 V. Allowing 8 V for the voltage drop in the connecting leads between the generators, and the generator and the load and a contact drop at 2 volts per brush, calculate the speed at which the generator should be driven. The flux per pole is 30 mWb and the armature and shunt field resistances are 0.065 Ω and 70 Ω, respectively.**

Solution Given data

Number of poles, $p = 4$

Type of winding = Wave winding (i.e., $A = 2$)

Number of conductors, $Z = 420$

Load particulars = 500 lamps × 75 = 37.5 kW

Terminal voltage, $V = 250$ V

Brush drop = 2 × 2 = 4 V

Load contact drop = 8 V

Flux per pole = 30 mWb

Armature resistance, $R_a = 0.065\ \Omega$

Shunt field resistance, $R_{sh} = 70\ \Omega$

Speed =?

Load current, $I = 500 \times 75/250 = 150$ A

Voltage across shunt = 250 + 8 = 258 V

Shunt current, $I_{sh} = 258 / 70 = 3.68$A

Armature current, $I_a = 150 + 3.68 = 153.68$ A

Armature drop, $I_aR_a = 153.68 \times 0.065 = 9.9$ V

Brush drop = 2 × 2 = 4 V

Drop in connecting leads = 8 V

Generated emf, $E_g = 250 + 4 + 8 + 9.98 = 271.98$ V

$$E_g = \frac{\phi ZN}{60}\left(\frac{p}{A}\right) \text{ volts}$$

$$271.98 = \frac{30\times10^{-3}\times420\times N}{60}\times\frac{4}{2}$$

Speed, $N = 647.57$ rpm

EXAMPLE 8.22 A 4-pole dc shunt generator with wave wound armature has 45 slots, each having 15 conductors. The armature resistance is 0.5 Ω and the shunt field resistance is 200 Ω. The flux per pole is 30 mWb. If a load of 15-Ω resistance is connected across the armature terminals, calculate the voltage across the load when the generator is driven at 1200 rpm.

Solution Given data

Number of poles, $p = 4$
Type of winding = Wave (i.e., $A = 2$)
Number of conductors, $Z = 4515 = 675$
Flux per pole = 30 mWb
Armature resistance, $R_a = 0.5\ \Omega$
Shunt field resistance, $R_{sh} = 200\ \Omega$
Load resistance, $R_L = 15\ \Omega$
Speed of the generator, $N = 1200$ rpm
Terminal voltage, $V = ?$

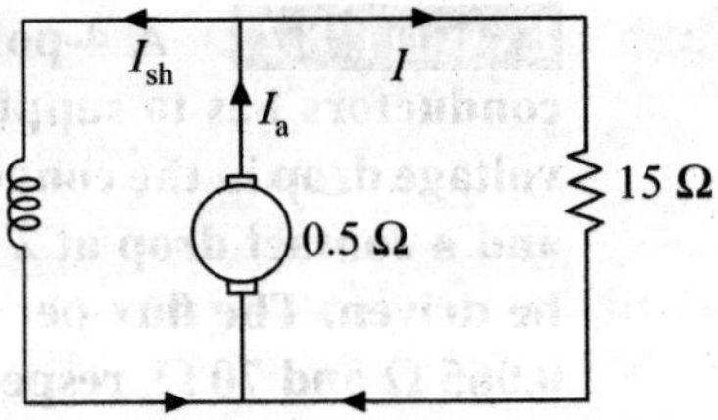

Figure 8.28

$$\text{Generated emf, } E_g = \frac{\phi ZN}{60}\left(\frac{p}{A}\right) \text{ volts}$$

Total conductors, $Z = 45 \times 15 = 675$
Speed, $N = 1200$ rpm
Flux, $\phi = 30 \times 10^{-3}$ Wb; poles, $p = 4$; parallel paths, $A = 2$

$$\text{Generated emf, } E_g = \frac{30\times10^{-3}\times675\times1200}{60}\times4/2 = 810\text{ V}$$

The shunt and load resistances are in parallel with each other.

$$\therefore \text{ Their combined resistance} = \frac{15\times200}{215} = 13.95\ \Omega = 13.95\ \Omega$$

Total circuit resistance = 0.5 + 13.95 = 14.45 Ω

$$\therefore \text{ Armature current} = \frac{810}{14.45} = 56.04\text{ A}$$

Armature drop = 56.04 × 0.5 = 28.02 V

∴ Terminal voltage, V = Generated emf – Armature drop = 810 – 28.02 = 781.97 V

EXAMPLE 8.23 A separately excited generator running at 1500 rpm supplies 225 A at 120 V to a current of constant resistance. What will be the current when the speed is dropped to 1000 rpm with the field current unaltered? The armature resistance is 0.06 Ω and the total drop at the brushes is 1.5 V. Ignore armature reaction.

Solution Given data

Speed, $N = 1500$ rpm

Load current, $I_L = 225$ A

Terminal voltage, $V = 120$ V

Speed is dropped, $N_2 = 1000$ rpm

Armature resistance, $R_a = 0.06\ \Omega$

Brush drop = 1.5 V

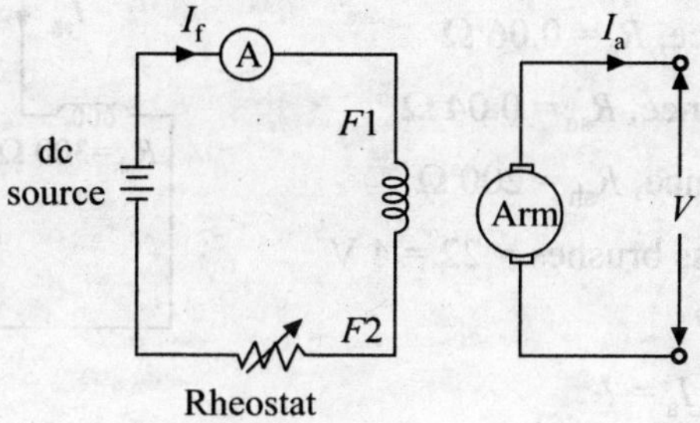

Figure 8.29

At 1500 rpm:

Load resistance, $R_L = 120/225 = 0.533\ \Omega$

Generated emf, $E_{g1} = 120 + 225 \times 0.06 + 1.5 = 135$ V

As the field current remains constant, the flux, ϕ, per pole is unchanged.

$$E \propto N$$

The generated emf at 1000 rpm,

$$E_{g2} = 135 \times \frac{1000}{1500} = 90 \text{ V}$$

Let V be the terminal voltage at 1000 rpm.

Then the armature current, $I_a = \dfrac{V}{R_L} = \dfrac{V}{0.533}$

But,

$$V = E_{g2} - I_a R_a - \text{(voltage drop across brushes)}$$

$$= 90 - \frac{V \times 0.06}{0.533} - 1.5$$

$$= 90 - 0.1125V - 1.5$$

$$= \frac{88.5}{1.1125} = 79.55 \text{ V}$$

$$I = \frac{79.55}{0.533} = 149.25 \text{ A}$$

EXAMPLE 8.24 A long-shunt compound generator delivers on load current of 45 A at 350 V and has armature, series field, and shunt field resistances of 0.06 Ω, 0.04 Ω, and 200 Ω, respectively. Calculate the emf of the generator and the armature current. Allow 2 V per brush for contact drop.

Solution Given data

Load current, $I_L = 45$ A

Terminal voltage, $V = 350$ V

Armature resistance, $R_a = 0.06\ \Omega$

Series field resistance, $R_{se} = 0.04\ \Omega$

Shunt field resistance, $R_{sh} = 200\ \Omega$

Voltage drop across brushes = 22 = 4 V

Generated emf = ?

Armature current, $I_a = ?$

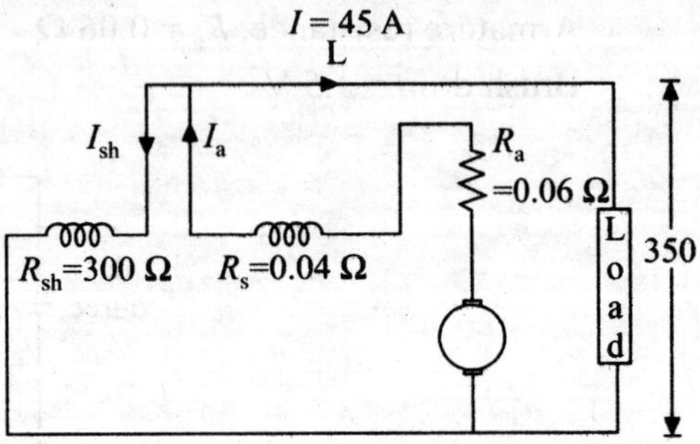

Figure 8.30

Shunt field current, $I_{sh} = \dfrac{350}{200} = 1.75$ A

Armature current, $I_a = 45 + 1.75 = 46.75$ A

Generated voltage, $E_g = V + I_a\,(R_a + R_s)$ + Voltage drop across brushes

$= 350 + 46.75\,(0.06 + 0.04) + 2 \times 2$

$= 358.67$ V

8.8 DC GENERATOR CHARACTERISTICS

A dc generator is an energy converter. The mechanical power input received from a prime mover is converted electromagnetically into electrical energy. This electrical energy can be changed into heat and is done by connecting load resistors across its armature terminals.

It is essential that the field windings of the generator be excited with dc current. The two main fields are shunt and series. If the shunt field alone is used then the generator is called a shunt generator, whereas if the series field alone is used, it is called a series generator. When both windings are used, we have the compound generators.

As the name indicates, the shunt field is connected in parallel with the armature, whereas the series field is connected in series with the armature or the load. The shunt field is of larger number of turns and of higher resistance than the series field. The shunt field can be excited from a separate source (separately excited generator) or from the armature of the same generator (self-excited shunt generator).

It is possible to have two types of compound generators: one is for the cumulative operation and the other for the differential operation: in the former, the series field assists the shunt field, whereas in the latter, it opposes. This will be seen from the results of the experiment. These different generators have very different V–I characteristics.

The behavior of a generator under operational conditions when plotted as a graph is called characteristics. In a generator the relation between the terminal voltage and the load current is a point of interest and the graph obtained between them is called performance characteristics or simply external characteristics. If, on the other hand, the relation between the actual induced emf and the armature current is found, then the graph is called internal characteristics. Performance characteristics also give an idea of voltage regulation, which is an important factor. The basic characteristics of a dc generator are broadly classified into three types:

(i) Open-circuit characteristic (OCC)

(ii) Internal or total characteristic (E/I_a)

(iii) External characteristic (V/I)

8.8.1 Separately Excited Generators

In a separately excited generator, the field flux is provided by a separate auxiliary source. This method of excitation has the obvious disadvantage of the need of a separate dc source.

Open-circuit characteristic

Open-circuit characteristics also known as *magnetization characteristics* depict the variation of generated voltage (or no-load voltage) with field current at a constant speed. Figure 8.31 shows the circuit diagram. The field winding is excited from a separate dc source to independent flux control. The generator is driven at a constant speed by a prime mover.

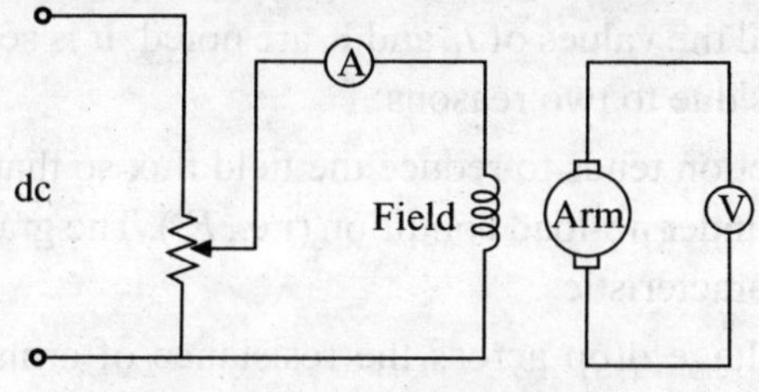

Figure 8.31 Circuit diagram for OCC

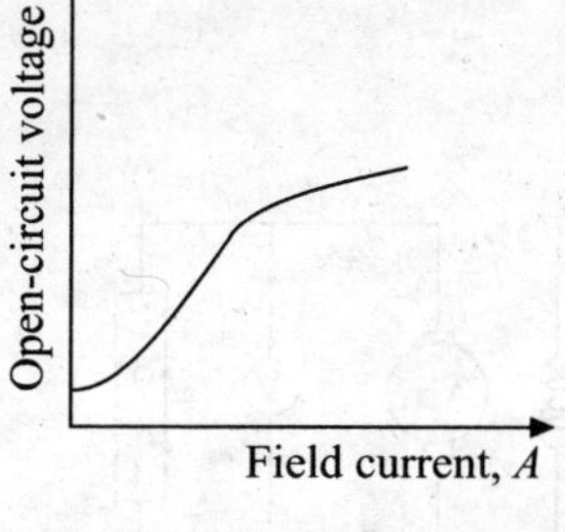

Figure 8.32 Open-circuit characteristics

Figure 8.32 shows the magnetization characteristic. The speed is constant, and the generated voltage is directly proportional to the field flux. When the field current is zero, a small residual flux is present in the machine. When the field current is zero, a small voltage is generated and is known as residual voltage. As the field current is increased, the flux increases proportionately and the voltage increases linearly with I_f. At some value of I_f, the magnetic circuit starts getting saturated. Therefore, a further increase in I_f does not increase the flux in the same proposition. Therefore, voltage increases slightly only in this region.

The generated voltage is directly proportional to speed. Therefore, the characteristics at any speed can be obtained from the test curve at one speed.

Internal characteristic

Let E_g be the no load generated voltage at no load conditions. If the generator is loaded, then an armature reaction occurs in the generator. Due to this effect, E_g drops to a voltage of E_a. The graph drawn between E_a and load current I_L is known as internal characteristic as shown in Figure 8.33.

External characteristic

External characteristic is also known as *technical characteristics*. It is a plot of the terminal voltage versus the line current (which is same as armature current in this case). Field current and speed are kept constant. Figure 8.33 depicts a test set up. At a constant speed the generator is driven by a prime mover. The field winding is excited from a separate dc source and I_f is kept constant at a value so that the terminal voltage, when $I_L = 0$, is equal to the rated value. Then the current is increased in steps, and the values of I_L and K are noted. It is seen that as I_L increases, the terminal voltage drops. This is due to two reasons:

(i) Armature reaction tends to reduce the field flux so that the actual generated voltage E_g is less than that under no-load condition (i.e., E_a). The graph of E_a versus I_L is usually known as internal characteristics.

(ii) There is a voltage drop across the resistance of armature winding ($=I_L R_a$). So there is further reduction in E_a to a voltage of V_t. The graph drawn between terminal voltage V_t versus load current I_L is known as external characteristics, as shown in Figure 8.33.

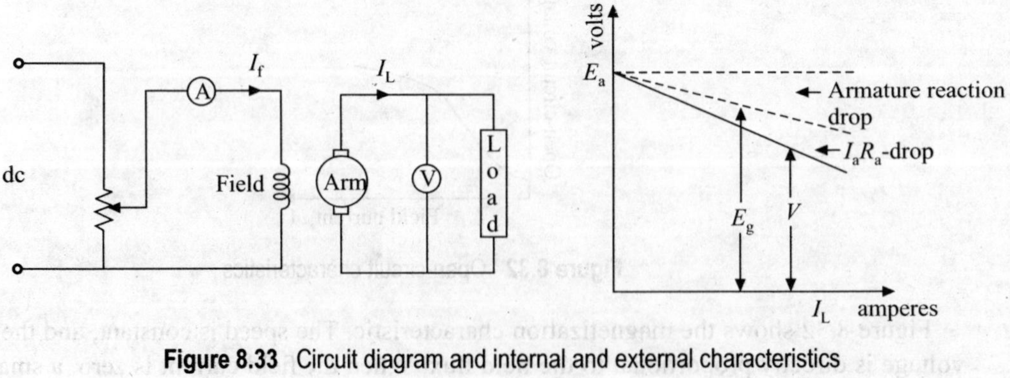

Figure 8.33 Circuit diagram and internal and external characteristics

8.8.2 Self-Excited Generators

In self-excited generators the residual magnetism in the generator is used to initiate the process of providing field flux. The self-excited generator can be a shunt generator, series generator, and compound generator.

Open-circuit characteristic

It is also known as magnetic characteristic or no-load saturation characteristic. It shows the relation between the no-load generated emf in the armature E_o and the field or exciting current, at a given fixed speed. In fact, it is just the magnetization curve for the material of the electromagnets. Its shape is practically the same for all generators whether separately excited or self-excited (Figure 8.34).

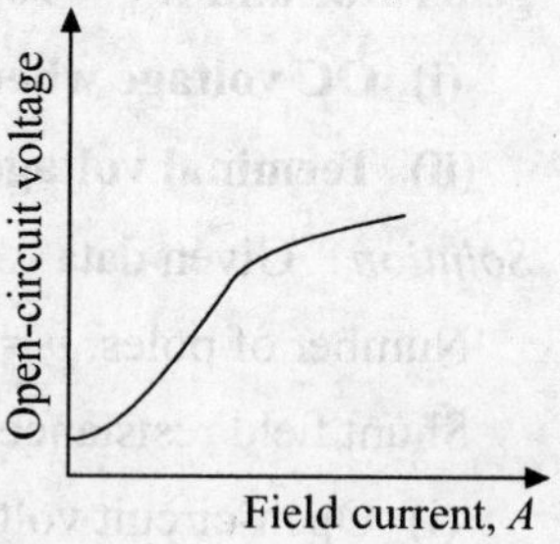

Figure 8.34 Open-circuit characteristics

Internal or total characteristic (E/I_a)

It gives the relation between emf E actually induced in the armature (after allowing for the demagnetizing effect of armature reaction) and the armature current I_a as shown in Figure 8.35. This characteristic is of interest mainly to the designer.

External characteristics (V/I)

It is also known as *performance characteristic* or sometimes *voltage regulation curve*. It gives a relation between the terminal voltage V and the load current I. This curve lies below the internal characteristics because it takes into account the voltage drop over the armature circuit resistance as shown in Figure 8.35. The values of V are obtained by subtracting $I_a R_a$ from the corresponding values of E. This characteristic is of great importance in judging the suitability of a generator for particular purpose.

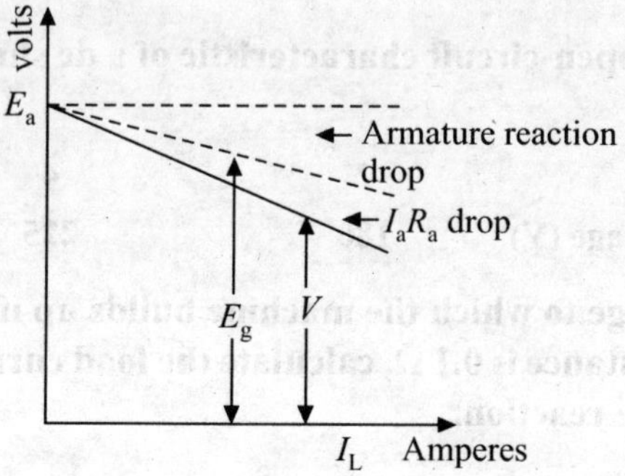

Figure 8.35 Internal characteristic (E/I_a) and external characteristics (V/I)

EXAMPLE 8.25 **The open-circuit characteristic of a 4-pole dc generator, when it is driven at normal speed, is as follows:**

Field current (A)	**0**	**0.5**	**1.0**	**1.5**	**2.0**	**2.5**	**3.0**
Open-circuit voltage (V)	**20**	**75**	**150**	**200**	**225**	**235**	**240**

The OCC of a 4-pole dc generator driven at normal speed. The machine as a shunt generator and R_{sh} = 100 Ω. Determine the following:

(i) OC voltage when the field resistance line meets the OCC at 232 V, OC voltage 232 V.

(ii) Terminal voltage when the load has resistance 10 Ω, no-load emf = 228 V.

Solution Given data

Number of poles, $p = 4$

Shunt field resistance, $R_{sh} = 100\ \Omega$

(i) Open-circuit voltage, $V_{oc} = 232$ V

Load resistance, $R_L = 10\ \Omega$

Total resistance = (100/110) + 0.25 = 9.34

Armature current, $I_a = 232/9.34 = 24.83$ A

(ii) Terminal voltage, $V = 232 - (24.83 \times 0.25)$
$= 225.8 \cong 226$ V.

Generated emf, $E_g = 50 \times 0.25 = 12.5$

$$= \frac{\phi ZN}{60}\left(\frac{p}{A}\right)$$

$$(Z \text{ is not given})\ \phi = \frac{E_g \times 60 \times A}{ZNp} = \frac{60 \times 2 \times 12.5}{252 \times 1000 \times 4}$$

$$= 1.488 \times 10^{-3}\ \text{Wb}$$

Flux, $\phi = 1.5$ mWb

EXAMPLE 8.26 **The open-circuit characteristic of a dc shunt generator running at 1000 rpm is as follows:**

Field current (A)	**2**	**4**	**6**
Open-circuit voltage (V)	**180**	**225**	**235**

Determine the voltage to which the machine builds up if the field circuit resistance is 50 Ω. If the armature resistance is 0.1 Ω, calculate the load current for a terminal voltage of 200 V. Ignore the armature reaction.

Solution Given data

Speed of the generator, $N = 1000$ rpm

Field resistance, $R_{sh} = 50\ \Omega$

Armature resistance, $R_a = 0.1\ \Omega$

Terminal voltage, $V = 200$ V

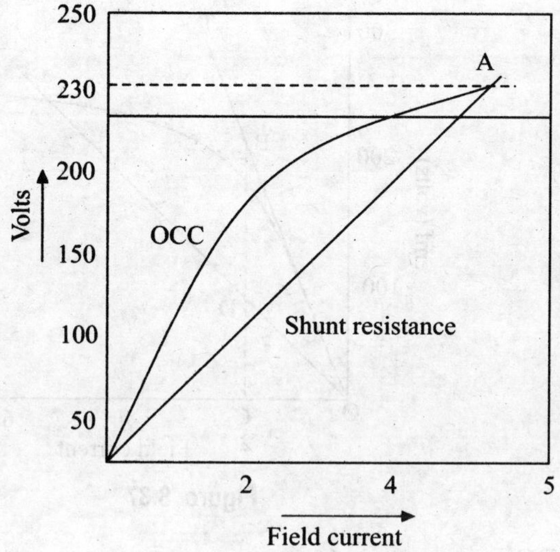

Figure 8.36

The OCC has been plotted from the given data. The shunt resistance line is drawn as usual. It cuts the OCC at point "A."

The voltage to which the machine will build up is 230 V. The terminal voltage is 200 V.

The shunt current, $I_{sh} = 200/50 = 4$ A

From OCC with shunt field current of 4 A, the total emf induced in the armature (neglecting armature reaction) is 225 V. As armature reaction is negligible, the difference of 225 – 200 = 25 V is the drop over the armature resistance alone. Therefore,

Armature drop, $I_a R_a = 25$

$$I_a = 25/0.1 = 250 \text{ A}$$

Load current, $I = I_a - I_{sh}$

$$= 250 - 4 = 246 \text{ A}$$

EXAMPLE 8.27 The magnetization curve at 800 rpm of a 4-pole, 220-V shunt generator with 576 conductors of lap-wound armature is as follows.

Field current (A)	**0**	**0.5**	**1**	**2**	**3**	**4**	**5**
Emf (V)	**10**	**50**	**100**	**175**	**220**	**245**	**262**

(i) If the field resistance is 75 Ω, calculate (a) the speed at which the machine just fails to excite and (b) the flux per pole if the open-circuit potential difference is 22 V.

(ii) What is the residual flux?

Solution The OCC is drawn from the given data as shown in Figure 8.37. If A is 75 Ω, then OT is tangential to the curve. Any line BC is drawn perpendicular to the base, cutting lines OA and OT at points D and B, respectively.

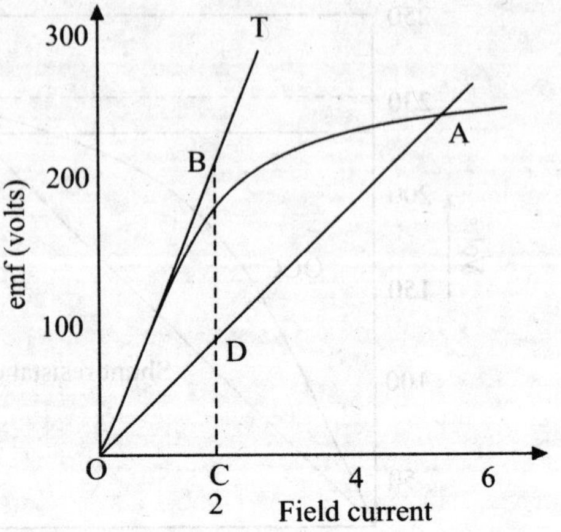

Figure 8.37

(i) (a) $\dfrac{N_e}{N} = \dfrac{DC}{BC}$

$$\frac{N_e}{800} = \frac{150}{194}$$

$$\therefore \quad N_e = 800 \times \frac{150}{194} = 60 \text{ rpm}$$

(b) Using the relation

$$E = \frac{\phi Z N}{60}\left(\frac{p}{A}\right) \text{volts} \qquad (1)$$

$$225 = \frac{\phi \times 576 \times 800}{60}\left(\frac{4}{A}\right)$$

$$\therefore \quad \phi = 29.3 \text{ mWb}$$

(ii) With no exciting current, the emf induced due to residual flux is 10 V. Using Eq. (1), we get

$$10 = \frac{\phi \times 576 \times 800}{60}\left(\frac{4}{4}\right)$$

$$\phi = 1.3 \text{ mWb}$$

EXAMPLE 8.28 The open-circuit characteristic of a separately excited generator driven at 900 rpm is given below. Find the voltage to which the machine will excite as a shunt generator on open circuit, when the field resistance is 600 Ω.

(i) At 600 rpm, (ii) at 750 rpm, and (iii) at what speed will the machine fail?

Field current I_f (A)	**0**	**2.4**	**3.6**	**6.7**	**8**	**10.4**
Generated emf (V)	**0**	**207**	**310**	**466**	**517**	**570**

Solution

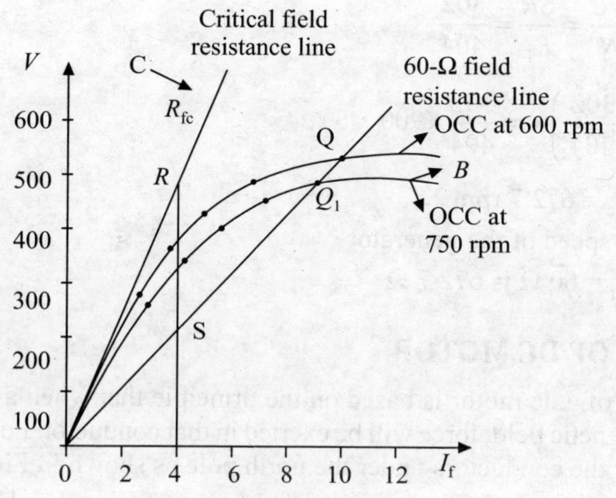

Figure 8.38

(i) From the given data the OCC at the speed of 900 rpm is drawn as shown in Figure 8.38. Then the field resistance line having a slope of 60 Ω is drawn by making the line OQ to pass through the point corresponding to our emf of 60 V (*y*-axis) and field current of P A (*x*-axis). The line OQ meets the OCC at point Q. The ordinate PQ represents the emf the machine can generate as a shunt generator. This emf PQ = 556 V.

(ii) The OCC at 750 rpm is obtained as follows. Let E_1 be the emf at 900 rpm and E_2 be the emf at 750 rpm. Then, since the generator emf is in proportion to the speed. We have,

$$\frac{E_1}{E_2} = \frac{900}{750}$$

i.e.,

$$E_2 = E_1 \times \frac{500}{600}$$

Thus, the emf values of the OCC at 900 rpm have to be multiplied by 5/6 to get the emf values of the OCC at 750 rpm. Then, the corresponding data are

Field current, I_f (A)	0	2.4	3.6	6.4	8	10.4
Generated emf (V)	0	172	258	388	430	475

The OCC at 750 rpm is plotted in the figure as shown by curve B. The field resistance line meets the OCC at point Q. The ordinate QP_1 = 426 V represents the emf the machine can generate at 750 rpm.

(iii) To estimate the critical speed when the field resistance is 60 Ω. Curve "C" represents the critical field resistance line, tangential to the OCC at 900 rpm. Let us take any point "R" on the tangent and from "R" draw a perpendicular R_T to the *x*-axis. This meets the field resistance line OQ at points. If N_c is the critical speed, we have

$$\frac{NC}{N} = \frac{SR}{R_T} = \frac{302}{404}$$

$$\left(\frac{302}{404}\right)N = \frac{302}{404} \times 900$$

$$N_c = 672.7 \text{ rpm}$$

The critical speed of the generator

$R_f = 60\ \Omega$ is $672.7\ \Omega$

8.9 PRINCIPLE OF DC MOTOR

The operation of a dc motor is based on the principle that when a current-carrying conductor is placed in a magnetic field, force will be exerted in that conductor. For the simplicity of the analysis, let us consider the conductors under the north pole as shown in Figure 8.39.

The current in the conductor is assumed to flow inward to the plane of the paper, indicated by a cross. According to the Fleming's right-hand rule, the flux around the conductor in the armature is in clockwise direction. So naturally by considering both the main field flux and the conductor flux, the resultant flux tends to the right side of the conductor, as shown in Figure 8.40. So this conductor tries to move towards the left; this twisting movement is known as *torque*. This developed torque makes the armature to rotate in the anticlockwise direction. Similarly, if we consider the conductor under the south pole, the current direction of the armature conductor will be outwards to the plane of the paper indicated by a dot, as shown in Figure 8.40. In this case, the flux set by the conductor is in anticlockwise direction, and the resultant of main field and conductor field is to the left side of the conductor. So the conductor tries to move towards the right. This torque again makes the armature to continue to rotate in the anticlockwise direction.

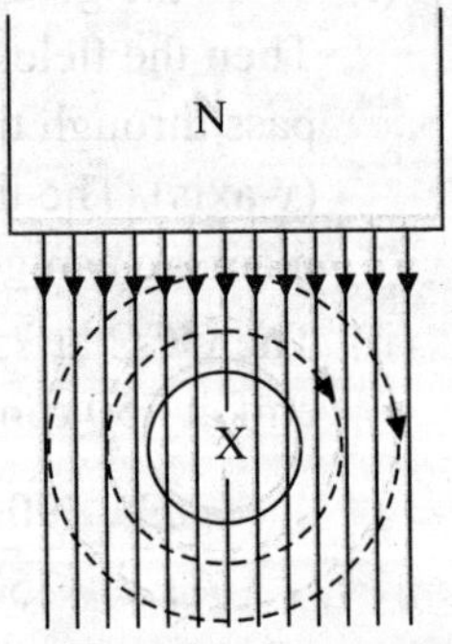

Figure 8.39 Conductor under north pole

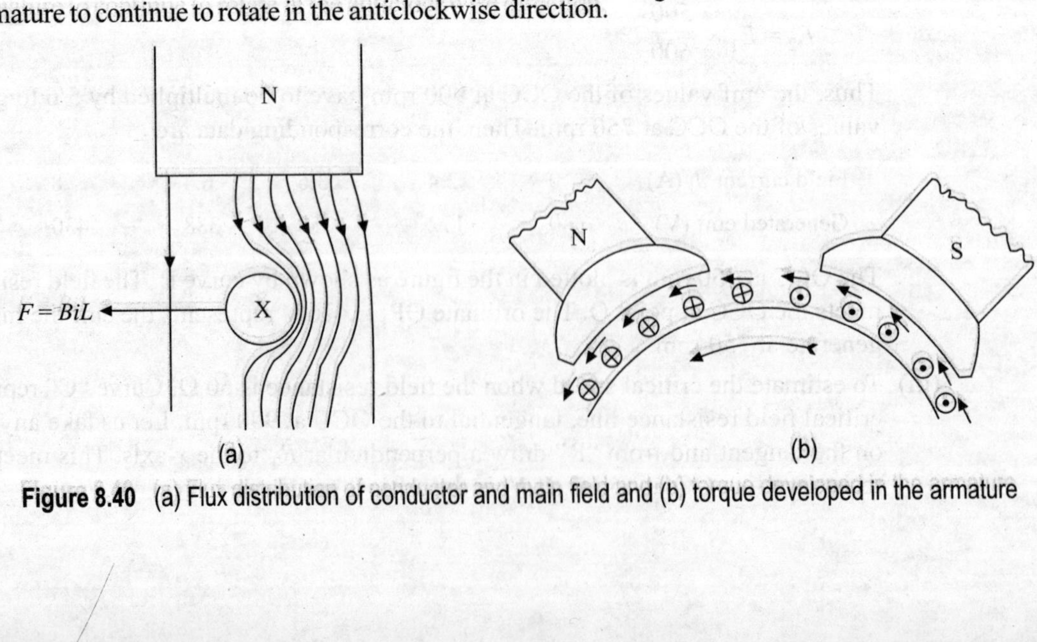

Figure 8.40 (a) Flux distribution of conductor and main field and (b) torque developed in the armature

Since, now the armature rotates in the magnetic field, an emf is induced in the armature. The direction of the induced emf is given by Fleming's right-hand rule and is found to be in opposition to the applied voltage. Therefore, this induced emf is termed as *back emf.*

8.10 SIGNIFICANCE OF BACK EMF

The applied voltage and back emf in a dc motor are in opposition, hence the net voltage across the armature is given as $V - E_b$. If the armature resistance is R_a, then the current drawn by the armature is I_a $[(V - E_b)/R_a]$ amperes.

If the load on the motor varies, the corresponding current drawn by the armature develops the required torque which means that the dc motor acts as a self-regulating machine.

When the motor is running on no-load, a small torque exists to overcome the friction and windage losses. Since the torque is small, the armature current is also small which makes the back emf nearly equal to the applied voltage.

When the motor is loaded, the armature slows down. So obviously the speed decreases and the back emf decreases (E_b). This in turn increases the armature current, which increases the driving torque.

Similarly, when the motor load is decreased, the momentary torque is in excess to the load required. The armature speed increases, thereby increasing the back emf (E_b) which in turn causes the armature current to decrease. The increase in the speed of the motor results in the acceleration of the armature, which stops when the armature current is sufficient to produce the torque required by the load.

EXAMPLE 8.29 A 230-V motor has an armature circuit resistance of 0.8 Ω. If the full-load armature current is 25 A and the no load armature current is 6 A, find the change in the back emf from no-load to full-load.

Solution Given data

Terminal voltage, $V = 230$ V

Armature circuit resistance, $R_a = 0.8\ \Omega$

Full-load armature current, $I_a = 25$ A

No-load armature current, $I_{ao} = 6$ A

Back emf, $E_b = V - I_a R_a$

At full load,

$E_b = 230 - (25 \times 0.8) = 210$ V

$E_{bo} = 230 - (6 \times 0.8) = 225.2$ V

Therefore, change in back emf $= 225.2 - 210 = 15.2$ V

EXAMPLE 8.30 A shunt motor runs at 450 rpm on a 200-V circuit. Its armature resistance is 0.6 Ω and the current taken is 25 A in addition to the field current. What resistance must be placed in series in order that the speed may be reduced to 250 rpm, the current in armature remaining the same.

Solution Given data

Terminal voltage = 200 V

Speed of the motor, $N_1 = 450$ rpm

Armature resistance, $R_a = 0.6\ \Omega$

Armature current, $I_a = 25$ A

Speed is reduced to 250 rpm, i.e., $N_2 = 250$ rpm

Field current remains the same, $\phi_2 = \phi$

$$\frac{N_2}{N_1} = \frac{E_{b2}}{E_{b1}}$$

$$E_{b1} = 200 - (0.6 \times 25) = 185 \text{ V}$$

$$E_{b2} = (200 - 25R_{se})$$

$$\frac{250}{450} = \frac{200 - 25R_{se}}{185}$$

$$185 \times 0.55 = 200 - 25R_{se}$$

$$102.77 = 200 - 25R_{se}$$

$$25R_{se} = 200 - 102.77$$

$$R_{se} = \frac{200 - 102.77}{25} = 3.88\ \Omega$$

EXAMPLE 8.31 A 450-V shunt motor runs at its normal speed of 225 rpm when the armature current is 150 A. If the resistance of armature is 0.12 Ω, calculate the speed when a resistance is inserted in the field, reducing the shunt field to 75% of the normal value and the armature current is 90 A.

Solution Given data

Terminal voltage, $V = 450$ V

Speed of the machine, $N = 225$ rpm

Armature current, $I_a = 150$ A

Armature resistance, $R_a = 0.12\ \Omega$

Reduced armature current, $I_{a2} = 90$ A (Since shunt field is reduced to 75%)

Back emf, $E_{b1} = V - I_{a1}R_a = 450 - 150 \times 0.12 = 432$ V

$E_{b2} = V - I_{a2}R_a = 450 - 90 \times 0.12 = 439.2$ V

$$\phi_2 = 0.75\ \phi_1$$

$$\frac{N_2}{N_1} = \frac{E_{b1}}{E_{b2}} \times \frac{\phi_1}{\phi_2}$$

$$\frac{N_2}{225} = \frac{439.2}{432} \times \frac{1}{0.75}$$

$$\frac{N_2}{225} = 1.0166 \times 1.333$$

$$N_2 = 305 \text{ rpm}$$

8.11 TORQUE OF DC MOTORS

A dc machine is loaded either as a motor or as a generator. The field of a machine is excited and a potential difference is impressed upon the machine terminals. The rotor conductors carry current. These conductors are the surface of the rotor at a common radius from its center. The torque is produced around the circumference of the rotor and the rotor starts rotating.

The meaning of torque is turning or twisting of a force about an axis. The torque is measured by the product of the force and radius at which this force acts.

In case of generators, the machine operates at constant speeds, and the torque is equal and opposite to that provided by the prime mover.

In case of motors, the torque is transferred to the shaft of the rotor and drives the mechanical load. The torque expression for the generator and the motor is same. It is derived in the following subsection.

8.11.1 Torque Equation

The voltage equation of a dc motor is

$$V = E + I_a R_a \qquad (8.19)$$

Multiplying both sides of Eq. (8.19) by I_a, we obtain

$$VI_a = EI_a + IR_a \qquad (8.20)$$

VI_a = Electrical power input to the armature

IR_a = Copper loss in the armature

We also know that, input = output + losses (8.21)

Comparing Eqs. (8.20) and (8.21), we get

EI_a = Electrical equivalent of gross mechanical power developed by the armature (electromagnetic power)

T = Average electromagnetic torque developed by the armature in newton meters (N-m)

At this value of torque the electromechanical power conversion takes place. The mechanical power developed by the armature,

$$P_m = \omega T = 2\pi NT$$

$$\therefore \quad P_m = EI_a = \omega T = 2\pi NT \qquad (8.22)$$

But $\quad E = \dfrac{\phi ZNP}{A}$

Therefore, from Eq. (8.22),

$$\frac{\phi ZNP}{A} I_a = 2\pi NT$$

$$T = \frac{PZ}{2\pi A}\phi I_a \tag{8.23}$$

Equation (8.23) is called the *torque equation of a dc motor*. For a given dc machine, Z and A are constants.

$$\therefore \quad \frac{PZ}{2\pi A} = K$$

Let $\quad \frac{PZ}{2\pi A}\phi I_a = K$

$$\therefore \quad T = K\phi I_a$$

$$T \propto \phi I_a$$

The torque developed by a dc motor is directly proportional to the flux per pole and the armature current.

(i) In case of shunt motors, ϕ is practically constant, hence $T \propto I_a$.

(ii) In case of series motors, ϕ is proportional to I_a before saturation (because field windings carry full armature current).

$$\therefore \quad T \propto I_a^2$$

8.11.2 Shaft Torque

The whole of the armature torque is not useful to calculate the above expression and is not available at the pulley because certain percentage of it is required for supplying iron in the motor. The torque which is available for useful work is known as shaft torque (T_{sh}). It is so called because it is available at the shaft. The horse power obtained by using the shaft torque is called brake horse power (bhp).

$$\text{bhp (metric)} = \frac{T_{sh} \times 2\pi N}{735.5}$$

$$T_{sh} = \frac{\text{bhp (metric)} \times 735.5}{2\pi N/60} \tag{8.24}$$

where N is the speed of armature in rpm.

The difference ($T_a - T_{sh}$) is known as lost torque (i.e., torque lost in iron and friction losses).

$$= 0.159 \times \frac{\text{Iron and friction losses}}{N/60} \text{ newton-meter} \tag{8.25}$$

The value of back emf (E_b) can be found from

(i) The equation, $E_b = V - I_a R_a$

(ii) The formula, $E_b = \phi ZN \times (p/A)$ volts

The shaft torque magnitude is always less than the armature torque, $T_{sh} < T_a$. The speed of the motor remains same all along the shaft say (N rpm). Then the product of shaft torque T_{sh} and the angular speed ω rad/s is called power available at the shaft, that is, the net output of the motor. The maximum power a motor can deliver to the load safely is called the output rating of a motor. Generally, it is expressed in hp. It is called hp rating of a motor.

$$\text{Net output of motor} = P_{out} = \omega T_{sh} \tag{8.26}$$

EXAMPLE 8.32 Determine the value of torque in N-m units established by the armature of a 4-pole motor having 760 conductors, two paths in parallel, 28 mWb per pole when the total armature current is 50 A.

Solution Given data

Number of poles, $p = 4$

Number of conductors, $Z = 760$

Flux per pole, $\phi = 28$ mWb

Number of parallel paths, $A = 2$

Total armature current, $I_a = 50$ A

Armature torque, $T_a = 0.159\phi Z I_a(p/A)$ newton-meter

Total conductors, Z = 760,

Armature current, $I_a = 50$ A

Flux, $\phi = 28 \times 10^{-3}$ Wb; poles, $p = 4$; parallel paths, $A = 2$

$$\therefore \quad T_a = 0.159 \times 28 \times 10^{-3} \times 760 \times 50 \times 2$$

$$= 338.35 \text{ N-m}$$

EXAMPLE 8.33 A 4-pole, 500-V, 75.6-kW wave-connected shunt motor has 600 conductors. The flux per pole is 45 mWb and the full-load efficiency is 90%. Its shunt field resistance is 250 Ω and armature circuit resistance is 0.2 Ω. Calculate for full-load, (i) the motor speed and (ii) the useful torque developed.

Solution Given data

Number of poles, $p = 4$

Terminal voltage, $V = 500$ V

Number of conductors, $Z = 760$

Flux per pole, $\phi = 45$ mWb

Full-load efficiency, $\eta = 90\%$

Shunt field resistance, $R_{sh} = 250\ \Omega$

Armature circuit resistance, $R_a = 0.2\ \Omega$

Motor input = 75.6 × 1000/0.90

$$\text{Motor input current} = \frac{75600}{0.90 \times 500} = 168 \text{ A}$$

Shunt current, $I_{sh} = 500/250 = 2$ A

Armature current, $I_a = 168 - 2 = 166$ A

Back emf, $E_b = 500 - (166 \times 0.2) = 466.8$ V

(i) $E_b = \phi ZN \times (p/A)$

$466.8 = 45 \times 10^{-3} \times 600 \times N \times 2$

Speed, $N = 8.64$ rps or 518.6 rpm

(ii) Useful torque or shaft torque, $T_{sh} = \dfrac{756 \times 100}{2\pi \times 8.64} = 1392$ N-m

EXAMPLE 8.34 A 6-pole dc motor has a wave-connected armature with 85 V, each slot containing 4 conductors. The flux per pole is 25 mWb and the armature has a resistance of 0.10 Ω. Calculate the speed when the motor is connected to a 250-V supply and taking an armature current of 75 A. Also calculate the torque in N-m, developed by the armature.

Solution Given data

Number of poles, $p = 6$

Since the type of connection is wave wound, $A = 2$

Total number of conductors, $Z = 85 \times 4$

Flux per pole, $\phi = 25$ mWb

Armature resistance, $R_a = 0.10\ \Omega$

Terminal voltage, $V = 250$ V

Armature current, $I_a = 75$ A

The motor back emf, $E_b = 250 - 75 \times 0.10 = 242$ V

Also,

$$E_b = \left(\frac{\phi NZ}{60}\right)\left(\frac{p}{A}\right)$$

Substituting the values, we get

$$242 = 2\left(\frac{85 \times 4}{2}\right)\left(\frac{25 \times 10^{-3} \times N \times 3}{60}\right)$$

$$N = \frac{242 \times 2 \times 60}{2 \times 85 \times 4 \times 25 \times 10^{-3} \times 3} = \frac{29040}{51} = 569.41 \text{ rpm}$$

$$E_b I_a = 2 \neq \frac{TN}{60}$$

$$T = \frac{E_b I_a \times 60}{2\pi N} = \frac{242 \times 75 \times 60}{2\pi \times 569.41}$$

$$= \frac{1089}{3577} = 304.38 \text{ N-m}$$

EXAMPLE 8.35 The armature of a 4-pole lap-wound dc machine has core length 20 cm, diameter 30 cm, 450 conductors, speed 1000 rpm, and current 15 A. Find the gross mechanical power developed and the internal torque. Consider the average flux density $B = 0.7$ T.

Solution Given data

Number of poles, $p = 4$

Since the type of connection is lap wound, $A = p$

Length of the core, $l = 20$ cm

Diameter of the core, $d = 30$ cm

Number of conductors, $Z = 450$

Speed of the machine, $N = 1000$ rpm

Current, $I = 15$ A

Average flux density, $B = 0.7$ T

Here, $l = 0.20$ m; $r = 0.3$ m; speed, $N = \dfrac{1000}{60} = 16.66$ rps

$$B_{ow} = \frac{P\phi}{2\pi rl}$$

Therefore, flux per pole,

$$\phi = \frac{0.7 \times 2\pi \times 0.3 \times 0.2}{4}$$

$$= 0.066 \text{ Wb}$$

Generated emf$_1$

$$E_a = \frac{\phi ZNp}{a}$$

$$= \frac{0.666 \times 450 \times 16.66 \times 4}{4}$$

$$= 494.60 \text{ V}$$

Gross mechanical power developed,

$$E_a I_a = 494.60 \times 15 = 7419 \text{ W}$$

Internal torque,

$$T_e = \frac{E_a I_a}{W_m} = \frac{7419}{2\pi \times 16.66} = 70.875 \text{ N-m}$$

EXAMPLE 8.36 **A 240-V, 40-A, 700-rpm dc shunt motor has armature circuit resistance of 0.3 Ω. If the load torque is reduced to 60% of its full-load value and a resistance of 4 Ω is inserted in series with the armature circuit, find the motor speed. Armature reaction weakens the field flux by 4% at full load and by 2% at 60% of full load.**

Solution Given data

Terminal voltage, $V = 240$ V

Current, $I = 40$ A

Armature resistance, $R_a = 0.3\ \Omega$

At rated load,

$$E_{a1} = V_t - I_{a1}R_a$$
$$= 240 - 40 \times 0.3 = 228 \text{ V}$$

Rated torque, $T_{L1} = K_a\phi_1 I_{a1} = K_a\phi_1 \times 40$

New torque, $T_{L2} = 0.6\, T_{L1} = K_a\phi_2 I_{a2}$

It is given that $\phi_1 = 0.96\phi$

$$\phi_2 = 0.98\phi$$

where ϕ is the main field flux at no load.

$$\frac{T_{L1}}{T_{L2}} = \frac{T_{L1}}{0.6T_{L1}} = \frac{K_a \times 0.96 \times 40}{K_a \times 0.98 \times I_{a2}}$$

$$\frac{1}{0.6} = \frac{38.4}{0.98 \times I_{a2}}$$

$$I_{a2} = \frac{38.4 \times 0.6}{0.98} = 23.51 \text{ A}$$

New, counter emf

$$E_{a2} = 240 - 23.51\,(4 + 0.2)$$
$$= 141.258 \text{ V}$$

$$\frac{E_{a1}}{E_{a2}} = \frac{N_1\phi_1}{N_2\phi_2} \text{ (or) } \frac{228}{141.258} = \frac{700 \times 0.96\phi}{N_2 \times 0.98\phi}$$

$$\text{New speed, } N_2 = \frac{700 \times 141.258 \times 0.96}{228 \times 0.98}$$
$$= 424.83 \text{ rpm}$$

EXAMPLE 8.37 A 400-V dc shunt motor takes a current of 4.5 A on no load and 58.2 A on full load. The armature reaction weakens the field by 3%. Calculate the ratio of full-load speed to no-load speed. Given $R_a = 0.2\ \Omega$, the brush voltage drop is 4 V, $R_F = 150\ \Omega$.

Solution Given data

Terminal voltage, $V = 400$ V

No-load current, $I_0 = 4.5$ A

Full-load current, $I_{fl} = 58.2$ A

Armature resistance, $R_a = 0.2\ \Omega$

Brush drop = 4 V

Shunt field resistance, $R_f = 150\ \Omega$

$$\text{Field current, } I_F = \frac{V}{R_F} = \frac{400}{150} = 2.66 \text{ A}$$

No load,

$$I_{a0} = 4.5 - 2.66 = 1.83 \text{ A}$$

$$E_{a0} = 400 - 0.2 \times 1.83$$

$$= 399.63 \text{ V}$$

Full load,

$$I_{a\,fl} = 58.2 - 2.66 = 55.54 \text{ A}$$

$$E_{a\,fl} = 400 - 0.2 \times 55.54 - 4 = 384.89 \text{ V}$$

$$\frac{N_{fl}}{N_0} = \frac{384.9}{399.63} \times \frac{1}{0.97} = 1.00$$

EXAMPLE 8.38 A 220-V shunt motor has an armature resistance of 0.5 Ω and takes a current of 40 A on full load. By how much percentage the main flux is reduced to raise the speed by 50% if the developed torque is constant.

Solution Given data

Terminal voltage, $V = 220$ V

Armature resistance, $R_a = 0.5\ \Omega$

Armature current, $I_a = 40$ A

We know that, $\dfrac{E_{b1}}{E_{b2}} = \dfrac{\phi_1 N_1}{\phi_2 N_2}$

$$\frac{N_2}{N_1} = \frac{E_{b1}}{E_{b2}} \frac{\phi_1}{\phi_2}$$

Since the torque is constant,

$$\therefore \quad \phi_1\, Ia_1 = \phi_2\, Ia_2$$

$$Ia_2 = Ia_1 \times \frac{\phi_1}{\phi_2} = 40X \quad \left(\text{where } X = \frac{\phi_1}{\phi_2}\right)$$

$$E_{b1} = 220 - (40 \times 0.5) = 200 \text{ V}$$

$$E_{b2} = 220 - (40 \times 0.5X) = 220 - 20X$$

$$\frac{N_2}{N_1} = \frac{3}{2}$$

$$\frac{3}{2} = \frac{200 - 20X}{200} \times X$$

$$X^2 - 11X + 15 = 0$$

$$X = 11 \pm \sqrt{\frac{121 - 60}{2}}$$

$$= \frac{11 \pm 7.81}{2} = 9.4 \text{ or } 1.6$$

$X = 9.4$ is rejected as it does not give the necessary increase in speed.

$$\therefore \quad \frac{\phi_1}{\phi_2} = 1.6 \text{ or } \frac{\phi_2}{\phi_1} = \frac{1}{1.6}$$

$$\frac{\phi_1 - \phi_2}{\phi_2} = \frac{1.6 - 1}{1.6} = \frac{3}{8}$$

Therefore, percentage change in the flux = 3/8 × 100 = 37.5%.

EXAMPLE 8.39 A series motor with an unsaturated field and negligible resistance when running at a certain speed on a given load takes 45 A at 450 V. If the load torque varies as the cube of speed, calculate the resistance required to reduce the speed by 25%.

Solution We know $T_a \propto \phi I_a$

As series field is unsaturated, therefore

$$\phi < I_a$$

$$\therefore \quad T_a \propto I \qquad (1)$$

and

$$T_a \propto N^3 \qquad (2)$$

From Eqs. (1) and (2), we get

$$I_a^2 \propto N^3$$

In the first case, $I_{a1}^2 < N_1^3$

In the second case, $I_{a2}^2 < N_2^3$

$$\left(\frac{I_{a2}}{I_{a1}}\right)^2 = \left(\frac{N_2}{N_1}\right)^3 = \left(\frac{3}{4}\right)^3 = \frac{27}{64}$$

$$I_{a2} = I_{a1}\sqrt{\frac{27}{64}} = 45\sqrt{\frac{27}{64}} = 29.22 \text{ A}$$

Hence, $E_{b1} = 460 - I_{a1} \times 0 = 460 - 45 \times 0 = 460$ V (Since resistance is neglected)

And $E_{b2} = 460 - 29.22\, R_t$

where R_t is the controller resistance in series with armature. Now,

$$\frac{N_2}{N_1} = \frac{E_{b1}}{E_{b2}} \times \frac{I_{a1}}{I_{a2}}$$

$$\frac{3}{4} = \frac{460 - 29.22R_t}{460} \times \frac{45}{29.22}$$

$$0.75 = 1 - 0.0635R_t \times 1.54$$

$$0.487 = 1 - 0.06352R_t$$
$$0.06352R_t = 1 - 0.487$$
$$R_t = \frac{0.513}{0.06352} = 8.076\ \Omega = 8\ \Omega \text{ (approximate)}$$

8.12 MECHANICAL POWER DEVELOPED BY MOTOR ARMATURE

In Figure 8.41, the voltage "V" is applied across the motor armature in order to

(i) overcome back emf, E_b

(ii) supply the armature ohmic drop, I_aR_a

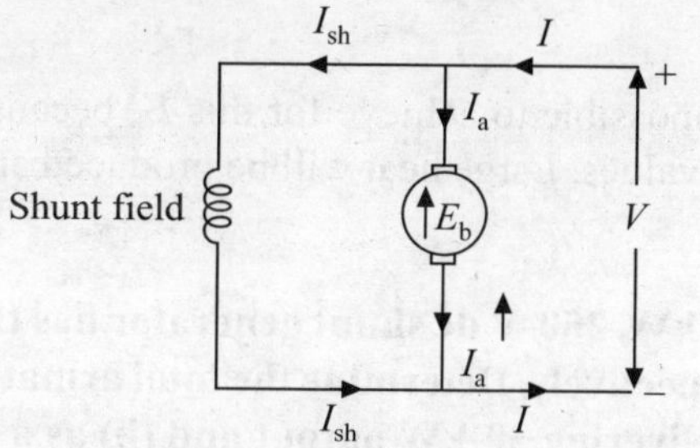

Figure 8.41

Therefore, the voltage equation of a motor is

$$V = E_b + I_aR_a \tag{8.27}$$

Multiplying both sides of Eq. (8.27) by I_a, we get

$$VI_a = E_bI_a + I_a^2R_a \tag{8.28}$$

Equation (8.28) is called the *power equation of a dc motor* in which VI_a is the electrical input to the armature, E_bI_a is the electrical equivalent of mechanical power P_m developed in the armature, and $I_a^2R_a$ is the copper loss in the armature.

The difference between VI_a and $I_a^2R_a$, that is, input-losses, gives the output of the armature. There are certain mechanical losses (such as bearing, friction, windage, and iron losses) that must be supplied by the driving power of the motor.

E_bI_a is called the electrical equivalent of gross mechanical power developed by the armature. This is denoted as P_m.

$\therefore$ Gross mechanical power developed in the armature = Power input to the armature – Armature copper loss

8.12.1 Condition for Maximum Power

The mechanical power developed by the motor,

$$P_m = VI_a - I_a^2R_a \tag{8.29}$$

Differentiating both sides of Eq. (8.29) with respect to I_a and equating the result to zero, we get

$$\frac{3(P_m)}{dt} = V - 2I_a R_a = 0$$

$$I_a R_a = V/2$$

Terminal voltage, $V = E_b + I_a R_a$

$$= E_b + \frac{V}{2}$$

$$\therefore \quad E_b = \frac{V}{2} \tag{8.30}$$

This is practically impossible to achieve for this E_b because the current required is much more than its normal rated values. Large heat will be produced and the efficiency of motor will be less than 50%.

EXAMPLE 8.40 A 30-kW, 250-V dc shunt generator has the armature and field resistances of 0.05 Ω and 90 Ω, respectively. Determine the total armature power developed when working (i) as a generator delivering 30-kW output and (ii) as a motor taking 30-kW input.

Solution Given data

Power output of the generator, $P = 30$ kW

Terminal voltage, $V = 250$ V

Armature resistance, $R_a = 0.05\ \Omega$

Shunt field resistance, $R_{sh} = 90\ \Omega$

As a generator,

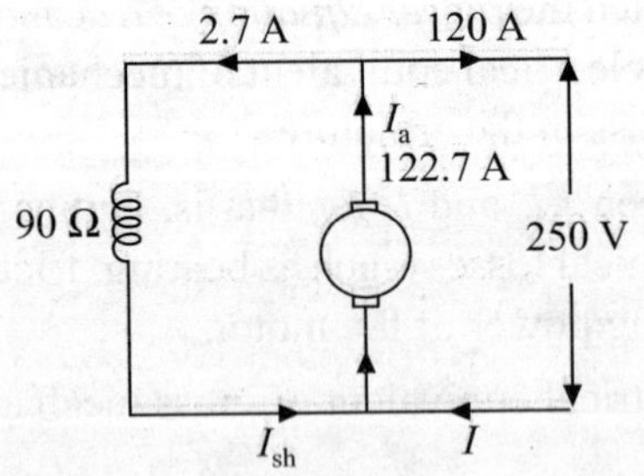

Figure 8.42

Output current, $I = 30000/250 = 120$ A

Shunt current, $I_{sh} = 250/90 = 2.7$ A

Armature current, $I_a = 120 + 2.7 = 122.7$ A

Generated emf, $E_g = V + I_a R_a$

$= 250 + 6.135 = 256.135$ V

Power developed in armature, $E_g I_a = \dfrac{256.135 \times 122.7}{1000} = 31.42$ kW

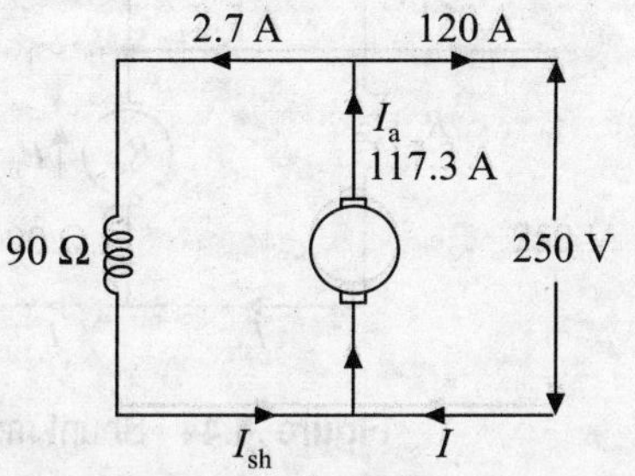

Figure 8.43

(ii) As a motor,

Motor input current, $I = 120$ A

$$I_{sh} = 250/90 = 2.7 \text{ A}$$

$$I_a = 117.3 \text{ A}$$

$$E_b = 250 - 117.3 \times 0.05 = 244.135 \text{ V}$$

Power developed in armature, $E_b I_a = \dfrac{244.135 \times 117.3}{1000} = 28.63$ kW

8.13 TYPES OF DC MOTOR

Similar to the dc generator the types of dc motor are also classified according to the manner in which the field and armature are connected:

(i) Series wound motor
(ii) Shunt wound motor
(iii) Compound wound motor
 (a) Long-shunt motor
 (b) Short-shunt motor

8.13.1 Shunt Wound Motor

The motor in which the field winding is parallel to the armature, as shown in Figure 8.44, is known as *shunt motor*. The voltage is same across the field and armature due to the shunt arrangement.

Armature current, $I_a = I_L - I_{sh}$

Load current, $I_L = I_a + I_{sh}$

$$I_{sh} = \frac{V}{R_{sh}}$$

There exists a drop in the brush contact, so the voltage equation becomes

$V = E_b + I_a R_a + V_{bc}$

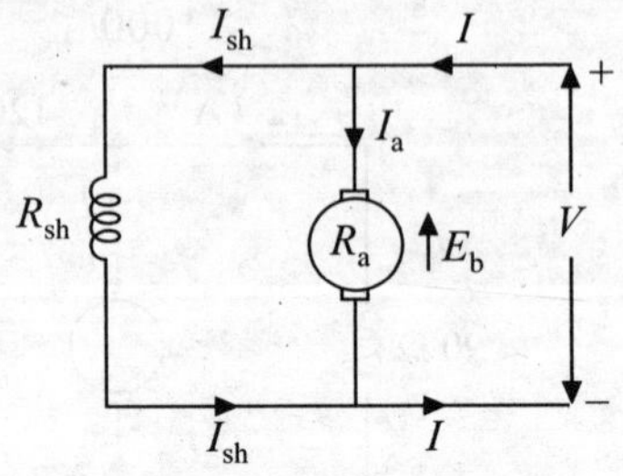

Figure 8.44 Shunt motor

EXAMPLE 8.41 **A shunt machine, connected to 200-V mains, has an armature resistance (including brushes) of 0.10 Ω, and the resistance of the field circuit is 90 Ω. Find the ratio of its speed as a generator to its speed as a motor, considering the line current in each case being 70 A.**

Solution Given data

Supply voltage, $V = 200$ V

Armature resistance, $R_a = 0.10\ \Omega$ (including brushes)

Field resistance, $R_{sh} = 90\ \Omega$

Line current, $I = 70$ A

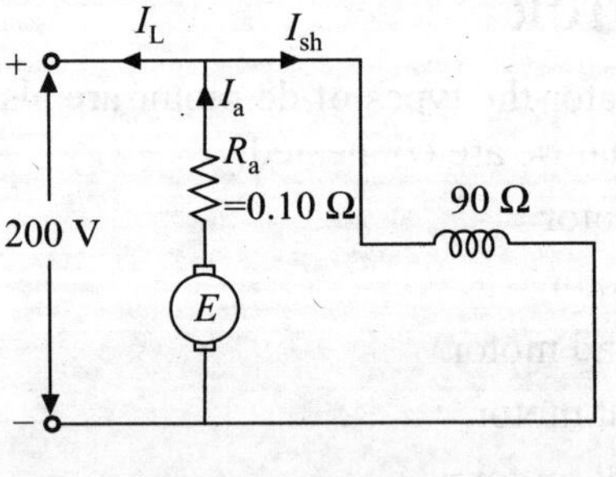

Figure 8.45 Generator

When dc machine is running as a generator,

Shunt current, $I_{sh} = \dfrac{200}{90} = 2.2$ A

The armature current, $I_a = 70 + 2.2 = 72.2$ A

Terminal voltage, $V = E_g - I_a R_a$

Generated emf, $E_g = V + I_a R_a$

$= 200 + 72.2 \times 0.10 = 207.22$ V

When a dc machine is running as a motor,

Armature current, $I_a = 70 - 2.2 = 67.8$ A

Back emf, $E_b = V - I_a R_a$

$= 200 - 67.8 \times 0.10 = 193.22$ V

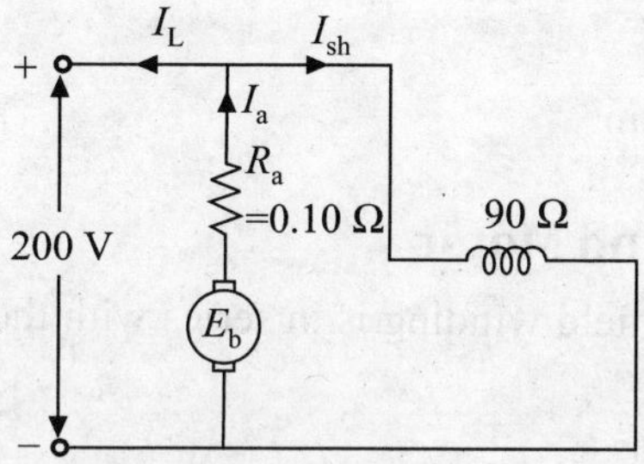

Figure 8.46

For a dc machine, the indirect emf E is proportional to the speed N,

i.e.,

$$E \propto N$$

Hence,

$$E_g = KN_1$$
$$E_b = KN_2$$

where N_1 and N_2 are the speeds at which the machine is operating as a generator and a motor, respectively.

$$\frac{N_1}{N_2} = \frac{207.22}{193.22} = 1.072$$

EXAMPLE 8.42 **A 200-V dc shunt motor has an armature resistance of 0.20 Ω and runs at 900 rpm, taking an armature current of 35 A. It is desired to reduce the speed to 800 rpm.**

(i) If the armature current remains the same, find the additional resistance to be connected in series with the armature circuit.

(ii) With the above additional resistance in the circuit, the armature current decreases to 20 A. Find the speed of the motor.

Solution Since, it is a shunt motor, its flux remains constant. The speed change is being made with the help of the armature control method.

(i) $N_1 = 900$ rpm, $E_{b1} = 200 - 35 \times 0.20 = 193$ V

$N_2 = 800$ rpm, $E_{b2} = 200 - 35R_t$

where R_t is the total armature circuit resistance.

$$\frac{800}{900} = \frac{200 - 35R_t}{193}$$

$R_t = 0.81$ Ω

If "R" is the additional resistance connected in the armature circuit, then

$R + 0.20 = 0.81$

$R = 0.61$ Ω

(ii) Here, $N_1 = 900$ rpm , $E_{b1} = 193$ V (as determined)

$N_2 = ?$, $E_{b2} = 200 - 20 \times 0.61 = 187.8$ V

$$\therefore \quad \frac{N_2}{900} = \frac{187.8}{193}$$

$$N_2 = 875 \text{ rpm}$$

8.13.2 Series Wound Motor

The motor in which the field winding is in series with the armature, as shown in Figure 8.47, is known as *series motor.*

Performance equations,

Terminal voltage, $V = E_b + I_{se}R_{se} + I_aR_a$ (8.31)

When $I_a = I_{se} = I_L$,

$V = E_b + I_L (R_a + R_{se})$ + Brush drop (8.32)

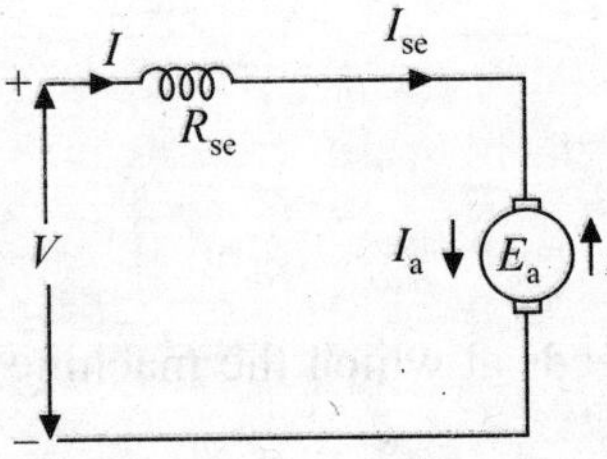

Figure 8.47

EXAMPLE 8.43 A 200-V dc series motor runs at 450 rpm when taking a line current of 20 A. The resistance of the armature is 0.4 Ω and that of the field 0.6 Ω. If the current taken remains constant, calculate the resistance necessary to reduce the speed to 325 rpm.

Solution Total armature circuit resistance = 0.4 + 0.6 = 1 Ω

For series motor, $I_{a1} = I_{a2} = 20$ A

$$E_{b1} = 200 - 20 \times 1 = 180 \text{ V}$$

$$E_{b2} = 200 - 20R_t$$

where R_t is the controller resistance = $R_a + R_{se} + R$

$$\frac{N_2}{N_1} = \frac{E_{b2}}{E_{b1}}$$

$$\therefore \quad \frac{325}{450} = \frac{200 - 20R_t}{180}$$

$$130 = 200 - 20R_t$$

$$20\,R_t - 70 = 0$$

$$\therefore \quad R_t = 3.5\ \Omega$$

$$R = 3.5 - 1 = 2.5\ \Omega$$

8.13.3 Compound Motor

If both the shunt and series fields are present in the motor, it is known as compound motor.

(i) In a long-shunt compound motor, the shunt field is across both the armature and the series field winding, as shown in Figure 8.48.

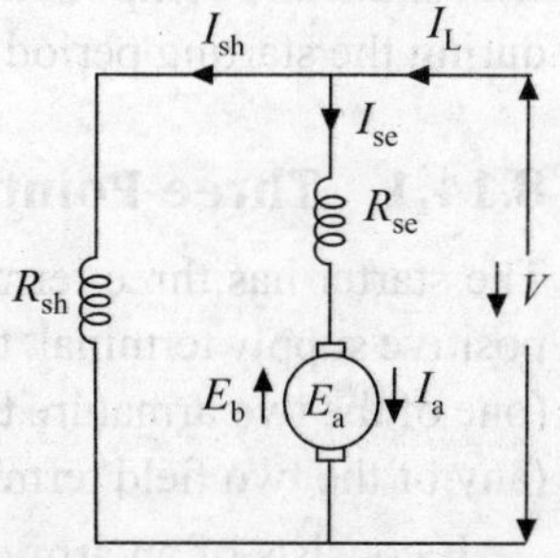

Figure 8.48 Long-shunt compound motor

$$I_a = I_{se}$$
$$I_a = I_{se} = I_L - I_{sh}$$
$$I_{sh} = \frac{V}{R_{sh}}\ ;\ I_L = I_a + I_{sh}$$
or $I_L = I_{se} + I_{sh}$
$$V = E_b + I_a R_a + I_{se} R_{se}$$
$$V = E_b + I_a (R_a + R_{se})$$

There exists a drop in the brush contact, so the voltage equation becomes

$$V = E_b + I_a (R_a + R_{se}) + V_{bc} \tag{8.33}$$

(ii) In a short-shunt compound motor, the shunt field winding is parallel only to the armature, as shown in Figure 8.49. Performance equations are

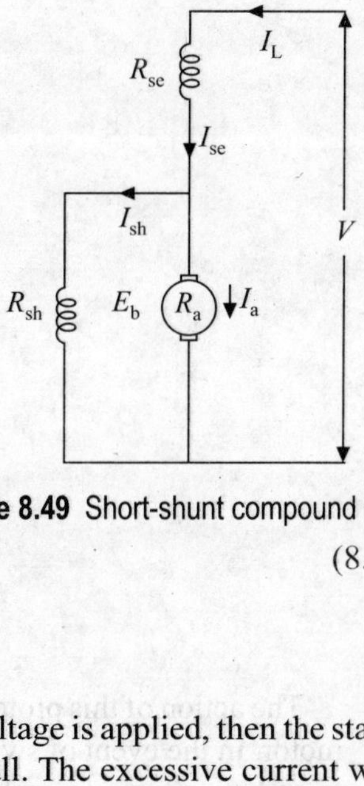

Figure 8.49 Short-shunt compound motor

$$I_L = I_{se}$$
$$I_a = I_L - I_{sh}$$
$$I_L = I_{sh} + I_a$$
$$I_{sh} = \frac{V - I_{se} R_{se}}{R_{sh}}$$
$$V = E_b + I_a R_a + I_{se} R_{se}$$

There exists a drop in the brush contact, so the voltage equation becomes,

$$V = E_b + I_a R_a + I_{se} R_{se} + V_{bc} \tag{8.34}$$

8.14 STARTERS

When a motor is at rest, there is no back emf, and if full supply voltage is applied, then the starting current is very high because the armature resistance is very small. The excessive current would be a many times of full-load value. This current will blow out the fuses and will damage the machine. For example, a 200-kW motor has a cold armature resistance of 0.28 Ω and a full-load current of 50 A. If this motor is directly switched on to a supply of 440 V, an armature current of 440/0.28 = 1572 A would flow which is 1572/50 = 31.4 times its full load current.

To avoid this excessive starting current, a resistance is introduced in series with the armature (for the duration of starting period only, say 5 to 10 seconds) and is gradually cut out as the motor gains speed and develops the back emf which then regulates its speed.

For series motors, the starter resistance is in series with both the armature and the field, but with shunt and compound motors, full shunt field is established on the first contact and maintained during the starting period while the starter resistance is progressively decreased.

8.14.1 Three-Point Starter

The starter has three terminals, L, A, and F. The starter terminal "L" (line) is connected to the positive supply terminal, the terminal "A" (armature) is connected to the motor armature terminal (one of the two armature terminals), and the terminal "F" (field) is connected to the field terminal (any of the two field terminal) as shown in Figure 8.50.

It consists of an arm or handle "H" which moves over the studs. When the arm touches the first stud, the field current is completed through brass "B" and the full resistance is placed in the armature which is gradually cut out as the handle is moved over. The handle moves against a strong spring as shown in Figure 8.50. It has a piece of soft iron "C" attached to it, which in the "full on" position is attracted and held by the electromagnet "E," which is energized by a shunt field current. This is known as *hold-on* coil or *low-voltage* (formerly no-voltage) release.

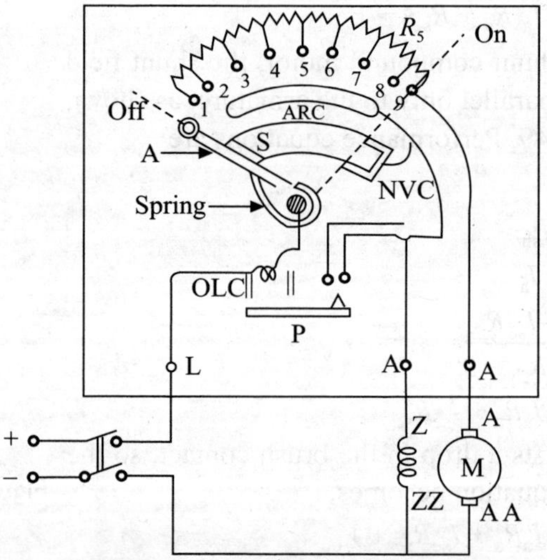

Figure 8.50 Three-point starter

The action of this protective device (hold on) is connected in series with the field winding of the motor. In the event of switching off, the supply voltage falls below a predetermined value, or the complete failure of supply while the motor is running. NVC is de-energized, this results to release the arm and allow the spring to bring it back to the OFF position. This prevents the fuses blowing, as they otherwise would, if the supply were stored with the handle in the "full on" position.

An overcurrent (or overload) release is also fitted in the starter. This consists of an electromagnetic "F," which is connected in the supply line. If the machine becomes overloaded beyond a certain predetermined value, then "D" is lifted and short circuited "E". Hence, the handle is released and returns to the "OFF" position.

8.14.2 Four-Point Starter

Figure 8.51 shows the schematic diagram of a four-point starter for a shunt motor. In a four-point starter, the no-volt release coil is connected directly across the supply line through a protective resistance R. Now, the no-volt release coil circuit is independent of the shunt-field circuit. Therefore, proper speed control can be exercised without affecting the operation of no-volt release coil. Note that the only difference between a three-point starter and a four-point starter is the manner in which the no-volt release coil is connected. However, the working of the two starters is the same. It may be noted that the three-point starter also provides protection against an open-field circuit. This protection is not provided by the four-point starter.

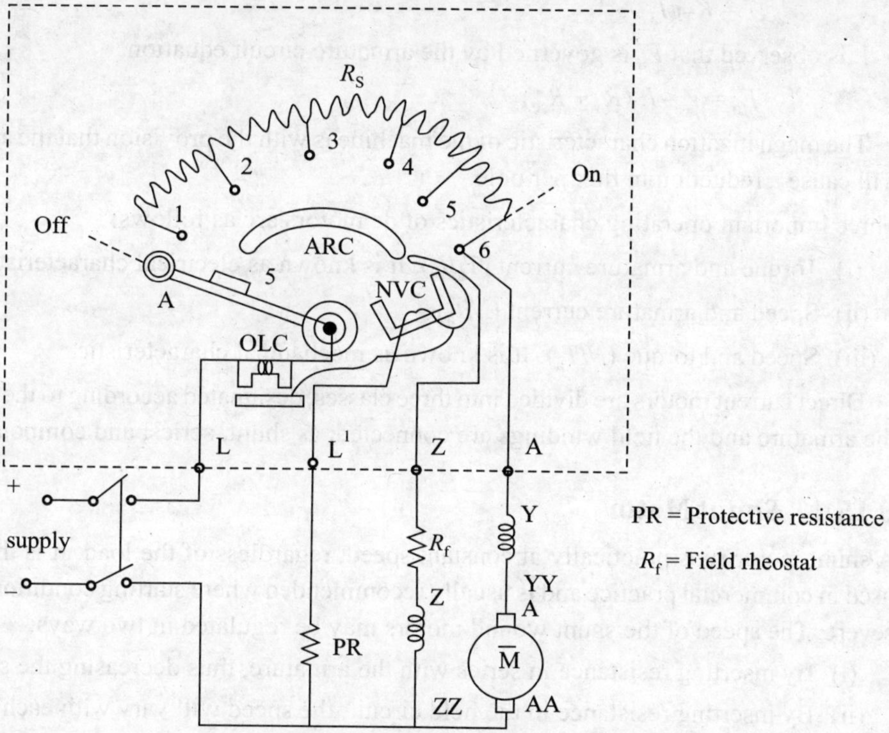

Figure 8.51 Four-point starter

8.15 CHARACTERISTICS OF DC MOTOR

In case of dc generator, the speed is fixed by the prime mover and remains nearly constant throughout the operating part of the characteristics, while the field conditions are adjusted to yield the desired terminal voltage at a given load.

On the other hand, when motor is considered the need is to match the speed–torque characteristic of the load and run the load at a specified speed by the adjustment of the field and by the adjustment of the armature voltage, in case speed control over a wide range is desired.

The fundamental emf and torque equations are

$$E_a = \frac{\phi NZ}{60}(p/A) = K_a\phi\omega \tag{8.35}$$

$$T = \frac{1}{2\pi}\phi I_a Z\,(p/A) = K_a\phi I_a \tag{8.36}$$

where N is the speed in rpm, while $\omega = 2\pi N/60$ is the speed in rad/s.

In motoring operation it is convenient to express these relationships in the form of armature speed and torque, that is,

$$N = K_N E_a/\phi$$

$$T = K_T\phi I_a \tag{8.37}$$

It is observed that E_a is governed by the armature circuit equation.

$$E_a = V - I_a\,(R_a + R_{se}) \tag{8.38}$$

The magnetization characteristic of the machine is with the provision that the armature reaction will cause a reduction in flux per pole.

Three important operating characteristics of dc motors are as follows:

(i) Torque and armature current (T_a/I_a): It is known as electrical characteristic.

(ii) Speed and armature current (N/I_a).

(iii) Speed and torque (N/T_a): It is known as mechanical characteristic.

Direct current motors are divided into three classes, designated according to the manner in which the armature and the field windings are connected, as shunt, series, and compound wound.

8.15.1 Shunt Motor

A shunt motor runs practically at constant speed, regardless of the load. It is the type generally used in commercial practice and is usually recommended where starting conditions are not usually severe. The speed of the shunt wound motors may be regulated in two ways:

(i) By inserting resistance in series with the armature, thus decreasing the speed.

(ii) By inserting resistance in the field circuit; the speed will vary with each change in load.

In the latter, the speed is practically constant for any setting of the controller. This is more generally used for adjustable speed curve, as in the case of machine tools.

For constant voltage supply the field current is constant. At small values of armature current, the demagnetizing effect of armature reaction is almost negligible, and therefore the air-gap flux is unaffected. For larger values of armature or load currents, the demagnetizing effect of armature reaction decreases the air-gap flux slightly.

The speed of a dc motor is

$$\omega_m = \frac{E_b}{K_a\phi}$$

We know that bach emf, $E_b = V_t - I_a R_a$

$$\omega_m = \frac{V_t - I_a R_a}{K_a \phi} \tag{8.39}$$

(i) T_a/I_a characteristic: Assuming ϕ to be practically constant (though at nearly loads, ϕ decreases somewhat due to increased armature reaction), we find that $T_a \propto I_a$.

Hence, the electrical characteristic is as shown in Figure 8.52(a). It is practically a straight line through the origin. The shaft torque is shown by dotted lines. The since a heavy starting load will need a nearly starting current, shunt motor should never be started on (heavy) load.

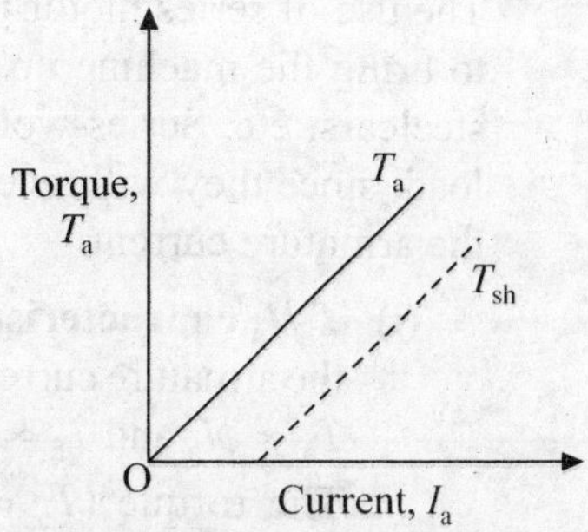

Figure 8.52 (a) *T_a/I_a* characteristic

When running on no load, a small armature current flow to supply the field and to drive the machine against the friction and other losses in it.

(ii) N/I_a characteristic: If ϕ is assumed constant, then $N \propto E_b$. As E_b is practically constant, speed is also constant.

$$\omega_{mo} = \frac{V_t}{K_a \phi}$$

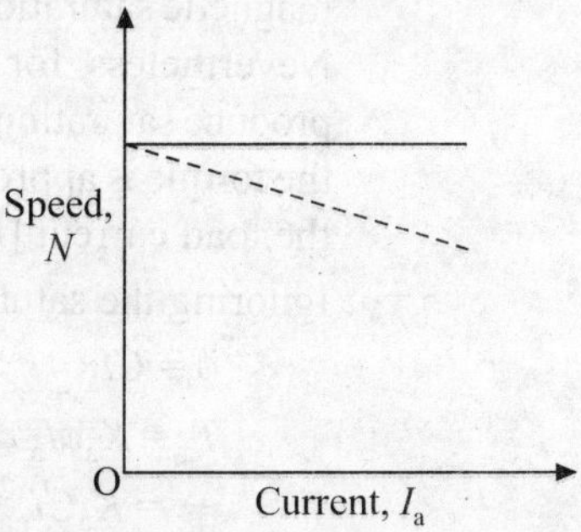

Figure 8.52 (b) *N/I_a* characteristic

As the load is applied to the motor and is increased, the torque rises almost proportionally to the increase in current. This is not quite true, because the flux has been assumed to be constant, whereas it decreases slightly owing to armature reaction. The effect of this is to cause the top of the curve connecting torque and line current to bend over as shown in Figure 8.52(b). There is no appreciable change in the speed of a shunt motor from no load to full load. It may be connected to loads which are totally and suddenly thrown off without any fear of excessive resulting speed. Due to the constancy of their speed, shunt motors are suitable for driving shafting, machine tools, lathes, wood-working machines, and for all other purposes where an approximately constant speed is required.

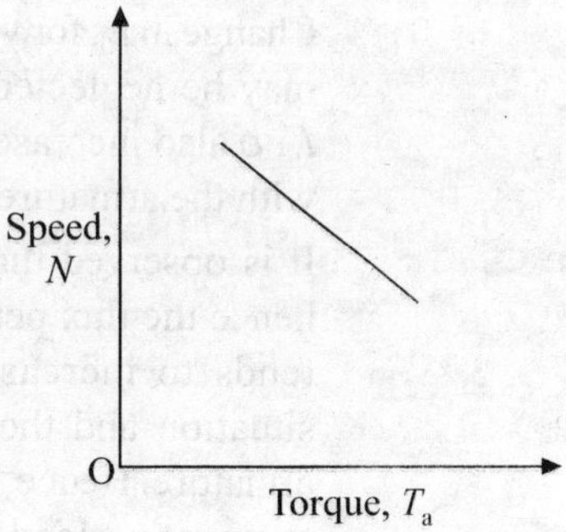

Figure 8.52 (c) *N/T* characteristic

(iii) *N/T* characteristic: The starting torque of a motor is determined by the starting resistance, which in turn, governs the initial current through the machine when the main switch is closed. At this moment the speed is zero, so that the back emf is zero and the starting current is given by $I = V/R$, where V is the supply voltage and R is the total resistance, which includes the armature and starting resistance.

8.15.2 Series Motor

The speed of series motor varies automatically with the load and increases as the load decreases. The use of series motor is generally limited to cases where a heavy power demand is necessary to bring the machine up to speed, as in the case of certain elevator and hoist installations, for steelcars, etc. Series-wound motors can never be used where the motor can be started without load, since they will race to a dangerous degree. For a series motor, the field current is equal to the armature current.

(i) T_a/I_a characteristic: We have seen that $T_a \propto \phi I_a$. In this case, as field windings also carry the armature current, $\phi \propto I_a$ up to the point of magnetic saturation. Hence, before saturation, $T_a \propto \phi I_a$ and $T_a \propto I_a^2$.

The torque ($T_a \propto I_a$) increases much than does the armature current. This is because the flux itself increases with the armature current, although owing to the magnetic saturation, the two are not strictly proportional. Nevertheless, for all but the heavy loads which tend to produce saturation of the field systems, it may be said that the torque is approximately proportional to the square of the load current [Figure 8.53(a)].

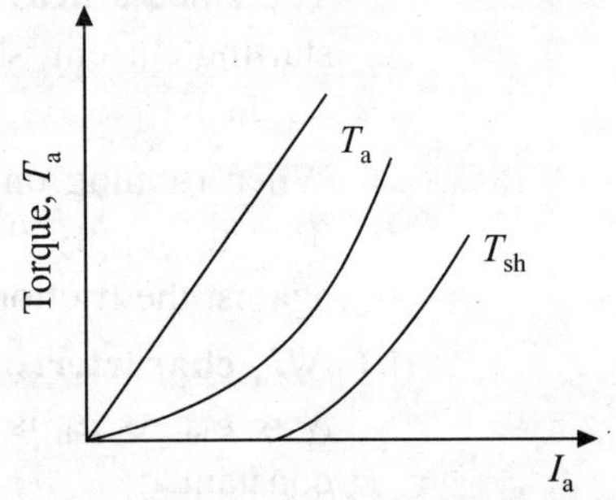

Figure 8.53 (a) T_a/I_a characteristic

Ignoring the saturation and armature currents,

$$\phi = CI_a$$

$$T_a = K_a \phi I_a$$

$$= K_a C I_a^2 = C_1 I_a^2$$

(ii) N/I_a characteristic: Variations of speed can be deduced from the formula $N \propto E_b/\phi$.

Change in E_b for various load currents is small and hence may be neglected for the time being. With increased I_a, ϕ also increases. Hence, the speed varies inversely with the armature current, as shown in Figure 8.53(b). It is observed that at no-load, the motor current and hence the flux per pole tends to zero as a motor speed tends to increase to infinity. This is a dangerous situation and the centrifugal forces will destroy the armature. Hence, a series motor must never be allowed to run at no-load, even accidentally.

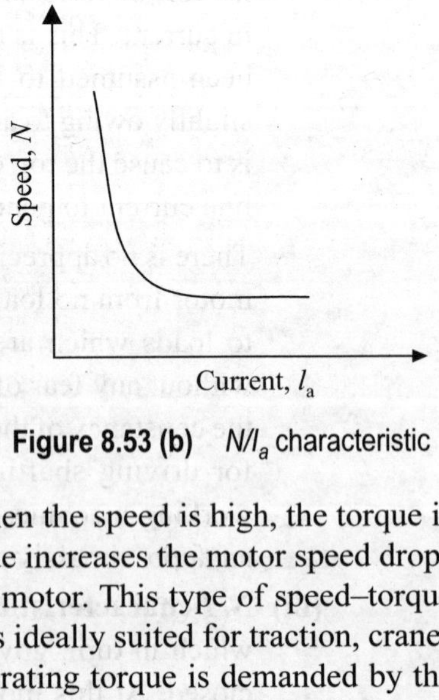

Figure 8.53 (b) N/I_a characteristic

(iii) N/T characteristic: It is found from the above that when the speed is high, the torque is low and vice-versa. It is observed that as the load torque increases the motor speed drops heavily, relieving thereby the kW load ($N \times T$) on the motor. This type of speed–torque characteristic is known as "*series*" *characteristic* and is ideally suited for traction, cranes applications, etc., where during starting a large accelerating torque is demanded by the loads. In two applications cited, the load is always on the motor. So there is no danger of underloading or no-loading.

Furthermore, if the supply voltage falls, the starting current may still be maintained at twice the full-load value by cutting out some of the starting resistance, so that the high value of starting torque may still be maintained.

This type of motor (series motor) is superior to shunt motor for drives in which machines have to be started and accelerated from rest when fully loaded, as is the case with traction equipment.

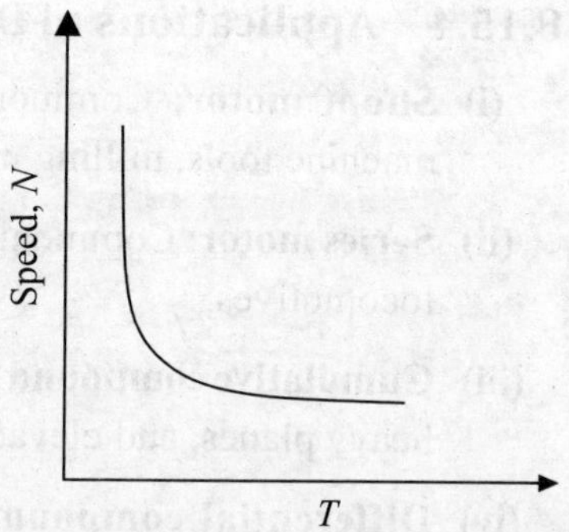

Figure 8.53 (c) *N/T* characteristic

8.15.3 Compound Motors

A combination of the shunt wound and series wound types combines the characteristics of both. Characteristics may be varied by the combination of the two windings. These motors are generally used where severe starting conditions are met and constant speed is required at the same time.

In a differential compound motor, the flux per pole decreases with increase in the armature current causing the motor speed to increase with load. This motor can be designed to give a full-load speed equal to no-load. Under accidental overload, the flux per pole reduces to almost zero and the motor can acquire high speeds similar to underloaded series motor.

In a cumulative compound motor, the series field aids the shunt field so that the flux pole increases with armature current and as a consequence (n vs I_a) lies between that of a shunt (ϕ = constant) and a series ($\phi \propto I_a$) motor. This motor finds application in rolling mills where the duty cycle comprises light- and heavy-load periods. Unlike the series motor, the cumulatively compounded motor has a finite no-load speed (no), while it is the load-receiving characteristic of a series motor under heavy-load conditions.

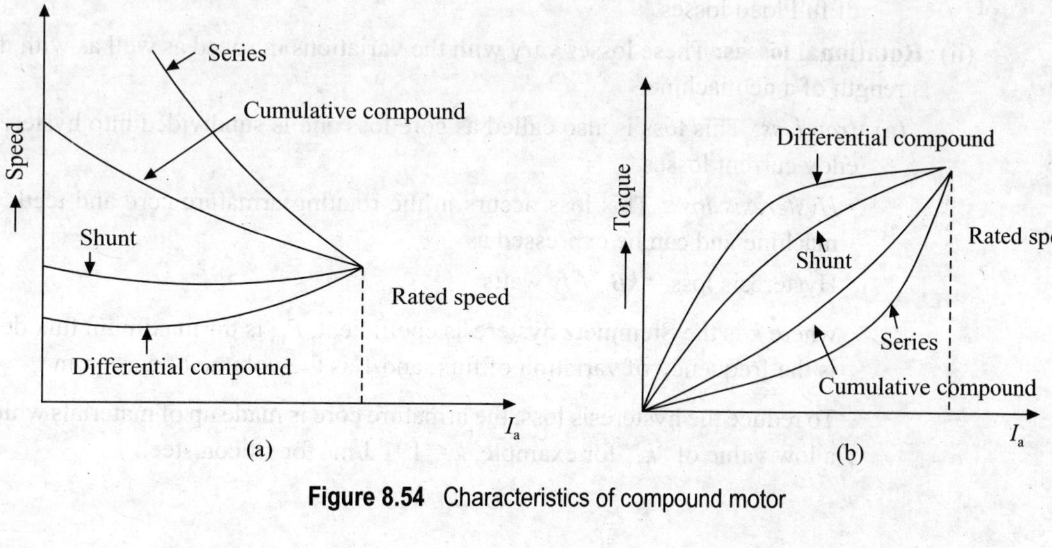

Figure 8.54 Characteristics of compound motor

8.15.4 Applications of DC Motors

(i) **Shunt motor:** Commonly used in blowers, fans, centrifugal pumps, lathe machines, machine tools, milling machines, and drilling machines.

(ii) **Series motor:** Commonly used in cranes, trolleys, conveyors, hoists, elevators, and electric locomotives.

(iii) **Cumulative compound motor:** These are commonly used in rolling mills, punches, shears, heavy planes, and elevators.

(iv) **Differential compound motor:** These are generally not suitable for any practical application.

8.16 LOSSES AND EFFICIENCY

The input energy of a machine is not available at the output because of various losses taking place in the machine. The difference between power input and power output under steady state conditions is called *power loss*. There is no use of this power loss, but it leads to heating of the machine. There are two general classifications of power losses: (i) the losses that are caused due to the flow of electric current in various parts of the machine and are termed as copper or electrical loss and (ii) the rotational losses developed due to the rotation of the armature.

(i) **Copper loss:** These losses are further subdivided into two types:

(a) *Armature copper loss:* Due to the current flow in the armature winding some amount of electrical energy is lost and is given as $I_a^2R_a$. This loss can also be called as variable loss, as it varies with the variation of load. (It varies with the square of the load current.) It is about 30–40% of full-load losses.

(b) *Field copper loss*: This loss also includes the field–rheostat loss. It is about 20–30% of full load losses.

(ii) **Rotational losses:** These losses vary with the variations in speed as well as with the field strength of a dc machine.

(a) *Iron loss:* This loss is also called as core loss and is subdivided into hysteresis and eddy current losses.

- *Hysteresis loss:* This loss occurs in the rotating armature core and teeth of a dc machine and can be expressed as

$$\text{Hysteresis loss} = kB_m^{1.6}fv \text{ watts} \tag{8.40}$$

where k is the Steinmetz hysteresis coefficient, B_m is the maximum flux density, f is the frequency of variation of flux, and V is the volume of core in m^3.

To reduce the hysteresis loss, the armature core is made up of materials which have a low value of "k," for example, $k = 191$ J/m^3 for silicon steel.

- *Eddy current loss:* When the armature core rotates in the magnetic field, an emf is induced in the core. This induced emf produces circulating current called as eddy currents and they produce some losses.

$$\text{Eddy current loss} = kB_m^2 f^2 t^2 v_c \text{ watts} \tag{8.41}$$

where k is the constant which depends on the core material, t is the thickness of the core laminations, B_m is the maximum flux density, F is the frequency of variation of flux, and V_c is the volume of core in m^3.

To reduce eddy current loss, magnetic core is made up of thin sheets of laminations. These losses are about 20–30% of full-load losses.

(b) *Mechanical loss:* This loss includes friction and windage losses. Friction losses are caused by the motion of moving parts and the friction between brushes and slip rings. Windage losses occur due to the rotation of armature in air. These losses vary with speed and are constant in a machine operating at constant speed. These losses are about 10–20% of full-load losses.

(c) *Stray load loss:* This loss is produced by (i) the distortion of air-gap flux due to the armature reaction and (ii) the currents in the commutated coil. It is difficult to measure this loss. For small machines, this loss can be neglected. For machines of 150 kW rating or more, they are assumed to be 1% of the rated output.

Efficiency of a dc machine

$$\text{Efficiency, } \eta = \frac{\text{Output}}{\text{Input}}$$

$$= \frac{\text{Output}}{\text{Input} + \text{Losses}}$$

Total output of a machine, $P_{out} = VI$

Copper losses, $P_{cu} = I_a^2 R_a$

Constant losses = P_o

Total input of a machine, $P_i = P_{out} + P_{cu} + P_o$

$$\%\eta = \frac{P_{out}}{P_{out} + P_{cu} + P_o} \times 100 \tag{8.42}$$

$$= \frac{VI}{VI + I_a^2 R_a + P_o} \times 100$$

$$= \frac{VI}{VI + I^2 R_a + P_o} \times 100 \quad \text{(Since } I_a = I \text{ neglecting shunt field current)}$$

$$= \frac{1}{1 + \frac{IR_a}{V} + \frac{P_o}{VI}} \times 100$$

Efficiency is maximum when the denominator is minimum. According to the maxima–minima theorem,

$$\frac{d}{dI}\left[1+\frac{IR_a}{V}+\frac{P_o}{VI}\right]=0$$

$$\therefore \quad P_{cu} = P_0 \tag{8.43}$$

8.17 TESTING OF DC MACHINES

8.17.1 Direct Test

Let V be the voltage applied, I the total line current drawn, R the radius of pulley in meters, N the speed in rpm, and W_1 the spring balance readings.

$$\text{Net pull} = W_1 - W_2 \text{ kg}$$

$$= (W - W_2)\, 9.81 \text{ N} \tag{8.44}$$

$$\text{Torque developed, } I_{sh} = \text{Net pull radius}$$

$$= 9.81\,(W_1 - W_2)\, R \text{ newtons/m} \tag{8.45}$$

$$\text{Output power obtained, } P_{out} = T_{sh} W$$

$$= 9.81 \times (W_1 - W_2) \times R \times \frac{2IIN}{60} \text{ watts} \tag{8.46}$$

$$\text{Power input, } P_{in} = VI \text{ watts} \tag{8.47}$$

$$\%\eta = \frac{P_{out}}{P_{in}} \times 100 \tag{8.48}$$

Advantages

(i) Actual efficiency of the motor under working conditions can be found.

(ii) This method is simple and easy.

(iii) This method can be performed on any type of dc motor.

Disadvantages

(i) The power developed gets wasted, hence this method is expensive.

(ii) Some type of cooling arrangement is necessary.

(iii) Overfriction and heat, and large dissipation of energy.

8.17.2 Swinburne Test

The machine is run as a motor on no-load. The input voltage and current are measured. The armature resistance and field resistance are measured by using battery, voltmeter, and ammeter. The obtained resistance is again calculated at required temperature. The stray load losses are assumed to be 1% of the full-load output.

Let V be the supply voltage; I_0 the input line current at no-load; R_a^{1}, R_f^{1} the measured values of armature and field resistances at ambient temperature, and R_a, R_f the armature and field resistances at operating temperature.

$R_a = R_a^1 (1 + \alpha_1 \times 40)$ ($\because$ At 40°C temperature)

$R_f = R_f^1 (1 + \alpha_1 \times 40)$

Input at no-load = VI_0

Field copper loss at no-load = $\dfrac{V^2}{R_f^1}$

Armature copper loss at no-load = $\left(I_0 - \dfrac{V}{R_f^1}\right)^2 R_a^1$

Mechanical and iron losses = $VI_0 - \dfrac{V_2}{R_f^1} - \left(I_0 - \dfrac{V}{R_f^1}\right) R_a^1$

Armature copper loss at load = $I_a^2 R_a$

Field copper loss at load = $\dfrac{V^2}{R_f}$

Total losses at load = Mechanical and iron losses + Armature copper losses + Field copper losses + Stray losses

To calculate the efficiency η at 1/4th load:

Let I_{fl} be the full-load current of motor, wf the field copper loss, and w the stray load losses.

Load current at 1/4th load = $\dfrac{I_{fl}}{4}$

Motor input at 1/4th full load = $V \times \dfrac{I_{fl}}{4}$ watts

Armature current at 1/4th load = $I_a^1 = \dfrac{I_{fl}}{4} - I_{sh}^1$

Armature copper loss at 1/4th load = $\left(\dfrac{I_{fl}}{4} - I_{sh}^1\right)^2 R_a$

Motor output at 1/4th load = Motor input at 1/4th load-losses

Motor output at 1/4th load = $\left(V \times \dfrac{I_{fl}}{4}\right) - \left(\dfrac{I_{fl}}{4} - I_{sh}^1\right)^2 R_a - wf - w$

$$\therefore \text{Efficiency, } \eta = \frac{\left(V \times \dfrac{I_{fl}}{4}\right) - \left(\dfrac{I_{fl}}{4} - I_{sh}^1\right)^2 R_a - wf - w}{V \times \dfrac{I_{fl}}{4}} \tag{8.49}$$

Advantages

(i) The efficiency can be estimated at any load because constant losses are known.

(ii) This method is convenient and economical.

Disadvantages

(i) This test cannot be performed for series motor.

(ii) Assumed temperature of 40°C at full load cannot be checked actually.

8.17.3 Hopkinson Test

It is a regenerative test and requires two identical dc shunt machines. The two machines are mechanically coupled. The field currents of the two machines are adjusted such that one works as a motor and drives the other machine, which works as a generator. The power input from the mains is only that needed for supplying the losses of the two machines. The two machines can be tested under full-load conditions. The connection diagram for the Hopkinson test is shown in Figure 8.55.

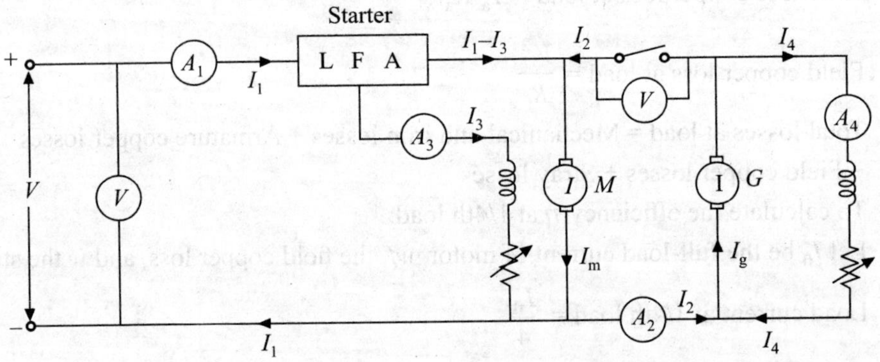

Figure 8.55 Connection diagram for Hopkinson test

Total input = VI_1

Armature current of generator = I_g

$I_g = I_2 + I_4$

Ammeter current motor = I_m

$I_m = I_1 - I_3 + I_2$

Armature copper loss of machine-I = $I_m^2 R_{a1}$

Armature copper loss of machine-II = $I_g^2 R_{a2}$

Field copper loss of machine-I = VI_3

Field copper loss of machine-II = VI_4

Total iron and mechanical losses of the two machines = $VI_1 - I_m^2 R_{a1} - I_g^2 R_{a2} - VI_3 - VI_4$

Iron and mechanical losses of each machine = $Pc = \frac{1}{2}(VI_1 - I_m^2 R_{a1} - I_g^2 R_{a2} - VI_3 - VI_4)$

Total losses of the motor = $I_m^2 R_{a1} + VI_3 + Pc$

Total losses of the generator = $I_g^2 R_{a2} + VI_4 + Pc$

Input of the motor = $V(I_1 + I_2)$

$$\therefore \quad \eta_{\text{motor}} = \frac{\text{Input} - \text{Losses}}{\text{Input}} = \frac{V(I_1 + I_2) - (I_n^2 Ra_1 + VI_3 + P_3)}{V(I_1 + I_2)} \tag{8.50}$$

Output of the generator = VI_2

$$\eta_{\text{gen}} = \frac{\text{Output}}{\text{Input} + \text{Losses}} = \frac{VI_2}{VI_2 + I_g^2 R_{a2} + VI_4 + Pc} \tag{8.51}$$

Advantages

(i) Two machines are tested under full-load conditions.
(ii) It can be used to determine efficiency as well as temperature rise.
(iii) The total power input is only to meet the losses.

Disadvantages

(i) The test requires two identical machines.
(ii) The iron and mechanical losses are assumed to be equal for the two machines, which are not correct for small machines.

8.17.4 Retardation Test or Running Down Test

The connection diagram for retardation test is shown in Figure 8.56. This method is generally employed to shunt generator and shunt motor. From this method we can get stray lasses. If the armature and field copper losses at any given load are known, then the efficiency of a machine can be calculated. The connections required for conducting this test is shown in the circuit below.

The machine whose test is to be taken is run at a speed which is slightly above the normal speed. The supply of the motor is to be cut-off, while the field is kept excited. The ammeter consequently slows down and its kinetic energy is used in supplying the rational or stray losses, which include iron, friction, and windage losses.

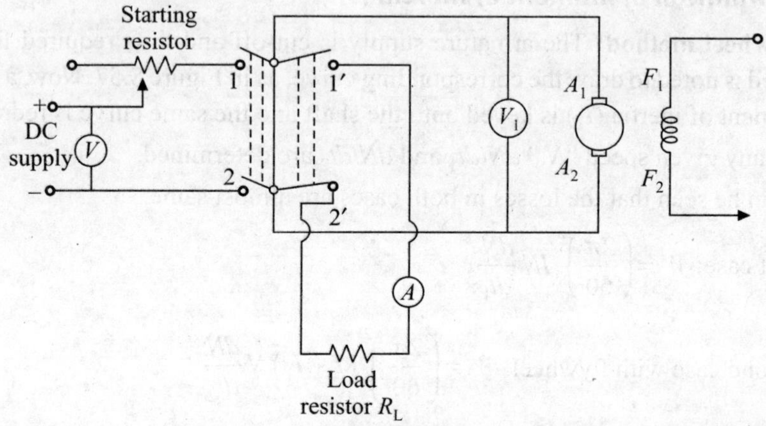

Figure 8.56

If I is the moment of inertia of the armature and W the angular velocity, then the kinetic energy of the armature is $(1/2)IW^2$.

Rotational losses, W = Rate of change of kinetic energy

$$= \frac{d}{dt}\frac{1}{2}\left(IW^2\right) = Iw\frac{dw}{dt}$$

$$W = I\left(\frac{2\pi N}{60}\right)\frac{d}{dt}\left(\frac{2\pi N}{60}\right)$$

$$= \left(\frac{2\pi}{60}\right)^2 IN\frac{dN}{dt} \quad (8.52)$$

So to calculate the stray losses, I and dN/dt must be known.

Determination of dN/dt

The voltmeter V_1, which is connected across the armature, will read the back emf of the motor. We know that the back emf is proportional to the speed, so the voltmeter is calibrated to read the speed directly. When the motor is cut-off from the supply, the speed decreases. The time required for definite decrease in the speed is noted with the help of a stop-watch.

A curve drawn between time and speed is shown in Figure 8.57.

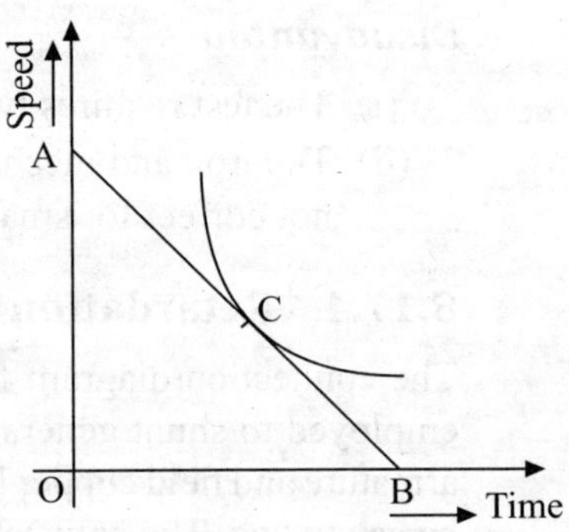

Figure 8.57

At any point C corresponding to the normal speed, a tangent AB is drawn. Then

$$\frac{dN}{dt} = \frac{\text{OA (in rpm)}}{\text{OB (in seconds)}}$$

Determination of moment of inertia (I)

(i) **Flywheel method:** The armature supply is cut-off and time required for definite change in speed is noted to draw the corresponding curve, as in Figure 8.57. Now, a flywheel with known moment of inertia (I_1) is keyed onto the shaft and the same curve is redrawn again.

For any given speed "N," dN/dt_1 and dN/dt_2 are determined.

It can be seen that the losses in both cases are almost same.

First case: $W = \left(\frac{2\pi}{60}\right)^2 IN\frac{dN}{dt_1}$

Second case with flywheel: $W = \left(\frac{2\pi}{60}\right)^2 (I + I_1)N\frac{dN}{dt_2}$

$$I_n\frac{dN}{dt_1} = (I + I_1)N\frac{dN}{dt_2}$$

$$\frac{I+I_1}{I} = N\frac{dN}{dt_1} \Big/ N\frac{dN}{dt_2}$$

$$\frac{I+I_1}{I} = \frac{dN/dt_1}{dN/dt_2}$$

$$I = \frac{(I+I_1)dN/dt_2}{dN/dt_1}$$

$$= \frac{I(dN/dt_2)}{dN/dt_1} + \frac{I_1(dN/dt_2)}{dN/dt_1}$$

$$I - \frac{I(dN/dt_2)}{dN/dt_1} = \frac{I_1(dN/dt_2)}{(dN/dt_1)}$$

$$\frac{I(dN/dt_1) - I(dN/dt_2)}{dN/dt_1} = \frac{I_1(dN/dt_2)}{dN/dt_1}$$

$$I[(dN/dt_1) - (dN/dt_2)] = I_1(dN/dt_2) \tag{8.53}$$

$$I = I_1\frac{dN/dt_2}{(dN/dt_1) - (dN/dt_2)}$$

$$= I_1\frac{dt_1}{dt_2 - dt_1}$$

$$= I_1\frac{t_1}{t_2 - t_1}$$

(ii) **Without flywheel:** In this method, a retarding torque, either mechanical or electrical, is applied. The method by which electrical torque can be provided is shown in Figure 8.56. If the switch "*S*" is connected to $1^1\ 2^1$, then the machine gets connected to a noninductive load resistance R_L. The power drawn by this resistance will act as a retarding torque on the armature, which will make it slow more quickly.

The additional loss in the resistance will be equal to the product of ammeter reading and the average reading of the voltmeter. Then the additional loss is $I_n^2\ (R_a+R_L)$. Let t_1 be the time when the armature is considered alone, t_2 the time when the armature is connected across the load resistance, V the average voltage across R_L, I_a the average current, and W^1 the retarding electrical torque supplied by the motor.

$$W = \left(\frac{2\pi}{60}\right)^2 IN\frac{dN}{dt_1}$$

$$W + W^1 = \left(\frac{2\pi}{60}\right)^2 IN\frac{dN}{dt^2}$$

$$\frac{W+W^1}{W} = \frac{1/dt_2}{1/dt_1} = \frac{t_1}{t_2}$$

$$(W + W^1)t_2 = Wt_1$$

$$W = W^1 \frac{t_2}{t_1 - t_2} \tag{8.54}$$

EXAMPLE 8.44 A 440-V shunt generator has a full-load current of 250 A, its armature resistance is 0.08 Ω, and field resistance 150 Ω, the stray losses are 2800 W. Find the horse power of the prime mover when it is delivering full load and also find the load for which the efficiency of the generator is maximum.

Solution Given data

Terminal voltage, $V = 440$ V

Full-load current, $I_L = 250$ A

Armature resistance, $R_a = 0.08\ \Omega$

Shunt field resistance, $R_{sh} = 150\ \Omega$

Stray losses = 2800 W

Full-load output = $VI = 440 \times 250 = 110000$ W

Shunt current, $I_{sh} = 440/150 = 2.93$ A

Armature current, $I_a = 250 + 2.93 = 252.933$ A

Armature copper loss $= 252.933^2 \times 0.08 = 5118$ W

Shunt copper loss = $440 \times 2.93 = 1289.2$ W

Stray losses = 2800 W

Total losses in the generator = 2800 + 5118 + 1289 = 9207 W

Total input = 110000 + 9207 = 119207 W

$$\text{hp of prime mover} = \frac{119207}{735.5} = 162.07$$

The total current for which efficiency is maximum is given by

$$I = \sqrt{\frac{W_c}{R_a}}$$

Constant loss = Stray losses + Shunt copper loss

= 2800 + 1289 = 4089 W

$$I = \sqrt{\frac{4089}{0.08}} = \sqrt{51112} = 226.08 \text{ A}$$

EXAMPLE 8.45 A shunt generator gives a full-load output of 10 kW at a terminal potential of 220 V. The armature and field resistances are 0.8 Ω and 60 Ω, respectively. If the mechanical and iron losses combined amount to 400 W, calculate the horse power required at the driving shaft on full load and the full-load efficiency.

Solution Given data

Generator output, $P = 10$ kW

Terminal voltage, $V = 220$ V

Armature resistance, $R_a = 0.8\ \Omega$

Shunt field resistance, $R_{sh} = 60\ \Omega$

Mechanical and iron losses = 400 W

Shunt current, $I_{sh} = 220/60 = 3.6$ A

Output current, $I = 10000/220 = 45.45$ A

Armature current, $I_a = 3.6 + 45.45 = 49.05$ A

Armature copper loss, $I_a^2 R_a = 2406 \times 0.8 = 1924$ W

Shunt field copper loss = 220 × 3.6 = 792 W

Total losses = 1924 + 792 + 400 = 3116 W

Input = 10000 + 3116 = 13116 W

$$\text{hp (metric)} = \frac{13116}{735.5} = 17.83$$

$$\text{Full-load efficiency} = \frac{10000}{13116} \times 100 = 76.24\%$$

EXAMPLE 8.46 A cutting tool exerts a tangential force of 350 N on a steel of diameter 15 cm, which is being turned in a simple latche. The latche is drawn by chain at 900 rpm from a 220-V dc motor running at 2000 rpm. Calculate the current taken by the motor if its efficiency is 85%. What is the size of the motor pulley if the latche pulley has a diameter of 26 cm?

Solution Given data

Tangential force, $F_T = 350$ N

Diameter of steel, $D_1 = 15$ cm

Speed of the machine, $N_1 = 2000$ rpm

Terminal voltage, $V = 220$ V

Speed of chain, $N_2 = 900$ rpm

Efficiency, $\eta = 85\%$

Diameter of the pulley, $D_2 = 26$ cm

Torque = Tangential force × Radius

= 350 × 0.075 = 26.25 N-m

$$\text{Output power} = \frac{T_a \times 2\pi N}{60} \text{ watts}$$

$$= 26.25 \times 2\pi \times 900 / 60$$

$$= 2474 \text{ W}$$

Motor efficiency = 0.85 or 85%

Motor input = 2474/0.85 = 2910 W

Current drawn by motor = 2910/220 = 13.22 A

Let N_1 and D_1 be the speed and diameter of the driver pulley, respectively, and N_2 and D_2 be the speed and diameter of the latche pulley, respectively.

$$N_1 \times D_1 = N_2 \times D_2$$

$$2000 \times D_1 = 900 \times 0.26$$

$$D_1 = \frac{900 \times 0.26}{2000} = 0.117 \text{ m or } 11.7 \text{ cm}$$

EXAMPLE 8.47 A 250-V dc shunt motor takes 5 A at rated voltage on no-load. The armature and field resistances are 0.35 Ω and 250 Ω, respectively. Calculate the efficiency of the motor for a load current of 45 A.

Solution Given data

Terminal voltage, $V = 250$ V

No-load current, $I_o = 5$ A

Armature resistance, $R_a = 0.35\ \Omega$

Shunt field resistance, $R_{sh} = 250\ \Omega$

Full-load current, $I_L = 45$ A

No-load motor input = 250 × 5 = 1250 W

This input goes to meet no-load losses, which consist of (i) armature copper losses and (ii) constant losses.

This no-load power can be used to find the constant losses of the motor.

Shunt current, $I_{sh} = 250/250 = 1$ A

Armature current, $I_a = 5 - 1 = 4$ A

Armature copper loss, $I R_a = 16 \times 0.35 = 5.6$ W

Constant losses of motor = 1250 – 5.6 = 1244.4 W

With load current of 45 A,

$$I_a = 45 - 1 = 44 \text{ A}$$

Armature copper loss = $44^2 \times 0.35 = 677.6$ W

Motor input = 250 × 45 = 11250 W

Total losses = 677.6 + 1244.4 = 1922 W

Motor output = 11250 – 1922 = 9328 W

Motor efficiency = 9328/11250

= 0.83 or 83%

EXAMPLE 8.48 A shunt generator delivers 198 A at a terminal potential difference of 250 V. The armature resistance and shunt field resistance are 0.04 Ω and 50 Ω, respectively. The iron and friction losses equal 960 W. Find

(i) The emf generated

(ii) The copper losses

(iii) The output of the prime mover in kW

(iv) The commercial, mechanical, and electrical efficiencies

Solution: Given data

Terminal voltage, $V = 250$ V

Load current, $I_L = 198$ A

Armature resistance, $R_a = 0.04\ \Omega$

Shunt field resistance, $R_{sh} = 50\ \Omega$

Iron and friction losses = 960 W

(i) Shunt current, $I_{sh} = 250/50 = 5$ A

Armature current, $I_a = 198 + 5 = 203$ A

Armature voltage drop, $I_aR_a = 203 \times 0.04 = 8.12$

∴ Generated emf = 250 + 8.12 = 258.12 V

(ii) Armature copper loss = $I^2_a\, R_a = 203^2 \times 0.04 = 1648$W

Shunt copper loss = $V\,I_{sh} = 250 \times 5 = 1250$ W

∴ Total copper loss = 1250 + 1648 = 2898 W

(iii) Stray losses = 960W

Total losses = 2898 + 960 = 3858 W

Output = 250 × 198 = 49500 W

Input = 49500 + 3858 = 53358 W

∴ Output of the prime mover = 53358 W = 53.35 kW

(iv) Input = 53358 W, Stray loss = 960 W

Electrical power produced in armature = 53358 − 960 = 52398 W

$$\text{Electrical efficiency} = \frac{52398}{53358}\times 100 = 98.2\%$$

Electrical or copper losses = 2898 W

$$\eta_c = \frac{49500}{49500+2898}\times 100 = 94.47\%$$

$$= \frac{7.46\times 10^3}{0.85}\times 100 = 92.77\%$$

EXAMPLE 8.49 **A 440-V shunt motor has an armature resistance of 0.4 Ω and a field resistance of 150 Ω. Determine the back emf when giving an output of 7.46 kW at 85% efficiency.**

Solution Given data

Terminal voltage, $V = 440$ V

Armature resistance, $R_a = 0.4\ \Omega$

Shunt field resistance, $R_{sh} = 150\ \Omega$

Output power, $P = 7.46$ kW

Efficiency = 85%

$$\text{Motor input power} = \frac{7.46 \times 10^3}{0.85} = 8776.47 \text{ W}$$

$$\text{Motor input current, } I = \frac{8776.47}{440} = 19.95 \text{ A}$$

Shunt current, $I_{sh} = 440/150 = 2.93$ A

Armature current, $I_a = I - I_{sh} = 19.95 - 2.93 = 17.01$ A

Back emf, $E_b = V - I_a R_a$

$$= 440 - 17.01 \times 0.4 = 433.2 \text{ V}$$

EXAMPLE 8.50 **A dc shunt motor takes 2 A at no-load when connected to a 250-V dc mains with an armature resistance of 1.5 Ω when the field current is 1 A. Determine the load current corresponding to the maximum efficiency.**

Solution Given data

No-load current, $I_o = 2$ A

Terminal voltage, $V = 250$ V

Armature resistance, $R_a = 1.5\ \Omega$

Shunt field current, $I_{sh} = 1$ A

No-load armature current, $I_{ao} = 2 - 1 = 1$ A

No-load input to the motor = $250 \times 2 = 500$ W

No-load copper loss = $I_{a0}^2 R_a = 1.5$ W

The total constant losses on no-load

$$W_c = 500 - 1.5 = 498.5 \text{ W}$$

If I_a is the current corresponding to maximum efficiency,

$$I_a^2 R_a = W_c$$

$$I_a^2 R_a = 498.5 \text{ W}$$

$$R_a = 1.5\ \Omega \quad \text{(Given)}$$

$$I_a^2 = \frac{498.5}{1.5} = 332.33$$

$$I_a = 18.23 \text{ A}$$

Therefore, load current corresponding to maximum efficiency

$$I = I_a + I_f$$
$$= 18.23 + 2$$
$$= 20.23 \text{ A}$$

Calculation of maximum efficiency

Input to the motor = 250 × 20.23 = 5057.5 W

Armature copper loss = $20.23^2 \times 1.5 = 613.88$ W

∴ Total loss = 613.88 × 2 = 1227.76 W

∴ Output of the motor = 5057.5 – 1227.76

= 3829.74 W

$$\therefore \text{ Maximum efficiency of the motor} = \frac{3829.74}{5057.5} \times 100 = 76\%$$

EXAMPLE 8.51 A 4-pole, 250-V wave-connected motor gives 9 kW when running at 900 rms and drawing armature and field currents of 60 A and 1 A, respectively. It has 560 conductors. Its armature resistance is 0.4 Ω. Assuming a drop of 1 V per brush, determine the (i) total torque, (ii) useful torque, (iii) useful flux per pole, (iv) rotational losses, and (v) efficiency.

Solution Given data

Number of poles, $p = 4$

$A = 2$ (since the type of connection is wave wound)

Terminal voltage, $V = 250$ V

Output power, $P = 9$ kW

Speed of the motor, $N = 900$ rpm

Armature current, $I_a = 60$ A

Shunt field current, $I_{sh} = 1$ A

Number of conductors, $Z = 560$

Armature resistance, $R_a = 0.4\ \Omega$

Voltage drop per brush, $V_b = 1$ V

$$E = V - I_a R_a - \text{brush drop}$$
$$= 250 - 60 \times 0.4 - 2\ 1$$
$$= 224 \text{ V}$$

(i) $$EI_a = T \times \frac{2\pi N}{60}$$

$$T = \frac{60 \times E \times I_a}{2\pi N}$$

$$= \frac{60 \times 224 \times 60}{2\pi \times 9000} = 142.6 \text{ N-m}$$

(ii) T useful = T shaft

$P_{out} = 9 \text{ kW} = 9 \times 10^3 \text{ W}$

$$T_{useful} \times \frac{2\pi N}{60} = P_{out}$$

$$T_{useful} = \frac{60 \times 9 \times 10^3}{2\pi \times 900}$$

$= 95.5$ N-m

(iii) $$E = \frac{NP\phi Z}{60A}$$

$$\phi = \frac{60EA}{NPZ} = \frac{60 \times 224 \times 2}{900 \times 4 \times 560}$$

$= 0.0133$ Wb

(iv) Armature drop $= VI_{a2} = 250 \times 60 = 15000$ W

Armature copper loss $= I_a^2 R_a$

$= (60)^2 \times 0.4 = 1440$ W

Brush contact loss $= V_b I_a$

$= 2 \times 60 = 120$ W

Power developed $= VI - I_a^2 R_a - V_b I_a$

$= 15000 - 1440 - 120$

$= 13440$ W

Total power output + Rotational losses = Power developed

$\therefore$ Rotational losses = Power developed – Total power output

$= 13440 - 9000 = 4440$ W

(v) Total input to motor,

$VI = V(I_a + I_{sh})$

$= 250\,(60 + 1)$

$= 15250$ W

$$\text{Motor efficiency } = \frac{\text{Motor output}}{\text{Motor input}} \times 100 = \frac{9 \times 10^3}{15250} \times 100 = 59\%$$

EXAMPLE 8.52 A 8-kW, 250-V dc shunt generator has total no-load rotational loss of 350 W. The armature circuit (including brushes) and shunt field resistances are 0.7 Ω and 250 Ω, respectively. Calculate the shaft power input and the efficiency at rated load. Also calculate the maximum efficiency and the corresponding power output.

Solution Given data

Terminal voltage, $V = 250$ V

Output power, $P = 8$ kW

No-load rotational loss = 350 W

Armature resistance, $R_a = 0.7\ \Omega$

Shunt field resistance, $R_{sh} = 250\ \Omega$

Constant shunt field current, $I_f = 250/250 = 1$ A

Constant losses = Shunt field losses + Rotational losses

= 250 × 1 + 350

= 600 W

Generator output current at rated output, $I_L = \dfrac{8000}{250} = 32$ A

Therefore, the generated armature current, $I_a = I_L + I_f$

$= 32 + 1 = 33$ A

Armature circuit loss, $I_a^2 R_a = (33)^2 \times 0.7$

= 762.3 W

Therefore, the generator shaft power input = 8000 + 600 + 762.3

= 9362.3 W

Therefore, at rated load,

$$\eta_g = \frac{\text{Output}}{\text{Input}} \times 100$$

$$= \frac{8000}{9362.3} \times 100$$

= 85.45%

Maximum efficiency occurs when,

Variable losses = Constant losses = 600 W

Therefore, the total losses at maximum efficiency = 600 + 600 = 1200 W

But variable losses $I_a^2 R_a = 600$ W; therefore, armature current at maximum efficiency is

$$I_a = \sqrt{\frac{600}{0.7}} = 29.27 \text{ A}$$

Generator output current, $I_a - I_f = 28.27$ A

$\therefore$ Generator output power = 250 × 28.27 = 7069 W.

Generator input = 7069 + 1200 = 8269 W

$$\therefore \text{Maximum, } \eta = \left(1 - \frac{1200}{8269}\right) \times 100$$

$$= 85.48\%$$

EXAMPLE 8.53 **A 250-V, 10-kW shunt motor has a maximum efficiency of 85% and a speed of 900 rpm, when delivering 80% of its rated output. The resistance of its shunt fields is 90 Ω. Determine the efficiency and speed when the motor draws a current of 75 A from the mains?**

Solution Given data,

Terminal voltage, $V = 250$ V

Output power, $P = 10$ kW

Maximum efficiency, $\eta_m = 85\%$

Speed of the motor, $N = 900$ rpm

Load current, $I_L = 75$ A

Efficiency at rated output, $\eta_{FL} = 80\%$

$$\text{Efficiency, } \eta = \frac{\text{Output}}{\text{Output} + \text{Losses}}$$

$$\text{Total losses} = \left(\frac{1}{\eta} - 1\right)\text{Output}$$

$$= \left(\frac{1}{0.85} - 1\right) \times 0.8 \times 10000$$

$$= 0.1764 \times 8000$$

$$= 1412 \text{ W}$$

At maximum efficiency

Constant losses = Rotational losses + Shunt field losses

= Variable armature circuit losses $I_a^2 R_a$

= 1412/2 = 706 W

Input to motor at maximum efficiency = 0.8 × 10000 + 1412

= 9412 W

$$\text{Input line current, } I_{L1} = \frac{9412}{250} = 37.64 \text{ A}$$

∴ Armature current,

$$I_{a1} = 37.64 - \frac{250}{90}$$

$$= 34.86 \text{ A}$$

$$I_{a1}^2 R_a = (34.86)^2 R_a = 706 \text{ W}$$

Armature circuit resistance, $R_a = \frac{706}{(34.86)^2} = 0.58\ \Omega$

Now, motor input current, $IL_2 = 75$ A

$$I_{a2} = 75 - 2.77 = 72.22 \text{ A}$$

$$I_{a2}^2 R_a \text{ losses} = (72.22)^2 \times 0.58$$

$$= 3025.3 \text{ W}$$

Total power input = 75 × 250 = 18750 W

Total losses = 3025.3 + 706 = 3731.3 W

∴ Motor efficiency for a line current of 72.22 A = $\left(1 - \frac{3731.3}{18750}\right) \times 100$

$$= 80\%$$

Counter emf, $E_{a1} = V - I_{a1} R_a$

$$= 250 - 34.86 \times 0.58$$

$$= 229.78 \text{ V}$$

$$E_{a2} = 250 - 72.22 \times 0.58$$

$$= 208.11 \text{ V}$$

$$\frac{E_{a1}}{E_{a2}} = \frac{228}{228} = \frac{N_1 \phi_1}{N_2 \phi_2} = \frac{900 \times \phi_1}{N_2 \times \phi_2}$$

Constant field current $\phi_1 - \phi_2$,

$$\eta_2 = \frac{900 \times 208.11}{229.73} = 815 \text{ rpm}$$

EXAMPLE 8.54 A dc shunt motor, with armature circuit resistance of 0.01 Ω runs at 1500 rpm, while taking an armature current of 90 A from a 225-V dc source. The friction and windage loss is 250 W, no-load core losses are 1200 W, and the total I^2R loss is 2000 W. The stray load loss equals 1.5% of the output. Find the shaft torque of the motor and its efficiency.

Solution Given data

Armature resistance, $R_a = 0.1\ \Omega$

Speed of the motor, $N = 1500$ rpm

Armature current, $I_a = 90$ A

Terminal voltage, $V = 225$ V

Friction and windage losses = 250 W

No-load core losses = 1200 W

I^2R losses = 2000 W

Motor counter emf, $E_a = 225 - 90 \times 0.1 = 216$ V

Electromagnetic power, $E_a I_a = 216 \times 90 = 19440$ W

No-load rotational loss = 250 + 1200 = 1450 W

Substituting no-load rotational loss from electromagnetic power, we get

$$= 19440 - 1450$$

$$= 17990 \text{ W}$$

Let P_{sh} be the shaft power output.

Therefore, stray load loss = $0.015P_{sh}$

$17990 - 0.015P_{sh} = 0$

$$\text{Shaft power,}\quad P_{sh} = \frac{17990}{0.015} = 17724.14$$

$$\text{Shaft torque, } T_{sh} = \frac{17724.14 \times 60}{2\pi \times 1500} = 112.83 \text{ N-m}$$

Power input to motor = Electromagnetic power + Total I^2R loss

$$= 19440 + 2000$$

$$= 214400 \text{ W}$$

$$\therefore \text{ Motor efficiency} = \frac{\text{Shaft power}}{\text{Output power}} = \frac{17724}{21440} \times 100 = 82.8\%$$

EXAMPLE 8.55 **A 90-kW, 500-V shunt generator runs as a motor on no-load at its rated voltage and speed. The total current taken was 9.0 A including a shunt field current of 2.0 A. The resistance of the armature circuit at normal working temperature is 0.2 Ω. Calculate the efficiency of the generator at full load and at half load.**

Solution Input at no-load = 9.0 × 500

$$= 4500 \text{ W}$$

This includes iron, friction, and windage losses, armature copper loss at no-load, and field copper loss.

Therefore, iron, friction, and windage losses $= 4500 - (7)^2 \times 0.2 - 500 \times 2.0$

$$= 4500 - 9.8 - 1000$$

$$= 3509.8 \text{ W}$$

Full-load output current of the generator = 200 + 2.0

= 202 A

Full-load armature copper loss = $(202)^2 \times 0.2$

= 8160.8 W

Shunt field copper loss = 500 × 2.0

= 1000 W

Efficiency of the generator at full load

$$= \frac{100\times1000}{100\times1000+3590.8+8160+1000}\times100$$

$$= \frac{100000}{100000+12562}\times100$$

= 88.69%

For calculating efficiency at half-load, the variable loss, the armature copper loss at half load is to be calculated.

Half-load output = 100 A

Armature current at half-load = 100 + 2.0 = 102 A

Armature copper-loss at half-load = $(102)^2 \times 0.2 = 2080.8$ W

$$\text{Efficiency of the generator at half-load} = \frac{50\times1000}{50\times1000+2080+1000+3509.8}\times100$$

$$= \frac{50000}{50000+6590.6}\times100$$

= 88.35%

EXAMPLE 8.56 A 8-kW, 220-V dc shunt motor draws a line current of 5.6 A, while running at a no-load speed of 1000 rpm. It has an armature resistance of 0.30 Ω and a field resistance of 150 Ω. Estimate the efficiency of the motor when it delivers rated load.

Solution Constant shunt field current, = 220/150 = 1.46 A

No-load armature current, $I_{a0} = 5.6 - 1.46 = 4.13$ A

Constant no-load rotational loss, $\omega_0 = V_t I_{a0} - I_a^2 r_a$

$= 220 \times 4.13 - 17.08 \times 0.30$

$= 908.6 - 5.125$

= 903.4 W

Electromagnetic power, $E_a I_a$ = Shaft power + No-load rotational loss

$(220 - I_a \times 0.30) I_a = 8000 + 903.4$

$0.30 I_a^2 - 220 I_a + 8903.4 = 0$

Its solution gives $I_a = 42.994$

Motor input = $E_a I_a$ + Full-load $I_a^2 R_a$ loss + Shunt field loss

Motor input = $8903.4 + (42.99)^2 \times 0.30 + 1.46 \times 240$

$= 8903.4 + 554.44 + 350.4$

$= 9808.24$ W

$$\text{Motor efficiency} = \frac{8000}{9808.24} \times 100$$

$= 0.8156 \times 100$

$= 81.56\%$

CHAPTER AT A GLANCE

- An electrical machine converts energy from mechanical to electrical form or electrical to mechanical form. This process is called electromechanical energy conversion.
- An electrical machine which converts mechanical energy into electrical energy is called electric generator, whereas an electrical machine which converts an electrical energy into mechanical energy is called an electric motor.
- Machines are of two types: dc machines and ac machines.
- The dc machines are classified as dc generators and dc motors.
- The direction of flux can be reversed either by changing the direction of current through the conductor or by reversing the polarities of the supply or by changing the direction of winding of the conductors around the core.
- The basic components of generating action are conductor, flux, and relative motion between the conductor and the flux.
- The main function of commutator is to collect current from the armature conductors.
- The emf equation of a dc generator is given by $E = \phi ZNP/60A$.
- The field winding of generator is also called as exciting winding, and the current carried by the field current is called an exciting current.
- There are two methods of excitation used for dc generators: (i) separate excitation and (ii) self-excitation.
- For shunt generator, $I_a = I_L + I_{sh}$.
- For series generator, $I_a = I_{se} = I_L$.
- While plotting the load characteristics, the speed of the generator is maintained constant.
- A motor is a device which converts an electrical energy into mechanical energy.
- The magnitude of the force experienced by the conductor in a motor is given by $F = BIl$ newtons.
- The emf induced in the motor is called back emf.
- The back emf regulates the flow of armature current and automatically alters the armature current to meet the load requirement. This is the practical significance of the back emf.

- The condition for maximum power of a motor is $E_b = V/2$.
- The torque equation of dc motor is given by $T_a = 0.159\phi I_a\ (ZP)/A$.
- The different types of motors are (i) shunt motor, (ii) series motor, and (iii) compound motor.
- Generally, a motor is switched on with normal voltage, and as armature resistance is very small, the armature current at start is very high.
- The various losses in a dc machine are copper losses, iron losses, and mechanical losses.
- The condition for maximum efficiency of a dc machine is that variable losses must be equal to constant losses.

MULTIPLE CHOICE QUESTIONS

1. The dc generator works on the application principle of

(a) Fleming's left-hand rule (b) Fleming's right-hand rule
(c) Lenz's law (d) None of these

2. A dc shunt motor is operating at no-load. If the armature is shunted by a resistance, its speed will

(a) increase (b) decrease (c) remain the same (d) be unpredictable

3. Fleming's left-hand rule is applicable to

(a) dc generator (b) dc motor (c) alternator (d) transformer

4. Consider the following statements regarding speed control of dc motors:

(i) Ward Leonard method is suitable for constant torque drive.
(ii) Field control methods facilitate speed control below the base speed.
(iii) Armature resistance control method is more efficient when compared to the Ward Leonard method.
(iv) Field control method is suitable for constant power drives.

Of the statements

(a) (i), (ii), and (iii) are correct (b) (i), (iii), and (iv) are correct
(c) (ii), (iii), and (iv) are correct (d) (i) and (iv) are correct

5. The emf induced in a conductor rotating in a bipolar field is

(a) dc (b) ac (c) both dc and ac (d) none of these

6. Consider the following statements regarding the starting of a dc shunt motor:

(i) The external field resistance should be minimum.
(ii) The external field resistance should be maximum.
(iii) The reduced voltage should be applied to the armature.
(iv) The increased voltage should be applied to the armature.

Of these statements

(a) (i) and (iii) are correct (b) (i) and (iv) are correct
(c) (ii) and (iii) are correct (d) (ii) and (iv) are correct

7. If a current-carrying coil is placed in a unique magnetic field with its plane perpendicular to the direction of magnetic induction, then

(a) both the net force and the torque on the coil are zero
(b) the net force is zero but the torque is finite
(c) the net force is finite but the torque is zero
(d) both the net force and the torque are finite

8. Which of the following tests can be conducted on other than shunt machines?

(a) Swinburne test (b) Retardation test (c) Field test (d) Back-to-back test

9. The commutator in a dc machine can convert

(a) ac to dc (b) dc to ac (c) both (a) and (b) (d) none of these

10. A 4-pole dc generator is running at 1500 rpm. The frequency of the current in the armature is

(a) 25 Hz (b) 50 Hz (c) 0 (d) 100 Hz

11. Pole shoe of a dc machine is laminated for the purpose of

(a) decreasing hysteresis loss
(b) decreasing eddy current loss
(c) decreasing both hysteresis and eddy current losses
(d) manufacturing case

12. Poles of dc machine are often laminated to

(a) reduce pulsation loss (b) reduce armature reaction
(c) reduce iron weight (d) dissipate more heat

13. Running the machine at no-load is inadvisable for

(a) dc shunt motor (b) dc series motor
(c) induction motor (d) synchronous motor

14. In dc machines, the armature windings are placed on the rotor because of the necessity for the

(a) electromechanical energy conversion (b) generation of voltage
(c) commutation (d) development of torque

15. A 4-pole lap-wound dc machine had an armature current I_a. The conductor current in the armature winding is

(a) I_a (b) I_a/P (c) PI_a (d) none of these

16. A 4-pole generator with 16 coils has a two-layer lap winding. The pole pitch is

(a) 32 (b) 16 (c) 8 (d) 4

17. The number of parallel paths for a 4-pole duplex lap winding will be

(a) 2 (b) 4 (c) 6 (d) 8

18. Minimum number of coils for 400-V, 6-pole lap-connected dc machine for a maximum voltage of 15 V between adjacent commutators segments should be

(a) 80 (b) 160 (c) 200 (d) 400

19. The purpose of providing dummy coils in a wave-wound dc machine armature is to

(a) enhance flux density (b) amplify voltage
(c) mechanical balance for the rotor (d) reduce eddy current loss

20. The armature resistance of a 6-pole lap-wound dc machine is 0.05 Ω. If the armature is rewound using a wave winding, then the armature resistance will be

(a) 0.45 (b) 0.30 (c) 0.15 (d) 0.10

21. A frog-leg winding is used in a dc machine to eliminate

(a) the use of equalizers in lap winding
(b) the use of dummy coils in a armature
(c) the use of compensating winding
(d) the reactance produced due to commutation

22. Approximate physical size of a 50-kW, 1800-rpm dynamo as a function of a 25-kW, 900 rpm dynamo will be

(a) 1 (b) 1/2 (c) 1/4 (d) none of these

23. Air gap at the pole tips of a dc machine is kept more than that at the center of the mains to reduce

(a) reactance voltage (b) effect of armature reaction
(c) losses of armature core (d) noise of the machine

24. If the armature current in a dc machine is increased to double its previous value and the time of commutation is halved, then the reactance voltage will

(a) be halved (b) remain the same (c) be doubled (d) become four times

25. Interpoles are provided in dc machines to

(a) neutralize the cross-magnetizing component of armature reaction
(b) neutralize the demagnetizing component of armature reaction
(c) reduce iron loss
(d) reduce copper losses

26. A dc shunt motor drives a load at a rated speed and supply voltage. If both the load and the voltage are halved, the speed of the motor will be

(a) almost doubled (b) almost halved
(c) the rated speed (d) slightly less than the rated speed

27. The voltage at the terminals of a dc series generator running at a rated rpm and no-load will be

(a) more than the rated voltage (b) full-rated voltage
(c) a very small voltage (d) zero

28. A dc generator beyond critical resistance will generate

(a) maximum power (b) maximum voltage
(c) maximum current (d) no voltage

29. If the circuit resistance of a dc shunt generator exceeds its critical value, the generator

(a) fails to build (b) builds up a very high voltage
(c) exceeds its current (d) produces power beyond its rating

30. The voltage regulation of a dc generator at full load being zero implies that the generator is

(a) shunt-connected (b) differentially compounded
(c) cumulatively compounded (d) series-connected

31. The type of generator used for welding purpose is a

(a) series generator (b) shunt generator
(c) cumulatively compound (d) differentially compound generator

32. Which motor should not be used for centrifugal pumps?

(a) Shunt (b) Series
(c) Cumulatively compound (d) Differentially compound

33. A 200-V dc machine has an armature resistance of 0.5. If the full-load armature current is 30 A, the induced emf when the machine run (i) as a generator and (ii) as a motor will be

(a) 230 V, 170 V (b) 225 V, 175 V (c) 185 V, 215 V (d) 215 V, 185 V

34. The direction of the rotation of a dc shunt motor can be reversed by interchanging

(a) the supply terminals (b) the field terminals only
(c) the armature terminals only (d) either field or armature terminals

35. When the supply terminals of a dc shunt motor are interchanged

(a) the motor will stop
(b) the motor will run at its normal speed in the same direction as before
(c) the direction of the rotation will be reversed
(d) the motor will run much faster in the same direction

36. If the load current and flux of a dc motor are held constant and voltage applied across its armature is increased by 5%, the speed of the motor will

(a) increase by 5% (b) reduce by 5%
(c) remain unalternated (d) depends on other factors

37. A dc shunt motor is running at light load, what will happen if the field windings get opened?

(a) The motor will stop (b) The motor will burn
(c) The motor will make noise (d) The motor will pick up high speed

38. If the applied voltage to a dc machine is 230 V, then the back emf, for maximum power, will be

(a) 115 V (b) 200 V (c) 230 V (d) 460 V

39. A dc series motor develops a torque of 20 N-m at 3 A of load current. If the current is increased to 6 A, the torque developed will be

(a) 10 N-m (b) 20 N-m (c) 80 N-m (d) 40 N-m

40. The speed of a dc shunt motor can be increased above its normal speed by

(a) increasing the field current (b) decreasing the field current
(c) decreasing the terminal voltage (d) increasing the armature resistance

41. A fly wheel is normally fitted to a

(a) dc series motor driving a constant torque load
(b) separately excited dc motor driving pulsed torque load
(c) cumulatively compound motor driving torque load
(d) differentially compound motor driving pulsed torque load

42. An electric train employing a dc series motor is running at a fixed speed, when a sudden slight drop in the mains voltage occurs. This would result in the
    (a) drop in speed and rise in current
    (b) rise in speed and drop in current
    (c) rise in speed and current
    (d) drop in speed and current

SHORT ANSWER QUESTIONS

1. Define dc generator.
2. List the basic requirements to produce emf.
3. State Fleming's right-hand rule.
4. Write a brief note on electromechanical energy conversion.
5. Draw the simple loop generator showing slip rings and brushes.
6. Explain the principle of a dc generator.
7. List the various parts of a dc generator under the heads stationary and rotating.
8. State the functions of yoke of a dc generator.
9. List the materials used for (i) yoke, (ii) brush, (iii) armature, and (iv) brush holder.
10. List the advantages and disadvantages of brushes.
11. Draw the sketch of a magnetic pole and label the parts.
12. What purpose is served by the pole shoe in a dc machine?
13. State the function of commutator in a dc generator.
14. Define (i) conductor, (ii) turn, (iii) coil, and (iv) coil span.
15. Define (i) pole pitch, (ii) full pitch, (iii) back pitch, and (iv) front pitch.
16. Compare lap and wave windings.
17. Derive an emf equation for a dc generator.
18. Classify dc generators based on their field excitation.
19. What is the basis for classification of dc generators?
20. Compare dc shunt and series generators.
21. List the various losses in a dc machine.
22. Draw the power flow diagram of a dc generator.
23. Derive the condition for maximum efficiency of a dc generator.
24. What is armature reaction?
25. Define MNA and GNA.
26. What is axis of commutation?
27. Derive an expression for AT_d/pole.
28. Derive an expression for AT_c/pole.
29. Define communication.
30. What do you understand by under communication?
31. Draw the interlope arrangement.
32. State the functions of interlopes.
33. List the types of generator characteristics.

34. Plot an OCC of a separately excited dc generator.
35. Plot an OCC of a dc shunt generator.
36. Plot an OCC of a dc series generator.
37. Plot the characteristic of compound generator.
38. Plot the internal and external characteristics of (a) shunt and (b) series generators.
39. Define level, under, and over compounding of a compound generator.
40. State the requirements of voltage build up in self-excited dc generator.
41. Define critical speed.
42. Define critical resistance.
43. List the advantages of parallel operation.
44. State the conditions for parallel operation.
45. What is the necessary requirement of equalizer bar?
46. Which generator requires equalizer bar?
47. State the applications of dc generators.
48. What is a dc motor?
49. State Fleming's left-hand rule.
50. What is the significance of back emf?
51. Write the formula for back emf and their units.
52. Classify dc motors.
53. Derive the torque equation of a dc motor.
54. Give voltage equation and power equation for a dc motor.
55. List the various losses in a dc motor.
56. Briefly explain the power stages in a dc motor.
57. Draw the electrical and mechanical characteristics of a dc shunt motor.
58. Draw the electrical and mechanical characteristics of a dc series motor.
59. Draw the electrical and mechanical characteristics of a dc compound motor.
60. Why series motor is having high-starting torque?
61. State the uses of dc motors.
62. What is the necessity of speed control of dc motors?
63. List the methods of speed control of a dc motor.
64. Draw the methods of speed control of a dc motor.
65. What are the various methods of speed control of a dc series motor?
66. What is the necessity of starter for a dc motor?
67. State the function of a no-volt coil in a 3-point starter.
68. State the function of an over-load coil in a 3-point starter and a 4-point starter.
69. List the various methods of motor testing.
70. State the merits and demerits of Swinburne test.
71. What is the main difference between brake test and Swinburne test?
72. Draw the connection diagram for brake test.

73. Draw the connection diagram for Swinburne test.
74. Draw the connection diagram for Hopkinson test.
75. State the merits and demerits of Hopkinson test.

LONG ANSWER QUESTIONS

1. Explain the simple loop generator.
2. Explain the construction of a dc generator with neat sketches wherever required.
3. Briefly explain the working of a dc generator.
4. Classify the dc generators along with voltage and current equations showing diagrammatically.
5. Explain the various losses in a dc generator.
6. Explain the armature reaction with diagrams.
7. Explain the commutation with neat sketches.
8. Describe the various methods to improve the commutation.
9. Explain an OCC of a separately excited dc generator.
10. Explain an OCC of a dc shunt generator.
11. Explain an OCC of a dc series generator.
12. Draw the load characteristics of dc shunt generator and explain.
13. Explain the load characteristics of dc series generator.
14. Explain the load characteristics of dc compound generator.
15. How critical resistance can be obtained? Explain.
16. Explain the parallel operation of dc shunt generators.
17. Describe the parallel operation of dc series generators.
18. Narrate the parallel operation of dc compound generators.
19. Explain the working of dc motor with a neat diagram.
20. Explain the OCC, internal, and external characteristics of a dc shunt motor.
21. Explain the OCC, internal, and external characteristics of a dc series motor.
22. Explain the OCC, internal, and external characteristics of a dc compound motor.
23. Explain with diagrams the speed control of a dc shunt motor.
24. Explain the various methods of speed control of a dc series motor.
25. Draw the neat sketch of a 3-point starter and explain.
26. What is the importance of NVL and OLC in starter?
27. Explain 4-point starter with a diagram.
28. With a neat diagram explain the working of a drum controlled stater.
29. Explain with diagram the brake test on a given dc shunt motor.
30. Draw a circuit diagram to perform a test and determine constant loss.
31. Describe with a diagram the back-to-back test.

Chapter 9

Induction Motors

Chapter Objectives

- To give constructional details and principle of operation
- To discuss slip–torque characteristics, losses, and efficiency of machine
- To explain testing and starting of induction motors
- To discuss speed control and applications of induction motors
- To give comparison between synchronous and induction motors

9.1 INTRODUCTION

The three-phase induction motors are the most widely used electric motors in industry. Almost more than 90% of the mechanical power used in industry is provided by three-phase induction motors. Induction motor has a simple and rugged construction. It is a self-starting motor. Moreover, it is a constant speed motor from no-load to full load. It is available in both single phase and three phases. These are also called *generalized transformers* because the principle of operation very much resembles that of a transformer. The main difference is that the induction motor has a rotating part. Further in the case of a transformer, there is no conversion of energy, whereas in an induction motor, the electrical energy input is converted into mechanical energy output. In this also, the energy is transferred by induction. The main advantages which make an induction motor so popular for industrial use are as follows:

(i) It is simple and extremely rugged construction
(ii) It is relatively cheap
(iii) High reliability
(iv) Less maintenance
(v) It has a self-starting torque
(vi) It has high efficiency

The disadvantages of induction motor are as follows:

(i) When the load is varied, speed is not constant

(ii) Low starting torque compared to dc shunt motor

(iii) Reduction in efficiency when the speed is varied

9.2 CONSTRUCTION OF THREE-PHASE INDUCTION MOTOR

The three-phase induction motor is very simple in construction compared to a dc motor or a synchronous motor. An induction motor essentially consists of two parts: (i) stator and (ii) rotor.

9.2.1 Stator

It consists of a steel frame which encloses a hollow, cylindrical core made up of thin laminations of silicon steel to reduce hysterisis and eddy-current loss. A number of slots are provided on the inner periphery of the laminations. The slots house the three-phase winding. The three-phase winding is called *stator winding*. It may be connected either in delta or in star. The stator winding is wound for definite number of poles. The requirement of poles will be determined by the requirement of speed. Greater the number of poles lesser the speed and vice versa. When the three-phase supply is given to stator winding a rotating magnetic field is produced. This rotating field induces an emf in the rotor by mutual induction hence the name is called induction motor. Figure 9.1 shows the stator lamination.

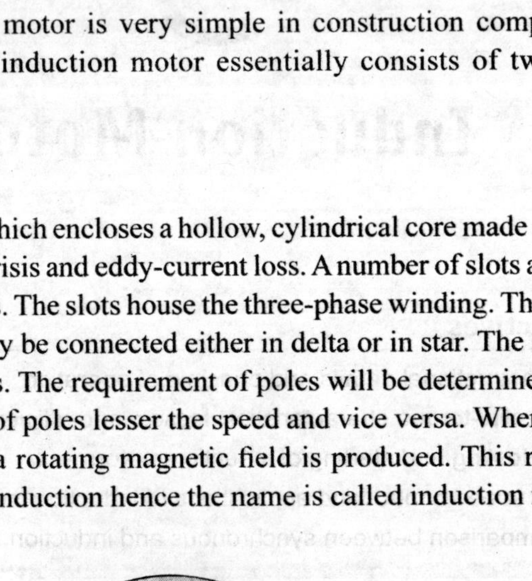

Figure 9.1 Stator lamination

9.2.2 Rotor

There are two types of rotors used in induction motors: (i) squirrel cage rotor and (ii) wound rotor or slip ring rotor.

Squirrel cage rotor

Rotors employing this type of rotor are known as squirrel cage induction motors. Figure 9.2 shows the squirrel cage rotor. This is made up of a cylindrical laminated core with slots to carry the rotor conductors. One copper or aluminum bar is placed in each slot. All these bars are joined at each end by using metal rings called end rings. This forms a permanently short-circuited winding. The rotor is not connected electrically to the supply but has current induced in it by induction from the stator.

Most of the three-phase induction motors are squirrel cage induction motors because of its simple and robust construction. But the disadvantage of this type of motor is low starting torque because the rotor bars are permanently short circuited and it is not possible to add any external resistance to the rotor circuit to have a large starting torque.

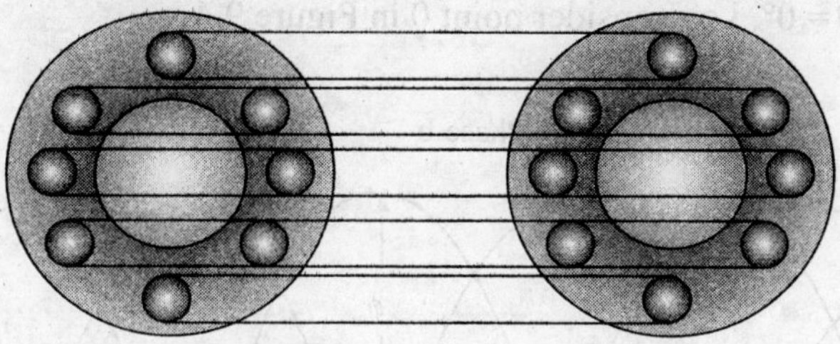

Figure 9.2 Squirrel cage rotor

Wound rotor or slip ring rotor

Figure 9.3 shows the slip ring or wound rotor. In this type of rotor, rotor windings are similar to stator windings. It consists of laminated steel core and carries three-phase windings. The rotor winding is uniformly distributed in the slots and is usually star connected. The open ends of the rotor winding are brought out and connected to slip rings mounted on the rotor shaft and a variable external resistance can be connected in the rotor circuit, with the help of brushes and slip ring arrangements. By varying external resistance in the rotor circuit, the motor speed and torque can be controlled.

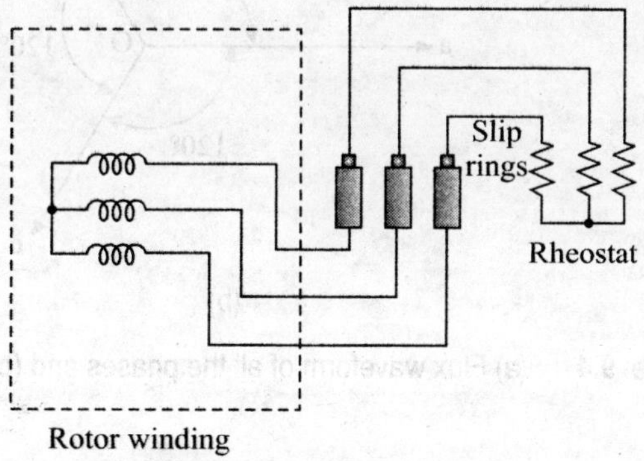

Figure 9.3 Slip ring rotor

9.3 PRODUCTION OF ROTATING MAGNETIC FIELD

When three-phase windings, displaced by 120° in space, are energized by three-phase currents, displaced in time by 120°, they produce a resultant magnetic flux which rotates in space. The field is such that its poles do not remain in a fixed position on the stator but go on shifting their position around the stator. Let the maximum value of flux due to any phase is ϕ_r, which at any instant is given by the vector sum of three fluxes ϕ_a ϕ_b ϕ_c due to three phases.

Now we will find out the value of ϕ_r in some instants, 0, 1, 2, 3 as shown in Figure 9.4. The assumed positive directions of flux are shown in Figure 9.4.

(i) When $\theta = 0°$, i.e., consider point 0 in Figure 9.4.

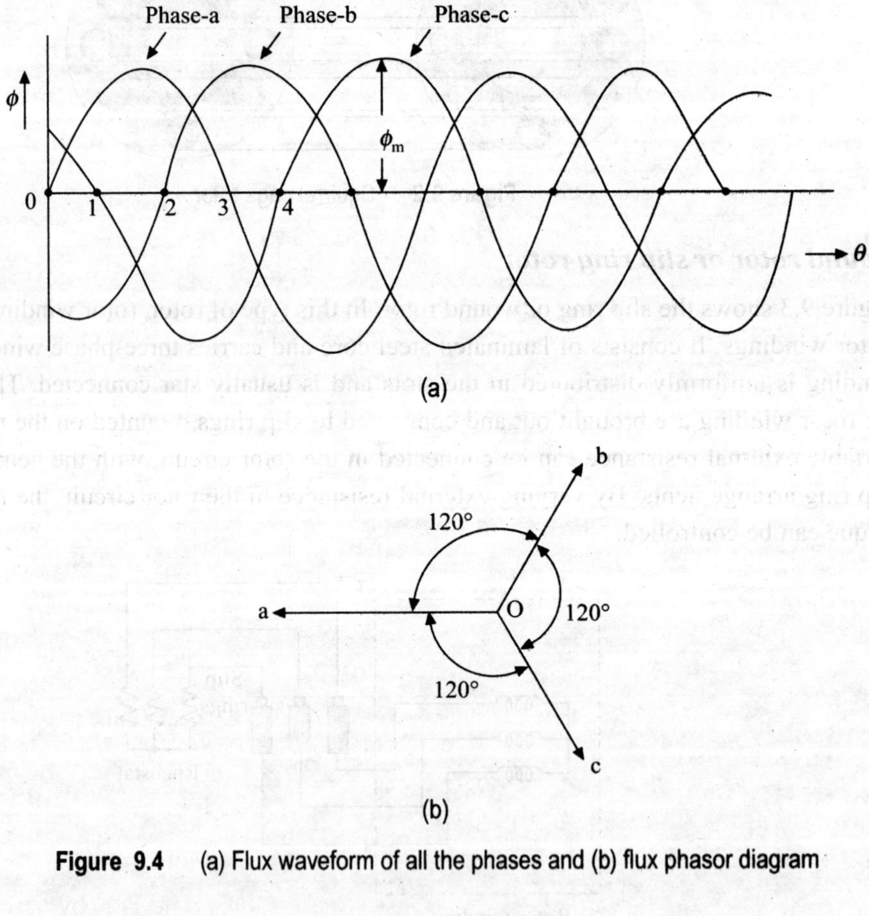

Figure 9.4 (a) Flux waveform of all the phases and (b) flux phasor diagram

At this case

$$\phi_a = 0,\ \phi_b = \frac{-\sqrt{3}}{2}\theta_m,\ \phi_c = \frac{\sqrt{3}}{2}\theta_m \tag{9.1}$$

Therefore, the resultant flux ϕ_r is

$$\phi_r = 2\times\frac{\sqrt{3}}{2}\phi_m \cos\frac{60}{2} = \frac{3}{2}\phi_m \quad (9.2)$$

(ii) $\theta = 60°$, i.e., consider point (i)

Here

$$\phi_a = \frac{\sqrt{3}}{2}\phi_m;\ \phi_b = \frac{-\sqrt{3}}{2}\phi_m;\ \phi_c = 0 \quad (9.3)$$

Resultant flux

$$\phi_r = 2\times\frac{\sqrt{3}}{2}\phi_m \cos\frac{60}{2} = \frac{3}{2}\phi_m \quad (9.4)$$

(iii) $\theta = 120°$, i.e., at point (ii)

Here

$$\phi_a = \frac{\sqrt{3}}{2}\phi_m;\ \phi_b = 0;\ \phi_c = \frac{-\sqrt{3}}{2}\phi_m \quad (9.5)$$

Resultant flux

$$\phi_r = 2\times\frac{\sqrt{3}}{2}\phi_m \cos\frac{60}{2} = \frac{3}{2}\phi_m \quad (9.6)$$

(iv) $\theta = 180°$, i.e., at point (iii)

Here

$$\phi_a = 0;\ \phi_b = \frac{\sqrt{3}}{2}\phi_m;\ \phi_c = \frac{-\sqrt{3}}{2}\phi_m \quad (9.7)$$

Resultant flux

$$\phi_r = 2\times\frac{\sqrt{3}}{2}\phi_m \cos\frac{60}{2} = \frac{3}{2}\phi_m \quad (9.8)$$

Therefore, the magnitude of this rotating magnetic field is constant and is equal to 1.5 times ϕ_m, where ϕ_m is the maximum flux due to any phase.

9.3.1 Synchronous Speed

The speed at which the revolving magnetic flux rotates is called synchronous speed, N_s. Its value depends on the number of poles and the supply frequency and is given by

$$N_S = \frac{120f}{P} \quad (9.9)$$

where P is the number of poles on stator and f is the supply frequency in Hz.

9.3.2 Slip

Slip is defined as the difference between the speeds of stator field, i.e., synchronous speed and the rotor expressed as a fraction of the synchronous speed. It is denoted by S.

It is usually expressed as a fraction of synchronous speed.

$$\text{Slip, } S = \frac{\text{Synchronous speed} - \text{Rotor speed}}{\text{Synchronous speed}} \tag{9.10}$$

$$\therefore \quad = \frac{N_S - N}{N_S} \times 100 \tag{9.11}$$

In terms of slip, the actual speed of the motor (N) is expressed as

$$N = N_S (1 - S) \tag{9.12}$$

When the motor is at standstill speed, $N = 0$, hence the slip $S = 1$. When the motor runs at synchronous speed the slip is equal to zero. But practically this case will not arise because an induction motor always runs at less than the synchronous speed.

EXAMPLE 9.1 **A 10-pole, 3-phase induction motor runs at a speed of 485 rpm at 50 Hz supply. Determine the (i) synchronous speed and (ii) slip.**

Solution Given data

Speed of motor = 485 rpm

Number of poles = 10

Frequency = 50 Hz

(i) Synchronous speed, $N_S = \dfrac{120f}{P}$

$$= \frac{120 \times 50}{10} = 600$$

(ii) Slip, $S = \dfrac{N_S - N}{N_S} \times 100$

$$= \frac{600 - 485}{600} \times 100$$

EXAMPLE 9.2 **A 6-pole, 100-hp, 3-phase, 440-V, 50-Hz induction motor has a slip of 5% on full load. Calculate the speed of the motor.**

Solution Given data

Number of poles = 6

Frequency = 50 Hz

Slip = 5%

Synchronous speed = $N_S = \dfrac{120f}{P} = \dfrac{120 \times 50}{6} = 1000$ rpm

Speed of motor $N = N_S (1 - S) = 1000 (1 - 0.05) = 950$ rpm

9.4 PRINCIPLE OF OPERATION OF INDUCTION MOTOR

When the three-phase stator windings are fed by a three-phase supply, a rotating magnetic field is set up which rotates round the stator at synchronous speed. The speed is given by $N_s = 120f/P$.

The rotating magnetic field produced passes through the air-gap and cuts the rotor conductors which are stationary. On the basis of Faraday's law of electromagnetic induction, an emf is induced due to the relative speed between the rotating magnetic field and the stationary conductor. As the rotor circuit is short circuited, the induced emf helps currents to flow through rotor conductors and the current-carrying conductors are placed in the magnetic field produced by the stator. Therefore, the mechanical force acts on the rotor conductors and produces the torque to rotate the rotor in the same direction as the rotating magnetic field. This can be explained by Lenz's law. According to Lenz's law, the direction of rotor currents is such that they tend to oppose the cause of producing them. Here, the cause of producing the rotor currents is the relative speed between the rotating field and the stationary rotor conductors. Therefore, to reduce this relative speed, the rotor starts running in the same direction as that of rotating magnetic field.

9.4.1 Relation between the Stator and Rotor Quantities

Let f be the frequency of the stator impressed voltage and N rpm be the speed of induction motor corresponding to a slip of S. Then the relative speed between the synchronously rotating magnetic field and the rotor conductors will be $S \cdot N_s$. So, the frequency of the induced emf in the rotor winding is given by

$$f' = \frac{(\text{Relative speed between rotating flux and the rotor in rpm}) \times P}{120} \tag{9.13}$$

$$= \frac{(N_S - N) \times P}{120}$$

$$= \frac{SN_S \times P}{120}$$

$$= S \times f \tag{9.14}$$

$f' = S \times f$, i.e., the frequency of the rotor induced emf is slip times the supply frequency. When the rotor is at standstill, i.e., $N = 0$, $S = 1$, then f and f' are equal.

EXAMPLE 9.3 **A 4-pole, 3-phase induction motor operates at supply frequency of 60 Hz. Calculate the frequency of the rotor currents when the slip is 0.04 and at standstill.**

Solution Frequency of rotor current when slip is $0.04 = f' = S \times f = 0.04 \times 60 = 2.4$ Hz

At standstill, the frequency of rotor current $= f' = 1 \times 60 = 60$ Hz.

EXAMPLE 9.4 **If the current in the stator of a four-pole, three-phase induction motor has a frequency of 50 Hz and that in the rotor is 2 Hz. Determine the slip and speed of the motor.**

Solution Slip, $S = \dfrac{\text{Frequency of rotor currents}}{\text{Freqency of stator currents}}$

$$= \frac{2}{50} = 0.04$$

Synchronous speed, $N_S = \dfrac{120 \times f}{P}$

$$= \frac{120 \times 50}{4} = 1500 \text{ rpm}$$

Speed of the motor, $N = N_S (1 - S)$

$$= 500(1 - 0.04)$$

$$= 1440 \text{ rpm}$$

9.5 TORQUE EQUATION

The torque of an induction motor which develops due to the interaction of the rotor and stator field depends on the strength of those fields and phase relation between them. Therefore, the torque developed by the induction motor depends on the

(i) Rotor current, I_2

(ii) Stator flux per pole, ϕ

(iii) Rotor circuit power factor, cos θ_2

$$\therefore \quad T \propto \phi I_2 \cos \theta_2 \tag{9.15}$$

If E_2 is the induced emf in the rotor at standstill, then

$$T \propto E_2 I_2 \cos \theta_2 \tag{9.16}$$

or

$$T = k_1 E_2 I_2 \cos \theta_2$$

where k_1 is a constant. From Eq. (9.16) it is clear that as the power factor increases torque increases and vice versa.

9.5.1 Starting Torque

The torque developed at the instant of starting of the motor is called starting torque. Let E_2 be the rotor emf per phase at standstill; R_2 be the rotor resistance per phase; and X_2 be the rotor reactance per phase at standstill.

Then, the rotor current

$$I_2 = \frac{E_2}{Z_2} = \frac{E_2}{\sqrt{R_2^2 + X_2^2}} \tag{9.17}$$

Power factor

$$\cos \theta_2 = \frac{R_2}{Z_2} = \frac{R_2}{\sqrt{R_2^2 + X_2^2}} \tag{9.18}$$

Substituting Eqs. (9.17) and (9.18) into Eq. (9.16), we get

$$T_{ST} = k_1 \times E_2 \frac{E_2}{\sqrt{R_2^2 + X_2^2}} \times \frac{R_2}{\sqrt{R_2^2 + X_2^2}}$$

$$= \frac{k_1 E_2^2 R_2}{R_2^2 + X_2^2} \tag{9.19}$$

From Eq. (9.19) it is observed that the starting torque is directly proportional to square of the supply voltage as the E_2 is directly proportional to supply voltage.

9.5.2 Condition for Maximum Starting Torque

Equation (9.19) gives the expression for starting torque. Keeping the supply voltage constant and in turn rotor emf E_2, Eq. (9.19) can be rewritten as

$$T_{ST} = \frac{k_2 R_2}{R_2^2 + X_2^2} \tag{9.20}$$

where k_2 is a constant and is equal to $k_1 E_2^2$. Differentiating Eq. (9.20) with respect to resistance R_2 and equating it to zero gives the condition for maximum starting torque.

$$\frac{dT_{ST}}{dt} = \frac{d}{dt}\left[\frac{k_2 R_2}{R_2^2 + X_2^2}\right] = 0 \tag{9.21}$$

$$k_2\left[\frac{1}{R_2^2 + X_2^2} - \frac{R_2(2R_2)}{(R_2^2 + X_2^2)^2}\right] = 0 \tag{9.22}$$

Simplifying Eq. (9.22), we get

$$R_2 = X_2 \tag{9.23}$$

Hence, the condition for maximum starting is $R_2 = X_2$ and the expression for maximum starting torque is given by

$$T_{ST} = \frac{k_2}{2X_2} \tag{9.24}$$

9.5.3 Torque Under Running Conditions

When the rotor is stationary or standstill the emf induced in the rotor is maximum and is equal to E_2. Under running conditions, the relative speed between the rotor and the rotating stator flux decreases and he ce the rotor-induced emf E_2 decreases. Hence for slip S the rotor-induced emf will be S times the standstill-induced emf E_2. Under running conditions the frequency of the rotor-induced emf is also S times the supply frequency. As frequency is varied, reactance also varies.

Under running conditions, let

Induce emf per phase in rotor = SE_2

Rotor winding resistance per phase = R_2

Rotor winding reactance per phase = $2\pi SfL_2$

$= S(2\pi fL_2) = SX_2$

Rotor winding impedance per phase is

$$\sqrt{R_2^2 + (SX_2)^2} \tag{9.25}$$

Rotor current per phase

$$I_2 = \frac{SE_2}{\sqrt{R_2^2 + (SX_2)^2}} = \frac{SE_2}{\sqrt{R_2^2 + S^2 X_2{}^2}} \tag{9.26}$$

Rotor power factor

$$\cos\theta_2 = \frac{R_2}{\sqrt{R_2^2 + S^2 X_2{}^2}} \tag{9.27}$$

Substituting Eqs. (9.26) and (9.27) in Eq. (9.16), we get

$$\therefore \quad \text{Torque, } T = k_1 E_2 \times \frac{SE_2}{\sqrt{R_2^2 + S^2 X_2{}^2}} \times \frac{R_2}{\sqrt{R_2^2 + S^2 X_2{}^2}}$$

$$= \frac{k_1 S R_2 E_2^2}{R_2^2 + S^2 X_2{}^2} \tag{9.28}$$

9.5.4 Condition for Maximum Torque Under Running Conditions

Similar to the standstill derivation, here also differentiating Eq. (9.28) with respect to "R_2" and equating it to zero, we get the condition for maximum torque.

$$\frac{dT}{dt} = \frac{d}{dt}\left[\frac{k_1 S E_2^2 R_2}{R_2^2 + S^2 X_2^2}\right] = 0 \tag{9.29}$$

$$k_1 S E_2^2 \left[\frac{R_2^2 + S^2 X_2^2 - 2R_2^2}{(R_2^2 + S^2 X_2^2)^2}\right] = 0$$

$$S^2 X_2^2 - R_2^2 = 0$$

$$R_2 = SX_2 \tag{9.30}$$

$$\therefore \quad S = \frac{R_2}{X_2} \tag{9.31}$$

Substituting Eq. (9.30) into Eq. (9.28) gives maximum torque,

$$T_{\max} = \frac{k_1 S E_2^2 \times SX_2}{(SX_2)^2 + S^2 X_2^2} \tag{9.32}$$

$$= \frac{k_1 E_2^2 X_2}{X_2^2 + X_2^2}$$

$$= \frac{k_1 E_2^2}{2X_2} = \frac{k_2}{2X_2} \tag{9.33}$$

It is found that the constant

$$k_1 = \frac{3}{w_s} = \frac{3}{2\pi N_S} \tag{9.34}$$

From Eq. (9.33) it is found that maximum torque is independent of rotor resistance and depends only on the rotor reactance.

9.5.5 Starting Torque, Full-Load Torque, and Maximum Torque Ratios

Here, we can obtain the ratios of starting torque to maximum torque as well as full-load torque to maximum torque.

$$\frac{T_{ST}}{T_{max}} = \frac{\left[\left(k_2 R_2 / R_2^2\right) + X_2^2\right]}{\frac{k_2}{2X_2}} \tag{9.35}$$

Dividing the numerator and denominator of Eq. (9.35) by X_2^2, we get

$$\frac{(2R_2 / X_2)}{1 + (R_2 / X_2)^2} = \frac{2a}{1 + a^2} \quad \left(\text{where } a = \frac{R_2}{X_2}\right) \tag{9.36}$$

$$\frac{T_{FL}}{T_{max}} = \frac{\left(k_2 S R_2 / R_2^2 + S^2 X_2^2\right)}{(k_2 / 2X_2)} \tag{9.37}$$

$$= \frac{2SR_2 X_2}{R_2^2 + S^2 X_2^2} \tag{9.38}$$

Dividing the numerator and denominator of Eq. (9.38) by X_2^2, we get

$$\frac{2S(R_2 / X_2)}{\left(R_2^2 / X_2^2\right) + S^2} = \frac{2aS}{a^2 + S^2} \quad \left(\text{where } a = \frac{R_2}{X_2}\right) \tag{9.39}$$

EXAMPLE 9.5 **A 3-phase induction motor having star-connected rotor has an induced emf of 100 V between slip rings at standstill on open circuit with normal voltage applied to the stator. The resistance and reactance of each rotor phase are 0.8 Ω and 5 Ω, respectively. Calculate the current and power factor (i) when slip rings are connected to a star-connected rheostat resistance of 2 Ω and reactance of 3 Ω per phase and (ii) when running short circuited with 3% slip.**

Solution Standstill rotor emf per phase, $E_2 = \dfrac{100}{\sqrt{3}} = 57.73$ V

(i) Rotor resistance per phase, $R_2 = 0.8 + 2 = 2.8\ \Omega$

Rotor reactance per phase, $X_2 = 5 + 3 = 8\ \Omega$

Rotor impedance per phase, $Z_2 = \sqrt{R_2^2 + X_2^2} = \sqrt{(2.8)^2 + 8^2} = 8.47\ \Omega$

Rotor current per phase, $I_2 = \dfrac{E_2}{Z_2} = \dfrac{57.73}{8.47} = 6.81$ A

Power factor, $\cos\theta_2 = \dfrac{R_2}{Z_2} = \dfrac{2.8}{8.47} = 0.33$

(ii) Rotor impedance at 3% slip $= 0.8 + 0.03 \times j5 = 0.8 + j0.15 = 0.814\ \Omega$

Rotor current per phase, $I_2 = \dfrac{SE_2}{|Z_2|} = \dfrac{0.03 \times 57.73}{0.814} = 2.127$ A

Power factor, $\cos\theta_2 = \dfrac{R_2}{Z_2} = \dfrac{0.8}{0.814} = 0.982$

EXAMPLE 9.6 **A 3-phase slip ring induction motor with star-connected rotor has an induced emf of 60 V between the slip rings at standstill with normal voltage applied to the stator. The rotor winding has a resistance per phase of 0.5 Ω and standstill reactance per phase of 3 Ω. Calculate the slip and rotor current per phase when the rotor is developing maximum torque.**

Solution The condition for developing maximum torque $R_2 = SX_2$

$$S = \frac{R_2}{X_2} = \frac{0.5}{3} = 0.166$$

Reactance X_2 at slip $S = 0.166 \times 3 = 0.5\ \Omega$

Impedance, $Z_2 = \sqrt{0.3^2 + 0.5^2} = 0.58\ \Omega$

Rotor current per phase $= \dfrac{S \times E_2}{|Z_2|} = \dfrac{0.166 \times \dfrac{60}{\sqrt{3}}}{0.58} = 9.91$ A

EXAMPLE 9.7 **A 2000-V, 3-phase, 50-Hz, 12-pole induction motor has a rotor impedance of (0.015 + *j*0.3) Ω at standstill. A full-load torque is obtained at a speed of 450 rpm. Calculate the (i) ratio of maximum to full-load torque and (ii) speed at which maximum torque occurs.**

Solution Synchronous speed, $N_S = \dfrac{120 \times f}{P}$

$$= \frac{120 \times 50}{12} = 500 \text{ rpm}$$

Full-load speed, $N = 450$ rpm

Full-load slip, $S = \dfrac{500 - 450}{500} = 0.1$

$$a = \frac{R_2}{X_2} = \frac{0.015}{0.3} = 0.05$$

(i) $\dfrac{T_{FL}}{T_{max}} = \dfrac{2aS}{a^2 + S^2} = \dfrac{2 \times 0.05 \times 0.1}{0.05^2 + 0.1^2} = 0.8$

(ii) Slip corresponding to maximum torque, $S = \dfrac{R_2}{X_2} = 0.05$

Speed at which maximum torque occurs, $N = (1 - S)\, N_S$

$= (1 - 0.05)500$

$= 475$ rpm

EXAMPLE 9.8 **An 8-pole, 50-Hz, 3-phase induction motor has rotor resistance and standstill reactance per phase of 0.06 Ω and 0.4 Ω, respectively. Calculate the value of the resistance to be added per phase in the rotor to obtain 80% of the maximum torque at starting.**

Solution The ratio of starting torque to maximum torque is given by $= \dfrac{T_{ST}}{T_{max}} = \dfrac{2a}{1 + a^2}$

Given that $= \dfrac{T_{ST}}{T_{max}} = 0.8$

$$\therefore \quad \frac{2a}{1 + a^2} = 0.8$$

Solving the above equation, $a = 0.5$

$$0.5 = \frac{R_2 + r}{X_2} = \frac{0.06 + r}{0.4}$$

Resistance to be added, $r = 0.14\ \Omega$

EXAMPLE 9.9 **A 3-phase induction motor runs at 1440 rpm at full load when supplied power from a 50-Hz, 3-phase line. Calculate the (i) slip at full load, (ii) frequency of rotor voltage, and (iii) speed of rotor at a slip of 10%.**

Solution Supply frequency = 50 Hz

Full-load speed of the motor, $N = 1440$ rpm

Synchronous speed N_S will be around the full-load speed. Hence, the synchronous speed will be 1500 rpm.

(i) Slip at full load, $S = \dfrac{N_S - N}{N_S} = \dfrac{1500 - 1440}{1500} = 4\%$

(ii) Frequency of rotor voltage, $f' = S_f = 0.04 \times 50 = 2$ Hz

(iii) Speed of rotor at a slip of 10%, $N = N_S(1 - S) = 1500\ (1 - 0.1) = 1350$ rpm

9.5.6 Slip Versus Torque Characteristics

The torque equation at full load is given by

$$T = \frac{k_1 S R_2 E_2^2}{R_2^2 + S^2 X_2^2} \tag{9.40}$$

(i) When the slip is zero, i.e., motor runs at synchronous speed the torque is zero.

(ii) When the speed is near to the synchronous speed (N_S) the slip is very small and is negligible compared to R_2.

$\therefore$ Torque, $T = \frac{k_1 S R_2 E_2^2}{R_2^2}$

Keeping R_2 and E_2 constant, torque is almost proportional to the slip S.

(iii) As the slip increases the torque increase and reaches maximum value when $S = R_2/X_2$.

(iv) If the slip is increased for further level, the term R_2 is negligible compared to SX_2.

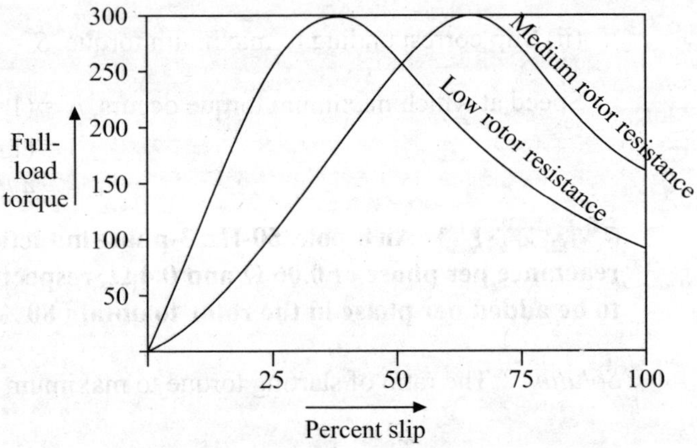

Figure 9.5 Slip–torque characteristics

Therefore, Eq. (9.40) becomes

$$T = \frac{k_1 S R_2 E_2^2}{S^2 X_2^2} = \frac{k_1 R_2 E_2^2}{S X_2^2} \Rightarrow T \propto \frac{1}{S}$$

keeping other parameters constant.

Therefore, the torque is inversely proportional to the slip, i.e., as the slip increases, the torque decreases as shown in Figure 9.5.

9.6 LOSSES IN AN INDUCTION MOTOR

The input electric power fed to the stator of the motor is converted into mechanical power at the shaft of the motor. During this process so many losses take place. There are electrical losses due to heat produced by the current in the stator and rotor windings, known as *copper losses*. Magnetic losses in the iron due to hysteresis and eddy currents are known as *iron* or *core losses*. Mechanical losses due to friction in the bearings and ventilation are known as *friction* and *windage losses*.

There are core losses and copper losses in the stator and copper losses and mechanical losses in the rotor. Core loss in the rotor is very small and is negligible. Therefore, total losses can be divided into two categories. One is fixed losses and the other is variable losses. The sum of the stator iron loss, friction, and windage loss is called fixed losses. The sum of stator copper losses and rotor copper losses is called variable losses.

Fixed losses = Stator iron loss + Friction and windage loss

Variable losses = Stator copper losses + Rotor copper losses

9.7 POWER STAGES

The following flowchart shows gives how the power given to a stator is being flown in the induction motor.

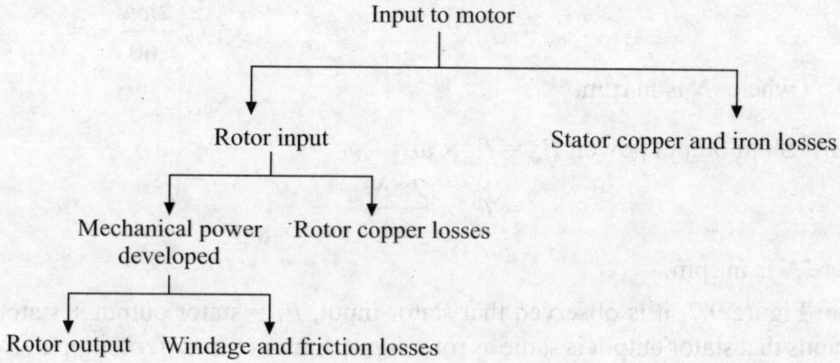

Figure 9.6 Power flowchart

From Figure 9.6 it is seen that power input to the motor will be divided into two, i.e., rotor input and stator copper and iron losses. The rotor input is again divided into mechanical power developed and rotor copper losses. The mechanical power developed is again divided into rotor shaft output and windage and friction losses. Figure 9.6 can be redrawn as shown in Figure 9.7.

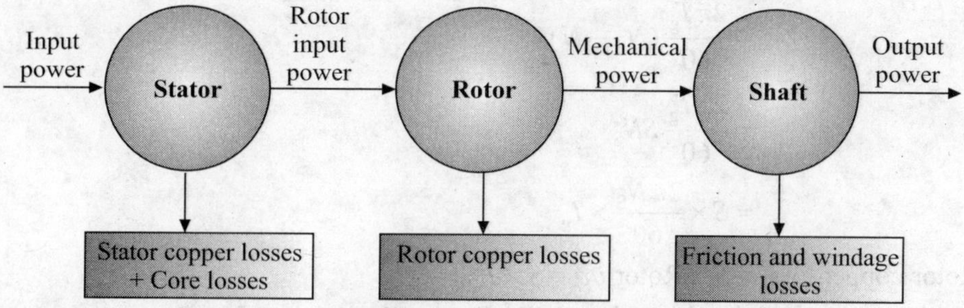

Figure 9.7 Power flow stages

9.7.1 Power Relations

In general, the relation between power and torque is given by

$$P = T \times w. \qquad (9.41)$$

Induction motor also develops gross torque due to mechanical power developed in the rotor. Its value can be expressed either in terms of rotor input or mechanical power on the rotor. Similarly, it also develops shaft torque due to shaft output power.

(i) Rotor input power

$$P_2 = T_g \times \omega_S \qquad (9.42)$$

where T_g = gross torque developed

ω_S = angular velocity = $2\pi N_S$ (N_S is in rps)

$$= \frac{2\pi N_S}{60} \ (N_S \text{ is in rpm})$$

$$P_2 = \frac{2\pi N_S}{60} \times T_g \tag{9.43}$$

(ii) Mechanical power developed in the rotor, $P_{mech} = T_g \times \omega$ (9.44)

$$= T_g \frac{2\pi N}{60}$$

where N is in rpm.

(iii) Shaft output power, $P_{sh} = T_{sh} \times \omega$ (9.45)

$$= T_{sh} \times \frac{2\pi N}{60} \tag{9.46}$$

where N is in rpm.

From Figure 9.7, it is observed that stator input, P_1 = stator output + stator copper losses. It is obvious that stator output is same as rotor input. Stator output = rotor input, P_2. Rotor gross output or mechanical power output, P_{mech} = rotor input P_2 – rotor copper losses.

Rotor copper loss = Rotor input, P_2 – Mechanical power, P_{mech} (9.47)

$$= T_g \times \omega_S - T_g \times \omega$$

$$= \frac{2\pi N_S}{60} T_g - \frac{2\pi N}{60} T_g$$

$$= \frac{2\pi T_g}{60}(N_S - N)$$

$$= \frac{2\pi T_g}{60} S N_S$$

$$= S \times \frac{2\pi N_S}{60} \times T_g$$

Rotor copper loss $= S \times$ Rotor $i/p = S \times P_2$ (9.48)

Mechanical power developed by motor, P_{mech} = Rotor input P_2 – Rotor copper loss

$$= P_2 - S \times P_2$$

$$P_{mech} = (1 - S)P_2 \tag{9.49}$$

$$\frac{\text{Copper loss in rotor}}{P_{mech}} = \frac{S \times P_2}{(1-S) \times P_2} = \frac{S}{1-S} \tag{9.50}$$

From Eqs. (9.48), (9.49), and (9.50), we can establish the following equation:

Rotor i/p : Mechanical power developed : Rotor copper loss, i.e., $P_2 : P_{mech} : P_{cu} = 1 : (1 - S) : S$

9.7.2 Efficiency

The efficiency of the induction motor is given by

$$\text{Efficiency} = \frac{\text{Output power}}{\text{Input power}} \times 100 \tag{9.51}$$

The efficiency of the rotor of induction motor is given by

$$\text{Efficiency} = \frac{\text{Rotor output power}}{\text{Rotor input power}} \times 100 \tag{9.52}$$

$$\eta_{mech} = \frac{P_{mech}}{P_2} \times 100 \tag{9.53}$$

EXAMPLE 9.10 **A 3-phase, 400-V, 6-pole induction motor drawing a line current of 78 A at 0.8 pf and at a slip of 4% is giving the efficiency of 85%. Calculate the shaft torque and shaft power output.**

Solution Given data

Line voltage, $V = 400$ V

Full-load current, $I = 78$ A

Number of poles, $P = 6$

Power factor, $\cos\phi = 0.8$

Slip, $S = 4\% = 0.04$

Efficiency, $\eta = 85\%$

Input power, $P_1 = \sqrt{3}\,VI\cos\phi$

$= \sqrt{3} \times 400 \times 78 \times 0.8$

$= 43.232$ kW

Shaft output power, $P_{sh} = 43.232 \times 0.85$

$= 36.74$ kW

$$\text{Synchronous speed, } N_S = \frac{120f}{p} = \frac{120 \times 50}{6} = 1000$$

Rotor speed, $N = N_S(1 - S)$

$= 1000\,(1 - 0.04) = 960$ rpm

$$\text{Angular speed of rotor, } \omega = \frac{2\pi N}{60} = \frac{2\pi \times 960}{60} = 100.5 \text{ rad/s}$$

$$\text{Shaft torque} = \frac{\text{Shaft output power}}{\text{Angular speed}} = \frac{36.7471 \times 1000}{100.5} = 365.64 \text{ N-m}$$

EXAMPLE 9.11 **A 4-pole, 50-Hz, 3-phase induction motor develops a useful torque of 90 N-m at full load and the motor emf generated at 120 cycles/min, take the friction and core loss as 6 N-m. Estimate the following:**

(i) Shaft output power

(ii) Rotor copper loss

(iii) Motor input

(iv) Motor efficiency

Assume the total core loss as 700 W.

Solution Given data

Number of poles, $P = 4$

Frequency, $f = 50$ Hz

Useful torque, $T = 90$ N-m

emf generated = 120 cycles/min

Torque loss on friction and core loss = 6 N-m

Total core loss = 700 W

$$\text{Rotor frequency } f_2 = \frac{120}{60} = 2 \text{ Hz}$$

$$f_2 = S_f$$

$$\therefore \quad S = \frac{2}{50} = 0.04$$

$$\text{Synchronous speed, } N_S = \frac{120f}{p} = \frac{120 \times 50}{4} = 1500 \text{ rpm}$$

Rotor speed, $N = (1 - S)N_S = (1 - 0.04) \times 1500 = 1440$ rpm

$$\text{Angular speed of rotor, } \omega = \frac{2\pi N}{60} = \frac{2\pi \times 1440}{60} = 150.7 \text{ rad/s}$$

(i) Shaft power output, $P_0 = T \times \omega = 90 \times 150.7$

$= 13.563$ kW

Mechanical power developed, $P_m = (90 + 6) \times 150.7 = 14.4672$ kW

$$\text{(ii) Rotor copper loss} = 3I_2^2R_2 = P_m\left[\frac{S}{1-S}\right]$$

$$= 14.4672\left[\frac{0.04}{1-0.04}\right]$$

$$= 0.6028 \text{ kW}$$

(iii) Motor input = $P_0 + P_m + 3I_2^2R_2 = 14.4672 + 0.6028 + 0.6 = 15.67$ kW

$$\text{(iv) Motor efficiency, } \eta = \frac{\text{Shaft power output}}{\text{Motor input}} = \frac{13.563}{15.67} \times 100$$

$$= 86.55\%$$

EXAMPLE 9.12 **A 20-kW, 4-pole, 50-Hz, 3-phase induction motor has friction and windage losses of 2% of the output. The full-load slip is 5%. Determine for full load (i) rotor copper loss, (ii) rotor input, (iii) shaft torque, and (iv) gross torque.**

Solution Motor output = $P_{sh} = 20000$ W

Friction and windage loss = 2% of 20000 = 400 W

Rotor gross output or mechanical power, $P_{mech} = 20000 + 400 = 20400$ W

$$\text{(i) Rotor copper loss} = \frac{S}{1-S} \times P_{mech} = \frac{0.05}{1-0.05} \times 20400 = 1073.68 \text{ W}$$

$$\text{(ii) Rotor input, } P_2 = \frac{\text{Rotor copper loss}}{S} = \frac{1073.68}{0.05} = 21473.68 \text{ W}$$

(iii) Shaft torque, $T_{sh} = \dfrac{60 \times P_{sh}}{2\pi N}$

Synchronous speed, $N_S = \dfrac{120 \times 50}{4} = 1500$ rpm

Rotor speed, $N = (1 - S)\, N_S = (1 - 0.05)1500 = 1425$ rpm

$$T_{sh} = \frac{60 \times 20000}{2 \times \pi \times 1425} = 134.03 \text{ N-m}$$

(iv) Mechanical torque, $T_{mech} = \dfrac{60 \times P_{mech}}{2\pi N} = \dfrac{60 \times 20400}{2 \times \pi \times 1425} = 136.72$ N-m

9.7.3 Synchronous Watt

It is another unit of torque and is defined as the torque which develops a power of 1 W at the synchronous speed of the motor. When we say that a motor is developing a torque of 100 synchronous watts, we mean that the rotor input is 100 W and that the torque is such that the power developed would be 100 W provided the rotor was running synchronously and developing the same torque.

Rotor input = $T_{SY} \times 2\pi N_S$

or

$$T_{SY} = \frac{\text{Rotor input in watt}}{2\pi N_S} \quad (9.54)$$

9.8 EQUIVALENT CIRCUIT DIAGRAM OF INDUCTION MOTOR

In case of induction motor, the energy is transferred from stator (primary) winding to rotor (secondary) winding entirely by induction. Hence, the induction motor can be considered as transformer. At standstill the induction motor is static transformer having its rotor (secondary) winding short circuited.

9.8.1 Rotor Circuit

In case of an induction motor the load is mechanical and it can be replaced by an equivalent electrical load of resistance R_L, i.e.,

$$R_L = R_2\left(\frac{1}{S} - 1\right) \quad (9.55)$$

The equivalent circuit of rotor is shown in Figure 9.8(a).

R_2 and X_2 are the resistance and reactance of rotor. At the starting, i.e., at zero speed the slip $S = 1$.

$$\therefore \quad \text{Load, } R_L = R_2\left(\frac{1}{1} - 1\right) = 0 \quad (9.56)$$

To find the equivalent circuit of the induction motor we can transfer the primary values to secondary or secondary values to primary. Therefore, while shifting resistance or impedance from secondary to primary, it should be divided by the transformation ratio K and current should by multiplied by K. The equivalent circuit of an induction motor where all the values have been referred to primary (stator) is shown in Figure 9.8(b). To make the calculations easier the approximate equivalent circuit is shown in Figure 9.8(c).

R_0 and X_0 are the no-load resistance and reactance per phase, respectively. R_1 and X_1 are the stator resistance and reactance per phase, respectively.

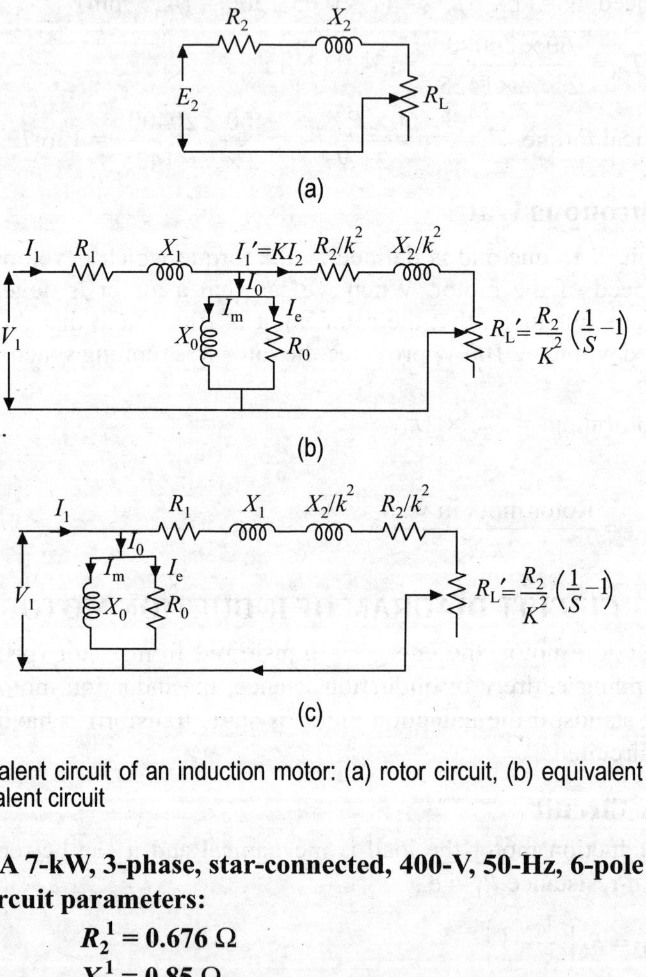

Figure 9.8 Equivalent circuit of an induction motor: (a) rotor circuit, (b) equivalent circuit, and (c) simplified equivalent circuit

EXAMPLE 9.13 A 7-kW, 3-phase, star-connected, 400-V, 50-Hz, 6-pole induction motor has the following circuit parameters:

$R_1 = 1.16\ \Omega$ $\qquad$ $R_2^1 = 0.676\ \Omega$

$S_1 = 1.78\ \Omega$ $\qquad$ $X_2^1 = 0.85\ \Omega$

$X_m = 44.3\ \Omega$

The total windage and friction loss, core losses may be assumed to be 400 W at any load. The motor runs at a speed of 985 rpm at rated voltage and frequency on a particular load. Calculate the input torque and power, stator current, power factor, and efficiency.

Solution Given data

Supply voltage, $V = 400$ V

Input power, $P_1 = 7$ kW

Frequency, $f = 50$ Hz

Number of poles, $p = 6$

$R_1 = 1.16\ \Omega, R_2^1 = 0.676\ \Omega,$

$X_1 = 1.78\ \Omega, X_2^1 = 0.85\ \Omega$

$X_m = 44.3\ \Omega$

Total windage and friction loss and core losses = 400 W

Rotor speed, $N_r = 985$ rpm

Synchronous speed, $N_S = \dfrac{120 \times 50}{6} = 1000\,\text{rpm}$

$$\omega_S = \frac{2\pi N_S}{60} = \frac{2\pi \times 1000}{60} = 104.7 \text{ rad/sec}$$

$$\text{Slip, } S = \frac{N_S - N}{N_S} = \frac{1000 - 985}{1000} = 0.015$$

$$\text{Impedance, } Z = (1.16) + \left(\frac{0.676}{0.015}\right) + (1.78 + 0.85)\ 46.26 + j2.63 = 46.30130 \angle 3.25\ \Omega$$

$$I_2^1 = \frac{\frac{400}{\sqrt{3}}}{46.3013} = \frac{231}{46.3013} = 4.98 \quad \angle -3.25 \text{ A}$$

$$I_0 = I_m = -j\frac{400/\sqrt{3}}{44.3} = \frac{231}{44.3} = -j\ 5.21 \text{ A}$$

$$\begin{aligned} I_1 = I_0 + I_2{}^1 &= 4.9877 \angle -3.25 - j\,5.2144 \\ &= 4.9796 - j\,0.2827 - j\,5.2144 \\ &= 4.9796 - j\,5.4971 \\ &= 7.4171 \angle -47.82 \text{ A} \end{aligned}$$

$pf = \cos(47.82°) = 0.67$ lag

$$\text{Input power} = \sqrt{3} \times 400 \times 7.4171 \times 0.67 = 3.4429 \text{ kW}$$

$$\begin{aligned} \text{Power output gross} &= 3I_2^{12}\,R_2^1\left(\frac{1}{S} - 1\right) \\ &= 3(4.9877)^2 \times (0.676) \times \left(\frac{1}{0.015} - 1\right) \\ &= 3.3129 \text{ kW} \end{aligned}$$

$$\begin{aligned} \text{Power output (net)} &= 3.3129 - 0.4 \\ &= 2.9129 \text{ kW} \end{aligned}$$

$$\text{Efficiency, } \eta = \frac{2.9129}{3.4429} \times 100 = 84.6\%$$

$$\text{Torque (net)} = \frac{P_{in}\,N_S}{\omega_S(1 - S)} = \frac{2.9129 \times 1000}{104.7(1 - 0.015)} = 28.24 \text{ N-m}$$

EXAMPLE 9.14 **A 200-V, 3-phase, 4-pole, star-connected induction motor has a stator impedance of (0.04 + *j* 0.4) Ω and an equivalent rotor impedance of (0.06 + *j*0.24) Ω. The motor draws a no-load current of 20 A at a pf of 0.2 lag. When the motor operates at a slip of 4%, calculate the (i) input line current and pf and (ii) electromagnetic torque developed in N-m.**

Solution:

$$\text{Applied voltage } V_1 = \frac{200}{\sqrt{3}} = 115.47 \text{ V}$$

$$I_2' = \frac{V_1}{\sqrt{(R_1 + \frac{R_2'}{S}) + j(X_1 + X_2)}}$$

$$= \frac{115.47}{\left(0.04 + \frac{0.06}{0.04}\right) + j(0.4 + 0.24)}$$

$$= \frac{115.47}{(1.54 + j0.64)} = \frac{115.47}{1.66\angle 22.56°} = 69.56 \angle -22.56° \text{A}$$

No-load current $I_0 = 20 \angle -78.46°$ A

(i) Primary current $I_1 = I_0 + I_2' = 20 \angle -78.46° + 69.56 \angle -22.56°$

$= 4.0 - j19.59 + 64.23 - j26.68$

$= 68.23 - j46.27$

$= 82.44 - 34.14°$A

Power factor $pf = \cos(34.14°) = 0.83$ lag

(ii) Torque developed $T_g = 9.55 \dfrac{P_2}{N_S}$

$$P_2 = (I_2')^2 \frac{R_2'}{S} = 3 \times (69.56)^2 \times \frac{0.06}{0.04} = 21.773 \text{ kW}$$

$N_S = 1500$ rpm

$$T_g = 9.55 \times \frac{21.773 \times 1000}{1500}$$

$= 138.62$ N-m

9.9 TESTING OF INDUCTION MOTORS

Various tests may be performed on an induction motor to determine its efficiency and other operating characteristics. These tests are (i) no-load test and (ii) blocked rotor test. These are also called as indirect tests. There are certain advantages of indirect testing. One is the power required for this test is small as compared to direct testing. Another advantage is the equivalent circuit parameters of induction motor which can be determined in addition to the performance parameters.

9.9.1 No-Load Test

No-load test is performed to determine the following:

(i) No-load current, I_0

(ii) No-load power factor cos ϕ_0

(iii) Windage and friction losses

(iv) No-load core losses

(v) No-load input

(vi) Core resistance R_0 and reactance X_0

The no-load test is performed with different values of applied voltage and below- and above-rated voltage, while the motor is running on no load. The connection diagram is shown in Figure 9.9. This test is conducted as follows.

The motor is uncoupled from its load, and rated voltage is applied to the stator. Since there is no power output, the power supplied to the stator furnishes its copper loss, its core loss, and the friction and windage loss in the rotor. Since slip at no load is often within one-tenth of 1%, the rotor current is practically zero, and hence the analogy to the no-load test of the transformer. However, the no-load stator current of the induction motor is comparatively large; therefore, copper losses in the stator may not be neglected as they were for the transformer primary. It is thus necessary to calculate this copper loss and subtract from the no-load input in order to obtain the sum of the core and friction and windage losses of the motor.

Speed and flux vary but little from no-load to full load, and these losses may be assumed constant for the operating range of the poly-phase induction motor. Knowing the total core losses, P_{core}, no-load current per phase I_0 and applied voltage per phase V, the values of magnetizing current I_m, wattful current I_w, no-load resistance R_0, and no-load reactance X_0 are determined as follows:

$$I_w = I_0 \cos \phi_0 = I_0 \times P_{core}/3V_{io} = P_{core}/3\ V$$

$$I_m = \sqrt{I_o^2 - I_w^2}$$

$$R_0 = V/I_w \text{ and } X_o = V/I_m$$

The power input is measured by two wattmeters, I_0 by an ammeter and V by a voltmeter, as shown in Figure 9.9. Since the motor is running on no-load, the power factor would be low, i.e., less than 0.5, hence total power input will be the difference of the two wattmeters readings W_1 and W_2.

At no load, the power input is equal to the sum of the following losses

(i) Core losses

(ii) Stator copper losses $3I_0^2R_1$

(iii) Friction and windage losses

To determine the friction and windage losses, a curve is drawn between the input power in watts and the applied voltage in volts. This curve is extended to meet the power line at same point as shown in Figure 9 10. OA indicates the friction and windage losses.

By subtracting friction and windage losses plus copper losses from no-load power input, we can determine the core losses. ϕ_0 can be found using the equation

$$W_0 = \sqrt{3}\, VI_0 \cos\phi_0 \tag{9.57}$$

$$\cos\varphi_0 = \frac{W_0}{\sqrt{3}\, VI_0} \tag{9.58}$$

where V is the line voltage and W_0 is the no-load stator input.

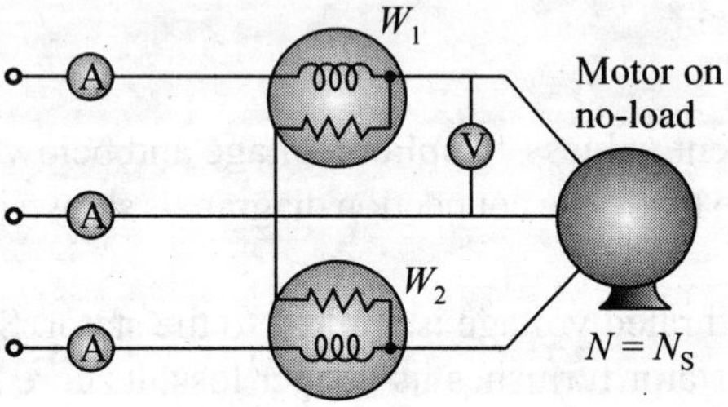

Figure 9.9 Circuit diagram for no-load test

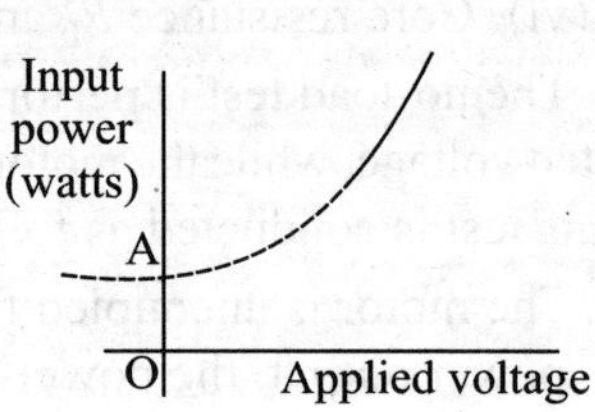

Figure 9.10 Graph of no-load test

9.9.2 Blocked Rotor Test

This test is also known as short-circuit test. The connection diagram for this test is shown in Figure 9.11. This test is performed to determine

(i) Short-circuit current, I_{sc} with low voltage applied to stator.

(ii) Power factor on short circuit.

We can find total equivalent resistance and reactance of the motor as referred to stator and full-load copper loss. The test is conducted as follows:

The rotor is prevented from turning by blocking it, and a low voltage of less than 10% of rated voltage is applied to the stator. This voltage is best applied through a variable autotransformer, or three-phase variac.

The voltage is then gradually increased until rated current is obtained in the stator. Since the voltage required is quite low, the power supplied to the stator will be dissipated as the copper loss in both the stator and rotor.

If at short circuit, V_s = applied voltage per phase, I_s = short-current per phase in the stator winding, and P_s = total input power, then $I_{sc} = I_s \times (V/V_s)$ and power factor,

$$\cos\phi_s = \frac{P_s}{3V_s I_s} \tag{9.59}$$

Since the input on short circuit meets with the stator and rotor copper losses and core losses being very small can be neglected, therefore, input on short circuit,

$$P_s = 3I_s^2 R_{01} \tag{9.60}$$

Equivalent resistance per phase referred to the stator,

$$R_{01} = (W_s)/3I_s^2 = P_s/3I_s^2 \tag{9.61}$$

Motor equivalent impedance per phase as referred to the stator,

$$Z_{01} = (V_s)/(I_s) \tag{9.62}$$

Motor equivalent reactance per phase as referred to the stator,

$$X_{01} = \sqrt{Z_{01}^2 - R_{01}^2} \tag{9.63}$$

In this test, the rotor is blocked. A reduced voltage, i.e., 15–20% of the normal value is applied to the stator terminals and is adjusted such that the full-load current flows in the stator. The values of current, voltage, and power are measured by a connecting ammeter, voltmeter, and wattmeter.

It can be found that relation between the short circuit and voltage is approximately a straight line. Hence, if V is normal stator voltage, V_S the short-circuit voltage, then short-circuit current corresponding to the normal volt is found using the relation

$$I_{sn} = I_s \times \frac{V}{V_s} \tag{9.64}$$

where I_{sn} is the short-circuit current with normal voltage and I_s is the short-circuit current with voltage V_s.

Power factor on short circuit is found from

$$W_s = \sqrt{3} V_{sl} I_{sl} \cos \phi_s$$

$$\therefore \quad \cos \phi_s = \frac{W_s}{\sqrt{3} V_{sl} I_{sl}}$$

where W_s is the total power on short circuit; V_{sl} is the line voltage on short circuit; and I_{sl} is the line current on short circuit.

Since the input on short circuit meets the stator and copper losses, input on short circuit is

$$P_s = 3 I_s^2 R_{01}$$

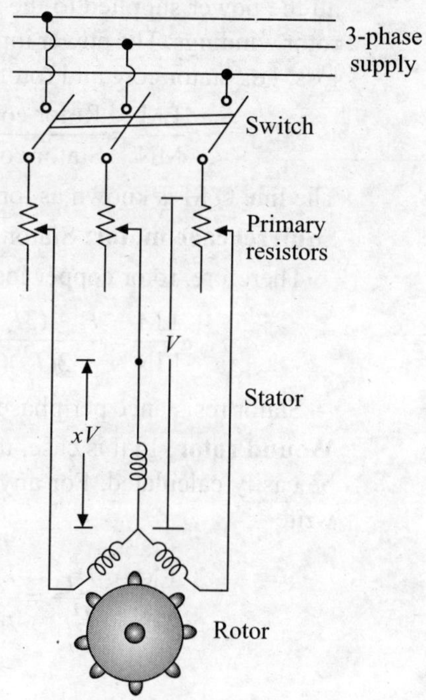

Figure 9.11 Blocked rotor test

9.9.3 Circle Diagram

For the purpose of studying the various characteristics of an induction motor, the circle diagram, which is a vector locus diagram of stator and rotor currents, is very useful and invaluable for estimating certain working conditions. A circuit diagram of an induction motor can be drawn by using the data obtained from the no-load test, short-circuit test, and stator resistance test. The procedure is as follows:

Calculate I_0 and no-load power factor ϕ_0 from the no-load test and lay off vector for I_0 lagging ϕ_0 behind the applied voltage V.

Find from the blocked rotor test or short-circuit test, the short-circuit current I_{sc} corresponding to the normal voltage and ϕ_S. The vector OL represents $I_{sn} = (I_{sx} V/V_s)$ in magnitude and phase. Vector OL represents rotor current I_2' as referred to stator. Clearly, the two points O′ and A lie on the required circle.

To find center G of this circle, bisect chord O′L at right angles, its bisector gives point G. Draw the diameter O′H perpendicular to voltage vector. With center G and radius GO′, draw the circle. The line O′L is known as the "output line" or mechanical power developed line. It may be noted that as the voltage vector is drawn vertically, all vertical distances represent the active power or energy components of currents.

Torque line

This is a line which separates the stator and rotor copper losses. When the rotor is locked, then all the power supplied to the motor goes to meet core losses and copper losses in the stator and rotor windings. The power input is proportional to LS. Out of this, NS = (O′F) represents the fixed loss, i.e., stator core and rotor core losses. The point M is such that

$$\frac{\text{LM}}{\text{MN}} = \frac{\text{Rotor copper loss}}{\text{Stator copper loss}}$$

The line O′M is known as torque line (or rotor input line). Point M is located as follows:

Squirrel cage motor: Stator copper loss = $3(I_s^2)(R_1)$.

Therefore, rotor copper loss = $(P_s) - (3I_s^2R_1)$

$$\frac{\text{LM}}{\text{MN}} = \frac{P_s - (3I_s^2 - R_1)}{3(I_s^2)(R_1)}$$

Stator resistance per phase, i.e., R_1 is found from the stator resistance test.

Wound rotor: In this case, the rotor and stator resistances per phase r_2 and r_1, respectively, can be easily calculated. For any values of stator and rotor currents I_1 and I_2, respectively, we can write

$$\frac{\text{LM}}{\text{MN}} = \frac{I_2^2 r_2}{I_1^2 r_1} = \frac{r_2}{r_1}\left(\frac{I_2}{I_1}\right)^2 \quad (I_1/I_2 = k = \text{transformation ratio}) \tag{9.65}$$

$$= \frac{r_2}{r_1} \times \frac{1}{k^2} = \frac{\left(\frac{r_2}{k^2}\right)}{r_1} = \frac{r_2'}{r_1} \tag{9.66}$$

r_2'/r_1 is the equivalent rotor resistance per phase/stator resistance per phase (value of k may be found from the short-circuit test itself by using two ammeters both in stator and in rotor circuits).

If the motor is running and taking a current O′A (see Figure 9.12), then the perpendicular DE represents fixed losses. DC is the stator copper losses, CA is the rotor input, CB is the rotor copper loss, BA is the rotor output, and AE is the motor input.

Now,

Motor input = $\sqrt{3}V_L$AE

Stator copper loss = $\sqrt{3}V_L$BE

Rotor input (torque) = $\sqrt{3}V_L$CA

Fixed losses = $\sqrt{3}V_L$DE

Rotor copper loss = $\sqrt{3}V_L$CB

Mechanical output = $\sqrt{3}V_L$BA

(i) Efficiency = Output/Input = BA/AE

(ii) Rotor copper loss/Rotor input = Slip = S = CB/CA

(iii) Rotor output/Rotor input = 1 – S = BA/CA

(iv) Power factor = AE/OA

Maximum quantities

Maximum output: The maximum output occurs at A where the tangent is parallel to the output line O′L. To locate point A draw a line from point G such that it is perpendicular to the output line O′L. Vertical line AB represents the maximum output.

Maximum torque or rotor input: The maximum torque or rotor input occurs at point C where the tangent is parallel to torque line O′M. Again, point C may be found by drawing GC perpendicular to the torque line.

Vertical line CD represents the maximum torque. The maximum torque is also known as pull out or stalling torque.

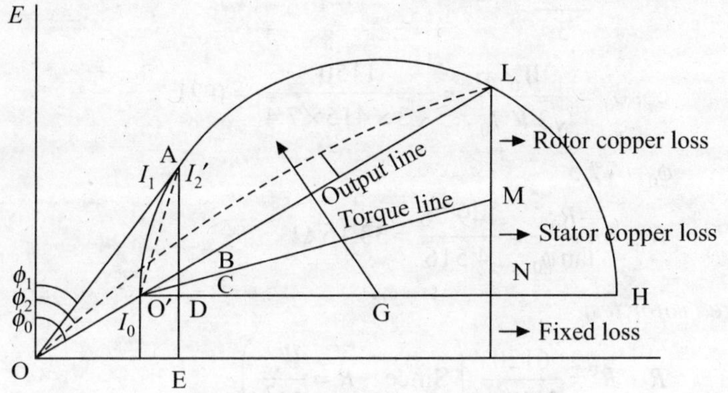

Figure 9.12 Circle diagram

Maximum input power: It occurs at the highest point of the circle, i.e., at point E where tangent to the circle is horizontal. It is proportional to ET. As point E is beyond the point of maximum torque, the induction motor will be unstable here.

However, the maximum input is a measure of the size of the circle and is an indication of the ability of the motor to carry short time over loads. Generally, ET is twice or thrice the motor input at rated load.

EXAMPLE 9.15 **The no-load and blocked rotor test reading of a 3-ϕ, 415-V, 22-kW, 50-Hz, 6-pole induction motor is given below**

No-load:	**415 V,**	**7.4 A,**	**1150 W**
Blocked rotor:	**220 V,**	**48 A,**	**6120 W**

Extimate the following:

(i) Parameter of the circuit model

(ii) Line current, power factor, torque developed, and motor efficiency at 960 rpm

(iii) Line current and torque developed at the starting

(iv) Maximum torque and slip

The stator resistance is 0.35 Ω per phase

Solution Given data

Supply voltage, $V = 415$ V

Rated power, $P = 22$ kW

Number of poles, $p = 6$

Rotor resistance, $R = 0.35\ \Omega$ per phase

Rotor speed, $N_r = 960$ rpm

(i) Motor parameters (equivalent star-bases)

No-load test

$$R_0 = \frac{\left(\frac{415}{\sqrt{3}}\right)^2}{\frac{1150}{3}} = \frac{(240)^2}{383.34} = 149.76\ \Omega$$

$$\cos\phi_0 = \frac{W_0}{\sqrt{3}\,V\,I_0} = \frac{1150}{\sqrt{3}\times 415\times 7.4} = 0.21$$

$$\phi_0 = 77.5$$

$$X_m = \frac{R_0}{\tan\phi_0} = \frac{149.76}{4.516} = 33.16\ \Omega$$

Blocked rotor test

$$R + R^1 = \frac{6120}{3(48)^2} \quad \left(\text{Since } R = \frac{W_{sc}}{3I_{sc}^2}\right)$$

$$= 0.89\ \Omega$$

$$R^1 = 0.89 - 0.35 = 0.54\ \Omega$$

$$Z = \frac{V}{I} = \frac{220/\sqrt{3}}{48} = \frac{127.01}{48} = 2.64\ \Omega$$

$$X + X^1 = \sqrt{2.64^2 - 0.89^2}$$

$$= 2.5\ \Omega$$

(ii) $$N_S = \frac{120f}{P} = \frac{120\times 50}{6} = 1000 \text{ rpm}$$

$$\omega_S = \frac{2\pi N_r}{60} = 100.5 \text{ rad/s}$$

$$S = \frac{N_S - N_r}{N_S} = \frac{1000 - 960}{1000} = 0.04$$

$$Z = R + \frac{R^1}{S} + j(X + X^1)$$

$$= 0.35 + \frac{0.54}{0.04} + j\,(2.5)$$

$$= 0.35 + j2.5$$

$$Z = 14.07 \angle 10.23\ \Omega$$

$$I^1 = \frac{\frac{415}{\sqrt{3}} \angle 0}{14.07 \angle 10.23}$$

$$= 17.03 \angle -10.23\ \text{A}$$

$$I_0 = 7.4 \angle -77.5\ \text{A}$$

$$I_1 = I_0 + I^1$$

$$= 7.4 \angle -77.5° + 17.03 \angle -10.23°$$

$$= 21.01 \angle -29.16°\ \text{A}$$

$$I_1 = 21.01, \quad \cos\phi = 0.87\ \text{(lag)}$$

Now,

$$T = \frac{3(I^1)^2 R^1}{S\omega_s} = \frac{3 \times 17.03^2 \times 0.54}{0.04 \times 100.5} = \frac{469.83}{4.02} = 116.87\ \text{N-m}$$

$$\text{Power output} = 3(I^1)^2 R^1 \left(\frac{1}{S} - 1\right)$$

$$= 3 \times 17.03^2 \times 0.54 \left(\frac{1}{0.04} - 1\right)$$

$$= 11.276\ \text{kW}$$

$$\text{Power input} = \sqrt{3} \times 415 \times 21.01 \times 0.87$$

$$= 13.1387\ \text{kW}$$

$$\text{Efficiency, } \eta = \frac{\text{Power output}}{\text{Power input}} \times 100 = \frac{11.276}{13.1387} \times 100 = 85.82\%$$

(iii) Starting, $S = 1$

$$Z_1 = R + \frac{R^1}{S} + j\,X + X^1$$

$$= 0.89 + j\,2.5 = 2.65 \angle 70.4°\ \Omega$$

$$I^1 = \frac{\frac{415}{\sqrt{3}} \angle 0^0}{2.65 \angle 70.4} = 90.41 \angle -70.4$$

$$I_1 = I_0 + I^1$$

$$= 7.4 \angle -77.5 + 90.41 \angle -70.4$$

$$= 97.74 \angle -70.94 \text{ A}$$

$$T_{start} = \frac{3I^{12}R^1}{\omega_s} = \frac{3(90.41)^2 \times 0.54}{100.5} = 131.75 \text{ N-m}$$

(iv) For maximum torque, P_G should be maximum

$$\sqrt{R^2 + (X + X^1)^2} = \frac{R^1}{S_{max}}$$

$$S_{max} = \frac{R^1}{\sqrt{R^2 + (X_0 + X^1)^2}}$$

$$= \frac{0.54^2}{\sqrt{0.35^2 + 2.5^2}} = \frac{0.2916}{2.5243} = 0.11$$

Now, for $S = 0.11$

$$Z = \left[0.35 + \frac{0.54}{0.11}\right] + j\,2.5\,\Omega$$

$$= 5.25 + j \angle 2.5\,\Omega$$

$$= 5.82 \angle 25.42$$

$$I^1 = \frac{\frac{415}{\sqrt{3}}}{5.82} = 41.16\,A$$

$$T_{max} = \frac{3 \times 41.16^2 \times 0.54}{100.5 \times 0.11} = 248.26 \text{ N-m}$$

EXAMPLE 9.16 **A 3-ϕ, 15.38-kW, 415-V, 50-Hz star-connected induction motor yielded the following data on testing.**

No-load test: 415 V, 7.4 A, cos $\phi_0 = 0.25$

Blocked rotor test: 220 V, 48 A, cos $\phi_S = 0.42$

From the circle diagram, find the (i) line current, (ii) power factor, and (iii) efficiency at rated load.

Solution Given data

Power = 15.38 kW

Supply voltage, $V = 415$ V

Frequency, $f = 50$ Hz

No-load test

$$I_0 = 7.4 \text{ A}, V_0 = 415 \text{ V}, \cos\phi_0 = 0.25$$

$$P = \sqrt{3}V_0 I_0 \cos\phi_0$$

$$= \sqrt{3} \times 415 \times 7.4 \times 0.25$$

$= 1330$ W

$\cos \phi_0 = 0.25 \Rightarrow \phi_0 = 75.52$

Short-circuit current with normal voltage applied to stator, $I_{sc} = I_s \times \dfrac{\text{OC voltage}}{\text{SC voltage}}$

$$= 48 \times \frac{415}{220} = 90.54 \text{ A}$$

$\cos\phi_S = 0.42 \Rightarrow \phi_S = 65.16$

$P_{sc} = \sqrt{3} \times 415 \times 90.54 \times 0.42$

$= 27.33$ kW $= 27333.7$ W

AF = 4 cm

$$\text{Power scale} = \frac{27333.7}{4} = 6833.42 \text{ W/cm}$$

Motor output is 15380 W.

$$\text{It is represent by AT} = \frac{15380}{6833.42} = 2.25 \text{ cm.}$$

From the circle diagram shown in Figure 9.13

(i) Line current on full-load "OH" = 4.5 cm

$= 4.7 \times 10 = 47$ A

(ii) Power factor at full load $= \dfrac{NH}{OH} = \dfrac{3.5}{4.7}$

$$\text{Slip} = \frac{KL}{LH} = \frac{0.35}{2.6} = 0.13$$

(iii) Efficicncy $= \dfrac{KH}{NH} \times 100 = \dfrac{2.2}{4.7} = 46\%$

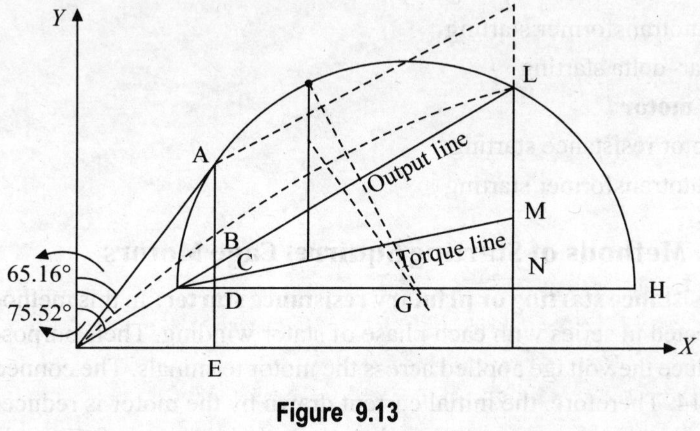

Figure 9.13

9.10 STARTING METHODS FOR INDUCTION MOTOR

Induction motors, when direct-switched, take five to seven times their full-load current and develop only 1.5 to 2.5 times full-load torque. This initial excessive current is objectionable because it will produce large line voltage drop, which will affect the operation of other electrical equipment connected to the same lines.

Direct online starting: In this method, as the name indicates the motor is started by connecting it directly to three-phase supply. When the motor is connected to the supply directly, the starting current will be high, i.e., 5 to 7 times full-load current. Hence, this method of starting is suitable for relatively small machines. If I_{st} and I_{fl} are the starting and full-load currents, respectively, we can obtain the relationship between starting torque T_{st} and full-load torque T_{fl}.

$$\frac{T_{st}}{T_{fl}} = \frac{\left(\frac{3I_{st}^2 R_2'}{1}\right)}{\left(\frac{3I_{fl}^2 R_2'}{S_{fl}}\right)} = \frac{I_{st}^2}{I_{fl}^2} \times S_{fl} \tag{9.67}$$

where S_{fl} is the full-load slip and at starting $S = 1$. If the motor is directly connected to the supply at starting I_{st} = short-circuit current I_{sc} at rated voltage.

$$\frac{T_{st}}{T_{fl}} = \frac{I_{sc}^2}{I_{fl}^2} \times S_{fl} \tag{9.68}$$

If I_{sc} is six times the full-load current and the full-load slip $S_{fl} = 0.04$, then

$$\frac{T_{st}}{T_{fl}} = 36 \times 0.04 = 1.44 \tag{9.69}$$

i.e., starting torque is 1.44 times the full-load torque but the starting current is six times the full-load current.

The methods of starting for an induction motor are:

Squirrel cage motors

(i) Stator resistance starting or primary resistance starter
(ii) Autotransformer starting
(iii) Star–delta starting

Slip ring motor

(i) Rotor resistance starting
(ii) Autotransformer starting

9.10.1 Methods of Starting Squirrel Cage Motors

Stator resistance starting or primary resistance starter: In this method, the external resistances are connected in series with each phase of stator winding. Their purpose is to drop some voltage hence reduce the voltage applied across the motor terminals. The connection diagram is shown in Figure 9.14. Therefore, the initial current drawn by the motor is reduced. The main drawback of this method is that the current varies directly with the square of the applied voltage. If the voltage applied across the motor is reduced by 50%, the starting current is reduced by 50%, but the torque is reduced by 25%. Another drawback of this method is that a lot of power is wasted in the starting resistance. Hence, this method is useful for smooth starting of small machines only.

Let V be the supply voltage per phase. A fraction of this voltage equal to $K V (K < 1)$ is applied to the stator at the starting. The starting current per phase will be $I_{st} = K I_{sc}$, where I_{sc} is the starting current (short-circuit current) with a voltage V applied.

$$\therefore \quad \frac{T_{st}}{T_{fl}} = \left(\frac{I_{st}}{I_{fl}}\right)^2 \times S_{fl} \tag{9.70}$$

$$= \left(\frac{KI_{sc}}{I_{fl}}\right)^2 \times S_{fl}$$

$$= K^2 \times \left(\frac{I_{sc}}{I_{fl}}\right)^2 \times S_{fl} \tag{9.71}$$

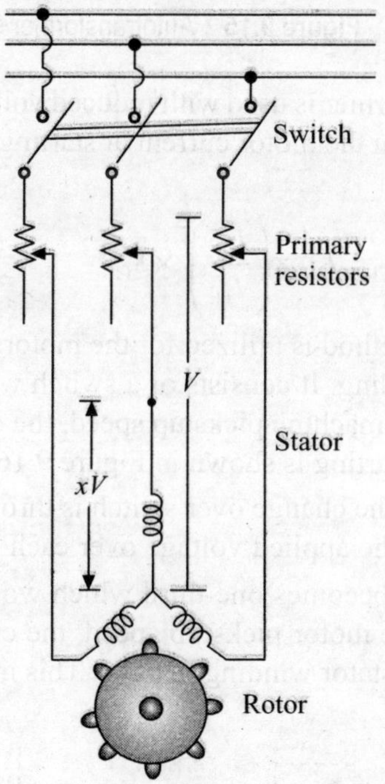

Figure 9.14 Connection diagram of primary resistors

Autotransformer starting: In this method also, the reduced voltage is applied to the induction motor at the starting and then the full voltage is applied as the motor picks up its speed. Figure 9.15 shows the arrangement for autotransformer starting.

Most of the autotransformers are provided with three set of taps, so as to reduce voltage to 80, 65, or 50% of the line voltage, to suit the local conditions of supply.

At the time of starting, the change over switch is thrown to "start" position. This connects the autotransformer in the circuit and the reduced voltage is applied to induction motor. When the motor attains about 80% of the speed the switch is thrown to "run" position. This eliminates the autotransformer from the circuit and hence full voltage is applied to the motor. The advantages of this method are low power loss, low starting current. This method of starting is used for large machines.

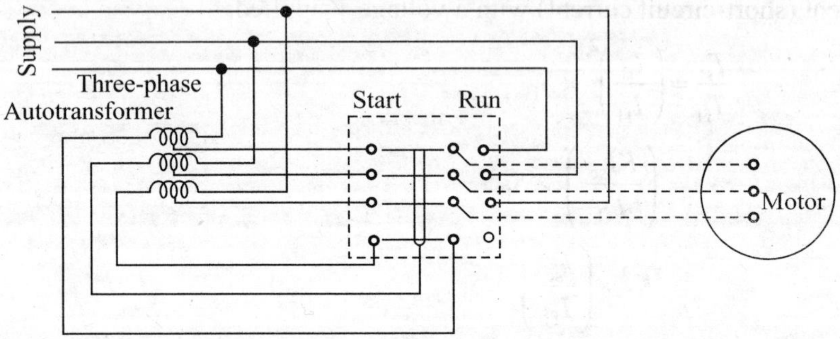

Figure 9.15 Autotransformer starting

Suppose, if the autotransformer is used with reduced voltage applied of magnitude K V (where $K < 1$) of normal voltage, then the motor current at starting is $I_{st} = KI_{sc}$ and the ratio of starting to full-load torque is given by

$$\frac{T_{st}}{T_{fl}} = \left(\frac{I_{st}}{I_{fl}}\right)^2 \times S_{fl} = K^2 \times \left(\frac{I_{sc}}{I_{fl}}\right)^2 \times S_{fl} \tag{9.72}$$

Star–delta starting: This method is utilized for the motors which are built to run normally with a delta-connected stator winding. It consists of a switch which connects the motor in star while starting the motor. When the machine picks up speed, the connections are changed to delta. The arrangement for star–delta starting is shown in Figure 9.16.

At the instant of starting, the change over switch is thrown to "start" position which connects the stator windings in star. The applied voltage over each motor phase is reduced by a factor of $1/\sqrt{3}$ and hence the torque becomes one-third which would have been developed if the stator is delta-connected. When the motor picks up speed, the change over switch is thrown to "run" position which connects the stator winding in delta. This method of starting is used for medium-sized machines.

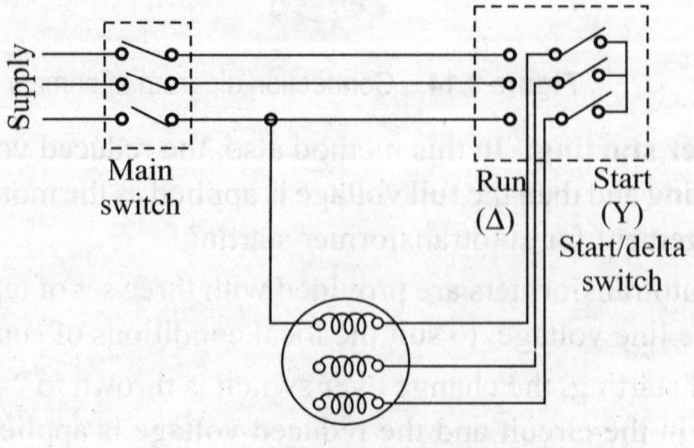

Figure 9.16 Arrangement for star–delta starting

$$I_{st} = \frac{1}{\sqrt{3}} I_{sc}$$

$$\frac{T_{st}}{T_{f1}} = \left(\frac{I_{st}}{I_{f1}}\right)^2 \times S_{f1} = \frac{1}{3} \times \left(\frac{I_{sc}}{I_{f1}}\right)^2 \times S_{f1} \tag{9.73}$$

This method is cheap and effective, as long as the starting torque is not required to be more than 50% of the full-load torque. The change over from star to delta should be done only when the speed of the motor is around 90% of the rated speed.

9.10.2 Methods of Starting Slip Ring Induction Motors

The three methods discussed above can be utilized to control the speed of slip ring motor in the same manner. In addition to those methods the speed of the slip ring induction motor can be controlled using rotor resistance starting. This method gives low starting current and high starting torque.

Rotor resistance starting: In this method, a variable star-connected rheostat is connected in the rotor circuit through slip rings, and full voltage is applied to the stator winding. The circuit arrangement is shown in Figure 9.17.

At the time of starting, the handle of rheostat is sent in the off position so that maximum resistance is placed in each phase of rotor circuit. This reduces the starting current and increases the starting torque. As the motor picks up speed the external resistance is gradually cut out in each phase of rotor circuit. When the motor attains rated, the whole external resistance is cut out.

Since this method of starting develops high starting torque, we can start the motor with load. The maximum torque can be obtained if at starting the total resistance of the motor winding including that of external resistance is made equal to that of rotor reactance, i.e., $R_2' + R = X_1 + X_2'$. This type of starting can be used even if the rotor is delta connected.

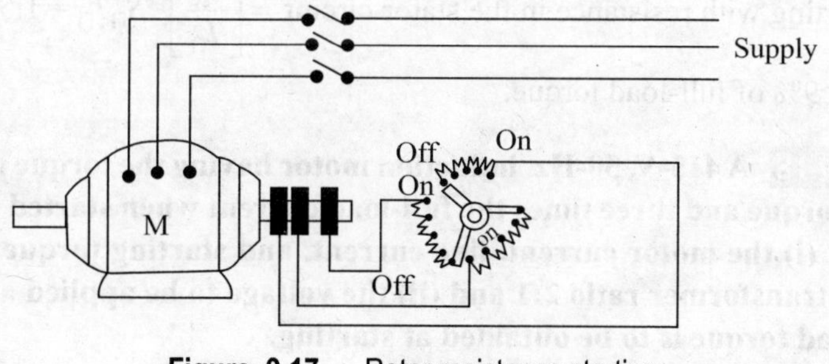

Figure 9.17 Rotor resistance starting

EXAMPLE 9.17 **A short-circuit current of a 3-ϕ squirrel cage induction motor is equal to three times the full-load current. Find the starting torque as a percentage of full-load torque if the motor is started by (i) direct switching to the supply, (ii) a star–delta starter, (iii) an out transformer, and (iv) a resistance in the stator circuit. The stator current in (iii) and (iv) is limited to 1.5 times the full-load current. Assume that the full-load slip is 4%.**

Solution Given data

Short-circuit current, $I_{sc} = 3I_{fl}$

Full-load slip, $S_{fl} = 0.04$

(i) Starting with the direct switch

$$T_{st} = \left[\frac{I_{sc}}{I_{fl}}\right]^2 S_{fl}T_{fl}$$

$$= 3^2 \times 0.04T_{fl}$$

$$= 0.36T_{fl}$$

i.e., 36% of full-load torque

(ii) Starting with star–delta starter

$$T_{st} = \frac{1}{3}\left[\frac{I_{sc}}{I_{fl}}\right]^2 S_{fl}T_{fl}$$

$$= \frac{1}{3} \times 3^2 \times 0.04T_{fl} = 0.12T_{fl}$$

i.e., 12% of full load torque

(iii) Starting with an out-transformer supply line current = $K^2 I_{sc} = 1.5I_{fl}$

K = Transformation ratio

$$= \sqrt{\frac{1.5I_{fl}}{I_{sc}}} = \sqrt{\frac{1.5I_{fl}}{1.5I_{fl}}} = 1$$

$$\therefore \quad l^2 \times 1.5^2 \times 0.04\ T_{fl}$$

$$= 0.09\ T_{fl}$$

i.e., 9% of full-load torque.

(iv) Starting with resistance in the stator circuit $= \left[\frac{I_{sc}}{I_{fl}}\right]^2 S_{fl}T_{fl} = 1.5^2 \times 0.04T_{fl} = 0.09T_{fl}$

i.e., 9% of full-load torque.

EXAMPLE 9.18 **A 415-V, 50-Hz, induction motor having the torque produced is 1.5 times the full-load torque and three times the full-load current when started directly from the mains. Calculate (i) the motor current, line current, and starting torque when started by means of an autotransformer ratio 2:1 and (ii) the voltage to be applied and the motor current, if the full-load torque is to be obtained at starting.**

Solution Given data

Supply voltage, V = 415 V

Frequency, f = 50 Hz

Transformation ration = 2:1

$$= \frac{1}{2} = 0.5$$

(i) Let us assume that the winding is star-connected,

$$\text{Phase voltage, } V_{ph} = \frac{415}{\sqrt{3}} = 240\text{ V}$$

Voltage across the motor terminal = $KV_L = 0.5 \times 415 = 207.5$ V

The phase voltage when started by an autotransformer = $\dfrac{207.5}{\sqrt{3}} = 120$ V

As the current is proportional to the voltage per phase

Therefore, new starting current per phase = $\dfrac{120}{240} \times 3I_{f1} = 1.5I_{fl}$

We know $\dfrac{V_2}{V_1} = \dfrac{N_2}{N_1} = \dfrac{I_1}{I_2}$

$$K = \frac{V_2}{V_1} = \frac{1}{2} = 0.5 \text{ and } I_2 = 1.5I_{fl}$$

$$\frac{I_1}{I_2} = K$$

or

$$I_1 = KI_2$$
$$= 0.5 \times 1.5\, I_{fl} = 0.75\, I_{fl}$$

Hence, line current = $0.75\, I_{fl}$

(ii) Let the voltage per phase be V to get full-load starting current,

$$T_f \propto V^2$$

From equations $1.5\, T_f \propto (240)^2$ and $T_f \propto V^2$, we get

$$\frac{T_f}{1.5T_f} = \frac{V^2}{(240)^2}$$

$$V^2 = \frac{(240)^2}{1.5}$$

$$V = 195.9 \text{ V}$$

Motor voltage = $195.9 \times \sqrt{3}$

$= 340$ V

Motor current = $\dfrac{195.6}{240} \times 3I_f$

$= 2.445I_f$ A

When the star–delta starter is used, the voltage applied is $1/\sqrt{3}$ times the normal voltage and the current is $3I_f$.

$\therefore$ Motor phase current with autotransformer $= 3I_f \times \dfrac{0.75\,V}{V/\sqrt{3}} = 3.9I_f$ A

Motor line current = $\sqrt{3} \times 3.9I_f$

$= 6.75I_f$ A

Line current from the mains I_1

$$\frac{V_{Ph}}{0.75\, V_{Ph}} = \frac{6.75I_f}{I_1}$$

$$I_1 = 0.75 \times 6.75 I_f$$
$$= 5.0625 I_f$$

Starting torque with 75% tapping

$$\frac{T_{st}}{T_f} = \left[\frac{3.9\, T_f}{I_f/\sqrt{3}}\right]^2 \times 0.02$$
$$= 0.91 \text{ or } 91\% \text{ full-load torque}$$

EXAMPLE 9.19 **A squirrel cage-type induction motor when started by means of a star–delta starter takes 300% of the full-load line current and develops 56% of the full-load torque at starting. Calculate the starting current and currents in terms of the full-load values, if an autotransformer with 75% tapping were employed.**

Solution Given data

Short-circuit current, $I_{sc} = 3I_f$

Starting torque $T_{st} = 0.56T_f$

Using the relation,

$$\frac{T_{st}}{T_f} = \left[\frac{I_{sc}}{I_f}\right]^2 \times S_f \qquad (1)$$

Hence, the full-load phase current $= \dfrac{I_f}{\sqrt{3}}$

When started with a star–delta starter

$$T_{st} = 0.56T_f$$
$$I_{sc} = 3I_f \quad \text{(Since line current = phase current)}$$
$$\frac{0.56T_f}{T_f} = \left[\frac{3I_f}{\frac{I_f}{\sqrt{3}}}\right]^2 \times S_f$$
$$0.56 = [5.2]^2 \times S_f$$
$$S_f = \frac{0.56}{27} = 0.02 \text{ or } 2\%$$

When started through an autotransformer with 75% tapping and winding in delta, 75% of the normal voltage is applied.

$$T_f = \text{Full-load torque}$$
$$T \propto V^2$$
$$1.5T_f \propto (240)^2 \qquad (2)$$

When the applied voltage is 120 V

Let the torque = T

$$T \propto (120)^2 \qquad (3)$$

Dividing Eq. (3) by Eq. (2), we get

$$\frac{T}{1.5T_f} = \left[\frac{120}{240}\right]^2$$

$$T = 0.375T_f$$

9.11 SPEED CONTROL OF INDUCTION MOTORS

The speed of an induction motor can be changed with various techniques. These speed control techniques are divided into two categories.

(i) **Control from stator side**

(a) *Changing the applied voltage:* By changing the applied voltage to the induction motor the speed of the motor can be changed. This method is very easy and cheapest. This method is rarely used because a large change in the voltage is required for a small change in the speed.

(b) *Changing the applied frequency:* The synchronous speed of the induction motor is given by $N_S = 120f/p$. From this equation it is clearly observed that the speed of induction motor can be changed by changing the frequency. This method is also rarely used.

(c) *Changing the number of poles:* From the above speed equation, it can be observed that the speed of induction motor can be changed by changing the number of poles. Only two or four speeds are possible by this method. This is achieved by utilizing independent stator windings wound for different number of poles.

(ii) **Control from rotor side**

(a) *Rotor rheostat control:* This method is applicable to slip ring induction motors only. The motor speed is reduced by introducing an external resistance in the rotor circuit to change the resistance, rotor starter may be used.

(b) The disadvantage of this method is that with increase in rotor resistance, I^2R losses will increase. Therefore, operating efficiency of the motor decreases.

(c) *Injecting an emf in the rotor circuit:* The speed of the induction motor is controlled by injecting a voltage in the rotor circuit. The injected voltage should have the same frequency as the slip frequency.

When we inject a voltage which is phase opposition to the rotor-induced emf, it decreases the speed of the motor. When we inject a voltage which is in phase with the induced rotor emf the speed will be increased.

EXAMPLE 9.20 **A 6-pole, 3-ϕ, 50-Hz, induction motor slip ring is short circuited and it rotates 960 rpm. The rotor resistor and standstill reactance are 0.2 and 0.8 Ω, respectively, if extra resistance of 0.15 Ω per phase is added to the rotor circuit. Calculate the new full-load speed.**

Solution Given data

Number of poles, $p = 6$

Frequency, $f = 50$ Hz

Rotor speed, $N_r = 960$ rpm

Rotor resistance, $R_2 = 0.2\ \Omega$

Rotor reactance, $X_2 = 0.8\ \Omega$

Extra resistance, $R_2 = 0.15$

Synchronous speed, $N_S = \dfrac{120f}{p} = \dfrac{120 \times 50}{6} = 1000$ rpm

Slip, $S_1 = \dfrac{1000 - 960}{1000} = 0.04$

We know $T_1 = \dfrac{KS_1R_2}{R_2{}^2 + S_1^2 X_2{}^2}$

$$T_2 = \frac{KS_2R_2'}{(R_2 + R_2')^2 + (S_2X_2)^2}$$

$$T_1 = T_2 = T_{fl}$$

i.e.,

$$\frac{S_1R_2}{R_2{}^2 + S_1^2X_2{}^2} = \frac{S_1R_2'}{(R_2 + R_2')^2 + S_2^2X_2{}^2}$$

$$\frac{0.04 \times 0.2}{0.2^2 + (0.04 \times 0.8)^2} = \frac{S_2(0.2 + 0.15)}{(0.35)^2 + S_2^2\ 0.8^2}$$

$$0.195 = \frac{0.35\,S_2}{0.1225 + 0.64\,S_2^2}$$

$$0.0238 + 0.1248\,S_2{}^2 = 0.35\,S_2$$

$$0.1248\,S_2{}^2 - 0.35\,S_2 + 0.0238 = 0$$

$$S_2 = 0.069 = \frac{N_S - N_r'}{N_S}$$

$\therefore$

$$0.069 = \frac{1000 - N_r'}{1000}$$

$$69 = 1000 - N_r'$$

or

$$N_r' = 931 \text{ rpm}$$

EXAMPLE 9.21 **A 6-pole, 50-Hz, 3-ϕ, slip ring induction motor runs at 960 rpm and has a resistance of 0.18 Ω per phase on full load. Determine how much external resistance must be added to lower the speed to 800 rpm. Assume the torque remaining constant in both the cages.**

Solution Given data

Number of poles, $p = 6$

Frequency, $f = 50$ Hz

Rotor speed, $N_r = 960$ rpm

Rotor resistance, $R_2 = 0.18\ \Omega$

Lower speed, $N_r = 800$ rpm

Synchronous speed, $N_S = \dfrac{120f}{p} = \dfrac{120 \times 50}{6} = 1000$ rpm

Slip, $S_1 = \dfrac{N_S - N_r}{N_S} = \dfrac{1000 - 960}{1000} = 0.04$

In the second case

$$S_2 = \frac{1000 - 800}{1000} = 0.2$$

Now let the rotor circuit resistance be "*R*."

The expression for torque is

$$T = \frac{KSR}{R^2 + S^2X^2}$$

For low value of slip, S_2 in small

$$\therefore \quad T \cong \frac{kSR}{R^2} = \frac{kS}{R}$$

In the first case, the torque

$$T_1 = \frac{KS_1}{R}$$

Similarly, the second case, $T_2 = \dfrac{KS_2}{R^1}$

Since torque remains constant, $T_1 = T_2$

$$\frac{kS_1}{R} = \frac{kS_2}{R^1}$$

$$\therefore \quad R^1 = R\frac{S_2}{S_1} = 0.18 \times \frac{0.2}{0.04} = 0.9\ \Omega$$

Extra resistance to be added in the motor = $R^1 - R = 0.90 - 0.18 = 0.72\ \Omega$

9.12 APPLICATIONS OF INDUCTION MOTORS

The squirrel cage induction motors are comparatively cheaper than the wound rotors and are rugged in construction. They require less conductor material than the wound rotor. Hence, the copper loss is less and such motors will be little more efficient than the wound rotor.

Since a cage motor is more rugged in construction hence it requires no slip rings, brushes, etc. The maintenance for cage motors is low.

The main disadvantages of squirrel cage induction motors are that it is not possible to insert any external resistance in the rotor. Speed control is not possible. Hence, squirrel cage induction motors are used for

- Constant speed industrial drives of small power
- Applications, where starting torque requirements are of medium or low value, i.e., printing machinery, flour mills

A slip ring induction motor has high starting torque, and its speed can be controlled very easily. Hence, the slip ring induction motors are used for loads requiring high starting torques or for loads which require speed control, i.e., for driving line shafts, lifts, pumps, generators, cranes, hoists, elevators, and compressors.

9.13 COMPARISON BETWEEN SYNCHRONOUS MOTOR AND INDUCTION MOTOR

Synchronous motor	Induction motor
1. It is not self-starting and some external means is required for its staring.	1. It is self-starting hence requires no external means for starting.
2. The synchronous motor runs at constant average speed whatever the load.	2. The speed of an induction motor falls somewhat with the increase in load.
3. It can be operated over a wide range of power factor both lagging and leading.	3. It operates only at lagging power factor.
4. No speed control is possible.	4. Speed can be controlled to some extent.
5. It requires dc excitation so it is a doubly excited device.	5. It does not require any dc excitation so it is a singly excited device.
6. Its torque is less sensitive to applied voltage changes.	6. Its torque is more sensitive to changes in applied voltage.
7. The break down torque is proportional to the supply voltage.	7. The break down torque is proportional to the square of the supply voltage.
8. It is more complicated and costs more.	8. It is simpler and costs less.

EXAMPLE 9.22 **A 3-ϕ star-connected induction motor is having the rotor emf between slip rings at standstill of 70 V. The rotor resistance and standstill reactance are 0.7 Ω and 5 Ω, respectively. Calculate the following:**

(i) Rotor current per phase at starting when the slip rings short circuited.

(ii) Rotor current per phase if a star-connected rheostat of resistance 8 Ω per phase is connected across the slip rings.

(iii) Full-load rotor current when the rotor power factor of the full-load slip is 0.035.

(iv) Rotor emf per phase under the full-load condition.

Solution Given data

Rotor emf at standstill (between slip rings) = 70 V

Rotor resistance = 0.7

Rotor reactance = 5 Ω

Slip = 3.5%

Rotor emf per phase at standstill, $E_2 = 70\sqrt{3} = 40.41$ V

(i) At starting

$$\text{Rotor current per phase at starting, } I_2 = \frac{E_2}{Z_2} = \frac{E_2}{\sqrt{R_2^2 + X_2^2}} = \frac{40.41}{\sqrt{0.7^2 + 5^2}}$$

$$= \frac{40.41}{5.0487} = 8.0039\text{A}$$

(ii) The total resistance, $R_2' = R_2 + 8 = 0.7 + 8 = 8.7$

Rotor current per phase with rheostat resistance, $I_2' = \frac{40.41}{\sqrt{8.7^2 + 52}} = \frac{40.41}{10.034} = 4.027$ A

(iii) Full-load rotor current slip, $S_{fl} = \frac{SE_2}{\sqrt{R_2^2 + (S_{fl}X_2)^2}}$

$$= \frac{0.035 \times 40.41}{\sqrt{0.7^2 + (0.035 \times 5)^2}}$$

$$= \frac{1.4143}{0.7215} = 1.9601 \text{ A}$$

(iv) Rotor power factor at full load, $\cos\phi = \frac{R_2}{\sqrt{R_2^2 + (S_{fl}X_2)^2}} = \frac{0.7}{\sqrt{0.7^2 + (0.035 \times 5)^2}}$

$$= \frac{0.7}{0.7215} = 0.97(\text{lag})$$

EXAMPLE 9.23 A 3-ϕ, 6-pole, 50-Hz, 400-V star-connected induction motor takes a current of 65 A and delivers useful power of 20 kW while running at a speed of 960 rpm. Its stator resistance per phase is 0.12 Ω and mechanical losses are 800 W. Calculate the (i) shaft torque, (ii) gross torque developed, (iii) rotor copper losses, (iv) stator copper losses, and (v) overall efficiency. The power factor of the motor is 0.8 lag.

Solution Given data

Supply voltage, $V = 400$ V

Useful power, $P = 20$ kW

Running speed, $N_r = 960$ rpm

Current, $I = 65$ A

Stator resistance per phase = 0.12

Mechanical losses = 800 W

Power factor, $\cos\phi = 0.8$

Number of poles, $p = 6$

Frequency, $f = 50$ Hz

Synchronous speed, $N_S = \frac{120f}{P} = \frac{120 \times 50}{6} = 1000$ rpm

Slip, $S = \frac{N_S - N_r}{N_S} = \frac{1000 - 960}{1000} = 0.04$

Mechanical power, $P_m = P_{out}$ + Mechanical losses

$$= 20 + 0.8 = 20.8 \text{ kW}$$

Rotor power, $P_r = \frac{SP_m}{1-S} = \frac{0.04 \times 20.8 \times 10^3}{1 - 0.04} = \frac{832}{0.96} = 866.67$ W

(i) Shaft torque $= \dfrac{P_{\text{out}}}{\omega_r} = \dfrac{20\times10^3}{\dfrac{2\pi N_r}{60}} = 199.04$ N-m

(ii) Gross torque $= \dfrac{P_{\text{m}}}{\omega_r} = \dfrac{20.8\times10^3}{\dfrac{2\pi N_r}{60}} = 207.006$ N-m

(iii) Copper losses (P_r) = 866.67 W

$$P_g = \frac{P_r}{S} = \frac{866.67}{0.04} = 21.6667 \text{ kW}$$

Total input (P_{in}) $= \sqrt{3}\, V_2 I_2 \cos\phi$

$= \sqrt{3}\times 400\times 65\times 0.8 = 36.0266$ kW

(iv) Stator copper losses $= 3I^2R$

$= 3\times 65^2\times 0.12 = 1.521$ kW

(v) Overall efficiency $= \dfrac{P_{\text{out}}}{P_{\text{in}}}\times100$

$$= \frac{20\times10^3}{36.0266\times10^3}\times100 = 55.51\%$$

EXAMPLE 9.24 **A 3-ϕ, 50-Hz, 6-pole, 400-V induction motor draws an input power to the rotor is 70 kW. The motor emf is observed to make 120 cycles/min. Calculate the (i) slip, (ii) rotor speed, (iii) mechanical power developed, (iv) rotor resistance loss per phase, and (v) rotor resistance per phase, if the rotor current is 65 A.**

Solution Given data

Supply voltage, $V = 400$ V
Number of poles, $p = 6$
Frequency, $f = 50$ Hz
Input power, $P = 70$ kW
Rotor current, $I = 65$ A
Rotor frequency = 120 cycles/min

(i) Synchronous speed, $N_S = \dfrac{120f}{p} = \dfrac{120\times50}{6} = 1000$ rpm

Given rotor frequency, $f_r = \dfrac{120}{60} = 2$ Hz

Rotor frequency, $f_r = Sf$

$$S = \frac{f_r}{f} = \frac{2}{50} = 0.04$$

(ii) Rotor speed, $N_r = (1-S)N_S$

$= (1-0.04)\times 1000 = 960$ rpm

(iii) Power input per phase, $P_{in} = \dfrac{70}{3} = 23.33$ kW

Mechanical power developed, $P_m = 3\,P_{in}(1 - S)$

$= 3 \times 23.33 \times 10^3\ (1 - 0.04) = 67.19$ kW

(iv) Rotor resistance loss per phase $= S \times P_{in}$

$= 0.04 \times 23.33 \times 10^3 = 933.2$ W

(v) If r_2 and i_2 per phase are the rotor resistance and current, respectively

$I^2{}_2 r_2$ = Rotor resistance loss = 933.2 W

= 933.2

$$r_2 = \frac{933.2}{I_2^2} = \frac{933.2}{65^2} = 0.2208\ \Omega$$

EXAMPLE 9.25 The efficiency of an induction motor is 0.85 when the load is 70 kW. At this load, the stator resistance and rotor resistance loss each equals the in loss. The machine losses are one-fourth of the no-load loss. Calculate the slip.

Solution Given data

Efficiency, $\eta = 0.85$

Output power, $p_{out} = 70$ kW

Mechanical losses = 1/4 of no-load loss

$$\text{Power input per phase} = \frac{\text{Power output per phase}}{\text{Efficiency}}$$

$$= \frac{70/3}{0.85} = 27.45 \text{ kW}$$

Let iron loss per phase be "X" kW, then mechanical loss = $0.25X$ kW

Stator resistance loss per phase = Rotor resistance per phase

= X kW air-gap power per phase

P_{ag} = Input – (Iron loss + Stator resistance loss + Rotor resistance loss)

$= 27.45 - 3X$

But P_{ag} = Power output per phase + Mechanical loss = $23.33 + 0.25X$

Then, $27.45 - 3X = 23.33 + 0.25X$

$4.12 = 3.25X$

$X = 1.2676$ kW

$\therefore P_{ag} = 23.33 + 0.251.2676 = 23.6469$ kW

Now rotor resistance loss = SP_{ag}

$$\therefore \text{Slip, } S = \frac{1.2676}{23.6469} = 0.05$$

EXAMPLE 9.26 **A 4-pole, 3-ϕ, 50-Hz, 415-V induction motor develops 22 kW including mechanical losses of 2 kW at a speed of 1440 rpm. Calculate the (i) slip, (ii) rotor resistance loss, (iii) total input if the stator losses are 1.8 kW, (iv) efficiency, (v) line current, and (vi) number of complete cycles of the rotor emf per minute. Take the power factor as 0.88.**

Solution Given data

Supply voltage, $V = 415$ V

Number of poles, $p = 4$

Frequency, $f = 50$ Hz

Output power $P_{out} = 22$ kW

Mechanical loss = 2 kW

Rotor speed, $N_r = 1440$ rpm

Stator losses = 1.8 kW

Power factor = 0.88

(i) Synchronous speed, $N_S = \dfrac{120f}{p} = \dfrac{120 \times 50}{4} = 1500$ rpm

Slip, $S = \dfrac{N_S - N_r}{N_S} = \dfrac{1500 - 1440}{1500} = 0.04$

(ii) The total mechanical power developed,

P_m = Power developed + Mechanical loss

$= 22 + 1.5 = 23.5$ kW

We know, $P_m = (1 - S)\,P_{ag}$

$$P_{ag} = \frac{P_m}{1-S} = \frac{23.5 \times 10^3}{1 - 0.04} = 24.48 \text{ kW}$$

Then rotor resistance loss $= SP_{ag}$

$= 0.04 \times 24.48 = 0.98$ kW

(iii) Total input $= P_{ag}$ + Stator loss $= 24.48 + 1.8 = 26.28$ kW

(iv) Efficiency $= \dfrac{\text{Output}}{\text{Input}} \times 100 = \dfrac{22}{26.28} \times 100 = 83.7\%$

(v) Input power $= \sqrt{3}\, VI \cos\phi$

$\therefore$ Current, $I = \dfrac{26.28 \times 10^3}{\sqrt{3} \times 415 \times 0.88} = 41.54$ A

(vi) Rotor frequency, $f_r = S_f = 0.04 \times 50 = 2$ Hz

The number of complete cycles of the rotor emf per minute is $2 \times 60 = 120$.

EXAMPLE 9.27 **Calculate the frequency of the rotor-induced emf, where the rotor of a 3-ϕ induction motor rotates at 960 rpm when a 50-Hz supply is connected across the stator terminals.**

Solution Given data

Rotor speed, $N_r = 960$ rpm

Frequency, $f = 50$ Hz

Synchronous speed, $N_S = \dfrac{120f}{p}$

Assume the number of poles be $p = 6$.

$$\therefore N_S = \frac{120 \times 50}{6} = 1000 \text{ rpm}$$

$$\text{Slip, } S = \frac{N_S - N_S}{N_S} = \frac{1000 - 960}{1000} = 0.04$$

Rotor frequency, $f_r = S_f = 0.04 = 2$ Hz

EXAMPLE 9.28 **The rotor resistance and standstill reactance of a 6.6-kV, 20-pole, 50-Hz, 3-ϕ induction motor is 0.15 Ω. The motor has a speed of 282.2 rpm at full load. Calculate the (i) slip at maximum torque, and (ii) ratio of maximum to the full-load torque. Neglect the stator impedance.**

Solution Given data

Line voltage, $V = 6.6$ kV

Number of poles, $p = 20$

Frequency, $f = 50$ Hz

Resistance, $R_2 = 0.15$

Reactance, $X_2 = 1.15$

Motor speed, $N_r = 282.5$ rpm

$$\text{Synchronous speed, } N_S = \frac{120f}{p} = \frac{120 \times 50}{20} = 300 \text{ rpm}$$

$$\text{Slip, } S = \frac{N_S - N_r}{N_S} = \frac{300 - 282.5}{300} = 0.058$$

$$S_{max} = \frac{R_2}{X_2} = \frac{0.15}{1.15} = 0.1304$$

$$\text{Torque at full load, } T_{fl} = \frac{3}{\omega_S}\frac{\left(\dfrac{V}{a}\right)^2 \dfrac{R_2}{S}}{\left(\dfrac{R_2}{S}\right)^2 + X_2^{\,2}} = \frac{3}{\omega_S}\frac{\left(\dfrac{V}{a}\right)^2 \times 0.15/0.058}{\left(\dfrac{0.15}{0.058}\right)^2 + (1.15)^2}$$

$$= \frac{3}{\omega_S}\left(\frac{V}{a}\right)^2 \times 0.3228 \qquad (1)$$

$$\text{(i) Maximum torque, } T_{max} = \frac{3}{\omega_S}\frac{0.5\left(\dfrac{V}{a}\right)^2}{1.15}$$

$$= \frac{3}{\omega_S}\left(\frac{V}{a}\right)^2 \times 0.4347 \qquad (2)$$

(ii) Dividing Eq. (2) by Eq. (1), we get

$$\therefore \quad \frac{T_{max}}{T_{f1}} = \frac{0.4347}{0.3228} = 1.347$$

EXAMPLE 9.29 **Prove that in a 3-ϕ induction motor the ratio of the maximum to starting torque is $(1 + K^2)/(2K)$, where K is the ratio of rotor resistance to rotor reactance. Neglect the stator impedance.**

Solution Given data

$$\text{Maximum torque, } T_{max} = K_T \frac{SR_2}{R_2^2 + (SX_2)^2}$$

$$S_{max} = \frac{R_2}{X_2}$$

$$\therefore \quad T_{max} = \frac{K_T}{2X_2} \qquad (1)$$

Slip at starting = 1

$$T_{start} = K_T \frac{R_2}{R_2^2 + X_2^2} \qquad (2)$$

Dividing Eq. (1) by Eq. (2), we get

$$\frac{T_{max}}{T_{start}} = \frac{R_2^2 + X_2^2}{2R_2X_2}$$

$$= \frac{1 + (R_2/X_2)^2}{2(R_2/X_2)}$$

$$= \frac{1 + k^2}{2k} \quad \left(\text{where } k = \frac{R_2}{x_2}\right)$$

EXAMPLE 9.30 **The rotor resistance and standstill reactance of an 8-pole, 50-Hz, 3-ϕ induction motor is 0.6 Ω, respectively. Calculate the (i) speed at which the torque is maximum, (ii) ratio of maximum torque to starting torque, and (iii) value of external resistance to be connected in series and the rotor circuit for the ratio in (ii) to be 2. Neglect the stator impedance.**

Solution Given data

Number of poles, $p = 8$

Frequency, $f = 50$ Hz

Rotor resistance, $R_2 = 0.6\ \Omega$

Rotor reactance, $X_2 = 6\ \Omega$

(i) Synchronous speed, $N_S = \dfrac{120f}{p} = \dfrac{120 \times 50}{8} = 750$ rpm

$$S_{max} = \frac{R_2}{X_2} = \frac{0.6}{6} = 0.1$$

Rotor speed, $N_r = (1 - S)\, N_S = (1 - 0.1)\, 750 = 675$ rpm

$$K = \frac{R_2}{X_2} = 0.1$$

(ii) $$\frac{T_{max}}{T_{start}} = \frac{1+k^2}{2k} \quad \left(\text{where } k = \frac{R_2}{X_2} = 0.1\right)$$

$$\therefore \quad \frac{T_{max}}{T_{start}} = \frac{1+(0.1)^2}{2\times 0.1} = 5.05$$

(iii) $$\frac{T_{max}}{T_{start}} = \frac{1+k^2}{2k} = 2$$

$$1 + k^2 = 4k$$

$$k^2 - 4k + 1 = 0$$

By solving the above equation, we get

$$k = 3.732 \text{ or } 0.268$$

Now,

$$R_2 = 6 \times 3.732 = 22.4\ \Omega \text{ or } 6 \times 0.268 = 1.608\ \Omega$$

$$R_{2\,(ext)} = 22.4 - 0.6 = 21.8\ \Omega \text{ or } R_{2\,(ext)} = 1.608 - 0.6 = 1.008$$

EXAMPLE 9.31 **A 3-ϕ, 33-kV, 4-pole, 50-Hz star-connected induction motor has slip ring rotor, the rotor resistance, and reactance per phase are 0.05 Ω and 0.7 Ω, respectively. Calculate the following:**

(i) Speed at maximum torque.

(ii) Ratio of full-load torque to maximum torque if the full-torque is obtained at 1440 rpm.

Solution Given data

Supply voltage, $V = 33$ kV

Number of poles, $p = 4$

Frequency, $f = 50$ Hz

Rotor resistance, $R_2 = 0.05\ \Omega$

Rotor reactance, $X_2 = 0.7\ \Omega$

Rotor speed, $N_r = 1440$ rpm

Let the maximum torque occur at $S = S_{max}$.

$$S_{max} = \frac{R_2}{X_2} = \frac{0.05}{0.7} = 0.071$$

(i) Synchronous speed, $$N_S = \frac{120f}{p} = \frac{120\times 50}{4} = 1500 \text{ rpm}$$

Rotor speed at $S = S_{max}$

$$N_r = N_S\,(1 - S_{max}) = 1500\,(1 - 0.071) = 1393.5 \text{ rpm}$$

(ii) Slip at full-load speed (S_{fl}) = $\dfrac{1500-1440}{1500} = 0.04$

$$\frac{T_{fl}}{T_{max}} = \frac{2S_{fl}S_{max}}{S^2_{fl}+S^2_{max}} = \frac{2\times 0.04\times 0.071}{(0.04)^2+(0.071)^2} = 0.855$$

EXAMPLE 9.32 **A 3-ϕ induction motor delivers 60 kW and the stator losses are 4 kW. Determine the (i) total mechanical power developed and rotor I^2R loss when the slip is 5%, (ii) output horse power of the motor if the friction and windage losses are 1.8 kW, and (iii) efficiency of the motor.**

Solution Given data

Input power, P_{in} = 60 kW

Stator losses = 4 kW

Slip, S = 5%

Friction and windage loss = 1.8 kW

Input to rotor = Input to stator – Stator loss = 60 – 4 = 56 kW

Power developed by the rotor = Rotor input – Rotor I^2R loss – Rotor core loss

Rotor I^2R loss = Slip rotor input

= 0.05 × 56 = 2.8 kW

(i) The rotor core loss at 5% slip is very small and can be neglected.

∴ Power developed by the rotor = 56 – 2.8 = 53.2 kW

$= \dfrac{53.2}{0.7355} = 72.529$ hp = (1 hp = 0.7355 kW)

(ii) Output power = Power developed by the rotor – Friction and windage loss

= 53.2 – 1.8

= 51.4 kW

(iii) Efficiency of the motor, $\eta = \dfrac{\text{Output power}}{\text{Input power}} \times 100 = \dfrac{51.4}{60} \times 100 = 85.66\%$

EXAMPLE 9.33 **Calculate the rotor I^2R loss and rotor speed of a 25-hp, 3-ϕ, 50-Hz, 4-pole induction motor and it has full-load slip of 4%. Assume the friction of windage loss 800 W.**

Solution Given data

Rated power, P = 25 hp

Frequency, f = 50 Hz

Number of poles, p = 4

Slip, S = 4%

Friction and windage loss = 800 W

Output of the motor = 25 hp

= 25735.5

= 18.3875 kW

Power developed by the rotor = Output of the motor + Friction of windage loss

$$= 18.3875 + 0.8$$

$$= 19.1875 \text{ kW}$$

Rotor input = Rotor I^2R loss + Power developed by the motor

= Slip × Rotor input + Power developed by the motor

(1–S) × Rotor input = Power developed by the motor

Rotor I^2R loss = S rotor input

$$\text{Rotor input} = \frac{\text{Rotor } I^2 R \text{ loss}}{S}$$

$$(1\text{-}S)\ \frac{\text{Rotor } I^2 R \text{ loss}}{S} = \text{Power developed by the motor}$$

$$\text{Rotor } I^2R \text{ loss} = \frac{S}{1-S} \times \text{ Power developed by the motor}$$

$$\text{In this case, rotor } I^2R \text{ loss} = \frac{0.04}{1-0.04} \times 19.1875 = 0.8 \text{ kW}$$

$$\text{Slip} = \frac{N_S - N_r}{N_S} \Rightarrow 0.04 = \frac{1500 - N_r}{1500} \left(\because \quad N_S = \frac{120 \times 50}{4} = 1500 \text{ rpm} \right)$$

$$\Rightarrow \quad 0.04 \times 1500 = 1500 - N_r$$

$$N_r = 1440 \text{ rpm}$$

EXAMPLE 9.34 A 940-rpm speed occurs at maximum torque of 3-ϕ, 6-pole, 50-Hz induction motor. The rotor resistance per phase is 0.3 Ω. What is the standstill rotor reactance?

Solution Given data

Rotor speed, $N_r = 940$ rpm

Number of poles, $p = 6$

Frequency, $f = 50$ Hz

Rotor resistance, $R_2 = 0.3\Omega$

$$\text{Synchronous speed, } N_S = \frac{120 f}{p} = \frac{120 \times 50}{6} = 1000 \text{ rpm}$$

$$\text{Slip,} \quad S = \frac{N_S - N_r}{N_S} = \frac{1000 - 940}{1000} = 0.06$$

$$\text{Maximum torque occurs at a slip, } S_{max} = \frac{R_2}{X_2}$$

$$\therefore \quad X_2 = \frac{R_2}{S} = \frac{0.3}{0.06} = 5 \ \Omega$$

EXAMPLE 9.35 A 6-pole, 50-Hz, 3-ϕ induction motor has a full-load slip of 0.04. What value of slip does the maximum torque obtain? The maximum torque is twice the full-load torque.

Solution Given data

Number of poles, $p = 6$

Frequency, $f = 50$ Hz

Full load slip, $S_{fl} = 0.04$

Given $T_{max} = 2\ T_{fl}$

Condition for maximum torque is

$$R_2 = S_{max} X_2$$

$$S_{max} = \frac{R_2}{X_2}$$

We know $T_m = \dfrac{k}{2X_2}$ and $T_{fl} = \dfrac{kSR_2}{R_2^2 + (SX_2)^2}$

Substituting above torques in the given torque condition, i.e., $T_{max} = 2T_{fl}$

$$\frac{k}{2X_2} = 2\frac{kSR_2}{R_2^2 + S^2 X_2^2}$$

$$k\left(R_2^2 + S^2 X_2^2\right) = 4kSR_2X_2$$

or

$$\left(R_2^2 + S^2 X_2^2\right) = 4SR_2X_2$$

Dividing both sides by "X_2," we get

$$\frac{R_2^2}{X_2^2} + S^2 = \frac{4SR_2}{X_2}$$

$$\left[\frac{R_2}{X_2}\right]^2 - 4S\left[\frac{R_2}{X_2}\right] + S^2 = 0$$

Substituting the values of $S = 0.04$ in the above equation,

$$\left[\frac{R_2}{X_2}\right]^2 - 4\times 0.04\left[\frac{R_2}{X_2}\right] + 0.04^2 = 0$$

$$\left[\frac{R_2}{X_2}\right]^2 - 0.16\left[\frac{R_2}{X_2}\right] + 0.0016 = 0$$

By solving the above equation, we get the value of slip at which the maximum torque occurs.

$$\frac{R_2}{X_2} = 0.0107$$

EXAMPLE 9.36 A 25-hp, 6-pole, 3-ϕ induction motor has a full-load slip of 5% and 2.5% of the output power and the friction of a windage loss, respectively. Determine the full-load (i) rotor I^2R losses, (ii) rotor input, and (iii) output torque.

Solution Given data

Rated power, $P = 25$ hp

Friction and windage = 2.5% of full load
Full load slip, $S_{fl} = 0.05$
Motor shaft output = 25 hp
$= 2573 \times 5.5$
$= 18.3875$ kW
Friction and windage loss = 2.5%
$= 0.025 \times 18.3875$
$= 0.46$ kW
Power developed by the rotor = 18.3875 + 0.46 = 18.847 kW

(i) Rotor I^2R loss $= \dfrac{S}{(1-S)}$ power developed by the rotor

$$= \frac{0.05}{(1-0.05)} \times 18.847$$

$= 0.9919$ kW

(ii) Rotor input = Power developed by the rotor + Rotor I^2R loss + Rotor iron loss
$= 18.847 + 0.9919$
$= 19.8389$ kW

(Rotor iron loss at 5% slips is very small and can be neglected)

(iii) Output torque, $T = \dfrac{\text{Shaft output}}{2\pi N_r}$

$$= \frac{18.3875}{2\pi \dfrac{950}{60}} = \frac{18.3875}{99.4837} = 184.823 \text{ N-m}$$

EXAMPLE 9.37 **A 3-ϕ induction motor having a small short-circuited current is three times the full-load currents. Calculate the starting torque as a percentage of full-load torque in which the load slip is 5%.**

Solution Given data

Short-circuit current, $I_{sc} = 3I_f$
Full-load slip, $S_{fl} = 0.05$

$$\text{Starting torque, } T_{st} = T_{fl}\left[\frac{I_{sc}^2}{I_{fl}^2}\right]S_{fl}$$

$= T_{fl}\,[3]^2 \times 0.05$
$= 0.45\, T_{fl}$

Hence, the starting torque is 45% of the full-load torque.

EXAMPLE 9.38 **In a 16-kW, 415-V, 1440-rpm, 3-ϕ, 50-Hz, 4-pole squirrel cage induction motor, the load current is 35 A but it takes five times the full-load current at standstill and develops two times the full-load running torque. The magnetizing current and stator impedance may be neglected.**

(i) At starting, what voltage must be reduced in the full-load torque, and calculate the current for this voltage?

(ii) Calculate the line current when the voltage is obtained by an autotransformer.

(iii) If the starting current is limited to the full-load current by an autotransformer, what will be the starting torque as a percentage of full-load torque?

Solution Given data

Motor output = 16 kW

Supply voltage = 415 V

Motor speed = 1440 rpm

Current at standstill, I_{sc} = 5 times full-load current = $5I_f$

Full-load current, I_f = 35 A

Standstill torque = 2 times of full-load running torque.

(i) We know that $T \propto V^2$ on the first case,

$$2T_f \propto 415^2 \qquad (1)$$

and

$$T_f \propto V^2 \qquad (2)$$

$$\left[\frac{V}{415}\right]^2 = \frac{T_f}{2\,T_f}$$

$$V = 293.45 \text{ V}$$

Since currents are proportional to the applied voltage

$$5I_f \propto 415 \qquad (3)$$

$$I \propto 293.45 \qquad (4)$$

$$\frac{5I_f}{I} = \frac{415}{293.45}$$

$$I = \frac{5\times 35\times 293.45}{415}$$

$$= 123.74 \text{ A}$$

(ii) Here, $K = \dfrac{293.45}{415}$

$$= 0.707$$

Line current = $K^2 I_{sc}$

$$= \left[\frac{293.45}{415}\right]^2 \times 5\times I_f$$

$$= 0.707^2 \times 5 \times 35$$

$$= 87.5 \text{ A}$$

(iii) We know that line current = $K^2 I_{sc}$.

Now, line current = full-load current, I_f

$$35 = K^2 \times 5 \times 35$$

$$K^2 = \frac{1}{5}$$

Now, using the relation, $\frac{T_{st}}{T_f} = K^2 \left(\frac{I_{sc}}{I_f}\right)^2 S_f$

$$S_f = \frac{N_S - N_r}{N_S} \times 100$$

$$= \frac{1500 - 1440}{1500} \times 100$$

$$= 4\% \text{ or } 0.04$$

$$\frac{T_{st}}{T_f} = \frac{1}{5}\left(\frac{5I_f}{I_f}\right)^2 \times 0.04$$

$$T_{st} = 0.2T_f$$

i.e., 20% full-load torque.

EXAMPLE 9.39 A 70-kW, 3-ϕ, induction motor is having eight notches. Design the sections of rotor starter; the rotor resistance per phase is 0.015. The upper current limit is to be full-load current, for which the slip is 3%.

Solution Given data

Motor output = 70 kW

Notches (or studs) = 8

Number of elements, $n = 7$

Rotor resistance per phase, $r_m = 0.015\ \Omega$

The upper current limit = full-load current

Slip, $S = 3\% = 0.03$

Using the relation, $\alpha = (S)^{1/n} = (0.03)^{1/7} = 0.605$

Also,

$$r_m = R_1\,(\alpha)^n = RS$$

$$R_1 = \frac{r_m}{S} = \frac{0.015}{0.03} = 0.5\ \Omega$$

The resistances of elements are

$r_1 = R_1\,(1 - \alpha) = 0.5\,(1 - \alpha)$

$= 0.5\,(1 - 0.605) = 0.1970\ \Omega$

$r_2 = \alpha r_1 = 0.605 \times 0.197 = 0.1192\ \Omega$

$r_3 = \alpha^2 r_1 = (0.605)^3 \times 0.197 = 0.07211\ \Omega$

$r_4 = \alpha^3 r_1 = (0.605)^4 \times 0.197 = 000436\ \Omega$

$r_5 = \alpha^4 r_1 = (0.605)^5 \times 0.197 = 0.02639\ \Omega$

$r_6 = \alpha^5 r_1 = (0.605)^5 \times 0.197 = 0.01596\ \Omega$

$r_7 = \alpha^6 r_1 = (0.605)^5 \times 0.197 = 0.00966\ \Omega$

$r_m = 0.015$

Total $= 0.5\ \Omega$

This resistance nearly equals to R_1.

EXAMPLE 9.40 **A 3-ϕ, wound round induction motor having the slip at full-load current is 4%. Design six sections of a seven-stud rotor starter. The resistance of rotor is 0.018 Ω per phase and the maximum starting current is 1.2 times full-load current.**

Solution Given data

Number of sections, $n = 6$

Slip at full-load current = 4%

Maximum starting current = 1.2 times full-load current

Rotor resistance per phase, $r_m = 0.018\ \Omega$

$$\text{Rotor current} = \frac{E}{\sqrt{\left(\frac{R}{S}\right)^2 + X^2}} = \frac{SE}{R} \quad \text{(if the rotor reactance is neglected)}$$

This means that slip changes directly in proportion with the trotor current. The full-load slip is 4%

$S_m = 1.2$ times full-load current,

$= 1.2 \times 4 = 4.8$ or 0.048

Using the relation, $\alpha = (S_m)^{1/n} = (0.048)^{1/6} = 0.6028$

$$R_1 = \frac{r_m}{S_m} = \frac{0.018}{0.048} = 0.375\ \Omega$$

The resistances of the sections are

$r_1 = R_1 (1 - \alpha) = 0.375 (1 - 0.6028) = 0.1489\ \Omega$

$r_2 = \alpha r_1 = 0.6028 \times 0.1489 = 0.0897\ \Omega$

$r_3 = \alpha^2 r_1 = (0.602)^2 \times 0.1489 = 0.0541\ \Omega$

$r_4 = \alpha^3 r_1 = (0.6028)^3 \times 0.1489 = 0.03261\ \Omega$

$r_5 = \alpha^4 r_1 = (0.6028)^4 \times 0.1489 = 0.0196\ \Omega$

$r_6 = \alpha^5 r_1 = (0.602)^5 \times 0.1489 = 0.0118\ \Omega$

$r_m = 0.018\ \Omega$

Total $= 0.3745\ \Omega$

This resistance nearly equals to R_1.

EXAMPLE 9.41 **A three-phase slip ring induction motor attains a 65-V across slip rings on open circuit. At normal voltage, it is on standstills and the impedance of the rotor is (0.5 + *j*6) Ω per phase and is stator connected. Determine the rotor current (i) at standstill with slip rings joined to a star-connected starter with a phase impedance of (8 + *j*6) Ω per phase, and (ii) running normally with a slip of 0.04.**

Solution Given data

Induced emf, $E = 65$ V

Impedance of the rotor = $(0.5 + j6)\ \Omega$ per phase

Impedance of the star-connected starter = $(8 + j6)\ \Omega$ per phase

Slip, $S = 0.04$

(i) At standstill

Induced emf in rotor winding/phase, $E_2 = \dfrac{65}{\sqrt{3}} = 37.52$ V

Resistance per phase in rotor circuit, $R_2 = 0.5 + 8 = 8.5\ \Omega$

Resistance per phase in rotor circuit, $X_2 = 6 + 6 = 12\ \Omega$

Impedance per phase in rotor circuit, $Z_2 = \sqrt{R_2^2 + X_2^2}$

$$= \sqrt{8.5^2 + 12^2} = 14.7054\ \Omega$$

Rotor current per phase, $\dfrac{E_2}{Z_2} = \dfrac{37.52}{14.7054} = 2.5514$ A

Rotor power factor, $(\cos\phi_2) = \dfrac{R_2}{Z_2} = \dfrac{8.5}{14.7054} = 0.578$ (lag)

(ii) Running at 0.04 slip

Induced emf in rotor winding per phase = $SE_2 = 0.04 \times 37.52 = 1.5008$ V

Rotor impedance per phase = $\sqrt{R_2^2 + (SX_2^2)}$

$$= \sqrt{0.5^2 + (0.04 \times 6^2)} = 0.5546$$

Rotor current per phase, $\dfrac{SE_2}{Z_2} = \dfrac{1.5008}{0.5546} = 2.706$ A

Rotor power factor, $(\cos\phi) = \dfrac{R_2}{Z_2} = \dfrac{0.5}{0.5546}$ 0.901 (lag)

CHAPTER AT A GLANCE

- An induction motor mainly consists of stator and rotor.
- The synchronous speed of an induction motor is given by $N_S = 120\,f/P$.
- Slip, $S = (N_S - N)/N_S \times 100$.
- Rotor frequency, $f_r = sf$.
- Torque, $T \propto E_2 I_2 \cos\phi_2$.
- The slip corresponding to maximum torque, $S = R_2/X_2$.
- Maximum torque, $T_{max} \propto E_2{}^2/X_2$.

- A wound rotor induction motor can be distinguished from squirrel cage induction motor by the presence of slip rings.
- A squirrel cage induction motor will have low starting torque and widely used motor.
- The operation of induction motor is based on the principle of mutual induction.
- The principle of operation of a three-phase induction motor closely resembles to that of a two-winding transformer with its secondary short circuited.
- The torque developed by an induction motor is directly proportional to the square of the supply voltage.
- The power factor of an induction motor at full load is likely to be 0.85 lag and at no-load 0.2 lag.
- The core losses and windage and friction losses of an induction motor are determined from the no-load test.
- The size of a high-speed motor is smaller in comparison to that of a low-speed motor for the same kW output.

MULTIPLE CHOICE QUESTIONS

1. The synchronous speed of a 12-pole, 3-phase induction motor will be
   (a) 500 rpm (b) 600 rpm (c) 700 rpm (d) 400 rpm
2. When the supply frequency is f Hz, what will be the frequency of rotor current?
   (a) f (b) sf (c) 0 (d) 1
3. If the synchronous speed and speed of a three-phase induction motor are N_s and N, respectively, then which of the following statements is true?
   (a) $N > N_s$ (b) $N = N_s$ (c) $N \geq N_s$ (d) $N < N_s$
4. What is the condition to obtain the maximum starting torque?
   (a) $r^2 = x^2$ (b) $2r^2 = x^2$ (c) $r^2 = 3x^2$ (d) $r^2 = 4x^2$
5. The maximum induced emf in an induction motor rotor is at
   (a) running (b) standstill
   (c) maximum torque (d) maximum efficiency
6. A four-pole induction motor and a six-pole induction motor are connected in cumulative cascade. If the six-pole induction motor rotor circuit frequency is 1 Hz and the four-pole induction motor is excited with 50 Hz, the combined speed of the set is ______ rpm?
   (a) 600 (b) 622 (c) 608 (d) 588
7. If the slip is 10% in an induction motor, the ratio of rotor copper loss to rotor output is
   (a) 1:8 (b) 1:9 (c) 1:10 (d) 1:11
8. The induced emf in a rotor of an induction motor is given by Fleming's ______ rule.
   (a) left hand (b) dot (c) right hand (d) right hand
9. At small loads, the power factor of induction motor is
   (a) low (b) large (c) medium (d) 0

10. If the slip is 10% in an induction motor, the ratio of rotor copper loss to rotor output is

(a) 1:8 (b) 1:9 (c) 1:10 (d) 1:11

11. A 5% supply voltage change to induction motor leads to a ______ % change in torque?

(a) 5% (b) 10% (c) 25% (d) 1.5%

12. The maximum torque in an induction motor of phase wound depend on

(a) rotor resistance (b) stator resistance

(c) stator reactance (d) stator leakage

13. Synchronous wattage of an induction motor is the power transferred across the

(a) air-gap to rotor (b) rotor shaft (c) supply to stator (d) none

14. If a motor has a slip of 2% at normal voltage, the approximate slip at the same torque by increasing the voltage to 10% is

(a) 2 (b) 1.65 (c) 1.25 (d) 1.02

15. The rotor output of three-phase induction motor is 15 kW. The rotor copper losses at a slip of 4% will be

(a) 600 W (b) 625 W (c) 650 W (d) 700 W

16. In an induction motor, if the air-gap is increased

(a) speed will reduce (b) efficiency will improve

(c) pf will be lowered (d) breakdown torque will reduce

17. The starting current of an induction motor is five times the full-load current while the full-load slip is 4%. The ratio of starting torque to full-load torque is

(a) 0.6 (b) 0.8 (c) 1.0 (d) 1.2

18. For a constant torque load the addition of a rotor resistance in a slip ring induction motor

(a) reduces the stator current as well as slip

(b) reduces the stator current but increases slip

(c) causes no change in the stator current but increases its slip

(d). causes no change in the stator current but reduces its slip

19. The blocked rotor test of squirrel cage induction motor determines its equivalent

(a) series resistance and reactance as seen from the stator

(b) shunt resistance and reactance as seen from the stator

(c) series resistance and reactance as seen from the rotor

(d) shunt resistance and reactance as seen from the rotor

20. A three-phase slip ring induction motor is fed from the rotor side with stator winding short circuited. The frequency of the currents flowing in the short-circuited stator is

(a) slip frequency (b) supply frequency

(c) frequency corresponding to rotor speed (d) 0

SHORT ANSWER QUESTIONS

1. How do you determine the number of poles in a three-phase induction motor?

2. What is the speed of the rotating field due to rotor currents relative to stator core in an induction motor?

3. A four-pole, three-phase induction motor is running at 4% slip at full load. If the speed of the motor is 720 rpm, what is the supply frequency?
4. What is the speed of rotating field due to rotor currents relative to the rotor surface in an induction motor?
5. The full-load slip of a 50-Hz, 12-pole squirrel cage induction motor is 5%. What is its full-load speed?
6. Define synchronous speed.
7. What will happen if an induction motor runs at synchronous speed?
8. Why the rotor of an induction motor rotates in the direction of the rotation of stator field?
9. How the direction of rotation of a three-phase induction motor is reversed?
10. What is the reason to keep the air-gap flux density low in a three-phase induction motor?
11. If the rotor winding of a three-phase slip-ring induction motor is connected in delta, how the stator must be connected?
12. What are the two important parts of an induction motor?
13. Why the stator core of a three-phase induction motor is laminated?
14. What is the frequency of rotor currents at standstill?
15. A three-phase, 400-V, four-pole induction motor is fed from a three-phase, 400-V, 50-Hz supply and runs at 1440 rpm. What is the frequency of rotor emf?
16. What are the losses that occur in an induction motor?
17. Why iron loss in the rotor of a three-phase induction motor is negligible?

REVIEW QUESTIONS

1. Explain the concept of rotating magnetic field.
2. Explain the construction details and principle of operation in detail.
3. Define the slip of an induction motor.
4. Derive the torque equation of an induction motor.
5. Derive the condition for maximum torque under running conditions and standstill.
6. Explain the power stages of induction motor in detail.
7. Explain the various losses in an induction motor.
8. Explain the speed control methods of a three-phase induction motor.
9. Draw and explain the torque–slip characteristics of a three-phase induction motor.
10. Explain the starting methods of a three-phase induction motor.
11. Give the various applications of a three-phase induction motor.
12. Obtain the equivalent circuit of a three-phase induction motor.
13. Explain the procedure to conduct the no-load test.
14. Explain the procedure to conduct the blocked rotor test.
15. Give the comparison of induction motor with synchronous motor.

EXERCISE PROBLEMS

1. A six-pole, 50-Hz, squirrel cage induction motor runs on no-load at 975 rpm. Calculate the percentage slip and frequency of the rotor current.
2. If the emf induced in a rotor of an eight-pole induction motor is 50 Hz and that in the rotor is 1.5 Hz. Find the speed and slip.
3. The rotor of a three-phase, 440-V, 50-Hz, four-pole induction motor has a resistance of 0.3 Ω and an inductance of 0.008 H per phase of rotor. The turns ratio of stator to rotor is 2. The stator is delta-connected and rotor is star-connected. Determine the voltage between the slip rings when the motor is a standstill. Calculate the standstill rotor current and rotor current when it is running at 3% slip.
4. A 3.3-kV, 50-Hz, 12-pole induction motor has a full-load slip of 0.018. if the stator copper loss is 2.4 kW, no-load loss is 3.5 kW and mechanical loss is 1.2 kW. Calculate the rotor copper loss, line current, and efficiency, assuming a power factor of 0.85.
5. Find the efficiency and percentage slip if the stator and rotor copper losses are 3.2 and 2.9 kW, respectively. The power input is 95 kW, stator iron loss is 1.1 kW, and mechanical loss is 1.3 kW.
6. The power output of an induction motor is 5 hp when running with 4% slip. The friction and windage losses are 500 W and stator loss is 300 W. Calculate the stator input and efficiency.
7. A 4-pole, 3-phase star-connected induction motor is connected to a 220-V, 50-Hz supply. The standstill reactance per phase and resistance per phase of stator are 0.9 and 0.1 Ω, respectively. Calculate the (a) rotor torque at 5% slip, (b) maximum torque, and (c) speed at maximum torque if the rotor to stator turns 0.67. Neglect the impedance of stator.
8. An induction motor having full-load torque of 60 N-m when delta-connected develops a starting torque of 120 N-m. For the same supply voltage, if the motor is changed to star connection, calculate the starting torque developed.
9. At a slip of 4%, calculate the maximum possible speed of a three-phase squirrel cage induction motor.
10. The power input to a 415-V, 50 Hz, six-pole, three-phase induction motor running at 975 rpm is 40 kW. The stator losses are 1 kW and friction and windage losses are 2 kW. Calculate the efficiency of the motor.

Chapter 10

Synchronous Machines

Chapter Objectives

- To explain the construction and principle of operation of synchronous machine
- To discuss the types of windings, emf equation, etc.
- To explain the behavior of synchronous machine both as generator and motor
- To discuss the regulation of an alternator
- To discuss the hunting phenomenon

10.1 INTRODUCTION

Nowadays three-phase ac system is extensively used for commercial as well as domestic purposes. The machines that are associated with an alternating supply are called *synchronous machines*. A synchronous machine can be operated either as an alternator or as a motor. Similar to dc machines, the machines associated with alternating quantities are classified as ac generators and motors.

When a synchronous machine is run at a constant speed by a prime mover, it generates an emf and supplies an electrical load, thus converting the mechanical energy to the electrical energy. The synchronous machine operating in this mode is called an alternator or a synchronous generator. These are also known as ac generators. All modern power stations consist of large-capacity alternators.

On the other hand, if the same synchronous machine is connected to an ac supply and if the rotor is made to run at synchronous speed and a mechanical load is connected to its shaft, the machine continues to run at synchronous speed thus converting the electrical energy to the mechanical energy. The synchronous machine operating in this mode is called *synchronous motor*.

10.2 CONSTRUCTION OF SYNCHRONOUS MACHINE

The armature winding either exports or imports ac power based on the mode of operation of synchronous machine. In both the modes, i.e., synchronous generator and synchronous motor, the field winding is always excited from dc source. In other words, the synchronous machines are doubly excited energy conversion devices. The synchronous machines are invariably constructed with high-power armature winding on the stator and low-power field winding on the rotor. There are some advantages of providing the field winding on the rotor and armature winding on the stator.

10.2.1 Construction of Stator

The stator is a stationary armature. It consists of a core and slots to hold the armature winding. The stator core uses laminated construction. To reduce the eddy current losses, laminated construction is preferred. It is built up of special steel stampings insulated from each other.

10.2.2 Construction of Rotor

The rotor carries a field winding which is supplied with dc. There are two types of rotors used in alternators:

(i) Salient pole type
(ii) Non-salient pole type or cylindrical rotors

Synchronous machines are of two types depending upon the geometrical structure of the rotor, i.e., salient pole synchronous machines and cylindrical synchronous machines.

Salient pole-type rotors

It is also called as projected pole type. Salient means standing out or projecting. The poles are mounted on a large circular steel frame which is fixed to the shaft of the alternator. The field winding is provided on the pole shoe as shown in Figure 10.1. The field winding on the salient poles is a concentrated winding.

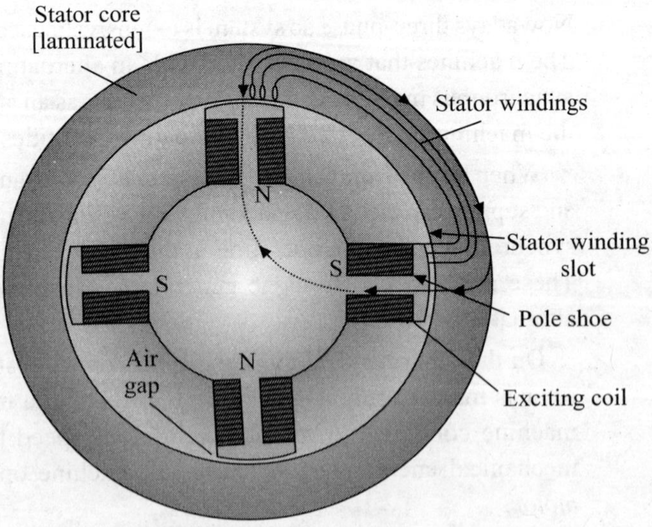

Figure 10.1 Salient pole-type rotor

The field pole windings are connected in series in such a way that when the field winding is excited by the dc source, adjacent poles develop opposite polarities [i.e., N, S, N, S]. These rotors have large diameters and small axial lengths. They are used for low-speed alternators ranging from

120 to 500 rpm. Since the mechanical strength of salient pole type is less, the prime movers used for these rotors are water turbines, diesel engines, etc.

Non-salient pole-type rotors

It is also known as smooth cylindrical rotor or non-projected pole-type rotor. Figure 10.2 shows a non-salient pole-type rotor or a cylindrical rotor in which the field winding is a distributed winding housed in rotor slots.

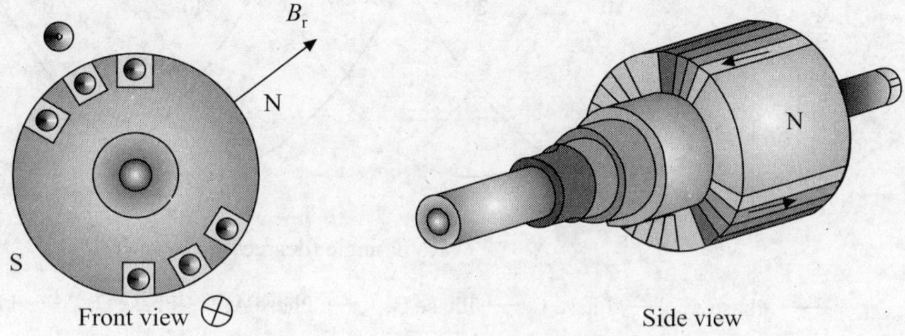

Figure 10.2 Non-salient pole-type rotor

The rotor consists of smooth solid steel cylinder having number of slots to accommodate the field coils. The field windings are embedded in these slots and are connected in series. The slots are covered at the top with the help of steel. The unslotted portions of the cylinder themselves act as the poles. It is clear that the poles formed are non-salient, i.e., the poles are not projecting out of the surface of the rotor. These rotors have small diameters and large axial length. These are mechanically strong and used for high-speed alternators ranging from 1500 to 3000 rpm. The prime movers used to drive these rotors are steam turbines and electric motors.

10.3 FLUX DENSITY DISTRIBUTION IN SYNCHRONOUS MACHINE

Figure 10.3 shows the flux density distribution for a four-pole salient synchronous machine with concentrated field coils. The flux density is constant in this region because of the uniform air gap over the pole faces. At the pole tips, the flux density begins to reduce and becomes zero in the interpolar region due to fringing. Since such a flat topped flux density distribution is not helpful to generate a sinusoidal emf, curving the pole faces increases the airgap outward. This scheme helps to make the flux density distribution almost sinusoidal.

In a non-salient pole machine, since the rotor is cylindrical, the airgap is uniform. The armature windings are uniformly distributed. With these two factors the number of ampere-turns reduces progressively due to a decrease in the flux linkages and also because the movement is away from the magnetic axis of the field. In this arrangement, the flux density distribution has a stepped waveform corresponding to the rotor slots, but it can be approximated to a sine wave for the purpose of analysis since the effect of high-frequency harmonics can be ignored without introducing significant errors in the results.

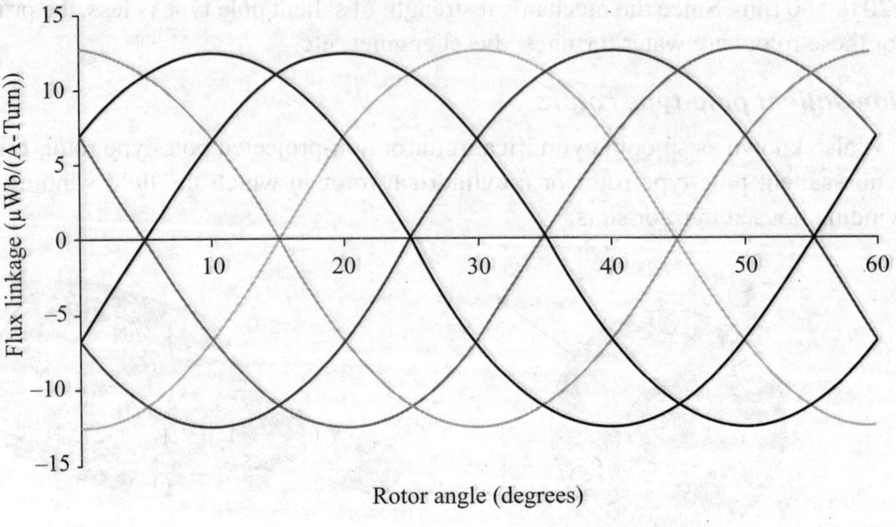

Figure 10.3 Flux density distribution in synchronous machine

10.4 WORKING PRINCIPLE OF SYNCHRONOUS GENERATOR

An alternator or a synchronous generator works on same fundamental principles of electromagnetic induction as dc generators. When the flux linking the conductor changes, an emf is induced in the conductor. It consists of an armature winding and a magnetic field. There is one important difference between dc generators and synchronous generators, i.e., ac generators. In case of dc generators, the armature rotates and the field system is stationary whereas it is vice versa in case of ac generators. Here, the armature windings are mounted on a stationary element called *stator* and field windings on a rotating element called *rotor*. The details of construction are shown in Figure 10.4.

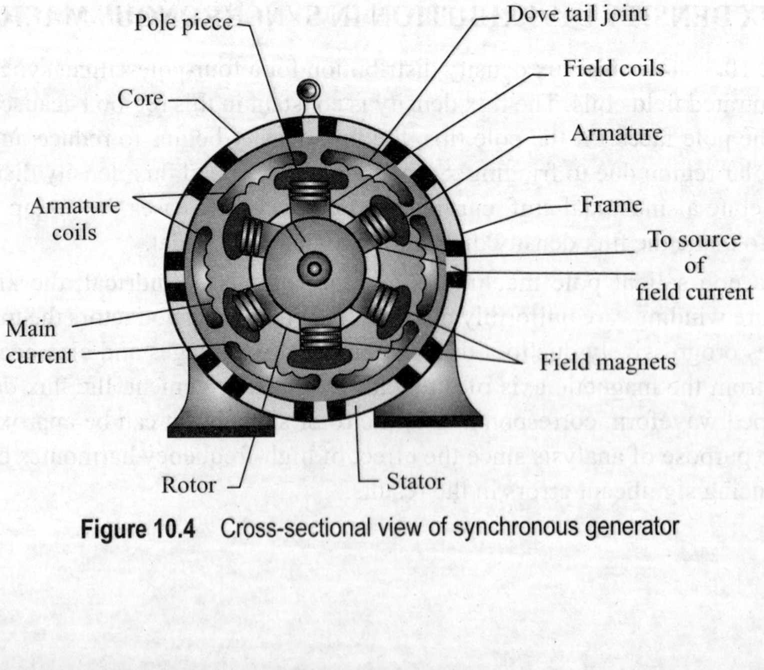

Figure 10.4 Cross-sectional view of synchronous generator

10.4.1 Advantages of Stationary Armature

The main advantages of stationary armature are as follows:

(i) It is very easy to insulate stationary winding for high voltages for which the alternators are basically designed. As the armature is stationary, it is not subjected to any centrifugal forces.

(ii) The output of stationary armature can be directly drawn into load circuit without going through large, unreliable slip rings and brushes.

(iii) Only two slip rings are required for dc supply to the field winding on the rotor.

(iv) Because of simple and robust construction of the rotor, higher speed of rotating dc field is possible, which increases the machine output of given dimensions.

10.4.2 Three-Phase Armature Winding

The armature windings of dc machines are usually closed circuit windings but alternators windings may be either closed or open. The armature winding of a three-phase alternator is similar to that of the stator winding of three-phase induction motor. The three windings are 120° electrical and are connected either in delta or star and the corresponding leads are brought out and connected to the terminals fixed on the stator. Single-phase alternator will have only one winding.

The armature windings can be classified into the following categories:

(i) Single-layer and double-layer winding

(ii) Full-pitch and short-pitch winding

Single-layer and double-layer winding

Single-layer winding is also known as concentric or chain winding. There will be one layer in each of the stator as shown in Figure 10.5. For this type of winding, the number of slots is equal to twice the number of coils. For the emf to be maximum, the two coil sides should be 180 electrical degrees apart. In this type of winding, polar group of each phase is apart.

(i) It is necessary to use two different shapes of coils to avoid fouling of end connections

(ii) The disadvantage is that short-pitched coils cannot be used

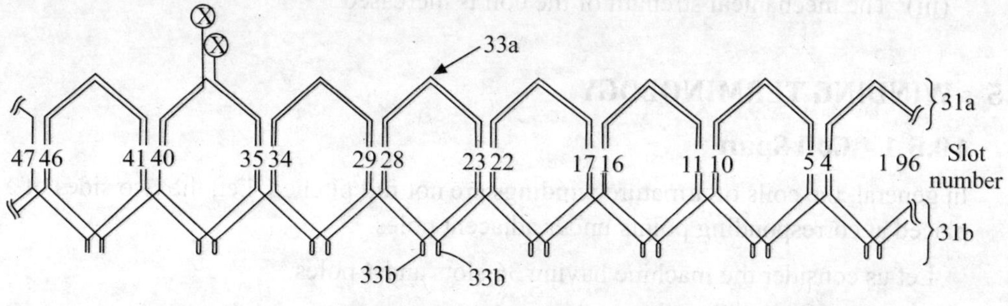

Figure 10.5 Single-layer winding

In the case of concentric or chain winding, each coil may be single turn or multi turn. The winding has one coil per pair of poles per phase, therefore the number of coils is half the number of slots. Generally, two different shapes of coils are used, which facilitate end connection.

Double-layer winding may be either wave wound type or lap wound type. It is the simplest and mostly used in all machines. In this winding there will be two coil sides in each slot; and the winding may be lap or wave wound.

A coil in a double-layer winding represents the entire set of conductors in one slot layer in association with the similar set in the other layer of another slot; as such the number of coils is equal to the number of slots. Basically, the connection can be lap connected or wave connected, with one turn or more than one turn. However, the choice depends upon number of turns per phase which in turn depends upon emf to be generated. Important points regarding this winding are:

(i) The number of slots in a stator is multiple of the number of poles and number of phases. Therefore, a stator of a four-pole, three-phase alternator may have 12, 24, 36, 48, ... , slots which are multiples of 12.
(ii) The number of stator slots is equal to the number of coils.

Full-pitch and short-pitch winding

When the two coil sides forming a complete coil of a winding are 180 electrical space degrees apart, the winding is known as *full-pitch winding*, i.e., it indicates that in this type of winding when one side of coil is under N pole, the other side will be in the corresponding position under S pole. The induced emfs differ by 180° in phase, but the coil is connected in such a way that the emfs will add together and give the resultant emf E.

When the two coil sides forming a complete coil of a winding are not 180 electrical space degree apart, it is known as *short-pitch winding* or *fractional pitch winding*. In this type of winding, the emfs induced in each coil are not in phase, so the resultant emf is phasor sum of two emfs. This emf is less than their arithmetic sum.

Therefore, the induced emf in the short pitch winding is less than that of full-pitch winding. The advantages of short-pitch winding are as follows:

(i) The waveform of the induced emf is improved.
(ii) Copper is saved in the coil ends due to less span.
(iii) The mechanical strength of the coil is increased.

10.5 WINDING TERMINOLOGY

10.5.1 Coil Span

In general, the coils of armature windings are not full pitched, i.e., the two sides of a coil are not placed at corresponding points under adjacent poles.

Let us consider the machine having 36 slots and 4 poles.

$$\therefore \qquad \frac{\text{Slots}}{\text{Pole}} = \frac{36}{4} = 9$$

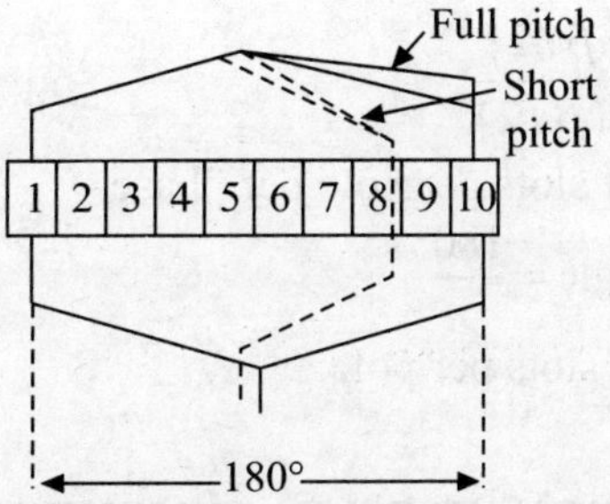

Figure 10.6 Coil span

Here, the dotted line indicates short-pitch winding. It falls short by two slots.

$$\therefore \quad \text{Coil span} = \frac{7}{9} \times 180° = 140°$$

If the other end of the coil is placed in slot 9, then the coil span $= \frac{8}{9} \times 180° = 160°$.

10.5.2 Pitch Factor

We utilize the short-pitch armature winding because of the following reasons:

(i) The short-pitch armature winding requires less copper.

(ii) The waveform of output voltage is improved.

When the coil is full pitched, the emfs in the coil sides will have phase difference. The resultant emf in the coil will be less than that of full-pitched coil. Therefore, the pitch factor (K_p) is defined as:

$$K_p = \frac{\text{emf with short-pitch coil}}{\text{emf with full-pitch coil}} \tag{10.1}$$

Pitch factor is also known as the chording factor or coil span factor. Therefore, for full-pitch coil, $K_p = 1$, while for short-pitch coil, $K_p < 1$.

$$K_p = \cos\frac{\alpha}{2} \quad (\text{where } \alpha = 180° - \text{Coil span}) \tag{10.2}$$

10.5.3 Distribution Factor

The armature winding of a synchronous generator is distributed over the entire armature. The distributed winding produces nearly a sine wave form and the heating is more uniform.

Once the conductors are spread over the entire armature in slots, their emfs differ in phase. Therefore, the total emf is the vector sum but not the arithmetic sum. Therefore, the distribution factor is defined as

$$K_d = \frac{\text{emf with distributed winding}}{\text{emf with concentrated winding}} \tag{10.3}$$

Distribution factor is also known as breadth factor or spread factor. Hence, for concentrated winding, $K_d < 1$.

$$K_d = \frac{\sin(m\beta/2)}{m\sin(\beta/2)} \tag{10.4}$$

where m is the number of slots per pole per phase.

$$\beta = \text{Slot angle} = \frac{180^\circ}{n}$$

where n is the number of slots per pole.

10.6 GENERATED EMF EQUATION IN SYNCHRONOUS GENERATOR

Let

Z = Number of conductors of coil side in series per phase

ϕ = Flux per pole in Wb

P = Number of rotor poles

N = Rotor speed in rpm

f = Frequency of induced emf in Hz

Therefore, in one revolution, total flux cut = $P\phi$ weber

Time taken for one revolution = $\frac{1}{N}$ min = $\frac{60}{N}$

Therefore, average emf induced in one conductor = $\frac{P\phi}{60/N} = \frac{P\phi N}{60}$ (10.5)

But there are Z conductors per phase.

$$\text{Average emf per phase} = \frac{P\phi N}{60} \times Z$$

But

$$N = \frac{120f}{P}$$

$$\text{Average emf per phase} = \frac{P\phi \times 120 \times f}{60 \times P} \times Z$$

$$= 2\phi f Z$$

Number of conductors $Z = 2T$, where T = number of coils or turns per phase

Average emf per phase = $4\phi fT$

rms value of voltage = Average value × Form factor

For sine wave, form factor = 1.11

rms emf per phase = $4\phi fT \times 1.11$

$$= 4.44\phi ft \text{ V} \tag{10.6}$$

Here, the derivation assumes that all the turns are concentrated and full pitched. In general, they are distributed over armature periphery and short pitched. To compensate this, the voltage is multiplied by distribution factor K_d and pitch factor K_p.

$$E_{rms} \text{ per phase} = 4.44 f\phi T K_p K_d \text{ V} \tag{10.7}$$

$$\text{Distribution factor, } K_d = \frac{\text{emf with distributed winding}}{\text{emf with concentrated winding}}$$

$$\text{Pitch factor, } K_p = \frac{\text{emf with short-pitch coil}}{\text{emf with full-pitch coil}} = \cos\frac{\alpha}{2}$$

10.7 FREQUENCY OF INDUCED EMF

The frequency of induced emf in the armature depends upon speed and the number of poles. Let N is the rotor speed in rmp, P is the number of poles, and f is the frequency of emf in Hz. One cycle of emf is induced in a conductor when one pair of poles, i.e., south and north, passes over it. Therefore, if a positive voltage gets induced when a N-pole passes over the conductor, a similar negative voltage will be induced when a S-pole passes over it.

Therefore, number of cycles per revolution = number of pair of poles

$$= P/2$$

$$\text{Therefore, number of cycles per second } \frac{P}{2} \times \frac{N}{60} = \frac{NP}{120}$$

Therefore, number of cycles per second is nothing but the frequency of emf

$$\therefore \quad f = \frac{NP}{120} \tag{10.8}$$

Here, N is the synchronous speed. Once the alternator is designed, the number of rotor poles is fixed and therefore the alternator must run at synchronous speed to give the desired frequency.

EXAMPLE 10.1 **A 3-phase, 8-pole alternator has a star connected winding with 144 slots and 8 conductors per slot. The flux per pole is 0.04 wb and speed is 400 rpm. Find the frequency and phase voltage.**

Solution Frequency, $f = \frac{NP}{120} = \frac{400 \times 8}{120} = 26.6 \text{ Hz}$

EXAMPLE 10.2 **A 415 V, 4-pole, 3-phase alternator speed is 1500 rpm. Calculate the frequency of operation.**

Solution Given

$V = 415$ V, $P = 4$, $N = 1500$ rpm

$$f = \frac{PN}{120} = \frac{4 \times 1500}{120} = 50 \text{ Hz}$$

EXAMPLE 10.3 **A synchronous generator has 9 slots per pole. If each coil spans 7 slot pitches, find the value of pitch factor?**

Solution Given number of slots per pole = 9, coil span = 7

Therefore, it falls short by (9 – 7) = 2

$$\therefore \quad \text{Coil span} = \frac{7}{9} \times 180° = 140°$$

$$\alpha = 180° - \text{coil span}$$
$$= 180° - 140° = 40°$$

$$\therefore \quad \text{Pitch factor, } K_p = \cos\frac{\alpha}{2}$$
$$= \cos\frac{40}{2} = 0.99$$

Example 10.4 **Calculate the pitch factor for the following cases:**

(i) 36 stator slots, 2 poles, coil span 1–17

(ii) 96 stator slots, 4 poles, coil span 1–21

Solution (i) Given 36 slots, 2 poles

$$\text{Number of slots per pole} = \frac{36}{2} = 18$$

Coil span 1 to 17

Therefore, it falls short by 2 slots (19–17).

$$\alpha = \frac{2}{18} \times 180° = 20°$$

$$K_p = \cos\frac{\alpha}{2} = \cos\frac{20}{2} = 0.984$$

$$\text{(ii) Number of slots per pole} = \frac{96}{4} = 24$$

Therefore, it falls short by 4 slots (25–21).

$$\alpha = \frac{4}{24} \times 180° = 30°$$

$$K_p = \cos\frac{\alpha}{2} = \cos\frac{30}{2} = 0.965$$

Example 10.5 **A 4-pole synchronous generator has a 3-phase winding wound in 36 slots with coil span of 140°. Find the (i) pitch factor and (ii) distribution factor.**

Solution Given

$P = 4,\ S = 36,\ \text{coil span} = 140°$

$$\text{Slot per pole per phase} = \frac{S}{P} = \frac{36}{4} = 9$$

$$\text{Slot per pole per phase} = \frac{9}{3} = 3$$

$$B = \frac{180°}{\text{Slots/pole}} = \frac{180°}{9} = 20°$$

$$\alpha = 180° - \text{Coil span}$$
$$= 180 - 140 = 40°$$

(i) Pitch factor, $K_p = \cos\frac{\alpha}{2} = \cos\frac{40}{2} = 0.939$

(ii) Distribution factor, $K_d = \frac{\sin(m\beta/2)}{m\sin(\beta/2)}$

$$= \frac{\sin(3\times 20/2)}{3\ \sin(20/2)} = 0.96$$

EXAMPLE 10.6 For a 3-phase, 6-pole synchronous generator with 72 slots, find the (i) pitch factor and (ii) distribution factor. The coil span is 1–10 slots.

Solution Given

$P = 6$, $S = 72$, coil span = 1–10 slots

Number of slots per pole = 72/6 = 12

$$m = \text{Number of slots per pole per phase } \frac{12}{3} = 4$$

$$B = \frac{180°}{\text{Number of slots per pole}} = \frac{180°}{12} = 15°$$

For full-pitch winding, coil span should be 1–13.

Therefore, the coil fall is short by three slots.

$$\alpha = \frac{3}{12}\times 180° = 45°$$

(i) Pitch factor, $K_p = \cos\frac{\alpha}{2} = \cos\frac{45}{2} = 0.924$

(ii) Distribution factor, $K_d = \frac{\sin(m\beta/2)}{m\sin\beta/2}$

$$= \frac{\sin(4\times 15/2)}{4\ \sin(15/2)} = 0.958$$

10.8 ARMATURE REACTION

It is evident that the induced emf remains constant if the field current and speed are kept constant. Under no-load conditions, the terminal voltage V becomes equal to the induced emf. When the alternator supplies the load, the load current passes through the phase windings and sets up its own flux which disturbs the airgap flux, due to which the terminal voltage does not remain the same as the induced voltage, in addition to the voltage drop in the resistance of the armature winding. The power factor of the load also influences the terminal voltage. Thus, the effect of armature magneto motive force (mmf) on the main flux is called the armature reaction.

When the alternator is loaded, it sets up flux of constant magnitude having the same number of poles as that of the rotor and rotates at constant speed (synchronous speed). The magnitude of the flux depends upon the load current magnitude and the load power factor. Now, we shall consider three cases: (i) unity power factor, (ii) zero power factor lag, and (iii) zero power factor load.

(i) **Unity power factor:** For simplicity, we assume concentrated winding. The alternator is loaded with unity power load, i.e., resistive load, with the result that the phase currents will be in phase with the corresponding phase voltages. With the rotor position as shown in Figure 10.7, the induced emf of phase (a) is maximum. Since the phase voltage is maximum in phase (a), the current is also maximum because of the unity power factor (upf). At this instant the currents in phases (b) and (c) also become half the current in phase (a). The resultant flux produced by the three phase currents also becomes maximum and gets along the axis of the coil (a). At this instant, the main flux is upwards and the armature flux is horizontal from right to left. The wave forms of these two fluxes are shown in Figure 10.7.

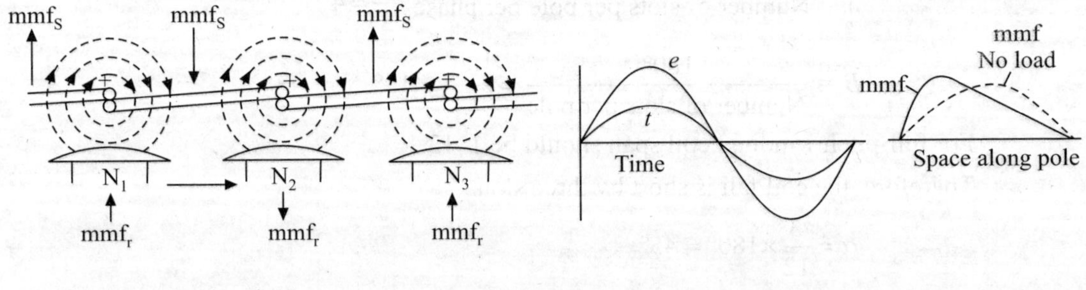

(a)

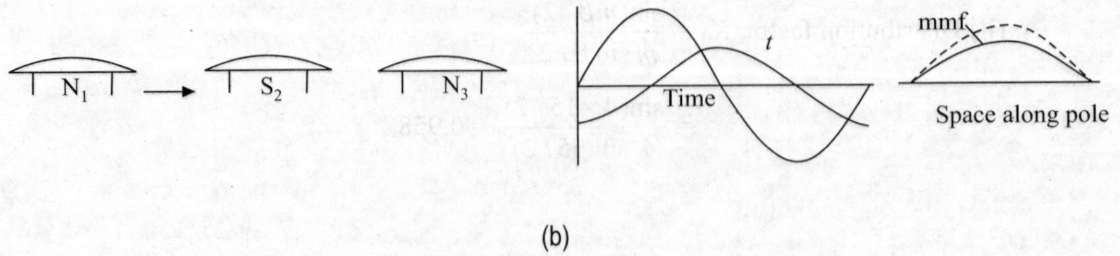

(b)

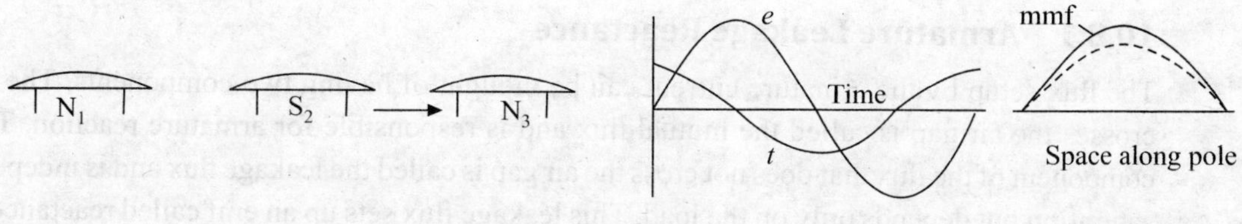

(c)

Figure 10.7

From Figure10.7(a), it is observed that the armature flux lags behind the main flux by an angle of 90° for the case of unity power factor load. As the time increases, the rotor position changes and the armature flux moves along the same direction and at the same speed. Thus, the armature flux will always be lagging behind the main flux by 90° (for upf loads). It results in the addition of half the main flux to one of the pole and opposes the other half. Neglecting saturation effects, the resultant flux is same as the main flux, but its waveform is distorted. Hence, when the alternator supplies resistive loads, the main flux gets distorted.

(ii) **Zero power factor lagging load:** With lagging loads, the armature currents lag behind the voltage by 90°, with the result that the armature flux ϕ_a lags behind the main flux by 180° as shown in Figure 10.7(b), resulting in demagnetization, and causes the reduction of the resultant flux, consequently leading to the reduction of the induced emf. Therefore, the effect of armature reaction with zero power factor lagging loads is demagnetization of the main flux.

(iii) **Zero power factor leading load:** With pure capacitive loads, i.e., zero power factor loads, the armature currents lead the corresponding armature voltages by 90°, and the armature flux ϕ_a will be in phase with the main flux ϕ. This results in more flux and so increased induced emf per phase [Figure10.7(c)].

In general, if the power factor is lagging (not zero power factor lagging) the result of armature flux on main flux, i.e., the effect of armature reaction is partly distortion and partly demagnetization, while for the leading power factor, the effect of armature reaction is partly distortion and partly cross magnetization.

From the above discussion, it can be concluded that the effect of armature reaction is to cause reduction in the terminal voltage if the power factor is lagging, and to cause increases in the terminal voltage if the power factor is leading; the change in the terminal voltage is proportional to the load current.

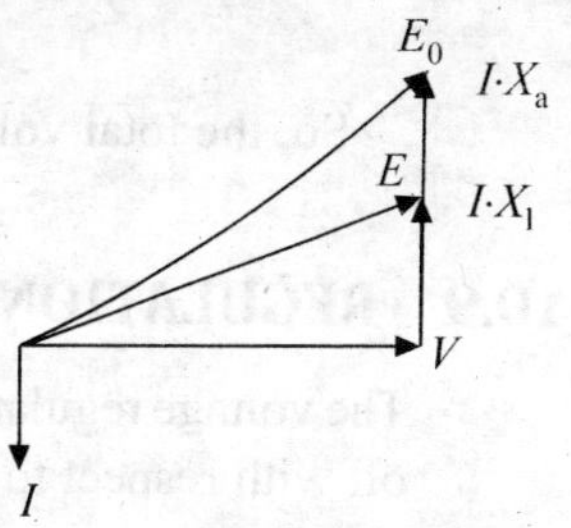

Figure 10.8

The change in the terminal voltage due to armature reaction may be attributed to a voltage drop in a fictitious reactance X_a in the armature winding, such that the voltage drop across X_a, i.e., $I{\cdot}X_a$, represents the change in terminal voltage due to armature reaction. The corresponding circuit diagram and vector diagram are shown in Figure 10.8.

10.8.1 Armature Leakage Reactance

The flux setup by the armature current can be thought of having two components. The flux that crosses the air gap is called the mutual flux and is responsible for armature reaction. The other component of the flux that does not cross the air gap is called the leakage flux and is independent of saturation but depends only on the load. This leakage flux sets up an emf called reactance voltage and leads the load current by 90°. this voltage can be attributed due to a voltage drop across a reactance called leakage reactance X_l. So, the generated emf has to overcome this voltage drop.

10.8.2 Synchronous Reactance

From the above discussion, we conclude that the terminal voltage changes from no load to full load due to the following reasons:

(i) Voltage drop due to the armature reaction
(ii) Voltage drop due to the armature leakage reactance
(iii) Voltage drop due to the armature resistance of windings

The voltage drops due to reasons (i) and (ii) are in line, so that the two reactances may be combined and are called synchronous reactance X_s. When combined with the resistance of the winding, they are called synchronous impedance Z_s. The relevant circuit diagram or vector diagram is shown in Figure 10.9.

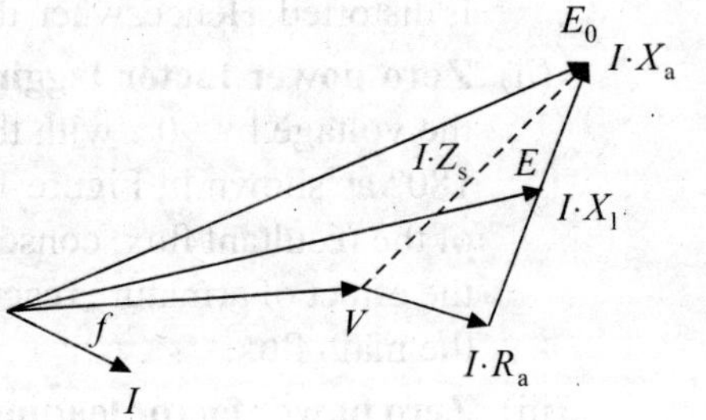

Figure 10.9

The sum of armature reactance X_a and armature leakage reactance X_l is called synchronous reactance X_s.

$$X_s = X_a + X_l$$

And the synchronous reactance combined with armature resistance R_a is called synchronous impedance Z_s. That is,

$$Z_s = R_a + jX_s = \sqrt{R_a^2 + X_s^2}$$

So, the total voltage drop under load conditions is $I_a Z_s$.

10.9 REGULATION OF AN ALTERNATOR

The voltage regulation of an alternator is defined as "the rise in voltage when the full load is thrown off with respect to the rated terminal voltage keeping excitation and speed constant."

$$\% \text{ Regulation } = \frac{E_0 - V}{V} \times 100 \tag{10.9}$$

The percentage regulation can be calculated from the vector diagram given in Figure 10.10. Here it is for lagging loads, the same can be derived for unity or leading power factor loads.

E_0 is the no-load emf and is due to the voltage induced in the absence of three factors, i.e., IR_a drop, IX_l drop, and drop due to armature reaction. Hence, it represents the maximum value of the induced emf. V is the load-induced emf. It is the induced emf after allowing for armature reaction. V is vectorially less than E_0.

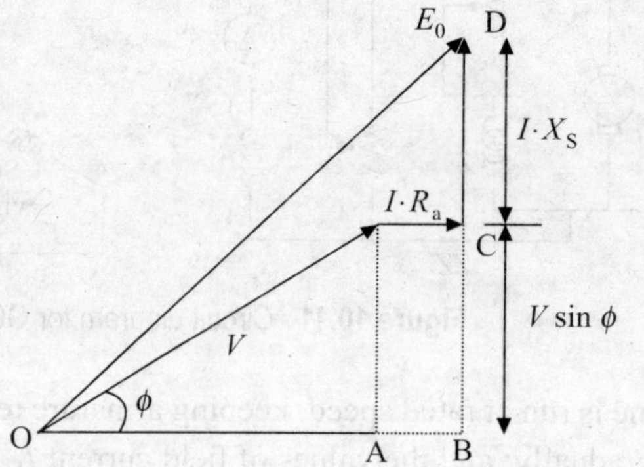

Figure 10.10 Vector diagram for lagging power factor load

Let R_a and X_s be armature resistance and synchronous reactance, respectively.

From vector diagram, OD = E_0

$$\therefore \quad E_0 = \sqrt{\left(\text{OB}^2 + \text{BD}^2\right)} \tag{10.10}$$

$$= \sqrt{\left[\left(V\cos\phi + IR_a\right)^2 + \left(V\sin\phi + IX_s\right)^2\right]} \tag{10.11}$$

In the case of leading power factors

$$E_0 = \sqrt{\left[\left(V\cos\phi + IR_a\right)^2 + \left(V\sin\phi - IX_s\right)^2\right]} \tag{10.12}$$

10.9.1 Tests to Determine Regulation

To determine the regulation of an alternator, generally the following methods are used:

(i) Synchronous impedance method

(ii) Ampere turn or mmf method

(iii) Zero power factor method

Methods (ii) and (iii) are not in the scope of this book and the synchronous method is explained in the following sections:

(i) **Synchronous impedance method:** For this method of calculations, the following three tests are carried out:

(a) Open-circuit test (OC test)
(b) Short-circuit test (SC test)
(c) Armature resistance (R_a) measurement test

(a) **Open-circuit test:** The connection of circuit diagram is shown in Figure 10.11.

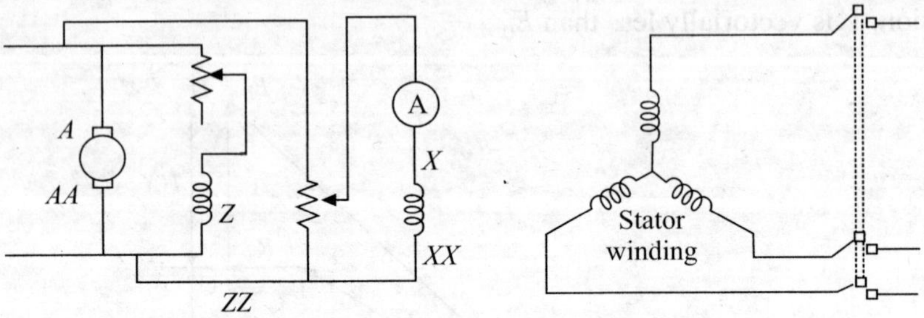

Figure 10.11 Circuit diagram for OC test

The machine is run at rated speed, keeping armature terminals open the field excitation is increased gradually; and the values of field current I_f and open-circuit voltage across the armature terminals are measured. The excitation is increased upto 125% of the rated voltage and a graph of I_f versus the voltage is drawn and the graph is shown in Figure 10.12.

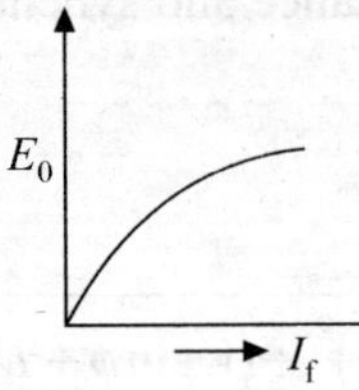

Figure 10.12 Open-circuit characteristic

(b) **Short-circuit test:** The connection of circuit diagram is shown in Figure 10.13.

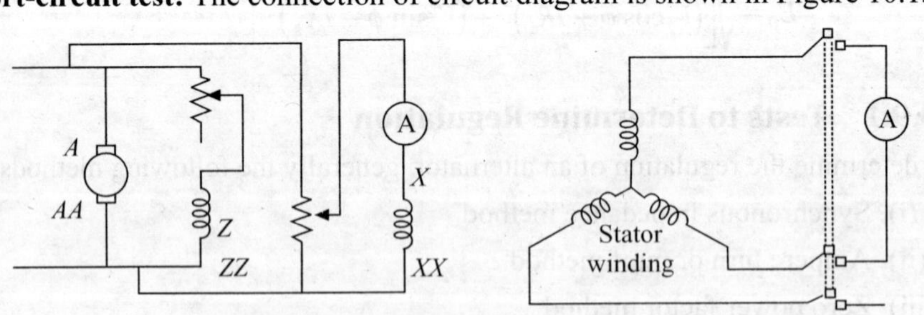

Figure 10.13 Circuit diagram for SC test

The armature terminals are short circuited through an ammeter, the machine is run at its rated speed, and the excitation is gradually increased till rated armature current flows under short-circuit condition, and the corresponding field current and short-circuit current are noted.

Then, a graph between I_f and I_a is drawn and it is called as the short-circuit characteristics (SCC) of the alternator. In this test full load reading is enough, because the other point is origin. When the field current $I_f = 0$, the armature current $I_a = 0$. But for calculations, both the graphs (OCC and SCC) are to be drawn on the same graph. The SCC is shown in Figure 10.14.

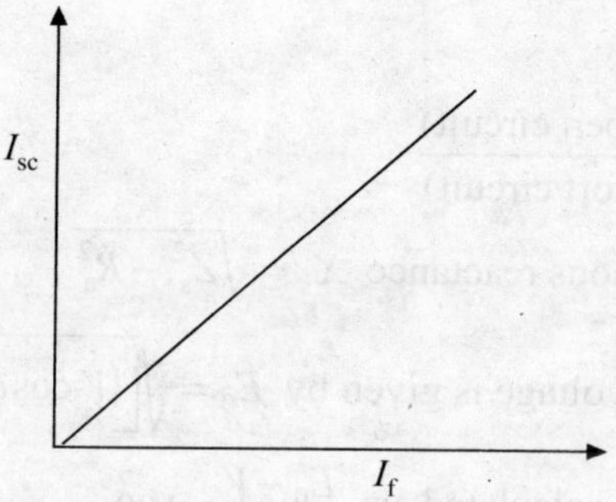

Figure 10.14 Short-circuit characteristic

(c) **Armature resistance test:** The armature resistance per phase is measured using the voltmeter, ammeter method, and the resulting value is multiplied by a factor equal to 1.6 (this factor is generally taken between 1.2 and 1.6) to take into account the equivalent ac resistance.

After the experimentation part, the following procedure is carried out in order to determine the regulation.

Step 1: Plot OCC by conducting an open-circuit test as shown in Figure 10.15 (curve A).

Step 2: Plot SCC by conducting a short-circuit test as shown in Figure 10.16 (curve B). Both these (OCC and SCC) curves are drawn on a common field current base.

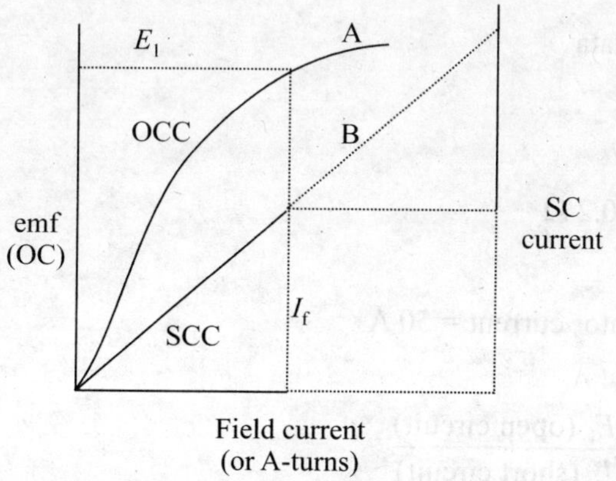

Figure 10.15

Step 3: Consider a field current I_f, and take the open-circuit voltage, E_1, corresponding to the field current I_f.

When the winding is short circuited, the terminal voltage is zero. Hence, it may be assumed that the whole of voltage E_1 is being used to circulate the armature short-circuit current I_{sc} against the synchronous impedance Z_s.

Therefore, the open-circuit voltage is

$$E_1 = I_{sc} Z_s$$

i.e.,

$$Z_s = \frac{E_1 \text{ (open circuit)}}{I_{sc} \text{ (short circuit)}} \tag{10.13}$$

Step 4: The synchronous reactance $X_s = \sqrt{Z_s^2 - R_a^2}$.

Step 5: The no-load voltage is given by $E_0 = \sqrt{\left[(V\cos\phi + IR_a)^2 + (V\sin\phi - IX_s)^2\right]}$.

Step 6: Regulation is calculated as: $\dfrac{E_0 - V}{V} \times 100$.

By using the above formula, we can determine the power factor and regulation characteristics for different power factors of loads.

The value of regulation calculated by this method is always greater than the actual value as determined from actual load tests. This method, known as emf method, is also called as pessimistic method.

EXAMPLE 10.7 **A 3-phase alternator is rated at 5 kVA, 110 V, 26.3 A, 50 Hz, and 1200 rpm. The stator resistance between terminals as measured with dc is 0.2 Ω. With no-load and rated speed the stator line voltage is 160 V for a field current of 4 A. At rated speed, the short-circuit stator current per terminal is 50 A for a field current of 4 A. Compute the voltage regulation of alternator at 0.8 power factor lagging using synchronous impedance method.**

Solution Given data

Rating = 5 kVA

Voltage = 110 V

dc resistance = 0.2 Ω

$\cos\phi = 0.8$

Short-circuit stator current = 50 A

Field current = 4 A

$$Z_s = \frac{E_1 \text{ (open circuit)}}{I_{sc} \text{ (short circuit)}}$$

$$X_s = \sqrt{Z_s^2 - R_a^2}$$

ac resistance, $R_a = 1.6 \times 0.2 = 0.32\ \Omega$

Generated volts corresponding to rated short circuit is 160 V.

$$Z_s = \frac{E_1 \text{ (open circuit)}}{I_{sc} \text{(short circuit)}}$$

$$= \frac{160/\sqrt{3}}{50}$$

$$= 5\ \Omega$$

$$X_s = \sqrt{Z_s^2 - R_a^2}$$

$$= \sqrt{1.847^2 - 0.32^2}$$

$$= 1.819\ \Omega$$

$$IR_a = 26.3 \times 0.32 = 8.416 \text{ V}$$

$$IX_s = 26.3 \times 1/819 = 47.839 \text{ V}$$

$$\cos\phi = 0.8, \sin\phi = 0.6$$

$$E_0 = \sqrt{\left[(V\cos\phi + IR_a)^2 + (V\sin\phi + IX_s)^2\right]}$$

$$= \left[\left(\frac{110}{\sqrt{3}} \times 0.8 + 8.416\right)^2 + \left(\frac{110}{\sqrt{3}} \times 0.6 + 47.839\right)^2\right]^{1/2}$$

$$= 104.374 \text{ V}$$

$$\% \text{ Regulation} = \frac{E_0 - V}{V}$$

$$= \frac{104.374 - 63.510}{63.510}$$

$$= 64.34\%$$

EXAMPLE 10.8 A 100-kVA, 3000-V, 50-Hz, 3-phase star-connected alternator has an effective armature resistance of 0.2 Ω. The field current of 40 A produces short-circuit current of 200 A and an open-circuit emf of 1100 V (line value). Calculate the full-load voltage regulation at 0.8 power factor lag.

Solution Given data

Rating = 100 kVA
Voltage, $V = 3000$ V
Frequency, $f = 50$ Hz
Field current, $I_f = 40$ A
Short-circuit current, $I_{sc} = 200$ A
Armature resistance, $R_a = 0.2\ \Omega$
Generated emf, $E_1 = 1100$ V

$$E_0 = \sqrt{\left[(V\cos\phi + IR_a)^2 (V\sin\phi + IX_s)^2\right]}$$

$$\% \text{ Regulation} = \frac{E_0 - V}{V} \times 100$$

$$Z_s = \frac{E_1(\text{open circuit})}{I_{sc}(\text{short circuit})}$$

$$= \frac{1100/\sqrt{3}}{200}$$

$$= 3.175\ \Omega$$

$$X_s = \sqrt{Z_s^2 - R_a^2}$$

$$= \sqrt{(3.175)^2 - 0.2^2}$$

$$= 3.168\ \Omega$$

$$\text{Full-load current, } I = \frac{100000}{\sqrt{3} \times 3000} = 19.2 \text{ A}$$

$$IR_a = 19.2 \times 0.2 = 3.84 \text{ V}$$

$$IX_s = 19.2 \times 3.168 = 60.838 \text{ V}$$

$$\text{Voltage per phase} = 3000/\sqrt{3} = 1730 \text{ V}$$

$$\cos\phi = 0.8,\ \sin\phi = 0.6$$

Power factor = 0.8 lagging

$$E_0 = \sqrt{\left[(V\cos\phi + IR_a)^2 + (V\sin\phi + IX_s)^2\right]}$$

$$= \left[(1730 \times 0.80 + 3.84)^2 + (1730 \times 0.6 + 60.838)^2\right]^{\frac{1}{2}} = 1770 \text{ V}$$

$$\% \text{ Regulation "up"} = \frac{E_0 - V}{V} \times 100$$

$$\% \text{ Regulation} = \frac{1770 - 1730}{1730} \times 100$$

$$= 2.324\%$$

EXAMPLE 10.9 **A 3-phase, 10-kVA, 400-V, 50-Hz alternator has per phase armature resistance and synchronous reactance of 0.5 and 2.5 Ω, respectively. Calculate the generated voltage corresponding to the (i) full load at unity power factor and (ii) full load at 0.85 lagging power factor.**

Solution Given data

Rating = 10 kVA

Voltage, $V = 400$ V

Frequency, $f = 50$ Hz

Armature resistance = 0.5 Ω

Synchronous reactance = 2.5 Ω

$$E_0 = \left[\left(V\cos\phi + IR_a\right)^2 + \left(V\sin\phi + IX_s\right)^2\right]^{\frac{1}{2}}$$

Full-load current, $I = 1000/\sqrt{3} \times 400 = 14.434$ A

$$IR_a = 14.434 \times 0.5 = 7.217 \text{ V}$$

$$IX_s = 14.434 \times 2.5 = 36.085 \text{ V}$$

Voltage per phase = $400/\sqrt{3} = 230.94$ V

(i) Unity power factor

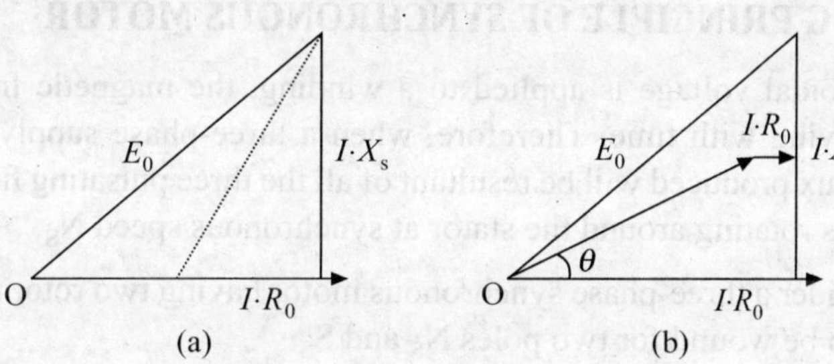

Figure 10.16

$$E_0 = \sqrt{[(V\cos\phi + IR_a)^2 + (IX_s)^2]}$$

$$= [(230.946 + 7.217)^2 + (36.085)^2]^{1/2}$$

$$= 240.88 \text{ V/phase} = 417.22 \text{ V (L-L)}$$

(ii) 0.85 power factor lagging

$$E_0 = \sqrt{[(V\cos\phi + IR_a)^2 + (V\sin\phi + IX_s)^2]}$$

$$= [(230.946 \times 0.85 + 7.217)^2 + (230.946 \times 0.526 + 36.085)^2]^{1/2}$$

$$= 257.384 \text{ V/phase} = 445.8 \text{ V (L-L)}$$

10.10 CONSTRUCTION OF THREE-PHASE SYNCHRONOUS MOTOR

A synchronous motor is a machine which operates at synchronous speed and converts electrical energy to mechanical energy. A synchronous generator may be operated as a motor by connecting its armature winding to a three-phase supply. As the name implies a synchronous motor runs at synchronous speed, i.e., in synchronism with the revolving field produced by the three-phase supply.

Like a synchronous generator, a synchronous motor also has two parts: (i) stator and (ii) rotor.

Stator: It consists of a three-phase armature winding in the slots of stator core and receives power from the three-phase supply.

Rotor: The rotor can be either salient pole type or non-salient pole type (smooth cylindrical). The rotor poles are normally excited by external dc source. The details of construction is shown in Figure 10.5.

The synchronous speed of the motor is given by

$$N_s = \frac{120f}{P}$$

where f is the frequency of supply in Hz and P is the number of poles.

10.11 WORKING PRINCIPLE OF SYNCHRONOUS MOTOR

When a sinusoidal voltage is applied to a winding, the magnetic field produced will also be sinusoidal varying with time. Therefore, when a three-phase supply is given to a three-phase winding, the flux produced will be resultant of all the three pulsating fields, so the stator develops a field which is rotating around the stator at synchronous speed N_S.

Let us consider a three-phase synchronous motor having two rotor poles N_R and S_R. Then, the stator will also be wound for two poles N_S and S_S.

Suppose at any instant the stator poles are at positions A and B as shown in Figure 10.17(a). The position of rotor poles is also shown in the figure. It is clear that poles N_S and N_R repel each other, and same is the case with S_S and N_R and S_S and S_R. So the rotor tends to move in anti-clock wise direction. After a period of half cycle, the polarities of the stator poles are reversed but the polarities of the rotor poles remain same as shown in Figure 10.17(b).

Now, the poles S_S and N_R, and N_S and S_R attract each other. The rotor tends to move in clockwise direction. Since stator poles change their polarities rapidly, they will pull the rotor first in one direction and then in other direction. Because of the high inertia of the rotor, the motor fails to start.

Suppose if the rotor is now rotated separately by a prime mover in the same direction as stator rotating field, with same speed, then the rotor will experience unidirectional torque.

Let the stator field is rotating in clockwise direction and the rotor is also rotated clockwise externally by some means at such a speed that the rotor poles interchange their positions along with the stator poles.

At some instant the position of stator and rotor poles is shown in Figure 10.17(a). It can be clearly observed that because of the attraction force between stator and rotor poles the rotor tends to rotate in clockwise direction. After the half cycle the stator poles reverse their polarities and same is the case with rotor poles. Positions of the rotor and stator poles is shown in Figure 10.17(b). Here, the rotor once again tends to rotate in clockwise direction.

Therefore, poles on the rotor always face poles of opposite polarity on the stator, and strong magnetic attraction is set up between them. This mutual attraction locks the rotor and stator together, and the rotor always rotates with the synchronous speed.

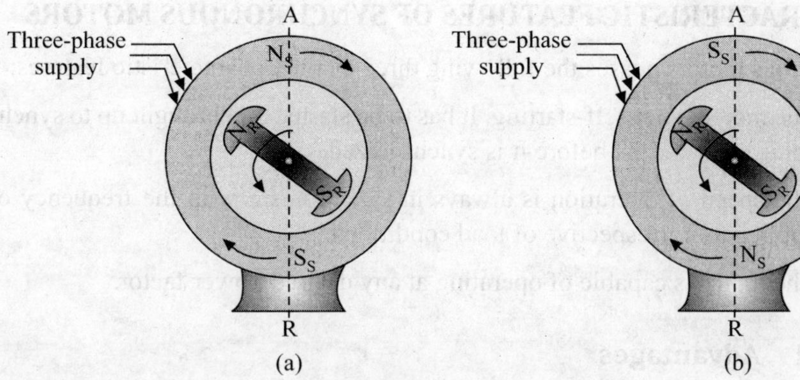

Figure 10.17

10.12 WHY SYNCHRONOUS MOTORS ARE NOT SELF-STARTING

Synchronous motors are not self-starting motors. This property is due to the inertia of the rotor. When the power supply is switched on, the armature winding and field winding are excited. Instantaneously, the armature winding creates a rotating magnetic field, which revolves at the designated motor speed. The rotor, due to inertia, does not follow the revolving magnetic field.

As per the operating principle of the synchronous motor, due to continuous and rapid rotation of stator poles, the rotor is subjected to a torque which is rapidly reversing, i.e., in quick succession, the rotor is subjected to a torque which tends to move it first in one direction and then in the opposite direction. Owing to its large inertia, the rotor cannot instantaneously respond to such quickly reversing torque, with the result that it remains stationary or in other words it is not self-starting.

In practice, the rotor should be rotated by some other means near to the motor's synchronous speed to overcome the inertia. Once the rotor nears the synchronous speed, the field winding is excited and the motor pulls into synchronization.

10.13 HUNTING

When we employ synchronous motor for driving varying load, the condition hunting is produced.

When the load on a synchronous motor is continuously increased, the rotor falls back to produce more torque to meet the required load. If the load is decreased suddenly the motor will be immediately pulled up to new value corresponding to the new load. At this stage the rotor may overshoot and oscillate around its new position of equilibrium corresponding to the new load.

To stop these oscillations, damper windings are employed. These windings consist of short circuited Cu bars placed in the faces of field poles. The oscillatory motion of the motor sets up eddy currents in the dampers which flow in such a way so as to suppress these oscillations.

10.14 CHARACTERISTIC FEATURES OF SYNCHRONOUS MOTORS

Synchronous motors posses the following three distinct characteristic features:

(i) The motor is not self-starting. It has to be started and brought up to synchronous speed by some other means before it is synchronized.

(ii) The speed of operation is always in synchronism with the frequency of the ac voltage applied to it irrespective of load conditions.

(iii) The motor is capable of operating at any desired power factor.

10.14.1 Advantages

(i) These motors can be used for power factor correction.

(ii) These are more efficient than induction motors when operated at unity power factor for same output (kW) and voltages rating.

(iii) They may be less costly for the same output, speed, and voltage ratings as compared to induction motors.

(iv) They give constant speed from no load to full load.

(v) Electromagnetic power varies linearly with the voltage.

10.14.2 Disadvantages

(i) These are not self-starting machines.

(ii) They require dc excitation, which must be supplied from external source.

(iii) They cannot be used for variable speed jobs.

(iv) They require collector rings and brushes.

(v) They cannot be started under load.

(vi) Their starting torque is zero.

(vii) They may fall out of synchronism and stop when overloaded.

10.14.3 Applications

(i) Used in power houses, and sub-stations in parallel are used to improve the power factor and the continuity of the supply.

(ii) Used to improve the power factor where in factories having large number of induction motors or other power apparatus, operating at lagging power factor.

(iii) Used in textile mills, rubber mills, mining industries, cement factories, and paper mills, where constant speed is required.

(iv) Used to drive continuously operating and constant speed equipment such as fans, blowers, centrifugal pumps, motor generator sets, and air compressors.

EXAMPLE 10.10 A 3-phase, 8-pole, 750-rpm, star-connected alternator has 72 slots on the armature. Each slot has 12 conductors, and the winding is short chorded by 2 slots. Find the phase and line emfs. Assume the flux per pole is 0.05 Wb.

Solution Given number of slots per pole $= \frac{72}{8} = 9$

Winding is short pitched by 2 slots.

$$\alpha = \frac{2}{9} \times 180° = 40°$$

Therefore, pitch factor, $K_p = \cos\frac{\alpha}{2} = \cos\frac{40°}{2} = 0.9396$

Number of slot angle, $B = \frac{180°}{\text{Slots per pole}} = \frac{180°}{9} = 20°$

Number of slots per pole per phase $= \frac{9}{3} = 3$

Distribution factor, $K_d = \frac{\sin(m\beta/2)}{3\ \sin(20/2)} = 0.9597$

Induced emf, $E_{ph} = 440\phi TK_p/K_d$

$$f = \frac{PN}{120} = \frac{8 \times 750}{120} = 50 \text{ Hz}$$

Number of conductors per slot = 12

Total conductors = 72 × 12

Total conductors per phase $= \frac{72 \times 12}{3} = 288$

Therefore, turns per phase, $T_{ph} = \frac{7}{2} = \frac{288}{2} = 144$

$$\therefore \quad E_{ph} = 4.44 \times 0.05 \times 50 \times 144 \times 0.9396 \times 0.9597$$
$$= 1441.33 \text{ V}$$

Line voltage $E_L = \sqrt{3}\, E_{ph}$

$$= \sqrt{3} \times 1441.33 = 2496.4 \text{ V}$$

EXAMPLE 10.11 A 3-phase star-connected alternator on open circuit is required to generate a line voltage of 3.4 kV, 50 Hz. When driven at 500 rpm, the stator has 3 slots per pole per phase and 10 conductors per slot. The coils are short pitched by slot. Calculate the (i) number of poles and (ii) useful flux per pole.

Solution m = slots per pole per phase = 3

∴ Slots per pole = 3 × 3 = 9

$$B = 180° \text{ slots per pole} = \frac{180°}{9} = 20°$$

$$K_d = \frac{\sin(m\beta/2)}{m[\sin(\beta/2)]} = \frac{\sin(3\times 20/2)}{3\ \sin(20/2)} = 0.96$$

$$\alpha = \frac{1}{9}\times 180° = 20°$$

$$K_p = \cos(\alpha/2) = \frac{120\times 50}{500} = 12$$

(i) $P = \frac{120f}{N} = \frac{120\times 50}{500} = 12$

Therefore, total number of slots = Slots per pole × Number of poles

$= 9 \times 2 = 108$

Therefore, total conductors per phase, $z = \frac{10\times 108}{3} = 360$

Total turns per phase, $T = \frac{360}{2} = 180$

Given, $E_L = 3.4 \times 10^3$ V

$$\therefore \quad E_{ph} = \frac{E_L}{\sqrt{3}} = \frac{3.4\times 10^3}{\sqrt{3}} = 1963 \text{ V}$$

$$\therefore \quad E_{ph} = 4.44\phi fTK_pK_d$$

$$\phi = \frac{E_{ph}}{4.44fTK_pK_d} = \frac{1963}{4.44\times 50\times 180\times 0.96\times 0985} = 0.052 \text{ Wb}$$

EXAMPLE 10.12 **A 3-phase, 6-pole alternator has a total of 144 turns. What is the induced emf per phase when the alternator runs at 1200 rpm with a flux of 0.055 Wb in each pole. Assume concentrated full-pitch winding.**

Solution Given concentrated full-pitch winding

$\therefore \quad K_d = 1$ and $K_p = 1$

Now $\quad E_{ph} = 4.44\phi fT$

$$f = \frac{NP}{120} = \frac{1200\times 6}{120} = 60 \text{ Hz}$$

Turns per phase, $T = \frac{144}{3}$

$$E_{ph} = 4.44\times 0.055\times 60\times \frac{144}{3}$$

$= 703.3$ V

EXAMPLE 10.13 A 3-phase star-connected alternator has 8 poles and runs at 750 rpm. It has 24 slots per phase and 10 conductors per slot, the flux being 0.055 Wb/pole. Calculate the line voltage. Assume the winding factor to be 0.96.

Solution Given data

Number of poles = 8

Slots per phase = 24

Conductor per slot = 10

Flux, $\phi = 0.055$ Wb/pole

Winding factor, $K_d = 0.96$

Speed, $N = 750$ rpm

Frequency, $f = PN/120$ Hz

The induced emf, $E_{ph} = 4.44K_pK_d\phi fT$ volts

Then,

$$Z = 24 \times 10 = 240$$

$$T = Z/2 = 240/2 = 120$$

Frequency, $f = PN/120 = 8 \times 750/120 = 50$ Hz

Induced emf, $E_{ph} = 4.44K_pK_d \times fT$

$$= 4.44 \times 1 \times 0.96 \times 0.55 \times 50 \times 120$$

$$= 1406 \text{ V/phase}$$

$$E_l = \sqrt{3} \times E_{ph}$$

$$= 2436 \text{ V (line to line)}$$

EXAMPLE 10.14 The stator of a 3-phase, 8-pole, 750-rpm alternator has 72 slots, each containing 10 conductors. Calculate the induced emf per phase, if the flux per pole is 0.1 wb, sinusoidally distributed. Assume full-pitched coil and a winding distribution factor of 0.96.

Solution Given data

Number of poles $P = 8$

Speed, $N = 750$ rpm

Number of slots = 72

Flux per pole, $\phi = 0.1$ Wb

Distribution factor, $K_d = 0.96$

$E_{ph} = 4.44 \times K_d \times K_p \times f \times N_{ph} \times \phi$ volts

Frequency of supply, $f = \dfrac{N_p}{120} = \dfrac{750 \times 8}{120} = 50$ Hz

Given, distribution factor, $K_d = 0.96$

Pitch factor, $K_p = 1$

Total number of conductors = 72 × 10 = 720

$$\text{Total number of turns} = \frac{720}{2} = 360$$

$$\text{Number of turns per phase, } N_{ph} = \frac{360}{3} = 120$$

$$\text{Induced emf per phase, } E_{ph} = 4.44 \times K_b \times K_p \times f \times N_{ph} \times \phi$$

$$= 4.44 \times 0.96 \times 1 \times 50 \times 120 \times 0.1$$

$$= 2557.44 \text{ V}$$

EXAMPLE 10.15 A 3-phase, 50-Hz, 20-pole salient pole alternator with star-connected stator winding has 180 slots on the stator. Each slot consists of 8 conductors. The flux per pole is 25 mWb and is sinusoidally distributed. The coils are full pitch. Calculate (i) the speed; (ii) the generated emf per phase; and (iii) the line emf.

Solution Given

Number of poles, $P = 20$

Number of slots = 180

Conductors per slot = 8

Flux, $\phi = 25$ mWb

$$\text{Distribution factor, } K_d = \frac{\sin(m\beta/2)}{m(\sin \beta/2)}$$

$$\text{Speed, } N = \frac{120 f}{P} \text{ rpm}$$

The generated emf per phase, $E_{ph} = 4.44\, K_p K_d\, \phi f T$ volts

Number of slots per pole = 180/20 = 9

$$\beta = \frac{180°}{9} = 20°$$

m = Number of slots per pole per phase

$= 180/(20 \times 3)$

$= 3$

$$\text{Distribution factor, } K_d = \frac{[\sin(\beta/2)]}{m[\sin(\beta/2)]}$$

$$= \frac{\sin\left(\frac{3 \times 20}{2}\right)}{3[\sin(20/2)]}$$

$$= 0.96$$

Z per phase = 180 × 8/3 = 1440/3 = 480

$T = Z/2 = 240$

(i) Speed, $N = \dfrac{120f}{p} = \dfrac{120 \times 50}{20}$

$= 300$ rpm

(ii) The generated emf per phase, $E_{ph} = 4.44K_pK_d\phi fT$

$= 4.44 \times 1 \times 0.96 \times 0.025 \times 50 \times 240$

$= 1278.72$ V

(iii) Generated emf $E_{(L-L)} = \sqrt{3} \times E_{ph}$

$= 2214.80$ V

EXAMPLE 10.16 A 6-pole, 3-phase, 60-cycles/s alternator has 12 slots per pole and 4 conductors per slot. The winding is 5/6-pole pitch. The flux per pole is 1.5 Wb. The armature coils are connected in series. The winding is star connected. Calculate the induced emf per phase.

Solution Given data

Number of pole, $P = 6$

Number of slot = 12

Frequency $f = 60$ Hz

Number of conductors per slot = 4

Pole pitch = 5/6

Flux, $\phi = 1.5$ Wb

The distribution factor, $K_d = \dfrac{\sin \dfrac{m\beta}{2}}{m\left(\sin \dfrac{\beta}{2}\right)}$

The pitch factor, $K_p = \cos \dfrac{\alpha}{2}$

Total number of slots around the entire armature = $12 \times 6 = 72$

Total number of conductors = $72 \times 4 = 288$

Hence, the number of turns connecting in series per phase = $\dfrac{288}{2 \times 3} = 48$

Number of slots per pole per phase, $m = \left(\dfrac{72}{6}\right)/3 = 4$

The short chorded angle, $\alpha = \dfrac{1}{6} \times 180° = 30°$

The slot pitch, $\beta = \dfrac{180°}{12} = 15°$

The distribution factor, $K_d = \dfrac{\sin\left(\dfrac{m\beta}{2}\right)}{m\left(\sin\dfrac{\beta}{2}\right)}$

$$= \frac{\sin\left(4\times\dfrac{15°}{2}\right)}{4\times\left(\sin\dfrac{15°}{2}\right)}$$

$$= 0.958$$

The pitch factor, $K_p = \cos\dfrac{\alpha}{2} = \cos\dfrac{30°}{2} = \cos 15° = 0.966$

∴ The induced emf per phase $= 4.44 \times \phi \times f \times N \times K_p \times K_d$

$$= 4.44 \times 1.5 \times 60 \times 48 \times 0.966 \times 0.958$$

$$= 17750 \text{ V}$$

EXAMPLE 10.17 A 3-phase, 1385-V star-connected synchronous motor having synchronous reactance of 2 Ω per phase and negligible resistance takes an input of 207.8 kW with an induced emf of 916.5 V/phase. Calculate the motor line current and its power factor.

Solution Given data

Synchronous reactance = 2 Ω

Input power = 207.8 kW

Induced emf per phase = 916.5V

Line voltage = 1385 V

Input power = $\sqrt{3}\, V_1 I_1 \cos\phi$

Voltage per phase, $V_{ph} = \dfrac{1385}{\sqrt{3}}$

$= 799.65$V

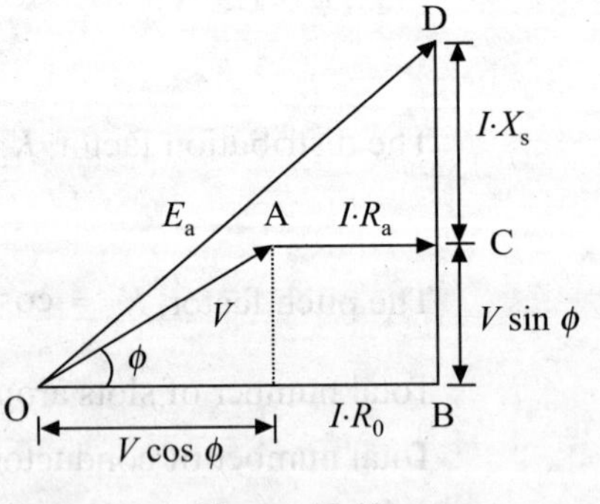

Figure 10.18

Input power = $\sqrt{3}\, V_1 I_1 \cos\phi$

207.8 kW = $\sqrt{3} \times 1385 \times I_1 \cos\phi$

$I_1 \cos\phi = 86.625$ A

From Figure 10.18

$OB = IX_s = 2I$

$BC = OB \cos\phi$

$= 2I \cos\phi$

$= 2 \times 86.625 = 173.25$

From Figure 10.18

$AB^2 = AC^2 + BC^2$

$916.5^2 = AC^2 + 173.25^2$

$$AC = 899.975 \text{ V}$$

$$OC = 899.975 - 799.65$$

$$= 100.32$$

$$\tan \phi = \frac{100.32}{86.625}$$

$$= 1.15$$

Power factor, $\cos \phi = \cos [\tan^{-1}(1.158)] = 0.653$

Current, $I = \dfrac{86.625}{0.653} = 132.53 \text{ A}$

EXAMPLE 10.18 The following test results are obtained on a 6600-V alternator.

Open-circuit voltage	**3100**	**4900**	**6600**	**7500**	**8300**
Field current (A)	**16**	**25**	**37.5**	**50**	**70**

A field current of 20 A is found necessary to circulate full-load current on short circuit of the armature. Calculate the full-load regulation at 0.8 power factor (lag) by the synchronous impedance method.

Solution Given data

Field current, $I_f = 20$ A

Voltage, $V = 6600$ V

Full-load regulation = 0.8 pf (lag)

$$I_f = \sqrt{V^2 + (IZ_s)^2 - 2VIZ_s \cos(90 + \theta)}$$

$$Z_s = \frac{\text{Open-circuit voltage}}{\text{Short-circuit current}}$$

Synchronous impedance method

Let the voltage of 6600 V be taken as 100% and also let the excitation of 37.5 A required to give 6600 V on open circuit be 100%.

Full load or 100% armature current is produced on short circuit by field current of 20 A.

If 100%, i.e., 37.5 A of field current were applied on the short circuit, the current would be

$$I_f = 100 \times \frac{37.5}{20} = 187.5 \text{ A}$$

Percentage impedance, $Z_s = \dfrac{\text{Open-circuit voltage}}{\text{Short-circuit current}}$

$$= \frac{100}{187.5} \times 100 = 53.3\%$$

Since armature resistance is negligible, therefore, impedance drop, $IZ_s = 53.3\%$ of the normal voltage will lead the load current by 90° or ∠OAB = 900 + θ = 90° + 36.86° = 126.87° as shown in Figure 10.19.

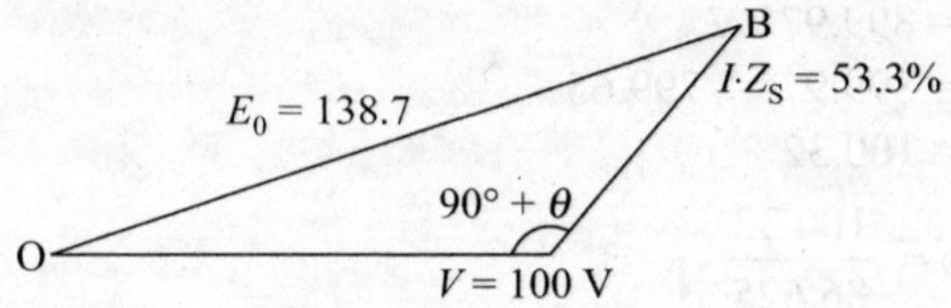

Figure 10.19

The phasor sum of normal voltage and impedance drop will give open-circuit voltage, E_0.

$$E_0 = \sqrt{(100)^2 + (53.3)^2 - 2\times100\times53.3\times\cos(90^0 + \phi)}$$

$$= \sqrt{(100)^2 + (53.3)^2 - 2\times100\times53.3(-0.6)}$$

$$= 138.7 \text{ V}$$

$$\text{Percentage of regulation} = \frac{138.7-100}{100}\times100 = 38.7\%$$

CHAPTER AT A GLANCE

- In alternators, the field system is rotating and the armature is stationary.
- The electric machine converts mechanical energy into electrical energy and is ac in nature.
- The rotor of an alternator may be of either salient pole type or smooth cylindrical type.
- Salient pole rotors are used in low- and medium-speed alternators.
- Non-salient pole rotors are used for alternators driven by steam turbines, i.e., turbo-alternators or turbo-generators, which run at very high speeds.
- The generated emf frequency, $f = (PN/120)$ Hz.
- In order to reduce eddy current loses, the stator of the alternator is laminated.
- The emf equation in an alternator is given by $4.44\, f\phi TK_pK_d$ volts.
- The distribution factor is given by $K_d = \dfrac{\sin\left(\dfrac{nm\beta}{2}\right)}{m\sin\left(\dfrac{n\beta}{2}\right)}$.
- The pitch factor is given by $K_p = \cos(\alpha/2)$.
- The synchronous impedance of a synchronous machine is given as phasor sum of armature resistance and synchronous reactance.
- With the use of damper winding, hunting can be minimized.
- The cylindrical-type synchronous motor is used for high-speed operation.
- The rating of a synchronous machine is usually governed by its temperature rise.

- The salient pole-type synchronous motor is used for low-speed operation, and in the pole faces damper bars are inserted and are used for starting purposes and also for minimizing the hunting.
- In the synchronous motor, the stator and the rotor magnetic fields rotate at the same speed.

MULTIPLE CHOICE QUESTIONS

1. In a synchronous machine, the voltage induced by armature reaction flux acts like the voltage drop in

(a) inductive reactance (b) resistance
(c) capacitive reactance (d) inductive impedance

2. In which coil the harmonic component of the generated emf will be more

(a) full-pitch coil (b) short-pitch coil
(c) long-pitch coil (d) same in coils

3. Which of the following has significant influence on the power factor of an alternator?

(a) Excitation (b) Speed of alternator
(c) HP of prime mover (d) Load

4. The purpose of damper winding in an alternator is

(a) to reduce hunting
(b) to reduce hunting and to help while starting
(c) to help in starting
(d) all of the above

5. In synchronous motors, hunting is reduced by

(a) damper winding (b) lap winding
(c) stator winding (d) field winding

6. Stationary armature windings are preferred in an alternator because

(a) they are cheaper (b) they require less insulation
(c) they are easy to make (d) all of the above

7. What is the purpose of airgap in a rotating machine?

(a) Reduce the mmf required
(b) Provide a path for flux
(c) Allows the rotor to rotate
(d) Improve the reluctance of the magnetic circuit

8. In a wave winding the number of paths is equal to

(a) 2 (b) number of poles (c) 4 (d) 6

9. Which of the following statement is true for synchronous motor?

(a) It is self-starting
(b) It is not self-starting
(c) It can run at a speed close to synchronous speed
(d) It can run at any speed

10. The advantage of using short-pitched coil in an alternator is
   (a) to reduce the volume of copper required
   (b) to reduce the efficiency
   (c) to reduce certain harmonics in the emf waveform
   (d) to reduce the synchronous impedance

11. A synchronous motor operates only at the synchronous speed which is determined by the
   (a) back emf and flux
   (b) number of poles and the supply frequency
   (c) supply voltage and the synchronous impedance
   (d) back emf and the supply frequency

12. A synchronous motor draws minimum current when the pf of the load is
   (a) zero (b) 0.6 leading (c) 0.6 lagging (d) unity

13. In which case the terminal voltage of alternator will be made
   (a) UPF (load) is thrown off (b) ZPF (lead) is thrown off
   (c) ZPF (lag) is thrown off (d) none of the above

14. The load taken up by the alternator directly depends upon the driving torque and
   (a) excitation (b) load current
   (c) load speed characteristics (d) angular advance of its rotor

15. The load angle of a synchronous generator is the angle between the excitation voltage and
   (a) bus bar voltage (b) bus bar current
   (c) bus bar impedance (d) none

16. An over excited synchronous motor can work at
   (a) UPF (b) leading pf (c) ZPF (d) lagging pf

17. The synchronizing power is maximum at
   (a) full load (b) slightly higher than full load
   (c) twice full load (d) no load

18. A dc current is passed through the stator of a synchronous machine with its rotor rotating with synchronous speed. Which of the following statements is true?
   (a) This induces huge currents in damper and field coils
   (b) Rotor decelerates
   (c) Alternator works as a dc motor
   (d) Governor restores the stability

19. The per unit synchronous impedance of an alternator is 2.0. Its short-circuit ratio is
   (a) 1 (b) 2 (c) 0.05 (d) 0.5

20. The maximum speed in rpm of an alternator at power frequency of 50 Hz will be
   (a) 1000 (b) 4000 (c) 3000 (d) 2000

21. The number of poles in turbo alternators is usually
   (a) 2 (b) 4 (c) 12 (d) 50

22. The machine that supplies dc to the rotor is called the
(a) exciter (b) rectifier (c) convertor (d) invertor

23. Which of the following prime-movers normally has high speed?
(a) Steam engine (b) Diesel engine
(c) Hydraulic turbine (d) Steam turbine

24. A three-phase synchronous motor needs
(a) 3-phase ac supply (b) dc supply
(c) 3-phase and 1-phase ac supply (d) 3-phase and dc supply

25. The speed of a synchronous motor can be varied by varying its
(a) excitation (b) supply voltage (c) load (d) frequency

SHORT ANSWER QUESTIONS

1. In modern alternators, what is the rotating part?
2. Why the stator core of a synchronous machine is laminated?
3. What is the purpose of the stator frame in a synchronous machine?
4. What is the purpose of dampers in a synchronous generator?
5. What is a turbo generator?
6. Why salient pole rotors are not suitable for high-speed turbo generators?
7. What is the maximum possible speed at which an alternator can be driven to generate 50 Hz and 3000 V?
8. Why distributed winding is preferred over concentrated winding?
9. Define pitch factor?
10. Define distribution factor?
11. For a 3-phase, 4-pole, 50-Hz synchronous motor, the frequency, pole number, and load torque are all halved. What is the motor speed?
12. How the power factor of a synchronous motor is varied?
13. What is the purpose of squirrel cage winding in a synchronous motor?
14. What is hunting?
15. What is synchronous capacitor?

REVIEW QUESTIONS

1. Explain the constructional features of an alternator?
2. Describe the working principle of synchronous motors?
3. How emf is induced in the armature of an alternator. What are the factors that cause change in the alternator terminal voltage?
4. Derive the expressions for the distribution factor and pitch factor?
5. Describe predetermination of regulation of an alternator from the open-circuit test and short-circuit test?

6. Explain what is hunting?
7. Explain why the synchronous motor is not self-starting?
8. Derive the emf equation of an alternator?
9. Explain the advantages of rotating field type of alternators?
10. Define (i) pitch factor and (ii) distribution factor.

EXERCISE PROBLEMS

1. A 4-pole, 3-phase, 50-Hz star-connected alternator has a single-layer winding in 36 slots with 30 conductors per slot. The flux per pole is 0.05 Wb and winding is full pitched. Find the synchronous speed and the emf induced on no-load?
2. A 4-pole alternator is run at a speed of 1500 rpm. What is the frequency of the armature emf? If the number of poles is 2, at what speed should they be run so that the armature emf has the same frequency?
3. If the coil span used in a particular winding is 150°, what is its coil span factor?
4. Calculate the value of distribution factor for a 6-pole, 3-phase, 36-slot alternator?
5. A single-phase alternator gives 200 V on open circuit and circulates 40 A on short circuit for the same field excitation. What is its synchronous impedance?
6. A synchronous motor is running at 1500 rpm when connected to a 3-phase, 50-Hz supply. At what speed will it run if the frequency is increased to 60 Hz?
7. The power factor of a synchronous motor is unity when the field current is 100 A. if the load on the motor is kept constant and the excitation is increased to150 A, what will be the nature of the power factor, will it be lagging or leading?
8. A 300-V, 10-kW, 1-phase alternator has the following open-circuit and short-circuit test results: open-circuit test—400 V, 1 A, short-circuit test—0 V, 1 A. If the excitation is such that the open-circuit voltage at a particular load when it is thrown off is 400 V, determine the (a) maximum power output of the generator, (b) current supplied at maximum output, and (c) pf at maximum output.

Chapter

11

Measuring Instruments

Chapter Objectives

- To explain the essential features of the indicating instruments
- To discuss the principle of operation of moving coil and moving iron type instruments
- To study the extension of meter ranges
- To explain the principle of the operation of single-phase energy meter and megger

11.1 INTRODUCTION

The fundamental quantities of electrical engineering such as current, voltage, power, energy, and frequency have to be measured with the help of instruments for the purpose of computing the system efficiency and stability. The instruments which are designed to measure these quantities are called *measuring instruments*.

11.2 CLASSIFICATION OF MEASURING INSTRUMENTS

The measuring instruments are classified as shown in Figure 11.1.

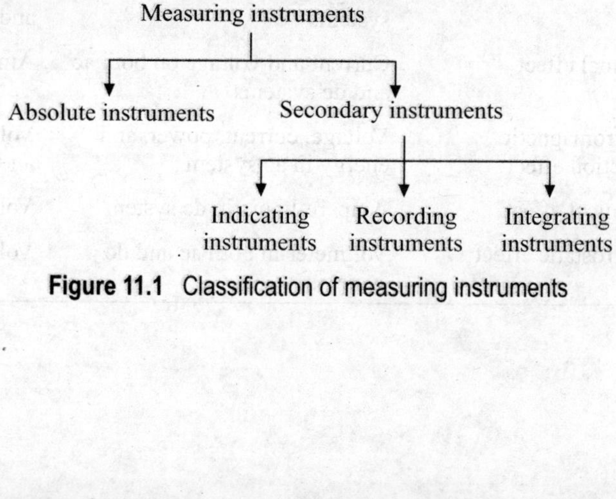

Figure 11.1 Classification of measuring instruments

11.2.1 Absolute Instruments

Absolute instruments measure the quantity to be measured in terms of physical constant or instrument constant, and its deflection requires no comparison with any other standard instrument. For example, tangent galvanometer and Rayleigh's current balance.

11.2.2 Secondary Instruments

The most generally used instruments are the secondary instruments. *Secondary instruments* measure the value of the quantity by the amount of deflection which is precalibrated by comparison with absolute instruments, for example, ammeter and voltmeter. According to the way of measuring the values the secondary instruments are further classified into the following types:

Indicating instruments: These instruments indicate the magnitude of the instantaneous value being measured by means of a pointer moving over a calibrated scale. If the magnitude of the quantity is being changed with respect to time, the indication of pointer also changes. Hence, there is no chance to know the previous values. For example, ammeter, voltmeter, and wattmeter.

Recording instruments: These instruments not only measure instantaneous values but also record continuously the magnitude of the quantity on a graph paper for a given period of time. In these instruments, the pointer is provided with a pen which touches lightly a sheet of graph paper wrapped round on a drum that rotates uniformly. These instruments are generally used in powerhouses where continuous information is required.

Integrating instruments: These instruments measure and register the total quantity of the electricity or the instrument measuring quantities or values and do the summation over a period of time and show or register the summation. For example, ampere-hour meter, energy meter, and speedometer.

Principle of operation

Table 11.1 shows the working principle of various measuring instruments.

Table 11.1 Principle of operation

S. No.	Effect used	Suitable for type of measurements	Instruments utilizing these effects
1.	Magnetic effect	Current, voltage, power, and energy on both ac and dc systems	Ammeter, voltmeters, wattmeter, integrating meter, and other electrical instruments
2.	Thermal effect	Current and voltage on both ac and dc systems	Ammeters and voltmeters
3.	Electromagnetic induction effect	Voltage, current, power, and energy in ac system	Voltmeter, ammeter, wattmeter, and energy meter
4.	Chemical effect	Ampere-hours in dc system	Voltmeter
5.	Electrostatic effect	Voltmeter in both ac and dc systems	Voltmeter

11.3 ESSENTIALS OF INDICATING INSTRUMENTS

For the operation of indicating instruments, three different torques are required to measure the quantity: (i) deflecting torque, (ii) controlling torque, and (iii) damping torque.

(i) Deflecting torque is used to move the pointer over a scale to the magnitude of the quantity being measured and is denoted by "T_d."

(ii) The pointer should be stopped at a value of the magnitude of the quantity; hence, there must be an equal and opposite torque on the pointer to bring the pointer to a stable position. This opposite torque to the deflection torque is known as the *controlling torque* and is denoted by "T_c."

(iii) Because of the deflecting and controlling torques the pointer gets oscillations, and due to this oscillation the pointer takes some time to set into a stable position. To reduce this oscillation and bring the pointer quickly to a stable position on the scale, a torque is required and is known as *damping torque*. This torque also acts in the opposite direction of the motion of the pointer at the deflected position.

11.3.1 Deflection Torque

The deflection torque moves the pointer of the instrument from its initial position (zero). The magnitude of the deflecting torque depends on the magnitude of the measuring quantity. This torque can be produced by utilizing any one of the following effects:

(i) Magnetic
(ii) Electromagnetic
(iii) Thermal
(iv) Electrostatic
(v) Chemical

The deflecting torque is constant and continuous. As long as the instrument is connected to the circuit, it reads or shows the value of the measuring quantity.

Derivation of deflecting torque

Let us consider a coil is placed inside a magnetic field as shown in Figure 11.2. When a current is passed through the coil, flux is set up around the coil by Faraday's law. This flux will interact with the magnetic field and produces the torque such as an electrical machine. This torque is continuous. The magnitude of the torque is proportional to the current passing through the coil.

Let B be the flux density in the air gap coil and main field in Wb/m^2, l the length of the moving coil in meters, b the breadth of the coil in meters, N the number of turns in moving coil, and I the current through the coil in amperes. Then the magnitude of the force experienced by the coil side for one turn is

$$F = BIl \text{ newton} \tag{11.1}$$

The force produced for N number of turns in coil side is

$$F = NBIl \text{ newton} \tag{11.2}$$

The total force for two sides of the coil is

$$F = 2NBIl \text{ newton} \tag{11.3}$$

Deflecting torque = Force × Radius

$$T_d = 2NBIl \times \frac{b}{2} \text{ newton-meter} \quad (11.4)$$

$$= NBI\,(lb)$$

$$\therefore \quad T_d = NBIA \text{ newton-meter} \quad (11.5)$$

where A is the cross-sectional area of the coil.

From Eq. (11.5), it is seen that N, A, and B are the constants.

$$\therefore \quad T_d \propto I \quad (11.6)$$

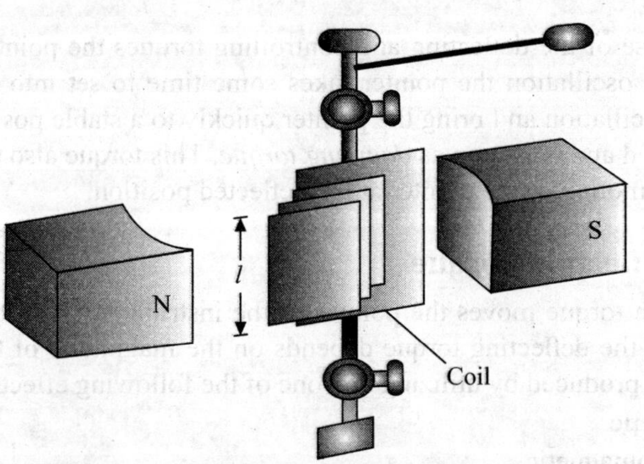

Figure 11.2 A coil placed inside a magnetic field

11.3.2 Controlling Torque

In the measuring instrument the deflection torque produced is constant and continuous. So the pointer moves continuously on the scale. The pointer should be stopped to know the corresponding measurable value. So there must be another torque in the opposite direction and in equal magnitude at the desired position as shown in Figure 11.3. Thus, force or torque increases along with the movement of the pointer until it becomes equal and opposite at its final deflected position.

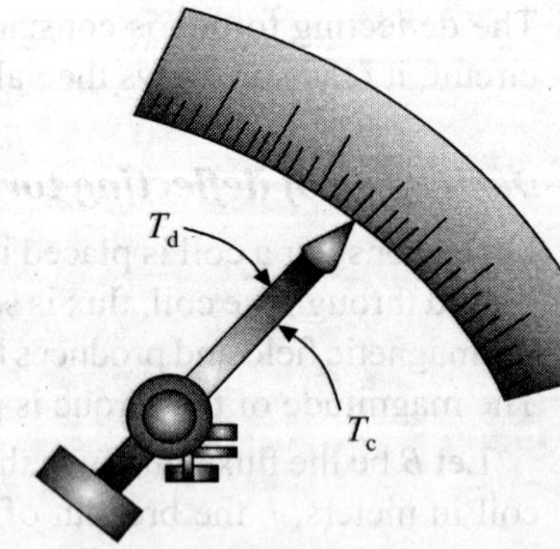

Figure 11.3

Normally, there are two methods adapted to get the controlling torque, namely,

(i) Spring control method

(ii) Gravity control method

Spring control method

In this method a helical spring is used in all the deflecting instruments to produce the controlling torque (Figure 11.4). To use spring control in an instrument, the following characteristics must be met:

(i) Non-magnetic
(ii) Low-resistance temperature coefficient
(iii) Free from mechanical stress

Generally, silicon bronze, hard-rolled silver or copper, platinum silver, platinum-iridium, or German silver has been used as spring materials, but for most applications phosphor bronze is used.

Let us consider a flat spiral spring having l meter length, d meter depth, t meter radial thickness, and Young's modulus of the spring material is E kg/m^2.

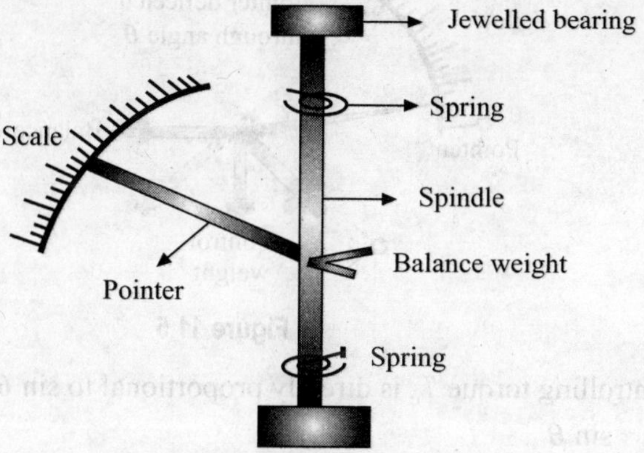

Figure 11.4

The torque developed due to deflection of a flat spiral spring is given by

$$T_c = \frac{Edt^3}{12l}\theta \text{ kg-m} \tag{11.7}$$

From the above equation for a particular spring, all the terms are constant except θ.

$$\therefore \quad T_c \propto \theta \tag{11.8}$$

Thus, the controlling torque is directly proportional to the angle of deflection. We know that the deflecting torque is proportional to the current, that is,

$$T_d \propto I \tag{11.9}$$

At the final deflection,

$$T_d = T_c$$

$$\therefore \quad I \propto \theta \tag{11.10}$$

So the scale of these types of instruments is uniform. The spring control is more popular and is employed widely in indicating instruments.

Gravity control method

In gravity control method, a small amount of weight is attached to the moving system as shown in Figure 11.5. If the pointer gets deflected, then only the controlling torque is produced in some proportion to the deflection. The magnitude of this controlling torque can be varied by adjusting the position of the weight upon the arm. In the zero position of the pointer, the control weight is vertical. When the pointer moves by an angel θ from the initial position, the control weight also changes its position by the same angle "θ" as shown in Figure 11.5 with dotted lines. At this position the controlling torque T_c can be written as

$$T_c = wl \sin \theta \tag{11.11}$$

where l is the distance from the axis of rotation, θ is the deflection, and w is the controlling weight.

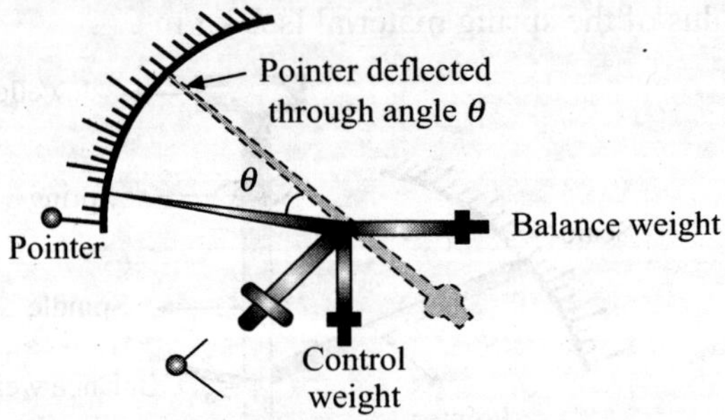

Figure 11.5

Thus, the controlling torque T_c is directly proportional to sin θ.

$$T_c \propto \sin \theta \tag{11.12}$$

$$T_d \propto I \tag{11.13}$$

At final deflection,

$$T_d = T_c \tag{11.14}$$

$$I \propto \sin \theta \tag{11.15}$$

$$\Rightarrow \quad I = k \sin \theta \tag{11.16}$$

$$\theta = \sin^{-1} \frac{I}{K} \tag{11.17}$$

So the scale of the gravity control is not uniform but crowded in the beginning.

The advantages of gravity control method are as follows:

(i) Simple and cheap
(ii) Control torque is not dependent on temperature
(iii) Performance is independent of time
(iv) By adjusting the position of the control weight, the control torque can be varied

The disadvantages of gravity control method are as follows:

(i) Problems in recording accurate reading because of non-uniform scale
(ii) The instrument must always be in vertical position only and must be properly leveled

The differences between the spring control method and gravity control method are explained in Table 11.2.

Table 11.2 Contrast of controlling systems

S. No.	Spring control	Gravity control
1.	For producing control torque, two hair springs are employed	For producing control torque, adjustable small weight is employed
2.	Control torque is constant	Control torque varies
3.	Scale is uniform	Scale is non-uniform
4.	Control torque is proportional to θ	Control torque is proportional to sin θ
5.	The performance is dependent on temperature	The performance is independent on temperature
6.	Very popularly employed in most of the instruments	Rarely employed in the instruments
7.	Simple and cheap	Simple, but costly

11.3.3 Damping Torque

In the indicating instruments the moving system is controlled by the two opposite torques such as deflection torque and controlling torque. At all the positions of final deflection these two torques are equal and in opposite direction to the pointer, so the pointer tends to oscillate because of the inertia and the two opposite torques about its final position before coming to rest. To bring the pointer to rest in the quickest possible time, a damping torque is required to avoid the oscillations. Without such damping force it is difficult to know the value of the reading at that instant. The oscillation at the deflected position is shown in Figure 11.6. The degree of damping should be adjusted to such a value which is sufficient enough to bring the pointer to rest about its deflected position without overshooting. Such type of damping is called *critical damping* and the instrument is said to have *dead boat*. Generally, the damping torque is adjusted to the value slightly less than the critical damping value.

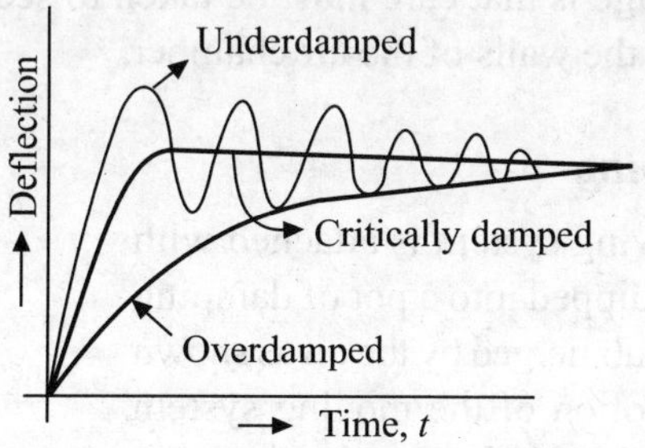

Figure 11.6

The various methods of damping mechanism employed to eliminate the oscillation are as follows:

(i) Air friction damping
(ii) Fluid friction damping
(iii) Eddy current damping

Air friction damping

Two methods are employed in air friction damping. The first method uses a light aluminum piston to the moving system as shown in Figure 11.7(a). The piston moves with a very small clearance in fixed air chamber closed at one end. If the piston moves rapidly into the chamber, the air inside the chamber gets compressed and thus the pressure inside the chamber opposes the motion of the piston, that is, of the moving system to which the piston is attached.

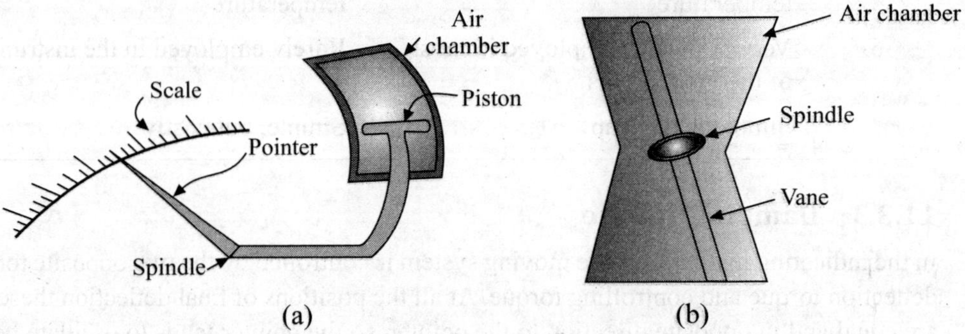

Figure 11.7

The second method employs a vane mounted on the spindle of the moving system. It is made of thin sheet of aluminum and moves in a closed sector shaped box as shown in Figure 11.7(b). This method is more popular.

The advantages of air friction damping are as follows:

(i) It is very simple and cheap.
(ii) It does not require the use of a permanent magnet.

The only disadvantage is that care must be taken to see that the piston is not bent or twisted, otherwise it will touch the walls of the air chamber.

Fluid friction damping

The spindle of the moving system is attached with light vanes, which are dipped into a pot of damping oil and are completely submerged by the oil as shown in Figure 11.8. The motion of the moving system is always opposed by the friction on the damping oil on the vanes. The damping force is increased by increasing the velocity of the vanes. There is no damping force when the vanes are stationary.

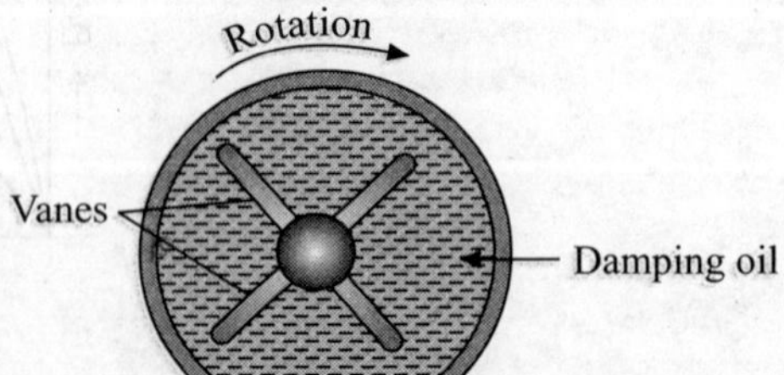

Figure 11.8

The advantages of fluid friction damping are as follows:

(i) Because of more viscosity of fluid, more damping is provided.

(ii) The fluid can also be used for insulation purposes.

(iii) The load on the bearing is reduced due to upthrust of oil, thus reducing the frictional errors.

The disadvantages of fluid friction damping are as follows:

(i) These are only used for instruments which are in vertical position.

(ii) The instruments cannot be kept clean because of the seeping of oil.

Eddy current damping

A thin disc of conducting material copper or aluminum is mounted on the spindle of the moving system as shown in Figure 11.9. This non-magnetic and conducting material moves in a magnetic field by Faraday's law and an eddy current is produced in a conducting material. Due to this eddy current, a force opposing the motion of the sheet is experienced between them and the magnetic field. This force is proportional to eddy currents and the strength of the magnetic field. But the eddy currents are proportional to the velocity of the moving system. Therefore, the damping force is caused by the velocity of the moving system because the magnetic field strength is constant.

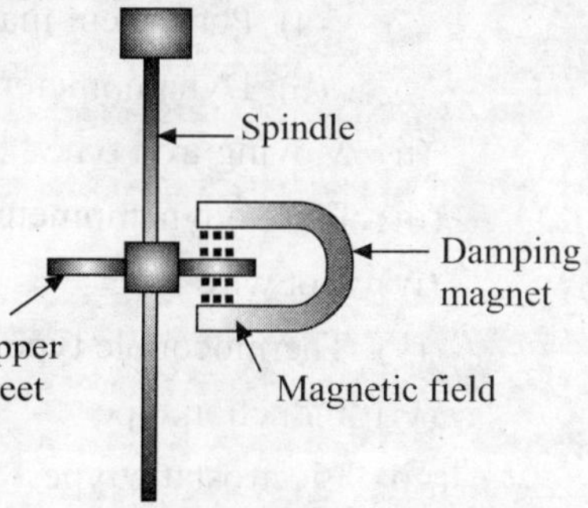

Figure 11.9

The advantages of eddy current damping are as follows:

(i) This is the most efficient form of damping.

(ii) It is very convenient to use in instruments where a metallic disc and permanent magnet already exist.

The only disadvantage of eddy current damping is that it cannot be used in instruments where eddy currents will distort the existing magnetic field.

11.4 VOLTMETER AND AMMETER

For any circuit, the characteristic cannot change when an instrument is connected to it. There is no difference in the principle of operation of an ammeter and a voltmeter. Both meters are basically current-measuring instruments, and the actuating or deflecting torque is directly proportional to the magnitude of the current ($T_d \propto I$) flowing through the operating coil. The ammeter is connected in series with the circuit and offers very low resistance, so that it does not affect the load current. A voltmeter is connected in parallel with the load (or across the circuit), so the voltmeter offers very high impedance, so that the current drawn by it is negligible. Therefore, ammeters have less number of turns and resistance, whereas voltmeters have more number of turns and high impedance to produce the required emf and deflecting torque.

11.4.1 Operation of Voltmeters and Ammeters in AC Circuits

The deflecting torque of a meter is proportional to the current flowing through the meter employed in an ac circuit; no deflection will be observed on it because the average torque over a cycle is zero. When the meter is employed in an ac circuit, if the operating torque is proportional to the square of voltage or current, the meter indicates some reading. Therefore, the deflection of the meter is proportional to the rms value of the voltage or current of the circuit.

11.4.2 Types of Voltmeters and Ammeters

The main types of instruments employed in voltmeters and ammeters are as follows:

(i) Moving coil type
   (a) Permanent magnet type
   (b) Dynamometer type
(ii) Moving iron type
(iii) Electrodynamometer type
(iv) Hot wire type
(v) Thermocouple type
(vi) Induction type
(vii) Electrostatic type

Permanent magnet moving coil type can be employed to measure dc quantity only and the induction type for ac only. The other types of meters can be used for both ac and dc.

Moving iron instruments are perhaps the cheapest and when used for measuring alternating currents and voltages, they give very accurate results. Both attraction type and repulsion type moving iron meters depend on the magnetic effect of current for their action.

Moving coil meters, especially the permanent magnet type, give accurate results when used in dc circuits and they may even be used as standard while calibrating other ammeters and voltmeters. The operation of the moving coil instruments is based on the production of mechanical force on a current-carrying conductor placed in a magnetic field.

Electrostatic instruments can be used as only voltmeters and such voltmeters find application in the measurements of high voltages.

Thermal instruments find use in both ac and dc circuits and the same calibration holds good because the operating torque depends on the heating effect of current. Thus, when used for measuring alternating current, the meter directly gives the rms value of the current to be measured. Also, its operation is independent of frequency and waveform.

Induction instruments are not very accurate as ammeters and voltmeters. Such instruments are mostly used as wattmeters and energy meters.

Only moving coil and moving iron instruments are discussed in detail in this chapter.

11.5 MOVING COIL INSTRUMENTS

These can be classified into the following two types:

(i) Permanent magnet moving coil type: only on dc

(ii) Dynamometer type: both ac and dc

11.5.1 Permanent Magnet Moving Coil-Type Instruments

The permanent magnet moving coil (PMMC) instruments consist of a powerful permanent magnet fitted with pole shoes to make the field radial. A cylindrical iron coil is mounted between the two poles giving a very narrow air gap. A rectangular coil of many turns wound on light aluminum. The former is located such that it freely rotates on the cylindrical core. The coil normally supported by a spindle on jewel bearing moves a pointer on the scale as shown in Figure 11.10.

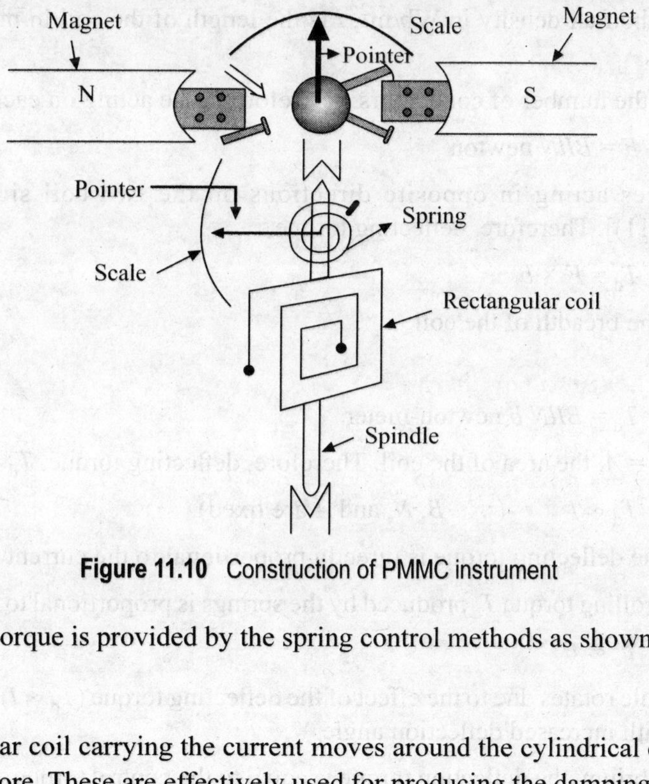

Figure 11.10 Construction of PMMC instrument

The controlling torque is provided by the spring control methods as shown in Figure 11.4.

Damping

When the rectangular coil carrying the current moves around the cylindrical coil, eddy currents are induced in the core. These are effectively used for producing the damping torque.

Expression for deflecting torque (T_d)

When the current (to be measured) is passed through the moving coil, the instrument is being used as ammeter. When this current-carrying coil is placed in a magnetic field (field due to the permanent magnet), it experiences mechanical forces.

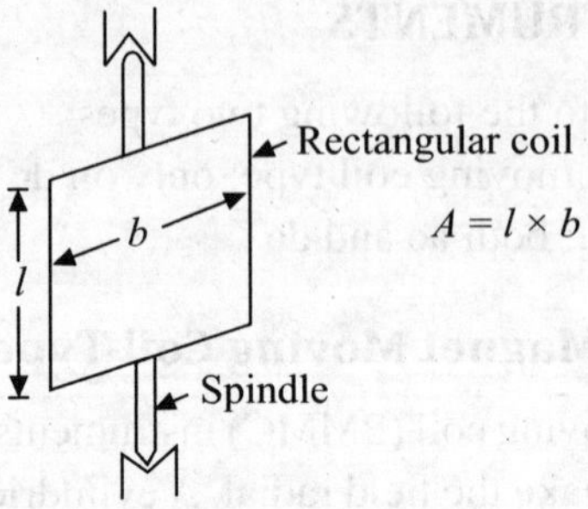

Figure 11.11

Force exerted on each conductor = BIl newton (11.18)

where B is the flux density in Wb/m^2, l is the length of the coil in m, and I is the current to be measured in A.

Let N be the number of conductors. Therefore, force acting on each coil side,

$$F = BIlN \text{ newton} \tag{11.19}$$

The forces acting in opposite directions on the two coil sides give rise to a torque [Figure (11.11)]. Therefore, deflecting torque,

$$T_d = F \times b \tag{11.20}$$

where b is the breadth of the coil.

or

$$T_d = BIlN\, b \text{ newton-meter}$$

But $l \times b = A$, the area of the coil. Therefore, deflecting torque, $T_d = BNAI$

$$T_d \propto I \qquad (\because \; B, N, \text{and } A \text{ are fixed}) \tag{11.21}$$

Hence, the deflecting torque is directly proportional to the current to be measured.

The controlling torque T_c produced by the springs is proportional to the angle of deflection "θ."

$$\therefore \quad T_c \propto \theta \tag{11.22}$$

The spindle rotates due to the effect of the deflecting torque ($T_d \propto I$), and the controlling torque increases with increased deflection angle.

At equilibrium, the deflection torque is equal to the control torque, and then the spindle comes to the rest position, i.e.,

$$T_d = T_c$$

$$\therefore \quad I \propto \theta. \tag{11.23}$$

Therefore, the deflection of the pointer is directly proportional to the current (I) to be measured. So the scale of PMMC instrument is uniform. If the current deflection in the coil reverses, then the torque experienced by the coil (dc motor action) also reverses and hence cannot be used on ac.

Advantages

(i) It has uniform scale.
(ii) It consumes low power.
(iii) High torque/weight ratio, as powerful permanent magnet is used which does not contribute to the weight of moving system.
(iv) The effect of stray magnetic field is negligible, as intense permanent magnetic field is employed.

Disadvantages

(i) It can be used only on dc.
(ii) The cost is high compared with moving iron instrument.
(iii) The field strength of a permanent magnet may be lost with time.

Errors

The main sources of errors in PMMC are as follows:

(i) Weakening of permanent magnet due to ageing at temperature effect.
(ii) Frictional error is generally very small with proper pivoting and balancing because of high torque–weight ratio.
(iii) Temperature error
(iv) Stray magnetic field error
(v) Thermocouple error

11.5.2 Dynamometer-Type Instruments

The permanent magnet moving coil instruments are only suitable on dc. With ac supply the deflecting torque produced will be of alternating nature. Owing to the inertia of the moving system, the instrument will not be able to indicate any reading. However, if the field is also of alternating nature so that its direction is also reversed in the moving coil, then the deflecting torque produced would act in one direction only and will cause the moving system to deflect. This principle is used in dynamometer type of instruments, thereby making them suitable on both ac and dc.

The essential features of dynamometer-type instruments are very similar to permanent magnet moving coil instruments except that the permanent magnetic field is replaced by a fixed coil which carries the current to be measured (or proportional voltage to be measured). These coils are usually air cored to avoid hysteresis, eddy currents, and other errors when the instrument is used on ac. In Figure 11.12, two air-cored fixed coils F_1 and F_2 are connected in series with the moving coil M and placed together and parallel to each other so as to provide a fairly uniform field within the range of the movement of the moving coil.

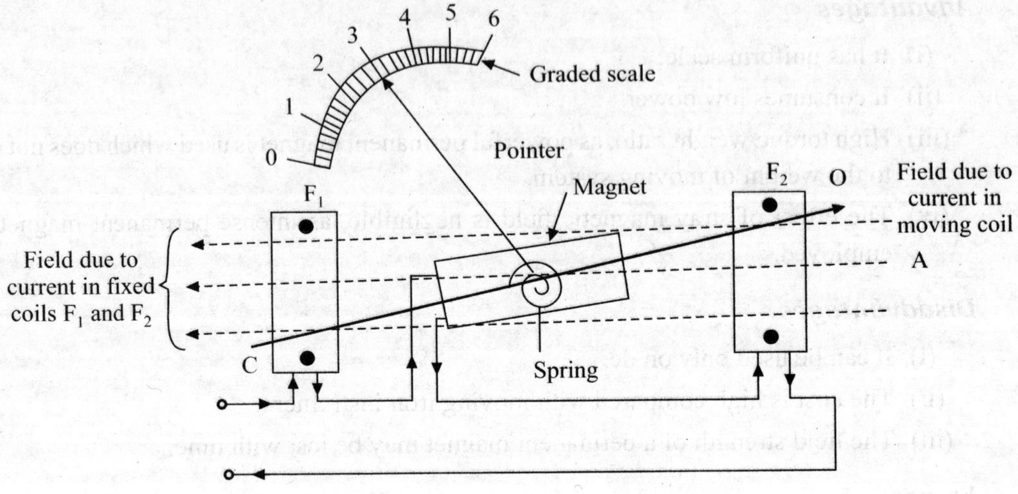

Figure 11.12 A dynamometer-type instrument

Damping: Air friction damping is used.

Control: Spring control method is used.

Arrangements of coils: They are connected as shown in Figure 11.13. Coils are used as the moving element should not carry heavy currents because of lead difficulties. As the current is small, the moving coil is connected in series with the fine coils.

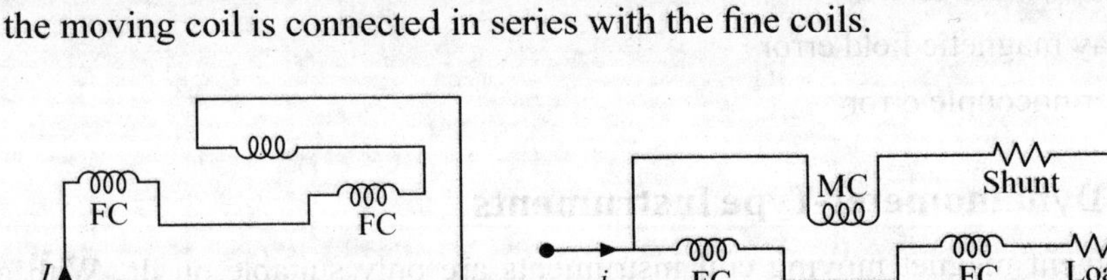

Figure 11.13 Arrangement for the measurement of (a) small current, (b) higher current, and (c) voltage. FC, fixed coil; MC, moving coil

As the current is high, the moving coil takes only a part of the current.

To increase the resistance of the instrument, all the coils are joined in series with an external resistance R_{se}.

Deflecting torque

Let I_f and I_m are the currents flowing through the fixed and moving coils, then the deflecting torque is given by

$$T_d \propto I_f I_m$$

In spring control instruments, control torque, $T_c \propto \theta$ (deflection). But, at steady deflection position,

$$T_c = T_d$$

Thus, $\theta \propto I_f I_m$

or

$$\theta \propto I^2$$

because both coils are in series.

An expression for the torque may also be obtained in terms of the mutual inductance M between the fixed and moving coils.

The energy stored in self-inductances and mutual inductors is given by

$$E = \frac{1}{2} L_1 I^2{}_1 + \frac{1}{2} L_2 I^2{}_2 + i_1 i_2 M \tag{11.24}$$

where i_1 is the current through fixed coils, i_2 is the current through moving coils, L_1 is the self-inductance of fixed coils, L_2 is the self-inductance of moving coil, and M is the mutual inductance between the fixed coils and moving coils.

As supply is given to the instrument, the moving coil deflects due to the interaction of fluxes produced by moving coils and fixed coils.

The rate of change of energy is torque,

$$\frac{dE}{d\theta} = i_1 i_2 \frac{dM}{d\theta} \tag{11.25}$$

This rate of change of energy is the torque produced on the moving system.

$$\therefore \quad i_1 i_2 \frac{dM}{d\theta} = T_d \tag{11.26}$$

Whatever may be the connection, i_1 and i_2 are proportional to the total current as shown in Figure 11.13(d).

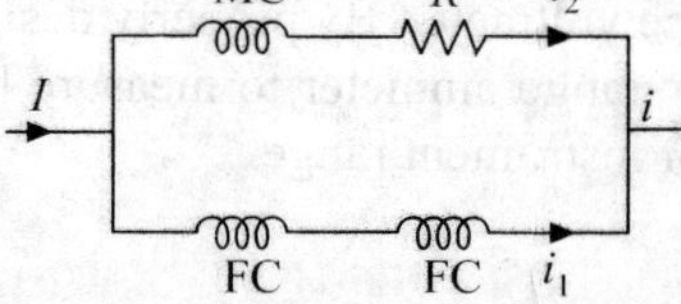

Figure 11.13(d)

$$\therefore \quad T_d \propto I^2 \frac{dM}{d\theta} \tag{11.27}$$

If $dM/d\theta$ is constant,

$$T_d \propto I^2 \tag{11.28}$$

$$T_c = K\theta$$

$$\therefore \quad I^2 \propto \theta$$

Hence, it has square law.

The working torques developed in these instruments are low and delicate. Dynamometer ammeters and voltmeters are not in common use. However, they can be constructed to substandard accuracy. They find much application as wattmeters.

Errors

(i) Affected by stray magnetic fields
(ii) Temperature errors
(iii) Frequency errors

Advantages of dynamometer-type instruments

(i) They can be used on both ac and dc.
(ii) They are free from hysteresis and eddy current errors.
(iii) These types of voltmeters are very useful for accuracy.
(iv) These instruments are used as transfer and for calibration purposes.

Disadvantages of dynamometer-type instruments

(i) The scale is not uniform.
(ii) Power consumption is high compared to other instruments.
(iii) Frictional losses are slightly more.
(iv) Adequate screening of the movements against stray magnetic fields is essential.

11.5.3 Extension of Instrument Range

Since the resistances of the moving coil of a PMMC instrument is quite small, a full-scale reading is obtained even with a small current. During the measurement, the current through the meter or voltage across the meter must be limited to a safe low value. This is achieved, in practice, by connecting a low-resistance shunt (i.e., a low resistance in parallel) across the ammeter and a high resistance in series with the voltmeter. By properly designing these shunt and series resistances, it is possible to use a low-range ammeter to measure high currents and high voltages. This is referred as an extension of instrument range.

Voltmeter

If the moving coil instrument is used to measure the voltage across a load in dc circuits, it has to control the currents in the instruments so that currents will be within limits. These are controlled with an extra high resistance connected in series with the instrument coil as shown in Figure 11.14. This series resistance is called a *multiplier*. The multiplier is used to limit the current through the basic meter so that meter current does not exceed the full-scale deflection value. Voltmeters are connected always in parallel to the load.

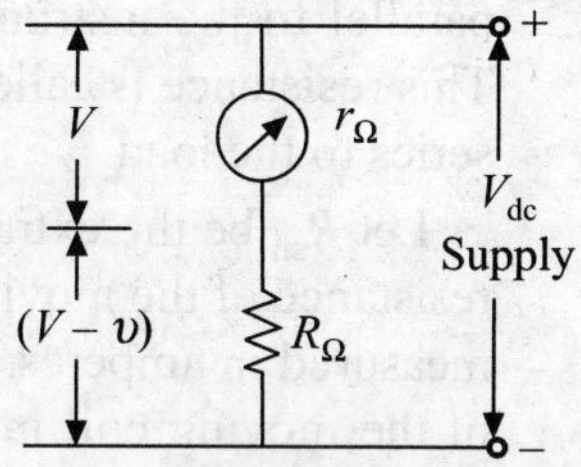

Figure 11.14

Let r be the resistance of the moving coil in ohms, R_{se} is the extra resistance connected in series in ohms, I_d is the full-scale deflection current in amperes, and v is the voltage across the coil.

$$v = I_d\, r \text{ volts} \tag{11.29}$$

V is the voltage to be measured in volts.

The voltage across the series R_{se} Ω resistors is $(V - v)$ volts.

The value of the extra resistance,

$$R_{se} = \frac{(V - v)}{I_d} \text{ ohms} \tag{11.30}$$

$$I_d\, R_{se} = V - v \tag{11.31}$$

Dividing on both sides of Eq. (11.31) by v,

$$\frac{I_d\, R_{se}}{v} = \frac{V - v}{v} \tag{11.32}$$

$$\frac{I_d\, R_{se}}{v} = \frac{V}{v} - 1$$

$$\frac{I_d\, R_{se}}{I_d r} = \frac{V}{v} - 1 \quad (\because I_d r = v)$$

$$\frac{V}{v} = \frac{R_{se}}{r} + 1 \tag{11.33}$$

$$m = \frac{R_{se}}{r} + 1 \tag{11.34}$$

where m is the multiplying factor and is defined by the ratio of full-range voltage to measured voltage.

The multiplying resistance is expressed as

$$R_{se} = (m - 1)\, r \text{ ohms} \tag{11.35}$$

From Eq. (11.35), it is observed that to increase the range of voltmeter by m times, the series resistance required is $(m - 1)$ times the basic meter resistance. It is nothing but the extension of the range of a voltmeter.

Ammeter

If the moving coil instrument is used to measure the current passing through the load circuit, it has to control voltage drop across the instrument. So an extra resistance has to be connected in parallel to the instrument coil as shown in Figure 11.15. This resistance is called *shunt.* Ammeters are connected in series to the load.

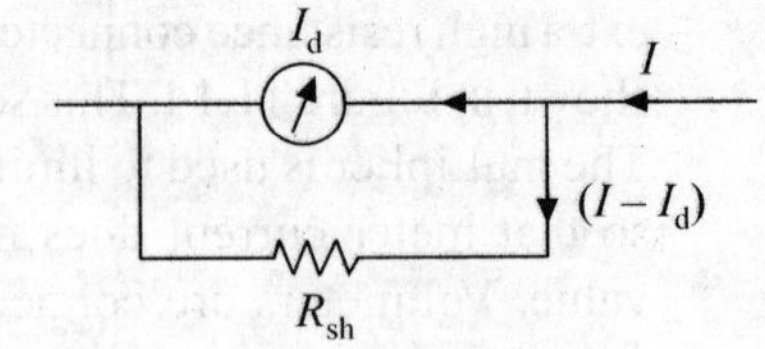

Figure 11.15

Let R_{sh} be the extra shunt resistance in ohms, "r" is the resistance of the moving coil in ohms, I is the current to be measured in amperes, I_d is the full-scale deflection current of the moving coil in amperes, and $(I - I_d)$ is the current passing through the shunt resistance (R_{sh}). Then

$$I_d r = R_{sh}(I - I_d) \quad (\because \text{ in parallel branches, voltage is same})$$

$$\frac{I_d r}{R_{sh}} = I - I_d \tag{11.36}$$

Dividing both sides of Eq. (11.36) by I_d, we get

$$\frac{I_d r}{I_d R_{sh}} = \frac{I - I_d}{I_d} \tag{11.37}$$

$$\frac{r}{R_{sh}} = \frac{I}{I_d} - 1 \tag{11.38}$$

$$\frac{I}{I_d} = \frac{r}{R_{sh}} + 1$$

or

$$m = \frac{r}{R_{sh}} + 1 \tag{11.39}$$

where m is called the multiplying power of the shunt. It is defined as the ratio of the total current to the current through the coil.

The shunt resistance required the following conditions:

(i) The coefficient of temperature of shunt and the meter should be small.

(ii) It should be stable.

(iii) It should not carry currents which will cause excessive rise in temperature.

(iv) The material used to join the shunts must have low thermo dielectric voltage drop, which means the soldering of joints should not cause voltage drop.

(v) The resistance value should not change due to soldering.

Manganin and constantan are generally employed for the shunts of dc instruments and ac instruments, respectively.

11.6 MOVING IRON INSTRUMENTS

These instruments are very widely used in laboratories and commercial purposes because these are cheap, robust, and accurate. The operating principle is simple. It consists of a vane of soft iron of high permeability which forms the moving system of the instrument along with the pointer, spring, pivots, and damping mechanisms. The iron vanes are so situated that they get attracted or repelled from the main magnetic field of force in a solenoid having the current-carrying coil. So the deflection of pointer is proportional to the quantity being measured. These instruments are generally of two categories:

(i) Attraction type
(ii) Repulsion type

11.6.1 Attraction-Type Moving Iron Instruments

In the attraction type, a piece of iron is drawn into the solenoid when the current flows in the coil. The moving iron is eccentrically pivoted and consists of a thick disc of soft iron as shown in Figure 11.16. This iron disc tends to move from the weaker magnetic field outside the solenoid to stronger magnetic field inside the solenoid. The shape of the disc is cut so to give a nearly uniform scale. Generally, these instruments have spring control to produce the controlling torque and air friction damping for damping torque.

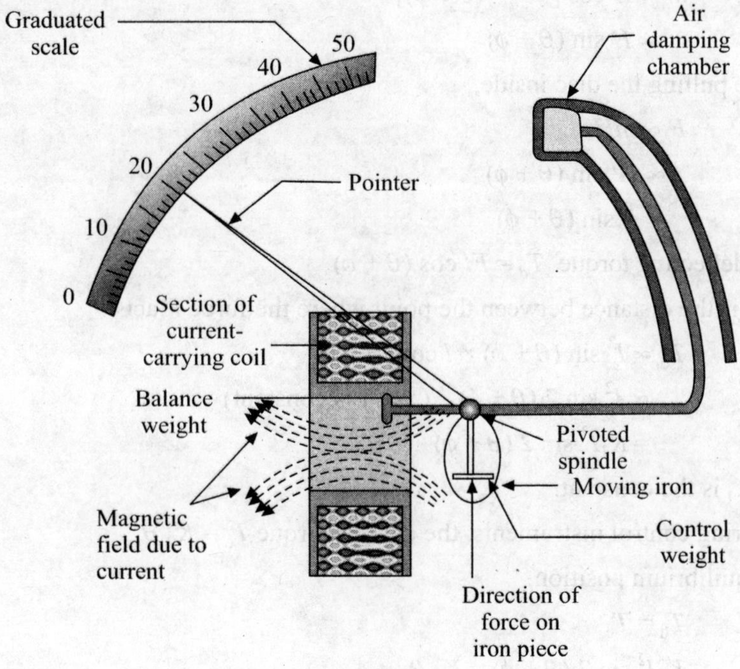

Figure 11.16

Deflecting torque

Consider that a soft iron disc makes an angle ϕ with the direction perpendicular to the field of the coil, when it is in zero position shown in Figure 11.17. The deflection produced due to flow of current I in the coil is θ. Current I in the coil set up a magnetic field of strength H parallel to the axis of the coil.

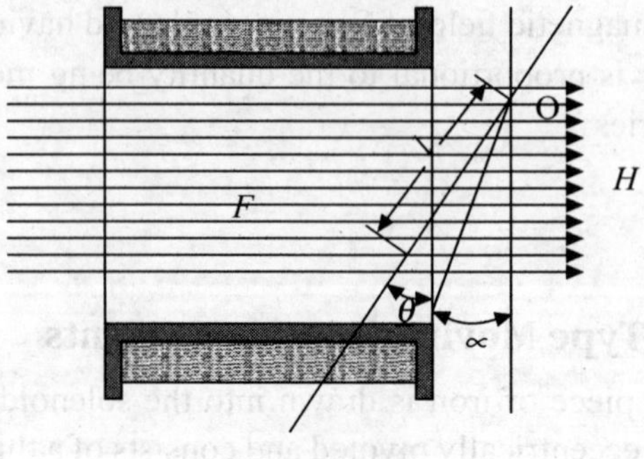

Figure 11.17

The disc is magnetized due to magnetic field set up by the coil, and in this position magnetization of the disc is proportional to H in the direction of the axis of the disc, that is, the pole strength of the disc,

$$m \propto H \cos [90° - (\theta + \phi)] \tag{11.40}$$

$$\propto H \sin (\theta + \phi) \tag{11.41}$$

Force pulling the disc inside,

$$F \propto mH \tag{11.42}$$

$$\propto H^2 \sin (\theta + \phi) \tag{11.43}$$

$$\propto I^2 \sin (\theta + \phi) \tag{11.44}$$

The deflecting torque, $T_d = Fl \cos (\theta + \phi)$ (11.45)

where l is the distance between the point where the force F acts

$$\therefore \quad T_d \propto I^2 \sin (\theta + \phi) \times l \cos (\theta + \phi) \tag{11.46}$$

$$\propto I^2 \sin 2 (\theta + \phi) \quad (\because \quad l \text{ is constant})$$

$$= K_1 I^2 \sin 2 (\theta + \phi) \tag{11.47}$$

where K_1 is the constant.

In spring control instruments, the control torque $T_c = K_2\, \theta$.

At equilibrium position,

$$T_d = T_c$$

$$K_1 I^2 \sin 2 (\theta + \phi) = K_2\, \theta \tag{11.48}$$

or

$$I = \sqrt{\frac{K_2 \theta}{K_1 \sin 2(\theta + \phi)}} \tag{11.49}$$

$$= K_3 \sqrt{\frac{\theta}{\sin 2(\theta + \phi)}} \tag{11.50}$$

In gravity control instruments, the control torque $T_c = K_4 \sin \theta$.

At equilibrium position,

$$T_d = T_c$$

$$K_1 I^2 \sin 2(\theta + \phi) = K_4 \sin \theta \tag{11.51}$$

or

$$I = \sqrt{\frac{K_4 \sin \theta}{K_1 \sin 2(\theta + \phi)}} \tag{11.52}$$

$$= K_5 \sqrt{\frac{\sin \theta}{\sin 2(\theta + \phi)}} \tag{11.53}$$

From Eqs. (11.50) and (11.53), it is observed that the range of the instrument will be maximum when ϕ is zero. The deflecting torque per unit of current will be maximum when $\sin 2(\theta + \phi) = 1$ or $(\theta + \phi) = 45°$, when and it will be zero when $(\theta + \phi) = 90°$. Therefore, the deflecting torque is proportional to the square of the current, so the instrument connected in ac circuit will indicate the rms value of voltage or current.

11.6.2 Repulsion-Type Moving Iron Instruments

In the repulsion-type instruments, two iron pieces are situated in the solenoid to produce a force of repulsion between each other when current flows in the coil as shown in Figure 11.18. There are two iron vanes within the coil: one stationary and the other moveable. The moveable vane is mounted on the spindle which carries the pointer spring/gravity control and damping mechanism. When the current passes through the coil of a solenoid, a magnetic field sets up. This magnetic field causes to magnetize the two vanes in the same direction. By the principle of magnetism the movable vane is repulsed from the fixed vane. This repulsive force is utilized to give the deflection force such that the pointer moves on the scale. If there is no current in the coil, the two vanes lie in phase to each other. The shape of the vanes is designed such that it gives nearly uniform scale. The instantaneous torque is proportional to the instantaneous square of the current due to which the direction of deflection is independent of the direction flow of current. So this instrument is suitable for both ac and dc measurements.

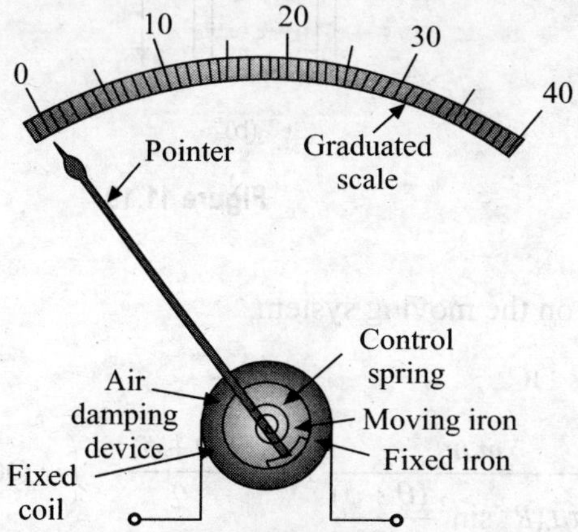

Figure 11.18

Deflecting torque

Torque can be obtained for straight bar instruments if it is assumed that the flux density B is proportional to H, the magnetizing force, that is, the saturation of iron is neglected.

Let the linear displacement between the two vanes in the deflected position is "d" for an angular displacement of "θ" while ϕ is the initial angle between them. The pole strength of these rods is m_1 and m_2 units when current I flows through the coil.

Assume that the distance "d" is very small in comparison with the length of the bar magnet "l," hence neglect the force of attraction between the poles at opposite ends.

The total repulsive force between the two rods,

$$F = \frac{2m_1 m_2}{4\pi\mu_0 d^2} \tag{11.54}$$

$$= \frac{2m_1 m_2}{4\pi\mu_0 \left(2R \sin\dfrac{(\theta+\phi)}{2}\right)^2} \tag{11.55}$$

where R is the distance of the rods from the axis of the rotation of moving rod.

$$F = \frac{m_1 m_2}{8\pi\mu_0 R^2 \sin^2\dfrac{(\theta+\phi)}{2}} \tag{11.56}$$

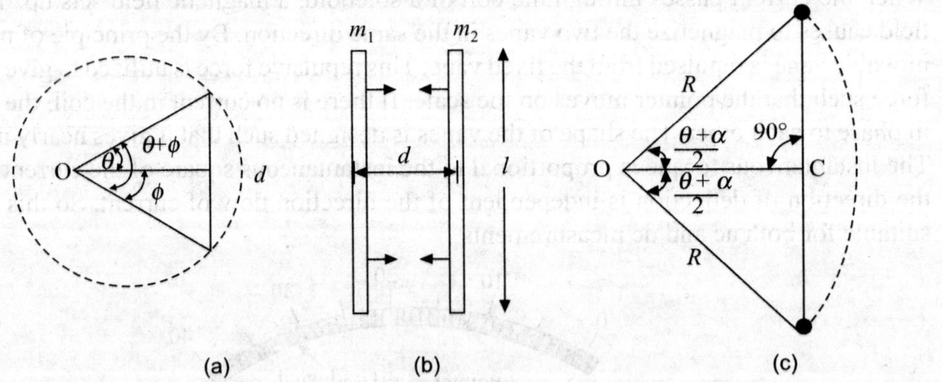

Figure 11.19

Deflecting torque on the moving system,

$$T_d = F \times OC$$

$$= \frac{m_1 m_2}{8\pi\mu_0 R^2 \sin^2\dfrac{(\theta+\phi)}{2}} R\cos\frac{(\theta+\phi)}{2} \quad \left(\because OC = R\cos\frac{(\theta+\phi)}{2}\right)$$

$$= \frac{m_1 m_2 \cos\dfrac{(\theta+\phi)}{2}}{8\pi\mu_0 R \sin^2\dfrac{(\theta+\phi)}{2}} \tag{11.57}$$

If the B–H curve is linear, both m_1 and m_2 are proportional to current I.

$$\therefore \qquad T_d \propto \frac{I^2 \cos\dfrac{(\theta+\phi)}{2}}{\sin^2\dfrac{(\theta+\phi)}{2}} \tag{11.58}$$

$$= K_1 \frac{I^2 \cos\dfrac{(\theta+\phi)}{2}}{\sin^2\dfrac{(\theta+\phi)}{2}} \tag{11.59}$$

For spring control,

$$T_c = K_2\,\theta$$

At equilibrium position,

$$T_d = T$$

$$K_1 \frac{I_2 \cos\dfrac{(\theta+\phi)}{2}}{\sin^2\dfrac{(\theta+\phi)}{2}} = K_2\theta \tag{11.60}$$

or

$$I = K_3 \sin\frac{(\theta+\phi)}{2}\sqrt{\frac{\theta}{\cos\dfrac{(\theta+\phi)}{2}}} \tag{11.61}$$

For gravity control,

$$T_c = K_4 \sin\theta$$

At equilibrium position,

$$T_d = T_c$$

$$I = K_3 \sin\frac{(\theta+\phi)}{2}\sqrt{\frac{\theta}{\cos\dfrac{(\theta+\phi)}{2}}} = K_4 \sin\theta$$

or

$$I = K_5 \sin\frac{(\theta+\phi)}{2}\sqrt{\frac{\sin\theta}{\cos\dfrac{(\theta+\phi)}{2}}} \tag{11.62}$$

From Eqs. (11.61) and (11.62), it is observed that the deflection is not proportional to θ, and therefore in the repulsion type of instruments, the scale is not uniform but is cramped. The meter reading is proportional to the rms value of voltage or current measured.

11.6.3 General Relation between Torque and Inductance of Moving Iron Instruments

If there is a small increase in the current magnitude dI, then the corresponding change in the angle of deflection is $d\theta$. Since the coil has inductance, it opposes change in the current magnitude; hence work must be done in order to overcome this opposition.

$$\therefore \quad \text{Work done} = T\,d\theta \text{ joules} \tag{11.63}$$

Due to stored energy in inductance, there will be a change in the stored energy in the magnetic field. This is because the vane tries to occupy the less reluctance position. Therefore, the force is always in such a direction so as to increase the coil inductance, since the inductance is inversely proportional to the reluctance of the magnetic coil.

Let T be the deflecting torque in newton-meters, I the current passing through the fixed coil in amperes, L the inductance of the coil in henry, and θ the angle of the deflection in radians.

The energy supplied would be stored in the magnetic field of the coil.

Let there be a small increase of the applied voltage in order to cause the current increment dI.

$$\text{Increase in voltage, } e = \frac{d}{dt}(LI)$$

$$= L\frac{di}{dt} + I\frac{dL}{dt} \tag{11.64}$$

The electrical energy supplied is $eI\,dt$ joules. (11.65)

Substituting the value of e from Eq. (11.64) into Eq. (11.65), we get

$$eI\,dt = \left(L\frac{di}{dt} + I\frac{dL}{dt}\right)(Idt)$$

$$= LI\,di + I^2 dL \tag{11.66}$$

The expression for energy stored is $(1/2)LI^2$ joules, and when the current has increased to $(I + dI)$ and the inductance has increased to $(L + dL)$, the stored energy after time dt is as follows:

$$\text{New energy stored} = \frac{1}{2}(L+dL)(I+dI)^2 \text{ joules}$$

$$\therefore \text{ Change in the stored energy} = \frac{1}{2}(L+dL)(I+dI)^2 - \frac{1}{2}LI^2$$

$$= \frac{1}{2}(L+dL)(I^2 + dI^2 + 2I\,dI) - \frac{1}{2}LI^2$$

If we ignore the second and higher order terms, the above expression approximately reduces to

Change in the stored energy = $LI\,dI + (1/2)\,I^2\,dL.$ (11.67)

From the law of conservation of energy, we have

Energy supplied = Work done + Energy stored (11.68)

From Eq. (11.63),

Work done = $T\,d\theta$ joules

Energy supplied = $LI\,DI + I^2dL$

Energy stored = $LI\,dI + \frac{1}{2}I^2dL$

From Eq. (11.63), we get

$$LI\,dI + I^2dL = (T\,d\theta) + \left(LI\,dI + \frac{1}{2}I^2dL\right) \tag{11.69}$$

or

$$T\,d\theta = \frac{1}{2}I^2dL \tag{11.70}$$

Hence,

$$T = \frac{1}{2}I^2\frac{dL}{d\theta} \tag{11.71}$$

From Eq. (11.71), it is noted that the torque is proportional to the square of the current being measured. Hence, the instrument can be used in both dc and ac circuits. But the scale tends to become non-uniform, unless the soft iron piece is properly designed.

11.6.4 Advantages and Disadvantages of Moving Iron Instruments

The various advantages of moving iron instruments are as follows:

(i) These instruments can be used to measure both ac and dc systems.
(ii) They have simple design and are robust in construction.
(iii) These instruments allows for maintenance-free working.
(iv) They have high staring torque.
(v) These are capable of giving good accuracy.
(vi) These are comparatively cheap .

The various disadvantages of moving iron instruments are as follows:

(i) The scale of these instruments is non-uniform.
(ii) The power consumption is higher for low-range instruments.
(iii) The stiffness of the spring decreases with increase in temperature.
(iv) The change in the frequency of operation causes serious error.

(v) Errors are caused due to hysteresis in the iron of the operating system and stray magnetic field.

(vi) Calibration is different for dc and ac usage.

11.6.5 Errors in Moving Iron Instruments

The errors usually occurring in moving iron instruments are as follows:

(i) **Friction error:** The friction of the pivot in the jewel produces a frictional torque which opposes the deflection of the moving element and as a result the pointer comes to rest at a short distance from its correct position and so there is an error in the instrument reading. This error is quite small due to high torque–weight ratio. This error is also reduced when the instrument is used on ac circuits because of the slight vibration caused by the rapid alternating flux which occurs due to the iron movement.

(ii) **Temperature error:** It arises due to the effect of temperature on the temperature coefficient of the spring. The error is due to the expansion of control spring, may be 0.04% per degree centigrade change in temperature. Errors occur due to self-heating of the coil which causes change in the resistance of the coil. Therefore, the coil and series resistance should have low temperature coefficient.

(iii) **Stray magnetic field error:** The error caused by the external field may be serious if not guarded against because of the weakness of the operating magnetic field in case of moving iron instruments. This error depends on the direction of the magnetic field with respect to the operating field of the instrument.

(iv) **Hysteresis error:** Due to the hysteresis in the iron of the working system, the readings are higher when descending values are measured than the ascending values (current or voltage). This error may be reduced by using small iron parts which can demagnetize quickly or by working with lower flux density.

(v) **Frequency error:** The change in frequency may cause errors due to the change of reactance of the operating coil and also due to the changes in the magnitudes of eddy currents set up in the metal parts of the instrument near the operating portion. This causes error in the instruments. This error is related to the ac operation of the instrument.

11.7 WATTMETER

A wattmeter is essentially an inherent combination of an ammeter and a voltmeter and, so, consists of two coils known as current coil and pressure coil. Due to the interaction of fluxes on account of currents in current coil and pressure coil, the operating torque is developed. The current coil is inserted in series with the line carrying current to be measured, and the pressure coil in series with a high resistance R is connected across the supply terminals.

There are four types of wattmeters, namely, (i) dynamometer type, (ii) induction type, (iii) electrostatic type, and (iv) thermal type. The first one, that is, the dynamometer type of wattmeter, is the most generally used wattmeter among all types and can be used in both ac and dc circuits.

11.7.1 Dynamometer-Type Wattmeter

This type of wattmeter consists of a fixed coil, which is divided into two equal portions in order to provide uniform field. One is used as current coil (fixed coil) and the other is used as a pressure coil (moving coil). A high resistance R is connected in series with the moving coil in order to limit the current in it. The magnetic fields of the fixed and the moving coils react on one another causing the moving coil to turn about its axis. In a wattmeter, the fixed coil (current coil) has two ends M and L called as current terminals and the moving coil (pressure coil) has two ends ± or C and V called as pressure terminals as shown in Figure 11.20. Damping is provided by light aluminum vanes moving in an air dash pot.

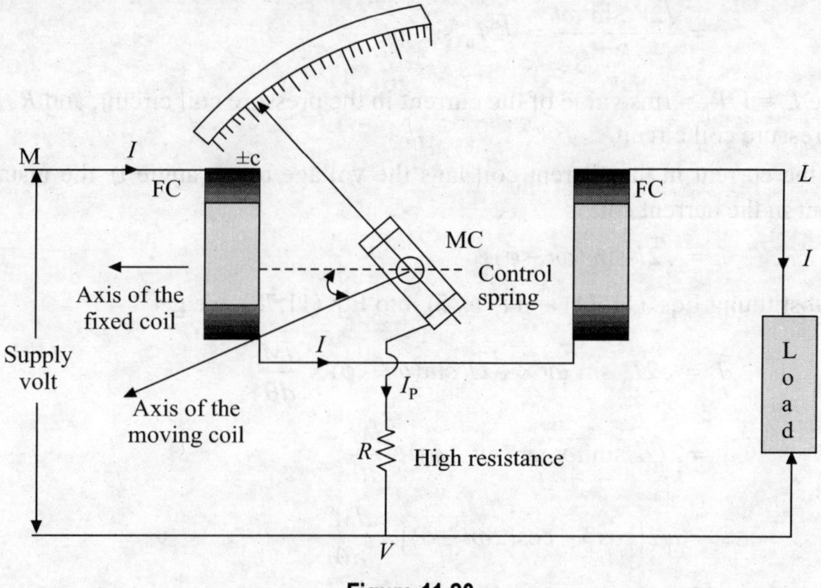

Figure 11.20

Torque equation

The instantaneous torque of a dynamometer instrument is given by

$$T_{\text{i}} = i_{\text{c}} i_{\text{p}} \frac{dM}{d\theta} \tag{11.72}$$

where i_c and i_p are the instantaneous current of current and pressure coils, respectively, and M is the mutual inductance between the current and pressure coils.

Let V and I be the rms values of voltage and current, respectively, to be measured.

The instantaneous value of voltage across the moving coil circuit is $v = \sqrt{2}V \sin\omega t$.

The pressure coil current i_p is in phase with the voltage, since, the pressure coil circuit has a very high resistance.

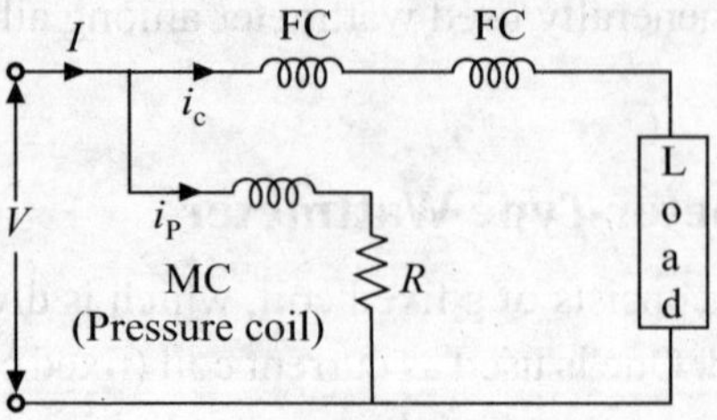

Figure 11.21

Instantaneous value of pressure coil current,

$$i_p = \frac{v}{R_p} \tag{11.73}$$

$$= \frac{\sqrt{2}V \sin \omega t}{R_p} = \sqrt{2}I_p \sin \omega t \tag{11.74}$$

where $I_p = V/R_p$ = rms value of the current in the pressure coil circuit, and R_p is the resistance of the pressure coil circuit.

If the current in the current coil lags the voltage by an angle ϕ, the instantaneous value of current in the current coil is

$$i_c = \sqrt{2}I \sin (\omega t - \phi) \tag{11.75}$$

Substituting Eqs. (11.74) and (11.75) into Eq. (11.72), we get

$$T_i = \sqrt{2}I_p \sin \omega t \times \sqrt{2}I \sin(\omega t - \phi) \times \frac{dM}{d\theta}$$

$$= 2I_p I \sin \omega t \sin(\omega t - \phi) \times \frac{dM}{d\theta}$$

$$= I_p I[\cos\phi - \cos(2\omega t - \phi)] \times \frac{dM}{d\theta} \tag{11.76}$$

From Eq. (11.76), it is clear that there is a component of power which varies as twice the supply frequency of the voltage and current. The average deflecting torque,

$$T_d = \frac{1}{T}\int_0^T T_i d(\omega t) \tag{11.77}$$

$$= \frac{1}{T}\int_0^T I_p I[\cos\phi - \cos(2\omega t - \phi)] \times \frac{dM}{d\theta} d(\omega t)$$

$$= I_p I \cos \phi \times \frac{dM}{d\theta}$$

$$= \frac{VI \cos \phi}{R_p} \times \frac{dM}{d\theta} \tag{11.78}$$

The control torque exerted by spring, $T_c = K\theta$

At equilibrium position,

$$T_d = T_c$$

$$\frac{VI\cos\phi}{R_p} \times \frac{dM}{d\theta} = K\theta$$

$$\text{Deflection, } \theta = \left(\frac{1}{KR_p} \times \frac{dM}{d\theta}\right) VI\cos\phi \tag{11.79}$$

$$= K_1 \frac{dM}{d\theta} P \tag{11.80}$$

where P is the power being measured is equal to $VI\cos\phi$

$$K_1 = \frac{1}{KR_p} = \text{constant}$$

From Eq. (11.80), it is observed that the deflection θ is proportional to the power under measurement, and the scale is essentially uniform over the range in which $dM/d\theta$ is constant.

Advantages

The advantages of dynamometer-type wattmeters are as follows:

(i) Scale is uniform
(ii) Used for both ac and dc circuits
(iii) Low power consumption
(iv) More accurate
(v) Used as a waveform harmonic analyzer

Disadvantages

The disadvantages of dynamometer-type wattmeters are as follows:

(i) Large frictional error due to heavy moving element
(ii) Limited scale range
(iii) Required large number of ampere-turns
(iv) Power loss is more

11.8 SINGLE-PHASE INDUCTION-TYPE ENERGY METER

The main components of a single-phase induction-type energy meter are shown in Figure 11.22. This energy meter consists of two electromagnets. The coil of one of the electromagnet is excited by the load current. This coil is called the current coil. Another coil of the electromagnet is connected across the supply terminals and is called as pressure coil. Consequently, the two electromagnets are known as series and shunt magnets, respectively. The pressure coil is mounted on the central limb of the shunt magnet. Copper shading bands are provided on the central limb of the shunt magnet so that phase displacement of 90° between the magnetic field set up by the shunt magnet and applied voltage is achieved by their adjustment. The copper shading bands are also called the pf compensator. A part of the leakage flux of the shunt magnet cuts the aluminum disc mounted on a spindle. The aluminum disc rotates in the air gap of a C-shaped brake magnet. The aluminum disc moves in the field of this magnet and thus provides a braking torque.

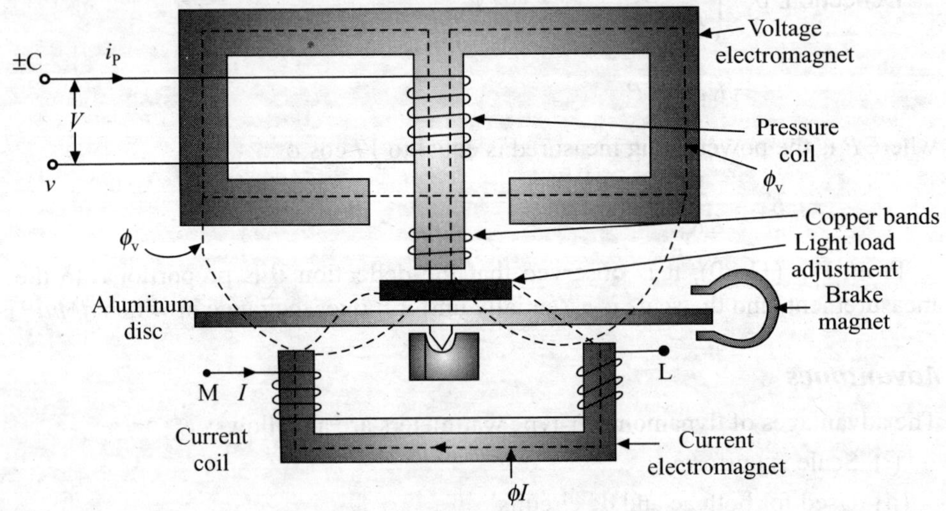

Figure 11.22

11.8.1 Theory of Operation

Figures 11.23(a) and (b) show the simplified functional diagram of the driving system of induction-type energy meter and the corresponding phasor diagram.

Let the load current lags the circuit voltage by an angle ϕ, as shown in Figure 11.23(c). The flux ϕ_{sh} of the shunt magnet is made to lag behind the supply voltage V by 90° by adjusting the position of copper shading bands.

Assume ϕ_{se} is proportional to and in phase with load current I. The emfs, E_{sh}, and E_{se} are induced in disc by shunt and series magnetic fluxes ϕ_{sh} and ϕ_{se}, respectively, and lags behind their respective fluxes by 90°. Eddy currents I_{sh} and I_{se} are set up by the induced emfs E_{sh} and E_{se}, respectively, and are in phase with their respective induced emfs.

Two opposite torques given by $\phi_{sh} I_{se}$ and $\phi_{se} I_{sh}$ will act on the disc. The instantaneous value of the resultant torque is the difference of the two, that is, since phase angle between ϕ_{sh} and I_{se} is ϕ and ϕ_{se} and I_{sh} is $(180° - \phi)$.

$$\text{Average torque} \propto \left[\phi_{sh} I_{se} \cos\phi - \phi_{se} I_{sh} \cos(180° - \phi)\right] \tag{11.81}$$

$$\propto \left[\phi_{sh} I_{se} \cos\phi + \phi_{se} I_{sh} \cos\phi\right]$$

$$\propto \left[\phi_{sh} I_{se} + \phi_{se} I_{sh}\right]\cos\phi$$

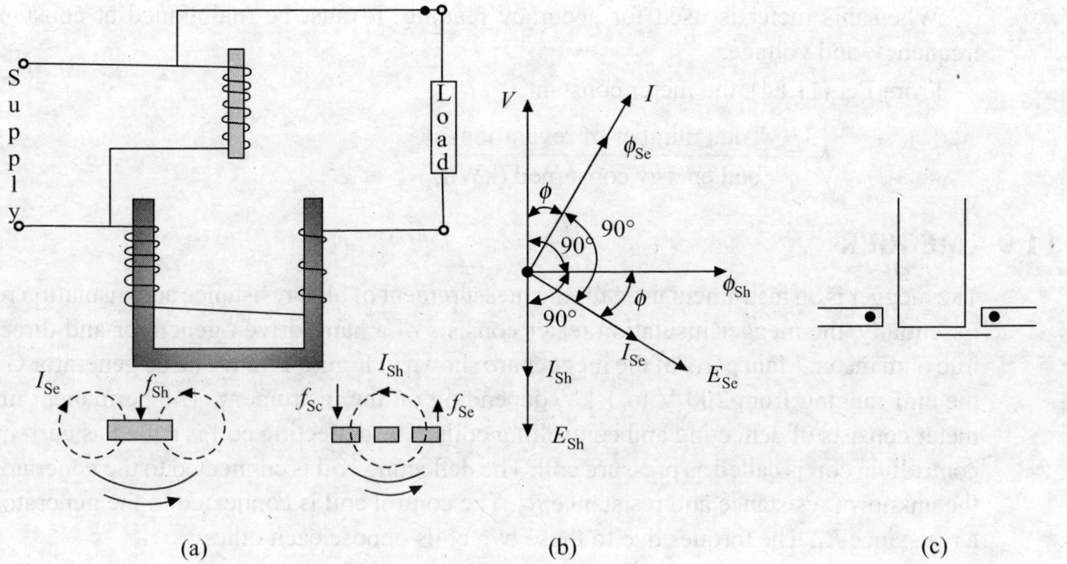

Figure 11.23

$$\phi_{sh} I_{se} \propto VI = k_1 VI$$

and

$$I_{se} I_{sh} \propto VI = k_2 VI$$

Since

$$\phi_{sh} \propto V, \phi_{se} \propto I, I_{se} \propto I, I_{sh} \propto V$$

$$\therefore \quad T_d \propto (K_1 VI + K_2 VI) \cos \phi$$

$$\propto VI \cos \phi \tag{11.82}$$

$$\propto \text{True power of circuit}$$

The braking torque is due to eddy currents induced in the aluminum disc. Since the magnitude of eddy currents is proportional to the speed of the disc, the braking torque T_b is proportional to the disc speed N. Thus,

$$T_b \propto N$$

For steady speed of rotation, $T_d = T_b$.

$$P \propto N \tag{11.83}$$

Hence, in a given period of time, the total number of revolutions, $\int_0^t N dt$, is proportional to $\int_0^t P dt$, that is,

$$\int_0^t N dt = K_3 \int_0^t P dt$$

Total number of revolutions mode = K_3 × Total energy consumed (11.84)

When this meter is used for accuracy reading, it must be maintained at constant supply, frequency, and voltage.

From Eq. (11.84), the meter constant,

$$K_3 = \frac{\text{Total number of revolutions}}{\text{Total energy consumed (kWh)}}$$

11.9 MEGGER

The megger is an instrument used for the measurement of high resistance and insulation resistance. Essentially, the megger insulation tester consists of a hand-driven generator and direct reading true ohm meter. Main parts of the megger are shown in Figure 11.24. The dc generator G produces the emf ranging from 200 V to 1 kV, depending on the instrument. The permanent magnet dC meter consists of deflecting and controlling coils. The deflecting coil is called as current coil and controlling coil is called as pressure coil. The deflecting coil is connected to the generator through the unknown resistance and resistance R_2. The control coil is connected to the generator through a resistance R_1. The torques due to these two coils oppose each other.

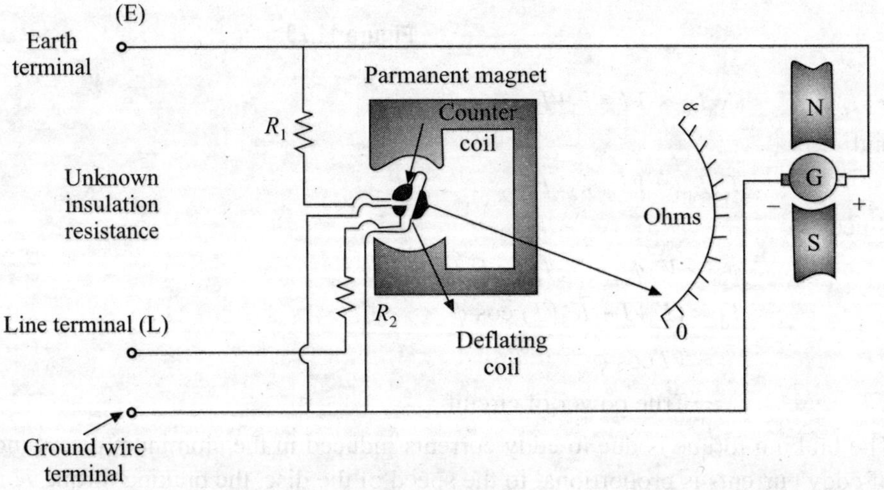

Figure 11.24

When the test terminals marked as L and E are open, no current flows through the deflecting coil and the torque due to the control coil moves the pointer to one end of scale (infinite resistance side). When the test terminals are short circuited, corresponding to zero external resistance, the current flowing through the deflecting coil is large enough to produce enough torque to overcome the counterclockwise torque on control coil. This moves the pointer to its extreme clockwise position. The point on the scale indicated by the pointer under this condition is marked zero resistance. When a resistance under test is connected between test terminals, the opposing torque of the coils balances each other so that the pointer comes to rest at some intermediate point on the scale. The scale is calibrated in ohms so that the pointer indicates directly the value of resistance under test.

The effective insulation resistance is the combination of insulation volume resistance and surface leakage resistance. The guard wire terminals are provided so that, if desired, only insulation volume resistance is measured and the surface leakage current bypasses the instrument. Usually, a terminal known as guard terminal is provided by means of which this guard ring may be connected to a guard wire on the insulation under test.

EXAMPLE 11.1 A moving coil instrument of resistance 25 Ω gives a full-scale deflection with a current of 1 mA. This instrument is to be used with a manganin shunt to extend its range to 100 mA. Calculate the error caused by a 10°C rise in temperature when

(i) The copper moving coil is connected directly across the manganin shunt.

(ii) A 75-Ω manganin resistance is used in series with the instrument moving coil. The temperature coefficient of copper is 0.0004 per degree centigrade and that of manganin is 0.00015 per degree centigrade.

Solution Given data

Resistance of instrument, $r = 25\ \Omega$

Full-scale deflecting current, $I_d = 1$ mA

New range with manganin resistance as a shunt, $I = 100$ mA

(i) When copper moving coil is connected directly across the manganin shunt, the multiplying factor of shunt is

$$N = \frac{I}{I_d} = \frac{100}{1} = 100$$

$$\therefore \text{ Resistance of shunt, } R = \frac{r}{N-1} = \frac{25}{100-1} = 0.2525\ \Omega$$

Given that the coefficient of resistance of copper is 0.0004 per degree centigrade and that of manganin is 0.00015 per degree centigrade. Therefore, for 10°C rise in temperature, instrument resistance,

$$r = 25\,(1 + 10 \times 0.0004)$$
$$= 25\ \Omega$$

(ii) When a 75-Ω manganin resistance is used in series with the moving coil of the instrument, the total resistance of meter circuit, $r_1 = r + R_{se}$, where R_{se} is the series resistance used and is equal to 75 Ω.

$$\therefore \quad r_1 = 25 + 75 = 100\ \Omega$$

$$\therefore \text{ Shunt resistance, } R_2 = \frac{r_1}{N-1} = \frac{100}{100-1} = 1.01\ \Omega$$

At 10°C rise in temperature,

Total resistance of the instrument circuit = $25(1 + 10 \times 0.0004) + 75(1 + 10 \times 0.00015)$

$= 101.1125\ \Omega$

Shunt resistance, $R_2 = 1.01(1 + 10 \times 0.00015) = 1.0155\ \Omega$

Current through meter for 100 mA measurement is given by

$$I_{d2} = 100 \times \frac{1.0115}{101.1125 + 1.0115} = 0.9905 \text{ mA}$$

$$\therefore \text{Percentage error} = \frac{I_{d2} - I_d}{I_d} \times 100 = \frac{0.9905 - 1}{1} \times 100$$

$$= -0.952\%$$

Hence, from the above, the percentage error decrease from –3.7% to 0.952% by the addition of a serious manganin resistor to the moving coil.

EXAMPLE 11.2 Explain how the PMMC instrument could be used to measure the (i) voltage of 300 V and (ii) current of 25 A.

Solution PMMC instruments are designed for very low values of currents which are in the order of microamperes or milliamperes, and the voltage drop across the instrument is of the order of millivolt. To make use of PMMC instrument for higher current and voltage ranges, shunts and multipliers are employed along with the instrument.

(i) To measure a voltage of 300 V, an external resistor (multiplier) is connected in series with the meter as shown in Figure 11.25.

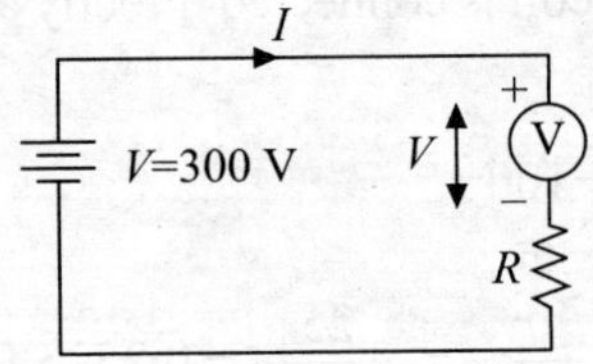

Figure 11.25

Let R be the resistance of the multiplier, V the voltage to be measured = 300 V, r the internal resistance of the instrument, and I_d the full-scale deflection current.

Applying the KVL, we get

$$V = I_r + IR$$

$$IR = V - Ir$$

$$R = \frac{V - Ir}{I} = \frac{V}{I} - r = \left(\frac{300}{I} - r\right) \Omega$$

Hence, to measure a voltage of 300 V, the minimum value of multiplier should be greater than

$$\left(\frac{300}{I_{max}} - r\right) \text{ ohms, i.e., } R_{min} > \left(\frac{300}{I_{max}} - r\right) \text{ ohms}$$

(ii) To measure a current of 25 A, a shunt is connected in parallel with the PMMC instrument as shown in Figure 11.26.

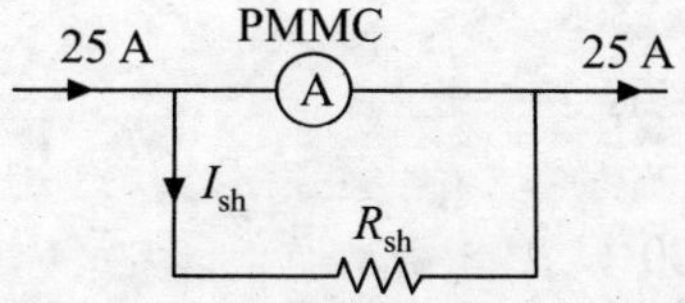

Figure 11.26

Let I_m be the current through the meter in amperes, r the internal resistance of the meter in ohms, R_{sh} the resistance of the shunt in ohms, I_{sh} the current through the shunt resistance in amperes, and I the current to be measured, which is equal to 25 A.

$$I = I_m + I_{sh} = 25 \text{ A}$$

$$I_{sh} = (25 - I_m) \text{ A} \tag{1}$$

Since the meter and the shunt resistor are connected in parallel, the voltage drop is connected in parallel is same.

$$\therefore \quad I_m r = I_{sh} R_{sh}$$

$$R_{sh} = \frac{I_m r}{I_{sh}} \tag{2}$$

From Eqs. (1) and (2)

$$R_{sh} = \frac{I_m r}{(25 - I_m)}$$

If I_m is the maximum current, that is, the full-scale deflection current of the instrument, then the shunt resistance should be less than $I_m r/(25 - I_m)$.

EXAMPLE 11.3 **A moving coil instrument gives a full-scale reading of 30 mA. When the potential difference across the terminals is 600 V, calculate the shunt resistance required for the full-scale deflection corresponding to 120 A.**

Solution Given data

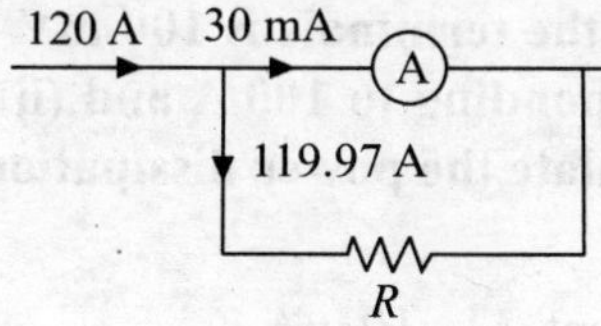

Figure 11.27

Voltage, $V = 600$ V

Current which causes full-scale deflection, $I_d = 30$ mA

Current to be measured, $I = 120$ A

Current through $R = I - I_d$

$= 120 - 30 \times 10^{-3}$

$= 119.97$ A

Voltage drop across "R" is

$V = 600$ V

$119.97R = 600$

$$R = \frac{600}{119.97} = 5.001\ \Omega$$

EXAMPLE 11.4 A moving coil instrument has a resistance of 25 Ω and gives a full-scale deflection when carrying 30 mA. Show how it can be adopted to measure currents up to 120 A.

Solution Given data

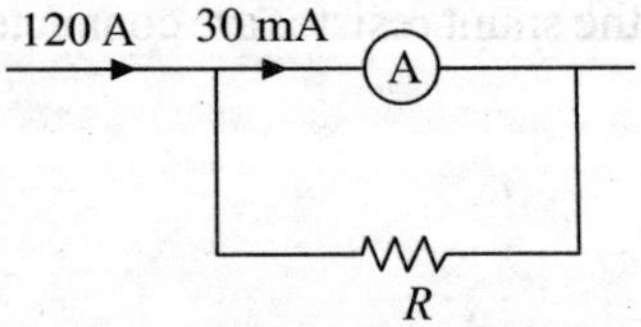

Figure 11.28

Resistance of the meter $r = 25\ \Omega$

Current which causes full-scale deflection $I_d = 30$ mA $= 0.03$ A

Full-line current $= 120$ A

Then the current to be bypassed $= I - I_d = 120 - 0.03 = 119.97$ A $=$ current through "R"

We have $119.97\,R = 0.03 \times 25$

$$\text{Shunt resistance, } R_{sh} = \frac{0.03 \times 25}{119.97} = 6.25\ \text{m}\Omega$$

EXAMPLE 11.5 A moving coil instrument gives a full-scale deflection of 10 mA. When the potential deference across the terminals is 100 mV, calculate (i) the shunt resistance for a full-scale deflection corresponding to 100 A and (ii) the series resistance for the full-scale reading with 1000 V. Calculate the power dissipation of the meter.

Solution Given data

Full-scale deflection current, $I_d = 10$ mA

Voltage applied, $V = 100$ mV

(i) Figure 11.29 shows the circuit connection to get a full-scale deflection corresponding to 100 A.

$$\text{Meter resistance, } r = \frac{V}{I_d} = \frac{100 \times 10^{-3}}{10 \times 10^{-3}} = 10\ \Omega$$

Voltage across shunt resistance (R_{sh}), $V_{sh} = V = 100$ mV

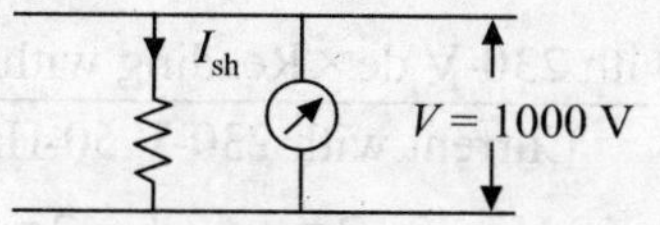

Figure 11.29

Voltage drop across shunt = $I_{sh}R_{sh} = 100$ V

$(I - I_d)R_{sh} = 100 \times 10^{-3}$

$$\text{Shunt resistance, } R_{sh} = \frac{100 \times 10^{-3}}{100 - 10 \times 10^{-3}}$$

$= 1.0001 \times 10^{-3}\ \Omega$

$= 1.0001$ mΩ

Power dissipation of the meter = $V \times I = 100 \times 10^{-3} \times 100 = 10$ W

(ii) Figure 11.30 shown the circuit arrangement for full-scale reading with 1000 V.

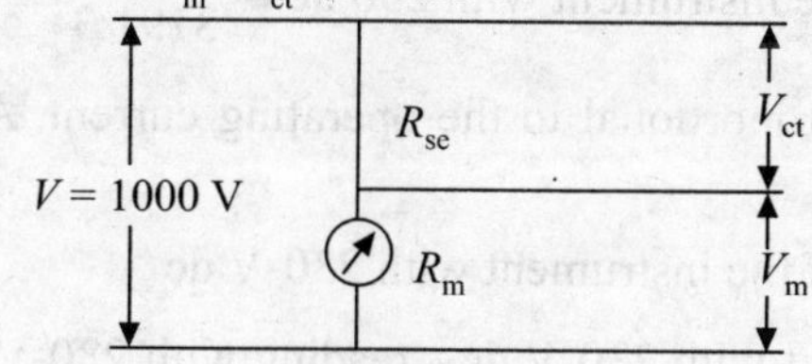

Figure 11.30

Voltage across series resistance, $R_{se} = V_{se} = V' - V$

$I R_{se} = 1000 - 100 \times 10^{-3}$

$10 \times 10^{-3} \times R_{se} = 999.9$

$$\text{Series resistance, } R_{se} = \frac{999.9}{10 \times 10^{-3}} = 99990\ \Omega = 99.99\ \text{k}\Omega$$

The power dissipation of the meter = $V' \times I = 1000 \times 10 \times 10^{-3} = 10$ W

EXAMPLE 11.6 The coil of a 320-V moving iron voltmeter has a resistance of 500 Ω and an inductance of 0.75. The instrument reads correctly at 50 Hz ac supply and takes 100 mA at full-scale deflection. What is the percentage error in the instrument rating when it is connected to a 230-V dc supply?

Solution Given data

Moving iron voltmeter resistance = 500 Ω

Moving iron voltmeter inductance = 0.75 H

At full-deflection current = 100 mA

The rating of the instrument with 230-V dc

$$= \frac{\text{Current with 230-V dc} \times \text{Reading with 230-V 50-Hz ac}}{\text{Current with 230-V 50-Hz ac}}$$

The reactance of coil at 50 Hz, $X = 2\pi \times f \times L = 2\pi \times 50 \times 0.75$

$$= 235.6\ \Omega$$

Current taken by the meter at 50 Hz and 320 V = 100 mA = 0.1 A

Impedance of the meter at 50 Hz, $Z = \dfrac{320}{0.1} = 3200\ \Omega$

Resistance of meter circuit = $\sqrt{Z^2 - X^2} = \sqrt{(3200)^2 - (235.6)^2}$

$$= 3191.3\ \Omega$$

Current taken by the meter at 230-V, 50-Hz supply = $\dfrac{230}{3200} = 0.0718$ A

Current taken by the instrument with 230 dc = $\dfrac{230}{3191.3} = 0.072$ A

The deflection is proportional to the operating current. Also the instrument reads correctly with 50-Hz ac.

Hence, the rating of the instrument with 230-V dc

$$= \frac{\text{Current with 230-V dc} \times \text{reading with 230-V 50-Hz ac}}{\text{Current with 230-V 50-Hz ac}}$$

$$= \frac{0.072}{0.0718} \times 230 = 230.64$$

EXAMPLE 11.7 **A 0- to 120-mA moving iron ammeter is converted to a 0–400-V, 50-Hz voltmeter by adding a series resistance and inductance $L = (0.01 + 0.2\theta)/4\pi$ H, where θ is the deflection in radians. The total angle of span of meter is 100°. Compute (i) the spring constant of the meter and (ii) the series resistance required.**

Solution Given data,

The coil has negligible resistance.

The coil has inductance $L = (0.01 + 0.2\theta)/4\pi$H, the total angle of span of meter is 100°.

Coil inductance, $L = \dfrac{0.01 + 0.2\theta}{4\pi}$

$$\frac{dL}{d\theta} = \frac{0.1}{2\pi} \text{ H/rad}$$

(i) Spring constant, $K = \dfrac{1}{2\theta} \times I^2 \times \dfrac{dL}{d\theta}$

$$= \frac{1\times(120\times10^{-3})^2}{2\times100^\circ\times\left(\dfrac{\pi}{180^\circ}\right)}\times\frac{0.1}{2\pi}$$

$$= 65.6 \times 10^{-6}\ \text{N-m/rad}$$

(ii) Inductance of the meter coil for full-scale deflection,

$$L = \frac{0.01+0.2\theta}{4\pi} = \frac{0.01+0.2\times100^\circ\times\dfrac{\pi}{180^\circ}}{4\pi}$$

$$= 0.0286\ \text{H}$$

Impedance required for converting the meter to 400-V voltmeter,

$$Z = \frac{400}{(120\times10^{-3})} = 3333.3\ \Omega$$

Reactance of the meter coil at 50 Hz = $2\pi \times 50 \times 0.0286 = 8.98\ \Omega$

Series resistance required,

$$R_{se} = \sqrt{(3333.3)^2-(8.98)^2} = 3333.3$$

EXAMPLE 11.8 The inductance of a moving iron ammeter is given by the expression $L = (20 + 10\theta - 3\theta^2)\ \mu\text{H}$, where θ is the angle of deflection in radians. Determine the deflection in the meter for a current of 10 A, if the spring constant is 10×10^{-6} N-m/rad.

Solution Given data

Inductance of the coil, $L = (20 + 10\theta - 3\theta^2)\ \mu\text{H}$

Spring constant, $K = 10 \times 10^{-6}$ N-m/rad

Current through the coil, $I = 10$ A

$$\therefore \quad \frac{dL}{d\theta} = (0+10-6\theta)\ \mu\text{H}$$

The deflecting torque of the instrument is given as

$$T_d = \frac{1}{2}I^2\frac{dL}{d\theta}$$

$$= \frac{1}{2}\times10^2\times(10-6\theta)\times10^{-6}$$

$$= 50(10\text{-}6\theta) \times 10^{-6}\ \text{N-m}$$

Control torque, $T_c = K\theta$, where K is the spring constant

$$T_c = (10 \times 10^{-6})\theta$$

$$= 10^{-5}\ \text{N-m}$$

At equilibrium, $T_c = T_d$.

$$\therefore \quad 10^{-5} = 50\,(10 - 6\theta)\,10^{-6}$$

or

$$10\theta = 50\,(10 - 6\theta)$$

$$\therefore \quad \theta = \frac{500}{310} = 1.1612 \text{ rad}$$

Angle of deflection, $\theta = 92.36°$ (since 1 rad = 57.3°)

EXAMPLE 11.9 A moving coil voltmeter gives a full-scale deflection with a current of 6 mA. The coil has 120 turns, effective depth of 3 cm, and width of 2.5 cm. The controlling torque of the spring is 0.5 N-cm for full-scale deflection. Estimate the flux density in the air gap.

Solution Given data

Number of turns in the coil, $N = 120$

Width of the coil, $w = 2.5$ cm

Depth of the coil, $d = 3$ cm

Controlling torque of the spring = 0.5 N-cm

Full-scale deflection current, $I = 6$ mA

Area, $A = w \times d$

$$= 2.5 \times 10^{-2} \times 3 \times 10^{-2} = 7.5 \times 10^{-4} \text{ m}^2$$

Deflecting torque, $T_d = BIAN$

$$= B \times 6 \times 10^{-3} \times 120 \times 7.5 \times 10^{4}$$

$$= 5.4 \times 10^{-4} \times B \text{ N-m}$$

Control torque, $T_c = 0.5$ N-cm

$$= 0.5 \times 10^{-2} \text{ N-m}$$

At balance, $T_c = T_d$

$$5.4 \times 10^{-4} \times B = 0.5 \times 10^{-2}$$

$$\text{Flux density, } B = \frac{0.5 \times 10^{-2}}{5.4 \times 10^{-4}}$$

$$= 9.26 \text{ Wb/m}^2$$

EXAMPLE 11.10 A moving coil mill voltmeter has a resistance of 200 Ω, and a full-scale deflection is reached when a potential difference of 100 mV is applied across the terminals. The moving coil has effective dimensions of 30 mm × 25 mm and is wound with 100 turns. The flux density in air gap is 0.2 Wb/m². Determine the control constant of the spring if the final deflection is 100° and has suitable diameter of copper wire for the coil winding if 20% of the total instrument resistance is due to the coil winding. Resistivity of the copper is 1.7 × 10⁻⁸ Ω-m.

Solution Given data

(i) Resistance of the meter = 200 Ω

Voltage measured = 100 mV

Full-scale deflection, $\theta = 100°$

Length of the coil, $l = 30$ mm

Breadth of the coil, $b = 25$ mm

Number of the turns in the coil, $N = 100$

Flux density in the gap, $B = 0.2$ Wb/m^2

Resistivity of the copper, $\rho = 1.7 \times 10^{-8}$ Ω-m

$$\text{Current through the meter, } I = \frac{\text{Voltage measured}}{\text{Meter resistance}}$$

$$= \frac{100 \times 10^{-3}}{200}$$

$$= 5 \times 10^{-4} \text{ A}$$

Area, $A = l \times b = 30 \times 10^{-3} \times 25 \times 10^{-3} = 7.5 \times 10^{-4}$ m^2

Deflecting torque, $T_d = BIAN$

$$= 0.2 \times 5 \times 10^{-4} \times 100 \times 7.5 \times 10^{-4}$$

$$= 7.5 \times 10^{-6} \text{ N-m}$$

Controlling torque, $T_c = K\theta$

$$= K \times 100 \text{ M-m}$$

At balance, $T_d = T_c$

$$7.5 \times 10^{-6} = 100 \times K$$

$$\text{Spring constant, } K = \frac{7.5 \times 10^{-6}}{100} = 7.5 \times 10^{-8}$$

(ii) Resistance of the coil, $r = 20\%$ of 200 Ω

$$= 0.2 \times 200$$

$$= 40\ \Omega$$

We have resistance,

$$r = \rho \frac{l'}{A}$$

where l' is the total length of the coil and A the cross-sectional area of the copper wire

$$l' = N \times \text{perimeter} = 2N(l + b) = 2 \times 100 \times (30 + 25) \times 10^{-3}$$

$$= 11 \text{ m}$$

$$40 = \frac{1.7 \times 10^{-8} \times 11}{A}$$

$$\therefore \quad A = 1.7 \times 10^{-8} \times 11 \times 40$$

$$= 4.675 \times 10^{-9} \text{ m}^2$$

But

$$A = \frac{\pi d^2}{4}$$

$$\therefore \quad d^2 = \frac{4A}{\pi}$$

Diameter of a copper wire $d = 2\sqrt{\frac{A}{\pi}} = 2\sqrt{\frac{4.675\times10^{-9}}{3.14}} = 7.7710^{-5}$ m

EXAMPLE 11.11 A 200-V moving iron voltmeter has an inductance of 0.8 H and a total resistance of 200 Ω. It is calibrated to read correctly on a 50-Hz circuit. What series resistance would be necessary to increase its range to 500 V?

Solution Given data

Resistance, $R = 2000\ \Omega$

Inductance, $L = 0.8$ H

Frequency, $f = 50$ Hz

Impedance, $Z = \sqrt{R^2 + (2\pi fL)^2}$

Impedance of the voltmeter, $Z = \sqrt{(200)^2 + (2\pi\times50\times0.8)^2}$

$= 2015.73\ \Omega$

Current through the voltmeter for full-scale deflection,

$$I = \frac{200}{2015.73} = 0.099 \text{ A}$$

Impedance required when used for 600 V, $Z' = \frac{500}{0.099} = 5050.5\ \Omega$

Hence, the total resistance of voltmeter,

$$R' = \sqrt{(Z')^2 - (2\pi fL)^2}$$

$$R' = \sqrt{(5050.5)^2 - (2\pi\times50\times0.8)^2} = 5044.24\ \Omega$$

$\therefore$ Additional series resistance required, $R_{se} = 5044.24 - 2000 = 3044.24\ \Omega$

Example 11.12 A moving coil instrument takes a current of 75 mA for full-scale deflection and has a resistance of 20 Ω. Calculate the value of resistance to be added in series to make it work as voltmeter to measure 500 V.

Solution Resistance, $r = 20\ \Omega$

Galvanometer current, $I_g = 75$ mA $= 75 \times 10^{-3}$ A

Voltage across the meter resistance, $v = I_g r = 75 \times 10^{-3} \times 20 = 1.5$ V

Extra resistance to be added in series to make it work as voltmeter to measure 500 V is

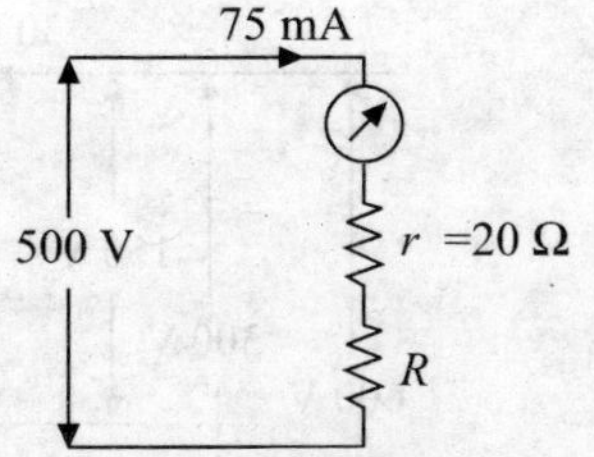

Figure 11.31

$$R = \left(\frac{V - v}{I_g}\right) = \frac{500 - 1.5}{75 \times 10^{-3}} = 6646.66\ \Omega$$

EXAMPLE 11.13 A moving coil instrument takes 50 mA at FSD (full-scale deflection) if the instrument is used as a voltmeter to measure 300 V at FSD by adding 5990 Ω in series. Calculate the resistance of the moving coil instrument.

Solution

$$I_g = 50\ \text{mA} = 50 \times 10^{-3}\ \text{A}$$

$$V = 300\ \text{V}$$

$$R = 5990$$

$$R = \frac{V - v}{I_g}$$

or

$$v = V - I_g R = 300 - 50 \times 10^{-3} \times 5990 = 0.5$$

$$r = \frac{v}{I_g} = \frac{0.5}{50 \times 10^{-3}} = 10\ \Omega$$

EXAMPLE 11.14 A moving coil instrument has a resistance of 25 Ω and takes a current of 30 mA at FSD if the instrument is to be used as voltmeter to measure 150 V, 300 V, and 600 V. Calculate the resistances values to be attached in series to the instrument.

Solution Voltmeter reading, $v = I_g\, r = 0.30 \times 10^{-3} \times 25 = 0.75$ V

For range of 150 V,

$$R_{150} = R_1 = \frac{V - v}{I_g} = \frac{150 - 0.75}{30 \times 10^{-3}} = 4975\ \Omega$$

For range of 300 V,

$$R_{300} = R_1 + R_2 = \frac{V - v}{I_g} = \frac{300 - 0.75}{30 \times 10^{-3}} = 9975\ \Omega$$

$$\therefore \quad R_2 = 9975 - R_1 = 9975 - 4975 = 5000\ \Omega$$

For range of 600 V,

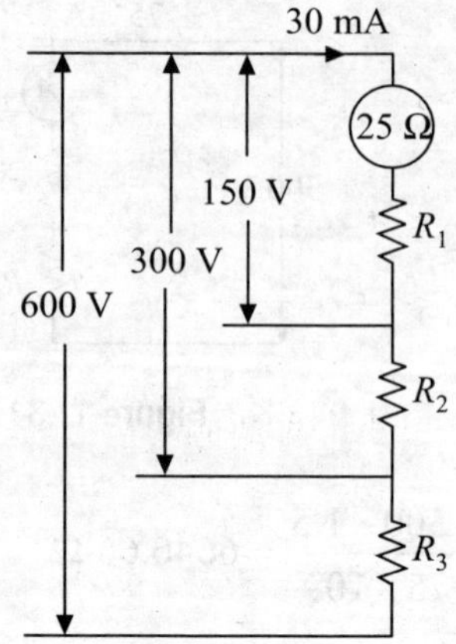

Figure 11.32

$$R_{600} = R_1 + R_2 + R_3 = \frac{V - v}{I_g} = \frac{600 - 0.75}{30 \times 10^{-3}} = 19975\ \Omega$$

$$\therefore \quad R_3 = 19975 - 4975 - 5000 = 10000\ \Omega$$

CHAPTER AT A GLANCE

- The three basic quantities in an electrical measurement are current, voltage, and power.
- The instrument which measures the current flowing in the circuit is called ammeter, whereas the instrument which measures the voltage across any two points of a circuit is called voltmeter.
- Instruments have been classified based on the type of work for which they are used. They are absolute instruments and secondary instruments.
- The examples of absolute instruments are tangent galvanometer and Rayleigh current balance.
- The examples of secondary instruments are ammeter and voltmeter.
- Electrical measuring instruments are mainly classified as indicating instruments, recording instruments, and integrating instruments.
- The examples of indicating instruments are ammeters, voltmeters, etc.
- An example of recording instrument is *X–Y* plotter.
- An example of integrating instrument is energy meter.
- For satisfactory operation of any indicating instrument, three essential systems are required. They are (i) deflecting system to produce deflecting torque, (ii) controlling system to produce controlling torque, and (iii) damping system to produce damping torque.
- Deflecting torque (T_d) can be obtained with any one of the following effects: magnetic effect, thermal effect, induction effect, electrostatic effect, and hall effect.
- Controlling torque (T_c) can be obtained either by gravity control or by spring control.
- Damping torque can be obtained by any one of the following methods: air friction damping, fluid friction damping, and eddy current damping.

- The permanent magnet moving coil instruments (PMMC) are most accurate type instruments for dc measurements.
- In gravity control method, the deflection is proportional to the current, that is, the quantity to be measured. But it is a function of sin θ, hence the scale for the instrument using gravity control is not uniform.
- In spring control method, the deflection is simply proportional to current. Hence the scale of the instrument using spring control is uniform.
- For PMMC instruments the deflecting torque is given by $T_d = NBAI$ N-m.
- The two types of moving iron instruments are moving iron attraction type and moving iron repulsion type.
- The deflecting torque equation for moving iron type instrument is given by $T_d = (1/2)I^2(dL/d\theta)$.
- The moving iron type can be used for both ac and dc measurements.
- The various errors in the moving iron instruments are hysteresis errors, temperature errors, stray magnetic field error, frequency error, and eddy current error.
- To increase the range of ammeter m times, the shunt resistance required is $1/(m - 1)$ times the basic meter resistance. This is nothing but the extension range of an ammeter.
- To increase the range of voltmeter m times, the resistance required is $(m - 1)$ times the basic meter resistance. This is extension range of voltmeter.
- Wattmeter is an instrument used for the measurement of power. The three types of instruments used for measurement of power are dynamometer type, induction type, and electrostatic type.
- Electrical energy measuring instruments are called energy or supply meters. The various types of energy meters are electrolytic type, motor type, and clock type.

MULTIPLE CHOICE QUESTIONS

1. A moving iron instrument can be used for current and voltage measurements in
   (a) ac circuits only
   (b) dc circuits only
   (c) both ac and dc circuits for any value of frequency.
   (d) both ac and dc circuits for frequencies up to about 125 Hz

2. The power consumption of PMMC instruments is typically about
   (a) 0.25 W to 2 W (b) 0.25 mW to 2 mW
   (c) 25 μW to 200 μW (d) none of these

3. In spring-controlled moving iron instruments, the scale is
   (a) uniform
   (b) cramped at the lower end and expanded at the upper end
   (c) expanded at the lower end and cramped at the upper end
   (d) cramped both at the lower and at the upper end

4. Moving iron instruments can be used as
   (a) standard instruments for calibration of other instruments
   (b) transfer type instruments
   (c) indicating type instruments
   (d) all of the above

5. The moving iron voltmeters indicate

(a) the same value for dc and ac voltages
(b) lower values for ac voltages than for corresponding dc voltages
(c) higher values for ac voltages than for corresponding dc voltages
(d) none of these

6. Moving iron instruments when measuring voltages or currents indicate

(a) the same values of the measurements for both ascending and descending values
(b) higher value of measured for ascending values
(c) higher value of measured for descending values
(d) none of the above

7. The frequency range of MI instruments is

(a) audio frequency band of 20 Hz to 20 kHz
(b) very low frequency band of 10 Hz to 30 kHz
(c) low frequency band of 30 Hz to 300 kHz
(d) power frequency of 0 to 125 Hz

8. Which instrument is the cheapest disregarding the accuracy

(a) PMMC (b) MI
(c) electrodynamometer (d) rectifier

9. Permanent magnets used in instruments are hardcore materials because

(a) they have broad hysteresis loop (b) their energy density is high
(c) they have a high $(BH)_{max}$ product (d) all of the above

10. Fluid friction damping can be used in

(a) horizontally mounted instruments
(b) vertically mounted instruments
(c) both in horizontally and in vertically mounted instruments
(d) none of the above

11. Moving iron instruments are used for

(a) ac (b) dc (c) both (d) none

12. Frictional errors are less in case of ____________ instruments.

(a) MC (b) MI (c) both (d) none

13. ____________ damping is similar to that of air friction damping.

(a) Electromagnetic (b) Eddy current (c) Fluid friction (d) None

14. ____________ damping method is used in hot wire and MI instruments.

(a) Air friction (b) Fluid friction (c) Eddy current (d) None

15. The methods for damping torque are

(a) air friction (b) fluid friction (c) eddy current (d) all of the above

16. Electrostatic type instruments are primarily used as

(a) ammeters (b) voltmeters (c) wattmeters (d) ohmmeters

17. Horizontally mounted MI instruments use ________ damping.
(a) electromagnetic (b) eddy current (c) fluid friction (d) air friction

18. Air friction damping should not be used where the deflecting torque in the instruments is produced due to
(a) magnetic field (b) electric field
(c) thermoelectric emf (d) none of the above

19. Electrostatic voltmeters draw a small value of current on dc
(a) under steady state condition response of the applied voltage
(b) when switched on irrespective of the applied volatage
(c) when measuring low voltage
(d) when measuring high voltage

SHORT ANSWER QUESTIONS

1. What is an ammeter and voltmeter?
2. Classify the measuring instruments?
3. What is an indicating instrument?
4. What are the essential torques of an indicating instrument?
5. What are the methods generally used to produce controlling torque?
6. What are the various methods generally used to produce damping torque?
7. Why controlling torque is necessary in an indicating instrument?
8. What is the material generally used for the control spring?
9. What is the purpose of anticreeping device in an energy meter?
10. Why induction-type instruments are less accurate compared to other types?
11. What are the various types of ammeters and voltmeters?
12. What are the advantages of permanent magnet moving coil instruments?
13. What are the disadvantages of induction-type instruments?
14. Why PMMC instruments cannot be used to read ac quantities?
15. What is the most efficient damping system of all the three?

REVIEW QUESTIONS

1. What are the different types of electrical measuring instruments?
2. Explain basic principle of indicating instruments.
3. Explain construction and working principle of permanent magnet moving coil instrument.
4. Explain the construction and working principle of dynamometer-type moving coil instrument.
5. Derive the expression for torque of a moving coil instruments.
6. Explain the construction and working principle of moving iron instruments.
7. Derive the torque equation for PMMC instrument?

8. Explain the working of moving iron attraction type instrument?

9. Derive the torque equation for moving iron instrument?

10. Explain the gravity control system used in instruments?

11. Explain the working of single-phase energy meter with a neat diagram?

12. Explain the extension range of ammeters and voltmeters.

13. Write the advantages and disadvantages of a moving coil and moving iron instruments.

EXERCISE PROBLEMS

1. A permanent magnet moving coil instrument has a coil of dimensions 14 mm and 12 mm. The flux density in the air gap is 1.8×10^{-3} Wb/m^2 and the spring constant is 0.14×10^{-6} N-m/rad. Determine the number of turns required to produce an angular deflection of 90° when a current of 5 mA is flowing through the coil.

2. The coil of moving coil voltmeter is 35 mm long and 45 mm wide and has 120 turns on it. The control spring exerts a torque of 240×10^{-6} N-m when the deflection is 100 divisions on full scale. If the flux density of the magnetic field in the air gap is 1.0 Wb/m^2. Estimate the resistance that must be put in series with the coil to give 1 V per division. The resistance of the voltmeter coil may be neglected.

3. A moving coil ammeter has a fixed shunt of 0.02 Ω with a coil resistance of 100 Ω and a potential difference of 500 mV across it, and full-scale deflection is also applied. (a) To what shunted current for thus correspond? (b) Calculate the value of R to give full-scale deflection when shunted current I is (a) 10 A and (b) 75 A. (c) With what value of R is 40% deflection is obtained with $I = 100$ A.

4. A moving coil instrument whose resistance is 30 Ω gives a full-scale deflection with a voltage of 35 mV. This instrument is to be used with a series multiplier to extend its range to 8 V. Calculate the error caused by 9° rise in temperature. The temperature coefficient of copper is 0.004 per degree centigrade and that of manganin is 0.0015 per degree centigrade.

5. A moving coil instrument has following data: number of turns = 130, width of the coil = 25 mm, depth of the coil = 35 mm, flux density in the gap = 0.1 Wb/m^2. Calculate the deflection torque when carrying a current of 12 mA. Also calculate the deflection if the control spring constant is 2×10^{-6} N-m/deg.

6. A moving coil instrument has resistance of 6 A between the terminals and a full-scale deflection is obtained with a current of 18 mA. This instrument is to be used with a manganin shunt to measure 110 A at full scale. Calculate the error caused by a 10° rise in temperature.

7. The inductance of a moving iron ammeter is given by $L = (0.01 + K\theta)^2$ mH, where θ is deflection from zero position in degree. The angular deflection of the instrument to 2 A and 2.5 A, are 90° and 120°, respectively. Find the value of K.

Answer Keys

Chapter 1

MULTIPLE CHOICE QUESTIONS

1. (d)	**2.** (b)	**3.** (a)	**4.** (b)	**5.** (d)
6. (a)	**7.** (b)	**8.** (a)	**9.** (b)	**10.** (d)
11. (d)	**12.** (c)	**13.** (a)	**14.** (d)	**15.** (d)
16. (c)	**17.** (d)	**18.** (c)	**19.** (b)	**20.** (b)

SHORT ANSWER QUESTIONS

1. 12 V
2. 12 Ω
3. 0.05 A
4. The resistance of a light bulb filament is effected by both length and cross-sectional area. Thicker wires have less resistance. A 120-W bulb has a greater current and a smaller resistance. Thus, a 120-W bulb must have a thicker filament than a 60-W bulb (assuming the lengths of the filaments are identical).
5. $P = 2.59$ W
6. Collisions between conduction electrons and atoms.
7. (i) Voltage-controlled voltage source, (ii) Voltage-controlled current source, (iii) Current-controlled current source, and (iv) Current-controlled voltage source.
8. It is defined as a group of network elements interconnected in a manner so as to perform a desired function.
9. The junction point of two or more branches is known as a node.
10. Any closed path formed by branches in a network is known as a loop. A mesh is a kind of loop which does not cross another loop.
11. 2 A

EXERCISE PROBLEMS

1. Two times greater
2. $R = 7053.33$, $I = 0.0326$ A
3. $R = 115.57\ \Omega$
4. $i_1 = 2.4$ A, $i_2 = 1.8$ A, $i_3 = 0.9$ A
5. $R_L = 10\ \Omega$
6. ₹ 3
7. 15 C
8. 1.4 μF, 120 J
9. $N = 4$, $b = 5$
10. $i_1 = 18$ A, $i_2 = -2$ A, $i_3 = 8$ A
11. $I = 4$ A

Chapter 2

MULTIPLE CHOICE QUESTIONS

1. (a)	**2.** (c)	**3.** (d)	**4.** (a)	**5.** (d)
6. (b)	**7.** (c)	**8.** (a)	**9.** (b)	**10.** (d)
11. (a)	**12.** (c)	**13.** (b)	**14.** (d)	**15.** (a)
16. (b)	**17.** (c)	**18.** (d)	**19.** (c)	**20.** (d)
21. (d)	**22.** (d)	**23.** (c)	**24.** (d)	**25.** (c)
26. (c)	**27.** (a)	**28.** (c)	**29.** (a)	**30.** (a)

SHORT ANSWER QUESTIONS

1. $R_1 + R_2$
2. All in parallel
3. Replacing all the sources by their internal resistances and determine the equivalent resistance looking from the open circuited terminals.
4. Series
5. Parallel
6. The condition for maximum power transfer is load resistance which must be equal to source resistance.
7. 50%
8. 25 H
9. 1 Ω
10. 16 μF

EXERCISE PROBLEMS

1. $V = 35$ V	**2.** $R = 3.57\ \Omega$
3. $i_1 = 2.4$ A, $i_2 = 0.6$A	**4.** 1.2 A, 2.8 A, 4.8 A
5. 1.6 A	**6.** $R = 2.5\ \Omega$
7. 1.481 A	**8.** 1.481 A
9. 2.389 A	**10.** 2.389 A

Chapter 3

MULTIPLE CHOICE QUESTIONS

1. (b)	**2.** (c)	**3.** (d)	**4.** (d)	**5.** (d)
6. (d)	**7.** (c)	**8.** (c)	**9.** (d)	**10.** (c)

11. (a)	**12.** (a)	**13.** (d)	**14.** (b)	**15.** (a)
16. (d)	**17.** (d)	**18.** (b)	**19.** (a)	**20.** (c)
21. (b)	**22.** (c)	**23.** (b)		

SHORT ANSWER QUESTIONS

1. Tesla
2. Leakage factor is defined as the ratio of the total flux to the useful flux.
3. $4\pi \times 10^{-7}$ H/m
4. The product of the number of turns and flux linking with the coil is known as flux linkage.
5. Using Fleming's right-hand rule.
6. Eddy current losses are reduced by using thin plates of high-resistivity material.
7. 0.15 J
8. An emf is induced in a conductor whenever it cuts the magnetic flux.
9. Number of turns, permeability of the core, and cross-sectional area of their common core.
10. The force experienced by a unit north pole at any point is called the magnetic flux strength.
11. Transformer
12. When a magnetizing force is removed from a material the kind of magnetism that remains is called the residual magnetism.

EXERCISE PROBLEMS

1. (a) 0.35 mWb (c) 2
2. (a) 1.3 (b) 0.58 (c) 2.9 Wb (d) 3
3. 1.8 Wb/m^2
4. (a) 12 N (b) 1.2 V (c) 120 W
5. 9.175 kg/m, Horizontally
6. 0.2 V
7. $L = 0.315$ H, $E = 3.94$ J
8. (a) 0.986 (b) 1.974 J
9. 39.8 N
10. (a) 750 AT (b) 7.073×10^5 AT/Wb (c) 1.0637 mWb
11. 47.183 mWb/m^2

Chapter 4

MULTIPLE CHOICE QUESTIONS

1. (a)	**2.** (c)	**3.** (b)	**4.** (b)	**5.** (c)
6. (a)	**7.** (d)	**8.** (b)	**9.** (a)	**10.** (a)
11. (b)	**12.** (c)	**13.** (d)	**14.** (a)	**15.** (b)
16. (d)	**17.** (a)	**18.** (b)	**19.** (a)	**20.** (c)

SHORT ANSWER QUESTIONS

1. Form factor is defined as the ratio of rms value to the average value.
2. Peak factor is defined as the ratio of peak value to the rms value.
3. Each of them pass through zero values at the same instant and rise in the same direction.
4. 3 A
5. $f = \dfrac{NP}{120}$
6. It is defined as time taken to complete one cycle for an alternating quantity.
7. 1.11
8. 1
9. 60 Hz, 12 A
10. Zero
11. $\sqrt{-1}$
12. –1
13. $-a + jb$
14. $(ac + bd) + j\,(bc - ad)$
15. $ab \angle \alpha + \beta$
16. $ab \angle \alpha - \beta$
17. It represents counterclockwise rotation of phasor by 90 degrees.
18. $a\,(\cos\theta + j\sin\theta)$
19. $\sqrt{a^2 + b^2}\ \angle \tan^{-1}\left(\dfrac{b}{a}\right)$
20. 75 degrees

EXERCISE PROBLEMS

1. (a) 5 Hz (b) 20 Hz (c) 2000 Hz
2. (a) 0.016 s (b) 2 ns (c) 1 ms
3. (a) 15.4 V (b) 26.57 V (c) 16.22 V (d) –16.22 V
4. (a) 108.92 mA (b) – 34.73 mA (c) – 68.4 mA (d) 0 A
5. 12.99 V, 12.99 V, 14.48 V, – 7.5 V, – 7.5 V
6. 64 V, 12 V
7. 14.14 mA, 0 A, 20 mA, 40 mA
8. 5 Ohms
9. 2 A, 200 V
10. 18.02 A
11. (a) 0.274 A (b) 1.37 A (c) 0.5 s
12. 2.258 V
13. 111.8 V

Chapter 5

MULTIPLE CHOICE QUESTIONS

1. (b)	2. (c)	3. (a)	4. (b)	5. (a)
6. (c)	7. (c)	8. (b)	9. (a)	10. (d)
11. (c)	12. (a)	13. (c)	14. (a)	15. (a)
16. (a)	17. (c)	18. (a)	19. (c)	20. (a)
21. (b)	22. (d)	23. (c)		

SHORT ANSWER QUESTIONS

1. The relationship between current and voltage in a pure inductor is the current lags behind the voltage by 90°.
2. Current will be half.
3. ohms
4. Unity
5. Triangular
6. 10 μF
7. 0 and 1
8. The current leads the voltage.
9. $VI \sin \theta$
10. j represents the rotation of phasor in counterclockwise direction by 90°.
11. –1
12. mho
13. Unity power factor and maximum current.
14. By varying the values L and C.
15. The range of frequencies between two half-power frequencies.
16. Minimum at unity power factor.
17. The condition for maximum power transfer in the load impedance is equal to complex conjugate of source impedance.
18. $I = V/Z$
19. 8 W
20. 0.992 lag

EXERCISE PROBLEMS

1. (a) 625 VA (b) 375 W (c) 0.6
2. (a) 0.925 (b) 1634.2 W (c) 673.08 VAR (d) 1767.8 VA
3. 316.23 VA
4. 40 V, 91.65 V
5. $14.57 \sin (1000\,t + 35.96°)$, –14.036°
6. 20 Ω, 56.27 Hz

7. 0.345 Hz

8. 1779.4 Hz

9. 0.832 lag

10. $-j5$ A; -7.5 A

11. $40 + j\,6282.2$

12. $1610.82 \angle 53°$ VA; $3221.65 \angle -53°$ VA

13. $8 + j9$ V; $14.8 + j\,14.4\ \Omega$

14. $5 + j10$ A; $4\ \Omega$

15. $14.06\ \Omega$; 66.66 V; 158.01 W

16. $15.18 \angle -132.87°$ V

Chapter 6

MULTIPLE CHOICE QUESTIONS

1. (d)	**2.** (c)	**3.** (d)	**4.** (b)	**5.** (c)
6. (d)	**7.** (d)	**8.** (d)	**9.** (a)	**10.** (d)
11. (d)	**12.** (c)	**13.** (a)	**14.** (b)	**15.** (a)
16. (c)	**17.** (c)	**18.** (d)	**19.** (a)	**20.** (d)
21. (a)	**22.** (d)	**23.** (a)	**24.** (c)	**25.** (a)

SHORT ANSWER QUESTIONS

1. 120°
2. Phase currents change in phase angle but not in magnitude.
3. (i) Line voltages must be equal, (ii) Impedances in each phase must be identical, and (iii) Phase difference between successive line voltages is equal.
4. zero
5. 0.8
6. 30 VAR
7. line voltage = phase voltage
8. 3 P
9. Power will be two-thirds of original.
10. Resistive load
11. 0.5

EXERCISE PROBLEMS

1. 30.12 A, 10 kW

2. (a) 46.18 Ω (b) 3.46 kW (c) 15 A (d) 10.39 kW

3. 8.79 A, 415 V

4. (a) 8.21 A (b) 240 V, 14.23 (c) 0.5144 (d) 3.038 kW

5. 0.168 H, 22.6 Ω

6. $13.8 \angle -66.86°$ A, 4.618 A

7. 24.58 A, 4.53 kW, 17.69 kVa, 17.07 kVAR

8. 17.21 A, 29.8 A

9. 59.1 A, 230 V

10. 29 A, 16.7 A

11. 0.725 $\angle 58.2°$
12. 25.4 $\angle -30°$ A, 25.4 $\angle -176.8°$ A, 25.4 $\angle 116.8°$ A,17.1 $\angle 150°$ A, 12.21 kW
13. 10.3 $\angle -3.43°$ A, 7.21 $\angle -150°$ A, 11.54 $\angle 53.13°$ A
14. 50 $\angle -30°$ A, 20 $\angle -120°$ A, 25 $\angle 150°$ A, 5.82 kW
15. (a) 3.097 kW (b) 9.29 kW
16. 10.5 Ω, 147.98 μF
17. (a) 214 V (b) 12 kW
18. 385 A, 516 A, 577.3 A
19. 384.9 $\angle -30°$ A, 516.3 $\angle -176.56°$ A, 577.3 $\angle 143.13°$ A
20. 8.98 $\angle -75°$ A, 3.816 kW, 1.022 kW

Chapter 7

MULTIPLE CHOICE QUESTIONS

1. (b)	2. (d)	3. (c)	4. (a)	5. (b)
6. (a)	7. (c)	8. (b)	9. (c)	10. (c)
11. (d)	12. (a)	13. (c)	14. (a)	15. (a)
16. (a)	17. (b)	18. (c)	19. (b)	20. (b)
21. (c)	22. (b)	23. (d)	24. (b)	25. (b)
26. (c)	27. (a)	28. (a)	29. (b)	30. (a)
31. (c)	32. (d)	33. (c)	34. (b)	35. (c)
36. (a)	37. (c)	38. (d)	39. (c)	40. (c)
41. (b)	42. (b)	43. (d)	44. (c)	45. (b)
46. (c)	47. (b)	48. (a)	49. (d)	50. (d)
51. (b)	52. (c)	53. (b)	54. (a)	55. (b)

SHORT ANSWER QUESTIONS

1. The eddy current losses are minimized in a transformer by laminating the core of the transformer.
2. The main purpose of using core in a transformer is to decrease the reluctance of the common magnetic circuit.
3. For leading power factors the transformers get negative regulation.
4. Copper loss varies inversely with power factor.
5. In order to get maximum efficiency copper loss must be equal to iron loss.
6. When transformer secondary is open circuited it will be under no-load condition. Hence, it draws no-load primary current.
7. Under no-load conditions, the magnitude of no-load primary current will be 3–5% of the full-load current.

8. It is defined as ratio of rise in voltage at the load end when the load is thrown off to the full-load voltage.
9. The all-day efficiency of a transformer mainly depends on amount of load and duration of load.
10. The essential conditions for parallel operation of transformer are same polarity, same voltage ratio, same phase sequence, and equal per unit impedances in magnitude and phase angle.
11. The function of transformer oil in a transformer is to provide cooling and insulation.
12. The desirable properties of transformer core material are high permeability and low hysteresis.
13. The function of breather in a transformer is to arrest the flow of moisture into the tank.
14. The good regulation of the transformer means output voltage variation from no-load to full-load must be minimum.
15. Transformer is a static device and there are no mechanical losses. Hence, efficiency of the transformer is high, whereas other electrical devices are rotating in nature.

EXERCISE PROBLEMS

1. (a) 0.286 Wb/m^2 (b) 924 turns
2. (a) 5.17 mWb (b) 345 W (c) 4.76 A
3. (a) 9 Ω, 23 Ω, 32 Ω primary (b) 9.5 Ω secondary
4. 0.026, 197.4
5. (a) 96.89 kVA (b) 96.43 kVA (c) 22.04 kVA
6. 309 A, 191.09 A
7. 1.5 kW

Chapter 8

MULTIPLE CHOICE QUESTIONS

1. (b)	2. (c)	3. (b)	4. (d)	5. (b)
6. (a)	7. (b)	8. (c)	9. (b)	10. (b)
11. (b)	12. (a)	13. (b)	14. (c)	15. (b)
16. (c)	17. (d)	18. (a)	19. (c)	20. (a)
21. (a)	22. (a)	23. (c)	24. (d)	25. (a)
26. (c)	27. (c)	28. (d)	29. (a)	30. (c)
31. (d)	32. (b)	33. (d)	34. (d)	35. (c)
36. (b)	37. (a)	38. (a)	39. (c)	40. (a)
41. (a)	42. (d)			

SHORT ANSWER QUESTIONS

1. A dc generator converts mechanical energy into electrical energy.
2. Whenever a conductor cuts the magnetic field, an emf is induced in that conductor and the magnitude of the induced emf depends on the rate of change of flux (df/dt).

3. According to Fleming's right-hand rule, if the forefinger, middle finger, and thumb of the right hand are stretched naturally perpendicular to each other, the forefinger indicates the direction of the magnetic field, the middle finger indicates the direction of the induced emf, and the thumb indicates the direction of motion of the conductor.
4. According to the principle of energy conversion, energy can neither be created nor be destroyed, it can merely be converted from one form to other from.
5.

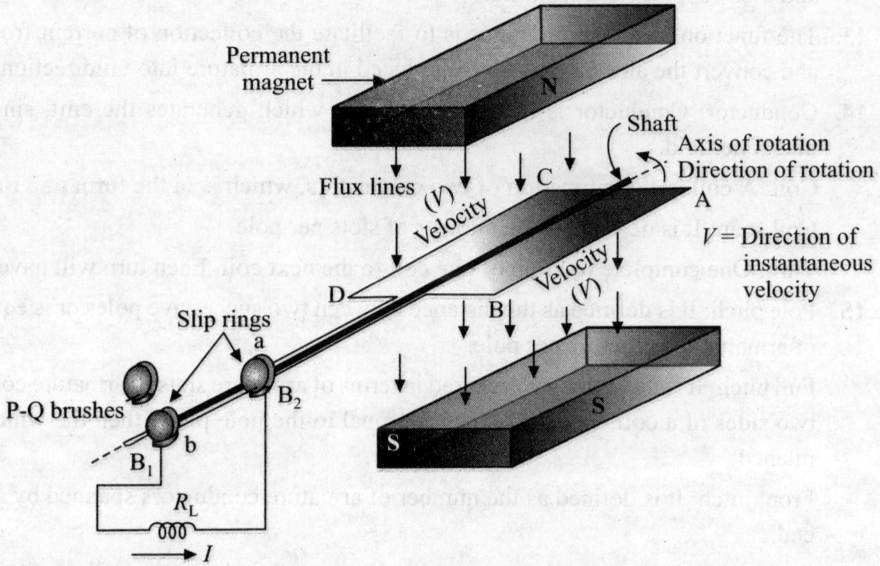

6. Whenever coil is placed in the magnetic field, the magnitude of the flux linked with the coil changes, according to Faraday's laws of electromagnetic induction, an emf induced in the coil is proportional to the rate of change of flux linkage.
7. The dc generator consists of three main parts: (i) magnetic field system, (ii) armature, and (iii) commutator and brushes. The magnetic field system consists of pole cores and pole shoes and rivet heads.
8. Important functions of yoke are as follows:
   (i) It provides the mechanical support for the poles.
   (ii) It provides the return path for magnetic flux produced by the field system.
   (iii) It acts as a protecting cover for the whole machine.
9. Yoke: It is a hollow cylinder made up of cast steel or rolled steel, and it is an unlamented ferromagnetic material.

   Brush: The brushes are made with graphite or composition of carbon and graphite to have good conductivity.

   Armature: Armature core made up of circular steel sheet disc.

   Brush holders: Brush holders are made up of carbon.
10. Advantage: The brushes are made with graphite and graphite to have good conductivity and low coefficient of friction to reduce the electrical loss and excessive wear.

Disadvantage: Without the use of brushes, the motor needs additional maintenance. Motors rely on the external power drive to perform the commutation of stationary copper winding on the stator.

11. Power lost as heat in the armature and field windings of a dc machine is called copper loss.

12. The pole shoes have a curved surface to distribute the flux uniformly throughout the air gap, and thus support field winding.

13. The function of the commutator is to facilitate the collection of current from armature coils and convert the alternating current induced in the armature into unidirectional current.

14. Conductor: Conductor is the active element which generates the emf, since it experiences magnetic field.

    Coil: A coil is a combination of two conductors, which is in the form of a ring.

    Coil span: It is defined as the number of slots per pole.

    Turn: One complete rotation of one coil to the next coil. Each turn will have two conductors.

15. Pole pitch: It is defined as the distance between two successive poles or is equal to the number of armature conductors per pole.

    Full pitch: It is the distance measured in terms of armature slots or armature conductors between two sides of a coil. If the coil pitch is equal to the pole pitch, then the winding is called full pitched.

    Front pitch: It is defined as the number of armature conductors spanned by a coil on the front end.

    Back pitch: It is generally measured in terms of the armature conductors and is equal to the number of difference of the conductors connected to a given signal of commutator.

16. Lap winding is used for high current and low voltage machines and wave winding is used for low current and high voltages. The lap winding is a two-layer winding in which each coil is connected in series to the adjacent coil. Wave type of connection winding always travels ahead avoiding overlapping.

    Lap winding, number of parallel paths = poles, i.e., $A = P = 4$

    Wave winding, number of parallel paths = 2

17. Let us consider a generator having "Z" number of conductors and the magnetic field system producing "f" flux per pole and running at a speed of "N" revolutions per minute. So

    f = Flux per pole in webers

    Z = Total number of conductors in armature

    p = Number of poles

    N = Speed of the armature in rpm

    A = Number of parallel paths in armature winding

    E_g = Induced emf

    Average emf generated per conductor $= N(d\phi/dt)$ volts (1)

    $= d\phi/dt$ (since $N = 1$) (2)

Flux cut by conductor in one revolution = ϕp.

$d\phi = \phi p$ (3)

Number of revolution per second = $N/60$

Time taken for one revolution = $60/N$.

$dt = 60/N$ seconds

$E_g = (d\phi/dt) = (\phi NP)/60$

The induced emf or generated emf for "Z" number of conductors is the emf generated in any one of the armature parallel path.

$E_g = \phi ZNp/60$ volts (4)

If the dc generator is having "A" number of parallel paths, then

$E_g = \phi ZNp/(60\ A)$ volts

18. The dc generators are broadly classified into two groups in which the field system is excited:

(i) Separately excited generators

(ii) Self-excited generators

19. Generators may be divided into (a) Separately-excited generators and (b) Self-excited generators Self-excited generators are classed according to the type of field connection they use. There are three general types of field connections—series-wound, shunt-wound (parallel), and compound-wound. Compound-wound generators are further classified as cumulative-compound and differential-compound.

20. (i) In shunt generators armature current is equal to sum of field current and load current whereas in series generators field current and load current is same. (ii) Shunt generators field winding has high resistance and large no of turns as compared to series generators. (iii) Shunt generator field winding has thin conductor and series generator has thick.

21. Losses in dc machine

1. Copper loss: (a) Armature copper loss

   (b) Field copper loss

2. Rotational losses: (a) Iron losses

   (i) Hysteresis losses

   (ii) Eddy current losses

   (b) Stray losses

22. Power flow diagram of dc generator

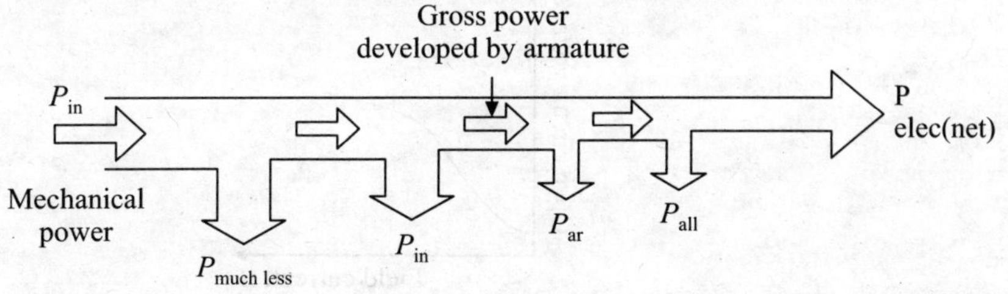

23. Efficiency is maximum when core losses equal to the copper losses

$$P_{cu} = P_0$$

24. The effect of armature flux on the main field flux is called the armature reaction.

25. The magnetic neutral axis (MNA): The magnetic neutral axis is the axis drawn perpendicular to the mean direction of the flux passing through the center of the armature.

Geometrical neutral axis (GNA): The geometrical neutral axis is the axis that bisects the angle between the center line of adjacent poles. Clearly, it is the axis of symmetry between two adjacent poles.

26. The axis of commutation is along with the brush axis.

27. Total number of turns in these angles = $(2\theta_m/360)\ Z$

Demagnetization ampere-turns per pair of poles = $(2\theta_m/360)\ ZI$

Demagnetization ampere-turns per pole, $AT_d = (\theta_m/360)\ ZI$.

28. Total armature ampere-turns per pole = $ZI/2p$

Demagnetizing conductors per pole = $(2\theta_m/360)\ Z$

Demagnetizing conductors per pole = $Z/P(2\theta_m/360)\ Z$

29. The process by which the current in the coil is reversed during the period in which it passes under the brushes called commutation.

30. The failure of the coil to get the current reversed to its full value during commutation is due to the production of the self-induced emf in the coil undergoing commutation. This induced emf is called the "reactance voltage" and opposes the reversal of current. To avoid the reactance voltage being induced in the coils undergoing commutation, the brushes are to be shifted to the MNA. The amount of brush shift depends on the load delivered by the armature.

31. In dc generator emf is induced in the armature coils.

32. Interpoles are similar to the main field poles and located on the yoke between the main field poles. They have windings in series with the armature winding. Interpoles have the function of reducing the armature reaction effect in the commutating zone. They eliminate the need to shift the brush assembly.

33. Characteristics of a dc generator are broadly classified into three types:

(i) Open-circuit characteristic (OCC)

(ii) Internal or total characteristic (E/I_a)

(iii) External characteristic (V/I)

34.

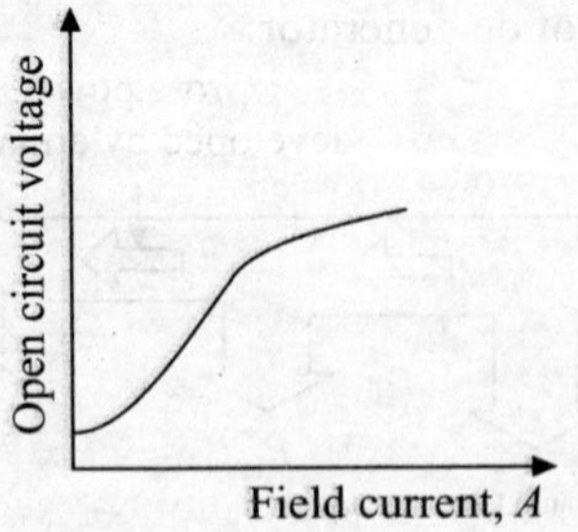

35.

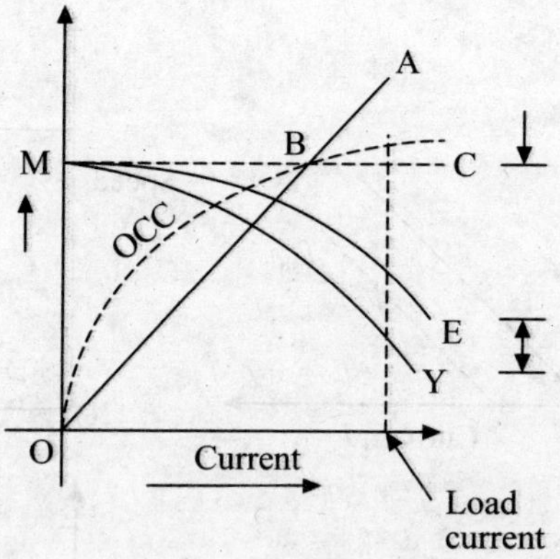
A
B
M
C
OCC
E
Y
O
Current
Load
current

36.

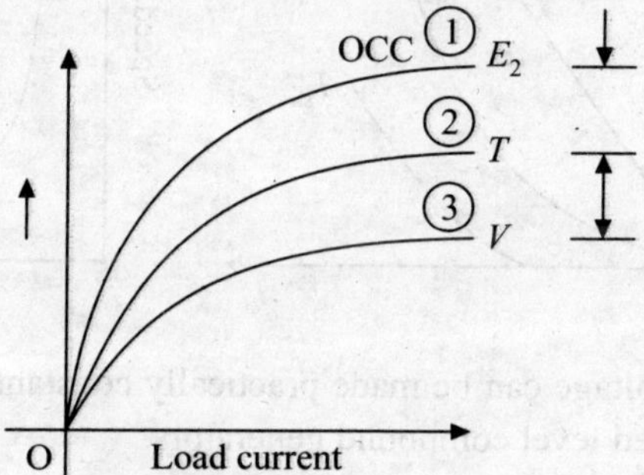
OCC
1
E_2
2
T
3
V
O
Load current

37.

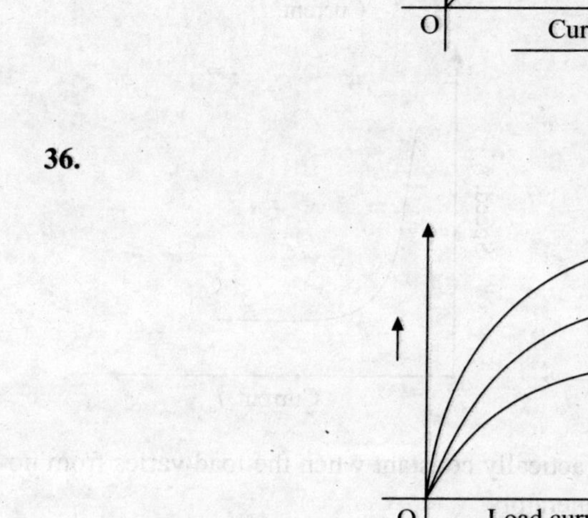
Series
Cumulative compound
Speed
Shunt
Rated speed
Differential compound
I_a

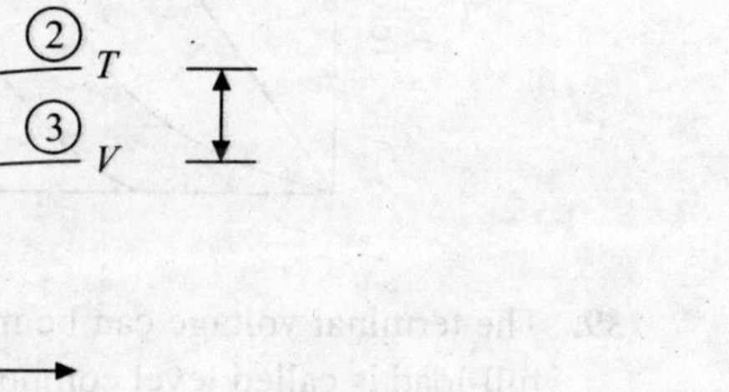
Differential compound
Torque
Rated speed
Shunt
Series
Cumulative compound
I_a

(a) (b)

38.

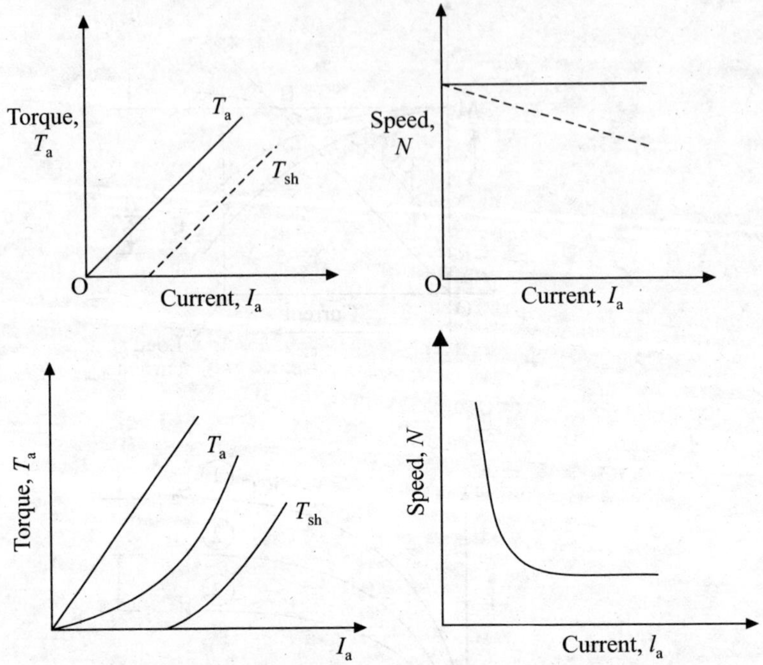

39. The terminal voltage can be made practically constant when the load varies from no-load to full-load is called level compound generators.

If a strong series field is used, the terminal voltage rises with an increase in the load current, and the generator is said to be over compound generator.

If a week series field is used, the terminal voltage decreases with the load current, and the generator is said to be under compound generator.

40. The requirements of voltage build up in self-excited dc generator: The field system retains the flux even if the magnetizing source is removed away. This small retained flux is known as the *residual flux*. This residual flux will result *voltage build up* in dc machine.

41. The critical speed of a shunt generator is the minimum speed below which it fails to excite.

42. Critical field resistance is defined as the resistance of the field circuit which will cause the shunt generator just to build up its emf at a specified field.

43. Shunt-wound generators will in general operate very well in parallel and will divide the load well if the machines have similar characteristics. If the machines do not have similar characteristics, one machine will take more than its share of the load and may tend to drive the other as a motor.

44. (i) Necessary conditions: For satisfactory parallel operation of dc generators without an equalizer connection:

(a) The prime movers must be free from speed pulsations of sufficient magnitude to cause objectionable fluctuations in generator voltage.

(b) At all loads the terminal voltage produced by each generating unit must decrease appreciably when its load current increases, and increase when its load current decreases.

(c) The curve which shows terminal voltage plotted against load current expressed as a percentage of full-load current must be substantially the same for each of the generating units to be paralleled.

(ii) Reasons for conditions: The first two of these conditions are to prevent surging of the load and ensure stable

Operation: The third is to provide for dividing the load between the generators in proportion to their ratings.

45. An equalizer, or equalizer connection, is used to connect two or more compound wound generators which are to operate in parallel.

46. Its use is necessary in paralleling compound wound generators when voltage regulators are not used. In order to be effective, its resistance must be considerably lower than that of the series field winding.

47. The terminal voltage of dc shunt generator is more or less constant from no load to full load. Therefore these generators are used where constant voltage is required.

(i) For electroplating

(ii) Battery charging

(iii) For excitation of alternators

48. A direct current (dc) motor is a fairly simple electric motor that uses electricity and a magnetic field to produce torque, which turns the motor.

49. According to Fleming's left-hand rule, if the forefinger, middle finger, and thumb of the left hand are stretched naturally perpendicular to each other, the forefinger indicates the direction of the magnetic field, the middle finger indicates the direction of the current, and the thumb indicates the direction of the force developed.

50. If the load on the motor varies, the corresponding current drawn by the armature develops the required torque which means that the dc motor acts as a self-regulating machine.

When the motor is running on no-load, a small torque exists to overcome the friction and windage losses. Since the torque is small, the armature current is also small which makes the back emf nearly equal to the applied voltage.

When the motor is loaded, the armature slows down. So obviously the speed decreases and the back emf decreases (E_b). This in turn increases the armature current, which increases the driving torque.

51. Let

V = applied voltage

E_b = back emf

R_a = armature resistance

I_a = armature current

Since back emf E_b acts in opposition to the applied voltage V, the net voltage across the armature circuit is $V - E_b$. The armature current I_a is given by

$$I_a = (V - E_b)/R_a$$

or $$V = E_b + I_a R_a$$

Back emf, $E_b = V - I_a R_a$

52. The types of dc motor are also classified according to the manner in which the field and armature are connected:

(i) Series wound motor
(ii) Shunt wound motor
(iii) Compound wound motor
  (a) Long-shunt motor
  (b) Short-shunt motor

53. The voltage equation of a dc motor is

$$V = E + I_a R_a \tag{1}$$

Multiplying both sides of Eq. (1) by I_a, we obtain

$$VI_a = EI_a + IR_a \tag{2}$$

VI_a = Electrical power input to the armature

IR_a = Copper loss in the armature

We also know that input = output + losses (3)

Comparing Eqs. (2) and (3), we get

EI_a = Electrical equivalent of gross mechanical power developed by the armature (electromagnetic power)

T = Average electromagnetic torque developed by the armature in Newton meters (N-m)

At this value of torque the electromechanical power conversion takes place. The mechanical power developed by the armature,

$$P_m = wT = 2\pi NT$$

$$P_m = EI_a = wT = 2\pi NT \tag{4}$$

But $E = (\phi ZNP)/A$

Therefore, from Eq. (4),

$$(\phi ZNP)/A = 2\pi NT$$

$$T = [PZ/(2\pi A)]\ \phi a \tag{5}$$

Equation (5) is called the *torque equation of a dc motor*. For a given dc machine, Z and A are constants.

54. The voltage equation of a dc motor is

$$V = E + I_a R_a$$

Power equation of a dc motor

$$P_m = E_b I_a + I^2{}_a R_a$$

55. Losses in dc motor

1. Copper loss: (i) Armature copper loss
   (ii) Field copper loss
2. Rotational losses: (i) Iron losses
   (a) Hysteresis losses
   (b) Eddy current losses
   (ii) Stray losses

56. The various power stages in a dc generator

Mechanical power input —Iron and fiction losses→ Electrical power developed in armature

↓ Cu losses

Electrical power output

57. Electrical and mechanical characteristics of shunt motor

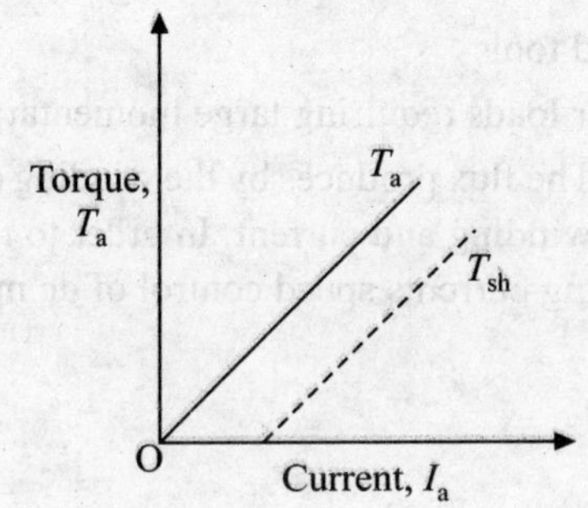

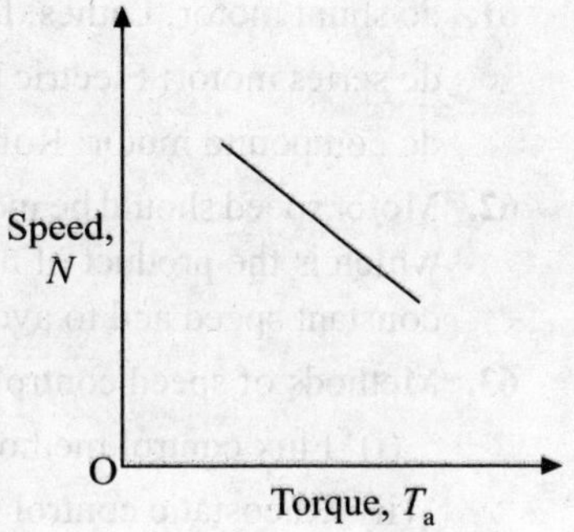

58. Electrical and mechanical characteristics of series motor

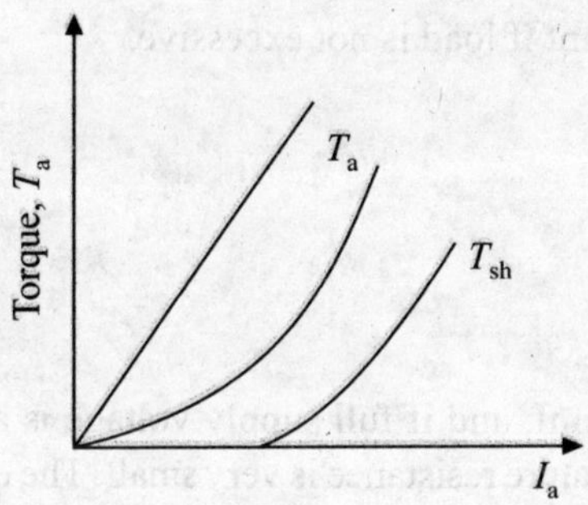

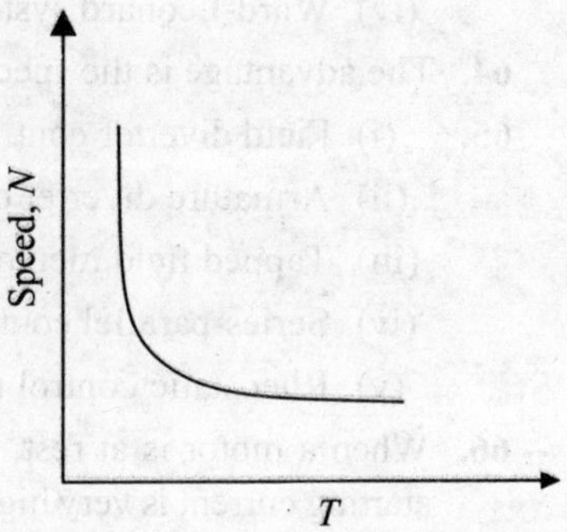

59. Electrical and mechanical characteristics of compound motor

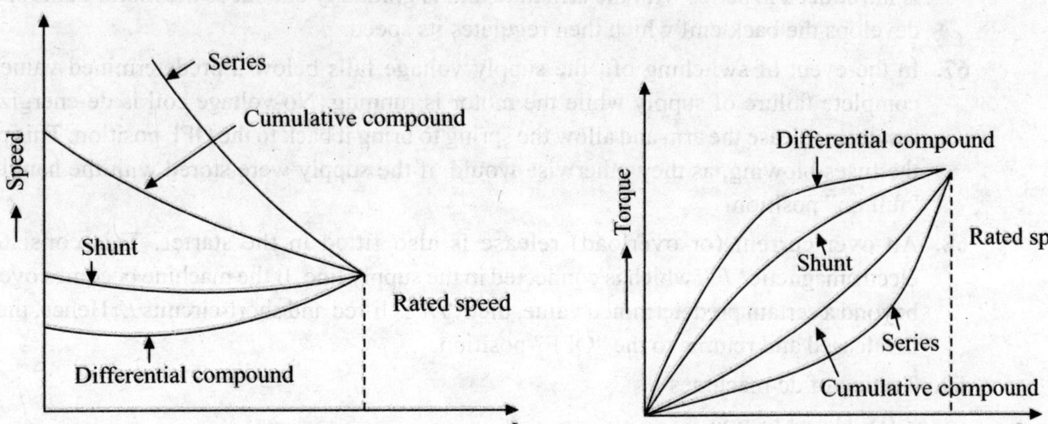

(a) (b)

60. Series-wound motors can never be used where the motor can be started without load. Since both field and armature are connected in series, the torque equation is

$$T = ki^2$$

The torque increases much than does the armature current. Furthermore, if the supply voltage falls, the starting current may still be maintained at twice the full-load value by cutting out some of the starting resistance, so that the high value of starting torque may still be maintained.

61. dc shunt motor: Lathes, fans, pumps disc and band saw drive requiring moderate torques.

dc series motor: Electric traction, high speed tools

dc compound motor: Rolling mills and other loads requiring large momentary toques.

62. Motor speed should be independent of load. The flux produces by the winding depends on mmf which is the product of number of turns of winding and current. In order to run the motor on constant speed and to avoid from high starting currents speed control of dc motors required.

63. Methods of speed control

(i) Flux control method

(ii) Rheostatic control method

(iii) Applied voltage control

(iv) Ward-Leonard system of speed control method

64. The advantage is the speed is almost constant if load is not excessive.

65. (i) Field diverter control method

(ii) Armature diverter control method

(iii) Tapped field method

(iv) Series-parallel connection of field

(v) Rheostatic control method

66. When a motor is at rest, there is no back emf, and if full supply voltage is applied, then the starting current is very high because the armature resistance is very small. The excessive current would be a many times of full-load value. To avoid this excessive starting current, a resistance is introduced in series with the armature and is gradually cut out as the motor gains speed and develops the back emf which then regulates its speed.

67. In the event of switching off, the supply voltage falls below a predetermined value, or the complete failure of supply while the motor is running. No-voltage coil is de-energized, this results to release the arm and allow the spring to bring it back to the OFF position. This prevents the fuses blowing, as they otherwise would, if the supply were stored with the handle in the "full on" position.

68. An over current (or overload) release is also fitted in the starter. This consists of an electromagnetic "*F*", which is connected in the supply line. If the machine becomes overloaded beyond a certain predetermined value, then "*D*" is lifted and short-circuits *E*. Hence, the handle is released and returns to the "OFF" position.

69. Testing of dc machines

(i) Direct testing

(ii) Swinburne test

(iii) Hopkinson test

(iv) Retardation or running down test

70. Merits:

(i) The efficiency can be estimated at any load because constant losses are known.

(ii) This method is convenient and economical.

Demerits:

(i) This test cannot be performed for series motor.

(ii) Assumed temperature of 40°C at full load cannot be checked actually.

71. Break test is a direct method of testing dc machine. It is a simple method of measuring motor output, speed, and efficiency.

Swinburne test is an indirect method of testing dc machine. In this method the constant losses of dc machine can be determined at no-load. This test cannot be performed for series motor.

72.

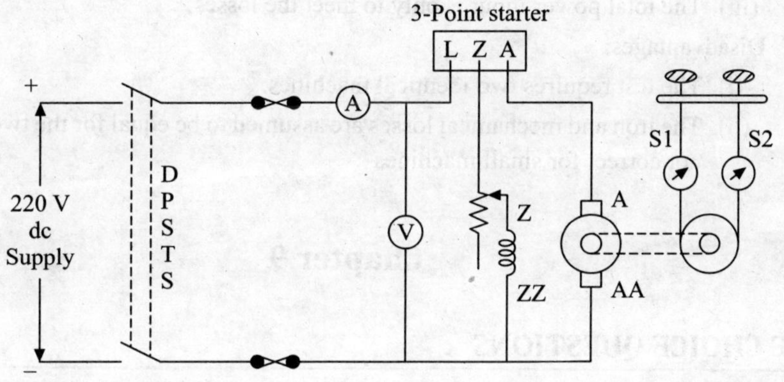

73.

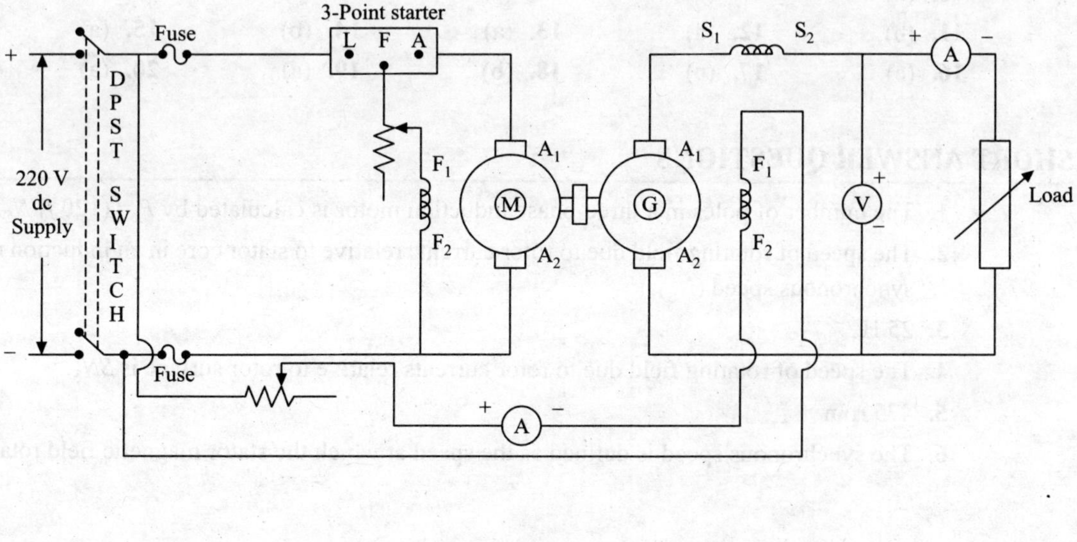

74.

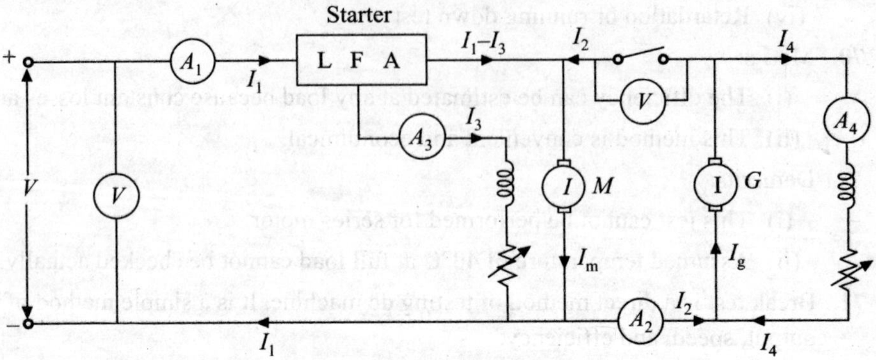

75. Advantages:

(i) Two machines are tested under full-load conditions.

(ii) It can be used to determine efficiency as well as temperature rise.

(iii) The total power input is only to meet the losses.

Disadvantages:

(i) The test requires two identical machines.

(ii) The iron and mechanical losses are assumed to be equal for the two machines, which are not correct for small machines

Chapter 9

MULTIPLE CHOICE QUESTIONS

1. (a)	**2.** (b)	**3.** (d)	**4.** (a)	**5.** (b)
6. (d)	**7.** (b)	**8.** (c)	**9.** (a)	**10.** (b)
11. (b)	**12.** (a)	**13.** (a)	**14.** (b)	**15.** (a)
16. (c)	**17.** (c)	**18.** (b)	**19.** (a)	**20.** (a)

SHORT ANSWER QUESTIONS

1. The number of poles in a three-phase induction motor is calculated by $P = (120\,f)/N_S$.
2. The speed of rotating field due to rotor currents relative to stator core in an induction motor is synchronous speed (N_S).
3. 25 Hz
4. The speed of rotating field due to rotor currents relative to rotor surface is SN_S.
5. 475 rpm
6. The synchronous speed is defined as the speed at which the stator magnetic field rotates.

7. When it runs at synchronous speed the relative speed between the rotating flux and rotor will be zero and so the emf induced in the rotor circuit and the torque developed will be zero.
8. The rotor of an induction motor rotates in the direction of the rotation of stator field in order to reduce the relative speed between the rotating stator field and the stationary rotor conductors.
9. The direction of rotation of three-phase induction motor is reversed by interchanging any two leads from the supply.
10. The reason to keep air-gap flux density low in an induction motor is to improve the power factor.
11. The stator must be connected in star.
12. The two important parts of an induction motor are stator and rotor.
13. The stator core of three-phase induction motor is laminated in order to reduce the eddy current loss.
14. The frequency of rotor currents at standstill is same as supply frequency.
15. 2 Hz
16. Losses that occur in an induction motor are stator copper loss, stator iron loss, rotor copper loss, and windage and friction loss.
17. The iron loss in the rotor of three-phase induction motor is negligible due to the very low frequency of emf induced in the rotor.

EXERCISE PROBLEMS

1. 2.5%, 1.25 Hz
2. 750 rpm, 0.03
3. 220 V, 87 A, 21.33 A
4. 1.85 kW, 22.4 A, 91.7% (assume $p = 100$ kW)
5. 91%, 0.03
6. 4.18 kW, 89.23%
7. (a) 47 N-m (b) 242 N-m (c) 1324 rpm
8. 69.28 N-m
9. 1500 rpm
10. 90%

Chapter 10

MULTIPLE CHOICE QUESTIONS

1. (a)	2. (a)	3. (a)	4. (a)	5. (a)
6. (d)	7. (c)	8. (a)	9. (b)	10. (c)
11. (b)	12. (d)	13. (c)	14. (d)	15. (a)
16. (b)	17. (d)	18. (a)	19. (d)	20. (c)
21. (a)	22. (a)	23. (d)	24. (d)	25. (d)

SHORT ANSWER QUESTIONS

1. The rotating part in modern alternators is field system.
2. Stator core of synchronous machine is laminated to reduce eddy current losses.
3. The stator frame serves the purpose of holding the armature stampings in position and also in some cases for circulating cold water for cooling.
4. Dampers are provided in order to increase stability and eliminate harmonic effects.
5. An alternator driven by steam turbine is known as turbo generator.
6. Salient pole rotors are not suitable for high speed turbo-generators due to high centrifugal force and windage loss.
7. 3000 rpm
8. Because it improves the generated emf waveform and adds rigidity and mechanical strength to the winding.
9. It is defined as the ratio of emf induced in a chorded coil to that in a full pitch coil.
10. It is defined as the ratio of the phasor sum of the emfs induced in all the coils distributed in a number of slots under one pole to the arithmetic sum of the emfs induced.
11. 1500 rpm
12. The power factor of a synchronous motor is varied by varying the excitation.
13. The purpose of squirrel cage winding in a synchronous machine is to make it self- starting.
14. The phenomenon of oscillation of the rotor of a synchronous motor about its equilibrium position corresponding to new load on sudden throwing off or increasing of load is called hunting.
15. An over-excited synchronous motor with no-load is called synchronous capacitor.

EXERCISE PROBLEMS

1. 1500 rpm, 1.9 kV
2. 50 Hz, 3000 rpm
3. 0.9659
4. 0.9659
5. 5 Ω
6. 1800 rpm
7. Leading
8. (a) 3.2 W (b) 20 A (c) 0.8

Chapter 11

MULTIPLE CHOICE QUESTIONS

1. (d)	2. (c)	3. (b)	4. (c)	5. (b)
6. (c)	7. (d)	8. (b)	9. (d)	10. (b)
11. (c)	12. (b)	13. (c)	14. (a)	15. (d)
16. (b)	17. (d)	18. (b)	19. (a)	

SHORT ANSWER QUESTIONS

1. The instrument which measures the current flowing in the circuit is called ammeter while the instrument which measures the voltage across any two points of a circuit is called voltmeter.
2. Electrical measuring instruments are mainly classified as: (i) Indicating instruments, (ii) Recording instruments, and (iii) Integrating instruments.
3. It is an instrument makes use of pointer and dial for showing or indicating magnitude of unknown quantity.
4. (i) Deflecting torque, (ii) Controlling torque, and (iii) Damping torque.
5. (i) Gravity control and (ii) Spring control.
6. (i) Air-friction damping, (ii) Fluid- friction damping, and (iii) Eddy-current damping.
7. To get the steady deflection, controlling torque is necessary.
8. German silver.
9. Its purpose is to prevent running the meter on no-load.
10. Induction-type instruments are less accurate compared to other types because of temperature and frequency errors.
11. (i) Moving coil type, (ii) Moving iron type, (iii) Hot wire type, (iv) Electrostatic type, and (v) Induction type.
12. (i) Power consumption is low, (ii) Uniform scale, (iii) It can be used as either ammeter or voltmeter, (iv) Torque to weight ratio is very high, and (v) It is also free from errors.
13. (i) Can be used only with ac voltages, (ii) Errors due to variation in frequency and temperature, and (iii) Consume large amount of power and are expensive compared to other types.
14. Because the deflecting torque is proportional to the current.
15. Eddy current damping.

EXERCISE PROBLEMS

1. 145
2. 2.4 ohms
3. (a) 25 A (b) 0.05 Ω (c) 6.66 mΩ
4. 2 MΩ, 0.09 V
5. 0.1365 milli N-m, 68.25°
6. 0.8
7. -3×10^{-5}

Index